THERMAL FOOD PROCESSING

New Technologies and Quality Issues

Second Edition

Contemporary Food Engineering

Series Editor

Professor Da-Wen Sun, Director
Food Refrigeration & Computerized Food Technology
National University of Ireland, Dublin
(University College Dublin)
Dublin, Ireland
http://www.ucd.ie/sun/

Thermal Food Processing: New Technologies and Quality Issues, Second Edition, *edited by Da-Wen Sun* (2012)

Physical Properties of Foods: Novel Measurement Techniques and Applications, *edited by Ignacio Arana* (2012)

Handbook of Frozen Food Processing and Packaging, Second Edition, *edited by Da-Wen Sun* (2011)

Advances in Food Extrusion Technology, *edited by Medeni Maskan and Aylin Altan* (2011)

Enhancing Extraction Processes in the Food Industry, *edited by Nikolai Lebovka, Eugene Vorobiev, and Farid Chemat* (2011)

Emerging Technologies for Food Quality and Food Safety Evaluation, *edited by Yong-Jin Cho and Sukwon Kang* (2011)

Food Process Engineering Operations, *edited by George D. Saravacos and Zacharias B. Maroulis* (2011)

Biosensors in Food Processing, Safety, and Quality Control, *edited by Mehmet Mutlu* (2011)

Physicochemical Aspects of Food Engineering and Processing, *edited by Sakamon Devahastin* (2010)

Infrared Heating for Food and Agricultural Processing, *edited by Zhongli Pan and Griffiths Gregory Atungulu* (2010)

Mathematical Modeling of Food Processing, *edited by Mohammed M. Farid* (2009)

Engineering Aspects of Milk and Dairy Products, *edited by Jane Sélia dos Reis Coimbra and José A. Teixeira* (2009)

Innovation in Food Engineering: New Techniques and Products, *edited by Maria Laura Passos and Claudio P. Ribeiro* (2009)

Processing Effects on Safety and Quality of Foods, *edited by Enrique Ortega-Rivas* (2009)

Engineering Aspects of Thermal Food Processing, *edited by Ricardo Simpson* (2009)

Ultraviolet Light in Food Technology: Principles and Applications, *Tatiana N. Koutchma, Larry J. Forney, and Carmen I. Moraru* (2009)

Advances in Deep-Fat Frying of Foods, *edited by Serpil Sahin and Servet Gülüm Sumnu* (2009)

Extracting Bioactive Compounds for Food Products: Theory and Applications, *edited by M. Angela A. Meireles* (2009)

Advances in Food Dehydration, *edited by Cristina Ratti* (2009)

Optimization in Food Engineering, *edited by Ferruh Erdoğdu* (2009)

Optical Monitoring of Fresh and Processed Agricultural Crops, *edited by Manuela Zude* (2009)

Food Engineering Aspects of Baking Sweet Goods, *edited by Servet Gülüm Sumnu and Serpil Sahin* (2008)

Computational Fluid Dynamics in Food Processing, *edited by Da-Wen Sun* (2007)

THERMAL FOOD PROCESSING

New Technologies and Quality Issues

Second Edition

Edited by Da-Wen Sun

CRC Press is an imprint of the
Taylor & Francis Group, an **informa** business

CRC Press
Taylor & Francis Group
6000 Broken Sound Parkway NW, Suite 300
Boca Raton, FL 33487-2742

First issued in paperback 2016

Version Date: 20120106

ISBN 13: 978-1-138-19963-7 (pbk)
ISBN 13: 978-1-4398-7678-7 (hbk)

Visit the Taylor & Francis Web site at
http://www.taylorandfrancis.com

and the CRC Press Web site at
http://www.crcpress.com

Contents

Series Preface

CONTEMPORARY FOOD ENGINEERING

Food engineering is the multidisciplinary field of applied physical sciences combined with the knowledge of product properties. Food engineers provide the technological knowledge transfer essential to the cost-effective production and commercialization of food products and services. In particular, food engineers develop and design processes and equipment to convert raw agricultural materials and ingredients into safe, convenient, and nutritious consumer food products. However, food engineering topics are continuously undergoing changes to meet diverse consumer demands, and the subject is being rapidly developed to reflect market needs.

In the development of food engineering, one of the many challenges is to employ modern tools and knowledge, such as computational materials science and nanotechnology, to develop new products and processes. Simultaneously, improving food quality, safety, and security continues to be a critical issue in food engineering study. New packaging materials and techniques are being developed to provide more protection to foods, and novel preservation technologies are emerging to enhance food security and defense. Additionally, process control and automation regularly appear among the top priorities identified in food engineering. Advanced monitoring and control systems are developed to facilitate automation and flexible food manufacturing. Furthermore, energy saving and minimization of environmental problems continue to be important issues in food engineering, and significant progress is being made in waste management, efficient utilization of energy, and reduction of effluents and emissions in food production.

The Contemporary Food Engineering Series, consisting of edited books, attempts to address some of the recent developments in food engineering. The series covers advances in classical unit operations in engineering applied to food manufacturing as well as such topics as progress in the transport and storage of liquid and solid foods; heating, chilling, and freezing of foods; mass transfer in foods; chemical and biochemical aspects of food engineering and the use of kinetic analysis; dehydration, thermal processing, nonthermal processing, extrusion, liquid food concentration, membrane processes, and applications of membranes in food processing; shelf-life and electronic indicators in inventory management; sustainable technologies in food processing; and packaging, cleaning, and sanitation. These books are aimed at professional food scientists, academics researching food engineering problems, and graduate-level students.

The editors of these books are leading engineers and scientists from many parts of the world. All the editors were asked to present their books to address the market's need and pinpoint the cutting-edge technologies in food engineering. All contributions have been written by internationally renowned experts who have both academic and professional credentials. All the authors have attempted to provide critical, comprehensive, and readily accessible information on the art and science of a relevant topic in each chapter, with reference lists for further information. Therefore, each book can serve as an essential reference source to students and researchers in universities and research institutions.

Da-Wen Sun
Series Editor

Series Preface

CONTEMPORARY FOOD ENGINEERING

[illegible]

Da-Wen Sun
Series Editor

Preface

Thermal processing is one of the most important processes in the food industry. The concept of thermal processing is based on heating foods for a certain length of time at a certain temperature. The challenge of developing advanced thermal processing for the food industry is continuing in line with the demand for enhanced food safety and quality as there is always some undesirable degradation of heat-sensitive quality attributes associated with thermal processing.

The first edition of this book was published in 2006, with the aim of providing a comprehensive review of the latest developments in thermal food processing technologies, of stressing topics vital to the food industry today, of pinpointing the trends in future research and development, and of assembling essential, authoritative, and complete references and data that can be used by the researcher in the university and research institution or can serve as a valuable reference source for undergraduate and postgraduate studies. This will continue to be the purpose of this second edition.

In the second edition, besides updating or rewriting individual chapters with the latest developments in each topic area, five new chapters have been added in order to enhance the contents of the book. In Part I, two new chapters, Thermal Effects in Food Microbiology, and Modeling Thermal Microbial Inactivation Kinetics, have been added to provide fundamental knowledge of the related food safety issues raised in subsequent chapters. In Part II, a new chapter, Thermal Processing of Fruits and Fruit Juices, has been added to provide a complete coverage of thermally processed food products. Finally, two new chapters, Aseptic Processing and Packaging, and Microwave Heating, have been added in Part III to provide a detailed description of two common thermal processing techniques.

Editor

Born in Southern China, **Professor Da-Wen Sun** is a world authority in food engineering research and education; he is a Member of Royal Irish Academy, which is the highest academic honor in Ireland, he is also a member of Academia Europaea (The Academy of Europe). His main research activities include cooling, drying, and refrigeration processes and systems; quality and safety of food products; bioprocess simulation and optimization; and computer vision technology. His innovative studies on vacuum cooling of cooked meats, pizza quality inspection by computer vision, and edible films for shelf-life extension of fruit and vegetables have been widely reported in national and international media. Results of his work have been published in about 600 papers, including 250 peer-reviewed journal papers (h-index = 35). He has also edited 12 authoritative books. According to Thomson Scientific's Essential Science Indicators[SM] updated as of July 1, 2010, based on data derived over a period of ten years plus four months (January 1, 2000–April 30, 2010) from ISI Web of Science, a total of 2554 scientists are among the top 1% of the most-cited scientists in the category of agriculture sciences, and Professor Sun tops the list with his ranking of 31.

Professor Sun received a first class BSc Honors and MSc in mechanical engineering and a PhD in chemical engineering in China before working in various universities in Europe. He became the first Chinese national to be permanently employed in an Irish university when he was appointed college lecturer at National University of Ireland, Dublin (University College Dublin), Ireland, in 1995, and was then continuously promoted in the shortest possible time to senior lecturer, associate professor, and full professor. Dr. Sun is now a professor of food and biosystems engineering and director of the Food Refrigeration and Computerised Food Technology Research Group at University College Dublin (UCD), Dublin, Ireland.

As a leading educator in food engineering, Professor Sun has significantly contributed to the field. He has trained many PhD students who have made their own contributions to the industry and academia. He has also given lectures on advances in food engineering on a regular basis in academic institutions internationally and delivered keynote speeches at international conferences. As a recognized authority in food engineering, he has been conferred adjunct, visiting, and consulting professorships by ten top universities in China, including Zhejiang University, Shanghai Jiaotong University, Harbin Institute of Technology, China Agricultural University, South China University of Technology, and Jiangnan University. In recognition of his significant contribution to food engineering worldwide and for his outstanding leadership in the field, the International Commission of Agricultural and Biosystems Engineering (CIGR) awarded him the "CIGR Merit Award" in 2000 and again in 2006, and the Institution of Mechanical Engineers (IMechE) based in the United Kingdom named him "Food Engineer of the Year 2004." In 2008, he was granted "CIGR Recognition Award" in honor of his distinguished achievements as the top 1% of agricultural engineering scientists in the world. In 2007, he was presented with the only "AFST(I) Fellow Award" in that year by the Association of Food Scientists and Technologists (India), and in 2010, he was presented with the "CIGR Fellow Award." The title of fellow is the highest honor in CIGR and is conferred to individuals who have made sustained, outstanding contributions worldwide.

Professor Sun is a fellow of the Institution of Agricultural Engineers and a fellow of Engineers Ireland (the Institution of Engineers of Ireland). He has also received numerous awards for teaching and research excellence, including the President's Research Fellowship, and has twice received the President's Research Award of University College Dublin, Dublin, Ireland. He is the editor in chief of *Food and Bioprocess Technology—An International Journal* (Springer) (2010 Impact Factor=3.576, ranked at the fourth position among 126 ISI-listed food science and technology journals), series editor of the Contemporary Food Engineering book series (CRC Press/Taylor & Francis), former editor of *Journal of Food Engineering* (Elsevier), and editorial board member of *Journal of Food Engineering* (Elsevier), *Journal of Food Process Engineering* (Blackwell), *Sensing and Instrumentation for Food Quality and Safety* (Springer), and *Czech Journal of Food Sciences.* He is also a chartered engineer.

On May 28, 2010, Professor Sun was awarded membership of the Royal Irish Academy (RIA), which is the highest honor that can be attained by scholars and scientists working in Ireland. At the 51st CIGR General Assembly held during the CIGR World Congress in Quebec City, Canada, on June 13–17, 2010, he was elected incoming president of CIGR and will become CIGR President in 2013–2014. The term of his CIGR presidency is six years, two years each for serving as incoming president, president, and past president. On September 20, 2011, he was elected to Academia Europaea (The Academy of Europe), which functions as European Academy of Humanities, Letters and Sciences and is one of the most prestigious academies in the world; election to the Academia Europaea represents the highest academic distinction.

Contributors

José Manuel Gallardo Abuín
Consejo Superior de Investigaciones Científicas
Instituto de Investigaciones Marinas
Vigo, Spain

Jasim Ahmed
Food and Nutrition Program
Kuwait Institute for Scientific Research
Safat, Kuwait

Mogessie Ashenafi
Institute of Pathobiology
Addis Ababa University
and
Ashraf Agricultural and Industrial Group
Addis Ababa, Ethiopia

Ursula Andrea Gonzales Barron
School of Biosystems Engineering
University College Dublin
Dublin, Ireland

Inês de Castro
Centro de Engenharia Biológica
Universidade do Minho
Braga, Portugal

Cuiren Chen
Campbell Soup Company
Camden, New Jersey

Xiao Dong Chen
Faculty of Health Engineering and Science
Department of Chemical Engineering
Monash University
Clayton, Victoria, Australia

Nivedita Datta
School of Biomedical and Health Sciences
Victoria University
Melbourne, Victoria, Australia

Paul L. Dawson
Department of Food Nutrition and Packaging Sciences
Clemson University
Clemson, South Carolina

Hilton C. Deeth
School of Agriculture and Food Sciences
University of Queensland
St. Lucia, Queensland, Australia

Adriana E. Delgado
School of Biosystems Engineering
University College Dublin
Dublin, Ireland

Pablo S. Fernández
Departamento de Ingeniería de Alimentos y del Equipamiento Agrícola
Universidad Politécnica de Cartagena
Cartagena, Spain

John D. Floros
Department of Food Science
The Pennsylvania State University
University Park, Pennsylvania

Alan L. Kelly
School of Food and Nutritional Sciences
University College Cork
Cork, Ireland

Qingyue Ling
Food Innovation Center
Oregon State University
Portland, Oregon

Min Liu
Department of Food Science
The Pennsylvania State University
University Park, Pennsylvania

Antonio Martínez Lopez
Consejo Superior de Investigaciones Científicas
Instituto de Agroquimica y Tecnologia de Alimentos
Valencia, Spain

Jean-François Maingonnat
Sécurité et Qualité des Produits d'Origine Végétale
Institut National de la Recherche Agronomique
Avignon, France

Sunil Mangalassary
School of Kinesiology and Nutritional Science
California State University
Los Angeles, California

Pamela Manzi
Department of Food Science
Istituto Nazionale di Ricerca per gli Alimenti e la Nutrizione
Rome, Italy

Weijie Mao
Department of Food Science and Technology
Guangdong Ocean University
Zhanjiang, People's Republic of China

María Isabel Medina Méndez
Consejo Superior de Investigaciones Científicas
Instituto de Investigaciones Marinas
Vigo, Spain

Colm O'Donnell
School of Biosystems Engineering
University College Dublin
Dublin, Ireland

Takashi Okazaki
Fisheries and Marine Technology Center
Hiroshima Prefectural Technology Research Institute
Hiroshima, Japan

Alfredo Palop
Departamento de Ingeniería de los Alimentos y del Equipamiento Agrícola
Universidad Politécnica de Cartagena
Cartagena, Spain

Ricardo Nuno Pereira
Centro de Engenharia Biológica
Universidade do Minho
Braga, Portugal

M. Consuelo Pina-Pérez
Consejo Superior de Investigaciones Científicas
Instituto de Agroquimica y Tecnologia de Alimentos
Valencia, Spain

Laura Pizzoferrato
Department of Food Science
Istituto Nazionale di Ricerca per gli Alimenti e la Nutrizione
Rome, Italy

Hosahalli S. Ramaswamy
Department of Food Science and Agricultural Chemistry
McGill University
Sainte Anne de Bellevue, Quebec, Canada

Catherine M.G.C. Renard
Sécurité et Qualité des Produits d'Origine Végétale
Institut National de la Recherche Agronomique
Avignon, France

Dolores Rodrigo
Consejo Superior de Investigaciones Científicas
Instituto de Agroquimica y Tecnologia de Alimentos
Valencia, Spain

Amelia C. Rubiolo
Consejo Nacional de Investigaciones Científicas y Técnicas
Instituto de Desarrollo Tecnológico para la Industria Química
Universidad Nacional del Litoral
Santa Fe, Argentina

Serpil Sahin
Department of Food Engineering
Middle East Technical University
Ankara, Turkey

Noboru Sakai
Department of Food Science and Technology
Tokyo University of Marine Science and Technology
Tokyo, Japan

Fernando Sampedro
Consejo Superior de Investigaciones Científicas
Instituto de Agroquimica y Tecnologia de Alimentos
Valencia, Spain

Brian W. Sheldon
Department of Poultry Science
North Carolina State University
Raleigh, North Carolina

Yujin Shigeta
Food Technology Center
Hiroshima Prefectural Technology Research Institute
Hiroshima, Japan

U.S. Shivhare
Department of Chemical Engineering & Technology
Panjab University
Chandigarh, India

Servet Gulum Sumnu
Department of Food Engineering
Middle East Technical University
Ankara, Turkey

Da-Wen Sun
Food Refrigeration and Computerised Food Technology
School of Biosystems Engineering
University College Dublin
Dublin, Ireland

Arthur A. Teixeira
Department of Agricultural and Biological Engineering
Institute of Food and Agricultural Sciences
University of Florida
Gainesville, Florida

José António Teixeira
Centro de Engenharia Biológica
Universidade do Minho
Braga, Portugal

Brijesh K. Tiwari
Department of Food and Consumer Technology
Manchester Metropolitan University
Manchester, United Kingdom

Gary Tucker
Department of Food Manufacturing Technologies
Campden and Chorleywood Food Research Association
and
Baking and Cereals Processing Department
Campden BRI
Gloucestershire, United Kingdom

António Augusto Vicente
Centro de Engenharia Biológica
Universidade do Minho
Braga, Portugal

Lijun Wang
North Carolina A&T State University
Greensboro, North Carolina

Z. Jun Weng
JBT FoodTech
Madera, California

Yanyun Zhao
Department of Food Science and Technology
Oregon State University
Corvallis, Oregon

MATLAB® Disclaimer

Part I

Modeling of Thermal Food Processes

Part I

Modeling of Thermal Food Processes

1 Thermal Physical Properties of Foods

Adriana E. Delgado, Da-Wen Sun, and Amelia C. Rubiolo

CONTENTS

1.1 INTRODUCTION

Thermal food processes, whether electrical or conventional, may be broadly classified as unit operations in blanching, cooking, drying, pasteurization, sterilization, and thawing, and involve raising the product to some final temperature that depends on the particular objective of the process [1]. The design, simulation, optimization, and control of any of these processes require the knowledge of basic engineering properties of foods.

Thermophysical properties, in particular, are within the more general group of engineering properties and primarily comprise specific heat and enthalpy, thermal conductivity and diffusivity, heat penetration coefficient. Other properties of interest are initial freezing point, freezing range, unfreezable water content, heat generation (and evaporation), and the more basic physical property: density [2].

The *thermal properties* depend on the chemical composition, structure of the product, and temperature; however, the processing of the food and the method of measurement are important as well. The temperature range of interest to food engineers, that is, –50°C to 150°C, covers two areas of food engineering, the applications of heat and cold. At low temperatures, where the conversion of water into ice takes place, the change in thermophysical properties is dramatic for all water-rich foods. However, the upper end, which is the temperature range considered here is less dramatic. Despite that, products rich in fat will also show phase-change effects [2].

In general, for each property, the following information is normally sought: (1) reliable method(s) of measurement, (2) key food component(s), and (3) widely applicable predictive relationships [3]. Sensitivity tests have demonstrated the significance of the thermophysical properties [4]. For example, two thermal properties, thermal conductivity and specific heat, and two mechanical properties, density and viscosity, determine how a food product heats after microwave energy has been deposited in it [5].

The thermal properties treated in this chapter are specific heat, thermal conductivity, thermal diffusivity, and density. Since alternative methods to supply heat are considered in this book (e.g., radio frequency [RF] and microwave heating), the dielectric constant and dielectric loss factor are also taken into account.

1.2 DEFINITION AND MEASUREMENT OF THERMOPHYSICAL PROPERTIES

The measurement of the *thermal properties* has been notably described previously by many authors in literature [6–11] and will not be detailed here. Thermal properties of food and their measurement, data availability, calculation, and prediction have also been well described as one of the subjects undertaken within European Cooperation in Science and Technology (*COST*)*90*, the first food technology project of COST [2,12–14]. Dielectric properties have also been discussed within electrical properties in the other European Union concerted project, COST90bis [15]. It was concluded from the COST research projects that the physical properties of foods depend not only on the specific food material but also on the processing of the food and the method of measurement [16]. Therefore, a brief comment on the recommended methods used to measure thermal properties is given later.

For most engineering heat transfer calculations performed in commercial heating or cooling applications, accuracies greater than 2%–5% are seldom needed, because errors due to variable or inaccurate operating conditions (e.g., air velocity and temperature) would overshadow errors caused by inaccurate thermal properties [9]. Thus, precision and accuracy of measurement with regard to the application of the data are important factors to consider when selecting a measurement method.

1.2.1 Specific Heat Capacity

The *heat capacity* (c) of a substance is defined as the amount of heat necessary to increase the temperature of 1 kg of material by 1°C at a given temperature. It is expressed as Joules per kilogram Kelvin in SI units and it is a measure of the amount of heat to be removed or introduced in order to change the temperature of a material. If ΔT is the increase in temperature of a given mass, m, as a consequence of the application of heat, Q, the calculated specific heat is the average, that is,

$$c_{aveg} = \frac{Q}{m\Delta T} \tag{1.1}$$

If ΔT is small and $q = Q/m$, Equation 1.1 gives the instantaneous value of c [11]:

$$c = \lim_{\Delta T \to 0} \left(\frac{q}{\Delta T} \right)_T = \left(\frac{dq}{dT} \right)_T \tag{1.2}$$

If both a temperature change and a thermal transition are included, this specific heat is then called the *apparent specific heat.*

There exist two specific heats, c_p and c_v; the former is for constant pressure process and the latter for constant volume process. The specific heat for solids and liquids is temperature dependant, but does not depend on pressure, unless very high pressures are applied. For the food industry, c_p is commonly used, as most of the food operations are at atmospheric pressure. Only with gases it is necessary to distinguish between c_p and c_v.

The methods often used to measure the *specific heat* and also the *enthalpy*, are the *method of mixing*, the *adiabatic calorimeter*, and the *differential scanning calorimeter* (DSC). Over the years, Riedel [17–20] has published extensively on both the specific heat and enthalpy of a wide range of food products [21]. He used the adiabatic calorimeter, which is a method that can provide high precision but involves long measuring times and difficulties for preparing the sample. Although DSC has the disadvantages such as necessity of calibration and requirement of small samples and good thermal contact, it is the method generally recommended for measuring specific heat.

1.2.2 Enthalpy

Enthalpy is the heat content or energy level in a system per unit mass and the unit is Joules per kilogram (J/kg). It can be written in terms of specific heat as follows [10]:

$$H = \int c\, dT \tag{1.3}$$

The specific heat and enthalpy are properties of state. Enthalpy has been used more for quantifying energy in steam than in foods. It is also convenient for frozen foods because it is difficult to separate *latent* and *sensible heats* in frozen foods, which often contain some unfrozen water even at very low temperatures [9].

1.2.3 Thermal Conductivity

Thermal conductivity (k) represents the basic thermal transport property and it is a measure of the ability of a material to conduct heat. It is defined by the basic transport equation written as *Fourier's law* for heat conduction in fluids or solids, which is integrated to give

$$\frac{q}{A} = \frac{k(T_1 - T_2)}{z} \tag{1.4}$$

where

q is the heat transfer rate in Watts (W)
A is the cross-sectional area normal to the direction of the heat flow in square meter (m^2)
z is the thickness of the material in meters (m)
T_1 and T_2 are the two surface temperatures of the material
k is the thermal conductivity in Watts per meter Kelvin (W/m K)

Thermal conductivity is an intrinsic property of the material. For hygroscopic moist porous materials k is a strong function of the porosity of the material [22]. Heat transfer in porous moist materials may occur by heat conduction and mass transfer simultaneously, so the *effective thermal conductivity* is used to precisely evaluate the coupled heat and moisture transfer through porous materials.

For measuring the *thermal conductivity* of foods the *line heat source probe* has frequently been used and it is the method recommended for most food applications. This technique is implemented in two designs: the hot-wire k apparatus and the k probe. The *hot-wire apparatus* is widely accepted as the most accurate method for measuring the k of liquids and gases, but is more complicated to adapt to instrumentation and more difficult to use in solid materials [23]. The k *probe* method is fast, uses small samples, and requires known and available instrumentation [24]. However, it is not well suited for nonviscous fluids due to convection currents that arise during probe heating [9]. The technique has also been used to measure k of moist porous foods materials at elevated temperatures [25]. Although the *thermal conductivity* probe is derived from an idealized heat transfer model, there are unavoidable differences between the real probe and the theoretical model, which cause errors in the application of the k probe and lead the researchers to corrective measures to either compensate or minimize these errors [23]. Various design parameters of the k probe have been analyzed and recommendations are given for applications to nonfrozen food materials [23]. As a result, it is recommended that users should design their thermal conductivity probes using the highest acceptable error for their intended application.

A dual-needle heat pulse (DNHP) probe containing two parallel needles spaced 6 mm apart was developed to determine simultaneously the thermal conductivity, thermal diffusivity, and specific heat [26]. One needle contains a line heat source and the other is a thermocouple. By applying a short duration pulse and monitoring the temperature response to the heat pulse measured by the thermocouple, the thermal conductivity and thermal diffusivity of the sample are determined simultaneously, and then the volumetric heat capacity is calculated, which is the product of the heat capacity and the density. The DNHP has successfully been used first to rapidly and accurately measure the thermal properties of soils, and its use was further evaluated with selected foods (e.g., apple, beef round, and white and yolk egg) [27]. The DNHP probe was found to be appropriate for the accurate measurement of thermal conductivity, thermal diffusivity, and specific heat of food products, with measured values within 7%–8% with respect to those reported in the literature. In addition to its economical advantage, measurements are rapid.

Recently, the feasibility to use the transient plane-source (TPS) method for simultaneously measuring k and α of foods was also evaluated [28]. The TPS sensor is made of a calibrated probe constructed with a resistance temperature detector (RTD). Results showed that TPS measured k and α values of standard materials (e.g., water, mineral oil, ethanol, ethylene glycol, and olive oil), were generally within 0.5%–10% from the standard values in literature. For food materials such as beef, chicken, flour, and apple, the deviation of experimental results with respect to published data were within 0.2%–15%. Results of this study suggested that the TPS method is an accurate method for simultaneously measuring thermal conductivity and thermal diffusivity of foods, though more studies are needed for improving the sensor construction and accuracy of the measurement.

1.2.4 Thermal Diffusivity

Thermal diffusivity (α) determines how rapidly a heat front moves or diffuses through a material and can be defined as

$$\alpha = \frac{k}{\rho c_p} \tag{1.5}$$

where

ρ is the density (kg/m^3)

c_p is the specific heat at constant pressure (J/kg K)

k the thermal conductivity (W/m K) of the material

The SI unit of α is square meters per second (m^2/s). Thermal diffusivity measures the ability of a material to conduct thermal energy relative to its ability to store thermal energy; products of large α will respond quickly to changes in their thermal environment, while materials of small α will respond more slowly.

The *thermal diffusivity of unfrozen foods* ranges from about 1.0×10^{-7} to 1.5×10^{-7} m^2/s and does not change substantially with moisture and temperature because any changes of k are compensated by changes of the density of the material [22]. In microwave heating, for example, the fact that thermal diffusivities of unfrozen foods are similar means that foods heat similarly for equivalent energy deposition [5].

The measurement of thermal diffusivity can be divided into two groups: *direct measurement* and *indirect prediction* [10]. The direct methods for experimental determination include the use of a cylindrical object and time-temperature data, the use of a spherical object and time-temperature data, and the use of a thermal conductivity probe. Indirect prediction, that is, the estimation from experimentally measured values of thermal conductivity, specific heat and density, is the recommended method to determine α [9]. Since specific heat can be estimated with sufficient accuracy from the product composition, the experimental determinations are the thermal conductivity and the mass density [9].

1.2.5 Density

Density (ρ) is a physical property widely used in process calculations. Density of food materials depends on temperature and composition and it is the unit mass per unit volume:

$$\rho = \frac{m}{V} \tag{1.6}$$

The SI unit of density is kilograms per cubic meter (kg/m^3). Different ways of density definition, measurement, and usage in process calculations is well discussed in literature [10]. In some cases the *apparent density* is used as *bulk density*. The apparent density is the density of a substance including all pores remaining in the material, while bulk density is the density of a material when packed or stacked in bulk [10]. Hence, it is necessary to mention the definition of density when presenting or using data in process calculations [10].

The density of a food product is measured by weighing a known volume of the product. Since food products are different in shape and size, the accurate measurement of volume can be challenging [29]. There exist different techniques for laboratory measurement of density, which involve the use of pycnometer (for fluids and solids), hydrostatic balance (for fluids and solids), Mohr–Westphal balance (for fluids), x-ray technique (for solids), and resonator frequency (for gases and fluids) [30]. An easy procedure recommended for measuring ρ in meat is to add a known mass (approximately 5 g) of sample to a calibrated 60 mL flask, and complete the volume with distilled water at 22°C [11]. Density is evaluated from the following equation:

$$\rho = \frac{m}{V_s} = \frac{m}{60 - V_w} \tag{1.7}$$

where m and V_s are the mass and volume of the sample respectively, calculated from the added water volume V_w.

The bulk and solid densities can be measured experimentally and they can be used to estimate the bulk porosity. *Porosity* is an important physical property, since its changes during processing may have significant effects on the heat and mass transport properties (e.g., thermal conductivity), and thus the quality (nutritive and sensory) of the food product [4].

1.2.6 Dielectric Constant and Dielectric Loss Factor

Electrical properties of foods are of general interest as they correlate the physical attributes of foods with their chemical ones, and they are of practical interest in optimization and control of *dielectric heating processes* [31]. The most intensively investigated electrical properties of foods have been the *relative dielectric constant* (ε') and *loss factor* (ε''). These dielectric properties determine the energy coupling and distribution in a material subjected to dielectric heating [32]. The dielectric constant or "capacitivity" is related to the material's capacitance and its ability to store electrical energy from an electromagnetic field and it is a constant for a material at a given frequency. The dielectric loss is related to a material's resistance and its ability to dissipate electrical energy from an electromagnetic field [1]. A material with high values of the dielectric loss factor absorbs energy at a faster rate than materials with lower loss factors [33]. It should always be remembered that *dielectric properties* in a time-varying electric field are complex, that is to say they have two components—real, ε', and imaginary, ε'' [34]. The dielectric loss factor in turn is the sum of two components: *ionic*, ε''_σ, and *dipole*, ε''_d *loss*. The ratio of the dielectric loss and dielectric constant is called the loss tangent or the dissipation (power) factor of the material (tan δ). The *permittivity*, which determines the dielectric constant, the dielectric loss factor, and dielectric loss angle, influences the *dielectric heating* [33].

The relative ionic loss, ε''_σ, is related to the electrical conductivity of a food material (σ) with the following relationship [35]:

$$\varepsilon''_\sigma = \frac{\sigma}{2\pi f \varepsilon_0} \tag{1.8}$$

where

ε_0 is the permittivity of free space (8.854×10^{-12} F/m)
f is the frequency of the electromagnetic waves (Hz)

Power *penetration depth* (d), one of the essential dielectric processing parameters, is defined as the distance that the incident power decreases to 1/e (e=2.718) of its value at the surface [35]. The penetration depth is calculated from the dielectric constant and dielectric loss data by using the following expression [36]:

$$d = \frac{c_0}{f} \frac{0.1125394}{\left\{\varepsilon'\left[\left(1+\tan^2(\varepsilon''/\varepsilon')\right)^{1/2}-1\right]\right\}^{1/2}} \tag{1.9}$$

where c_0 is the speed of light in vacuum (3×10^8 m/s). An approximation for determining the penetration depth that holds for virtually all foods is given by [5]

$$d = \frac{\lambda_0 \sqrt{\varepsilon'}}{2\pi\varepsilon''} \tag{1.10}$$

where λ_0 is the free space microwave wavelength, which can be in any units of length. For 2450 MHz, λ_0 is equal to 122 mm. Knowledge of the penetration depth can help in selecting a correct sample thickness to guide the microwave or RF *heating* processes [35]. It has been reported that for mashed potatoes, for example, after calculating the penetration depth, *microwave heating* is advisable for packages with relatively smaller thickness (e.g., 10–20 mm for two-sided heating), and *RF heating* should be applied for packages and trays with large institutional sizes (e.g., 40–80 mm depth) [35].

The *food map* plot for ε'' vs. ε' with constant penetration depth lines (d) is a recommended way to illustrate the dielectric properties [5]. The dielectric properties of liquid and semisolid food products depend primarily on their moisture, salts, and solid contents. However, the extent to which each of these constituents affects food dielectric behavior depends very much on the processing frequency and the temperature history of the product [1]. In an experiment about the effect of sample heating procedures (temperature being raised in 10°C intervals or being raised directly to a set point, 121°C) on the results of measurements for whey protein gel, cooked macaroni noodles, cheese sauce, macaroni and cheese, it was found that the heating procedures did not affect the results of the dielectric property measurements for the materials tested [37].

Dielectric properties can be measured by the methods reviewed within the collaborative research project COST90bis [34]. The measuring methods can vary even in a given frequency range. Four groups of measurement methods can be considered: lumped circuit, resonator, transmission line, and free space methods [38]. One of the most commonly used measuring methods employs resonant cavities, since they are very accurate but can also be sensitive to low-loss tangents [33]. The method can be easily adapted to high (up to 140°C) or low temperatures (–20°C). Another popular technique is the open-ended coaxial probe method [33], because it requires no particular sample shapes and offers broad-brand measurement [35]. Venkatesh and Raghavan [39] summarized the status of the research in this area, and provided an extensive review of the literature on measuring techniques and comparison, and potential applications of dielectric properties.

1.3 DATA SOURCES ON THERMOPHYSICAL PROPERTIES

Thermal property data have been measured since the late 1800s, with almost two-thirds of that being published in the 1950s and 1960s [9]. A problem that industrial users normally face is that the data available are often of limited value because information about composition, temperature, error in measurement, etc., is not reported. Furthermore, moisture and air content ranges tend to cover a narrow band and thermophysical data at both elevated and low temperatures are sparse [40]. Though information available is only partial, such data are very useful for preliminary design, heat transfer calculations, and food quality assessment.

Different ways can be recognized to obtain information on thermal properties, namely, (1) original publications, (2) summarizing publications such as articles, monographs, and books, (3) bibliographies and compilations of literature references, (4) handbooks and data books, and finally (5) computerized data banks [2]. The recommendation of the *COST90 project* in 1983 was to replace the first four choices mentioned earlier with one compilation of basic data and thermophysical properties by product, containing calculated values and experimental data as references, accompanied also by a reference to their source in case more information was needed. Computer programs such as *COSTHERM* and *FoodProp* were developed for the thermal properties of foods. In turn, a computer program based on the computer program COSTHERM was developed at the Budapest University of Technology and Economics [16]. This program [16] considered properties of new food products, especially those of liquid products (e.g., milk products, oils, and fruit juices), and it was also improved to predict further characteristics, mainly engineering properties. The data were organized and processed to make them suitable for use in computer aided design packages [16]. *Food Properties Database* 2.0 for Windows [41] was the first database assembled in the United States [40]. This database includes over 2400 food property combinations and over 2450 food materials; it also features a collection of *mathematical models* that have been proposed for predicting food property values. In the European Union, there is an online database available for physical properties of agro-food materials (www.nelfood.com) [40]. The database contains five main categories of data: thermal, mechanical (rheological and textural), electrical and dielectric, sorption and diffusional, and optical (spectral and color) properties of foods. The future work of *NELFOOD database* will improve the predictive features of the database. The novelty of the database is that it specifies both the experimental method and the descriptions of the food, and also provides a score (four-point scale) indicating the quality of the method specification and food definition. This characteristic in particular is very helpful when selecting appropriate values or models from many sources available, since it is not only the data but also the interpretation and application that are equally important.

References to important sources of information on thermophysical properties have been published in literature [2]; Table 1.1 presents additional information on data recently available. In particular, the *Food Properties Handbook*, second edition [10], contains exhaustive and extensive data on physical, thermal, and thermodynamic properties of foods, including measurement, predictions, and applications. The database of the Food Research Institute of Prague contains more than 16,000 manuscripts, which can be accessed in part through the NELFOOD database. Average values and variation ranges of thermal conductivity of more than 100 food materials, classified into 11 food categories were also compiled [42]. More than 95% of these data are in the ranges of 0.03–2 W/m K for thermal conductivity, 0.01–65 kg/kg db for moisture content and −43°C to 160°C for temperature range.

Very few thermal diffusivity data are available; however, thermal diffusivity can be calculated from specific heat, thermal conductivity, and mass density [9] if they are available as shown in Equation 1.5.

A compilation of dielectric property data has been presented (dielectric constant ε', dielectric loss ε'', and penetration depth d) for a wide range of fruits, vegetables, meat, and fish for the frequency range of 2000–3000 MHz [36]. The references for each set of data and the type of measurement used are provided as well. In some cases, the composition data are from sources other than those from which the data were taken. The amount of information available on the dielectric properties of foods in the RF range is limited in comparison with data at microwave frequencies. Dielectric properties of selected foods in the RF range 1–200 MHz have been reported along with information related to data sources [33]. Except for one work [46], very few studies provide the dielectric properties above 65°C [47]. Dielectric properties of whey protein gel, cooked macaroni

TABLE 1.1
Literature on Thermal Physical Properties of Foods

Source	Information
Nesvadba et al. [40], http://www.nelfood.com database	Thermal, mechanical, electrical, diffusional, and optical properties. Data available in tables and equations as function of temperature, pressure, composition, etc.
http://www.vupp.cz/envupp/research.htm database	Physical properties data at the Institute of Food Research Prague
Krokida et al. [42], Article	Compilation of thermal conductivity data with range of material moisture content and temperature
Singh [41], Database	Experimental values, mathematical models of food properties along with literature citations
Rahman [10], Chapters 13 through 20, Textbook	Density, specific heat, enthalpy and latent heat, thermal conductivity, and thermal diffusivity Measurement, experimental values, and prediction models
Datta et al. [36], Chapter 9	Dielectric property data of fruits, vegetables, meat, and fish
Sosa-Morales et al. [43], Article	Dielectric property data of fruits, vegetables, meat and seafood, nuts, bread, egg, and liquid foods
ASHRAE [44], Chapter 30	Specific heat, thermal diffusivity, and thermal conductivity
Singh [45], Chapter 5	Specific heat, thermal diffusivity, and thermal conductivity

noodles, cheese sauce, and macaroni and cheese, at both microwave and radio frequencies (27, 40, 915, and 1800 MHz) over a temperature range from 20°C to 121.1°C were reported [37]. Recently, a comprehensive review on dielectric properties for foods after the year 2000, along with the methods for their determination and the factors that influence on dielectric properties have been published [43]. This review [43] provides information of different foods such as (1) fruits and vegetables; (2) flour, dough, and bread; (3) nuts; (4) coffee grains; (5) meat, fish, and seafood; (6) dairy products; (7) eggs and egg products; and (8) liquid fluids.

It is important to note that the data file for specific heat capacity published in ASHARE (American Society of Heating, Air Conditioning and Refrigeration Engineers) Handbook of Fundamentals [44] is not experimentally measured values, instead they are calculated from equations based on water content, which can result in considerable error for calculations.

In summary, considerable data on thermal properties have been published to the present, though in many cases the information available is only partial. Therefore, when reporting thermal property data, researchers should provide a detailed and informative description of the product tested (variety, chemical composition, pretreatment, etc.), the experimental procedures (process variables), and the data obtained [9].

1.4 PREDICTIVE EQUATIONS

Thermal processing was the first food process to which *mathematical modeling* was applied, because of its great importance to the public health safety and the economics of food processing [4]. Modeling requires the information of the mean or effective values of the components together with the representation of the physical structure [11]. Because of the large variety of foods and formulations, it is almost impossible to experimentally measure the thermal properties for all possible conditions and compositions. Therefore, the most viable option is to predict the thermophysical properties of foods using mathematical models. However, if more accuracy is required, a good solution is the experimental determination.

Water as a major component in foods affects safety, stability, quality, and physical properties of food. Analysis of published data shows that the less water there is in the material the more discrepancies between predicted and measured values that exist [32]. It seems that discrepancies arise from the treatment of whole water in food as bulk water, without taking into account the interactions between water and food components, which must affect thermal properties.

Most of the thermal property models are empirical rather than theoretical; that is, they are based on statistical curve fitting rather than a theoretical derivation involving heat transfer analyses [9]. Artificial neural networks (ANNs) in particular are promising tools for application to process identification and controls owing to the ability to model functions with accuracy. They also offer a cost-effective method of developing useful relationships between variables, when the experimental data of these variables are available [48]. ANNs are optimization algorithms, which attempt to mathematically model the learning process by using basic foundations and concepts inherent in the learning processes of humans and animals [49]. Neural network modeling has generated increasing acceptance in the estimation and prediction of food properties and process related parameters (e.g., thermal conductivity of fruits and vegetables, bakery products, milk, and specific heat and density of milk) [48–50].

A comprehensive compilation of predictive equations of thermal physical properties of foods is provided in literature [10–12,51]. From the many published equations, some examples of commonly used correlations are given later.

1.4.1 Specific Heat

Water has a high specific heat in comparison to other food components; hence, even small amounts of water in foods affect its specific heat substantially [32]. The simplest *specific heat model* for low-fat foods has the following form [32]:

$$c_p = a + bx_w \tag{1.11}$$

where

a and b are constants that depend on the product and temperature
x_w is the water content in decimals
c_p is in Joules per kilogram degrees Centigrade (J/kg°C)

Table 1.2 lists the constants a and b, the moisture content and temperature range for a great variety of foods [10].

It is generally accepted that specific heat obeys the rules of additivity. This means that the specific heat of a product is equal to the sum of the fractional specific heats of the main constituents [32]. Using additivity principle specific heat can be calculated as follows:

$$c_p = \sum c_{pi} x_i \tag{1.12}$$

where

c_{pi} is the specific heat at constant pressure of the food component i
x_i is the mass fraction of the ith food component (water, x_w; protein, x_p; fat, x_f; carbohydrate, x_c; and ash, x_{as})

The thermal properties of the major food components as a function of temperature can be found in literature [52]. When the food contains a large amount of fat, the specific heat is made up from the contribution of the fat fraction and also from the phase transition of the fat.

TABLE 1.2
Linear Models for Specific Heat of Foods

Material	a	b	x_w Range	T Range (°C)
Foods	837	3349		
Fish and meat	1670	2500	≤0.25	
Fruits and vegetables	1670	2500	≥0.25	
Orange (navel)	1452	2515	0.00–0.89	
Lentil	1030	4080	0.02–0.26	10–80
Potato	904	3266	≥0.5	
Potato	1645	1830	0.20–0.50	
Milk products				
Cheese (processed)	1918	2258	0.425–0.684	40
Dulce de leche	1790	2640	0.28–0.60	30–50
Sorghum and cereals	1400	3200	Low water	
Sorghum	1396	3222	0.00–0.30	
Wheat (hard red spring)	1090	4046	0.00–0.40	0.6–21.1
Soybeans	1637	1927		
Soy flour (defatted)	1748	3363	0.092–0.391	130

Source: Rahman, S., *Food Properties Handbook*, 2nd edn., CRC Press, Boca Raton, FL, 2009, pp. 398–697.

The specific heat above the *initial freezing point* can be calculated if the c_p of the fat is assumed to be half of the c_p of water, and the c_p of the solids, which have similar specific heats, is assumed to be 0.3 times that of water c_p [21,53]:

$$c_p = 4180\,(0.5x_f + 0.3x_s + x_w) \tag{1.13}$$

Equation 1.13 gives a rough estimate of the specific heat above the freezing point of the product.

An empirical equation for the calculation of c_p of some different foods is given as follows [54]:

$$c_p = 4187[x_w + (\gamma + 0.001T)(1 - x_w) - \beta\exp(-43x_w^{2.3})] \tag{1.14}$$

where the temperature T is in degrees Centigrade (°C), and the numerical values of the coefficients in Equation 1.14 for some foods are

$\gamma_{beef} = 0.385$	$\beta_{beef} = 0.08$
$\gamma_{white\ bread} = 0.350$	$\beta_{white\ bread} = 0.09$
$\gamma_{sea\ fish} = 0.410$	$\beta_{sea\ fish} = 0.12$
$\gamma_{low\text{-}fat\ cheese} = 0.390$	$\beta_{low\text{-}fat\ cheese} = 0.10$

If detailed composition data are not available, the following simpler model can be used [55]:

$$c_p = 4190 - 2300x_s - 628x_s^3 \tag{1.15}$$

where

x_s is the mass fraction of solids

c_p is in Joules per kilogram degrees Centigrade

Gupta [56] developed the following correlation to predict the specific heat of foods as a function of moisture content and temperature considering 15 types of foods [10]:

$$c_p = 2476.56 + 2356x_w - 3.79T \tag{1.16}$$

where

T is in Kelvin (K)

c_p in Joules per kilogram (J/kg K)

x_w ranges from 0.001 to 0.80 and T from 303 to 336 K

Equation 1.16 gives fairly good values for substances like sugar, wheat flour, starch, dry milk, rice, etc. For substances containing higher moisture (more than 80%), Equation 1.16 shows higher deviations from reported values.

The specific heat is related to the dielectric properties and the temperature increase (ΔT) through the following equation [33]:

$$\Delta T = \frac{2\pi t f \varepsilon_0 \varepsilon' \tan\delta V^2}{c_p \rho} \tag{1.17}$$

where

t is the temperature rise time (s)
ε_0 is the dielectric constant of free space
V is the electric field strength, equal to voltage/distance between plates (V/cm)

Equation 1.17 shows that the specific heat affects the resulting ΔT. A material with greater specific heat will undergo a smaller temperature change since more energy is required to increase the temperature of 1 g of the material by 1°C [33]. In a multicomponent product, where the components have wide differences in dielectric and thermal properties, it is often necessary to balance both sets of properties in order to approach equal heating for each component. It is usually more fruitful to adjust specific heat rather than dielectric properties to obtain such a balance [5].

1.4.1.1 Specific Heat of Juices

The specific heat for fruit juices with water content greater than 50% can be calculated as follows [57]:

$$c_p = 1674.7 + 25.12x_w \tag{1.18}$$

The specific heat (J/kg °C) of clarified apple juice as a function of concentration (6°Brix–75°Brix) and temperature (30°C–90°C) can be estimated from [11,58]

$$c_p = 3384.57 - 18.1774Bx + 2.3472\,T \tag{1.19}$$

Equation 1.19 gives a good fit (correlation coefficient $R^2 = 0.99$) of the experimental data in the entire range of concentrations and temperatures under consideration.

Alvarado [59] developed a general correlation using 140 data for fruit pulps, with moisture content ranging from 0.012 to 0.945 and temperatures from 20°C to 40°C, which is given as follows [10]:

$$c_p = 1560\,[\exp(0.9446x_w)] \tag{1.20}$$

1.4.1.2 Specific Heat of Meat

Sanz et al. [51,60] presented a list with experimental values and the most appropriate equations to calculate the specific heat, the thermal conductivity, thermal diffusivity, and density of meat and meat products. The following general correlation for meat products for temperatures above the initial freezing point is proposed:

$$c_p = 1448\,(1 - x_w) + 4187x_w \tag{1.21}$$

c_p in lamb meat can be estimated with the following expression [61]:

$$c_p = 979 + 3175.4x_w \tag{1.22}$$

where

x_w is the moisture content in % wet basis
c_p is in Joules per kilogram degrees Centigrade

AbuDagga and Kolbe [62] measured and modeled the apparent specific heat of salt-solubilized surimi paste with 74%, 78%, 80%, and 84% moisture content in the temperature range 25°C–90°C. The following linear model was fitted to the experimental data as function of the temperature and moisture content:

$$c_p = 2330 + 6\,T + 14.9x_w \quad (1.23)$$

where

c_p is in Joules per kilogram degrees Centigrade

The moisture content, x_w, is in percent wet basis

Equation 1.23 can be considered as a workable engineering model in most design circumstances.

1.4.1.3 Specific Heat of Fruits and Vegetables

The specific heat (J/kg °C) of Golden Delicious apples for the temperature range from −1°C to 60°C can be estimated with the following correlation [63]:

$$c_p = 3360 + 7.5T \quad (1.24)$$

and for Granny Smith,

$$c_p = 3400 + 4.9T \quad (1.25)$$

Hsu et al. [64] proposed the following equation to predict c_p (J/kg °C) for pistachio with water contents ranging from 5% to 40% on wet basis [11]:

$$c_p = 1074 + 27.79x_w \quad (1.26)$$

c_p for potatoes (Desiree variety) can be estimated with the following correlation, which was generated from data obtained with DSC measurements, for a temperature range from 40°C to 70°C and moisture content from 0% to 80% on wet basis [65]:

$$c_p = 4180(0.406 + 1.46\times10^{-3}T + 0.203x_w - 2.49\times10^{-2}x_w^2) \quad (1.27)$$

The main relative percentage deviation of Equation 1.27 is equal to 3.36%, which indicates a reasonably good fit for practical purposes.

1.4.1.4 Specific Heat of Miscellaneous Products

For milk, the following expression is proposed [66] at temperatures above freezing [9]:

$$c_p = 4190\,x_w + [(1370 + 11.3T)(1 - x_w)] \quad (1.28)$$

where

T is in degrees Celsius

c_p is in Joules per kilogram degrees Centigrade

c_p (J/kg °C) for processed cheese can be estimated from the general correlation [67]:

$$c_p = 4101 + 1.2T - (1673 + 0.27T)\,x_f - (2716 - 1.1T)\,x_{ns} \quad (1.29)$$

Equation 1.29 is applicable to a temperature range from 40°C to 100°C, from 0.316 to 0.575 mass fraction of solutes, and from 0.135 to 0.405 mass fraction of nonfat solids (NFS), x_{ns} [10].

Christenson et al. [68] assumed that the dependence of the specific heat of bread with moisture follows a mass fraction model [11]:

$$c_p = c_{pw}\,x_w + c_{pdry\,solid}\,(1 - x_w) \quad (1.30)$$

where c_p of dry solid is given by

$$c_{pdry\,solid} = 98 + 4.9T \quad (1.31)$$

T is in Kelvin (K) for 298–358 K temperature range and c_p is in J/kg K.

The following empirical equation that incorporates temperature, moisture content, and protein content was developed to predict the specific heat of cereal flours and starches between 20°C and 110°C, moisture content between 0% and 70% dry basis and protein content from 7.5% to 16% dry basis [69]:

$$c_p = 1.056 + 0.0058T + 3.71x_w - 2.34(x_w)^2 + 0.62x_p \quad (1.32)$$

where

c_p is in J/g°C

x_w and x_p are the moisture and protein mass fraction in wet basis

Equation 1.32 predicts the specific heat of flours (wheat, corn, and rice) with an average absolute error of 4.3% with respect to the specific heat of cereal flours and starches reported in literature. Inclusion of a protein dependent term becomes significant for predicting the specific heat at low moisture content and low protein content, while the specific heat of water dominates at high moisture levels.

1.4.2 Enthalpy

The enthalpy content is a relative property. For temperatures above the initial freezing point, it can be evaluated with the following general expression [12]:

$$H = \sum x_i \int_0^T c_{pi}\,dT \quad (1.33)$$

which is valid at atmospheric pressure. If the specific heats, c_{pi}, are independent of temperature, then the following equation should be used:

$$H = T \sum x_i c_{pi} \quad (1.34)$$

1.4.3 Thermal Conductivity and Thermal Diffusivity

Thermal conductivity and thermal diffusivity strongly depend on moisture content, temperature, composition, and structure or physical arrangement of the material (e.g., voids and nonhomogeneities).

The thermal conductivity of fluid foods is a weak function of their composition, and simple empirical models can be used for its estimation. However, to model the thermal conductivity of solid foods, structural models are needed, due to differences in micro- and macrostructure of the heterogeneous materials [22,70]. For example, for meat the thermal conductivity along the meat fibers is different from what it is across the fibers. These differences were considered by Kopelman [71]. His models for thermal conductivity are presented by Heldman and Singh [72].

Porous foods are difficult to model because of the added complexity of the void spaces. The *effective thermal conductivity* depends on the heat flow path through solids and voids; it may be affected by pore size, pore shape, percent porosity, particle-to-particle resistance, convection within pores, and radiation across pores [9]. At low moistures, the thermal conductivity and thermal diffusivity of porous foods are nonlinear functions of the moisture content, due to significant changes of bulk porosity; at moistures higher than 30%, k increases linearly with the moisture content [4]. A review of thermal conductivity values and mathematical models for porous foods was given by Wallapapan et al. [73]. Despite the attempts in developing structural models to predict the thermal conductivity of foods, a generic model does not exist at the moment [42].

Since theoretical models have a number of limitations for application in food material, empirical models are popular and widely used for food process design and control, even though they are valid only for a specific product and experimental conditions [74].

Similar to specific heat, most of the models used to calculate the thermal conductivity of foods with high moisture content have the following form [32]:

$$k = c_1 + c_2 x_w \quad (1.35)$$

where c_1 and c_2 are constants. At $x_w = 1$ most of the equations converge on the thermal conductivity of water. Predictions agree at high water contents, and discrepancies between experimental and predicted values are marked at low water contents. Table 1.3 lists some simple equations, which

TABLE 1.3
Simple Thermal Conductivity Equations for Foods

Material	Model	Source
Fruits and vegetables	$k = 0.148 + 0.493x_w$ $0 < x_w < 0.60$	[9]
Tomato paste	$k = 0.029 + 0.793x_w$ $0.538 < x_w < 0.708$, $T = 30°C$	[10]
Tomato paste	$k = -0.066 + 0.978x_w$ $0.538 < x_w < 0.708$, $T = 40°C$	[10]
Tomato paste	$k = -0.079 + 1.035x_w$ $0.538 < x_w < 0.708$, $T = 50°C$	[10]
Spring wheat (hard red)	$k = 0.129 + 0.274x_w$ $0.014 < x_w < 0.148$, $32°C < T < 60°C$	[10]
Dairy products and margarine	$k = 0.141 + 0.412x_w$	[11]
Meat and fish	$k = 0.080 + 0.52x_w$ $0.60 < x_w < 0.80$, $0°C < T < 60°C$	[12]
Fish	$k = 0.0324 + 0.3294x_w$	[12]
Minced meat	$k = 0.096 + 0.34x_w$	[12]
Fruit juice	$k = 0.140 + 0.42x_w$	[12]
Sorghum	$k = 0.564 + 0.0858x_w$	[12]
Pistachio	$k = 0.0866 - 0.2817 \times 10^{-3}x_w$ $5\% < x_w < 40\%$	[64]

only takes into account the water content of the food (x_w is in decimal form). Linear models, similar to Equation 1.3, are also commonly used to estimate the thermal diffusivity with the moisture content or the temperature as variables [10]. Riedel [75] proposed the following equation for estimating α [12]:

$$\alpha = 0.088\times10^{-6} + \left(\alpha_w - 0.088\times10^{-6}\right)x_w \tag{1.36}$$

where α_w is the thermal diffusivity of water.

Several composition models have been proposed and discussed in literature [9,10,12]. Additivity principles can also be used to calculate the thermal conductivity of many liquids and solid foods, by taking into account the water, protein, carbohydrate, fat, and sometimes also the ash content. The following equations are recommended for estimating k of foods that are not porous [9]:

$$k = 0.61x_w + 0.20x_p + 0.205x_c + 0.175x_f + 0.135x_{as} \tag{1.37}$$

and

$$k = 0.58x_w + 0.155x_p + 0.25x_c + 0.16x_f + 0.135x_{as} \tag{1.38}$$

For the thermal diffusivity Hermans [76] proposed the following equation, which is slightly dependent on temperature (K) [12]:

$$\alpha = \left(0.0572x_w + 0.0138x_f + 0.0003T\right)\times10^{-6} \tag{1.39}$$

Although Equation 1.37 is for liquids foods [77], it is relatively accurate for solid foods [9]. Equation 1.38 was fitted to more than 430 liquids and solids foods with satisfactory results; but it is not accurate for porous foods containing air (e.g., apples). The temperature effect is not included in Equations 1.37 and 1.38, so they are valid at the fitting region approximately at 25°C.

For heterogeneous foods, the effect of geometry must be considered using structural models [22]. In general, the models assume that foods consist of two different components physically orientated so that heat travels either in parallel or in perpendicular through each of them [32,70]. In dispersed systems, the volume fraction of the dispersed or discontinuous component and the thermal conductivity of the dispersed or continuous component are considered [32]. These models are not widely applicable since foods tend to have more than two components and they are not arranged in simple configurations [9]. The continuous dispersed components model may be expanded to more than two components, and then appears to have application for some food systems [9].

Saravacos and Maroulis [22] presented another approach to estimate the thermal conductivity as a function of moisture content and temperature, which is given in the following. To develop the model, it was assumed that a material of intermediate moisture content consists of a uniform mixture of two different materials—a dried material and a wet material with infinite moisture—and that k can be estimated using a two-phase structural model. The temperature dependence of the thermal conductivity is then modeled by an Arrhenius-type model, and thus, the proposed mathematical model has the following form:

$$k = \frac{1}{1+X}k_0 \exp\left[-\frac{E_0}{R}\left(\frac{1}{T}-\frac{1}{T_r}\right)\right] + \frac{X}{1+X}k_i \exp\left[-\frac{E_i}{R}\left(\frac{1}{T}-\frac{1}{T_r}\right)\right] \tag{1.40}$$

where

X (kg/kg db) is the moisture content
T (°C) is the material temperature
T_r is the reference temperature at 60°C
R is the ideal gas constant (0.0083143 kJ/mol K)

The adjustable parameters are as follows: k_0 (W/m K) is the thermal conductivity at $X=0$, $T=T_r$, k_i (W/m K) is the thermal conductivity at $X=\infty$, $T=T_r$, E_0 (kJ/mol) is the activation energy for heat conduction in dry material at $X=0$, and E_i (kJ/mol) is the activation energy for heat conduction in wet material at $X=\infty$.

Table 1.4 shows the results of the parameter estimation by applying the model to all the data of each material, regardless of the data sources. Since the results are not based on the data of only one source, the accuracy is very high [22].

1.4.3.1 Thermal Conductivity of Meat

Few data are available on the thermal conductivity of meat in the cooking temperature range. For predictive purposes Baghe-Khandan et al. [78] developed models to estimate thermal conductivities at temperatures up to 90°C and heating rates of <0.5°C/min based on the initial water (x_{w0}) and fat (x_f) contents at 30°C [79]:

TABLE 1.4
Parameter Estimates of Equation 1.40

Material	k_i (W/m K)	k_0 (W/m K)	E_i (kJ/mol)	E_0 (kJ/mol)	SD (W/m K)
Cereal products					
Corn	1.580	0.070	7.2	5.0	0.047
Fruits					
Apple	0.589	0.287	2.4	11.7	0.114
Orange	0.642	0.106	1.3	0.0	0.007
Pear	0.658	0.270	2.4	1.9	0.016
Vegetables					
Potato	0.611	0.049	0.0	47.0	0.059
Tomato	0.680	0.220	0.2	5.0	0.047
Dairy					
Milk	0.665	0.212	1.7	1.9	0.005
Meat					
Beef	0.568	0.280	2.2	3.2	0.017
Baked products					
Dough	0.800	0.273	2.7	0.0	0.183
Model foods					
Amioca	0.718	0.120	3.2	14.4	0.037
Starch	0.623	0.243	0.3	0.4	0.006
Hylon	0.800	0.180	9.9		0.072
Other					
Rapeseed	0.239	0.088	3.6	0.6	0.023

Source: Saravacos, G.D. and Maroulis, Z.B., Thermal conductivity and diffusivity of foods, in: *Transport Properties of Foods*, Marcel Dekker, New York, pp. 269–358, 2001.

SD, standard deviation.

For whole beef,

$$k = 10^{-3}\left(732 - 4.32x_{f0} - 3.56x_{w0} + 0.636T\right) \tag{1.41}$$

For minced meat,

$$k = 10^{-3}\left(400 - 4.49x_{f0} + 0.147x_{w0} + 1.74T\right) \tag{1.42}$$

For lamb meat [11,51],

$$k = 0.48534 + 1.0627 \times 10^{-3}T; \quad T \geq -0.9113°C \tag{1.43}$$

Thermal conductivity for meat products and for *heat transfer parallel* or *perpendicular* to muscle fiber can be correlated as follows [78]:

$$k = 0.1075 + 0.501x_w + 5.052 \times 10^{-4}x_wT; \quad \text{Parallel to fiber} \tag{1.44}$$

$$k = 0.0866 + 0.501x_w + 5.052 \times 10^{-4}x_wT; \quad \text{Perpendicular to fiber} \tag{1.45}$$

The correlation given in the following is proposed for surimi paste for temperatures in the cooking range of 25°C–90°C [62]:

$$k = 1.33 - 4.82 \times 10^{-3}T + 5 \times 10^{-5}T^2 - 2.45 \times 10^{-2}x_w + 1.7 \times 10^{-4}x_w^2 + 2.4 \times 10^{-5}x_wT \tag{1.46}$$

where

x_w is in percent wet basis
T is in degrees Centigrade (°C)

The standard deviations of Equation 1.46 ranged from 0.1% to 5%, with higher deviations found at higher temperatures.

1.4.3.2 Thermal Conductivity and Thermal Diffusivity of Juices, Fruits and Vegetables, and Others

The thermal conductivity of clarified apple juice as a function of the concentration (Bx) in °Brix and temperature (T) in k can be expressed as follows [58]:

$$k = 0.27928 - 3.5722 \times 10^{-3}Bx + 1.1357 \times 10^{3}T \tag{1.47}$$

Thermal conductivity of Golden Delicious and Granny Smith apples can be evaluated as follows [11]:

$$k = 0.0159x_w + 0.0025T - 0.994 \tag{1.48}$$

and for the thermal diffusivity,

$$\alpha = (-0.00278T - 1.39) \times 10^{-7}; \quad \text{Golden Delicious} \tag{1.49}$$

$$\alpha = (-0.00556T - 1.31) \times 10^{-7}; \quad \text{Granny Smith} \tag{1.50}$$

for a temperature range from the initial freezing point up to 60°C, x_w and T are the water content on percent wet basis and temperature in °C, respectively. Simplified analytical equations are very useful for practical industrial calculations [80].

Quadratic and multiple form correlations are also used to model k in an attempt to cover a wide range of moisture contents [10]. Rahman et al. [74] improved a general model developed previously for fruits and vegetables, and obtained the following general power law correlation which is valid for a wide range of temperature (from 5°C to 100°C), water content (from 14% to 88%, wet basis), and porosity (from 0 to 0.56):

$$\frac{\alpha_0}{1-\varepsilon_a+\left(k_a/\left(k_w\right)_r\right)}=0.996\left(\frac{T}{T_r}\right)^{0.713} x_w^{0.285} \quad (1.51)$$

where α_0 is the Rahman–Chen structural factor, which includes the effective value of the thermal conductivity (k_e) and accounts for the effect of temperature and structure of a food item as

$$\alpha_0=\frac{k_e-\varepsilon_a k_a}{\left(1-\varepsilon_a-\varepsilon_w\right)k_s+\varepsilon_w k_w} \quad (1.52)$$

Equation 1.51 gives a mean percent deviation from 6.8% to 15.1%, so the model can be applied in process design and control purposes where around 15% maximum allowable error in data is permitted [74].

Califano and Calvelo [81] measured the thermal conductivity of potatoes between 50°C and 100°C, and obtained a quadratic equation, which fitted experimental values within a mean absolute deviation of 2.3%, the fitted equation is given as follows [11]:

$$k=1.05-1.96\times10^{-2}T+1.90\times10^{-4}T^2 \quad (1.53)$$

For extruded Durum wheat pasta, k for the range 20°C–60°C can be predicted from [82]:

$$k=0.305+0.285x_w \quad (1.54)$$

with an error in the order of 1% in the extreme temperatures [11]. For the thermal diffusivity, the following general relationship is proposed, involving moisture content and temperature with an error of 4% [11]:

$$\alpha=\left(1.73-0.9x_w-0.003T\right)\times10^{-7} \quad (1.55)$$

1.4.4 Density

In most engineering design, it is assumed that *density* moderately changes with temperature and pressure [10]. Pretreatments (heating, cooking, drying, etc.) and product preparation are factors that can also influence the value of density. Rahman [10] presented one of the most comprehensive books related to food properties. Experimental values and models to predict density are reviewed and discussed by the author [10].

The density of materials also influences their electrical properties [83]. This can be seen in Equation 1.17, which shows that density is inversely proportional to the temperature increase (ΔT).

Some of the many equations that exist in literature for density are given here. The most commonly used model to predict density is

$$\rho = \frac{1}{\sum x_i/\rho_i} \tag{1.56}$$

Since the *porosity* (ε) of a food material can strongly influence its density, Equation 1.55 should be modified to incorporate porosity (ε):

$$\rho = (1-\varepsilon)\frac{1}{\sum x_i/\rho_i} \tag{1.57}$$

and i denotes the ith component of the food system. A rule of thumb, regarding densities of major food components, states that values of fats, proteins, and carbohydrates fall within narrow ranges; consequently average values of 920, 1250, and 1550 kg/m^3 can be taken for most of the calculations [11].

A general correlation for density of fruit juices, which is claimed to agree with values calculated from the Choi and Okos model and fits experimental values obtained for 30 different juices, is presented in the following [84]. The following equation is valid for temperatures above freezing and for soluble solid concentration up to 30°Brix [11]:

$$\rho = (1002 + 4.61Bx) - 0.460T + 7.001\times10^{-3}T^2 - 9.175\times10^{-6}T^3 \tag{1.58}$$

For sour cherry, apple, and grape juices, correlations as a function of temperature (T) in K and concentration (Bx) are obtained as follows [11,85]:

$$\rho = 0.79 + 0.35\exp(0.0108\,Bx) - 5.41\times10^{-4}\,T;\quad \text{Sour cherry juice} \tag{1.59}$$

$$\rho = 0.83 + 0.35\exp(0.01Bx) - 5.64\times10^{-4}T;\quad \text{Apple juice} \tag{1.60}$$

$$\rho = 0.74 + 0.43\exp(0.01Bx) - 5.55\times10^{-4}T;\quad \text{Grape juice} \tag{1.61}$$

For fruit juices, Riedel [86] suggested measuring the index of refraction of the juice, s, and to use the following relationship [45]:

$$\rho = \frac{s^2-1}{s^2+2}\frac{62.4}{0.206}16.0185 \tag{1.62}$$

In general, density of food materials varies nonlinearly with moisture content. The density of fruits and vegetables during drying can be correlated as follows [87]:

$$\rho = g + h\frac{X_w}{X_{w0}} + p\left[\exp\left(-u\frac{X_w}{X_{w0}}\right)\right] \tag{1.63}$$

where g, h, p, and u are constants for each specific product considered (e.g., carrot, garlic, pear, potato, sweet potato, etc.).

For meat products, a density equal to 1053 kg/m^3 for temperatures above freezing point can be used [51]. AbuDagga and Kolbe [62] reported the following straight line regression in two variables for the density of Pacific whiting surimi paste:

$$\rho = 1511.2 - 1.16T - 5.4x_w \tag{1.64}$$

The moisture content for Equation 1.64 ranges from 74% to 84% and T from 30°C to 90°C, x_w is in percent wet basis. Equation 1.64 indicates that density decreases with increased moisture content and temperature. Protein denaturation during cooking is likely to be responsible for the decrease in ρ with temperature.

Pasta density as a function of moisture content (%) can be correlated as follows [11,82]:

$$\frac{1}{\rho} = (3.02x_w + 6.46) \times 10^{-4} \tag{1.65}$$

For milk it is often sufficient to use only three components, water, fat, and NFS. The following equation was suggested for the component densities in kg/m³ in terms of temperature (°C) [88]:

$$\rho_{fat} = 966.665 - 1.334T \tag{1.66}$$

$$\rho_{NFS} = 1635 - 2.6T + 0.01T^2 \tag{1.67}$$

The density of water as function of temperature and for the 5°C–100°C range can be calculated with good accuracy as follows:

$$\rho_{water} = 1000.35 + 0.004085T - 0.0057504T^2 + 1.50673 \times 10^{-5}T^3 \tag{1.68}$$

1.4.5 Dielectric Properties

Electrical properties of foods have generally been of interest for two reasons. One relates to the possibility of using the electrical property as a means for determining moisture content or some other quality factor. The other has to do with the absorption of energy in high-frequency dielectric heating or microwave heating applications in food processing [83]. Primary determinants of electrical properties of foods are frequency, temperature, chemical composition, and physical structure [31]. The effect of composition on *dielectric properties* is complex. For example, the *dielectric constant* increases with moisture content for most foods, but studies report different trends for the *dielectric loss factor* [89].

Extensive experimentally obtained data for the dielectric properties of various foods are reported in literature; however values of dielectric properties above 100°C for both *microwave* and *RF* ranges are scarce [35]. Furthermore, data extrapolation from low temperature up to *sterilization* region should not be done [46].

Efforts have been made to relate dielectric properties to food composition, temperature, and frequency [35]. The responses of different components in a system (e.g., salt or ash content, and carbohydrates) depend upon the manner in which they are bound to the other components in a food matrix; the particle geometry in a food mixture is also an important parameter [90]. Therefore, for food mixtures, the prediction of dielectric properties from data for the individual components is difficult [90]. On the other hand, prediction from composition can potentially avoid costly measurements and can provide valuable insight into the behavior of individual components in the food [36].

A series of polynomial equations have been developed to estimate ε′ and ε″ for cereal grains, fatty/low moisture content, fruits and vegetables, and meat products [90]. The predictive equations include the influence of food composition and temperature (below 70°C in general) at microwave frequencies between 0.9 and 3 GHz for all the products, and up to 10 GHz for grains. Table 1.5 shows the predictive equations which give coefficients of determination $R^2 \geq 0.9$.

TABLE 1.5
Predictive Equations for the Dielectric Constant and Loss Factor of Foods

Food Type	f (GHz)	T (°C)	x_w (%)	x_a (%)	x_f (%)	R^2
Fatty/low-moisture-content foods						
$\varepsilon' = 2.63 - 0.0015T + 0.162x_w$	1–3	0–100	0–30	0–5	70–100	0.98
Fruits/vegetables						
$\varepsilon' = 2.14 - 0.104T + 0.808x_w$	≈2.45	0–70	50–90	—	—	0.98
$\varepsilon'' = 3.09 - 0.0638T + 0.213x_w$	≈2.45	0–70	50–90	—	—	0.90
Meat products						
Pork						
$\varepsilon' = -70 - 0.1T + 1.7x_w + (1.5 + 0.02T)x_{as}$	2–3	0–70	60–80	0–6	0–20	0.89
Poultry						
$\varepsilon' = -87.3 - 0.051T + 1.91x_w + (2 + 0.02T)x_{as}$	2–3	0–70	60–80	0–5	≈10	0.90

Source: Calay, R.J., *Int. J. Food Sci. Technol.*, 29, 699, 1995.

Datta et al. [36] performed a regression analysis for the dielectric constant and loss data set of about 100 experimental points in the frequency range of 2400–2500 MHz and the temperature range of 5°C–65°C. The most important factors proposed to use in the predictive equations were moisture (moisture content greater than 60%), salt and temperature. After fitting to the reported data in literature, the authors [36] concluded that it was difficult to develop a generic composition-based model for all products; then, they separated meat products from those of fruits and vegetables. The different methods of measurement and the lack of composition data could have probably led to the significant variability observed. For a restricted data set containing about 30 data points for raw beef, beef juice, raw turkey, and turkey juice, the best correlation for the dielectric constant contains two components—water and ash [36]:

$$\varepsilon' = x_w(1.0707 - 0.0018485T) + x_{as}(4.7947) + 8.5452 \tag{1.69}$$

Equation 1.69 has a correlation coefficient of 0.94 and the maximum error in the prediction is 5%. The dielectric loss for this restricted data set was modeled as a function of the mass percentage of water and temperature. The addition of a temperature-dependent ash term resulted in the final correlation, which has a high correlation coefficient ($R^2 = 0.989$) and an error lower than 10%:

$$\varepsilon'' = x_w(3.447 - 0.0187T + 0.000025T^2) + x_{as}(-57.093 + 0.231T) - 3.599 \tag{1.70}$$

ε' and ε'' of ham as a function of temperature (0°C–70°C), moisture (38.2%–68.9%), and ash contents (1.78%–6.80%) at 2450 MHz can be predicted from [89]

$$\varepsilon' = -25.49 + 1.063x_w - 1.041x_{as} + 0.03452T \tag{1.71}$$

$$\varepsilon'' = -150.2 + 5.243x_w + 6.220x_{as} - 0.2845T - 0.04322x_w^2 - 0.4732x_{as}^2 + 0.002245T^2 + 0.1090x_{as}T \tag{1.72}$$

where

x_w and x_{as} are the percentages of moisture and ash, respectively

T is in °C

The adjusted coefficients of determination, R^2, are equal to 0.817 and 0.852, respectively.

TABLE 1.6
Predictive Equations for Vegetables and Fruits at 2450 MHz

Dielectric constant	
Overall	$\varepsilon' = 38.57 + 0.1255T + 0.4546x_w - 14.54x_{as} - 0.0037x_wT + 0.07327x_{as}T$
Vegetables	$\varepsilon' = -243.6 + 1.342T + 4.593x_w - 426.9x_{as} + 376.5x_{as}^2 - 0.01415x_wT - 0.3151x_{as}T$
Fruits	$\varepsilon' = 22.12 + 0.2379T + 0.5532x_w - 0.0005134T^2 - 0.003866x_wT$
Dielectric loss factor	
Overall	$\varepsilon'' = 17.72 - 0.4519T + 0.001382T^2 - 0.07448x_w + 22.93x_{as} - 13.44x_{as}^2 + 0.002206x_wT + 0.1505x_{as}T$
Vegetables	$\varepsilon'' = -100.02 - 0.1611T + 0.001415T^2 + 2.429x_w - 378.9x_{as} + 316.2x_{as}^2$
Fruits	$\varepsilon'' = 33.41 - 0.4415T + 0.001400T^2 - 0.1746x_w + 1.438x_{as} + 0.001578x_wT + 0.2289x_{as}T$

Source: Sipahioglu, O. and Barringer, S.A., *J. Food Sci.*, 68, 234, 2003.

Sipahioglu et al. [47] obtained dielectric constant and loss factor predictive equations for turkey meat with the addition of sodium chloride, sodium tripolyphosphate, glycerol, lactic acid, and sodium lactate as a function of temperature (5°C–130°C), moisture, water activity, and ash content at both 915 and 2450 MHz.

Sipahioglu and Barringer [91] developed predictive equations at 2450 MHz for 15 vegetables and fruits at 5°C–130°C as a function of temperature, ash content (0.26%–1.56%), and moisture content (57.30%–95.89%). It was found that separating vegetables from fruits increased the correlation between dielectric properties and food composition and temperature. Predictive equations for each vegetable and fruit as a function of temperature were also reported. The average percent error of prediction was 6.20% for the dielectric constant and 13.22% for the dielectric loss. Table 1.6 lists the predictive equations at 2450 MHz. The units for temperature, and moisture and ash contents are °C and percent wet basis, respectively.

Dielectric properties of mashed potatoes relevant to microwave and radio-frequency pasteurization and sterilization processes have been measured over 1–1800 MHz and 20°C–120°C, and regressed with polynomial relationships as a function of moisture content (81.6%–87.8% wet basis), and salt content (0.8%–2.8% wet basis) [35]. The regression equations are shown in Table 1.7 at four frequencies: 27, 40, 433, and 915 MHz. All the regression equations have R^2 values equal to or above 0.91. The calculated data for ε' differ from measured values by less than 10%, and the calculated data for ε'' differ by 1%–25%, except that the discrepancies are up to 40% at 20°C. Literature shows that it is more difficult to predict loss factor than dielectric constant [91].

ε' and ε'' for skim milk, green pea puree, and carrot puree were measured under continuous flow conditions, and were correlated as a function of temperature for a temperature range of 20°C–130°C. The second-order polynomial correlations obtained at 915 MHz are as follows [92]:
For skim milk,

$$\varepsilon' = 73.4 - 0.1615T - 0.001T^2 \quad (1.73)$$

$$\varepsilon'' = 13.15 + 0.0489T + 0.0009T^2 \quad (1.74)$$

For green pea puree,

$$\varepsilon' = 74.8 - 0.2957T - 0.0007T^2 \quad (1.75)$$

TABLE 1.7
Predictive Equations for the Dielectric Properties of Mashed Potatoes

Dielectric constant		$\mathbf{R^2}$
27 MHz	$\varepsilon' = 54 + 98.5x_{as} - 81.2x_{as}^2 + 0.000019T^3 + 0.00121T^2x_{as} - 0.000026T^2x_w + 18.6x_{as}^3$	0.95
40 MHz	$\varepsilon' = 37.5 + 114x_{as} - 86.4x_{as}^2 + 0.000020T^3 + 0.000683T^2x_{as} - 0.000035T^2x_w + 18.6x_{as}^3$	0.91
433 MHz	$\varepsilon' = -59.2 + 0.940x_w + 115x_{as} - 0.00138Tx_w - 82.4x_{as}^2 + 16.8x_{as}^3$	0.92
915 MHz	$\varepsilon' = -85.5 + 1.26x_w + 105x_{as} - 76.3x_{as}^2 + 0.000012T^3 - 0.000025T^2x_w + 15.7x_{as}^3$	0.91
Dielectric loss factor		
27 MHz	$\varepsilon'' = -285 + (636 + 0.0893T^2)x_{as}$	0.98
40 MHz	$\varepsilon'' = -187 + (426 + 0.0601T^2)x_{as}$	0.98
433 MHz	$\varepsilon'' = -9.51 + (39.1 + 0.0053T^2)x_{as}$	0.97
915 MHz	$\varepsilon'' = 0.12 + (19.3 + 0.00234T^2)x_{as}$	0.96

Source: Guan, D. et al., *J. Food Sci.*, 69, FEP30, 2004.

$$\varepsilon'' = 13.2 + 0.0829T + 0.0005T^2 \tag{1.76}$$

For carrot puree,

$$\varepsilon' = 77.5 - 0.2022T - 0.00001T^2 \tag{1.77}$$

$$\varepsilon'' = 18.4 + 0.0359T + 0.0019T^2 \tag{1.78}$$

The correlation coefficients for Equations 1.73 through 1.78 were higher than 0.97.

ANN was found to be a suitable method to express dielectric constant and dielectric loss factor of cakes [93]. The ANN model developed can be used in different formulations, with or without emulsifiers and/or fat replacers, and under a wide range of moisture content and porosity values. Such models are useful for the estimation of dielectric properties of bakery products, which are needed for modeling the heat transfer during microwave baking.

1.5 CONCLUSIONS

Extensive data on experimental values and predictive equations for thermal properties of foods have been reported in literature; in some areas, for example, dielectric properties, the data are very limited. Although the data are sometimes incomplete, they are of great value for preliminary design, heat transfer calculations, and food quality assessment. There are few databases available at present, and in some cases (e.g., Nelfood) the databases available give a score indicating the quality of the method specification and food definition. These databases can be very helpful tools when there is no enough expertise. In general, data or predictive equations for the upper temperature range of interest to food engineers are rather scarce. Mathematical models for predicting properties are very useful, but if more accuracy is desired, experimental determination should be carried out. The data analysis shows that when experimental values are compared with the ones obtained from predictive models, the less the water there is in the material, the higher are the discrepancies between predicted and measured values. The reason seems to be the treatment of water as bulk water, without taking into account the interactions with the food components. Future work, as suggested by many researchers, should be addressed to study the influence of the state of water on thermal and electrical properties.

NOMENCLATURE

Symbols

a,b	Constants in Equation 1.11
A	Area (m^2)
Bx	Concentration (°Brix)
c	Specific heat (J/kg °C)
c_0	Speed of light in vacuum (m/s)
c_1,c_2	Constants in Equation 1.35
d	Penetration depth (cm)
E_0	Activation energy in dry material (kJ/mol)
E_i	Activation energy in wet material (kJ/mol)
f	Frequency (Hz)
g,h	Constants in Equation 1.63
H	Enthalpy (J/kg)
k	Thermal conductivity (W/m K)
m	Mass (kg)
p	Constant in Equation 1.63
q	Heat transfer rate (W)
Q	Heat (kJ)
t	Time (s)
u	Constant in Equation 1.63
V	Electric field (V/cm)
V_S	Sample volume (m^3)
V_w	Water volume (m^3)
x	Mass fraction
z	Thickness (m)

Greek letters

α	Thermal diffusivity (m^2/s)
α_0	Rahman–Chen structural factor
β	Constant in Equation 1.14
δ	Dielectric loss angle (°)
ΔT	Temperature increase (°C)
ε	Porosity or volume fraction
ε_0	Dielectric constant of vacuum (8.85419×10^{-12} F/m)
ε'	Dielectric constant of the material (F/m)
ε''	Dielectric loss factor of the material (F/m)
ε''_d	Dipole loss component (F/m)
ε''_σ	Ionic loss component (F/m)
γ	Constant in Equation 1.14
λ_0	Free space microwave wavelength (m)
ρ	Density (kg/m^3)
σ	Electrical conductivity (S/m)

Subscripts

A	Air
As	Ash
C	Carbohydrate
E	Effective value
F	Fat

i ith component in a food system
ns Nonfat solids
P Protein
R Reference temperature (°C)
S Solids
W Water
W_0 Initial water content

REFERENCES

1. RE Mudgett. Electrical properties of foods. In: MA Rao, SSH Rizvi, eds. *Engineering Properties of Foods*. New York: Marcel Dekker, 1995, pp. 389–455.
2. HFTh Meffert. History, aims, results and future of thermophysical properties work within COST 90. In: R Jowitt, F Escher, B Hallström, HFTh Meffert, WEL Spiess, G Vos, eds. *Physical Properties of Foods*. Essex, England: Applied Science Publishers Ltd, 1983, pp. 229–267.
3. MA Rao. Engineering properties of foods current status. In: P Fito, E Ortega-Rodriguez, GV Barbosa-Cánovas, eds. *Food Engineering 2000*. New York: Chapman & Hall, 1997, pp. 39–54.
4. GD Saravacos, AE Kostaropoulos. Engineering properties in food processing simulation. *Computers & Chemical Engineering* 20: S461–S466, 1996.
5. CR Buffler, MA Stanford. Effects of dielectric and thermal properties on the microwave heating of foods. *Microwave World* 12: 5–10, 1995.
6. NN Moshenin. *Thermal Properties of Foods and Agricultural Products*. New York: Gordon and Breach, 1980, pp. 25–196.
7. P Nesvadba. Methods for the measurement of thermal conductivity and diffusivity of foodstuffs. *Journal of Food Engineering* 1: 93–113, 1982.
8. M Kent, K Christiansen, IA van Hanehem, E Holtz, MJ Morley, P Nesvadba, KP Poulsen. Cost90 collaborative measurements of thermal properties of foods. *Journal of Food Engineering* 3: 117–150, 1984.
9. VE Sweat. Thermal properties of foods. In: MA Rao, SSH Rizvi, eds. *Engineering Properties of Foods*. New York: Marcel Dekker, 1995, pp. 99–138.
10. S Rahman. *Food Properties Handbook*, 2nd edn. Boca Raton, FL: CRC Press, 2009, pp. 398–697.
11. MJ Urbicain, JE Lozano. Thermal and rheological properties of foodstuffs. In: KJ Vanlentas, E Rostein, RP Singh, eds. *Handbook of Food Engineering Practice*. Boca Raton, FL: CRC Press, 1997, pp. 425–486.
12. CA Miles, G van Beek, CH Veerkamp. Calculation of thermophysical properties of foods. In: R Jowitt, F Escher, B Hallström, HFTh Meffert, WEL Spiess, G Vos, eds. *Physical Properties of Foods*. Essex, England: Applied Science Publishers Ltd, 1983, pp. 269–312.
13. T Ohlsson. The measurement of thermal properties. In: R Jowitt, F Escher, B Hallström, HFTh Meffert, WEL Spiess, G Vos, eds. *Physical Properties of Foods*. Essex, England: Applied Science Publishers Ltd, 1983, pp. 313–330.
14. JD Mellor. Critical evaluation of thermophysical properties of foodstuffs and outline of future developments. In: R Jowitt, F Escher, B Hallström, HFTh Meffert, WEL Spiess, G Vos, eds. *Physical Properties of Foods*. Essex, England: Applied Science Publishers Ltd, 1983, pp. 331–353.
15. R Jowitt, F Escher, M Kent, B McKenna, M Roques, eds. *Physical Properties of Foods-2*. Amsterdam, the Netherlands: Elsevier Applied Science, 1987, pp. 159–233.
16. A Bálint. Prediction of physical properties of foods for unit operations. *Periodica Polytechnica Ser Chemical Engineering* 45(1): 35–40, 2002.
17. L Riedel. Kalorimetrische untersuchungen über das schmelzverhalten von fetten un ölen. *Fette, Seifen, Anstrichmittel* 57: 771–783, 1955.
18. L Riedel. Kalorimetrische untersuchungen über das gefrieren von seefischen. *Kältetechnik* 8: 374–377, 1956.
19. L Riedel. Kalorimetrische untersuchungen über das gefrieren von fleisch. *Kältetechnik* 9: 38–40, 1957.
20. L Riedel. Enthalpimessungen en lebensmitteln. *Chemie, Mikrobiologie und Technologie von Lebensmitteln* 5: 118–127, 1977.
21. I Lind. The measurement and prediction of thermal properties of food during freezing and thawing—A review with particular reference to meat and dough. *Journal of Food Engineering* 13, 285–319, 1991.
22. GD Saravacos, ZB Maroulis, eds. Thermal conductivity and diffusivity of foods. In: *Transport Properties of Foods*. New York: Marcel Dekker, 2001, pp. 269–358.

23. EG Murakami, VE Sweat, SK Sastry, E Kolbe, K Hayakawa, A Datta. Recommended design parameters for thermal conductivity probes for nonfrozen food materials. *Journal of Food Engineering* 27: 109–123, 1996.
24. AE Delgado, A Gallo, D de Piante, AC Rubiolo. Thermal conductivity of unfrozen and frozen strawberry and spinach. *Journal of Food Engineering* 31: 137–146, 1997.
25. DL Goedeken, KK Shah, CH Tong. True thermal conductivity determination of moist porous food material at elevated temperatures. *Journal of Food Science* 63: 1062–1066, 1998.
26. GS Campbell, C Calisssendorff, JH Williams. Probe for measuring soil specific heat using a het-pulse method. *Journal of Soil Science Society of America* 55: 291–293, 1991.
27. AJ Fontana, J Varith, J Ikediala, J Reyes, B Wacker. Thermal properties of selected foods using a dual needle heat-pulse sensor. *ASAE Meeting Presentation*, Toronto, Ontario, Canada, July 18–22, 1999, Paper No 996063, 1999.
28. L Huang, L-S Liu. Simultaneous determination of thermal conductivity and thermal diffusivity of food and agricultural materials using a transient plane-source method. *Journal of Food Engineering* 95: 179–185, 2009.
29. RP Singh. Thermal properties of frozen foods. In: MA Rao, SSH Rizvi, eds. *Engineering Properties of Foods*. New York: Marcel Dekker, 1995, pp. 139–167.
30. LO Figura, AA Teixeira. *Food Physics—Physical Properties—Measurement and Applications*. Berlin, Germany: Springer-Verlag, 2007, pp. 41–72.
31. RE Mudgett. Electrical properties of foods: A general review. In: R Jowitt, F Escher, M Kent, B McKenna, M Roques, eds. *Physical Properties of Foods-2*. Amsterdam, the Netherlands: Elsevier Applied Science, 1987, pp. 159–170.
32. PP Lewicki. Water as the determinant of food engineering properties: A review. *Journal of Food Engineering*, 61: 483–495, 2004.
33. P Piyasena, C Dussault, T Koutchma, HS Ramaswamy, GB Awuah. Radio frequency heating of foods: Principles, applications and related properties—A review. *Critical Reviews in Food Science and Nutrition* 43: 587–606, 2003.
34. M Kent, E Kress-Rogers. The COST90bis collaborative work on the dielectric properties of foods. In: R Jowitt, F Escher, M Kent, B McKenna, M Roques, eds. *Physical Properties of Foods-2*. Amsterdam, the Netherlands: Elsevier Applied Science, 1987, pp. 171–197.
35. D Guan, M Cheng, Y Wang, J Tang. Dielectric properties of mashed potatoes relevant to microwave and radio-frequency pasteurisation and sterilization processes. *Journal of Food Science* 69: FEP30–FEP37, 2004.
36. AK Datta, E Sun, A Solis. Food dielectric property data and their composition-based prediction. In: MA Rao, SSH Rizvi, eds. *Engineering Properties of Foods*. New York: Marcel Dekker, 1995, pp. 457–494.
37. Y Wang, TD Wig, J Tang, LM Hallberg. Dielectric properties of foods relevant to RF and microwave pasteurisation and sterilization. *Journal of Food Engineering* 57: 257–268, 2003.
38. S Ryynänen. The electromagnetic properties of food materials: A review of the basic principles. *Journal of Food Engineering* 26: 409–429, 1995.
39. MS Venkatesh, GSV Raghavan. An overview of dielectric properties measuring techniques. *Canadian Biosystems Engineering Journal* 47: 7.15–7.30, 2005.
40. P Nesvadba, M Houška, W Wolf, V Gekas, D Jarvis, PA Saad, AI Johns. Database of physical properties of agro-food materials. *Journal of Food Engineering* 61: 497–503, 2004.
41. RP Singh. *Food Properties Database. Version 2.0 for Windows*. Boca Raton, FL: CRC Press, 1995.
42. MK Krokida, NM Panagiotou, ZB Maroulis, GD Saravacos. Thermal conductivity: Literature data compilation for foodstuffs. *International Journal of Food Properties* 4: 111–137, 2001.
43. ME Sosa-Morales, L Valerio-Junco, A López-Malo, HS García. Dielectric properties of foods: Reported data in the 21st Century and their potential applications. *LWT-Food Science and Technology* 43: 1169–1179, 2010.
44. ASHRAE. Thermal properties of foods. In: RA Parsons ed. *ASHRAE Handbook: Fundamentals*. New York: American Society of Heating, Air Conditioning and Refrigeration Engineers, 1993, pp. 30.1–30.26.
45. RP Singh. Heating and cooling processes of foods. In: DR Heldman, DB Lund, eds. *Handbook of Food Engineering*. Boca Raton, FL: CRC Press, 2007, pp. 398–426.
46. T Ohlsson, NE Bengtsson. Dielectric food data for microwave sterilization processing. *Journal of Microwave Power* 10: 93–108, 1975.
47. O Sipahioglu, SA Barringer, I Taub, APP Yang. Characterization and modelling of dielectric properties of turkey meat. *Journal of Food Science* 68: 521–527, 2003.

48. M Hussain, MS Rahman. Thermal conductivity prediction of fruits and vegetables using neural networks. *International Journal of Food Properties* 2(2): 121–137, 1999.
49. S Sablani, MS Rahman. Using neural networks to predict thermal conductivity of food as a function of moisture content, temperature and apparent porosity. *Food Research International* 36: 617–623, 2003.
50. HL Mattar, LA Minim, JSR Coimbra, VPR Minim, SH Saraiva, J Telis-Romero. Modeling thermal conductivity, specific heat, and density of milk: A neural network approach. *International Journal of Food Properties* 7(3): 531–539, 2004.
51. PD Sanz, MD Alonso, RH Mascheroni. Equations for the prediction of thermophysical properties of meat products. *Latin American Applied Research* 19: 155–163, 1989.
52. Y Choi, MR Okos. Effects of temperature and composition on the thermal properties of foods. In: M LeMaguer, P Jelen, eds. *Food Engineering and Process Applications, Vol. 1, Transport Phenomena*. London, U.K.: Elsevier, 1986, pp. 93–101.
53. HA Leniger, WA Beverloo. *Food Process Engineering*. Dordrecht, the Netherlands: D Reidel, 1975.
54. L Riedel. Eine formel zur berechnung der enthalpie fettarmer lebensmitteln in abhängigkeit von vassergehalt und temperatur. *Chemie, Mikrobiologie und Technologie von Lebensmitteln* 5: 129–133, 1978.
55. CS Chen. Thermodynamic analysis of the freezing and thawing of foods: Enthalpy and apparent specific heat. *Journal of Food Science* 50: 1152–1162, 1985.
56. TR Gupta. Specific heat of Indian unleavened flat bread (chaphati) at various stages of cooking. *Journal of Food Process Engineering* 13: 217–227, 1990.
57. RW Dickerson. Thermal properties of foods. In: KK Tressler, WB van Arsdel, MJ Copely, eds. *The Freezing Preservation of Foods*. Westport, CT: AVI, 1968, pp. 26–51.
58. DM Constela, JE Lozano, GH Crapiste. Thermophysical properties of clarified apple juice as a function of concentration and temperature. *Journal of Food Science* 54: 663–668, 1989.
59. JDD Alvarado. Specific heat of dehydrated pulps of fruits. *Journal of Food Process Engineering* 14: 185–189, 1991.
60. PD Sanz, MD Alonso, RH Mascheroni. Thermophysical properties of meat products: General bibliography and experimental values. *Transactions of the ASAE* 30: 283–289, 296, 1987.
61. HC Bazán, RH Mascheroni. Transferencia de calor con simultáneo cambio de fase en la congelación de carnes ovinas. *Revista Latinoamericana de Transferencia de Calor y Materia* 8: 55–76, 1984.
62. Y AbuDagga, E Kolbe. Thermophysical properties of surimi paste at cooking temperature. *Journal of Food Engineering* 32: 325–337, 1997.
63. HS Ramaswamy, MA Tung. Thermophysical properties of apples in relation to freezing. *Journal of Food Science* 46: 724–728, 1981.
64. MH Hsu, JD Mannapperuma, RP Singh. Physical and thermal properties of pistachios. *Journal of Agricultural Engineering Research* 49: 311–321, 1991.
65. N Wang, JG Brennan. The influence of moisture content and temperature on the specific heat of potato measured by DSC. *Journal of Food Engineering* 19: 303–310, 1993.
66. F Fernández-Martín, F Montes. Influence of temperature and composition on some physical properties of milk and milk concentrates. III, Thermal conductivity. *Milchwissenschaft* 27: 772–776, 1972.
67. AS Thomareis, J Hardy. Evolution de la chaleour specifique apparente des fromages fondues entre 40 et 100°C. Influence de leur composition. *Journal of Food Engineering* 4: 117, 1985.
68. ME Christenson, CH Tong, DB Lund. Physical properties of baked products as functions of moisture and temperature. *Journal of Food Process and Preservation* 13: 201–217, 1989.
69. G Kaletunç. Prediction of specific heat of cereal flours: A quantitative empirical correlation. *Journal of Food Engineering* 82: 589–594, 2007.
70. DW Sun, X Zhu. Effect of heat transfer direction on the numerical prediction of beef freezing processes. *Journal of Food Engineering* 42: 45–50, 1999.
71. IJ Kopelman. Transient heat transfer and thermal properties in food systems. PhD thesis. Michigan State University, East Lansing, MI, 1966.
72. DR Heldman, RP Singh. *Food Process Engineering*. Westport, CT: AVI Publishers, 1981.
73. K Wallapapan, VE Sweat, KC Diehl, CR Engler. Thermal properties of porous foods. ASAE Paper 83-6575, American Society of Agricultural Engineering, St. Joseph, MI, 1983.
74. MS Rahman, XD Chen, CO Perera. An improved thermal conductivity prediction model for fruits and vegetables as a function of temperature, water content and porosity. *Journal of Food Engineering* 31: 163–170, 1997.
75. L Riedel. Temperaturleitfähigkeitsmessungen an wasserreichen lebensmitteln. *Kältetechnik* 21: 315–316, 1969.

76. F Hermans. The thermal diffusivity of foods. PhD thesis. University of Leuven, Leuven, the Netherlands, 1979.
77. Y Choi, MR Okos. Thermal properties of liquid foods—Review. *Proceedings of the Winter Meeting of the American Society of Agricultural Engineers*, December 13–16, Chicago, IL, Paper No 83-6516, 1983.
78. MS Baghe-Khandan, MR Okos, VE Sweat. The thermal conductivity of beef as affected by temperature and composition. *Transactions of the ASAE* 25: 1118–1122, 1982.
79. SJ James, J Stephen. Thermophysical properties of meat. In: SJ James, C James, eds. *Meat Refrigeration*. Cambridge, U.K.: Woodhead Publishing, 2002, pp. 273–281.
80. VO Salvadori, RH Mascheroni. Prediction of freezing and thawing times of foods by means of a simplified analytical method. *Journal of Food Engineering* 13: 67–78, 1991.
81. AN Califano, A Calvelo. Thermal conductivity of potato between 50 and 100°C. *Journal of Food Science* 56: 586–587, 589, 1991.
82. J Andrieu, E Gonnet, M Laurent. Thermal conductivity and diffusivity of extruded Durum wheat pasta. *Lebensmitteln-Wissenschaft und Technologie* 22: 6–10, 1986.
83. SO Nelson. Electrical properties of agricultural products—A critical review. *Transactions of the ASAE* 16: 384–400, 1973.
84. JD Alvarado, CH Romero. Physical properties of fruits: Density and viscosity of juices as functions of soluble solids content and temperature. *Latin American Applied Research* 19: 15–21, 1989.
85. L Bayindirli, O Özsan. Modeling the thermophysical properties of sour cherry juice. *Gida* 17: 405–407, 1992.
86. L Riedel. 1949. Measurements of the thermal conductivity of sugar solutions, fruit juices and milk. *Chemical Engineering & Technology* 21: 340, 1949 (in German).
87. JE Lozano, MJ Urbicain, E Rotstein. Shrinkage, porosity and bulk density of foodstuffs at changing moisture content. *Journal of Food Science* 48: 1497–1502, 1553, 1983.
88. K Morison, RW Hartel. Evaporation and freeze concentration. In: DR Heldman, DB Lund, eds. *Handbook of Food Engineering*. Boca Raton, FL: CRC Press, 2007, pp. 495–552.
89. O Sipahioglu, SA Barringer, I Taub, A Prakash. Modeling the dielectric properties of ham as a function of temperature and composition. *Journal of Food Science* 68: 904–909, 2003.
90. RJ Calay, M Newborough, D Probert, PS Calay. Predictive equations for the dielectric properties of foods. *International Journal of Food Science and Technology* 29: 699–713, 1995.
91. O Sipahioglu, SA Barringer. Dielectric properties of vegetables and fruits as a function of temperature, ash, and moisture content. *Journal of Food Science* 68: 234–239, 2003.
92. P Kumar, P Coronel, J Simunovic, VD Truong, KP Sandeep. Measurement of dielectric properties of pumpable food materials under static and continuous flow conditions. *Journal of Food Science* 72(4): E177–E183, 2007.
93. IH Boyacı, G Sumnu, O Sakiyan. Estimation of dielectric properties of cakes based on porosity, moisture content, and formulations using statistical methods and artificial neural networks. *Food Bioprocess Technology: An International Journal* 2: 353–360, 2009.

2 Heat and Mass Transfer in Thermal Food Processing

Lijun Wang and Da-Wen Sun

CONTENTS

2.1 INTRODUCTION

Thermal processing techniques are widely used to improve eating quality and safety of food products, and to extend shelf life of the products. These thermal processing techniques involve the production, transformation, and preservation of foods. Sterilization and pasteurization are heating processes to inactivate or destroy enzymatic and microbiological activity in foods. Cooking (including baking, roasting, and frying) is a heating process to alter the eating quality of foods and to destroy microorganisms and enzymes for food safety. Dehydration and drying are heating processes to remove the majority of water in foods by evaporation (or by sublimation for freeze drying) for extending the shelf life of foods due to a reduction in water activity.

When a food is placed in contact with a liquid or solid medium of different temperatures or concentrations, a potential for a flux of energy and/or mass appears. The principles of many food thermal processes are based on heat and mass exchanges between the food and processing medium. There is a need for qualitative and quantitative understanding of the *heat and mass transfer* mechanisms underlying various unit operations of food thermal processes. This is important for the development of new food sources and food products, for more economical and efficient processing of foods, and for better food quality and safety. If the mechanism of a process is well understood, *mathematical models* can be developed to present the process. Experiments can virtually be carried out on mathematical models under broad experimental conditions in an economical and time-saving manner. With process models, quantitative calculations and predictions can be made for more reliable design, optimization of design and operating conditions, and evaluation of process performance. Therefore, advances of food thermal processes may become possible on the basis of improved understanding of heat and mass transfer mechanisms.

This chapter first briefly presents the fundamental mechanisms and *physical laws* of heat and mass transfer. A review is then conducted on the applications of the *engineering principles* and physical laws of heat and mass transfer for analyzing the unit operations of thermal processes in the food industry. Finally, future improvements in understanding of heat and mass transfer mechanisms in food thermal processes and development of heat and mass transfer models for describing food thermal processes are discussed.

2.2 FUNDAMENTALS OF HEAT AND MASS TRANSFER

2.2.1 HEAT TRANSFER

2.2.1.1 Basic Heat Transfer Modes

Many *unit operations* of *thermal food processes* involve the transfer of heat into or out of a food. Heat may be transferred by one or more of the three mechanisms of *conduction*, *convection*, and *radiation*. Most of industrial heat transfer operations involve a combination of these but it is often the case that one mechanism is dominant.

Conduction is the transfer of heat through solids or stationary fluids due to lattice vibration and/or particle collision. The heat flux due to conduction in the x direction through a uniform homogeneous slab of materials as shown in Figure 2.1a is given by *Fourier*'s first law of conduction:

$$q = -kA\frac{\mathrm{d}T}{\mathrm{d}x} \tag{2.1}$$

Fourier's law of heat conduction may be solved for a rectangular, cylindrical, or spherical coordinate system, depending on the geometrical shape of the object being studied.

Convection uses the movement of fluids to transfer heat. The movement, which causes heat transfer, may occur in natural or forced form. *Natural convection* creates the fluid movement by the

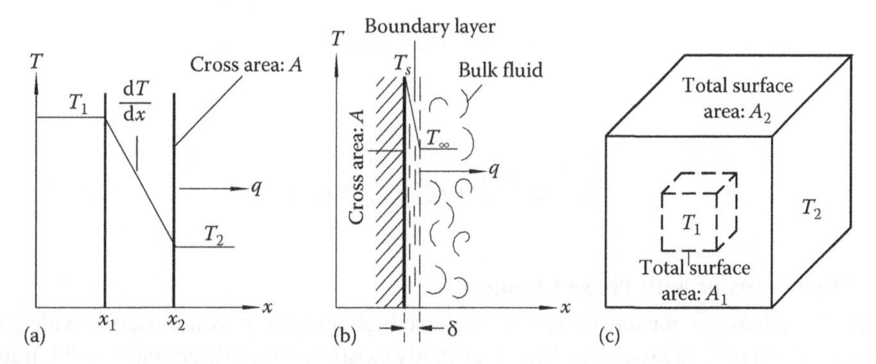

FIGURE 2.1 Schematic of three basic heat transfer modes: (a) Heat conduction through a uniform slab. (b) Heat convection through a vertical wall. (c) Heat radiation between a surface (1) of a body and the surroundings (2).

difference between fluid densities due to the temperature difference. *Forced convection* uses external means such as agitators and pumps to produce fluid movement. Convection heat transfer is the major mode of heat transfer between the surface of a solid material and the surrounding fluid. For analyzing convection heat transfer, a *boundary layer* is normally assumed near the surface of the solid material as shown in Figure 2.1b. Heat is transferred by conduction through this layer. The layer contains almost all of the resistance to heat transfer because of relatively low thermal conductivities and rapid heat transfer from the outer edge of the boundary layer into the bulk of the fluid. Using the boundary layer concept, the rate of convective heat transfer may be written as

$$q = kA\frac{(T_s - T_\infty)}{\delta} = A\frac{(T_s - T_\infty)}{\delta / k} \tag{2.2}$$

However, as the thickness of the boundary layer, δ, can neither be predicted nor measured easily, the thermal resistance of the boundary layer cannot be determined. δ/k is thus replaced with the term $1/h_c$, in which h_c is a film heat transfer coefficient. Equation 2.2 can then be rewritten as

$$q = h_c A(T_s - T_\infty) \tag{2.3}$$

Radiation does not require a medium for transferring heat but uses electromagnetic waves emitted by an object for exchanging heat. The energy emitted from a surface depends on the temperature of the surface, which can be described using the Stefan–Boltzmann law:

$$q = \sigma A \varepsilon T_K^4 \tag{2.4}$$

When the energy exchanges between two bodies by radiation, the energy emitted by one is not completely absorbed by the other as it can only absorb the portion that it intercepts. Therefore, a shape factor, F, is defined. The radiative energy exchange between a surface, 1, of a body and the surroundings, 2, can be determined by

$$q = \sigma F_{12} A_1 e\left(T_{K1}^4 - T_{K2}^4\right) \tag{2.5}$$

If the surface, 1, is enclosed by the surroundings, 2, as shown in Figure 2.1c, then $F_{12} = 1$. Similar to convective heat transfer coefficient, a radiative heat transfer coefficient may be expressed as

$$q = h_r A\left(T_{K1} - T_{K2}\right) \tag{2.6}$$

where

$$h_r = \sigma e\left(T_{K1}^2 + T_{K2}^2\right)\left(T_{K1} + T_{K2}\right) \tag{2.7}$$

2.2.1.2 Heat Transfer with Phase Changes

Most of foods such as raw meats and vegetables have high moistures. Water itself is widely used as processing medium in food processes. Water normally exists in one of three states: solid, liquid, and gas. The transition between two states is called a *phase transformation or phase change*. During phase transformation, the temperature of pure water keeps constant with added energy because all energy is used to transform water from one state to another. As water is widely present in foods and is used as processing medium, it is necessary to discuss the heat transfer with phase changes of water in the food industry.

During food thermal processes, the water in the food may experience *phase changes*. Frying and grilling of foods involve phase change from liquid water to vapor. There is an evaporation front, which divides the food body into two parts of the outer crust and the internal core regions as shown in Figure 2.2. The evaporation front moves toward the center as frying and grilling processes proceed. If a frozen food is used during frying and grilling, there will be two moving boundaries: the thawing front and the evaporation front.

The heat transfer mechanisms across the *moving boundary* must account for the latent heat of phase change of water. The moving front of phase change in the food can be tracked by the energy balance on the front, which is given by [1]

$$-k_1\left(\frac{\partial T}{\partial x}\right)_1 + k_2\left(\frac{\partial T}{\partial x}\right)_2 = \lambda \rho X_w \frac{\mathrm{d}S(t)}{\mathrm{d}t} \tag{2.8}$$

$$t > 0, \quad x = S(t) \tag{2.9}$$

Water is also widely used as processing medium. *Boiling* and *condensation* involve phase change between liquid water and vapor. Boiling heat transfer is particularly important in processing

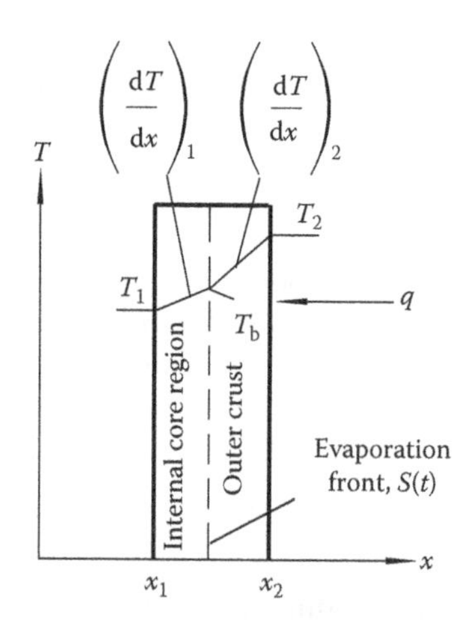

FIGURE 2.2 Schematic of heat transfer with phase changes (frying).

operations such as evaporation in which the boiling of liquids takes place either at submerged surfaces or on the inside of vertical tubes as in a climbing film evaporator. The heat flux changes dramatically as a function of the temperature difference between the surface and the boiling liquid, rising to a peak value and falling away sharply. This is caused by the strong dependence between the heat transfer coefficient and the temperature difference, which is shown in Figure 2.3. In order to avoid the danger of overheating and damaging the walls of the heater, equipment should ideally be operated in the nucleate boiling zone, just below the critical temperature difference as shown in Figure 2.3 [2]. Vapor condensation is also used in food thermal processes. Consider the food sterilization process used in canned foods, if steam is used as a heating medium, the condensing vapors on the metal surface of containers result in a significantly higher heat transfer than if hot water is used to heat the cans. A vapor condenses on a cold surface in one of two distinct ways: film condensation and drop condensation. Presence of noncondensable gases affects the rate of condensation and the film heat transfer coefficient may be reduced considerably [2].

The heat flux due to phase change of boiling and condensation can be expressed as

$$q = hA\left(T_s - T_\infty\right) \tag{2.10}$$

The heat transfer coefficients experienced when a liquid is vaporized or when a vapor is condensed are considerably greater than that for heat transfer without a phase change. However, it is rather more difficult to measure heat transfer coefficients of phase changes.

The heat transfer coefficient in nucleate boiling may be calculated by a correlation. Kutateladze's correlation is a commonly used one, which is given by [2]

$$\left(\frac{h_b}{k}\right)\psi^{0.5} = 0.0007\left[\frac{q_{max}}{\alpha\lambda\rho_v}\frac{P}{\sigma}\Psi\right]^{0.7}\mathrm{Pr}^{-0.35} \tag{2.11}$$

where

$$\Psi = \frac{\sigma}{g\left(\rho_l - \rho_v\right)} \tag{2.12}$$

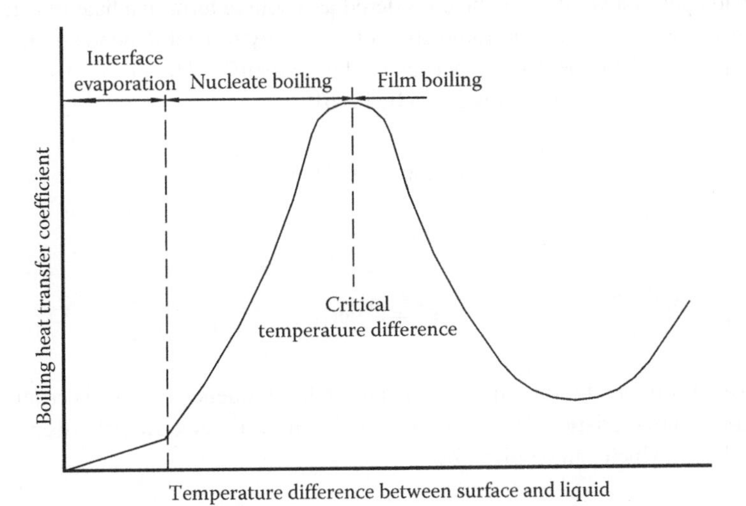

FIGURE 2.3 Relationship between boiling heat transfer coefficient and temperature difference.

$$q_{max} = 0.16\lambda\rho_v\left(\frac{\sigma g(\rho_l - \rho_v)}{\rho_v^2}\right)^{0.25} \quad (2.13)$$

The film heat transfer coefficient for condensation can be predicted from the Nusselt theory, which gives the mean film coefficient by [2]

$$h_{cd} = 0.943\left(\frac{\rho^2 k^3 g\lambda}{\mu L\Delta T}\right)^{0.25} \quad \text{for a vertical surface} \quad (2.14)$$

$$h_{cd} = 0.725\left(\frac{\rho^2 k^3 g\lambda}{\mu d\Delta T}\right)^{0.25} \quad \text{for a horizontal tube} \quad (2.15)$$

2.2.1.3 Heat Transfer with Electromagnetic Waves

Microwave and *radio frequency* are widely used in the food industry. Microwave energy is a type of electromagnetic waves with frequencies between 300 MHz and 300 GHz. Radio frequency is a rate of oscillations in the range of about 3 kHz–300 GHz, which are alternating currents to carry radio signals. Microwaves are nonionizing electromagnetic waves and commercial microwave heating applications use frequencies of 2450 MHz, sometimes 915 MHz in the United States and 896 MHz in Europe. During conventional heating, the energy is transferred to the material through convection, conduction, and radiation of heat from the surfaces of the materials. However, during microwave and radio frequency heating, the energy is delivered directly to the product through molecular interaction with an electromagnetic and electric field.

Microwaves and radio frequency are transmitted as waves, which can penetrate foods and interact with the polar molecules such as water in the foods to be converted to heat. Electromagnetic spectrum is normally characterized by wavelength (λ) and frequency (ν). The depth of penetration into a food is directly related to frequency and the lower frequency waves penetrate more deeply. As microwave and radio wave can penetrate into foods, they can heat foods quicker than the food heated through conduction from the outer surface. Once microwave and radio wave energy have been absorbed by foodstuffs, heat is transferred throughout the food mass by conduction or convection. The rate of energy conversion per unit volume can be considered as a source term in a heat transfer model.

The conversion of microwave and radio frequency energy to heat depends on the properties of the energy source and the dielectric properties of the foodstuffs. The power dissipation or rate of energy conversion per unit volume, S, is given by [2–5]

$$S = 5.56 \times 10^{-15} E^2 \nu \varepsilon'' \quad (2.16)$$

where

$$\varepsilon'' = \varepsilon' \tan\beta \quad (2.17)$$

However, the suitability of a food for microwave and radio frequency heating is crucially dependant on the penetration characteristics. The microwave and radio wave electric field strength is a function of penetration depth, which can be given by

$$E = E_0 e^{-2\alpha' x} \quad (2.18)$$

where

$$\alpha' = \frac{2\pi}{\lambda}\left[\frac{\varepsilon'}{2}\left(\sqrt{1+\tan^2\beta}-1\right)\right]^{0.5} \tag{2.19}$$

2.2.1.4 Heat Transfer during Ohmic Heating

Ohmic heating is a rapid and relatively uniform heating method. The basic principle of ohmic heating is that electric energy is converted to thermal energy within a conductor. Like microwave heating, the increase of food temperature during ohmic heating is caused by the heat generated inside an electrically conductive food material when a current is applied across the material. The internal heat generation rate on a volume basis, S, is given by [6,7]

$$S = \sigma\left|\nabla V\right|^2 \tag{2.20}$$

where

σ is the electric conductivity
V is the voltage
$|\nabla V|$ is the gradient of electric potential

The electric field distribution within an ohmic heating cell can be calculated using the following Laplace equation [6,7]:

$$\nabla(\sigma\nabla V) = 0 \tag{2.21}$$

The electric conductivity is a function of temperature. For most aqueous materials, the electric conductivity increases linearly with temperature, which can be calculated by [8]

$$\sigma = \sigma_0(1 - mT) \tag{2.22}$$

where

σ_0 is the electric conductivity at a reference temperature
m is a constant

2.2.1.5 Heat Transfer during Infrared Radiation

Infrared radiation can achieve contactless heating. Advantages of infrared over convective heating with hot air or water are higher heat transfer coefficients [9]. Infrared power absorption by foods has been treated by two formulations: zero penetration and finite penetration depth [10]. With the assumption of zero penetration, the infrared energy is absorbed by the surface of a food object and converted into heat on the surface, which is further transferred into the food by conduction. Therefore, on the boundary of a heat transfer model, there is [9,11]

$$k\nabla T = q_{inf,s} = \sigma\varepsilon\left(T_\infty^4 - T_s^4\right) \tag{2.23}$$

where

k is thermal conductivity of the food
σ is the Stefan–Boltzmann constant (5.669×10^{-8} W/m^2 K^4)
ε is emissivity
T is temperature in Kelvin

The formulation of finite penetration depth assumed that the infrared energy from heaters suddenly impinges upon a food surface and directly penetrates into the food by approximately 1 mm under the surface. Therefore, all of the infrared energy is completely absorbed from the food surface into the depth of 1 mm, which is called the penetrating layer. The interior of the food from the depth of 1 mm through the core of the food is called the conductive layer [12]. The heat transfer in the penetrating layer can be described by Fourier's equation of heat conduction with an inner heat generation term covering the infrared heat generation [12]. The inner heat generation in the penetrating layer can be calculated by an exponential decay model [13]:

$$S(x) = \frac{q_{inf,s}}{\delta_{inf}(x)} \exp\left(-\int_0^x \frac{dx}{\delta_{inf}}\right) \quad (2.24)$$

where

$q_{inf,s}$ is the surface infrared flux
δ_{inf} is infrared penetration depth

2.2.1.6 Heat Transfer during High-Pressure Processing

High-pressure processing technology has been used to destroy microorganisms and extend the shelf of foods. The compression during high-pressure processing of food generates heat, resulting in the increase of temperature in the pressure vessel. At an initial temperature of 25°C and a pressure from 150 to 600 MPa, the temperature increases were measured at 2.6°C–2.9°C/100 MPa for water, 2.9°C–3.5°C/100 MPa for honey, 3.2°C/100 MPa for ground beef and whole milk, and 6.6°C–9.2°C/100 MPa for vegetable oil [14]. Therefore, for a high-pressure sterilization process, the heat generated by the compression can provide additional synergetic sterilization effects. The heat generated by compression during high-pressure processing is dissipated by a combination of conduction and convection within the pressuring fluid in the chamber [15]. High-pressure compression or expansion is usually assumed to be adiabatic. The thermodynamic equation governing the adiabatic high-pressure compression or expansion is given by [16]

$$T\,dS = c_p\,dT - T\left(\frac{\partial V}{\partial T}\right)_P dP = 0 \quad (2.25)$$

where

T is the temperature (K)
P is pressure (MPa)
V is the specific volume (m^3/kg)
c_p is the specific heat of the product (kJ/kg K)

The temperature variation produced by an adiabatic pressure change is thus expressed as [16,17]

$$\frac{dT}{dP} = \frac{TV\beta}{c_p} \quad (2.26)$$

where β is thermal expansivity (1/K).

The internal heat generation rate on a volume basis, S, during high-pressure compression or expansion is given by [15]

$$S = \beta T \frac{dP}{dt} \tag{2.27}$$

Due to the internal heat generation, temperature gradients are established in food products during compression. Modeling heat transfer in high-pressure food processes is a useful tool to optimize the processes [17–19]. The main challenge in modeling heat transfer during high-pressure processing is the lack of appropriate thermophysical properties of food materials under pressure [17]. Accurate thermophysical properties including density, specific heat, thermal conductivity, and thermal expansivity of a food product under pressure are needed for modeling heat transfer during high-pressure processing. The methods for measuring those thermophysical properties of foods under pressure can be found in the literature [16,20].

2.2.2 Mass Transfer

Mass transfer is concerned with the movement of materials in fluid systems including gases or liquids. Mass transfer may take place according to two mechanisms: *molecular diffusion* or *convective mass transfer*. When there is a concentration gradient of the considered component between two points of the system, mass transfer is produced by molecular diffusion. When the entire mass moves from one point to another, the transfer is produced by convection.

2.2.2.1 Molecular Diffusion

Diffusion is the process by which matter is transported from one part of a system to another as a result of random molecular motion. Although no molecule has a preferred direction of motion, observation indicated that molecules transfer from a region of higher concentration to a region of lower concentration. The mass flux can be described by *Fick's first law*:

$$J_A = -D_A \frac{dC_A}{dx} \tag{2.28}$$

where D is diffusivity, which is defined as the constant of proportionality in Fick's law.

The diffusion of fluids within the pore spaces of a porous solid is of some interest to food processing such as drying. It is possible to quantify an effective diffusivity which describes the transfer of gas or liquid within a solid porous food. Effective diffusivities of moisture in some solid foods are listed in Table 2.1 [21].

In Equation 2.28, concentrations can be expressed in a number of ways such as molar concentration, partial pressure, mass concentration, and mass and mole fractions. The corresponding units of different concentrations differ considerably.

2.2.2.2 Convective Mass Transfer

During food processes, mass may be transferred between distinct phases across a phase boundary. Whitman assumed that two laminar films exist on each side of the interface when a gaseous solute transfers from a gas phase to a liquid phase or mass transfers between two liquid phases as shown in Figure 2.4 [2]. According to Whitman's theory, the resistance to mass transfer is contained within the two films and the concentration gradients across each film are linear. Whitman's two-film theory is the earliest and most generally one to account for interphase mass transfer. Mass transfer within the films is assumed to be due solely to molecular diffusion and thus Fick's law can be applied directly to each film. Integrating over the linear concentration gradient, the Fick's law can be expressed as [2]

TABLE 2.1
Effective Diffusivities in Solid Foods

	Moisture Content (%, By Dry Base)	Temperature (K)	Diffusivity $\times 10^{11}$ (m^2/s)
Starch gel	10	298	0.1
Starch gel	30	298	2.3
Blanched potato	60	327	26.0
Air-dried apple	12	303	0.65
Freeze-dried apple	12	303	12.0
Fish muscle	30	303	34.0
Raw minced beef	60	333	10.0
Cooked minced beef	60	333	12.0

Source: Saravacos, G.D., Mass transfer properties of foods, in: M.A. Rao, S.S.H. Rizvi, eds., *Engineering Properties of Foods*, 2nd edn., Marcel Dekker, New York, pp. 169–221, 1994.

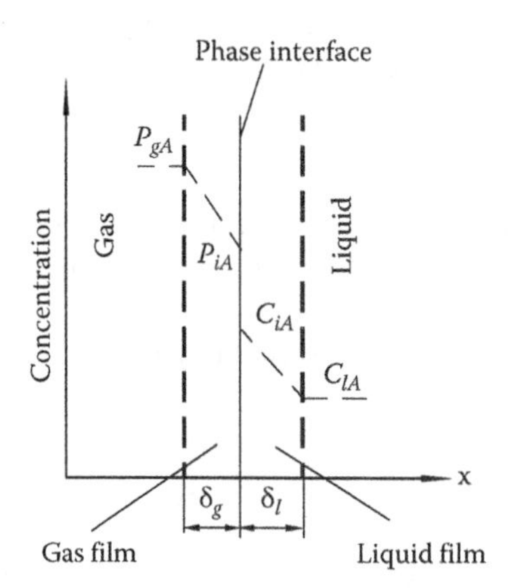

FIGURE 2.4 Whitman's two-film diffusion theory. (From Smith, P.G., *Introduction to Food Process Engineering*, Kluwer Academic/Plenum Publishers, New York, pp. 163–170, 179–186, 191–217, 2003.)

$$J_A = -\frac{D_{gA}}{\delta_g}\left(P_{iA} - P_{gA}\right) = -\frac{D_{lA}}{\delta_l}\left(C_{lA} - C_{iA}\right) \tag{2.29}$$

However, the thickness δ_g and δ_l cannot be measured or predicted independently. In order to overcome this difficulty, the terms of D_{gA}/δ_g and D_{lA}/δ_l are replaced by a gas film mass transfer coefficient, k_g, and a liquid film mass transfer coefficient, k_l, respectively.

The partial pressure and molar concentration on the interface cannot be determined independently because of the uncertainty of the interface position and the impossibility of measurement of interfacial concentration. Therefore, an overall gas and liquid mass transfer coefficients are introduced based on the concentration differences which can be determined. Equation 2.29 can be rewritten as

$$J_A = -K_g\left(P_A^* - P_{gA}\right) = -K_l\left(C_{lA} - C_A^*\right) \tag{2.30}$$

where

P_A^* is the partial pressure of A in the gas phase, which is in equilibrium with the bulk liquid concentration

C_{lA} and C_A^* are molar concentrations of A in the liquid phase which is in equilibrium with the bulk partial pressure, P_{gA}

The equilibrium relationships are determined by *Henry's law*, which are expressed as

$$P_A^* = HC_{lA} \tag{2.31}$$

$$C_A^* = \frac{P_{gA}}{H} \tag{2.32}$$

2.2.3 Unsteady-State Heat and Mass Transfer

For steady state heat and mass transfer, there is no change in temperature or concentrate of the material with time. However, in the majority of food thermal processing applications, the food temperature or concentrate of a food component is constantly changing, and unsteady-state heat and mass transfer is more commonly found. Heat and mass transfer follows the same pattern, which can be described in a generalized manner. The generalized governing equation of *unsteady-sate* heat and mass transfer can be expressed as

$$\frac{\partial \phi}{\partial t} + u_x \frac{\partial \phi}{\partial x} + u_y \frac{\partial \phi}{\partial y} + u_z \frac{\partial \phi}{\partial z} = \frac{\partial}{\partial x}\left(\alpha \frac{\partial \phi}{\partial x}\right) + \frac{\partial}{\partial y}\left(\alpha \frac{\partial \phi}{\partial y}\right) + \frac{\partial}{\partial z}\left(\alpha \frac{\partial \phi}{\partial z}\right) + S \tag{2.33}$$

where

ϕ is temperature for heat transfer and concentration for mass transfer

α is diffusivity (for heat transfer $\alpha = k/(\rho c)$ and for diffusion $\alpha = D$)

In order to find the solution of Equation 2.33, it is necessary to know the initial and boundary conditions. The *initial conditions* give what happens at the start. The initial conditions may be the same initial temperature or concentration, $\phi|_{t=0} = \phi_0$. The initial conditions may also be an initial profile of temperature or concentration, $\phi|_{t=0} = \phi_0(x,y,z)$. The *boundary conditions* give what happens at the boundaries of the phase to be investigated. The boundary conditions may be (1) a constant, $\phi|_\Gamma = \phi_s$; (2) a flux, $\phi|_\Gamma = q_s$; (3) a convection, $\phi|_\Gamma = h(\phi_s - \phi_\infty)$; or (4) a combination of flux and convection, $\phi|_\Gamma = q_s + h(\phi_s - \phi_\infty)$.

Sometimes, depending on the geometry of the product to be studied, it is useful to consider alternative coordinate systems such as the cylindrical coordinate and spherical coordinate systems. However, whichever system is used, the intrinsic mechanisms and physical laws of heat and mass transfer remain the same.

2.2.4 Overview of Solution Methods

Variables such as temperature and moisture used in modeling the unsteady-state thermal processes depend on time and position. The equations governing the physical mechanism of unsteady-state heat and mass transfer are thus of a partial differential type as shown in Equation 2.33. An *analytical solution* of the partial differential equation is continuous. However, the possibility of analytical solution is restricted to rather simple forms of the governing equations, boundary and initial conditions. *Numerical methods* have been widely used to solve the *partial differential equations* governing the heat and mass transfer. Numerical methods can generate discretized solutions for the partial differential equations.

Finite difference (FD) method is simple to formulate a set of discretized equations from the transport differential equations in a differential manner [22]. The FD method is normally used for simple geometries such as sphere, slab, and cylinder (Table 2.2). The FD method has been widely used to solve heat and mass transfer models of many food processes [23–41]. There are a number of important publications which have improved the knowledge of the FD scheme for predicting the heat and mass transfer during food processes [42,43]. Additional information can be found in a review paper by Wang and Sun [44]. Table 2.2 gives a summary of recent development of FD models for simulating food thermal processes. However, for foods with irregular shapes, the surface temperature predictions by the FD method are less satisfactory due to geometric simplification.

Finite element (FE) method may perform better than the FD method for irregular geometries, complex boundary conditions, and heterogeneous materials. The FE method involves discretizing a large domain into a large number of small elements, developing element equations, assembling the element equations for the whole domain, and solving the assembled equations. The FE discretization of the governing differential equations is based on the use of interpolating polynomials to describe the variation of a field variable within an element. Although the spatial discretization is different for the FE method compared with FD method, it is usual to employ an FD method for the time progression in a transient problem [45,46]. The FE method has been successfully used to solve the heat and mass transfer models of food processes [47–73]. Additional information can be found in two review papers by Wang and Sun [44] and Puri and Anantheswaran [74]. A summary of various FE methods developed recently for analyzing food thermal processes is listed in Table 2.3. However, the FE method is complex and computationally expensive than the FD method.

Computational fluid dynamics (CFD) is a simulation tool for the solution of fluid flow and heat transfer problems. In CFD calculation, the continuity equation, momentum conservation equation (also known as the Navier–Stokes transport equations), and energy conservation equation are numerically solved to give predictions of velocity, temperature, shear, pressure profiles, and other parameters in a fluid flow system [75]. In the last few years, there has been continuous progress in the development of CFD codes. Some of the common commercial codes include CFX (http://www.ansys.com/products/fluid-dynamics/cfx/), Fluent (http://www.fluent.com/), Phoenics (http://www.cham.co.uk/), and Star-CD (http://www.cd.co.uk). The computational procedure of most of commercial CFD packages is based on finite volume (FV) numerical method. In fact, the FV method was derived from the FD method. In the FV method, the domain is divided into discrete control volumes. The key step of the FV scheme is the integration of the transport equations over a control volume to yield a discretized equation at its nodal points [76]. Although CFD has been applied to industries such as aerospace, automotive, and nuclear for several decades, it has only recently been applied to the food processing industry due to the rapid development in computer and commercial software packages. A review of CFD in the food industry has been given by Scott and Richardson [77] and Xia and Sun [78]. Langrish and Fletcher reviewed the applications of CFD in spray drying [79]. Applications of CFD in the food industry include analyses of air flow in ovens and chillers, fluid flow of particle foods in processing systems, convection flow patterns in containers during thermal processing such as sterilization, and modeling of vacuum cooling process [80–96]. The transport equations of CFD can be applied to both laminar and turbulent flow conditions. The eddy viscosity models such as κ–ε approach, and second-order closure models are used to describe the flow turbulence if the effects of turbulence on the effective viscosity need to be considered. A summary of various CFD models developed recently for analyzing food thermal processes is presented in Table 2.4.

TABLE 2.2
Summary of FD Method in Thermal Food Processing

Processes	References	Affiliation	Heat Model	Mass Model	Dimension	Temperature-Dependent Properties	Foods
Dehydration and drying	[100]	Ciudad Universitaria, Argentina		√	1D	√	Potato
	[102]	University of Illes Balears, Spain		√	3D	√	Aloe vera
	[101]	University of Wisconsin-Madison, USA		√	3D	√	Sugar film
	[24]	University of Reading, United Kingdom	√	√	1D	√	Solid foods
	[39]	Swedish Institute for Food and Biotechnology	√	√	1D	√	Bread
	[38]	Hong Kong University of Science and Technology	√	√	1D	√	Vegetables
Pasteurization and sterilization	[26,30]	Higher Institute of Food and Flavour Industries, Bulgaria	√		1D		Mushroom
	[23]	Memorial University of Newfoundland and McGill University, Canada	√		3D		Various
	[41]	North Carolina State University, USA	√		2D-axi		Cucumber
Cooking and frying	[35,36]	University of Florida, USA	√		2D-axi	√	Shrimp
	[28,29]	North Carolina State University and University of California, Davis, USA	√	√	1D		Potato
	[40]	Archer Daniels Midland Co., USA, and University of California, Davis, USA	√	√	1D-axi	√	Hamburger patty

(continued)

TABLE 2.2 (continued)
Summary of FD Method in Thermal Food Processing

Processes	References	Affiliation	Heat Model	Mass Model	Dimension	Temperature-Dependent Properties	Foods
Others	[140]	Universidad Nacional de Misiones, Argentina	√		1D-axi	√	Twigs of yerba mate
	[128]	Middle East Technical University, Turkey, and Ohio State University, USA	√		1D	√	Potato

Source: Adapted from Wang, L.J. and Sun, D.-W., *Trends Food Sci. Technol.*, 14, 408, 2003.

TABLE 2.3
Summary of FE Method in Thermal Food Processing

Processes	References	Affiliation	Heat Model	Mass Model	Dimension	Temperature-Dependent Properties	Foods
Dehydration and drying	[103]	University of Saskatchewan, Canada, and Utah State University, USA	√	√	2D	√	Starch
	[60–64]	University College Dublin, Ireland, and China Agricultural University	√	√	2D-axi	√	Grain
	[65]	Purdue University, USA	√	√	2D-axi	√	Biscuits
Pasteurization and sterilization	[124,126,129]	Universidad Católica Portuguesa, Portugal, and University College Cork, Ireland	√		2D-axi		Various
Cooking and frying	[58]	University of Arkansas, USA	√		2D-axi	√	Chicken

(*continued*)

TABLE 2.3 (continued)
Summary of FE Method in Thermal Food Processing

Processes	References	Affiliation	Heat Model	Mass Model	Dimension	Temperature-Dependent Properties	Foods
	[54]	Technical University of Nova Scotia, Canada	√		2D-axi	√	Meat
	[59]	Cornell University, USA	√		2D-axi	√	Various
	[115]						
	[51,52,111,112]	Pennsylvania State University, USA	√	√	2D and 3D		Solid foods
	[119,126, 120–123]	Katholieke Universiteit Leuven, Belgium	√		2D-axi		Various
	[73]	University of California, Davis, USA	√	√	2D-axi	√	Hamburger patty
Others	[130]	University of Tennessee, USA	√		2D		Carrot
	[138]	Katholieke Universiteit Leuven, Belgium	√		3D	√	Various

Source: Adapted from Wang, L.J. and Sun, D.-W., *Trends Food Sci. Technol.*, 14, 408, 2003.

TABLE 2.4
Summary of CFD in Thermal Food Processing

Processes	References	Affiliation	Foods
Drying	[108]	NIZO Food Research, the Netherlands	Particle foods
	[106]	National Centre for Scientific Research, Greece	Fruits and vegetables
	[84]	INRA, France	Sausage
Pasteurization and sterilization	[80,81,86]	University of Auckland, New Zealand	Canned liquid foods
	[82]	University of Birmingham, United Kingdom	Various
	[96]	Indian Institute of Technology, India	Canned liquid foods
Heating	[109,110]	Katholieke Universiteit Leuven, Belgium	Various
Others	[132]	Katholieke Universiteit Leuven, Belgium	Various
	[131]	Institut National de la Recherche Agronomique and Institut de Mecanique des Fluides de Toulouse, France	Various

Source: Adapted from Wang, L.J. and Sun, D.-W., *Trends Food Sci. Technol.*, 14, 408, 2003.

2.3 HEAT AND MASS TRANSFER APPLIED TO THERMAL FOOD PROCESSING

During thermal food processing, heat must be transferred between a heat source or sink and the inside zone of the food usually through an interface such as food surface or container wall. External heat transfer between the source or sink and the interface may occur by any of the heat transfer mechanisms (including conduction, convection, radiation, and phase changes). Internal heat transfer from the interface to the inside zone of foods is usually by conduction of solid foods or by conduction and convection of liquid foods. If microwave and radio frequency are used, heat can also travel to the inside zone by penetrating radiation. Moisture, water vapor, nutrients, and flavor must first travel to the food surface by any of internal mass transfer mechanisms such as diffusion. Then they must travel from the food surface to the ambient by external mass transfer processes such as convective mass transfer. For a series of the mechanisms of external heat transfer, internal heat transfer, internal mass transfer, and external mass transfer, the step with the greatest effect on the rate will be the slowest one, which is the rate-determining step [97].

Heat transfer through solid foods is normally modeled by Fourier's equation of heat conduction, and mass transfer is generally described by Fick's law of diffusion [98]. For thermal processes of fluid foods, the conservation of mass, momentum, and energy in a fluid should be considered together. The *continuity equation* and *Navier–Stokes equations* are used to describe fluid flow [76]. The actual conditions imposed by the processing equipment are considered as the boundary conditions of the governing equations. Most heat and mass transfer models can only be solved analytically for simple cases. Numerical methods are useful for estimating the thermal behavior of foods under complex but realistic conditions such as variation in initial temperature, nonlinear, and nonisotropic thermal properties, irregular-shaped bodies and time-dependent boundary conditions. In solving the models, the FD and FE methods are widely used. In recent years, the FV method was the main computational scheme used in commercial CFD software packages. CFD has been increasingly used to simulate thermal processes of foods for analyzing complex flow behavior [44,75].

2.3.1 PASTEURIZATION AND STERILIZATION

Pasteurization and *sterilization* are widely used in the food industry to inactivate microorganisms present in foods for ensuring food safety and extending the shelf life of foods. In aseptic processing, the products are first thermally treated, then carried to a previously sterilized container, and sealed under sterile environment conditions. The thermal processing of packed products is carried out in equipment that uses steam or hot water as the heating fluid. The pasteurization and sterilization techniques are initially used in liquid foods such as milk and fruit juices. Recently it has also been applied to particulate food products [99].

2.3.1.1 Pasteurization and Sterilization of Liquid Foods

Sterilizing process of canned liquid foods is a typical example of fluid flow with heat transfer. CFD model can thus be used to predict transient flow patterns and temperature profiles in a can filled with liquid foods. For simulating the sterilizing process of canned liquid foods, the energy equation needs to be solved simultaneously with the continuity and momentum equations in a CFD model [80,81,86]. Continuous sterilization processes of single-phase mixtures such as milks and fruit juices have become more and more common. The continuous process is called the HTST (high-temperature-short-time) sterilization process, which gives the same level of sterility but a reduced quality loss as compared to batch sterilization process. For optimizing the quality of foods during continuous sterilization, the laminar flow of liquid foods in circular pipes with uniform wall temperature can be described by a CFD model [82].

2.3.1.2 Pasteurization and Sterilization of Particle–Liquid Foods

Sterilization of canned solid particle foods with a brine solution in a container is a typical liquid–solid thermal process. Blanching of fresh vegetables and sous vide processing of particulate foods are also heating practices in a liquid–solid systems. In this system, the low-viscosity brine liquid is heated by convection and the solid particle foods by conduction. A heat conduction model can be used to simply determine the temperature distribution in a canned particle body. Meanwhile, the temperature of brine liquid in the heated cans, which is variable with the temperature outside of the cans, can be simply described by the regular regime differential equation:

$$\frac{dT_l}{dt} = \frac{T_m - T_l}{\Phi} \tag{2.34}$$

where the thermal inertia, Φ, which characterizes the temperature lag of the brine liquid from the temperature of heating medium, is experimentally determined by monitoring the temperature of the brine with linearly increasing, holding, and linearly decreasing the temperature of the medium [26,30].

For liquid–solid thermal process, as heat transfer coefficient of surface convection is normally very large due to good circulation of brine liquid in the container, the effect of the coefficient on the temperature profiles of foods is normally assumed to be negligible. This means that if the coefficient is big enough, the total heat transfer rate is controlled by conduction through the particle food body. For this reason, the heat transfer coefficient can arbitrarily be set at a very high value in a simulation, for example, 5000 W/m^2 K [23,41].

2.3.2 Dehydration and Drying

Dehydration, or *drying*, is a unit operation of food thermal processing most commonly used for food preservation. Reduction of water in foods during drying can achieve better microbiological preservation and retard many undesirable reactions. Drying can also decrease packaging, handling, storage, and transport costs due to the decrease of food weight. Drying process is mainly characterized by moisture loss in foods. In most cases, the removal of water from a food is achieved by blowing a dry air flow, which transports water from the surface of the product to the air stream. However, spray drying, freeze drying, microwave drying, far-infrared drying, and other methods are also used for drying some special products. Drying of food materials is normally a complex process involving simultaneous coupled heat and mass transfer in the materials. It is important to know the mechanisms related to the movements of water inside and outside the food [99].

2.3.2.1 Air Drying

Air drying is the most popular drying method in the food industry. For a drying process with a small Biot number, a uniform temperature profile in foods can be assumed in simulation. This uniform temperature can be determined by a heat balance between the dried food body and drying medium [100,101], or be assumed to be the air temperature [102]. The moisture transfer through the foods is normally described by the differential equation of Fick's law of diffusion, which is expressed as

$$\frac{\partial X_w}{\partial t} = \nabla\left(D\nabla X_w\right) \tag{2.35}$$

The diffusion coefficient is important for the accuracy of model prediction. The diffusion coefficient can be regressed as a function of temperature and concentration by using data in the literature [101]. Alternatively, the diffusion coefficient can be determined by Arrhenius law as [24,100,102]

$$D = D_0 \exp\left(-\frac{E_a}{RT_K}\right) \tag{2.36}$$

and E_a and D_0 are varied during simulation until a reasonable agreement between predicted and experimental results is obtained.

However, for a drying process with a big Biot number, a coupled mass and heat transfer should be taken into account in the simulation. For drying of a composite food system, simulation found that the predicted temperature, moisture, and pressure distributions in the composite food system by the coupled model agreed with experimental data. However, there was a big difference between the predicted values by the uncoupled model and experimental data [103].

In most cases, it is often assumed that moisture diffuses to the outer boundaries in a liquid form and evaporation takes place only on the surface. The diffusion models do not separate liquid water and water vapor diffusion [24]. However, in some cases, inner water evaporation during drying is significant and therefore simultaneous heat, water and vapor diffusion should be considered in simulation [39]. For example, for predicting the drying process of breads, simultaneous heat, water and vapor diffusion through breads was described by using three governing equations, respectively. The three governing equations of heat, moisture, and vapor were connected by the equilibrium of local moisture evaporation and vapor condensation, which is determined by the relationship between saturated vapor pressure and local temperature [39]. Simulations on drying process of vegetables and fruits using the coupled heat, water and vapor diffusion model confirmed that the assumption of evaporation–condensation front in the drying model was valid for drying of porous moisture materials with big permeability such as banana. However, the assumption of evaporation–condensation front was invalid and more comprehensive analysis was necessary if the permeability of dehydrated foods and vegetables was below 10^{-19} m^2 [38]. To model coupled heat and mass transfer in *porous foods*, two types of formulations are usually developed: one involving distributed evaporation through the whole food object and the other involving a sharp moving interface where evaporation occurs [104,105].

On the surface of a food body, external mass transfer is normally assumed to be proportional to the vapor pressure difference between the surface and the drying media [24]. The surface mass transfer coefficients are affected by the properties of air, operating conditions, design of the dryer, and the product. Pressure profiles and velocity of heated air above products in an air dryer can be determined by a CFD model [106]. In this case, the turbulent flow, which is characterized by relatively high velocity and the presence of many obstacles in the air dryer, can be described by the Chen–Kim κ–ε model [107].

2.3.2.2 Spray Drying

During *spray drying*, coupled heat, mass, and pressure transfer phenomenon occurs. The drying of droplets is influenced by external and internal transport phenomena alike. For simulating gas flow in a spray dryer and calculating the trajectories and the course of the atomized particles, CFD is widely used [79]. The κ–ε turbulence model is used to calculate the gas flow field. The differential equation that describes the diffusion process in spherical particles is then solved simultaneously with equations for external heat and mass transfer [108].

2.3.2.3 Microwave and Radio Frequency Drying

Microwave is used in drying of some heat-sensitive foods [56,65]. The heat and moisture transfer during microwave can be described by Fourier's equation of heat conduction with inner heat generation and Fick's law of diffusion, respectively [65]. In modeling the coupled heat and moisture transfer through porous materials during microwave-assisted vacuum drying, a combination of liquid water and vapor transfer should be taken into account in the equation of mass transfer.

Meanwhile, heat transfer can be described by Fourier's equation of heat conduction with an inner heat generation term covering latent heat of water evaporation and source heat of microwave power. However, as moisture transfer is caused by the temperature gradient in foods, the equation of moisture transfer can even be simplified into an isothermal equation if the temperature gradient is too small [56].

2.3.3 Cooking and Frying

2.3.3.1 Air Convection Cooking

Air convection-heating oven is popular cooking equipment. For predicting transient temperature and moisture distribution in chicken patties of regular shapes in a cooking oven, a coupled heat and mass transfer model was found to give better prediction than that of single heat transfer model [58]. In some cases, if it is difficult to find data for mass diffusivity and mass transfer coefficient, a volumetric moisture loss rate due to evaporation can be experimentally determined and the heat removed due to moisture loss can then be incorporated into Fourier's equation of heat conduction as inner heat generation [54].

With powerful computers available, heating and cooking of solid foods in an industrial convection-type oven can be modeled as a fluid flow and heat transfer problem. CFD offers an efficient and effective tool to analyze the performance of industrial convection-type oven such as hot-air electric forced convection ovens. In the CFD models, the electric heating coils and the fan can be modeled in the momentum equation (the Navier–Stokes equations) as a distributed resistance and a distributed body force in the region of the flow domain where the coils and fan are positioned. The value of turbulent viscosity in the momentum equation can be obtained by using the standard and renormalization group version of the κ–ε turbulence model [109,110].

2.3.3.2 Microwave and Radio Frequency Cooking

Microwave- and radio wave-heated and cooked foods are becoming increasingly popular in the food market and at home. For modeling microwave and radio wave heating process, the heat transfer through a solid food body can also be described by Fourier's equation of heat conduction with inner heat generation due to the microwave and radio wave energy absorbed by the food components. The microwave and radio wave power density absorbed at any location in foodstuffs can be derived as a function of dielectric properties and geometry of the food given by Equation 2.16. Meanwhile, heat losses on the surface of food body by convention and evaporation can be included in the boundary conditions. For simulating microwave and radio wave heating of solid food with rectangular and cylindrical shapes, FE analysis may be a powerful tool to numerically solve the model [52]. During microwave heating, a big moisture loss sometimes occurs. In this case, a coupled heat and mass transfer model should be developed and additional moisture transfer through a solid food body can be modeled by the diffusion equation of Fick's law [51]. The moisture evaporation rate on surface can be obtained by using drying experiment and regressed as a function of temperature [111,112].

2.3.3.3 Ohmic Heating

For modeling ohmic heating process, the heat transfer through a solid food body can also be described by Fourier's equation of heat conduction with inner heat generation due to the conversion of electric energy into heat to be absorbed by the food components. The general governing equation, which is derived from Equation 2.33, is given by

$$\rho c_p \frac{\partial T}{\partial t} = \nabla(k\nabla T) + S \tag{2.37}$$

The inner heat generation rate can be calculated by Equation 2.20. A heat transfer model was developed to describe the ohmic heating of liquid foods such as chicken noodle soup in a flexible pouch package. The model was solved using commercial CFD software Fluent to optimize the design of electrodes for uniform heating of the material [6]. A heat transfer model based on Equation 2.37 can also be developed to quantify the temperature distribution in a solid food during ohmic heating [7].

2.3.3.4 Frying

When foods are fried, crust formation is easily observed in many foods. The crust layer increases in thickness as the frying process proceeds, and the interface between the crust and the core region becomes a moving boundary. For a phase change problem in frying, one side of the interface is crust and the other is core region. Fourier's law of heat conduction can be used to describe the heat transfer on both sides of the interface:

$$\rho_1 c_{p1} \frac{\partial T_1}{\partial t} = \nabla\left(k_1 \nabla T_1\right) \quad \text{(for frozen)} \tag{2.38}$$

$$\rho_2 c_{p2} \frac{\partial T_2}{\partial t} = \nabla\left(k_2 \nabla T_2\right) \quad \text{(for unfrozen)} \tag{2.39}$$

The interface between two phases is tracked by Equation 2.8. It should be stressed that the crust and core regions have significantly different thermophysical properties. Because the phase change in foods occurs over a range of temperature, the thermophysical properties of foods experience extreme discontinuities at the phase change temperatures. These discontinuities cause instability in the numerical solutions. Alternatively, the *enthalpy formulation* technique based on the relationship between enthalpy and temperature is used to model the phase change problem. One advantage of the enthalpy formulation is that it is not necessary to track the moving interface. Other advantages include the relative stability and simplicity of the method. Using the enthalpy method, Equations 2.38 and 2.39 can be replaced by one single equation as [113]

$$\rho \frac{\partial H}{\partial t} = \nabla\left(k \nabla T\right) \tag{2.40}$$

During frying, there occurs significant mass transfer as the movement of fat/oil and moisture into or out of the food. A set of mass transfer model based on Fick's law of diffusion is widely used to describe the moisture and oil/fat movement during frying. Both mass and heat transfer models are coupled for simulating the frying process of foods [73,113].

2.4 CHALLENGES IN MODELING HEAT AND MASS TRANSFER

Although continuous progress has been made in recent years in improving the accuracy of the modeling, much research work still needs to be carried out. The following identifies several possible areas where further research could be performed in order to further improve the accuracy of model prediction.

2.4.1 Mechanisms in Heat and Mass Transfer

Drying of moisture and porous foods is widely used in the food industry. Drying involves coupled heat and mass transfer through a porous media. It is still difficult to predict the moisture transfer rate through a porous media because the mechanisms involved are complex and not completely understood [104,105]. As a result, the design of drying process remains largely an art based on experience gained from trial and error testing. Often, the controlling resistance is from internal mass transfer and the internal mass transfer may occur through the solid phase or within the void spaces. Several mechanisms of internal mass transfer including vapor diffusion, moisture diffusion and then surface evaporation, hydrodynamic flow, and capillary flow have been proposed. The fundamental transport modes of molecular diffusion, capillary diffusion, and pressure-driven Darcy flow were given in the literature [104]. However, modeling of drying processes is complicated because there is nearly always more than one mechanism to the total flow [114].

During microwave heating, the heating patterns can be uneven. Food factors such as dielectric properties, size, and shape play more important role as compared to conventional heating because they affect not only the magnitude of heat generation but also its spatial distribution [115]. Modeling of microwave heating process involves solutions of electromagnetic equation and the energy equation. Lambert's law is a simple and commonly used power formulation, according to which the microwave power is attenuated exponentially as a function of distance of penetration into the sample [116,117]. Although *Lambert's law* is valid for samples thick enough to be treated as infinitely thick, it is a poor approximation in many practical situations. In such cases, a rigorous formulation of the heating problem requires solving *Maxwell's equations*, which govern the propagation of electromagnetic radiation in a dielectric medium [116,117]. During microwave heating, a large temperature change may cause a significant variation in dielectric properties, resulting in big change in the heating pattern. Therefore, a coupled Maxwell's equation with the heat transfer model is necessary to describe the microwave heating process. Besides, the potential for nonuniformity in the microwave heating process should be comprehensively described. Also, there occurs moisture accumulation at the food surface during microwave heating [13]. Therefore, the challenge is to understand the mechanism of microwave heating, gain insight into the changes in heating patterns, and verify the temperature distribution during microwave heating, and to develop a coupled heat, moisture, and electromagnetic transfer model.

Turbulence is a phenomenon of great complexity and has puzzled theoreticians for over 100 years. What makes turbulence so difficult to tackle mathematically is the wide range of length and time scales of motion even in flows with very simple boundary conditions. No single turbulence model is universally accepted as being superior for all classes of problems. The standard κ–ε model is still highly recommended for general purpose CFD computation. The mechanism of κ–ε models is derived for equilibrium flows in which the rates of production and destruction of turbulence are nearly balanced [76]. This assumption has been proven to be valid only in flows with a high Reynolds number and relatively far from the wall in the boundary layer. At low Reynolds numbers (lesser than 30,000), it was known that simplified turbulence models, such as κ–ε models or even the modified κ–ε models by the near-wall treatment based either on a wall function or on Wolfshtein's low Reynolds number, are rough approximations of reality. In many cases, these semiempirical models will fail to predict the correct near-wall limiting behavior near the product surface. However, κ–ε models remain popular because of their availability in user-friendly codes, which allows a straightforward implementation of the models, and because they are cheap in terms of computation time. Predictions by general codes based on κ–ε model are often very different from experimental data. As the shape of many food products is very complex, the experimental determination of heat transfer coefficients remains at the time quicker and much more reliable than predictions. The calculation based on the current CFD codes has to be used with caution and more research is needed to improve near-wall modeling particularly around blunt bodies placed in a turbulent flow. A full treatment of turbulence would require more complex models such as large eddy simulations (LESs) and Reynolds stress models (RSMs). However, LES models require large computing resources and not of use as general-purpose tools. As the RSM accounts for the effects of streamline curvature, swirl, rotation, and rapid changes in strain rate in a more rigorous manner compared to the κ–ε models, it has greater potential to give accurate predictions for complex flows. However, the fidelity of RSM predictions is still limited by the closure assumptions used to model various terms in the exact transport equations for the Reynolds stresses. The modeling of the pressure–strain and dissipation-rate terms is particularly challenging. Therefore, the RSM with additional computational expense might not always yield results that are clearly superior to simpler models in all cases of flows. The mathematical expressions of turbulence models may be quite complicated and they contain adjustable constants that need to be determined as best-fit values from experimental data. Therefore, any application of a turbulence model should not be beyond the data range. Besides, the current turbulence models can be used to guide the development of other models through comparative studies.

2.4.2 Judgment of Assumptions in Models

Accurate modeling of a thermal process of foods is complex. For simplification and saving of computational time, some assumptions made in the modeling are necessary. Most of assumptions come from the geometrical dimension and shape, surface heat and mass transfer coefficients, food materials properties, and volume change during thermal processes. Before simulation, whether or not to use a model of coupled heat and mass transfer or coupled heat transfer and fluid flow should also be determined.

Sensitivity analysis can make judgment for the acceptability of an assumption in the modeling. Some research has been carried out to investigate the sensitivity of variables in interest such as temperature on operating conditions of a thermal system and thermal properties of foods [118–126]. Findings from the research include that the time and location-dependent variations in operating conditions such as variable temperature and surface heat transfer coefficient cause the detachment of the thermal and geometric centers during processing of foods [118]. For simulating a thermal process with low heat transfer coefficient, small deviations in the coefficient may result in large deviations in the core temperature of foods [119,126]. The disturbances of different means but with the same scale of fluctuation in processing medium temperature resulted in comparable center temperature variation [122]. For a typical sterilization process, it was found that thermal–physical properties were the most important sources of variability [120,121,123]. It is stressed from the findings of sensitivity analyses in the publications that more efforts should be made to judge the acceptability of an assumption in the modeling.

2.4.3 Surface Heat and Mass Transfer Coefficients

Heat and mass transfer coefficients are important parameters in modeling heat and mass transfer during food thermal processes. The heat transfer coefficients of surface convection are mostly calculated using a correlation between a set of dimensionless numbers: *Nusselt number* ($Nu = h_c L/k$), *Prandtl number* ($Pr = c_p \mu/k$), *Reynolds number* ($Re = \rho L u/\mu$), and *Grashof number* ($Gr = L^3 \rho^2 g \beta \Delta T/\mu^2$) for flow across a body [31,49,66–68]. The surface mass transfer coefficient can be determined by using the Lewis relationship of heat and mass transfer coefficients, which is expressed as [127]

$$\frac{h}{K_p \lambda} = 64.7\,\text{Pa/K} \tag{2.41}$$

It should be noted that such correlations are normally restricted to a given range of operating conditions and reasonable accuracy can only be ensured under the given range of operating conditions. More attention should be paid to select a suitable correlation for a given case.

The heat transfer coefficients of surface convection can also be determined by fitting predicted temperatures to experimental data. The coefficient is determined by a trial and error method until the predicted model gives a good fit with experimental data [128,129]. For an aseptic system of fluid-particle foods, the coefficient of each particle can be determined by a trial and error matching of predicted temperature contours from a numerical heat transfer model with magnetic resonance imaging (MRI) images [130].

For simplicity, an average heat transfer coefficient of surface convection is used in most of the simulations. However, with the advance of CFD technology, a CFD model can offer an effective and efficient tool to calculate the average and local heat transfer coefficients of surface convection with an acceptable cost [96,131]. Verboven et al. used a 2D CFD model (CFX package) to investigate the variation in heat transfer coefficient around the surface of foods. Their simulations found that around the rectangular-shaped foods, there was a large variation in the local surface heat transfer coefficients. Using the local coefficients instead of the average surface coefficient caused changes in temperature in the foods to be considerably slower especially for slab-shaped foods and the coldest point was also no longer at the geometric center [132].

2.4.4 Food Properties

The food is introduced into a model through its properties. These properties including thermal conductivity, density, specific heat capacity, diffusivity, and porosity can identify the uniqueness of the food to the model. Food products are complicated materials. Their properties vary with species, process treatment, temperature, concentration, etc. All these factors will increase difficulties in describing and predicting the properties of a product in modeling heat and mass transfer.

Thermal properties are another one of the most important factors determining the accuracy of model predictions. Part of thermal properties of food products can be found in publications available [127,133–137].

The thermal properties of foods can be directly measured by experiments [23,41]. For measuring physical properties, heat transfer models can be used to optimize the experimental design [138]. The prediction accuracy of a model can be significantly improved by including temperature- and composition-dependent thermal properties [58,139]. However, it is difficult for experimental measurement to obtain a detailed description of the relationship between thermal properties and temperature and compositions of foods. Alternatively, thermal properties of foods can be calculated from the compositions of foods and the thermal properties of each composition [27,31,66–68]. The compositions of foods can be measured before and/or after processing and the variation in the compositions during processing can be determined by mass transfer models. The main compositions of foods usually are water, protein, and fat and other compositions such as salt and ash are very small. The temperature-dependent thermal properties of these compositions can be measured or found in literatures [127]. It should be noted that the calculations for thermal properties from food compositions are based on empirical or semiempirical relationship. More attention should be paid to select suitable correlation equations for a given case.

Thermal properties of foods can also be inversely found by using analytical or numerical heat transfer models and experimental temperature history. For determining a thermal property, an assumed value of the thermal property is first used to solve the numerical model. The predicted temperatures for given locations are compared with their corresponding measured values. The value of the thermal property is acceptable if the minimum difference between the predicted and measured temperatures is achieved [140].

2.4.5 Shrinkage of Solid Foods during Thermal Processes

Shrinkage in foods occurs due to moisture loss during thermal processes. Effects of shrinkage on the accuracy of models are sometimes significant [141]. Shrinkage is normally taken into account in models of drying processes [24,100,102]. Shrinkage can be expressed as functions of moisture and the functions are determined by experiments [24,141].

2.5 CONCLUSIONS

Changes in temperature and concentration during thermal processes are always initiated by a transfer to or from the surface of the product. The transfer rate may be controlled by internal resistance, external resistance, or both. Transfer of both heat and mass takes place according to several mechanisms. In most cases, more than one of these mechanisms are involved. In some processes, transfer of both heat and mass occurs simultaneously. It is important to understand heat and mass transfer mechanism for the improvement of existing food thermal processes and for the development of new and better processes.

The physical laws of heat and mass transfer have widely been used to describe food thermal processes, producing a large number of mathematical models. Some assumptions such as simplified geometrical shape, constant thermal–physical properties, constant surface heat and mass transfer coefficients, and no volume change during processing were widely used in modeling. However, more research should be conducted to justify the acceptability of those assumptions and to improve the accuracy of models by finding more information on surface heat and mass transfer coefficients, food properties, and shrinkage during processing. Before heat and mass transfer models can become a quantitative tool for correctly analyzing thermal processes, determination of thermal–physical propensities and surface mass and heat transfer coefficients remains an important area to be studied.

NOMENCLATURE

A Area (m^2)
c_p Specific heat capacity (kJ/kg K)
C Concentration (kg/m^3 or $kmol/m^3$)
C^* Concentration in equilibrium with the bulk gas partial pressure (kg/m^3 or $kmol/m^3$)
d Diameter (m)
D Mass diffusivity (m^2/s)
D_0 Pre-exponential factor in Arrhenius equation (m^2/s)
D_g Mass diffusivity in the gas phase (kmol/N m s)
D_l Mass diffusivity in the liquid phase (m^2/s)
e Emissivity (–)
E Electric field strength (V/m)
E_a Activation energy (J/kg·mol)
F_{12} View factor, fraction of radiation leaving surface 1 and arriving at surface 2 (–)
g Acceleration due to gravity (m/s^2)
h Heat transfer coefficient (W/m^2 K)
h_b Boiling heat transfer coefficient (W/m^2 K)
h_c Convection heat transfer coefficient (W/m^2 K)
h_{cd} Condensation heat transfer coefficient (W/m^2 K)
h_r Radiation heat transfer coefficient (W/m^2 K)
H Enthalpy (J/kg) or Henry's constant (J/kmol)
J Diffusive flow rate (kg/m^2 s or $kmol/m^2$ s)
k Thermal conductivity (W/m K)
K_g Gas mass transfer coefficient (kmol/N s)
K_l Liquid mass transfer coefficient (m/s)
K_p Surface mass transfer coefficient related to pressure (kg/Pa m^2 s)
L Length (m)
P Pressure (Pa)
P^* Partial pressure in equilibrium with the bulk liquid concentration (Pa)
Pr Prandtl number (–)

q	Heat rate (W)
q_{max}	Peak heat flux (W/m^2)
q_s	Heat flux (W/m^2) or mass flux on the surface (kg/m^2 s or kmol/m^2 s)
R	Gas constant (8.314 J/mol K)
S	Source term (W/m^3) or entropy (J/K)
$S(t)$	Position of moving boundary at time, t (m)
t	Time (s)
T	Temperature (°C)
T_K	Absolute temperature (K)
ΔT	Temperature difference (°C)
u	Velocity (m/s)
V	Specific volume (m^3/kg)
x, y, z	Orthogonal coordinates (m)
X	Moisture content (%)

Greek symbols

α	Thermal diffusivity (m^2/s)
α'	Attenuation factor (1/m)
β	Thermal expansion (1/K) or loss angle (–)
δ	Thickness (m)
δ_{inf}	Infrared penetration depth (m)
ε'	Resistive part of permittivity (–)
ε''	Capacitive part of permittivity (–)
λ	Latent heat (J/kg) or wavelength (m)
μ	Viscosity (Pa s)
$\nabla\cdot$	Divergence of a vector
∇	Vector operator
ν	Frequency (Hz)
ϕ	Temperature or concentration (–)
Φ	Thermal inertia (s)
ψ	Group defined by Equation 2.12
ρ	Density (kg/m^3)
σ	Stefan–Boltzmann constant (W/m^2 K^4) or surface tension (N/m)

Subscripts

Γ	Boundary
∞	Processing medium
0	initial
1	Surface or phase 1
2	Surface or phase 2
A	Component A
b	Boiling
c	Convection
cd	Condensation
g	Gas
i	Interface
inf	Infrared
K	Temperature in Kelvin
l	Liquid
m	Medium

max	Maximum
P	Pressure
s	Surface
v	Vapor
w	Water
x, y, z	Orthogonal coordinates

REFERENCES

1. RP Singh. Moving boundaries in food engineering. *Food Technology* 54: 44–53, 2000.
2. PG Smith. *Introduction to Food Process Engineering*. New York: Kluwer Academic/Plenum Publishers, 2003, pp. 163–170, 179–186, 191–217.
3. V Gekas. *Transport Phenomena of Foods and Biological Materials*. Boca Raton, FL: CRC Press, 1992, pp. 176–188.
4. G Tiwari, S Wang, J Tang, SL Birla. Computer simulation model development and validation for radio frequency (RF) heating of dry food materials. *Journal of Food Engineering* 105: 48–55, 2011.
5. V Romano, F Marra. A numerical analysis of radio frequency heating of regular shaped foodstuff. *Journal of Food Engineering* 84: 449–457, 2008.
6. S Jun, S Sastry. Modeling and optimization of ohmic heating of foods inside a flexible package. *Journal of Food Process Engineering* 28: 417–436, 2005.
7. F Marra, M Zell, JG Lyng, DJ Morgan, DA Cronin. Analysis of heat transfer during ohmic processing of a solid food. *Journal of Food Engineering* 91: 56–63, 2009.
8. S Palaniappan, S Sastry. Electrical conductivity of selected solid foods during ohmic heating. *Journal of Food Process Engineering* 14: 221–236, 1991.
9. S Jaturonglumlert, T Kiatsiriroat. Heat and mass transfer in combined convective and far-infrared drying of fruit leather. *Journal of Food Engineering* 100: 254–260, 2010.
10. F Tanaka, P Verboven, N Scheerlinck, K Morita, K Iwasaki, B Nicolai. Investigation of far infrared radiation heating as an alternative technique for surface decontamination of strawberry. *Journal of Food Engineering* 79: 445–452, 2007.
11. N Shilton, P Mallikarjunan, P Sheridan. Modeling of heat transfer and evaporative mass losses during the cooking of beef patties using far-infrared radiation. *Journal of Food Engineering* 55: 217–222, 2002.
12. N Meeso, A Nathakaranakule, T Madhiyanon, S Soponronnarit. Modeling of far-infrared irradiation in paddy drying process. *Journal of Food Engineering* 78: 1248–1258, 2007.
13. A Datta, H Ni. Infrared and hot-air-assisted microwave heating of foods for control of surface moisture. *Journal of Food Engineering* 51: 355–364, 2002.
14. E Patazca, T Koutchma, VM Balasubramaniam. Quasi-adiabatic temperature increase during high pressure processing of selected foods. *Journal of Food Engineering* 80: 199–205, 2007.
15. T Carroll, P Chen, A Fletcher. A method to characterise heat transfer during high-pressure processing. *Journal of Food Engineering* 60: 131–135, 2003.
16. S Denys, AM Van Loey, ME Hendrickx. A modelling approach for evaluating process uniformity during batch high hydrostatic pressure processing: Combination of a numerical heat transfer model and enzyme inactivation kinetics. *Innovative Food Science & Emerging Technologies* 1: 5–19, 2000.
17. L Otero, P Sanz. Modeling heat transfer in high pressure food processing: A review. *Innovative Food Science & Emerging Technologies* 4: 121–134, 2003.
18. AG Abdul Ghani, MM Farid. Numerical simulation of solid–liquid food mixture in a high pressure processing unit using computational fluid dynamics. *Journal of Food Engineering* 80: 1031–1042, 2007.
19. A Delgado, C Rauh, W Kowalczyk, A Baars. Review of modelling and simulation of high pressure treatment of materials of biological origin. *Trends in Food Science & Technology* 19: 329–336, 2008.
20. S Denys, ME Hendrickx. Measurement of the thermal conductivity of foods at high pressure. *Journal of Food Science* 64: 709–713, 1999.
21. GD Saravacos. Mass transfer properties of foods. In: MA Rao, SSH Rizvi, eds. *Engineering Properties of Foods*, 2nd edn. New York: Marcel Dekker, 1994, pp. 169–221.
22. PK Chandra, RP Singh. *Applied Numerical Methods for Food and Agricultural Engineers*. Boca Raton, FL: CRC Press, 1994.
23. S Ghazala, HS Ramaswamy, JP Smith, MV Simpson. Thermal process simulation for sous vide processing of fish and meat foods. *Food Research International* 28: 117–122, 1995.

24. N Wang, JG Brennan. A mathematical model of simultaneous heat and moisture transfer during drying of potato. *Journal of Food Engineering* 24: 47–60, 1995.
25. S Coulter, QT Pham, I McNeil, NG McPhail. Geometry, cooling rates and weight losses during pig chilling. *International Journal of Refrigeration* 18: 456–464, 1995.
26. SG Akterian. Numerical simulation of unsteady heat transfer in canned mushrooms in brine during sterilisation processes. *Journal of Food Engineering* 25: 45–53, 1995.
27. J Evans, S Russell, S James. Chilling of recipe dish meals to meet cook-chill guidelines. *International Journal of Refrigeration* 19: 79–86, 1996.
28. BE Farkas, RP Singh, TR Rumsey. Modelling heat and mass transfer in immersion frying. I. Model development. *Journal of Food Engineering* 29: 211–226, 1996.
29. BE Farkas, RP Singh, TR Rumsey. Modelling heat and mass transfer in immersion frying. II, model solution and verification. *Journal of Food Engineering* 29: 227–248, 1996.
30. SG Akterian. Control strategy using functions of sensitivity for thermal processing of sausages. *Journal of Food Engineering* 31: 449–455, 1997.
31. LM Davey, QT Pham. Predicting the dynamic product heat load and weight loss during beef chilling using a multi-region finite difference approach. *International Journal of Refrigeration* 20: 470–482, 1997.
32. S Chuntranuluck, CM Wells, AC Cleland. Prediction of chilling times of foods in situations where evaporative cooling is significant—Part 1. Method development. *Journal of Food Engineering* 37: 111–125, 1998.
33. S Chuntranuluck, CM Wells, AC Cleland. Prediction of chilling times of foods in situations where evaporative cooling is significant—Part 2. Experimental testing. *Journal of Food Engineering* 37: 127–141, 1998.
34. S Chuntranuluck, CM Wells, AC Cleland. Prediction of chilling times of foods in situations where evaporative cooling is significant—Part 3. Applications. *Journal of Food Engineering* 37: 143–157, 1998.
35. F Erdogdu, MO Balaban, KV Chau. Modelling of heat conduction in elliptical cross section: I. Development and testing of the model. *Journal of Food Engineering* 38: 223–239, 1998.
36. F Erdogdu, MO Balaban, KV Chau. Modelling of heat conduction in elliptical cross section: II. Adaption to thermal processing of shrimp. *Journal of Food Engineering* 38: 241–258, 1998.
37. FA Ansari. Finite difference solution of heat and mass transfer problems related to precooling of food. *Energy Conversion and Management* 40: 795–802, 1999.
38. ZH Wang, GH Chen. Heat and mass transfer during low intensity convection drying. *Chemical Engineering Science* 54: 3899–3908, 1999.
39. K Thorvaldsson, H Janestad. A model for simultaneous heat, water and vapour diffusion. *Journal of Food Engineering* 40: 167–172, 1999.
40. Z Pan, RP Singh, TR Rumsey. Predictive modelling of contact-heating process for cooking a hamburger patty. *Journal of Food Engineering* 46: 9–19, 2000.
41. OO Fasina, HP Fleming. Heat transfer characteristics of cucumbers during blanching. *Journal of Food Engineering* 47: 203–210, 2001.
42. RD Radford, LS Herbert, DA Lovett. Chilling of meat—A mathematical model for heat and mass transfer. *Refrigeration Science and Technology* 1: 323–330, 1976.
43. KV Chau, JJ Gaffney. A finite difference model for heat and mass transfer in products with internal heat generation and transpiration. *Journal of Food Science* 55: 484–487, 1990.
44. LJ Wang, D-W Sun. Numerical modeling of heating and cooling processes in the food industry: A review. *Trends in Food Science & Technology* 14: 408–423, 2003.
45. FL Stasa. *Applied Finite Element Analysis for Engineers*. New York: Dryden Press, 1985.
46. SS Rao. *The Finite Element Method in Engineering*, 2nd edn. New York: Pergamon Press, 1989.
47. JA Arce, PL Potluri, KC Schneider, VE Sweat, TR Dutson. Modelling beef carcass cooling using a finite element technique. *Transactions of the ASAE* 26: 950–954, 960, 1983.
48. SM Van Der Sluis, W Rouwen. TNO develops a model for refrigeration technology calculations. *Voedingsmiddelentechnologie* 26: 63–64, 1994.
49. P Mallikarjunan, GS Mittal. Heat and mass transfer during beef carcass chilling—Modelling and simulation. *Journal of Food Engineering* 23: 277–292, 1994.
50. P Mallikarjunan, GS Mittal. Prediction of beef carcass chilling time and mass loss. *Journal of Food Process Engineering* 18: 1–15, 1995.
51. L Zhou, VM Puri, RC Anantheswaran, G Yeh. Finite element modelling of heat and mass transfer in food materials during microwave heating—Model development and validation. *Journal of Food Engineering* 25: 509–529, 1995.

52. YE Lin, RC Anantheswaran, VM Puri. Finite element analysis of microwave heating of solid foods. *Journal of Food Engineering* 25: 85–112, 1995.
53. G Comini, G Cortella, O Saro. Finite element analysis of coupled conduction and convection in refrigerated transport. *International Journal of Refrigeration* 18: 123–131, 1995.
54. JN Ikediala, LR Correia, GA Fenton, NB Abdallah. Finite element modelling of heat transfer in meat patties during single-sided pan-frying. *Journal of Food Science* 61: 796–802, 1996.
55. N Carroll, R Mohtar, LJ Segerlind. Predicting the cooling time for irregular shaped food products. *Journal of Food Process Engineering* 19: 385–401, 1996.
56. G Lian, CS Harris, R Evans, M Warboys. Coupled heat and moisture transfer during microwave vacuum drying. *Journal of Microwave Power and Electromagnetic Energy* 32: 34–44, 1997.
57. Y Zhao, E Kolbe, C Craven. Computer simulation on onboard chilling and freezing of albacore tuna. *Journal of Food Science* 63: 751–755, 1998.
58. HQ Chen, BP Marks, RY Murphy. Modelling coupled heat and mass transfer for convection cooking of chicken patties. *Journal of Food Engineering* 42: 139–146, 1999.
59. H Zhang, AK Datta. Coupled electromagnetic and thermal modeling of microwave oven heating of foods. *Journal of Microwave Power and Electromagnetic Energy* 35: 71–85, 2000.
60. CC Jia, D-W Sun, CW Cao. Mathematical simulation of stresses within a corn kernel during drying. *Drying Technology* 18: 887–906, 2000.
61. CC Jia, D-W Sun, CW Cao. Mathematical simulation of temperature and moisture fields within a grain kernel during drying. *Drying Technology* 18: 1305–1325, 2000.
62. CC Jia, D-W Sun, CW Cao. Mathematical simulation of temperature fields in a stored grain bin due to internal heat generation. *Journal of Food Engineering* 43: 227–233, 2000.
63. CC Jia, D-W Sun, CW Cao. Finite element prediction of transient temperature distribution in a grain storage bin. *Journal of Agricultural Engineering Research* 76: 323–330, 2000.
64. CC Jia, D-W Sun, CW Cao. Computer simulation of temperature changes in a wheat storage bin. *Journal of Stored Products Research* 37: 165–177, 2001.
65. SS Ahmad, MT Morgan, MR Okos. Effects of microwave on the drying, checking and mechanical strength of baked biscuits. *Journal of Food Engineering* 50: 63–75, 2001.
66. LJ Wang, D-W Sun. Modelling three conventional cooling processes of cooked meat by finite element method. *International Journal of Refrigeration* 25: 100–110, 2002.
67. LJ Wang, D-W Sun. Evaluation of performance of slow air, air blast and water immersion cooling methods in cooked meat industry by finite element method. *Journal of Food Engineering* 51: 329–340, 2002.
68. LJ Wang, D-W Sun. Modelling three dimensional transient heat transfer of roasted meat during air blast cooling process by finite element method. *Journal of Food Engineering* 51: 319–328, 2002.
69. LJ Wang, D-W Sun. Modelling vacuum cooling process of cooked meat—Part 1: Analysis of vacuum cooling system. *International Journal of Refrigeration* 25: 852–860, 2002.
70. LJ Wang, D-W Sun. Modelling vacuum cooling process of cooked meat—Part 2: Mass and heat transfer of cooked meat under vacuum pressure. *International Journal of Refrigeration* 25: 861–872, 2002.
71. LJ Wang, D-W Sun. Numerical analysis of the three dimensional mass and heat transfer with inner moisture evaporation in porous cooked meat joints during vacuum cooling process. *Transactions of the ASAE* 46: 107–115, 2003.
72. LJ Wang, D-W Sun. Effect of operating conditions of a vacuum cooler on cooling performance for large cooked meat joints. *Journal of Food Engineering* 61: 231–240, 2003.
73. LJ Wang, RP Singh. Mathematical modeling and sensitivity analysis of double-sided contact-cooking process for initially frozen hamburger patties. *Transactions of the ASAE* 47: 147–157, 2004.
74. VM Puri, RC Anantheswaran. The finite-element method in food processing: A review. *Journal of Food Engineering* 19: 247–274, 1993.
75. D-W Sun. CFD applications in the agri-food industry. *Computers and Electronics in Agriculture* 34: 1–236, 2002 [Special issue].
76. HK Versteeg, W Malalsekera. *An Introduction to Computational Fluid Dynamics: The Finite Volume Method*. New York: Wiley, 1995.
77. G Scott, P Richardson. The application of computational fluid dynamics in the food industry. *Trends in Food Science & Technology* 8: 119–124, 1997.
78. B Xia, D-W Sun. Applications of computational fluid dynamics (CFD) in the food industry: A review. *Computers and Electronics in Agriculture* 34: 5–24, 2002.
79. TAG Langrish, DF Fletcher. Spray drying of food ingredients and applications of CFD in spray drying. *Chemical Engineering and Processing* 40: 345–354, 2001.

80. AGA Ghani, MM Farid, XD Chen, P Richards. Numerical simulation of natural convection heating of canned food by computational fluid dynamics. *Journal of Food Engineering* 41: 55–64, 1999.
81. AGA Ghani, MM Farid, XD Chen, P Richards. An investigation of deactivation of bacteria in a canned liquid food during sterilisation using computational fluid dynamics (CFD). *Journal of Food Engineering* 42: 207–214, 1999.
82. A Jung, PJ Fryer. Optimising the quality of safe food: Computational modelling of a continuous sterilisation process. *Chemical Engineering Science* 54: 717–730, 1999.
83. ZH Hu, D-W Sun. CFD simulation of heat and moisture transfer for predicting cooling rate and weight loss of cooked ham during air-blast chilling process. *Journal of Food Engineering* 46: 189–197, 2000.
84. PS Mirade, JD Daudin. Numerical study of the airflow patterns in a sausage dryer. *Drying Technology* 18: 81–97, 2000.
85. PS Mirade, L Picgirard. Assessment of airflow patterns inside six industrial beef carcass chillers. *International Journal of Food Science & Technology* 36: 463–475, 2001.
86. AGA Ghani, MM Farid, XD Chen, P Richards. Thermal sterilisation of canned food in a 3-D pouch using computational fluid dynamics. *Journal of Food Engineering* 48: 147–156, 2001.
87. G Cortella, M Manzan, G Comini. CFD simulation of refrigerated display cabinets. *International Journal of Refrigeration* 24: 250–260, 2001.
88. ZH Hu, D-W Sun. Effect of fluctuation in inlet airflow temperature on CFD simulation of air-blast chilling process. *Journal of Food Engineering* 48: 311–316, 2001.
89. ZH Hu, D-W Sun. Predicting local surface heat transfer coefficients by different turbulent κ–ε models to simulate heat and moisture transfer during air-blast chilling. *International Journal of Refrigeration* 24: 702–717, 2001.
90. G Cortella. CFD—Aided retail cabinet design. *Computers and Electronics in Agriculture* 34: 43–66, 2002.
91. AM Foster, R Barrett, SJ James, MJ Swain. Measurement and prediction of air movement through doorways in refrigerated rooms. *International Journal of Refrigeration* 25: 1102–1109, 2002.
92. PS Mirade, A Kondjoyan, JD Daudin. Three-dimensional CFD calculations for designing large food chillers. *Computers and Electronics in Agriculture* 34: 67–88, 2002.
93. ZH Hu, D-W Sun. CFD evaluating the influence of airflow on the thermocouple-measured temperature data during air-blast chilling. *International Journal of Refrigeration* 25: 546–551, 2002.
94. D-W Sun, Z Hu. CFD predicting the effects of various parameters on core temperature and weight loss profiles of cooked meat during vacuum cooling. *Computers and Electronics in Agriculture* 34: 111–127, 2002.
95. D-W Sun, Z Hu. CFD simulation of coupled heat and mass transfer through porous foods during vacuum cooling process. *International Journal of Refrigeration* 26: 19–27, 2003.
96. A Kannan, PCG Sandaka. Heat transfer analysis of canned food sterilization in a still retort. *Journal of Food Engineering* 88: 213–228, 2008.
97. CJ King. Heat and mass transfer fundamentals applied to food engineering. *Journal of Food Process Engineering* 1: 3–14, 1977.
98. NP Cheremisinoff. *Handbook of Heat and Mass Transfer*. Houston, TX: Gulf Publishing Co., 1986.
99. A Ibarz. *Unit Operations in Food Engineering*. Boca Raton, FL: CRC Press, 2003, pp. 491, 573.
100. CO Rovedo, C Suarez, PE Viollaz. Drying of foods: Evaluation of a drying model. *Journal of Food Engineering* 26: 1–12, 1995.
101. E Ben-Yoseph, RW Hartel, D Howling. Three-dimensional model of phase transition of thin sucrose films during drying. *Journal of Food Engineering* 44: 13–22, 2000.
102. S Simal, A Femenia, P Llull, C Rossello. Dehydration of aloe vera: Simulation of drying curves and evaluation of functional properties. *Journal of Food Engineering* 43: 109–114, 2000.
103. Y Wu, J Irudayaraj. Analysis of heat, mass and pressure transfer in starch based food systems. *Journal of Food Engineering* 29: 399–414, 1996.
104. AK Datta. Porous media approaches to studying simultaneous heat and mass transfer in food processes. I: Problem formulations. *Journal of Food Engineering* 80: 80–95, 2007.
105. AK Datta. Porous media approaches to studying simultaneous heat and mass transfer in food processes. II: Property data and representative results. *Journal of Food Engineering* 80: 96–110, 2007.
106. E Mathioulakis, VT Karathanos, VG Belessiotis. Simulation of air movement in a dryer by computational fluid dynamics: Application for the drying of fruits. *Journal of Food Engineering* 36: 183–200, 1998.
107. YS Chen, SW Kim. Computational of turbulent flows using an extended κ–ε turbulence closure model, NASA CR-179204, USA, 1987.

108. J Straatsma, GV Houwelingen, AE Steenbergen, P De Jong. Spray drying of food products: 1. Simulation model. *Journal of Food Engineering* 42: 67–72, 1999.
109. P Verboven, N Scheerlinck, J De Baerdemaeker, BM Nicolaï. Computational fluid dynamics modelling and validation of the isothermal airflow in a forced convection oven. *Journal of Food Engineering* 43: 41–53, 2000.
110. P Verboven, N Scheerlinck, J De Baerdemaeker, BM Nicolaï. Computational fluid dynamics modelling and validation of the temperature distribution in a forced convection oven. *Journal of Food Engineering* 43: 61–73, 2000.
111. RS Vilayannur, VM Puri, RC Anantheswaran. Size and shape effect on nonuniformity of temperature and moisture distributions in microwave heated food materials: Part I. Simulation. *Journal of Food Process Engineering* 21: 209–233, 1998.
112. RS Vilayannur, VM Puri, RC Anantheswaran. Size and shape effect on nonuniformity of temperature and moisture distributions in microwave heated food materials: Part II. Experimental validation. *Journal of Food Process Engineering* 21: 235–248, 1998.
113. RP Singh. Phase transition and transport phenomena in frying of foods. In: MA Rao, RW Hartel, eds. *Phase/State Transitions in Foods*. New York: Marcel Dekker, 1998, pp. 369–390.
114. S Bruin, KC Luyben. Drying of food materials: A review of recent developments. In: AS Mujumdar, ed. *Advances in Drying*. New York: Hemisphere, 1980.
115. H Zhang, AK Datta, IA Taub, C Doona. Electromagnetics, heat transfer, and thermokinetics in microwave sterilization. *AIChE Journal* 47: 1957–1968, 2001.
116. KG Ayappa, HT Davis, G Crapiste, J Gordon. Microwave heating: An evaluation of power formulations. *Chemical Engineering Science* 46: 1005–1016, 1991.
117. D Burfoot, CJ Railton, AM Foster, SR Reavell. Modelling the pasteurisation of prepared meals with microwaves at 896 MHz. *Journal of Food Engineering* 30: 117–133, 1996.
118. C De Elvira, PD Sanz, JA Carrasco. Characterising the detachment of thermal and geometric centres in a parallelepipedic frozen food subjected to a fluctuation in storage temperature. *Journal of Food Engineering* 29: 257–268, 1996.
119. BM Nicolaï, J De Baerdemaeker. Sensitivity analysis with respect to the surface heat transfer coefficient as applied to thermal process calculations. *Journal of Food Engineering* 28: 21–33, 1996.
120. BM Nicolaï, P Verboven, N Scheerlinck, J De Baerdemaeker. Numerical analysis of the propagation of random parameter fluctuations in time and space during thermal food processes. *Journal of Food Engineering* 38: 259–278, 1998.
121. BM Nicolaï, J De Baerdemaeker. A variance propagation algorithm for the computation of heat conduction under stochastic conditions. *International Journal of Heat and Mass Transfer* 42: 1513–1520, 1999.
122. BM Nicolaï, B Verlinden, A Beuselinck, P Jancsok, V Quenon, N Scheerlinck, P Verboven, J De Baerdemaeker. Propagation of stochastic temperature fluctuations in refrigerated fruits. *International Journal of Refrigeration* 22: 81–90, 1999.
123. BM Nicolaï, N Scheerlinck, P Verboven, J De Baerdemaeker. Stochastic perturbation analysis of thermal food processes with random field parameters. *Transactions of the ASAE* 43: 131–138, 2000.
124. S Varga, JC Oliveira, FAR Oliveira. Influence of the variability of processing factors on the *F*-value distribution in batch retorts. *Journal of Food Engineering* 44: 155–161, 2000.
125. S Varga, JC Oliveira, C Smout, ME Hendrickx. Modelling temperature variability in batch retorts and its impact on lethality distribution. *Journal of Food Engineering* 44: 163–174, 2000.
126. P Verboven, N Scheerlinck, J De Baerdemaeker, BM Nicolaï. Sensitivity of the food center temperature with respect to the air velocity and the turbulence kinetic energy. *Journal of Food Engineering* 48: 53–60, 2001.
127. MJ Lewis. *Physical Properties of Foods and Food Processing Systems*. Chichester, U.K.: Ellis Horwood, 1987.
128. S Sahin, SK Sastry, L Bayindirli. The determination of convective heat transfer coefficient during frying. *Journal of Food Engineering* 39: 307–311, 1999.
129. S Varga, JC Oliveira. Determination of the heat transfer coefficient between bulk medium and packed containers in a batch retort. *Journal of Food Engineering* 44: 191–198, 2000.
130. GJ Hulbert, JB Litchfield, SJ Schmidt. Determination of convective heat transfer coefficients using 2D MRI temperature mapping and finite element modelling. *Journal of Food Engineering* 34: 193–201, 1997.
131. A Kondjoyan, HC Boisson. Comparison of calculated and experimental heat transfer coefficients at the surface of circular cylinders placed in a turbulent cross-flow of air. *Journal of Food Engineering* 34: 123–143, 1997.

132. P Verboven, BM Nicolaï, N Scheerlinck, J De Baerdemaeker. The local surface heat transfer coefficient in thermal food process calculations: A CFD approach. *Journal of Food Engineering* 33: 15–35, 1997.
133. JD Mellor, AH Seppings. Thermophysical data for designing a refrigerated food chain. *Refrigeration Science and Technology* 1: 349–359, 1976.
134. JD Mellor. Critical evaluation of thermophysical properties of foodstuffs and outline of future developments. In: R Jowitt, ed. *Physical Properties of Food*. London, U.K.: Applied Science Publishers, 1983, pp. 331–353.
135. CA Miles, G VAN Beek, CH Veerkamp. Calculation of thermophysical properties of foods. In: R Jowitt, ed. *Physical Properties of Food*. London, U.K.: Applied Science Publishers, 1983, pp. 269–313.
136. VE Sweat. Thermal conductivity of food: Present state of the data. *ASHRAE Transactions* 91(part 2B): 299–311, 1985.
137. S Rahman. *Food Properties Handbook*. New York: CRC Press, 1995.
138. HB Nahor, N Scheerlinck, R Verniest, J De Baerdemaeker, BM Nicolaï. Optimal experimental design for the parameter estimation of conduction heated foods. *Journal of Food Engineering* 48: 109–119, 2001.
139. H Tewkesbury, AGF Stapley, PJ Fryer. Modelling temperature distributions in cooling chocolate moulds. *Chemical Engineering Science* 55: 3123–3132, 2000.
140. ME Schmalko, RO Morawicki, LA Ramallo. Simultaneous determination of specific heat capacity and thermal conductivity using the finite difference method. *Journal of Food Engineering* 31: 531–540, 1997.
141. M Balaban. Effect of volume change in foods on the temperature and moisture content predictions of simultaneous heat and moisture transfer models. *Journal of Food Process Engineering* 12: 67–88, 1989.

3 Thermal Effects in Food Microbiology

Mogessie Ashenafi

CONTENTS

3.1 INTRODUCTION

Thermal processing is the commonest food processing operation usually applied to processes such as canning, baking, and pasteurization of various food items. The major objective of thermal processing is to guarantee food safety by killing bacteria and inactivating their enzymes or other metabolites in foods. Heat also brings about physical changes to food that may assist to develop taste and flavor.

The amount of heat required to destroy microorganisms in a given product can be determined in a laboratory through thermal death time tests, which involve heating a known amount of microorganisms in food at several temperatures and for several time intervals at each temperature. The data are used to calculate *D*- and *z*-values. The *D*-value (also termed the decimal reduction time) is defined as the time at a particular temperature required to reduce a known number of microorganisms by 90% (1-log reduction). The *z*-value can be determined from the slope of the line that results from plotting the log of *D*-values versus temperature and indicates the change in the death rate based on temperature. Both *D*- and *z*-values are indirectly used to establish thermal processes.

The successful adaptation of microorganisms to changes in food production, processing, and preservation techniques has resulted in an increasing number of emerging and reemerging foodborne pathogens [1]. This emergence, in certain cases, is associated with very low infectious doses [2]. The food industry uses various techniques to reduce or prevent the introduction and survival of microorganisms in food. Physical methods, such as heat treatment, and many other chemical methods are traditionally used to exert pressure on microorganisms, eventually leading to growth inhibition or death. Heat is the most effective measure used to control microbial activity in foods because it eliminates vegetative cells, toxins, and spores of pathogenic and spoilage forms.

Microorganisms, however, have remarkable adaptive mechanisms, by which they respond to physical and chemical stresses used in the food industry [3].

3.2 SOME PATHOGENS OF CONCERN IN THE FOOD INDUSTRY

Escherichia coli O157:H7 is an emerging food-borne and waterborne pathogen of major public health concern. Meat, milk, fruit juices, and vegetables, among others, are implicated as vehicles of *E. coli* O157:H7, and its survival in acidic food environments has been well documented [4]. It causes diarrheal illness, hemorrhagic colitis, hemolytic uremic syndrome, and, in some cases, death [5]. The hurdle technology or the application of a combination of suboptimal growth factors could control the growth of pathogens in processed foods [6]. However, studies showed that there was lower reduction of *E. coli* O157:H7 in certain combined treatments [7]. Although hurdle concept for preservation of food may inhibit outgrowth, they may induce prolonged survival of *E. coli* O157:H7 in foods [8].

Enterococcus faecium is a gram-positive nonspore-forming coccus that has important technological properties, such as the improvement of flavor of several fermented foods [9]. It, however, is associated with the spoilage of thermally processed foods [10]. It is a heat-resistant bacterium and capable of growing at a wide range of temperatures [11]. A study showed that previous adaptation to a low pH increased the bacterial heat resistance of *E. faecium* [12].

Bacillus cereus is a spore-forming gram-positive rod. It may cause illness through the production of a heat-stable emetic toxin that causes vomiting and another heat-labile enterotoxin that causes diarrhea [13]. A strain of *B. cereus* produced a novel cytotoxin, CytK, which caused fatal necrotic enteritis [14]. *B. cereus* is ubiquitous in nature and can easily contaminate food production or processing systems [15]. *B. cereus* has an important role in food safety issues because of the ability of its spores to survive and the vegetative cells to colonize diverse ecological niches including food production and processing environments and human gastrointestinal tract [16].

Listeria monocytogenes is a gram-positive rod that has been linked to sporadic episodes and large outbreaks of human illness worldwide [17] showing a mortality rate of about 20% [18]. It has been isolated from a wide variety of raw and processed foods including milk and dairy products, meat and meat products, fresh produce as well as seafood [19,20]. Incidence of listeriosis is affected, among others, by consumption of minimally processed ready-to-eat and refrigerated foods [21].

Cronobacter sakazakii, formerly *Enterobacter sakazakii*, is a gram-negative, nonspore-forming rod. It has been implicated as a cause of meningitis and enterocolitis in infants, mostly under 1 year old, with mortality rates ranging from 40% to 80% [22]. Microbiological, genetic typing and epidemiological methods have confirmed that dried infant milk formula harbors this pathogen that causes most severe neonatal infections [22]. *C. sakazakii* is also isolated from a wide spectrum of other food and food ingredients of plant and animal origins [23,24]. Because it is an emerging food-borne pathogen, and due to its potential impact on human health, *E. sakazakii* has increasingly gained the interest and concern of regulatory agencies, health care providers, the scientific community, and the food industry [25].

Clostridium botulinum is a rod-shaped obligate anaerobe and forms the highly potent botulinum neurotoxin that is responsible for botulism, a severe disease with a high fatality rate. *C. botulinum* is ubiquitous in nature, and its spores are naturally present in soil and water. It is a heterogeneous species consisting of four distinct groups known as *C. botulinum* Groups I–IV. Most cases of food-borne botulism are caused by proteolytic *C. botulinum* (*C. botulinum* Group I) and nonproteolytic *C. botulinum* (*C. botulinum* Group II). The spores survive adverse conditions (e.g., heat treatment, high pressure, UV light, and desiccation) that vegetative forms would not survive. Proteolytic *C. botulinum* forms spores of high heat resistance and produces type A or type B neurotoxin. A standard minimum heat treatment of 121°C for 3 min is adopted by the canning industry as the "botulinum cook" for low-acid canned foods [26]. This has ensured the safe production of low-acid canned foods. Failure to achieve this temperature–time combination or postprocessing contamination leads to food-borne botulism outbreaks [27]. Spores formed by strains of nonproteolytic *C. botulinum* are of moderate heat resistance [28]. In the absence of lysozyme, heat treatment of 90°C for 10 min was sufficient to prevent growth and neurotoxin formation [27].

Salmonella is a gram-negative rod and is one of the most prevalent enteric pathogens causing a diversity of illnesses that include typhoid fever, gastroenteritis, and septicemia [29]. There are over 1500 *Salmonella* serovars, and the most frequently isolated serotype is *Salmonella* Typhimurium [30]. *S.* Typhimurium has developed responses to combat adverse conditions [31]. *Salmonella,* in general, has developed stress responses such as acid tolerance response (ATR), which protects it against severe acid stress [32]. This ATR is also reported to provide protection against heat [33]. This cross-protection is of great importance in food preservation, which depends on multiple stresses to guarantee microbiological safety of processed foods [34].

Staphylococcus aureus is a gram-positive coccus. An estimated 50% of the human population carry this pathogen on the skin and in the nasal passage [35] and, thus, is the most prevalent pathogenic bacteria found in foods handled by bare hands. As a food-borne pathogen, *S. aureus* is most important as the source of various heat-stable enterotoxins produced in contaminated foods. Ingestion of foods containing staphylococcal enterotoxins leads to food poisoning with severe gastroenteritis. *S. aureus* does not compete well with background flora normally found in unheated foods. However, mild pasteurization treatments designed to destroy heat-sensitive bacterial pathogens may reduce the competing microflora, allowing the more heat-resistant *S. aureus* to survive, multiply, and produce toxin in treated foods [36].

3.3 THERMAL TREATMENT AND MICROORGANISMS

Many food processing systems use high temperatures to preserve food by reducing the number of bacteria in a product. In food microbiology, high temperature means all temperatures above ambient [37]. This enhances food safety and increases the shelf life of a product.

To achieve these goals, the required time–temperature combinations are set. The most effective measure used to defend the food supply against microbial contamination is heat. The principal goal of heat treatment is to destroy vegetative cells, toxins, enzymes, and spores of microorganisms that may either be pathogenic to humans or cause the spoilage of food. In addition to cooking, two basic types of commercial heat processing methods exist: pasteurization and sterilization. Pasteurization is basically a mild heat treatment aimed at destroying roughly 99%–99.9% of vegetative cells. The purpose of pasteurization is to destroy most nonspore-forming bacteria that may be present. Sterilization involves a more intense treatment, which completely destroys a population of microorganisms as may be measured by an appropriate plating or enumerating technique [37].

Pasteurization of milk is achieved by heating as follows:

63°C for 30 min [low temperature, long time (LTLT)]
72°C for 15 s [primary high temperature, short time (HTST)]
89°C for 1 s
90°C for 0.5 s
94°C for 0.1 s
100°C for 0.01 s

These treatments are equivalent and are sufficient to destroy the most heat resistant of the nonspore-forming pathogenic organisms—*Mycobacteria tuberculosis* and *Coxiella burnetti.* These temperatures are also sufficient to destroy all yeasts, molds, gram-negative bacteria, and many gram positives. Organisms that survive, but do not grow at pasteurization temperatures are known as thermodurics. Many lactic acid bacteria belong to this group. Thermophiles not only survive high temperatures but also require high temperatures for their growth. Many thermophilic bacteria are found within the spore-forming groups—*Bacillus* and *Clostridium* spp.

Sterilization means the destruction of all viable organisms, vegetative, or spores, as may be measured by an appropriate plating or enumerating technique. Canned foods are sometimes called "commercially sterile" to indicate that no viable organisms can be detected by the usual cultural

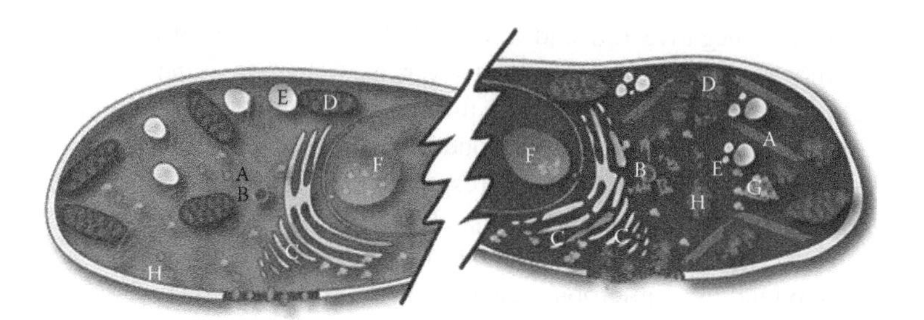

FIGURE 3.1 Effects of heat shock on the organization of the eukaryotic cell. An unstressed eukaryotic cell (left) is compared to a cell under heat stress (right). Heat stress leads to damage to the cytoskeleton, including the reorganization of actin filaments (A) into stress fibers and the aggregation of other filaments (microtubuli, B). Organelles like the Golgi and the endoplasmic reticulum (C) become fragmented and disassemble. The number and integrity of mitochondria (D) and lysosomes (E) decrease. The nucleoli, sites of ribosome (F) assembly, swell, and large granular depositions consisting of ribosomal proteins become visible. Large depositions, the stress granula (G), resulting from assemblies of proteins and RNA, are found in the cytosol in addition to protein aggregates (hexagonal versus spaghetti style, H). Finally, there are changes in the membrane morphology, aggregation of membrane proteins, and an increase in membrane fluidity. Together, all these effects stop growth and lead to cell-cycle arrest as indicated by the noncondensed chromosomes in the nucleus (From Richter, K. et al., *Mol. Cell*, 40, 253, 2010.)

methods employed or that the number of survivors is so low as to be of no significance under the conditions of canning and storage. Microorganisms may be present in canned foods but cannot grow in the product because of undesirable pH, oxidation–reduction potential (Eh), or temperature of storage.

Most cells die in response to high temperatures. Heat denatures nucleic acids, structural proteins, and enzymes, and results in the loss of vital cell functions [38] (Figure 3.1). Adequate cooking procedures during processing are effective in killing food-borne microbes.

A number of factors influence a microorganism's susceptibility to heat. These factors include:

1. *Water activity of the food*—This represents the amount of unbound water molecules in the food and, thus, are available for the microorganisms. Microorganisms in foods with high water activity are more susceptible to thermal treatment than those in foods with low water activity. This is because protein denaturation occurs faster in microbial cells heated in water than in air [37] or in low water activity foods. It is suggested that the heating of wet proteins causes the formation of free SH groups and increase in the water-binding capacity of proteins. Thermal breaking of peptide bonds requires less energy in the presence of water [37].
2. *Fat content of food*—Some microorganisms develop increased heat resistance in the presence of fat due to the protection offered by fat by directly affecting cell moisture [39].
3. *Type and concentration of salt in food*—The effect of salt on thermal resistance of microorganisms is related to water activity. Some salts bind water molecules and decrease water activity, thereby increasing heat resistance. Other salts, such as Ca^{2+} and Mg^{2+}, which may increase water activity, increase sensitivity of cells to heat [40].
4. *Presence of carbohydrates in food*—Microbial cells are more susceptible to heat in foods with less amount of sugars in them. The presence of sugars at higher concentrations in foods increases the heat resistance partly due to decreased water activity. However, wide differences in heat sensitivity occurred with different types of sugars [41].
5. *pH of food*—Heat sensitivity in microbial cells increases when the pH is lowered or raised from neutral (pH 7) where they are the most resistant to heat. Thermal processing of high-acid foods, thus, requires considerably less heat. It is well known that the

microbial susceptibility to high temperature increases as the pH of the food decreases [42,43].

6. *Proteins and other particles in food*—Microbial cells are susceptible to heat when the protein content of the food is low. Proteins have a protective effect on microorganisms when they are heated in high protein medium.
7. *Number of organisms in food*—Microbial cells are susceptible to heat when they are found in low numbers in the food. Degree of heat resistance increases with rise in the number of microorganisms in the particular food. A high number of microorganisms in food excrete proportionally higher amounts of extracellular compounds, which are predominantly proteins that offer some protection against heat.
8. *Age of organisms*—Bacteria are more susceptible to heat when they are in the exponential phase, and resistance increases as the bacterial population gets older and goes through the stationary phase. The mechanism why less active bacterial cells are resistant to heat is not well understood [37].
9. *Microbial growth temperature*—Bacterial cells are sensitive to heat when the temperature of incubation is within their optimum range. As this temperature increases, the heat resistance of microorganisms in the food also increases, particularly in the case of spore formers. This may, possibly, be due to genetic selection for more heat-resistant cells.
10. *Inhibitory compounds*—Most microorganisms are more susceptible to heat treatment when they are heated in food that contains microbial inhibitors. Adding inhibitory substances (such as nitrite) to foods prior to heat treatment reduces the amount of heat necessary to eliminate the microorganisms.

3.4 MICROBIAL THERMAL RESISTANCE AND ITS DEVELOPMENT

Currently, thermal processing is the most commonly used method for food preservation. It is an efficient and reliable process reducing the risks associated with food-borne pathogens and diminishing the activity of several enzymes, and it is also an economical technology for the food industry in terms of energy [44]. Since 1977, the frequency of discovering emerging food-borne pathogens has been quite high. These pathogens cause infection at relatively low doses and are found in thermally untreated foods [45]. Therefore, accurate prediction of heat resistance and thermal death rates of target pathogens is crucial to produce safe food [44].

In general, the following can be said about microbial heat resistance. Psychrophiles are the most heat sensitive followed by mesophiles and thermophiles, in that order. Spore formers are more resistant than nonspore formers. Among the nonspore formers, cocci are generally more resistant than rods. gram-positive bacteria are relatively more heat resistant than gram-negative bacteria. Yeasts and molds are fairly sensitive to heat. Among the fungal spores, sclerotia are the most heat resistant [37].

To eliminate contaminating microorganisms from food, the most effective measure taken by the food industry is heat treatment. The purpose of heat treatment, depending on its intensity, is to destroy vegetative cells, microbial toxins, or microbial spores, which are important either as human pathogens or as spoilage microorganisms.

High temperature invariably kills food-borne microbial cells mainly because it denatures nucleic acids, structural proteins, and other functional proteins such as enzymes. However, when conditions are created for microorganisms to survive a heat treatment just short of a temperature that normally eliminates them, various microorganisms develop a mechanism to allow them withstand even a higher temperature. This is usually achieved by producing heat shock proteins (HSPs) [46].

Spores of certain bacteria belonging to the genera *Clostridium* and *Bacillus* can survive at temperatures of >100°C. More severe processing parameters should be considered to control such microorganisms [37]. On the other hand, there are nonspore-forming bacterial species that have developed resistance to high temperatures. For example, *E. faecalis* and *E. faecium* are among the vegetative bacteria that are capable of withstanding heat treatment and spoil pasteurized canned meat products [47].

During microbial inactivation by heat, the right combination of temperature and time is of paramount importance to achieve two goals in a process: eliminate or reduce pathogenic or spoilage microorganisms to an acceptable level and, at the same time, cause minimum effect on product quality and acceptability [48,49].

Although heat has its own detrimental effect on microorganisms, the presence of other factors in the food may have influence on the thermal inactivation of microorganisms [7]. Hansen and Riemann [50] reported that combination of salt and heat did not show any significant reduction of *E. coli* O157:H7, and they considered this as additive effect. In the case of acid–heat combination, however, a higher reduction of *E. coli* O157:H7 was noted, and such effect was considered as synergistic. A combination of salt and acid resulted in a lower reduction of *E. coli* O157:H7 than acid alone treatment. Salt was believed to give protection against acid treatment (antagonistic effect) [7].

Microbial heat resistance is influenced by several environmental factors [51]. These factors can affect heat resistance when applied prior to heat treatment (as growth conditions and exposure to various sublethal stresses), simultaneously with heat treatment (as composition of the heating media, its pH and water activity), or subsequent to heat treatment (as recovery conditions that can affect the detection of survivors). Accordingly, microbial resistance to heat may vary [7,50].

In culture medium, it was observed that clostridia were more strongly inhibited by nitrite when the medium was autoclaved after, and not before, the addition of nitrite to the medium [52,53].

It was assumed that heating the medium in the presence of nitrite resulted in the formation of a substance called the Perigo factor [37].

The exposure of microorganisms to a previous heat shock before the pasteurization or sterilization treatment is another relevant point to consider [54]. The heat resistance of several microbial species increases when cells are exposed for a short time to moderately elevated temperatures, normally above their maximum for growth, before the actual heat treatment is applied [55–58].

L. monocytogenes cells develop enhanced resistance to heat when they grow in media containing fat or quaternary ammonia compounds. It is, thus, advised to consider the various growth conditions that *L. monocytogenes* cells could encounter before establishing and verifying adequacy of a thermal challenge [59].

According to a study by Gliemmo et al. [60], during thermal inactivation of an osmophilic yeast, *Zygosaccharomyces bailii*, it was observed that addition of potassium sorbate at a concentration of 0.025% increased the rate of heat inactivation while lowering the a_w to 0.985 using polyols did not have any effect on rate of heat inactivation. The use of glucose enhanced it. A synergistic effect on the rate of heat inactivation of *Z. bailii* was, however, observed by the combined use of potassium sorbate and sorbitol, xylitol, or glucose to depress a_w to 0.985–0.988. The authors [60] suggested that this behavior of *Z. bailii* might help to decrease the severity of the thermal treatment with no detrimental effect on sterility.

Among the various environmental stresses, starvation of microbial cells may increase resistance to other lethal stresses. Starvation, for example, increased the heat resistance of *L. monocytogenes* [61–63]. When evaluating thermal effects, it is important to consider the nutritional status of target bacteria. Thermal treatment studies [64] on starved cells on bologna showed that increased temperature alone reduced the thermal death time, whereas addition of sodium lactate showed protective effect. Other combination treatments either reduced or increased the thermal death time. They concluded that starvation rendered *L. monocytogenes* more susceptible to heat and additives.

In *E. faecium*, previous adaptation to a low pH increased the bacterial heat resistance, whereas the subsequent cold storage of cells reduced its thermal tolerance. The extent of increased heat tolerance varied with the acid used to lower the pH. In contrast, cold storage progressively decreased *E. faecium* thermal resistance. These findings highlight the need for a better understanding of microbial response to various preservation stresses in order to increase the efficiency of thermal processes [65].

In another study [66] on the effect of defrosting of ground beef in refrigerator, at room temperature or in microwave, it was observed that the thermal inactivation of *Salmonella* Enteritidis was not, overall, affected by defrosting practices. In contrast, defrosting at room temperature resulted in an increased heat tolerance of *L. monocytogenes* compared to the rest of the tested defrosting practices.

Byrne et al. [67] studied the effects of commercial beef burger production and product formulation on the heat resistance of *E. coli* O157:H7 and found out that in frozen tempered trimmings, the cells were less heat resistant than those found in fresh trimmings. Those found in frozen trimmings made of beef (70%), and other ingredients were more heat resistant than those in frozen trimmings made of 100% beef. Higher heat resistance was also observed in *E. coli* O157:H7 in unfrozen trimmings with 70% beef. The study concluded that commercial processing and product formulation have profound effects on the heat resistance of *E. coli* O157:H7 in beef burgers.

In *C. sakazakii* (formerly *E. sakazakii*), an emerging opportunistic pathogen in infants associated with infant formula [22] and other foods [23], it was shown that heat shock treatments given at and around its maximum growth temperature enhanced its thermal tolerance. It was also shown that heat shock treatment caused damage and disruption in *C. sakazakii* cells resulting in increased leakage of nucleic acid and protein as the temperature and duration of heat shock increased [25] (Figure 3.2).

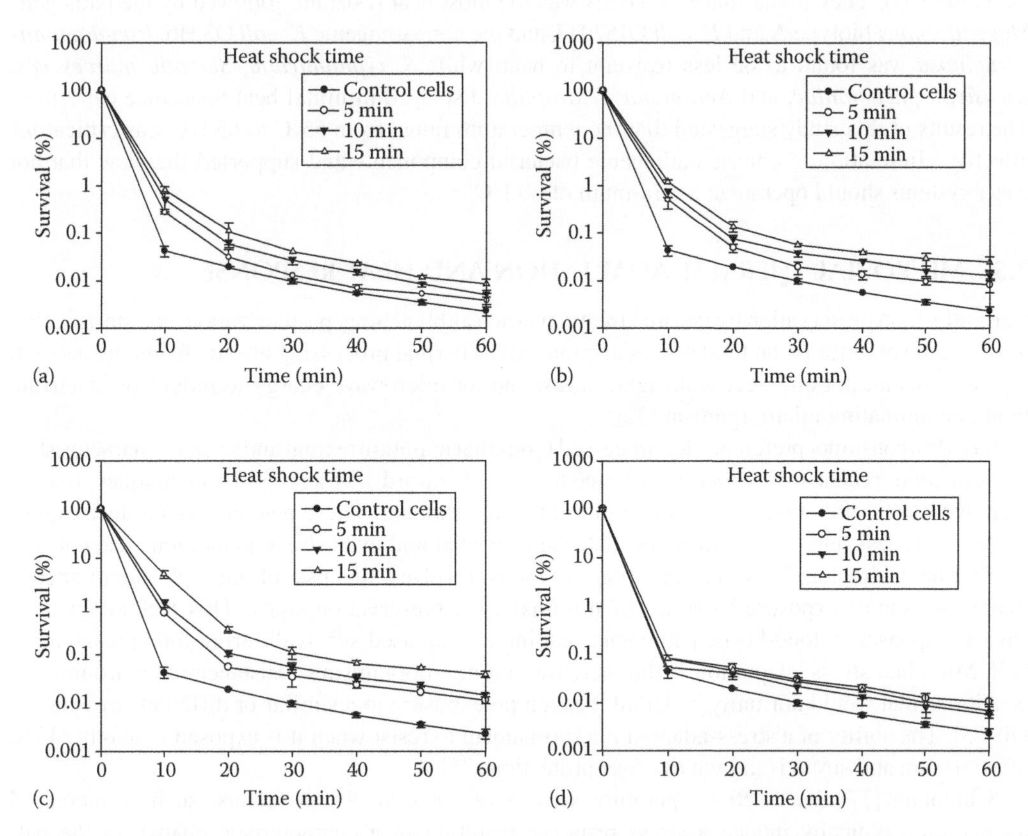

FIGURE 3.2 Effect of heat shock time on the survival of *C. sakazakii* BCRC 13988 at 51°C. The initial population of the control and heat shock *C. sakazakii* BCRC 13988 were about 10^6 CFU mL^{-1}. Surviving percentage was obtained by dividing the survival population by the initial population, which corresponds to 100%. Data were expressed as mean ± standard deviations from the three separate experiments. (a) Heat shock at 42°C, (b) heat shock at 45°C, (c) heat shock at 47°C, and (d) heat shock at 48°C. (From Chang, H. et al., *Int. J. Food Microbiol.*, 134, 184, 2009.)

In another study [44], it was demonstrated that the heat resistance of *C. sakazakii* depended on the strain studied, the growth phase, growth conditions, the characteristics of treatment medium, and the recovery conditions.

Byrne et al. [68] designed a thermal treatment for pork luncheon roll, which would destroy *B. cereus* and *Clostridium perfringens* vegetative cells and spores and determined that the decimal reduction time for *B. cereus* vegetative cells ranged from 1 min to 33.2 min at 60.1°C and 50.1°C, respectively. The decimal reduction time of *C. perfringens* vegetative cells also ranged from 0.9 min to 16.3 min at 65.1°C and 55.1°C, respectively. In case of *B. cereus* spores, the range was 2–32.1 min at 95.1°C and 85.1°C, respectively, whereas it ranged from 2.2 to 34.2 min at 100.1°C and 90.1°C, respectively, for *C. perfringens*.

Oteiza et al. [69] determined the thermal resistance of *E. coli* O157:H7 and *E. coli* in Morcilla, a link sausage, and observed that higher added fat and starch levels resulted in higher thermal resistance for both strains and concluded that the composition of product affected heat lethality of the two strains of *E. coli*. Velliou et al. [70] observed that addition of various acids to achieve a specific pH of the medium increased the heat resistance of stationary-phase *E. coli* K12 cells. The induced resistance depended on the type of acid and on the quantity added.

Spinks et al. [71] investigated water quality and health risks of domestic hot water systems from harvested rainwater by performing thermal inactivation analysis on eight species of non-spore-forming bacteria in a water medium at temperatures relevant to domestic hot water systems (55°C–65.1°C). They found that *E. faecalis* was the most heat resistant, followed by the pathogens *Shigella sonnei* biotype A and *E. coli* O157:H7, and the nonpathogenic *E. coli* O3:H6. *Pseudomonas aeruginosa* was found to be less resistant to heat, while *S. typhimurium*, *Serratia marcescens*, *Klebsiella pneumoniae*, and *Aeromonas hydrophila* displayed minimal heat resistance capacities. The results of this study suggested that the temperature range from 55°C to 65.1°C was critical for effective elimination of enteric/pathogenic bacterial components and supported the view that hot water systems should operate at a minimum of 60.1°C.

3.5 MICROBIAL THERMAL ADAPTATION AND HEAT RESPONSE

Currently, food preservation by thermal treatment such as blanching, pasteurization, and sterilization is a common practice in the food processing industry. Thermal processing uses different means such as water, steam, hot air, electrical, light, ultrasound, or microwave energy to reduce or inactivate food-contaminating microorganism [72].

Based on consumer preference for processed foods that maintain textual and sensory characteristics of fresh food, recent food processing procedures tend toward less aggressive techniques, that is, high hydrostatic pressure, irradiation, and mild heat treatment. These new processing techniques, however, are less harsh and may favor microbial survival and even stress adaptation. The concept of "hurdle technology" for preservation of foods involves the use of multiple simultaneous treatments and/or exposure to series of minimal food preservation steps. This technology may trigger responses in food-borne pathogens, leading to increased survival during food preservation [73]. Microbial stress adaptation is the increase in a microorganism's resistance to environmental conditions that would normally be lethal through preexposure to a similar or different stress factor [70–75]. The ability of a stress-adapted microorganism to resist when it is exposed to another kind of environmental stress is known as cross-protection [76].

Schumann [77] discussed temperature sensors of bacteria. Stress factors, such as increased temperature, typically induce a stress response resulting in a characteristic change in the pattern of gene expression. This stress response helps the bacterial cells to protect vital processes, to restore cellular homeostasis, and to increase the cellular resistance against subsequent stress challenges. Temperature sets physiological limits for a wide variety of biological activities and, therefore, the ability to sense and respond to changes in temperature is of vital importance. Three different thermosensors have been described so far: (1) DNA as nucleoid modulator, (2) RNA, and (3) proteins. Bacteria need thermosensors to prevent denaturation of their proteins caused by sudden increases in temperature. Denatured proteins form aggregates, which kill the cell. Bacteria show two different responses to high temperature: the high-temperature response (HTR), which is a constitutive expression of high-temperature genes at high temperatures and strongly reduced expression at low temperatures, and the heat shock response (HSR), which is a transient expression of heat shock genes after a sudden temperature up-shift (Table 3.1).

There are different targets in microbial cells upon which heat can act. Thus, heat resistance manifested by microorganisms is basically due to intrinsic stability of macromolecules, such as

TABLE 3.1
Thermosensors of Some Bacteria

Temperature Response	Thermosensors	Examples
	Nucleoid modulator (DNA)	*virF* promoter of *Shigella flexneri*
		Promoter of the *E. coli* hemolysin gene
		Synthesis of pili (fimbriae)
		Virulence genes encoded by *Yersinia* plasmids
HTR	RNA	*rpoH* transcript of *E. coli*
		ROSE (Repression of heat shock gene expression) element
		FourU element
		lcrF transcript of *Yersinia pestis*
		prfA transcript of *L. monocytogenes*
		hspA of *Spindasis vulcanus*
		hspA transcript of *Scaphirhynchus albus*
		Group II intron of *Azotobacter vinelandii*
	Proteins	TlpA repressor protein of *S. Enterica*
		RheA protein of *S. albus*
		ClpXP and Lon proteins of *Y. pestis* translational repressor of *Thermosynechococcus elongates*
		Sensor kinases
		Replication initiation protein
HSR	Molecular chaperones	σ^{32}—DnaK system of *E. coli*
		HrcA–GroEL system of *Bacillus subtilis*
	Proteases	HspR–DnaK system of *Streptomyces coelicolor*
		DegS of *E. coli*
Low-temperature response (LTR)	*cis*-acting RNA	Lysis–lysogeny decision of phage l
	trans-acting RNA protein	DsrA RNA
		VirA sensor kinase of *Agrobacterium tumefaciens*
		NifA activator of *Klebsiella pneumonia*
		Response regulator DegU of *L. monocytogenes*
		Conjugation
Cold shock response (CSR)	RNA	*cspA* transcript of *E. coli*
		Pap transcript of *E. coli*
	Protein	DesK sensor kinase of *B. subtilis*

Source: From Juneja, V.K. and Novak, J.S., Adaptation of foodborne pathogens to stress from exposure to physical intervention strategies, in: A.E. Yousef, V.K. Juneja, eds. *Microbial Stress Adaptation and Food Safety*, CRC Press, Boca Raton, FL, pp. 159–211, 2003.

ribosomes, nucleic acids, enzymes, and proteins, inside the microbial cell and membranes [72]. Elevated heat can destroy specific secondary and tertiary structures of ribosomal subunits, coagulated proteins, damage RNA, and single-stranded DNA and also change the fluidity of microbial membranes by affecting saturation and length of fatty acids [73].

However, in most microorganisms, thermotolerance is induced by slow heating or heating for short periods of time at temperatures above the optimum temperature [78]. Such temperatures trigger a transient physiological cellular response to stressful stimuli where various proteins, known as heat shock proteins (HSPs), are synthesized [79]. These are cytoplasmic proteins that play an important role in protein–protein interactions such as refolding of aberrant proteins, which have been denatured during exposure to heat. They assist in the establishment of proper protein conformation and proteolysis of unwanted protein aggregation [80,81].

There are practical consequences during the exposure of bacterial cells to heat shocks. During the thermal processing of solid foods, the slow heat penetration across the pieces exposes the microorganisms to a temperature gradient for a given period of time. In a similar way, during the processing of some liquid foods, the temperature increases slowly and gradually due to the thermal sensitivity of some components, which cannot withstand contact with hot surfaces in heat exchangers. The gradual increase in temperature, in both cases, may act as a heat shock, which would provoke the increase of heat resistance of the microorganisms to the final treatment temperature [82,83]. Many modern food preservation processes apply one or more environmental stresses, which may retard or inhibit bacterial growth. However, stress-inhibited bacteria make phenotypic variations such as expression of HSPs [84] and cross-protection against different challenges including resistance to antibiotics [85]. Genotypic variations are manifested in the form of stress-induced genetic plasticity where rates of microbial random mutagenesis and intra- and intercellular transfer of genes are increased [86]. These alterations increase population diversity, increasing the chances that at least some cells may survive and grow under the conditions of stress. Through such variations, population diversity is increased leading to increased chances for some cells to survive and grow under the conditions of stress [87]. Production of HSPs by a stressed cell is induced by heat shock factor (HSF), and their concentration is regulated at transcription [81].

Bacteria recognize and respond to a variety of environmental alterations through signal transduction mechanisms [88]. To adapt to rapidly changing conditions, bacteria produce several alternative sigma factors that play key roles in regulating bacterial gene expression [89]. The cytoplasm gets signals about the extracytoplasmic conditions through extracytoplasmic function (ECF) sigma factors [90]. Sigma B, for example, is an alternative sigma factor, which is involved in stress response of gram-positive bacteria [91].

One of the widespread responses that bacteria use to adapt to harsh environments is the SOS [92]. Under harsh environments, the SOS response plays an active role in stress resistance, in the induction of genetic diversity that may result in stress-resistant subpopulations. Spoilage and pathogenic bacteria likely survive food preservation steps because of their SOS response [93].

A number of stresses encountered in the food processing facilities activate the SOS response in a range of bacteria (Figure 3.3). The SOS response is regulated by repressor and activator proteins and is essential for DNA repair and restart of stalled or collapsed replication forks [94]. It generally consists of proteins involved in DNA repair, such as excinucleases, helicases, and recombinases, or proteins involved in translesion DNA synthesis, such as translesion DNA polymerases, and is typically induced by stresses and agents that cause damage to DNA or the collapse of replication forks resulting in exposure of ssDNA [93].

When bacteria are exposed to food preservation stresses in food processing facilities, DNA damage or replication fork stalling may occur consequently activating the SOS response. This may subsequently result in increased mutagenesis due to the activity of error-prone translesion polymerases [93]. Various stress responses have been described to be involved in adaptive mutagenesis as well [95]. Activation of the SOS and other mutagenesis mechanisms during food preservation

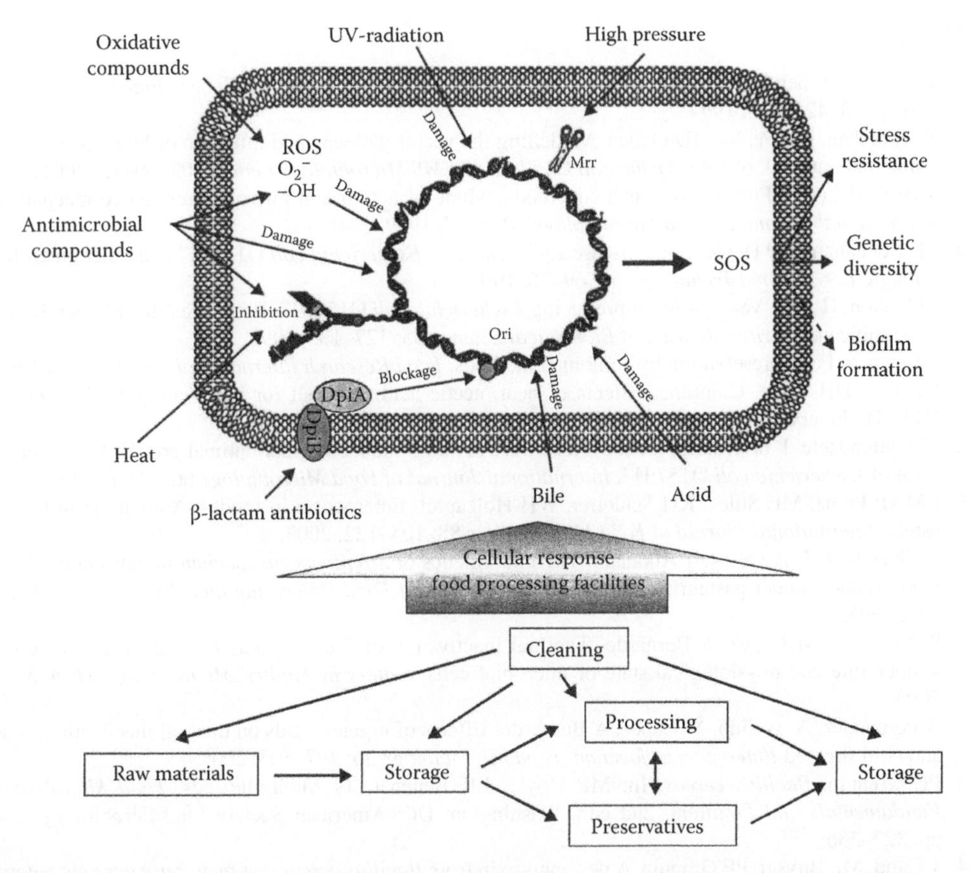

FIGURE 3.3 Stresses/triggers and their mechanisms of SOS response activation in bacteria. (From van der Veen, S. and Abee, T., *Curr. Opin. Biotechnol.*, 22, 136, 2011.)

could result in the occurrence of adaptive mutations in bacteria. These adapted bacteria can subsequently become persistent in the food processing environment and repeatedly contaminate food products [93].

3.6 CONCLUSIONS

In food microbiology, techniques to preserve food focus on lengthening the lag phase of microbial growth, retarding the growth rate, and/or reducing the maximum population. Microorganisms in foods interact with one another and their food environment. This interaction limits whether they are eliminated from the food, survive, or grow. It has encouraged microorganisms to sense and develop strategies to withstand environmental effects—formation of protective structures (spores), adaptive phenotypic responses (HSRs), or even genetic responses (SOS responses). Bacterial survival strategy mainly involves phenotypic variability in a genetically homologous population. In the food industry, each subsequent preservation treatment will play a role as natural selection agent resulting in more resistant strains and their subsequent progeny. Thermal treatment is the most dependable technique to eliminated pathogenic bacteria from food. Although product quality is a consumer sensory demand, microbial evolution has reached a point where product quality compromises food safety. Further research on ecophysiology of food-borne pathogens may bring us closer to resolving this dilemma.

REFERENCES

1. RV Tauxe. Emerging foodborne diseases: An evolving public health challenge. *Emerging Infectious Diseases* 3: 425–433, 1997.
2. NJ Strachan, DR Fenlon, ID Ogden. Modelling the vector pathway and infection of humans in an environmental outbreak of *Escherichia coli* O157:H7. *FEMS Microbiology Letters* 203: 69–73, 2001.
3. S Brul, P Coote. Preservative agents in foods: Mode of action and microbial resistance mechanisms. *International Journal of Food Microbiology* 50: 1–17, 1999.
4. RL Buchanan, MP Doyle. Food disease significance of *Escherichia coli* O157:H7 and other enterohemorrhagic *E. coli. Food Technology* 51: 69–76. 1997.
5. FJ Bolton, H Aird. Verocytotoxin-producing *Escherichia coli* O157:H7: Public health and microbiological significance. *British Journal of Biomedical Science* 55: 127–135, 1998.
6. L Leistner. Food preservation by combined methods. *Food Research International* 25: 151–158, 1992.
7. SY Lee, DH Kang. Combined effects of heat, acetic acid, and salt for inactivating *Escherichia coli* O157:H7 in laboratory media. *Food Control* 20: 1006–1012, 2009.
8. M Uyttendaele, I Taverniers, J Debevere. Effect of stress induced by suboptimal growth factors on survival of *Escherichia coli* O157:H7. *International Journal of Food Microbiology* 66: 31–37, 2001.
9. CMAP Franz, ME Stiles, KH Schleifer, WH Holzapfel. Enterococci in foods—A conundrum for food safety. *International Journal of Food Microbiology* 88: 105–122, 2003.
10. S Ghazala, D Coxworthy, T Alkanani. Thermal kinetics of *Streptococcus faecium* in nutrient broth/sous vide products under pasteurization conditions. *Journal of Food Processing and Preservation* 19: 243–257, 1995.
11. S. Martınez, M Lopez, A Bernardo. Thermal inactivation of *Enterococcus faecium*: Effect of growth temperature and physiological state of microbial cells. *Letters in Applied Microbiology* 37: 475–481, 2003.
12. A Fernandez, A Avelino, M Lopez, A Bernardo. Effects of organic acids on thermal inactivation of acid and cold stressed *Enterococcus faecium. Food Microbiology* 26: 497–503, 2009.
13. PE Granum. *Bacillus cereus*. In: MP Doyle, LR Beuchat, TJ Montville, eds. *Food Microbiology: Fundamentals and Frontiers*. 2nd edn. Washington, DC: American Society for Microbiology, 2001, pp. 327–336.
14. T Lund, ML Buyser, PE Granum. A new cytotoxin from *Bacillus cereus* that may cause necrotic enteritis. *Molecular Microbiology* 38: 254–261, 2000.
15. A. Kotiranta, K Lounatmaa, M Haapasalo. Epidemiology and pathogenesis of *Bacillus cereus* infections. *Microbes and Infection* 2: 189–198, 2000.
16. W van Schaik, MH Tempelaars, JA Wouters, WM de Vos, T Abee. The alternative sigma factor σ^B of *Bacillus cereus*: Response to stress and role in heat adaptation. *Journal of Bacteriology* 186: 316–325, 2004.
17. PS Mead, L Slutsker, V Dietz, LF McCaig, JS Bresee, C Shapiro, PM Griffin, RV Tauxe. Food-related illness and death in the United States. *Emerging Infectious Diseases* 5: 607–625, 1999.
18. BG Gellin, CV Broome. Listeriosis. *Journal of the American Medical Association* 261: 1313–1320, 1989.
19. B Gudbjonsdottir, ML Suihiko, P Gustavsson, G Thorkelsson, S Salo, AM Sjoberg. The incidence of *Listeria monocytogenes* in meat, poultry and seafood plants in Nordic countries. *Food Microbiology* 21: 217–225, 2004.
20. J Rocourt, P Cossart. Listeria monocytogenes. In: MP Doyle, LR Beuchat, TJ Montville, eds. *Food Microbiology: Fundamentals and Frontiers*. 2nd edn. Washington, DC: American Society for Microbiology, 1997, pp. 237–352.
21. J Rocourt, J Bille. Food-borne listeriosis. *World Health Statistics Quarterly* 50: 67–73, 1997.
22. J van Acker, F de Smet, G Muyldermans, A Bougatef, A Naessens, S Lauwers. Outbreak of necrotizing enterocolitis associated with *Enterobacter sakazakii* in powdered milk formula. *Journal of Clinical Microbiology* 39: 293–297, 2001.
23. M Friedemann. *Enterobacter sakazakii* in food and beverages (other than infant formula and milk powder). *International Journal of Food Microbiology* 116: 1–10, 2007.
24. LR Beuchat, H Kim, JB Gurtler, LC Lin, JH Ryu, GM Richards. *Cronobacter sakazakii* in foods and factors affecting its survival, growth, and inactivation. *International Journal of Food Microbiology* 136: 204–213, 2009.
25. H Chang, ML Chiang, CC Chou. The effect of temperature and length of heat shock treatment on the thermal tolerance and cell leakage of *Cronobacter sakazakii* BCRC 13988. *International Journal of Food Microbiology* 134: 184–189, 2009.

26. CR Stumbo, KS Purohit, TV Ramakrishna. Thermal process lethality guide for low-acid foods in metal containers. *Journal of Food Science* 40: 1316–1323, 1975.
27. MW Peck, SC Stringer, AT Carter. Review *Clostridium botulinum* in the post-genomic era. *Food Microbiology* 28: 183–191, 2011.
28. MW Peck. Biology and genomic analysis of *Clostridium botulinum. Advances in Microbial Physiology* 55: 183–265, 2009.
29. JY D'Aoust. *Salmonella*. In: BM Lund, TC Baird-Parker, GW Gould, eds. *The Microbiological Safety and Quality of Foods*. Gaithersburg, MD: Aspen, 2000, pp. 1233–1299.
30. MR Wilmes-Riesenberg, B Bearson, JW Foster, R Curtiss. Role of the acid tolerance response in virulence of *Salmonella typhimurium. Infection and Immunity* 64: 1085–1092, 1996.
31. JW Foster, MP Spector. How *Salmonella* survive against the odds. *Annual Review of Microbiology* 49: 145–174, 1995.
32. HG Yuk, KR Schneider. Adaptation of *Salmonella* spp. in juice stored under refrigerated and room temperature enhances acid resistance to simulated gastric fluid. *Food Microbiology* 23: 694–700, 2006.
33. AS Mazzotta. Thermal inactivation of stationary-phase and acid-adapted *E. coli* O157:H7, *Salmonella* and *Listeria monocytogenes* in fruit juices. *Journal of Food Protection* 64: 315–320, 2001.
34. A Álvarez-Ordóñez, A Fernández, M López, R Arenas, A Bernardo. Modifications in membrane fatty acid composition of *Salmonella typhimurium* in response to growth conditions and their effect on heat resistance. *International Journal of Food Microbiology* 123: 212–219, 2008.
35. JP Arbuthnott, DC Coleman, JS de Azevedo. Staphylococcal toxins in human disease. *Journal of Applied Bacteriology (Supplement)* 69: 101S–107S, 1990.
36. J Kennedy, IS Blair, DA McDowell, DJ Bolton. An investigation of the thermal inactivation of *Staphylococcus aureus* and the potential for increased thermotolerance as a result of chilled storage. *Journal of Applied Microbiology* 99: 1229–1235, 2005.
37. JM Jay, MJ Loessner, DA Golden. *Modern Food Microbiology*. 7th edn. New York: Springer Science and Business Media, 2005, pp. 415–441.
38. K Richter, M Haslbeck, J Buchner. The heat shock response: Life on the verge of death. *Molecular Cell* 40: 253–266, 2010.
39. H Sugiyama. Studies on factors affecting the heat resistance of spores of *Clostridium botulinum. Journal of Bacteriology* 62: 81–96, 1951.
40. HS Levinson, MT Hyatt. Effect of sporulation medium on heat resistance, chemical composition, and germination of *Bacillus megaterium* spores. *Journal of Bacteriology* 87: 876–886, 1964.
41. JEL Corry. The effect of sugars and polyols on the heat resistance of salmonellae. *Journal of Applied Bacteriology* 37: 31–43, 1974.
42. MA Casadei, R Ingram, E Hitchings, J Archer, JE Gaze. Heat resistance of *Bacillus cereus, Salmonella typhimurium* and *Lactobacillus delbrueckii* in relation to pH and ethanol. *International Journal of Food Microbiology* 63: 125–34, 2001.
43. R Rodrigo, C Rodrigo, PS Fernandez, M Rodrigo, A Martınez. Effect of acidification and oil on the thermal resistance of *Bacillus stearothermophilus* spores heated in food substrate. *International Journal of Food Microbiology* 52: 197–201, 1999.
44. C Arroyo, S Condón, R Pagan. Thermobacteriological characterization of *Enterobacter sakazakii. International Journal of Food Microbiology* 136: 110–118, 2009.
45. RV Tauxe. Emerging food-borne pathogens. *International Journal of Food Microbiology* 78: 31–41, 2002.
46. N Chotyakul, JG Velazquez, JA Torres. Assessment of the uncertainty in thermal food processing decisions based on microbial safety objectives. *Journal of Food Engineering* 102: 247–256, 2011.
47. CM Franz, WH Holzapfel, ME Stiles. Enterococci at the crossroads of food safety. *International Journal of Food Microbiology* 47: 1–24, 1999.
48. M Peck. *Clostridium botulinum* and the safety of minimally heated, chilled foods: An emerging issue? *Journal of Applied Microbiology* 101: 556–570, 2006.
49. DS Smith, JN Cash, WK Nip, YH Hui. *Processing Vegetables: Science and Technology*. New York: CRC Press, 1997, p. 434.
50. NH Hansen, H Riemann. Factors affecting the heat resistance of nonsporing organisms. *Journal of Applied Bacteriology* 26: 314–333, 1963.
51. RI Tomlins, ZJ Ordal. Thermal injury and inactivation in vegetative bacteria. In: FA Skinner, WB Hugo, eds. *Inhibition and Inactivation of Vegetative Microbes*. London, U.K.: Academic Press, 1976, pp. 153–190.
52. JA Perigo, TA Roberts. Inhibition of clostridia by nitrite. *Journal of Food Technology* 3: 91–94, 1968.

53. JA Perigo, E Whiting, TE Bashford. Observations on the inhibition of vegetative cells of *Clostridium sporogenes* by nitrite which has been autoclaved in a laboratory medium, discussed in the context of sublethally processed meats. *Journal of Food Technology* 2: 377–397, 1967.
54. M Hassani, P Manas, R Pagan, S Condon. Effect of a previous heat shock on the thermal resistance of *Listeria monocytogenes* and *Pseudomonas aeruginosa* at different pHs. *International Journal of Food Microbiology* 116: 228–238, 2007.
55. F Jorgensen, B Panaretou, PJ Stephens, S Knochel. Effect of pre- and post-heat shock temperature on the persistence of thermotolerance and heat shock-induced proteins in *Listeria monocytogenes. Journal of Applied Bacteriology* 80: 216–224, 1996.
56. F Jorgensen, TB Hansen, S Knochel. Heat shock induced thermotolerance in *Listeria monocytogenes* 13–249 is dependent on growth phase, pH and lactic acid. *Food Microbiology* 16: 185–194, 1999.
57. R Pagan, S Condón, FJ Sala. Effect of several factors on the heat shock-induced thermotolerance of *Listeria monocytogenes. Applied Environmental Microbiology* 63: 3225–3232, 1997.
58. YD Lin, CC Chou. Effect of heat shock on thermal tolerance and susceptibility of *Listeria monocytogenes* to other environmental stresses. *Food Microbiology* 21: 605–610, 2004.
59. KK Schultze, RH Linton, MA Cousin, JB Luchansky, ML Tamplin. Effect of preinoculation growth media and fat levels on thermal inactivation of a serotype 4b strain of *Listeria monocytogenes* in frankfurter slurries. *Food Microbiology* 24: 352–361, 2007.
60. MF Gliemmo, CA Campos, LN Gerschenson. Effect of sweet solutes and potassium sorbate on the thermal inactivation of *Z. bailii* in model aqueous systems. *Food Research International* 39: 480–485, 2006.
61. Y Lou, AE Yousef. Adaptation to sublethal environmental stresses protects *Listeria monocytogenes* against lethal preservation factors. *Applied and Environmental Microbiology* 63: 1252–1255, 1997.
62. AS Mazzotta, DE Gombas. Heat resistance of an outbreak strain of *Listeria monocyogenes* in hot dog batter. *Journal of Food Protection* 64: 321–324, 2001.
63. MA Lihono, AF Mendonca, JS Dickson, PM Dixon. A predictive model to determine the effects of temperature, sodium pyrophosphate, and sodium chloride on thermal inactivation of starved *Listeria monocytogenes* in pork slurry. *Journal of Food Protection* 66: 1216–1221, 2003.
64. C Grosulescu, VK Juneja, S Ravishankar. Effects and interactions of sodium lactate, sodium diacetate, and pediocin on the thermal inactivation of starved *Listeria monocytogenes* on bologna. *Food Microbiology* 28: 440–446, 2011.
65. S Mormann, M Dabisch, B Becker. Effects of technological processes on the tenacity and inactivation of norovirus genogroup II in experimentally contaminated foods. *Applied and Environmental Microbiology* 76: 536–545, 2010.
66. A Lianou, KP Koutsoumanis. Evaluation of the effect of defrosting practices of ground beef on the heat tolerance of *Listeria monocytogenes* and *Salmonella* Enteritidis. *Meat Science* 82: 461–468, 2009.
67. CM Byrne, DJ Bolton, JJ Sheridan, IS Blair, DA McDowell. The effect of commercial production and product formulation stresses on the heat resistance of *Escherichia coli* O157:H7 (NCTC 12900) in beef burgers. *International Journal of Food Microbiology* 79: 183–192, 2002.
68. B Byrne, G Dunne, DJ Bolton. Thermal inactivation of *Bacillus cereus* and *Clostridium perfringens* vegetative cells and spores in pork luncheon roll. *Food Microbiology* 23: 803–808, 2006.
69. JM Oteiza, L Giannuzzi, AN Califano. Thermal inactivation of *Escherichia coli* O157:H7 and *Escherichia coli* isolated from morcilla as affected by composition of the product. *Food Research International* 36: 703–712, 2003.
70. EG Velliou, E Van Derlinden, AM Cappuyns, E Nikolaidou, AH Geeraerd, F Devlieghere, JF Van Impe. Towards the quantification of the effect of acid treatment on the heat tolerance of *Escherichia coli* K12 at lethal temperatures. *Food Microbiology* 28: 702–711, 2003.
71. AT Spinks, RH Dunstan, T Harrison, B Coombes, G Kuczera. Thermal inactivation of water-borne pathogenic and indicator bacteria at sub-boiling temperatures. *Water Research* 40: 1326–1332, 2006.
72. DR Heldman, DB Lund. *Handbook of Food Engineering*, London, U.K.: CRC Press, 2006. p. 1040.
73. T Abee, JA Wouters. Microbial stress response in minimal processing. *International Journal of Food Microbiology* 50: 65–91, 1999.
74. G Cebrian, S Condon, P Manas. Heat-adaptation induced thermotolerance in *Staphylococcus aureus*: Influence of the alternative factor sigma (B). *International Journal of Food Microbiology* 135: 274–280, 2009.
75. PN Skandamis, JD Stopforth, Y Yoon, PA Kendall, JN Sofos. Heat and acid tolerance responses of *Listeria monocytogenes* as affected by sequential exposure to hurdles during growth. *Journal of Food Protection* 72: 1412–1418, 2009.

76. VK Juneja, JS Novak. Adaptation of foodborne pathogens to stress from exposure to physical intervention strategies. In: AE Yousef, VK Juneja, eds. *Microbial Stress Adaptation and Food Safety*. Boca Raton, FL: CRC Press, 2003, pp. 159–211.
77. W Schumann. Temperature sensors of eubacteria. *Advances in Applied Microbiology* 67: 213–256, 2009.
78. BM Mackey, MD Derrick. The effect of prior shock on the thermoresistance of *Salmonella thompson* in foods. *Letters in Applied Microbiology* 5: 115–118, 1987.
79. S Lindquist. The heat-shock response. *Annual Review of Biochemistry* 55: 1151–1191, 1986.
80. DM Katchinski. On heat and cells and proteins. *News in Physiological Sciences* 19: 11–15, 2004.
81. D Sergelidis, A Abrahim. Adaptive response of *Listeria monocytogenes* to heat and its impact on food safety. *Food Control* 20: 1–10, 2009.
82. BM Mackey, MD Derrick. Changes in the heat resistance of *Salmonella typhimurium* during heating at rising temperatures. *Letters in Applied Microbiology* 4: 13–16, 1987.
83. P Manas, R Pagán, I Alvarez, S Condón. Survival of *Salmonella senftenberg* 775W to current liquid whole egg pasteurization treatments. *Food Microbiology* 20: 593–600, 2003.
84. NJ Rowan. Evidence that inimical food-preservation barriers alter microbial resistance, cell morphology and virulence. *Trends in Food Science and Technology* 10: 261–270, 1999.
85. MN Alekshun, SB Levy. Regulation of chromosomally mediated multiple antibiotic resistance: The *mar* regulon. *Antimicrobial Agents and Chemotherapy* 41: 2067–2075, 1997.
86. VV Velkov. How environmental factors regulate mutagenesis and gene transfer in microorganisms. *Journal of Biosciences* 24: 529–559, 1999.
87. MAS McMahon, J Xu, JE Moore, IS Blair, DA McDowell. Environmental stress and antibiotic resistance in food-related pathogens. *Applied and Environmental Microbiology* 73: 211–217, 2007,
88. J Marles-Wright, RJ Lewis. Stress responses of bacteria. *Current Opinion in Structural Biology* 17: 755–760, 2007.
89. S Chaturongakul, S Raengpradub, M Wiedmann, KJ Boor. Modulation of stress and virulence in *Listeria monocytogenes*. *Trends in Microbiology* 16: 388–396, 2008.
90. JD Helmann. The extracytoplasmic function (ECF) sigma factors. *Advances in Microbial Physiology* 46: 47–110, 2002.
91. W van Schaik, T Abee. The role of sigma(B) in the stress response of Gram-positive bacteria—Targets for food preservation and safety. *Current Opinion in Biotechnology* 16: 218–224, 2005.
92. I Erill, S Campoy, J Barbe. Aeons of distress: An evolutionary perspective on the bacterial SOS response. *FEMS Microbiology Review* 31: 637–656, 2007.
93. S van der Veen, T Abee. Bacterial SOS response: A food safety perspective. *Current Opinion in Biotechnology* 22: 136–142, 2011.
94. SL Lusetti, MM Cox. The bacterial RecA protein and the recombinational DNA repair of stalled replication forks. *Annual Review of Biochemistry* 71: 71–100, 2002.
95. SJ Goldfless, AS Morag, KA Belisle, VA Jr Sutera, ST Lovett. DNA repeat rearrangements mediated by DnaK-dependent replication fork repair. *Molecular Cell* 21: 595–604, 2006.

4 Simulating Thermal Food Processes Using Deterministic Models

Arthur A. Teixeira

CONTENTS

4.1 INTRODUCTION

Thermal processing of canned foods has been one of the most widely used methods of food preservation during the twentieth century and has contributed significantly to the nutritional well-being of much of the world's population. Thermal processing consists of heating food containers in pressurized retorts at specified temperatures for prescribed lengths of time. These process times are calculated on the basis of achieving sufficient bacterial inactivation in each container to comply with public health standards and to ensure that the probability of spoilage will be less than some minimum. Associated with each thermal process is always some degradation of heat-sensitive vitamins and other quality factors that is undesirable. Because of these quality and safety factors, great care is taken in the calculation of these process times and in the control of time and temperature during processing to avoid either underprocessing or overprocessing. The heat transfer considerations that govern the temperature profiles achieved within the container of food are critical factors in the determination of time and temperature requirements for sterilization. This chapter will focus on the development and application of deterministic heat transfer models capable of accurately predicting internal product temperature at any location in response to retort operating conditions.

The topics to be covered in this chapter are organized into sections listed as follows:

1. Thermal death time relationships, which describe how thermal inactivation of bacterial spore populations can be quantified as a function of time and temperature
2. Process lethality and sterilizing value, which defines the concepts for specifying the process requirements with respect to public health considerations and spoilage probability
3. Heat transfer considerations, which describes the methods of temperature measurement and recording, and how these data are treated to obtain important heat penetration parameters for subsequent use in various methods of thermal process calculation
4. Process calculations, which describes the general method for calculating thermal processes, including the process lethality delivered by a specific process, as well as the process time required at a given temperature to deliver a specified lethality value
5. Deterministic model for heat transfer, which describes the development of deterministic heat transfer models for thermal process simulation and application to process design
6. Process optimization, which shows the use of deterministic models to find optimum process conditions that maximize quality retention without compromise of sterility assurance
7. Process deviations, which describes the application of deterministic models to the rapid evaluation of unexpected process deviations
8. Online real-time computer control, which shows the use of deterministic models for online correction of unexpected process deviations
9. Measuring and predicting headspace pressure, which introduces simplistic mathematical models to predict pressure buildup within a rigid container of food undergoing a retort process

4.2 THERMAL DEATH TIME RELATIONSHIPS

An understanding of two distinct bodies of knowledge is required to appreciate the basic principles involved in thermal process calculation. The first of these is an understanding of the thermal inactivation kinetics (heat resistance) of food-spoilage-causing organisms. The second body of knowledge is an understanding of heat transfer considerations that govern the temperature profiles achieved within the food container during the process, commonly referred to in the canning industry as *heat penetration*.

Figure 4.1 conceptually illustrates the interdependence between the thermal inactivation kinetics of bacterial spores and the heat transfer considerations in the food product. Thermal inactivation of bacteria generally follows first-order kinetics and can be described by logarithmic reduction in the concentration of bacterial spores with time for any given lethal temperature, as shown in the upper family of curves in Figure 4.1. These are known as *survivor curves*. The decimal reduction time, D, is expressed as the time in minutes to achieve one log cycle of reduction in concentration, C. As suggested by the family of curves shown, D is temperature dependent and varies logarithmically with temperature, as shown in the second graph. This is known as a thermal death time (TDT) curve and is essentially a straight line over the range of temperatures employed in food sterilization. The slope of the curve that describes this relationship is expressed as the temperature difference, Z, required for the curve to transverse one log cycle. The temperature in the food product, in turn, is a function

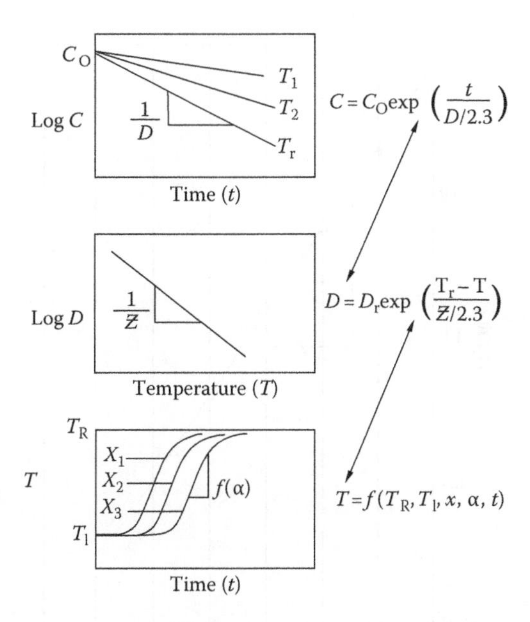

FIGURE 4.1 Time and temperature dependence of the thermal inactivation kinetics of bacterial spores in thermal processing of canned foods.

of the retort temperature (T_R), initial product temperature (T_I), location within the container (x), thermal diffusivity of the product (α), and time (t) in the case of a conduction-heating food.

Thus, the concentration of viable bacterial spores during thermal processing decreases with time in accordance with the inactivation kinetics, which are a function of temperature. The temperature, in turn, is a function of the heat transfer considerations, involving time.

4.3 PROCESS LETHALITY AND STERILIZING VALUE

4.3.1 Time at Temperature for Isothermal Process

Once the TDT curve (second graph in Figure 4.1) has been established for a given microorganism, it can be used to calculate the time–temperature requirements for any idealized thermal process (isothermal process in which the product is heated instantly and uniformly to the treatment temperature, held there for a specified time, and likewise cooled instantly and uniformly). For example, assume a process is required that will achieve a six-log-cycle reduction in the population of bacterial spores whose kinetics are described by the TDT curve in Figure 4.2, and that a temperature of 235°F has been chosen for the process. The TDT curve shows that the D value at 235°F is 10 min. This means that, at that temperature, 10 min will be required for each log-cycle reduction in population. If a six-log-cycle reduction is required, a total of 60 min is needed for the process. If a temperature of 270°F had been chosen for the process, the D value at that temperature is ~0.1 min, and only 0.6 min (or 36 s) would be required at that temperature to accomplish the same six-log-cycle reduction.

Since the TDT curve is a straight line on a semilog plot, all that is needed to specify such a curve is its slope and a single reference point on the curve. The slope of the curve is specified by the Z value, and the reference point is the D value at a reference temperature. For sterilization of low-acid foods (pH > 4.5), in which thermophilic spores of relatively high heat resistance are of concern, this reference temperature is usually taken to be 250°F. For high-acid foods or pasteurization processes in which microorganisms of much lower heat resistance are of concern, lower reference temperatures are used, such as 212°F or 150°F. In specifying a reference D value for a microorganism,

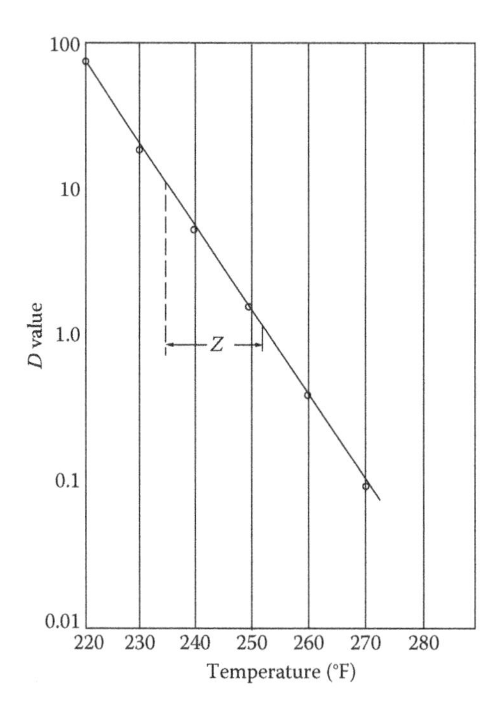

FIGURE 4.2 TDT curve showing temperature dependency of D value in terms of temperature change (Z) required for 10-fold change in D value.

the reference temperature is shown as a subscript, such as D_{250}. For example, the TDT curve in Figure 4.2 can be specified by a Z value of 18°F and a D_{250} value close to 1.5 min.

The ranges of D values for different classifications of bacteria are given in Table 4.1, and D_{250} values for specific organisms in selected food products are given in Table 4.2.

4.3.2 Process Lethality

The example process calculations carried out in the preceding subsection using the TDT curve in Figure 4.2 showed clearly how two widely different processes (60 min at 235°F and 0.6 min at 270°F) were equivalent with respect to their ability to achieve the same log-cycle reduction in spore population (sterilizing value). In fact, the straight line drawn between these two points plotted on the TDT graph would lie parallel to the TDT curve and would represent all possible combinations of time and temperature that would accomplish a six-log-cycle reduction for that microorganism. Therefore, for a given Z value, the specification of any one point on this line is sufficient to specify the sterilizing value of any process combination of time and temperature on that line. The reference point that has been adopted for this purpose is the time in minutes at the reference temperature of 250°F, or the point in time where the equivalent process curve crosses the vertical axis drawn at 250°F, and is known as the F value for the process. F is often referred to as the *lethality* of a process, and since it is expressed in minutes at 250°F, the *unit of lethality* is 1 min at 250°F. Thus, if a process is assigned an F value of 6, it means that the integrated lethality achieved by whatever time–temperature history is employed by the process must be equivalent to the lethality achieved from 6 min exposure to 250°F, assuming an idealized process of instantaneous heating to 250°F and instantaneous cooling from 250°F.

To illustrate, the example process calculation using the TDT curve in Figure 4.2 will be repeated by specifying the F value for the required process. Recall from that example that the

TABLE 4.1
***D* Values for Different Classifications of Food-Borne Bacteria**

Bacterial Groups	*D* Value
Low-acid and semiacid foods (pH above 4.5)	D_{250}
Thermophiles	
Flat-sour group (*Bacillus stearothermophilus*)	4.0–5.0
Gaseous-spoilage group (*Clostridium thermosaccharolyticum*)	3.0–4.0
Sulfide stinkers (*Clostridium nigrigicans*)	2.0–3.0
Mesophiles	
Putrefactive anaerobes	
C. botulinum (types A and B)	0.10–1.20
Clostridium sporogenes group (including P.A. 3679)	0.10–1.5
Acid foods (pH 4.0–4.5)	
Thermophiles	
Bacillus coagulans (facultatively mesophilic)	0.01–0.07
Mesophiles	D_{212}
Bacillus polymyxa and *Bacillus macerans*	0.10–0.50
Butyric anaerobes (*Clostridium pasteurianum*)	0.10–0.50
High-acid foods	D_{150}
Lactobacillus spp., *Leuconostoc* spp., yeast and molds	0.50–1.00

Source: From Stumbo, C.R., *Thermobacteriology in Food Processing*, Academic Press, New York, 1965. With permission.

TABLE 4.2
Comparison of D_{250} Values for Specific Microorganisms in Selected Food Substrates

Organism Substrate	TDT Methods	D_{250}
P.A. 3679	Cream-style corn	Can 2.47
P.A. 3679	Whole-kernel corn (1)	Can 1.52
P.A. 3679	Whole-kernel corn (2)	Can 1.82
P.A. 3679	Phosphate buffer	Tube 1.31
F.S. 5010	Cream-style corn	Can 1.14
F.S. 5010	Whole-kernel corn	Can 1.35
F.S. 1518	Phosphate buffer	Tube 3.01
F.S. 617	Whole milk	Can 0.84
F.S. 617	Evaporated milk	Tube 1.05

Source: From Stumbo, C.R., *Thermobacteriology in Food Processing*, Academic Press, New York, 1965. With permission.

process was required to accomplish a six-log-cycle reduction in spore population. All that is required to specify the F value is to determine how many minutes at 250°F will be required to achieve that level of log-cycle reduction. The D_{250} value is used for this purpose, since it represents the number of minutes at 250°F to accomplish one log-cycle reduction. Thus, the F value is equal to D_{250} multiplied by the sterilizing value (number of log cycles required in population reduction):

$$F = D_{250}(\log a - \log b), \tag{4.1}$$

where

a is the initial number of viable spores
b is the final number of viable spores (or survivors)

In this example, $D_{250}=1.5$ min as taken from the TDT curve in Figure 4.2 and multiplied by the required sterilizing value (six log cycles). Thus, $F=1.5$ (6)=9 min, and the lethality for this process has been specified as $F=9$ min. This is normally the way in which a thermal process is specified for subsequent calculation of a process time at some other temperature. In this way, proprietary information regarding specific microorganisms of concern and/or numbers of log-cycle reduction can be kept confidential and replaced by the F value (lethality) as a process specification.

Note also that this F value serves as the reference point to specify the equivalent process design curve discussed earlier. By plotting a point at 9 min on the vertical line passing through 250°F in Figure 4.2, and drawing a curve parallel to the TDT curve through this point, the line will pass through the two equivalent process points that were calculated earlier (60 min at 235°F and 0.6 min at 270°F). Alternatively, the equation of this straight line can be used to calculate the process time (t) at some other constant temperature (T) when F is specified:

$$F = 10^{[(T-250)/Z]t}. \tag{4.2}$$

Equation 4.3 becomes important in the general case when the product temperature varies with time during a process, and the F value delivered by the process must be integrated mathematically:

$$F = \int_0^t 10^{[(T-250)/Z]t}. \tag{4.3}$$

At this point, Equations 4.1 and 4.3 have been presented as two clearly different mathematical expressions for the process lethality, F. It is most important that the distinction between these two expressions be clearly understood. Equation 4.1 is used to determine the F value that should be *specified* for a process and is determined from the log-cycle reduction in spore population required for the process (sterilizing value) by considering factors related to safety and wholesomeness of the processed food, as discussed in the following section. Equation 4.3 is used to determine the F value *delivered* by a process because of the time–temperature history experienced by the product during the process. Another observation is that Equation 4.1 makes use of the D_{250} value in converting log cycles of reduction into minutes at 250°F, while Equation 4.3 makes use of the Z value in converting temperature–time history into minutes at 250°F. Because a Z value of 18°F (10°C) is so commonly observed or assumed for most thermal processing calculations, F values calculated with a Z of 18°F and reference temperature of 250°F are designated F_O.

4.3.3 Specification of Process Lethality

Establishing the sterilizing value to be specified for a low-acid canned food is undoubtedly one of the most critical responsibilities taken on by a food scientist or engineer acting on behalf of a food company in the role of a competent thermal processing authority. In this section, we outline briefly the steps normally taken for this purpose [1,2].

There are two types of bacterial populations of concern in canned food sterilization. First is the population of organisms of public health significance. In low-acid foods with pH above 4.5, the chief organism of concern is *Clostridium botulinum*. A safe level of survival probability that has been accepted for this organism is 10^{-12} or one survivor in 10^{12} cans processed. This is known as the $12D$ concept for botulinum cook. Since the highest D_{250} value known for this organism in foods is 0.21 min, the minimum lethality value for a botulinum cook, assuming an initial spore load of one organism per container, is

$$F = 0.21 \times 12 = 2.52.$$

Essentially all low-acid foods are processed far beyond the minimum botulinum cook in order to avoid economic losses from spoilage-causing bacteria of much greater heat resistance. For these organisms, the acceptable levels of spoilage probability are usually dictated by economic considerations. Most food companies accept a spoilage probability of 10^{-5} from mesophilic spore formers (organisms that can grow and spoil food at room temperature). The organism most frequently used to characterize this classification of food spoilage is a strain of *C. sporogenes*, known as P.A. 3679, with a maximum D_{250} value of 1 min. Thus, a minimum lethality value for a mesophilic spoilage cook assuming an initial spore of load of one spore per container is

$$F = 1.00 \times 5 = 5.00,$$

where thermophilic spoilage is a problem, more severe processes may be necessary because of the high heat resistance of thermophilic spores. Fortunately, most thermophiles do not grow readily at room temperature and require incubation at unusually high storage temperatures (110°F–130°F) to cause food spoilage. Generally, foods with no more than 1% spoilage (spoilage probability of 10^{-2}) upon incubation after processing will show less than the accepted 10^{-5} spoilage probability in normal commerce. Therefore, when thermophilic spoilage is a concern, the target value for the final number of survivors is usually taken as 10^{-2}, and the initial spore load needs to be determined through a microbiological analysis since contamination from these organisms varies greatly. For a situation with an initial thermophilic spore load of 100 spores per can, and an average D_{250} value of 4.00, the process lethality required would be

$$F = 4.00\,(\log 100 - \log 0.01) = 4.00(4) = 16.$$

The aforementioned procedural steps are only preliminary guidelines for average conditions and often need to be adjusted up or down in view of the types of contaminating bacteria that may be present, the initial level of contamination or *bioburden* of the most resistant types, the spoilage risk accepted, and the nature of the food product from the standpoint of its ability to support the growth of the different types of contaminating bacteria that are found. Table 4.3 contains a listing of process lethalities (F_O) specified for the commercial processing of selected canned foods.

TABLE 4.3
Lethality Values (F_O) for Commercial Sterilization of Selected Canned Foods

Products	Can Sizes	Lethality Value, F_O
Asparagus		2–4
Green beans, brine	No. 2	3.5
packed	No. 10	3.5
Chicken, boned	All	6–8
Corn, whole kernel,	All	9
brine packed	No. 10	15
Cream-style corn	No. 2	5–6
	No. 10	2.3
Dog food	No. 2	12
	No. 10	6
Mackerel in brine	301×411	2.9–3.6
Meat loaf	No. 2	6
Peas, brine packed	No. 2	7
	No. 10	11
Sausage, Vienna, in brine	Various	5
Chili con carne	Various	6

Source: Lopez, A.A., *Complete Course in Canning, Book 1, Basic Information on Canning*, 11th ed., The Canning Trade, Baltimore, MD, 1987.

4.4 HEAT TRANSFER CONSIDERATIONS

4.4.1 Unsteady (Nonisothermal) Heat Transfer

In the previous sections on the thermal inactivation kinetics of bacterial spores, frequent reference was made to an idealized process in which the food product was assumed to be heated instantaneously to a lethal temperature and then cooled instantaneously after the required process time. These idealized processes are important to gain an understanding of how the kinetic data can be used directly to determine the process time at any given lethal temperature. There are, in fact, commercial sterilization processes for which this method of process-time determination is applicable. These are high-temperature/short-time (HTST) pasteurization or ultrahigh-temperature (UHT) sterilization processes for liquid foods that make use of flow-through heat exchangers and/or steam injection heaters and flash cooling chambers for instantaneous heating and cooling. The process time is accomplished through the residence time in the holding tube between the heater and cooler as the product flows continuously through the system. This method of product sterilization is most often used with aseptic filling systems, discussed in other chapters.

In traditional thermal processing of most canned foods, the situation is quite different from the idealized processes described earlier. The cans are filled with relatively cool unsterile product, sealed after headspace evacuation, and placed in steam retorts, which apply heat to the outside can wall. The product temperature can then only respond in accordance with the physical laws of heat transfer and will gradually rise in an effort to approach the temperature at the wall followed by a gradual fall in response to cooling at the wall. In this situation, the lethality delivered by the process will be the result of the transient time–temperature history experienced by the product at the slowest heating location in the can; this is usually the geometric center. Therefore, the ability to determine this time–temperature history accurately is of paramount importance in the calculation of thermal processes. In this section, we review the various modes of heat transfer found in canned foods, and describe the methods of temperature measurement and recording and how these data are treated for subsequent use in thermal process calculation.

4.4.2 Modes of Heat Transfer

Solid-packed foods in which there is essentially no product movement within the container, even when agitated, heat largely by conduction heat transfer. Because of the lack of product movement and the low thermal diffusivity of most foods, these products heat very slowly and exhibit a nonuniform temperature distribution during heating and cooling caused by the temperature gradient that is set up between the can wall and geometric center. For conduction-heating products, the geometric center is the slowest heating point in the container. Therefore, process calculations are based on the temperature history experienced by the product at the can center. Solid-packed foods such as canned fish and meats, baby foods, pet foods, pumpkin, and squash fall into this category. These foods are usually processed in still cook or continuous hydrostatic retorts that provide no mechanical agitation.

Thin-bodied liquid products packed in cans such as, milk, soups, sauces, and gravies will heat by either natural or forced convection heat transfer, depending on the use of mechanical agitation during processing. In a still cook retort that provides no agitation, product movement will still occur within the container because of natural convective currents induced by density differences between the warmer liquid near the hot can wall and the cooler liquid near the can center [3,4]. The rate of heat transfer in nearly all convection-heating products can be increased substantially by inducing forced convection through mechanical agitation. For this reason, most convection-heating foods are processed in agitating retorts designed to provide either axial or end-over-end can rotation. Normally, end-over-end rotation is preferred and can be provided in batch retorts, while continuous agitating retorts can provide only limited axial rotation.

Unlike conduction-heating products, because of product movement in forced convection-heating products, the temperature distribution throughout the product is reasonably uniform under mechanical agitation. In natural convection, the slowest heating point is somewhat below the geometric center and should be located experimentally in each new case. The two basic mechanisms of conduction and convection heat transfer in canned foods are illustrated schematically in Figure 4.3.

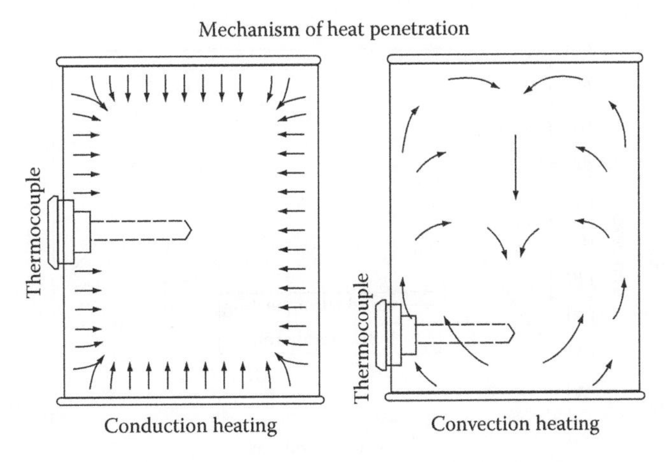

FIGURE 4.3 Conduction and convection heat transfer in solid and liquid canned foods, respectively. (From Lopez, A.A., *Complete Course in Canning, Book 1, Basic Information on Canning*, 11th ed., The Canning Trade, Baltimore, MD, 1987.)

4.4.3 Heat Penetration Measurement

The primary objective of heat penetration measurements is to obtain an accurate recording of the product temperature at the can cold spot over time while the container is being treated under a controlled set of retort processing conditions. This is normally accomplished through the use of copper-constantan thermocouples inserted through the can wall so as to have the junction located at the can geometric center. Thermocouple lead wires pass through a packing gland in the wall of the retort for connection to an appropriate data acquisition system in the case of a still cook retort. For agitating retorts, the thermocouple lead wires are connected to a rotating shaft for an electrical signal pickup from the rotating armature outside the retort. Specially designed thermocouple fittings are commercially available for these purposes [2,5,6].

The precise temperature–time profile experienced by the product at the can center will depend on the physical and thermal properties of the product, size, and shape of the container, and retort-operating conditions. Therefore, it is imperative that test cans of product used in heat penetration tests be truly representative of the commercial product with respect to ingredient formulation, fill weight, headspace, can size, and so on. In addition, the laboratory or pilot plant retort being used must accurately simulate the operating conditions that will be experienced by the product during commercial processing on the production-scale retort systems intended for the product. If this is not possible, heat penetration tests should be carried out using the actual production retort during scheduled breaks in production operations.

During a heat penetration test, both the retort temperature history and product temperature history at the can center are measured and recorded over time. A typical test process will include venting of the retort with live steam to remove all atmospheric air and then closing the vents to bring the retort up to operating pressure and temperature. This is the point at which "process time" begins, and the retort temperature is held constant over this period of time. At the end of the prescribed process time, the steam is shut off, and cooling water is introduced under overriding air pressure to prevent a sudden pressure drop in the retort. This begins the cooling phase of the process, which ends when the retort pressure returns to atmosphere, and the product temperature in the can has reached a safe low level for removal from the retort. A typical temperature–time plot of these data is shown in Figure 4.4 and illustrates the degree to which the product center temperature in the can lags behind the retort temperature during both heating and cooling.

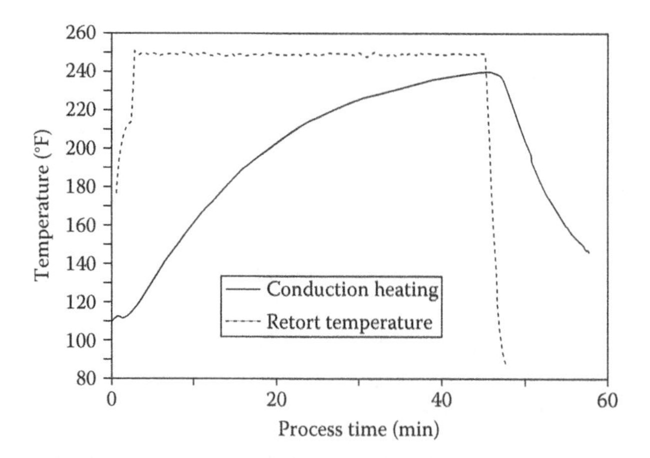

FIGURE 4.4 Generic heat penetration curve for a conduction heating food during a thermal process.

4.4.4 Heat Penetration Curves and Thermal Diffusivity

The response of the product temperature at the can center to the steam retort temperature applied at the can wall is governed by the physical laws of heat transfer and can be expressed mathematically. This mathematical expression serves as a basis for obtaining effective values for thermal properties of canned foods in order to use numerical computer (deterministic) models that are capable of simulating the heat transfer in thermal processing of canned foods.

A heat balance between the heat absorbed by the product and the heat transferred across the can wall from the steam retort could be expressed as follows for an element of food volume facing the can wall of surface area A and thickness L:

$$\rho LAC_p \frac{dT}{dt} = \frac{k}{L} A(T_r - T), \tag{4.4}$$

where

T is product temperature
T_r is retort temperature
ρ, C_p, and k are density, specific heat, and thermal conductivity of the product, respectively

Because of high surface heat transfer coefficient of condensing steam at the can wall and high thermal conductivity of the metal can, overall surface resistance to heat transfer can be assumed negligible, in contrast to the product's resistance to heat transfer. After rearranging the terms, Equation 4.4 can be written in the form of an ordinary differential equation:

$$\frac{dT}{dt} = \frac{k}{\rho C_p} L^2 (T_r - T). \tag{4.5}$$

By letting the thermal diffusivity (α) represent the combination of thermal and physical properties ($k/\rho C_p$), and letting T_O represent the initial product temperature, the solution to Equation 4.5 becomes

$$\frac{T_r - T}{T_r - T_O} = \exp\left(\frac{\alpha}{L^2} t\right). \tag{4.6}$$

Thus, the product center temperature can be seen to be an exponential function of time; a semilog plot of the temperature difference ($T_r - T$) against time would produce a straight line sloping downward, having a slope related to the product's thermal diffusivity and can dimensions (Figure 4.5). The heat penetration rate factor (f_h) is the reciprocal slope of the heat penetration curve (time required for one log-cycle temperature change). Therefore, it can be related to the overall apparent thermal diffusivity of the product and container dimensions for a given container shape. For a finite cylinder, with all parameters expressed in English units, the following relationship can be used to obtain the thermal diffusivity, α, from the heating rate factor taken from a heat penetration curve [2]:

$$\alpha = \frac{0.398}{1/R^2 + (0.427/H^2) f_h}, \tag{4.7}$$

where

R is the can radius in inches
H is one-half the can height in inches
f_h is the heating curve slope factor in minutes
α is the product thermal diffusivity in square inches per minute

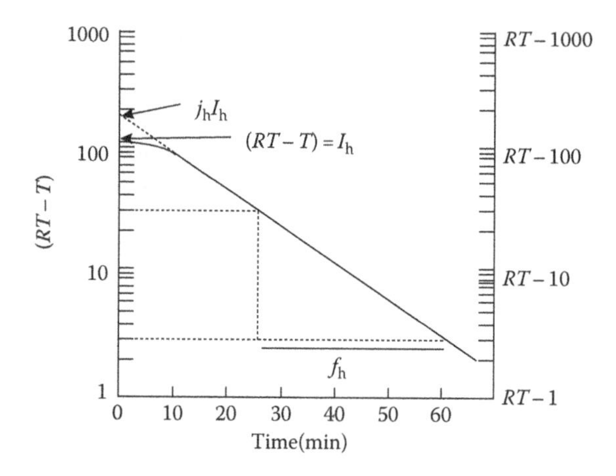

FIGURE 4.5 Semilog heat penetration curve showing unaccomplished temperature difference (on log scale) versus time, from which heating rate (f_h) and heating lag (j_h) factors can be estimated.

This relationship is also useful to determine the heating rate factor for the same product in a different size container, since thermal diffusivity is a combination of thermal and physical properties that characterize the product and its ingredient formulation and remains unaffected by container size or shape.

Another important heat penetration parameter obtained from the semilog heat penetration curve is the heating curve lag factor, j_{ch}, which is taken as the ratio of the difference between the retort temperature (T_r) and pseudoinitial temperature (T_O), the temperature at which an extension of the straight-line portion of the heating curve intersects the ordinate axis ($T_r - T_O$) over the difference between retort temperature and actual initial product temperature ($T_r - T_i$). The heating lag factor is used with deterministic conduction heat transfer models to account for heat transfer mechanisms other than pure conduction that often take place in most canned foods.

4.5 PROCESS CALCULATION

Once a heat penetration curve has been obtained from laboratory heat penetration data or predicted by a computer model, there are essentially two widely accepted methods for using these data to perform thermal process calculations. The first of these is the general method of process calculation [7], and the second is the Ball formula method of process calculation [8,9]. Only the general method is described in this chapter because of its connected use with deterministic heat transfer models.

As the name implies, the general method is the most versatile method of process calculation because it is universally applicable to essentially any type of thermal processing situation. It makes direct use of the product temperature history at the can center obtained from a heat penetration test (or predicted by a mathematical model) to evaluate the integral shown in Equation 4.3 for calculating the process lethality delivered by a given temperature–time history. A straightforward numerical integration of Equation 4.3 can be expressed as follows with reference to Figure 4.6:

$$F_O = \sum_{i=1}^{n} \Delta F_i = \sum_{i=1}^{n} 10^{[(T_i-250)/Z]\Delta t}. \tag{4.8}$$

Figure 4.6 is a direct plot of the can center temperature experienced during a heat penetration test. Since no appreciable lethality can occur until the product temperature has reached the lethal

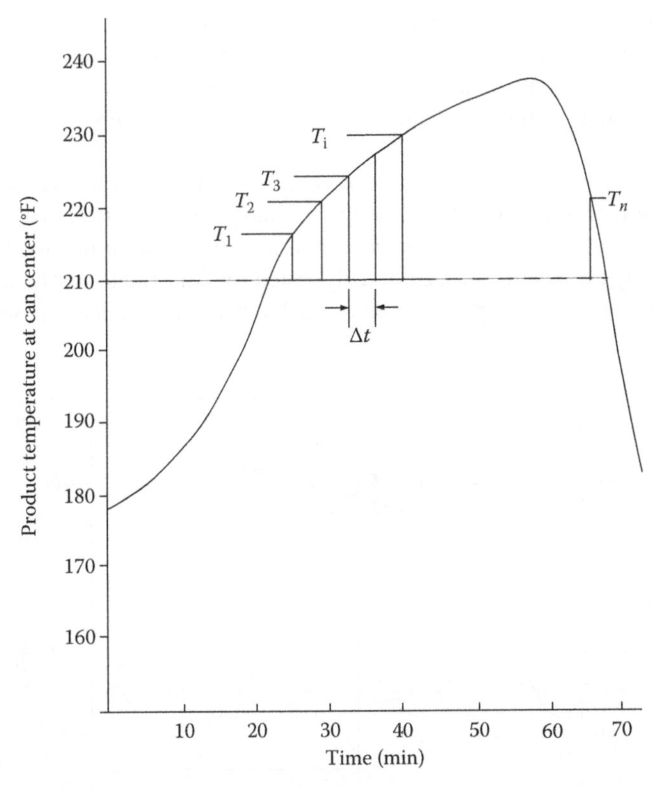

FIGURE 4.6 Temperature history at center of canned food during thermal process for calculation of process lethality by general method.

temperature range (above 220°F), Equation 4.8 need only be evaluated over the time period during which the product temperature remains above 220°F. By dividing this time period into small time intervals (Δt) of short duration as shown in Figure 4.6, the temperature T_i at each time interval can be read from the curve and used to calculate the incremental lethality (ΔF_i) accomplished during that time interval. Then the sum of all these incremental sterilizing values equals the total lethality, F_O, delivered by the test process. To determine the process time required to deliver a specified lethality, the cooling portion of the curve in Figure 4.6 is shifted to the right or left, and the integration is repeated until the delivered sterilizing value so calculated agrees with the value specified for the process.

When first introduced in 1920, this method was sometimes referred to as the graphical trial-and-error method because the integration was performed on specially designed graph paper to ease the tedious calculations that were required. The method was also time-consuming, and soon gave way in popularity to the historically more convenient (but less accurate) Ball formula method. With the current widespread availability of low-cost programmable calculators and desktop computers, these limitations are no longer of any consequence, and the general method is currently the method of choice because of its accuracy and versatility.

In fact, the general method is particularly useful in taking maximum advantage of computer-based data logging systems used in connection with heat penetration tests. Such systems are capable of reading temperature signals received directly from thermocouples monitoring both retort and product center temperature, and processing these signals through the computer. Through programming instructions, both retort temperature and product center temperature are plotted against time without any data transformation. This allows the operator to see what has actually happened throughout the duration of the process test. As the data are being read by the computer, additional programming instructions call for calculation of the incremental process lethality (ΔF_i) at each time interval between temperature readings and summing these over time as the process is under way. As a result, the accumulated lethality (F) is known at any time during the process and can be plotted on the graph along with the temperature histories to show the final value reached at the end of the process.

An example of the computer printout from such a heat penetration test is shown in Figure 4.7. Another test can be repeated quickly for a longer or shorter process time with instant results on the F_O achieved. By examining the results from both tests, the desired process time for the target F value can be closely estimated and then quickly tested for confirmation. The results of two such heat penetration tests are shown superimposed on each other in Figure 4.8. These results show that

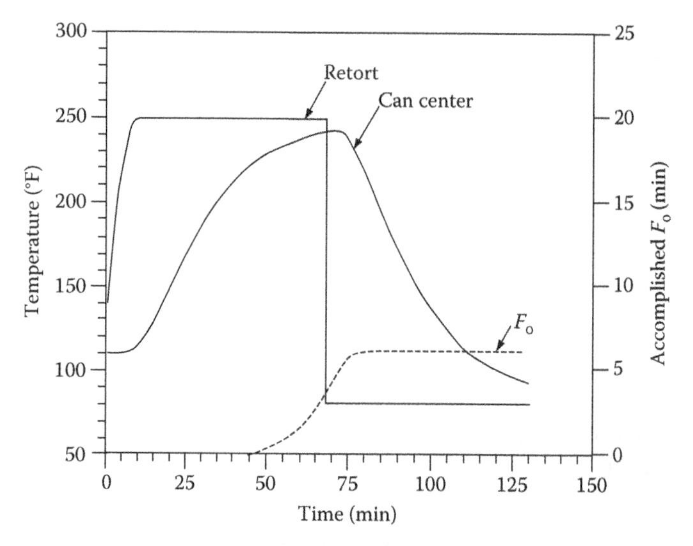

FIGURE 4.7 Computer-generated plot of retort temperature, can center temperature, and accomplished lethality (F_O) over time for thermal processing of a conduction-heated food. (From Datta, A.K. et al., *J. Food Sci.*, 51(2), 480, 1986. With permission.)

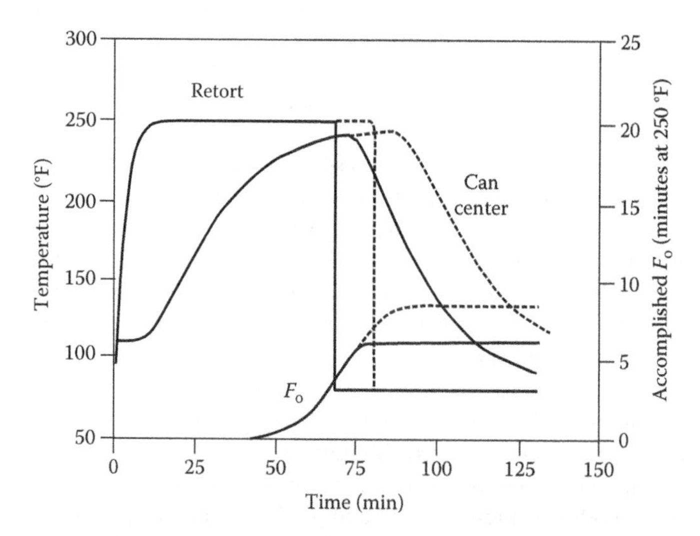

FIGURE 4.8 Computer-generated plot of retort temperature, can center temperature, and accomplished lethality (F_O) over time for two different process times superimposed on each other.

test 1, with a process time of 68 min, produced an F value of 6; test 2, with a process time of 80 min, produced an F value of 8, suggesting that a target F value of 7 will be achieved by an intermediate process time. This can be confirmed by running a test at the suggested process time and examining the resulting F value.

4.6 DETERMINISTIC MODEL FOR HEAT TRANSFER

A "deterministic" model of a process is a mathematical equation derived from mathematical expressions describing of fundamental scientific principles known to be responsible for the observed process behavior. The scientific principles governing heat transfer are well established and understood. Therefore, deterministic models are an appropriate means for mathematically simulating heat transfer in canned foods. One of the primary advantages of these models is that once the apparent thermal diffusivity has been determined from heat penetration tests, the model can be used to predict the product temperature history at any specified location within the can for any set of processing conditions and container size/shape specified. Thus, with the use of such models, it is unnecessary to carry out repeated heat penetration tests in the laboratory or pilot plant to determine the heat penetration curve for a different retort temperature or can size.

A second advantage of even greater importance is that the retort temperature need not be held constant but can vary in any prescribed manner throughout the process, and the model will predict the correct product temperature history at the can center in response to such dynamic (or transient) boundary condition. The use of these models has become invaluable for simulating the process conditions experienced in continuous sterilizer systems, in which containers pass from one chamber to another, experiencing a changing boundary temperature as they pass through the system. Another important application of these models is in the rapid evaluation of an unscheduled process deviation, such as when an unexpected drop in retort temperature occurs during the course of the process. The model can quickly predict the product center temperature profile in response to such a deviation and calculate the delivered lethality value, F_O, for comparison with the lethality value specified for the product.

The development and use of such a deterministic model for stimulating the thermal processing of canned foods have been well documented in published scientific literature [10–14]. The model makes use of a numerical solution by finite differences of the 2D partial differential equation that describes conduction heat transfer in a finite cylinder. During conduction heating, heat is applied only at the can surface, temperatures will first rise only in regions near the can walls, while temperature near the can center will begin to respond only after a considerable time lag. Mathematically, the temperature is a distributed parameter in that at any point in time during heating, the temperature takes on a different value with location in the can; and in any one location, the temperature changes with time as heat gradually penetrates the product from the can walls toward the center.

The mathematical expression that describes this temperature distribution pattern over time is shown in Figure 4.9 and lies at the heart of the numerical computer model. This expression is the classic partial differential equation for 2D unsteady heat conduction in a finite cylinder and can be written in the form of finite differences for numerical solution by digital computer, as shown in Figure 4.10. The finite differences are discrete increments of time and space defined as small intervals of process time and small increments of container height and radius (Δt, Δh, and Δr, respectively).

As a framework for computer iterations, the cylindrical container is imagined to be subdivided into volume elements that appear as layers of concentric rings having rectangular cross sections, as illustrated in Figure 4.11 for the upper half of the container. Temperature nodes are assigned at the corners of each volume element on a vertical plane, as shown in Figure 4.12, where I and J are used to denote the sequence of radial and vertical volume elements, respectively. By assigning appropriate boundary and initial conditions to all the temperature nodes (interior nodes set at initial product temperature and surface nodes set at retort temperature), the new temperature reached at each node

$$\frac{\partial T}{\partial t} = \alpha \left[\frac{\partial^2 T}{\partial r^2} + \frac{1}{r}\frac{\partial T}{\partial r} + \frac{\partial^2 T}{\partial h^2} \right]$$

Where

T is the temperature
t is the time
α is the thermal diffusivity
r is the radial position in cylinder
h is the vertical position in cylinder

FIGURE 4.9 Two-dimensional second-order partial differential equation for conduction heat transfer (heat conduction equation) in a finite cylinder. (From Teixeira, A.A. and Manson, J.E., *Food Technol.*, 36(4), 85, 1982. With permission.)

$$T_{(ij)}^{(t+\Delta t)} = T_{(ij)}^{(t)} + \frac{\alpha \Delta t}{\Delta r^2}\left[T_{(i-1,j)} - 2T_{(ij)} + T_{(i+1,j)} \right]^{(t)} + \frac{\alpha \Delta t}{2r\Delta r}\left[T_{(i-1,j)} - T_{(i+1,j)} \right]^{(t)} + \frac{\alpha \Delta t}{\Delta h^2}\left[T_{(i,j-1)} - 2T_{(i,j)} + T_{(i,j+1)} \right]^{(t)}$$

Where

Δt, Δr, Δh are the discrete increments of time, radius, and height,
i and j denote sequence of radial and vertical increments away from can wall and mid plane

FIGURE 4.10 Heat conduction equation for a finite cylinder expressed in the form of finite differences for numerical solution by computer iterations. (From Teixeira, A.A. and Manson, J.E., *Food Technol.*, 36(4), 85, 1982. With permission.)

can be calculated after a short time interval (Δt) that would be consistent with the thermal diffusivity of the product obtained from heat penetration data (f_h). This new temperature distribution is then taken to replace the initial one, and the procedure repeated to calculate the temperature distribution after another time interval. In this way, the temperature at any point in the container at any instant in time is obtained. At the end of the process time, when steam is shut off and cooling water is admitted to the retort, the cooling process is simulated by simply changing the boundary conditions from retort temperature T_R to cooling temperature T_C at the surface nodes and continuing with the computer iterations described earlier.

The temperature at the can center can be calculated after each time interval to produce a predicted heat penetration curve upon which the process lethality, F, can be calculated. When the numerical computer model is used to calculate the process time required at a given retort temperature to achieve a specified lethality, F, the computer follows a programmed search routine of assumed process times that quickly converges on the precise time at which cooling should begin in order to achieve the specified F value. Thus, the model can be used to determine the process time required for any given set of constant or variable retort temperature conditions.

4.7 PROCESS OPTIMIZATION

4.7.1 Objective Functions

The principal objective of thermal process optimization is to maximize product quality, minimize undesirable changes, minimize cost, and maximize profits. At all times, a minimal process must be

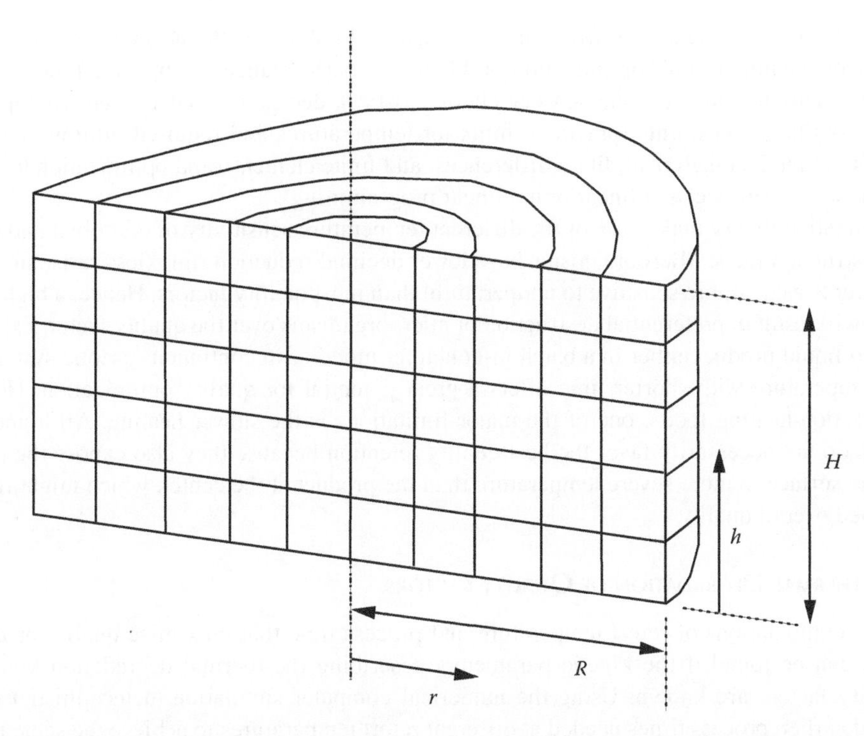

FIGURE 4.11 Subdivision of a cylindrical container for application of finite differences. (From Teixeira, A.A. et al., *Food Technol.*, 23(6), 137, 1969. With permission.)

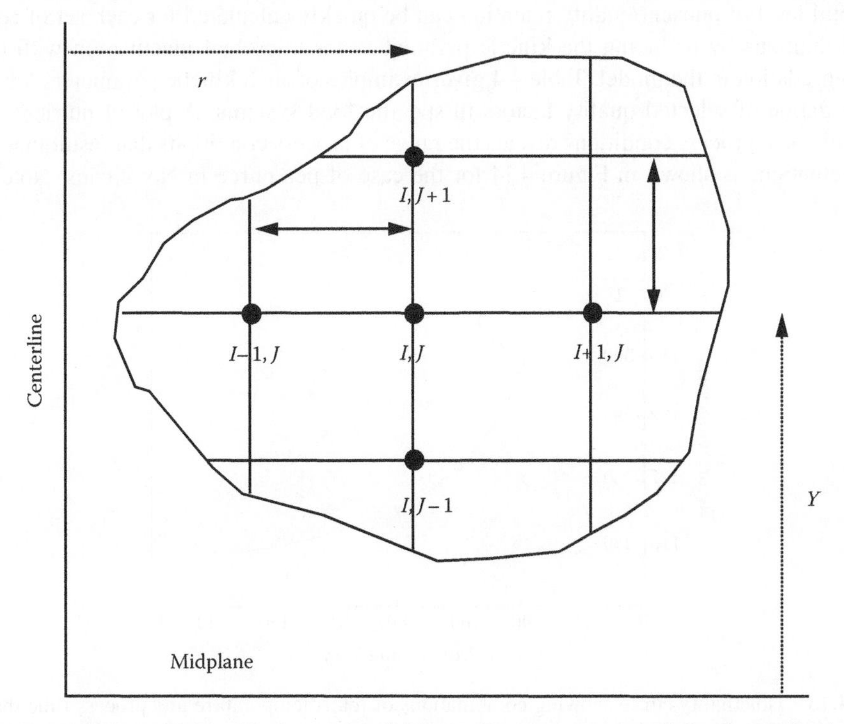

FIGURE 4.12 Labeling of grid nodes in matrix of volume elements on a vertical plan for application of finite differences. (From Teixeira, A.A. et al., *Food Technol.*, 23(6), 137, 1969. With permission.)

maintained to exclude the danger from microorganisms of public health and spoilage concern. The five elements common to all optimization problems are performance or objective function (quality factors, nutrients, texture, and sensory characteristics), decision variables (retort temperature and process time), constraints (practical limits for temperatures and required minimal lethality), mathematical model (analytical, finite differences, and finite element), and optimization technique (search, response surface, and linear or nonlinear programming).

Optimization theory makes use of the different temperature sensitivity of microbial and quality factor destruction rates. Microorganisms have lower decimal reduction time (less resistant to heat) and a lower Z value (more sensitive to temperature) than most quality factors. Hence, a higher temperature will result in preferential destruction of microorganisms over the quality factor. Especially applied to liquid product either in a batch in-container mode or in continuous aseptic systems, the higher temperature with shorter time offers a great potential for quality optimization. However, for conduction-heating foods, one of the major limitations is the slower heating. All higher temperatures do not necessarily favor the best quality retention because they also expose the product nearer the surface to more severe temperature than the product at the center, which might result in diminished overall quality.

4.7.2 Thermal Degradation of Quality Factors

Optimum combinations of retort temperature and process time that maximize quality or nutrient retention can be found if the kinetic parameters describing the thermal degradation kinetics of the quality factors are known. Using the numerical computer simulation (deterministic) models described earlier, process times needed at different retort temperatures to achieve the same process lethality can be quickly calculated over a range of retort temperatures that falls within the operating performance limitations of the retort. A plot of these equivalent retort temperature-process time combinations produces an "isolethality" curve such as the one shown in Figure 4.13 for the case of pea puree in No. 2 cans [10].

The total level of nutrient/quality retention can be quickly calculated for each set of equivalent process conditions by replacing the kinetic parameters for microbial inactivation with those for quality degradation in the model. Table 4.4 gives examples of such kinetic parameters for the thermal degradation of selected quality factors in specific food systems. A plot of nutrient retention versus equivalent process conditions reveals the range of process conditions that result in maximum nutrient retention, as shown in Figure 4.14 for the case of pea puree in No. 2 cans. Note that the

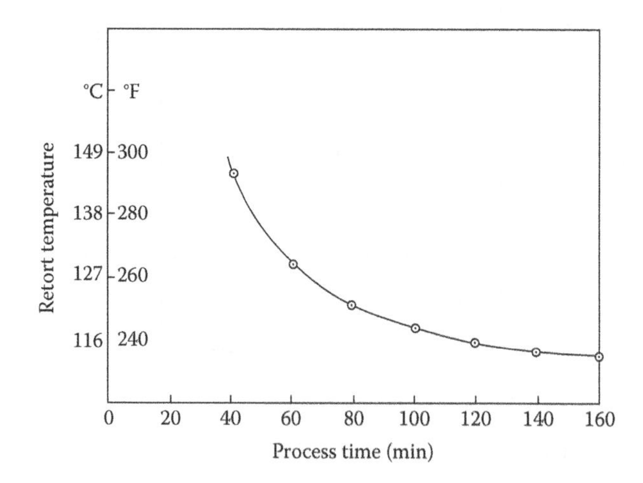

FIGURE 4.13 Isolethality curve showing combinations of retort temperature and process time that deliver the same level of lethality for pea puree in No. 2 cans. (From Teixeira, A.A. et al., *Food Technol.*, 23(6), 137, 1969. With permission.)

TABLE 4.4
Kinetic Parameters for Thermal Degradation of Quality Factors in Selected Thermally Processes Foods

Quality Factor in Food Systems	$D_{121°C}$ (min)	$K_{121°C}$ (/min)	Z (°C)	E_a (kcal/mol)
Thiamine in beans	329.77	6.9837×10^{-3}	27.95	25.416
Lysine in beans	178.28	9.051×10^{-2}	25.44	27.32
Texture in beans	101.68	2.260×10^{-2}	20.62	35.44

Source: Massaguer, P., Food Science Department, Food Engineering Faculty, UNICAMP, Campinas, Sao Paulo, Brazil, Personal communication, 1997.

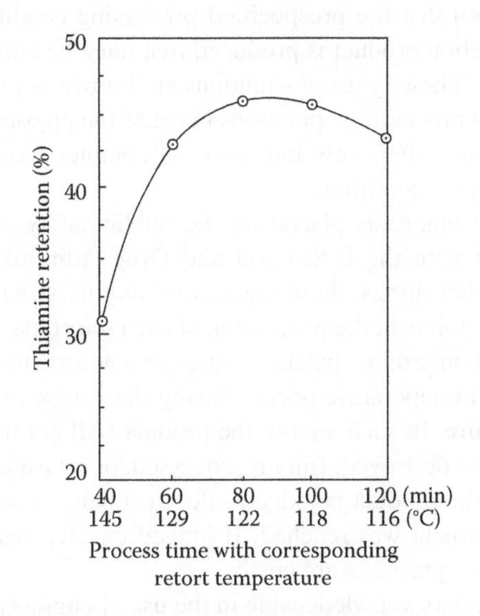

FIGURE 4.14 Optimization curve showing percentage thiamine retention for pea puree in No. 2 cans after various retort temperature and process time combinations that deliver the same level of lethality. (From Teixeira, A.A. et al., *Food Technol.*, 23(6), 137, 1969. With permission.)

same exercise is also useful when seeking to minimize process time, because these results reveal the price that is paid in lower quality retention caused by the higher surface temperatures needed to allow for shorter process time.

4.7.3 Volume Average Determination of Quality Retention

Quality retention in thermally processed conduction-heated foods is a nonuniformly distributed parameter. Relatively long exposure to the higher temperatures near the product surface causes much more quality degradation in product near the surface than will occur in product near the cold spot or center. This is because temperature distribution throughout the food container is nonuniformly distributed as heating and cooling proceed during the process. For this reason, quality retention must be calculated by volume integration of the different levels of retention at different locations. This is done by taking advantage of the finite element feature of the numerical simulation model. As the computer iterations make each sweep across the finite element nodes in carrying out the heat transfer calculations, the small change in nutrient concentration that occurs in that time interval can be calculated from the momentary value of rate constant that prevails at the local temperature at that time. When the process simulation is ended, a different final nutrient/quality concentration will exist within each volume element. Recall that the volume elements are in the shape of concentric rings with known dimensions from which the volume of each ring of different sizes can be calculated. Total nutrient retention within each ring is calculated by multiplying the final nutrient concentration within the ring by the volume of that ring. Total nutrient retention in the product is the summation of final retention in all the rings.

4.8 PROCESS DEVIATIONS

Control of thermal process operations in food-canning factories has consisted of maintaining specified operating conditions that have been predetermined from product and process heat penetration tests, such as the process calculations for the time and temperature of a batch cook. Sometimes unexpected changes can occur during the course of the process operation or at some point upstream in a processing sequence such that the prespecified processing conditions are no longer valid or appropriate, and off-specification product is produced that must be either reprocessed or destroyed at appreciable economic loss. These types of situations are known as process deviations and can be of critical importance in food processing operations because the physical process variables that can be measured and controlled are often only indicators of complex biochemical reactions that take place under the specified process conditions.

Because of the important emphasis placed on the public safety of canned foods, processors operate in strict compliance with the U.S. Food and Drug Administration's Low-Acid Canned Food Regulations. Among other things, these regulations require strict documentation and record-keeping of all critical control points in the processing of each retort load or batch of canned product. Particular emphasis is placed on product batches that experience an unscheduled process deviation, such as when a drop in retort temperature occurs during the course of the process that may result from the loss of steam pressure. In such a case, the product will not have received the established scheduled process and must be destroyed, fully reprocessed, or set aside for evaluation by a competent processing authority. If the product is judged safe, then batch records must contain documentation showing how that judgment was reached. If judged unsafe, then the product must be fully reprocessed or destroyed. Such practices are costly.

In recent years, food engineers knowledgeable in the use of engineering mathematics and scientific principles of heat transfer have developed deterministic computer models capable of simulating thermal processing of conduction-heated canned foods such as described in this chapter. These models make use of numerical solutions to mathematical heat transfer equations capable of predicting accurately the internal product cold spot temperature in response to any dynamic temperature experienced by the retort during the process. As such, they are very useful in the rapid evaluation of deviations that may unexpectedly occur.

Accuracy of such models is of paramount importance, and the models must work equally as well for any mode of heat transfer or size and shape container. Recall that the deterministic model described earlier in this chapter was derived for the case of pure conduction heat transfer in a solid body of finite cylinder shape. Recent work reported in the literature has described effective modification and simplification of the model to overcome these limitations [14,15]. These reports confirmed that food containers need not be of the same shape as the solid body assumed by the heat transfer model. They could be of any shape so long as temperature predictions were required only at the single cold spot location within the container from which heat penetration data were determined.

The improved model assumed the product was a pure conduction-heating solid in the form of a sphere. An "apparent" thermal diffusivity was obtained for the solid sphere that would produce the same heating rate as that experienced by the product cold spot. Similarly, the precise radial location where the heating lag factor (j_h) was the same as that at the product cold spot, would be used as the location at which temperature is calculated by the model (Figures 4.15 and 4.16). Thus, for any product with empirical parameters (f_h and j_h) known from heat penetration tests, it would be possible to simulate the thermal response at the product cold spot to any dynamic boundary condition (time-varying retort temperature) regardless of shape or process conditions.

Recall that heat penetration test data normally produce straight-line semilog heat penetration curves from which the empirical heat penetration parameters (f_h and j_h) can be determined. Incorporation of the parameters into the heat transfer model is accomplished by the relationship between thermal diffusivity (α) and heating rate factor (f_h) for a sphere (Equation 4.9) and the relationship between heating lag factor (j_h) and radial location (r) within the sphere (Equation 4.10). These and similar relationships for other regular solid body shapes can be found in the literature [9,15]:

$$f_h = 0.233\left(\frac{R^2}{\alpha}\right) \tag{4.9}$$

$$j(r) = 0.637\left(\frac{R^2}{r}\right)\sin\left(\frac{\pi r}{R}\right). \tag{4.10}$$

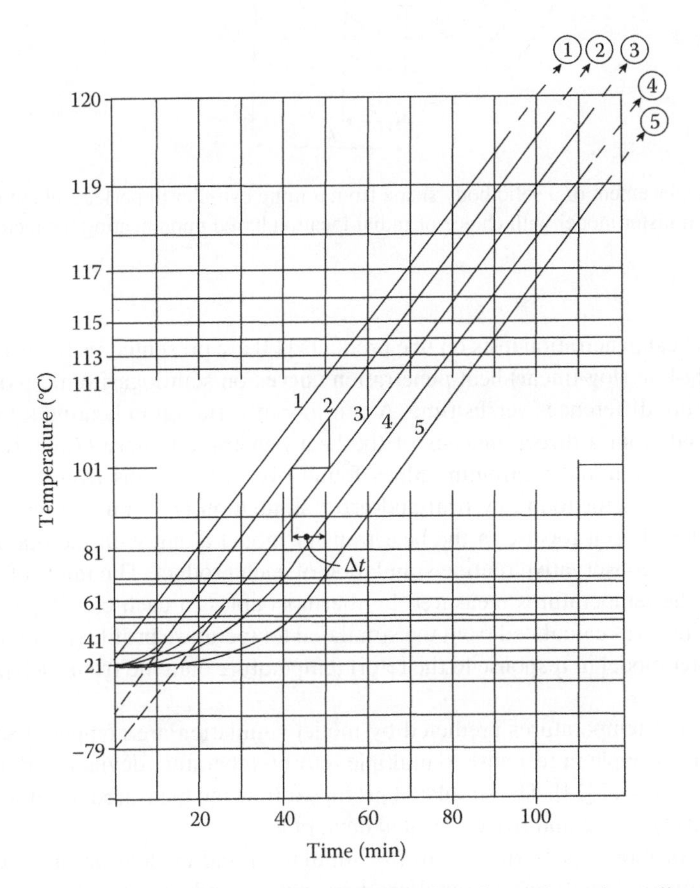

FIGURE 4.15 Heat penetration curves for five different locations along the radius on the midplane of a cylindrical container (see Figure 4.16), illustrating a relationship between location and heating lag factor (j_h).

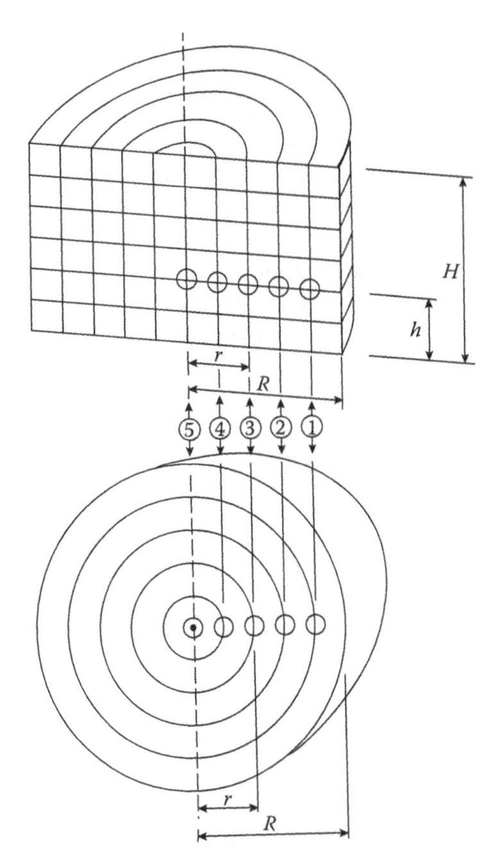

FIGURE 4.16 Replacement of a solid body shape from a finite cylinder to perfect sphere for simplification of numerical heat transfer model with choice of radial location based upon heating lag factor from heat penetration tests.

The results from heat penetration tests on five products [14] are presented in Table 4.5. All products exhibited straight-line (log-linear) heat penetration curves on semilogarithmic plots of unaccomplished temperature differences versus time. A can-to-can variation in heating rate factor (f_h) and lag factors derived from a direct analysis of the heat penetration curve (j_h analyze) were determined by the maximum and minimum values found over all six cans from two replicate tests. The true heating lag factor found by trial-and-error simulation (j_h simulate) was also compared. This was the value chosen for use in the heat transfer model along with the maximum f_h values (slowest heating) for conservative routine simulation of each product. The range of lethality values calculated from the temperatures measured by thermocouples in each can (F_O actual) were also compared. Lethality was calculated from the simulated temperature profile (F_O simulate) predicted by the heat transfer model in response to the retort temperature data file from each heat penetration test as input.

Internal cold spot temperatures predicted by model simulation are compared with the profiles measured by thermocouple in response to multiple retort temperature deviations during a heat penetration test in Figure 4.17 [14]. The simulated profiles follow the measured profiles quite closely in response to relatively severe and twice repeated deviations.

The final test of model performance in the simulation and evaluation of process deviations was a comparison of lethalities accomplished by actual and simulated temperature profiles (Table 4.6). Recall that the accomplished lethality (F_O) for any thermal process is easily calculated by numerical integration of the measured or predicted cold spot temperature over time as

TABLE 4.5
Heat Penetration Results on Products Using Two Replicated Heat Penetration Tests with Six Instrumented Cans for Each Product

Product and Process	f_h (min) (Range)	Analyze, j_h (Range)	Simulate, j_h	Actual, F_O (Range)	Simulate, F_O
5% Bentonite, 1 kg cans (98 × 110 mm), static cook	70.4–73.0	1.9–2.0	2.0	6.0–7.0	6.2
5% Bentonite, tuna cans (86 × 45 mm), static cook	20.0–22.0	1.4–1.6	1.4	7.5–9.8	7.4
Water, 1 kg cans (98 × 110 mm), static cook	3.0–3.1	1.8–2.3	1.0	9.8–10.8	9.9
Water, tuna cans (86 × 45 mm), static cook	1.7–1.9	2.5–3.9	1.0	7.9–10.6	7.7
Peas in brine, ½ kg cans (74 × 88 mm), agitated cook	2.5–3.0	2.6–3.4	1.0	10.8–12.0	11.0

Source: From Teixeira, A.A. et al., *J. Food Sci.*, 64(3), 488, 1999. With permission.

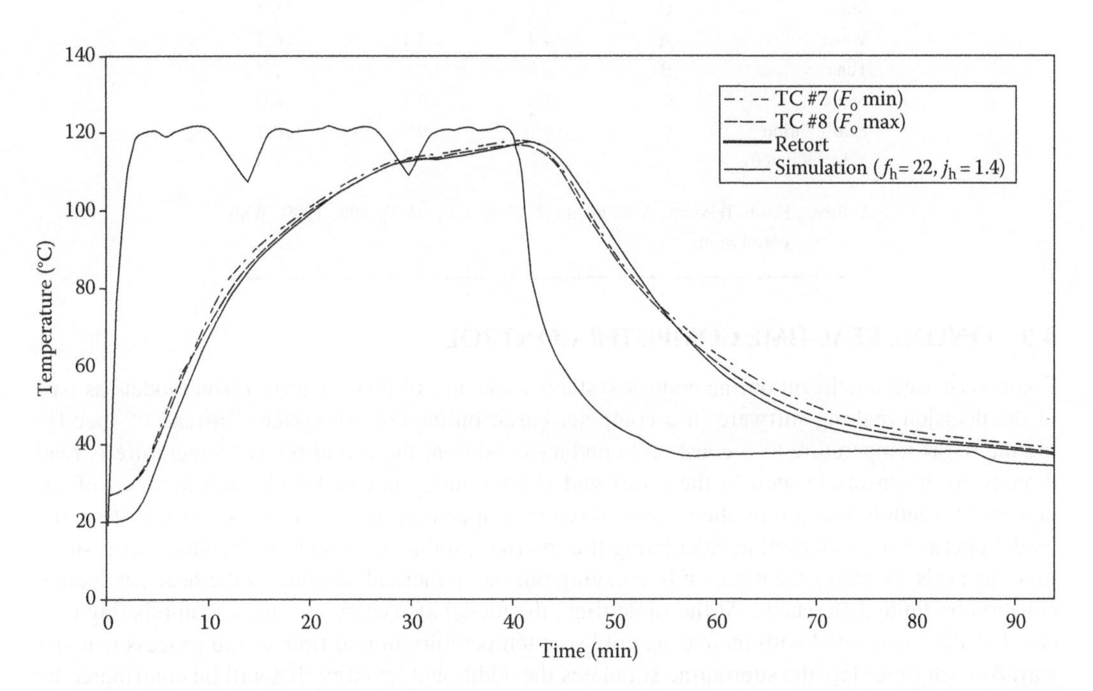

FIGURE 4.17 Comparison of internal cold spot temperatures predicted by model simulation with those measured by thermocouples in response to multiple retort temperature deviations during a heat penetration test with 5% bentonite suspension in 6-ounce tuna can. (From Teixeira, A.A. et al., *J. Food Sci.*, 64(3), 488, 1999. With permission.)

explained previously. Thus, if the cold spot temperature can be accurately predicted over time, so can accumulated process lethality. In all cases, the simulated lethality predicted agreed most closely with the minimum actual lethality calculated from measured temperature profiles. Model predictions that tend toward the minimum side of the range are always desirable for conservative decision making.

TABLE 4.6
Process Deviation Test Results Showing Lethalities Calculated from Temperatures Predicted by Model Simulation (F_O, Simulated), and Those Calculated from Actual Measured Temperatures (F_O, Actual) in Slowest and Fastest Heating Cans in Response to Different Types of Retort Temperature Deviations during Processing

Product and Process	Deviation (A, B, C)	Simulated, F_O	Actual, F_O	
			Minimum	Maximum
5% Bentonite	A	5.5	5.5	6.4
1 kg	B	3.0	3.3	5.3
Static	C	1.7	1.6	2.4
5% Bentonite	A	6.5	6.6	7.6
Tuna	B	5.6	5.7	7.0
Static	C	4.8	4.7	5.7
Water	A	7.4	7.4	8.2
1 kg	B	7.8	8.8	10.3
Static	C	7.1	7.4	8.8
Water	A	4.4	5.4	6.2
Tuna	B	6.0	6.6	7.7
Static	C	5.5	6.7	8.0
Peas in brine	C	9.1	9.2	10.0
½ kg (agitated)				

Source: From Teixeira, A.A. et al., *J. Food Sci.*, 64(3), 488, 1999. With permission.

4.9 ONLINE REAL-TIME COMPUTER CONTROL

Computer-based intelligent online control systems make use of these deterministic models as part of the decision-making software in a computer-based online control system. Instead of specifying the retort temperature as a constant boundary condition, the actual retort temperature is read directly from sensors located in the retort and is continually updated with each iteration of the numerical solution. Using only the measured retort temperature as input to the control system, the model operates as a subroutine calculating the internal product cold spot temperature after small time intervals for computer iteration in carrying out the numerical solution to the heat conduction equation by finite differences. At the same time, the model also calculates the accomplishing process lethality associated with increasing cold spot temperature in real time as the process is under way. At each time step, the subroutine simulates the additional lethality that will be contributed by the cooling phase if cooling were to begin at that time. In this way, the decision of when to end heating and begin cooling is withheld until the model has determined that final target process lethality will be reached at the end of cooling.

By programming the control logic to continue heating until the accumulated lethality has reached some designated target value, the process will always end with the desired level of lethality (F_O) regardless of an unscheduled process temperature deviation. At the end of the process, complete documentation of measured retort temperature history, calculated center temperature history, and accomplished lethality (F_O) can be generated in compliance with regulatory record-keeping requirements. Such documents are shown in Figure 4.18 for a normal process (above), and for the same intended process with an unexpected deviation (below) [16,17].

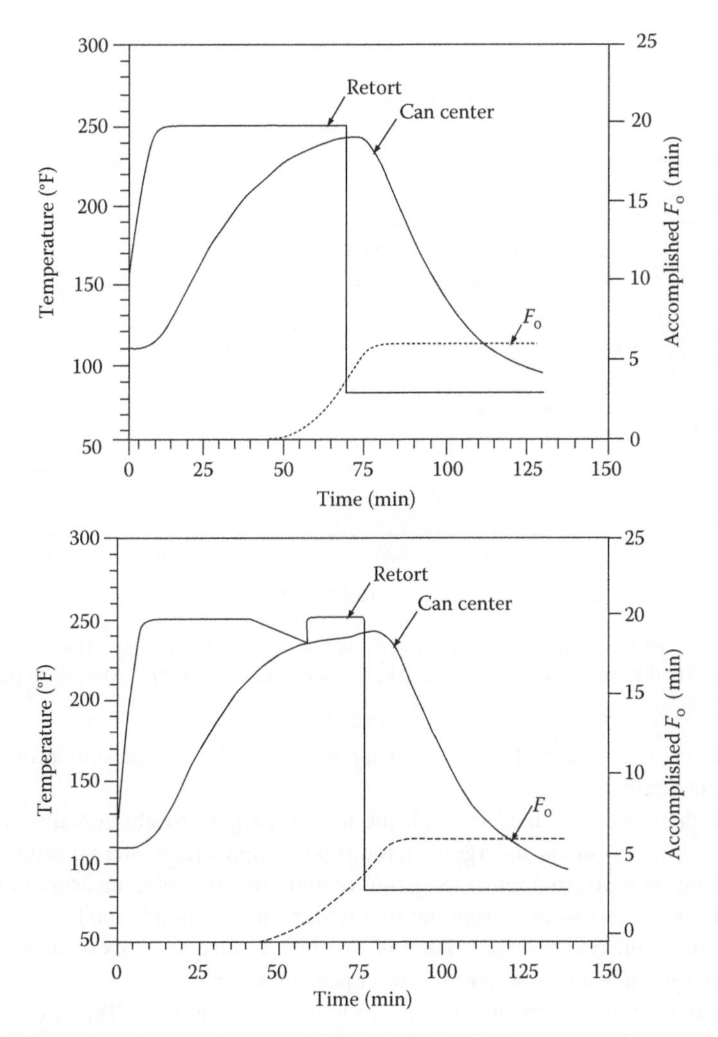

FIGURE 4.18 Output documentation of computer-based online control system showing scheduled heating time of 68 min for normal process (above) and heating time extended automatically to 76 min in compensation for unscheduled temporary loss of retort temperature (process deviation) (below).

4.10 MEASURING AND PREDICTING HEADSPACE PRESSURE

This last section attempts to address the technical challenges facing thermal processing of shelf-stable foods in flexible and semirigid packaging systems. Traditional metal cans and glass jars are now sharing shelf space with increasingly popular flexible pouches and semirigid bowls and trays. These flexible packages lack the strength of metal cans and glass jars and need greater control of external retort pressure during processing. Increasing internal package pressure without counter pressure causes volumetric expansion, putting excessive strain on package seals that may lead to serious container deformation and compromised seal integrity.

Recent work reported in the literature describes efforts to develop mathematical models that will predict internal headspace pressure profiles during retorting in response to known internal temperature and initial and boundary conditions [18]. The models were based on the use of an equation recommended by the International Association for the Properties of Water and Steam (IAPWS) [19] for the case of pure water, along with the Ideal gas law for noncondensable gases and Raoult's law for predicting vapor pressure of water in aqueous solutions with dissolved solids.

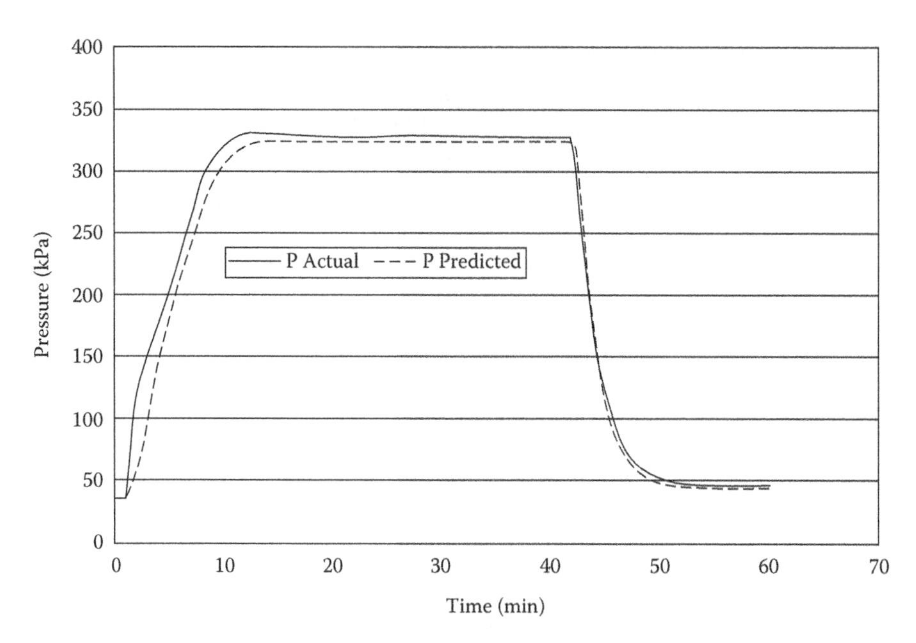

FIGURE 4.19 Comparison of model-predicted internal pressure profile with experimentally measured pressure profiles for distilled water. (From Ghai, G. et al., *J. Food Sci.*, 76(3), E298, 2011. With permission.)

Model performance was assessed by comparing model-predicted pressure profiles with those measured experimentally.

Experimental data were obtained through the use of a rigid airtight container (module) filled with various model food systems undergoing a retort process in which internal product temperature and pressure, along with external retort temperature and pressure, were measured and recorded at the same time. Experiments were carried out first with relatively simple model food systems (pure water and aqueous saline and sucrose solutions), as well as food systems of increasing compositional complexity (green beans in water and sweet peas in water).

Selected results from that work are shown in Figures 4.19 and 4.20 [18]. A comparison of the model-predicted internal pressure profile with the experimentally measured profile for the case of distilled water is shown in Figure 4.19, while the same comparison in the case of a more complex food system (sweet peas in distilled water) is shown in Figure 4.20. These results show clearly that the model could perform reasonably well in the simple case of pure distilled water but was incapable of predicting correct pressure in the case of a more complex food system, such as sweet peas in water. Sweet peas contain soluble sugars and starches that will impart noticeably lower vapor pressures to the system as compared to pure water, and the first-generation model developed in that work was incapable of taking such factors into account.

4.11 CONCLUSIONS

This chapter has focused on the development and application of deterministic heat transfer models for simulation of thermal processing in the process development and manufacture of heat-sterilized canned foods. The models are capable of accurately predicting internal product temperature over time during heating in response to dynamic retort operating conditions for any degree of combined conduction–convection mode of heat transfer and for any size or shape container. Applications of these models to process optimization in process design, rapid off-line evaluation of process deviations, and real-time online computer control of retort operations were described in detail.

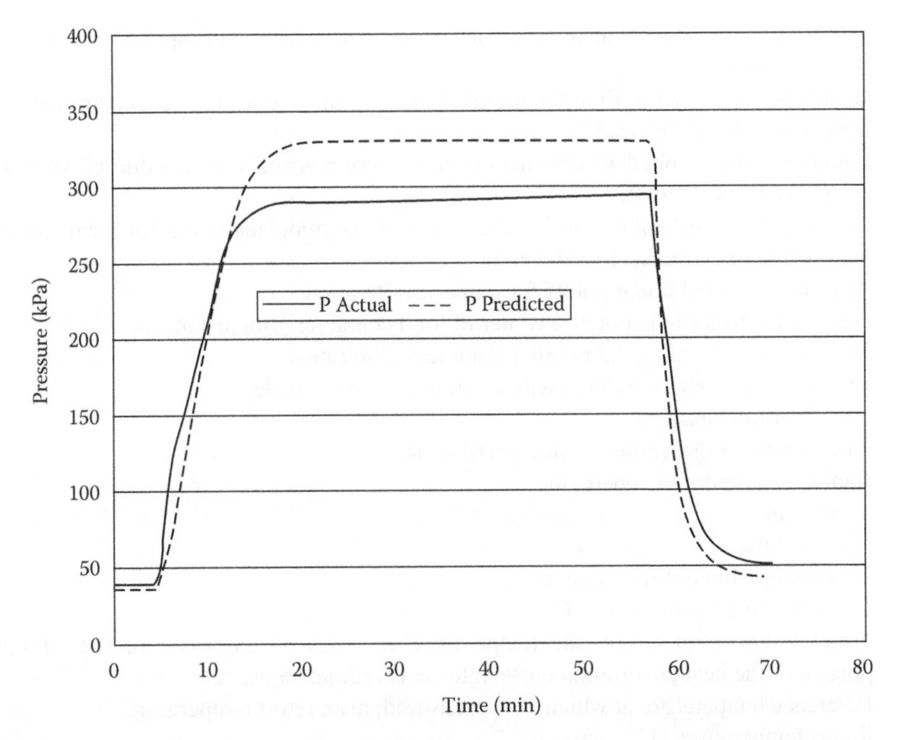

FIGURE 4.20 Comparison of model-predicted internal pressure profile with experimentally measured pressure profiles for sweet peas in distilled water. (From Ghai, G. et al., *J. Food Sci.*, 76(3), E298, 2011. With permission.)

The final chapter section describes efforts to develop mathematical models for predicting internal headspace pressure profiles during retorting in response to known internal temperature, as well as initial and boundary conditions. However, the relative simplicity of those models (Perfect gas law and Raoult's laws) renders them inadequate for food products with complex composition and chemistry. Improved models are needed that will likely require the inclusion of factors related to the chemical and physical composition of the food product.

The chapter also presented appropriate review of underlying principles and concepts of thermal processing important to understanding model development, applications, and limitations.

NOMENCLATURE

a	Initial number of viable bacterial spores at the beginning of a thermal process
A	Area through which heat transfer occurs, m^2
b	Final number of viable bacterial spores (survivors) at the end of a thermal process
C	Concentration of primary component in a first-order reaction, quantity per unit mass or volume (e.g., spores/mL)
C_O	Initial concentration of primary component at the beginning of a reaction, spores/mL
C_p	Specific heat or heat capacity, kJ/kg·°C
D	Decimal reduction time, time for one log-cycle reduction in population during exposure to constant lethal temperature, min
D_r	Decimal reduction time at a specified reference temperature (T_r), min
D_{121}	Decimal reduction time at temperature of 121°C (250°F), min
D_{100}	Decimal reduction time at temperature of 100°C (212°F), min
D_{65}	Decimal reduction time at temperature of 65°C (150°F), min

F	Process lethality, minute at any specified temperature and z value, applied to destruction of microorganisms
f_h	Heat penetration factor, time for straight-line portion of semilog heat penetration curve to traverse one log cycle, min
F_O	Lethality value applied to destruction of microorganisms with z value of 10°C (18°F), minute at 121°C (250°F)
h	Distance along vertical dimension (height) from the midplane in a cylindrical can, m
H	Half height of cylindrical food can, m
I	Sequence of radial nodal points for numerical iteration
j_h	Heating lag factor at geometric center of food container, dimensionless
J	Sequence of vertical nodal points for numerical iteration
k	Thermal conductivity, W/m·k, with reference to heat transfer
L	Length or thickness, m
r	Any distance along radius from centerline, m
R	Radius of cylinder or sphere, m
t	Time, min
T	Temperature, °C
T_c	Temperature of cooling medium, °C
T_I	Initial product temperature, °C
T_O	Pseudoinitial temperature, the temperature at which an extension of the straight-line portion of the heat penetration curve intersects ordinate axis, °C
T_r	Reference temperature at which D_r is measured; also, retort temperature, °C
T_R	Retort temperature, °C
T_{ih}	Initial product temperature at the beginning of heating, °C
T_{pih}	Pseudoinitial product temperature at the beginning of heating (see also T_O), °C
$T_{(ij)}$	Temperature at any grid node (i,j) in finite difference solution to heat transfer equation, °C
$T_{(ij)}^{(t)}$	Temperature at any grid node (ij) at time (t), °C
$T_{(ij)}^{(t+\Delta t)}$	Temperature at any grid node (i,j) at time $(t+\Delta t)$, or one time interval (Δt) later, °C
T_w	Cooling water temperature, °C
x	Spatial location within a food container, m
Z	Temperature-dependency factor in the thermal inactivation kinetics, temperature difference required for 10-fold change in decimal reduction time (D value), °C
α	Thermal diffusivity, m^2/s
ΔF_i	Incremental lethality accomplished over time interval (Δt), min
Δh	Vertical height of incremental volume element ring in finite difference solution to heat transfer equation, m
Δr	Radial width of incremental volume element ring in finite difference solution to heat transfer equation, m
Δt	Time interval between computational iterations in finite difference solution to heat transfer equation, min
ρ	Product density, kg/m^3

REFERENCES

1. R Graves. Determination of the sterilizing value (F_o) requirement for low acid canned foods. Presentation to Institute of Thermal Process Specialists, *Annual Conference*, November 4–6, 1987, Washington, DC.
2. CR Stumbo. *Thermobacteriology in Food Processing*. New York: Academic Press, 1965.
3. A K Datta, AA Teixeira. Numerical modeling of natural convection heating in canned foods. *Transactions of the ASAE* 30(5):1542–1551, 1987.
4. AK Datta, AA Teixeira. Numerically predicted transient temperature and velocity profiles during natural convection heating of canned liquid foods. *Journal of Food Science* 53(1):191–196, 1988.

5. NFPA. *Laboratory Manual for Food Canners and Processors*, Vol. 1. Westport, CT: AVI Publishing Company, 1980.
6. AA Lopez. *Complete Course in Canning, Book 1, Basic Information on Canning*, 11th edn. Baltimore, MD: The Canning Trade, 1987.
7. WD Bigelow, GS Bohart, AC Richardson, CO Ball. *Heat Penetration in Processing of Canned Foods*. Bulletin 16L, 1st edn. Washington, DC: National Canners Association. 1920.
8. CO Ball. Thermal process time for canned foods. *Bulletin of the National Research Council* 7(Part 1, No. 37):76, 1923.
9. CO Ball, FCW Olson. *Sterilization in Food Technology*. New York: McGraw-Hill, 1957.
10. AA Teixeira, JR Dixon, JW Zahradnik, GE Zinsmeister. Computer optimization of nutrient retention in thermal processing of conduction-heated foods. *Food Technology* 23(6):137–142, 1969.
11. AA Teixeira, GE Zinsmeister, JW Zahradnik. Computer simulation of variable retort control and container geometry as a possible means of improving thiamine retention in thermally processed foods. *Journal of Food Science* 40(3):656–659, 1975.
12. AA Teixeira, JE Manson. Computer control of batch retort operations with on-line correction of process deviations. *Food Technology* 36(4):85–90, 1982.
13. AK Datta, AA Teixeira, JE Manson. Computer-based retort control logic for on-line correction of process deviations. *Journal of Food Science* 51(2):480–483, 507, 1986.
14. AA Teixeira, MO Balaban, SPM Germer, MS Sadahira, RO Teixeira-Neto, A Vitali. Heat transfer model performance in simulation of process deviations. *Journal of Food Science* 64(3):488–493, 1999.
15. J Noronha, M Hendrickx, A Van Loeyand, P Tobback. New semi-empirical approach to handle time-variable boundary conditions during sterilization of non-conductive heating foods. *Journal of Food Engineering* 24:249–268, 1995.
16. AA Teixeira. Thermal processing of canned foods. In: DR Heldman and DB Lund, eds. *Handbook of Food Engineering*. 2nd edn. Boca Raton, FL: CRC Press, Taylor & Francis Group, 2007, Chapter 11, pp. 592–659.
17. AA Teixeira, CF Shoemaker. *Computerized Food Processing Operations*. New York: Van Nostrand Reinhold Company, 1988, pp. 169–185.
18. G Ghai, AA Teixeira, BA Welt, R Goodrich-Schneider, W Yang, S Almonacid. Measuring and predicting head space pressure during retorting of thermally processed foods. *Journal of Food Science* 76(3):E298–E308, 2011.
19. W Wagner, A Pruss. The IAPWS formulation for the thermodynamic properties of ordinary water substance for general and scientific use. *Journal of Physical Chemistry Reference Data* 31:387–535, 2002.

[illegible]

5 Modeling Food Thermal Processes Using Artificial Neural Networks

Cuiren Chen and Hosahalli S. Ramaswamy

CONTENTS

5.1 INTRODUCTION

Artificial neural networks (ANNs) are being successfully applied for a wide range of problem domains in diverse areas including engineering, physics, finance, medicine, and others related to purposes of prediction, classification, or control. This extensive success can be attributed to many factors: (1) Power of modeling: Neural networks (NNs) are very sophisticated techniques capable of modeling extremely complex functions. A priori knowledge of the system is not needed for constructing the ANN because the ANN will learn its internal representation from the input/output data of its environment and response. (2) Ease of use: NNs learn by example. The user of NNs gathers representative data and then invokes training algorithms to automatically learn the structure of the data. Although the user does need to have some heuristic knowledge of how to select and prepare data, how to select an appropriate NN, and how to interpret the results, the level of user knowledge needed to successfully apply NNs is much lower than to use some more traditional nonlinear statistical methods. (3) High computational speed: The ANN is an inherently parallel architecture. The result comes from the collective behavior of a large number of simple parallel processing units. Therefore, once trained, ANN can calculate results from a given input very quickly. Because of this feature, ANNs have a greater potential to be used for the online control system than conventional modeling methods.

The concept of NNs was based on the research in artificial intelligence, which was specifically intended to mimic the fault tolerance and capacity of biological neural systems by modeling the low-level structure of the brain. Warren McCulloch and Walter Pitts in 1943 were the first to open the idea on how neurons might work and they modeled simple NN using electrical circuits [1]. As computers became more advanced in the 1950s, it was finally possible to simulate a hypothetical NN. In 1959, Bernard Widrow and Marcian Hoff developed models called "ADALINE" and "MADALINE" [2,3]. In 1962, the same authors developed a learning procedure that examined the value before the weight adjustment (i.e., 0 or 1), which was one of the important fundamentals for the following success of NNs [4]. However, the NN concepts did not result in practical applications until 1980s when several new approaches such as bidirectional lines, hybrid network and multilayer NNs were developed [2–6]. In addition to these advances in algorithms, the rapid development of computer technologies including both hardware and software became an important driving force for NNs as a computing technique to be used in not only computing science but also in other areas as a tool for prediction, classification, and optimization.

5.2 INSPIRATION FROM BIOLOGICAL NEURONS

The human brain principally consists of over 10 billion neurons, each of which is connected to about 10,000 other neurons. A typical biological neuron, as shown in Figure 5.1, contains neuronal cell bodies (soma), dendrites, and axons. Each neuron receives electrochemical inputs from other neurons at the dendrites. If the sum of these electrical inputs is sufficiently powerful to activate the neuron, it transmits an electrochemical signal along the axon, and passes this signal to the other neurons whose dendrites are attached at any of the axon terminals. These attached neurons may

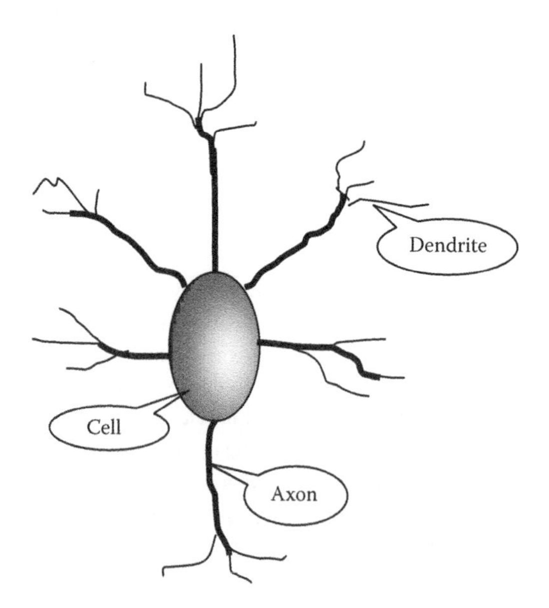

FIGURE 5.1 The structure of a typical biological neuron.

then fire. It is important to note that a neuron fires only if the total signal received at the cell body exceeds a certain level. The entire brain is composed of these interconnected electrochemical transmitting neurons. From a very large number of extremely simple processing units (each performing a weighted sum of its inputs and then firing a binary signal if the total input exceeds a certain level), the brain manages to perform extremely complex tasks.

This is the model on which ANNs are based. However, it should be noted that ANNs only represent extremely simplified formal models of biological neurons and their interconnections without making any attempt to model the biological system itself. Their importance lies in the fact that artificial networks are brain-inspired computational tools for solving complex problems.

5.3 PRINCIPLES OF A BASIC ARTIFICIAL MODEL

5.3.1 NN Architecture

NNs consist of a set of neurons or processing units, which are arranged in several parallel layers. The most commonly used NN architecture is the multilayer feed-forward network using backpropagation of error in the learning mechanism, which is shown in Figure 5.2. This NN has an input layer, two hidden layers, and one output layer. Each layer is essential to the operation of the network. A NN can be viewed as a "black box" into which a specific input to each node in the input layer is sent. The network processes this information through the interconnections between nodes, but this entire processing step is hidden. Finally, the network gives an output from the nodes on the output layer. The function of each layer is described as follows:

- *Input layer*—receives information from an external source and passes this information to the network for processing.
- *Hidden layers*—receives information from the input layer and does all of the information processing, which is hidden from view. The number of hidden layers can be one to three, dependent on the problem being investigated.
- *Output layer*—receives processed information from the network and sends the results out to an external receptor.

When the input layer receives the information from an external source, it will be activated and emit signals to their neighbors. The neighbors which receive excitations from an input layer, in turn, emit signals to their neighbors. Depending on the strength of the interconnections (i.e., the magnitude of

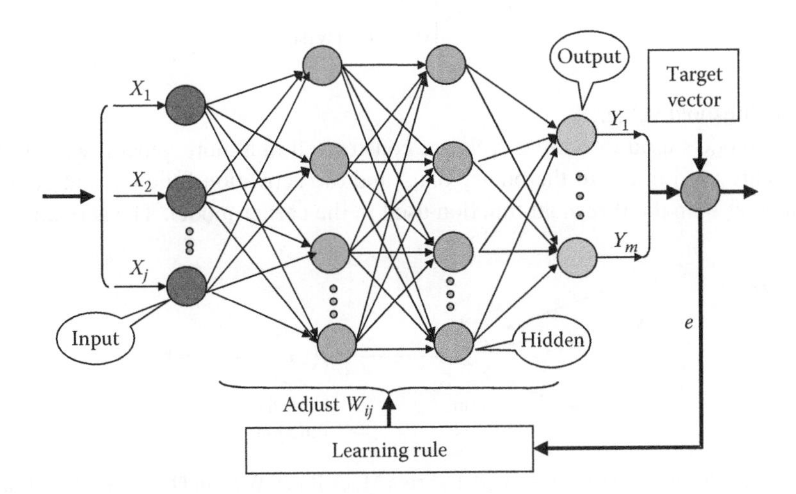

FIGURE 5.2 A typical multilayer NN with one hidden layer.

the so-called weight factor that adjusts the strength of the input signal), these signals can excite or inhibit the nodes. The result is a pattern of activation that eventually manifests itself in the output layer. Finally, the values from the output layer will be compared with the desired values. If the difference between output and desired values is larger than the set error range, then the weight factors are adjusted through the repeated training until the error is within the set error range or the number of learning runs is larger than the set number.

There are other ways of interconnections between neurons to construct different NN architectures such as feedback connections, lateral connections, time-delayed connections, and recurrent network.

5.3.2 Artificial Neurons

The artificial neurons are simple processing units similar to the biological neurons: they receive multiple inputs from other neurons but generate only a single output. The generated output may be propagated to several other neurons.

Each neuron has two basic functions: gathering information from the other neurons in the prior layer and sending the signals to the neurons in next layers. The first artificial neuron model proposed in 1943 by McCulloch and Pitts (Figure 5.3) is based on the simplified consideration of the biological model [1]. In this model, $x_1, x_2, \ldots, x_n$ are the binary inputs of the neurons. Zero represents absence, and one represents existence. The weight of connection between the ith input x_i and the neuron is represented by w_i. When $w_i > 1$, the input is excitatory and when $w_i < 0$, it is inhibitory.

The net summation, X, of inputs weighted by the synaptic strength w_i at connection i is

$$X = \sum_{i-1}^{n} w_i x_i \tag{5.1}$$

The net value is then mapped through an activation function of neuron output. The activation function used in the model is a threshold function:

$$y = f(X) \tag{5.2}$$

$$f(x) = \begin{cases} 1, & x > \theta \\ 0, & \text{otherwise} \end{cases} \tag{5.3}$$

where θ is the threshold value.

The neuron models used in current NNs are constructed in a more general way. The input and output signals are not limited to the binary data, and the activation function can be any continuous function other than the threshold function used in the earlier model. The activation function is

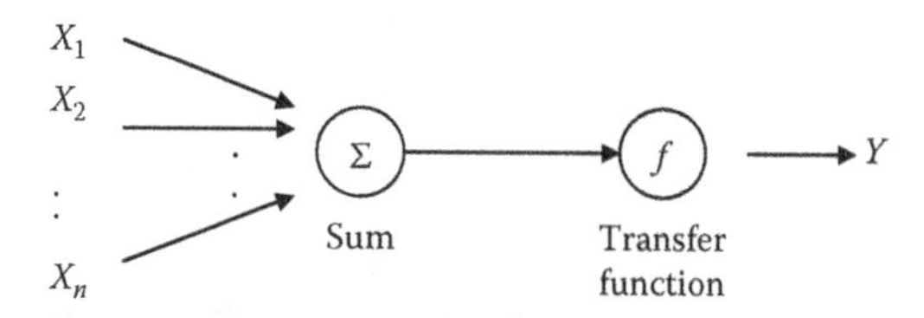

FIGURE 5.3 McCulloch and Pitts neural model. (From McCulloch, W.S. and Pitts, W., *Bull. Math. Biophys.*, 5, 115, 1943.)

typically a monotonic nondecreasing nonlinear function. Some of the often used activation functions are (where α denotes the parameter, and θ denotes the threshold value):

$$\text{Sigmoid function}: f(x) = \frac{1}{1+e^{-\alpha x}} \tag{5.4}$$

$$\text{Hyperbolic function}: f(x) = \tanh(\alpha x) = \frac{e^{\alpha x} - e^{-\alpha x}}{e^{\alpha x} + e^{-\alpha x}} \tag{5.5}$$

$$\text{Linear threshold}: f(x) = \begin{cases} 1 & x \geq \theta \\ \dfrac{x}{\theta} & 0 < x < \theta \\ 0 & x \leq \theta \end{cases} \tag{5.6}$$

$$\text{Gaussian function}: f(x) = e^{-\alpha x^2} \tag{5.7}$$

5.3.3 Learning Rules

There are primarily two learning methods used for NNs: supervised learning and unsupervised learning. For supervised learning, the training data set consists of pairs of input and desired output data. The error signal is generated as difference between the actual output and the desired output and then used to adjust weights of networks. For unsupervised learning, only input data are fed into network, because the desired output is not known and thus no explicit error information is given. The supervised learning networks are the most often used NNs for the modeling purpose. Therefore, only learning rules used in supervised learning networks are discussed.

Learning rule is a method to adjust the weight factors based on trial and error. Many learning rules have been developed to train NNs. The main training method is Error-Correction Learning [1] which uses the data to adjust network's weights and thresholds so as to minimize the error in its prediction on the training set, mathematically defined as follows:

$$\varepsilon_i = d_i - c_i \tag{5.8}$$

where

ε_i is the output error
d_i is the desired output
c_i is the calculated output

for the ith neuron on the output layer only. The total square error on the output layer can be calculated as follows:

$$E = \sum_i \varepsilon^2 = \sum_i (d_i - c_i)^2 \tag{5.9}$$

The change in the weight factor for the *j*th connection to the *i*th neuron is obtained by

$$\Delta w_{ji} = -\eta\left(\frac{\partial E}{\partial w_{ji}}\right) = \eta a_j \varepsilon_i \tag{5.10}$$

where

η is a linear proportionality constant, called the learning rate (typically, $0<\eta \ll 1$)
a_j is the *j*th input to neuron *i*

In order to speed up the learning during training, the following modification is often implored:

$$\Delta w_{ji}(\text{new}) = \eta a_j \varepsilon_i + \lambda_i \Delta w_{ji}(\text{old}) \tag{5.11}$$

where λ is the momentum.

There are a number of other approaches developed as learning rules [1]. *Reinforcement Learning* is a type of supervised learning closely related to error-correction learning. *Stochastic learning* utilizes statistics, probability, and/or random processes to adjust connection weights. *Hardwired NNs* have all connections and weights predetermined (hardwired). These networks have a speed advantage, and are used with additional a priori information in speech recognition, language processing vision, and robotics. *Hebbian learning* adjusts weights based on a correlation between the two nodes associated with the weight factor.

5.4 DEVELOPING NNs

NNs can be developed either based on the principles of NNs using a specific computer language or by the use of one of the commercial NNs software. Developing NN codes, which means to turn the theory of a particular network model into a computer simulation implementation, can be a challenging task for most of applied scientists and engineers who do not have both programming and related knowledge of NNs; so the use of commercial software has been the most popular method to develop a NN model. With the rapid development of computer software, several NN software have been developed which can be used for developing NN models for specific purposes, such as NeuralWare Profession, NeuralShell, Neurosolution (NeuroDimension, Inc., Gainesville, FL), MATLAB®, etc. Developing a NN by using a commercial NN software consists of following steps: (1) Selection of inputs and outputs, (2) data collection, (3) optimization of configurations, (4) training or learning, and (5) testing or generation.

The number of outputs can be determined by the problem being investigated while the number of inputs should be those influencing outputs significantly which can be decided by comparison of results from different experiments with different inputs. The size of data required for NN training is dependent on the complexity of the underlying function which the network is trying to model and the variance of the additive noise. Normally, NNs can only process numeric data in a limited range but it is also possible for NNs to handle different types of data such as an unusual range, missing data or nonnumeric through some methods. For example, numeric data can be scaled into an appropriate range for the network, missing values can be substituted for using the mean value, and nonnumeric data can be represented by a set of numeric values.

For each specific problem, in order to develop a NN model with the best performance, the configuration parameters of the NN being developed must be determined by trial and error. These parameters include transfer functions, learning rules, learning rate, momentum coefficient, number of hidden layers, number of neurons in each hidden layer, and learning runs.

In *the training or learning step*, a set of known input–output data are repeatedly presented to train the network. During this repetition process, the weight factors between nodes are adjusted until the specified input yields the desired output. Through these adjustments, the NN "learns" the

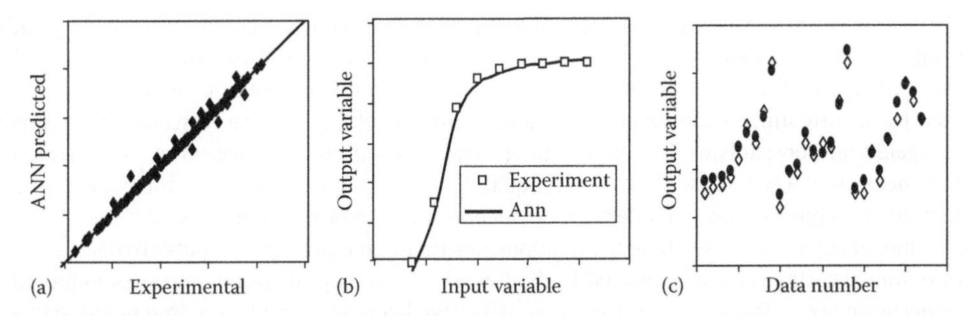

FIGURE 5.4 Graphs for performance of NN modeling. (a) Evaluation of NN modeling performance. (b) Matchability of NN model with specifically input variable. (c) Pair comparision of NN model predicted and desired outputs.

correct input–output response behavior. In NN development, this phase is typically the longest and most time-consuming, and it is critical to the success of the network.

After the training step, *the recall and generalization step* will be carried on. In the recall step, the network will be subjected to a wide array of input patterns used in training, and adjustments introduced to make the system more reliable and robust. During the generalization step, the network will be subjected to input patterns that it has not seen before, but whose outputs are known, and the system's performance will be monitored. The performance of NNs can be evaluated visually by graphs or in quantity by different statistic measures. Figure 5.4 shows three kinds of graphs used for the performance of NN modeling. In Figure 5.4a, *X* and *Y* axes represent desired and predicted results, respectively. The diagonal line is the ideal line and can be easily used for evaluation of the modeling performance. If NN predicted values are equal or close to desired results, then the points should be located on or close to the idea line. In Figure 5.4b, *x* axis is one of the input variables while *y* axis is used for outputs including desired and predicted results. This graph can be used to determine the matchability of the NN model with the specifically input variable. In Figure 5.4c, *x* axis is just representing the number order of data used for training or testing, while *y* axis is used for outputs as Figure 5.4b. Its emphasis is on comparison of the agreement between each pair of predicted and desired values.

Often used statistical calculations include the *Regression coefficient* R^2 of relationship between predicted and experimental results and *average relative error* (E_r), which are given as follows:

$$R^2 = 1 - \frac{\sum_{i=1}^{n} (y_i - y_{di})^2}{\sum_{i=1}^{n} (y_i - y_m)^2} \tag{5.12}$$

$$E_r = \frac{\sum_{i=1}^{n} (y_i - y_{di})/n}{y_{max} - y_{min}} \tag{5.13}$$

where

y_i is predicted by ANN model
y_{di} is the actual desired value (actual values)
n is the number of data
y_m is the average of actual values

5.5 APPLICATIONS IN THE FOOD THERMAL PROCESSING

Thermal processing is one of the major operations in the food processing and preservation system. It has generally been viewed as an energy-intensive preservation technique, but persists as the most widely used method of preservation. One major objective of thermal processing is to destroy pathogenic and spoilage microorganisms present in foods being processed so that they can be stored for extended periods of time and consumed with no safety concerns. Quality factors in foods are also affected by heat treatments, however, they usually show much higher heat resistance as compared with targeted microorganisms. Therefore, an optimal thermal process procedure for a given food product means that it will result in a minimal quality destruction while being sufficient to make the product safe for consumption. In order to solve an optimization problem, the key step is to develop a suitable model capable of describing the relationships between inputs and outputs. To date, a variety of conventional methods have been used for both modeling and optimization purposes in food thermal processing areas, from which different models have been developed including numerical, analytical, and experimental. These conventional methods applied for optimizing thermal processing operations need specific inputs. First, the models need a knowledge and understanding of relationships between the input and output variables. Second, it is necessary that information of physical and thermal properties of food products being modeled be available. Unlike other disciplines, the food processing deals with biomaterials which show much more complicated thermophysical properties and uncertainties during the processing period. This also results in more complicated relationships between input and output variables. Thus, often it is difficult to use a simple partial differential equation or model to accurately describe the phenomena occurring in food processing operations. In addition, the lower calculation speed of most conventional methods under complex situations of process optimization and control limits them to optimally applied for online application in industrial processes. NN models offer an attractive alternative in such instances. Hence, NNs are being continually extended to food thermal processing area as a modeling and optimization technique.

Since 1980s, NNs have received more and more interests in food processing areas. So far, the applications of NNs in food processing have covered various areas such as drying [7–14], fermentation [15,16], extrusion [17–20], freezing [21], baking [22,23], post-harvest [24,25], experimental design [26], etc. The application in food thermal processing began in the middle of 1990s. Sablani et al. [27] published one of the early reports on the application of NNs in food thermal processing. A four layer NN was developed with three inputs and three outputs to predict optimal sterilization temperatures under different processing conditions. Sablani et al. [28] developed ANN models for predictions of overall heat transfer coefficient and the fluid to particle heat transfer associated with liquid particle mixtures, in cans subjected to end-over-end rotation. Similarly, Y. Meng and H.S. Ramaswamy [29] developed ANN models for apparent heat transfer coefficients associated with canned particulates in high viscous Newtonian and non-Newtonian fluids during end-over-end thermal processing in a pilot-scale rotary retort. Manickaraj et al. [30] applied ANN method to estimate the fluid–particle heat transfer coefficient in a liquid–solid fluidized system. The application of the neural computing approach for prediction of the *residence time distribution* (RTD) under aseptic processing conditions was reported by Chen and Ramaswamy [31]. In this study, NNs were explored for modeling two RTD functions—the time specific (E-type distribution) and the cumulative particle concentration function (F-type distribution) of carrot cubes in starch solutions in a vertical scraped surface heat exchanger (SSHE) of a pilot-scale aseptic processing system. The NNs were also applied for prediction of thermophysical properties of food materials, such as physical property changes of dried carrot as a function of fractal dimension and moisture content [32], thermal conductivity of ethylene glycol–water solutions [33], thermal conductivity of bakery products [23], thermal conductivity of food as a function of moisture content, temperature, and apparent porosity [34], and thermal conductivity of pure gases [35]. Modeling of thermal kinetic change of microorganisms and quality factors using NNs was reported by several researchers including enzyme reaction rate estimation [36], thermal inactivation of *Escherichia coli.* [37], heat resistance of *Bacillus stearothermophilus* [38], and online quality assessment of food [39]. NNs have been used as alternative tool to Ball and Stumbo methods, which are the most often used for retort process calculation to predict process time (PT) and or process lethality in cans during thermal processing [40,41]. NN based Dynamic modeling of real retort processing was investigated by Gonçalves et al. [42]. Two NN models: the backpropagation through time and Jordan were trained and their generalization performance was compared. Results indicated that a better generalization capacity were obtained using the backpropagation through time network, which presented an average relative error of 2.2% between the calculated and predicted F values. A more systematic and in-depth application of NNs in food thermal processing areas was carried out by Chen and Ramaswamy [43–48]. In these studies, separate ANN prediction models were developed involving main input parameters such as retort temperature (RT) profile, thermophysical properties of food products, kinetics of microorganisms, and quality factors and outputs such as the PT, cumulative lethality value, quality retention, unit energy consumption, and the transient temperature at the can center. These ANN models not only could be directly used for the process establishment and validation for a given food product but also could be combined with a search technique to build optimal thermal process conditions in order to meet different optimization objectives. Kupongsak and Tan [49] applied NN combined with fuzzy techniques to determine food process control set points for producing products of certain desirable sensory quality. ANN technique was also utilized for high pressure processing to model the thermal behavior of foods during high pressure treatments [50,51]. Some details of selected studies in thermal processing based on ANN approach are given in the following sections.

5.5.1 NN Modeling of Heat Transfer to Liquid Particle Mixtures in Cans Subjected to End-over-End Processing

Overall heat transfer coefficient (U) and *fluid to particle heat transfer coefficient* (h_{fp}) are fundamental data needed to develop prediction models for transient temperature of canned foods undergoing agitation thermal processing which is necessary to establish and optimize the thermal process schedule for canned liquid/particle food system. Traditionally, the dimensionless correlations (DCs) are used for the development of an experimental model of U and h_{fp} involving other influencing parameters by the use of multiple regression analysis. However, selection of appropriate dimensionless groups requires prior knowledge of the phenomena under investigation. Sablani et al. [28] developed an ANN model for the overall heat transfer coefficient and the fluid to particle heat transfer coefficient associated with liquid particle mixtures, in cans subjected to end-over-end rotation. Experimental data obtained for U and h_{fp} under various test conditions (shown in Table 5.1) were used for both training and evaluation. Multilayer NNs with seven inputs and two output neurons (for a single particle in a can), and six inputs and two outputs neuron (for multiple particles in a can) were trained. The optimal network was obtained by initial trials as number of hidden layers = 2, number of neurons in each hidden layer = 10, and learning runs = 50,000. By the use of trained NN models with optimal configurations, the prediction performance of all NN models for both U and h_{fp} was found to be higher than 0.98, meaning that the developed NN models could be safely used for prediction of U and h_{fp} under the given experimental conditions. The comparison of NN models and dimensionless regression model using the same experimental data is summarized in Table 5.2. Prediction errors using ANN were less than 3% and 5%, respectively, for U and h_{fp}, which were about 50% better than those associated with dimensionless number models, indicating that the predictive performance of the ANN was far superior to that of DCs.

5.5.2 Neuro-Computing Approach for Modeling of RTD of Carrot Cubes in a Vertical SSHE

The *RTD* is one of the important parameters for establishing the aseptic processing of particulate liquids. Although a lot of different models have been developed for describing the RTD characteristics

TABLE 5.1
Range of System and Product Parameters Used in the Determination of Heat Transfer Coefficients (U and h_{fp})

No.	Parameter	Experimental Range
1	Retort temperature	110°C, 120°C, and 130°C
2	Radius of rotation	0, 0.09, 0.19, and 0.27 m
3	Rotation speed	10, 15, and 20 rpm
4	Can headspace	0.0064 and 0.01
5	Test fluid	Water and oil
6	Test particle	Polypropylene and nylon
7	Particle concentration	Single particle, 20%, 30%, and 40% (v/v)
8	Particle shape and size	
	Cube	0.01905 m
	Cylinder	0.01905 × 0.01905 m
	Sphere	0.01905, 0.02225, and 0.025 m
9	Can dimension	307 × 409 (8.73 × 11.6)

TABLE 5.2
Comparison of Error Parameters for NN Models and DC Models

	Single Particle				Multiple Particle			
Error	U		h_{fp}		U		h_{fp}	
Parameters	DC	NN	DC	NN	DC	NN	DC	NN
MAE1	17.1	5.11	31.3	17.2	25.1	9.85	75.4	48.1
SDE	25.4	4.76	43.3	16.0	32.0	11.0	63.4	40.7
MRE (%)	5.00	2.46	16.9	5.82	5.70	2.57	8.26	4.52
SRE (%)	3.76	2.51	11.9	7.00	4.65	1.96	7.12	3.90
R^2	0.99	0.99	0.83	0.98	0.98	0.99	0.96	0.98

MAE, mean absolute error; SDE, standard deviation of error; MRE, mean relative error; SRE, standard deviation of relative error.

using conventional mathematical methods, none of them give a fully satisfactory solution for the RTD covering the wide range of processing conditions. A neuro-computing approach was used by Chen and Ramaswamy [31] for modeling two RTD functions—the time specific (E-type distribution) and the cumulative particle concentration function (F-type distribution) of carrot cubes in starch solutions in a vertical SSHE of a pilot-scale aseptic processing system. In this study, 356 experimental data pairs obtained for $E(t)$ and $F(t)$ under various test conditions, including the concentration of particles, flow rate, particle dimension, and test time, were used for both training and evaluation. The optimal configurations of the NN model were determined by adjusting the number of hidden layers, the number of neurons in each hidden layer and learning runs, and a combination of learning rule and transfer functions. The results showed that the trained ANN model can accurately map experimental results with R^2 value = 0.98 and 0.99 for E and F functions, respectively. The prediction performance of ANN model under several typical processing conditions is shown in Figure 5.5. The ANN models were also compared with conventional models developed based on multiple variable regression techniques. The comparison indicated that average modeling errors associated for ANN model were 5.7% and 3.0%, respectively, for E and F, while those for the

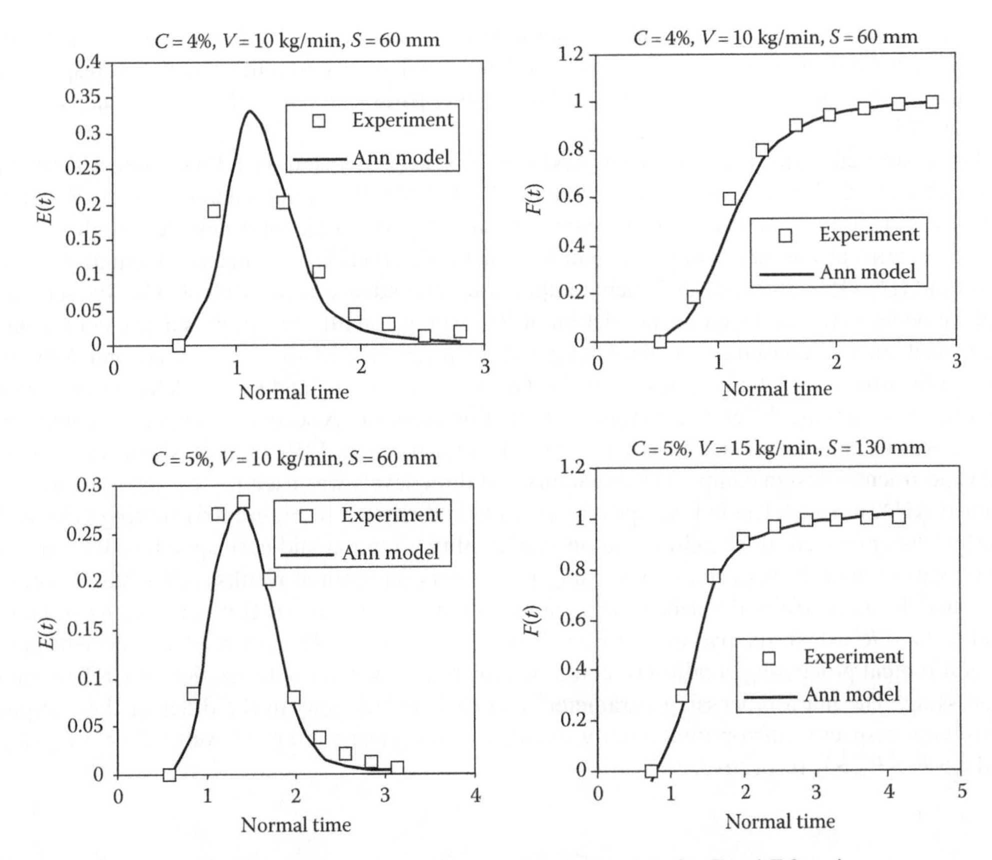

FIGURE 5.5 Comparison of ANN model predictions and experiments for *E* and *F* functions.

multiple regression models were 15.5% and 12.3%, meaning that the ANN model had higher precision for predicting *E* and *F* functions.

5.5.3 Modeling and Optimization of Constant Retort Temperature Thermal Processing Using Coupled NNs and Genetic Algorithms

Modeling and optimization of the thermal processing are of considerable interest, and are widely based on conventional mathematical methods. Traditionally, solving an optimization problem consists of two steps. First, different objective function models are developed using mathematical approaches which include regression methods, theoretical analysis models, and differential equations; and then the optimal conditions are sought using one of several search methods such as direct search, grid search, and gold section method, for single variable, and alternating variable search, pattern search, and Powell's method, for multi-variables. Like traditional modeling methods, NNs cannot provide direct answers for optimization problems. In order to be used for optimization purposes, NN models have to be combined with one of search techniques. *Genetic algorithms* (GAs) are a combinatorial optimization technique, searching for an optimal value of a complex objective function by simulation of the biological evolutionary process, based on crossover and mutation as in genetics. Chen and Ramaswamy [44] found that the combination of GA and ANN models can become an effective tool for optimization problems. This study might be the first report on application of ANN and GA for the thermal processing optimization. The focuses of this study were on (1) developing ANN models for predicting PT, average quality retention (Qv), surface cook value (Fs), equivalent unit energy consumption (En), temperature difference (*g*), and ratio of *F* value from heating to total desired F value (ρ) under constant retort temperature (*CRT*) processing conditions; (2) coupling ANN models and GA to search for the optimal quality retention and the corresponding RT; and (3) investigating the effects of main processing parameters on both optimal quality retention and RT.

Processing conditions as inputs for ANN models were selected as follows: retort temperature (RT = 110°C–140°C), thermal diffusivity ($\alpha = 1.1–2.14 \times 10^{-7}$ m^2/s), volume of can ($V = 1.64–6.55 \times 10^{-4}$ m^3), ratio of height to diameter of can ($R_{dh} = 0.2–1.8$), total desired lethality value ($F_o = 5–10$ min) at the can center and quality kinetic destruction parameters: decimal destruction time ($D_q = 150–300$ min) and their temperature dependence ($z_q = 15$°C–40°C). Six separate ANN models were developed for prediction of PT, average quality retention, surface cook value, equivalent energy consumption, final temperature difference at the can center, and lethality ratio, ρ (heating/total lethality), respectively. The data for training and testing ANN models were obtained from a finite difference computer simulation program. A second order central composite design was used for constructing experimental data for training ANN models, while an orthogonal experimental design composing six factors and three levels was used for the generalization of trained ANN models. The hybrid optimization method (shown in Figure 5.6) linking GAs with ANNs was employed for searching the optimal quality retention and corresponding RT, and for investigating the effects of main processing parameters on optimal results. ANN-based prediction models successfully described the various outputs of CRT thermal processing (correlation coefficients: $R^2 > 0.98$; relative errors: $E_r \leq 3\%$). The coupled ANN–GA models, verified under several typical processing conditions, could be effectively used for optimization of CRT thermal processing. The main processing parameters and their interactions in the order of their importance with respect to the optimal quality retention and corresponding RT were: $V > z_q > F_o > R_{dh}$; and $z_q > F_o > R_{dh} > V$, respectively.

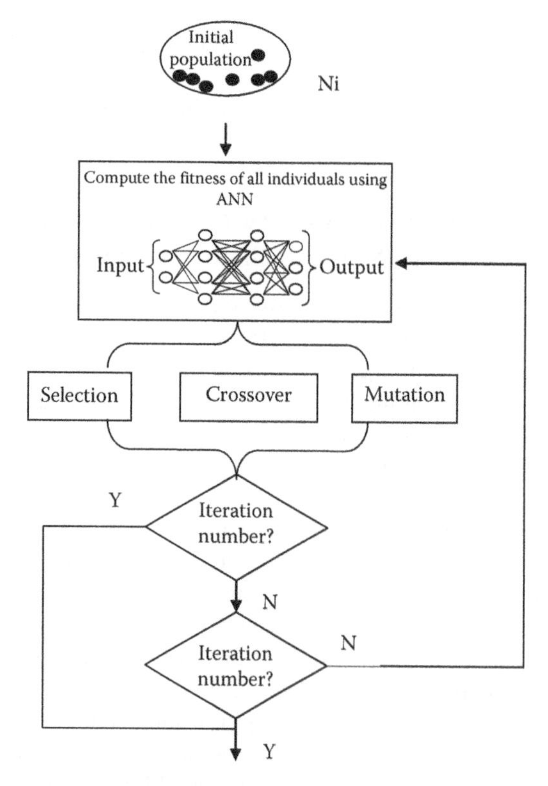

FIGURE 5.6 The procedure of the hybrid optimization method using GAs and ANNs.

5.5.4 ANN Model-Based Multiple Ramp-Variable Retort Temperature Control for Optimization of Thermal Processing

Variable retort temperature (VRT) thermal processing has been recognized as an innovative method to improve food product quality and save PTs. The key to designing a VRT thermal process is to choose a reasonable (optimal) VRT temperature profile for a given food product and package being thermally processed. The selection of optimal RT profiles with a multistage ramp function involving multiple variables is complex and difficult to be handled by conventional optimization methods [47]. The study consisted of three parts: Developing associated prediction models using ANN, sensitivity analysis of VRT parameters, investigating the sensitivity of VRT parameters to processing results, and searching for the optimal VRT temperature profile using a hybrid optimization technique coupling ANN with GA.

For the first part, three separate ANN models were developed for predictions of PT, average quality retention, and surface cook value, respectively, each as a function of five input variables: ramp time, *t*, and four step temperatures: *T*1, *T*2, *T*3, *T*4. ANN models were trained and tested by two data sets, respectively, which were generated by a computer simulation program of VRT thermal processing. The statistical results of the modeling performance for all ANN models were correlation coefficient >0.95 and average relative error <2.05%, indicating that these ANN models can be safely used for prediction purposes of the VRT thermal processing with a multiple ramp temperature profile.

In the second part, ANN model-based sensitivity analysis was for investigation of effects of five multiple ramp variable (MRV) parameters on PT, average quality retention (Qv), and surface cook value (Fs). For example, Figure 5.7 shows the effects of individual variable including four step temperatures and ramp time on process outputs: PT, Qv, and Fs for the small size. As expected normally, the increase of step temperatures resulted in the decrease of PTs (Figure 5.7a), but the decrease rate was dependent on the number of steps and temperature values. If the temperature was less than 119°C, *T*3 showed the most sensitive to PT; if larger than 119°C, then *T*2 was the most sensitive factor to PT. The effects on the quality retention Qv were illustrated in Figure 5.7b. It showed that *T*1 and *T*4 had no effect on Qv while effects of *T*2 and *T*3 were related to the temperature value. Qv increased with the temperature value of *T*2 or *T*3 until the temperature reached around 127°C and then decreased with the temperatures. The effects of step temperatures on the surface cook value Fs are presented in Figure 5.7c. It can be found that increasing temperatures *T*1 or *T*4 caused Fs increase, especially for *T*4; and for *T*2 or *T*3, the best temperature was about 119°C which had minimum surface cook value. Effects of the ramp time on process outputs are shown in Figure 5.7d through f. Basically, the increase of ramp time made the PT and the quality retention increase and the surface cook value decrease. However, the sensitivity of ramp time was dependent on the base temperature.

Coupled ANN models and GA were then used for searching the optimal RT profiles to meet the requirements of optimization objectives and constraint conditions. The typical optimal MRV profiles for the middle size (4.92×10^{-4} m^3) achieved by GA–ANN protocol were illustrated in Figure 5.8. It could be found that there were different specific MRV values for different optimization objectives and constraint conditions. By comparison of optimization objectives, it was indicated that the minimum PT used as the optimization objective needed much more ramp time than the minimum Fs as the optimization objective. For example, the ramp time was 70 min for the minimum PT with Qv $\geq$ 62%, while it was only 30 min for the minimum Fs with PT $\leq$ 74 or Qv $\geq$ 62%. From the step temperature point of view, the order of step temperatures was *T*4>*T*3>*T*2>*T*1 if the PT or Fs was used as the constraint condition while *T*4 was less than *T*3 if Qv was used as the constraint condition.

5.5.5 Analysis of Critical Control Points in Deviant Thermal Processes Using ANNs

The basic objective of thermal processing is to meet the safety requirements while trying to reduce quality degradation to a minimum. Theoretically, it is possible to design an optimal processing

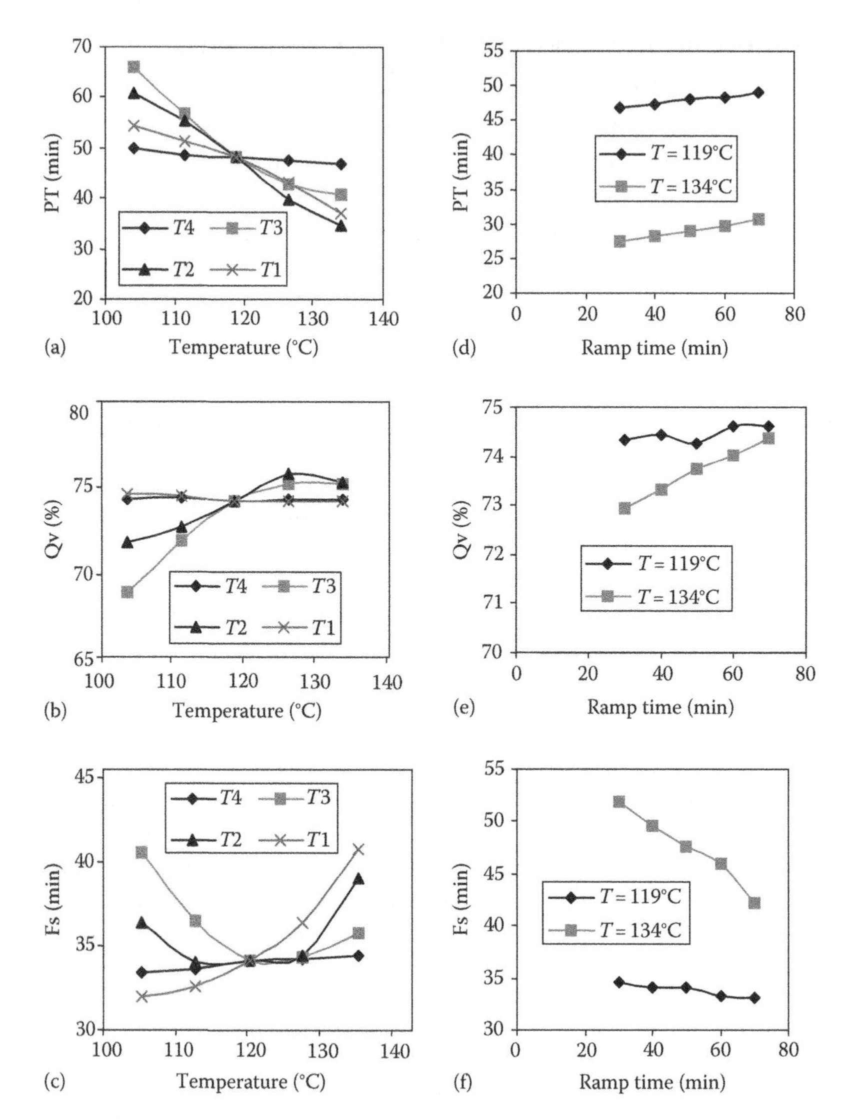

FIGURE 5.7 Effects of MRV parameters on process outputs for the small size ($V = 1.64 \times 10^{-4}$ m^3). (a) Step temperatures vs. process time. (b) Step temperatures vs. quality retention. (c) Step temperatures vs. surface cook value. (d) Ramp time vs. process time. (e) Ramp time vs. quality retention. (f) Ramp time vs. surface cook value.

protocol for any food product, but, in practice, it is difficult to obtain truly optimal results since considerable deviations exist in process parameters. In cases where deviations go beyond a certain critical level, there can be under-processing or over-processing. The former indicates that processed products cannot meet the sterility requirements for safety and consumption, while the latter means that quality destruction is more than optimal. Therefore, it is important to identify critical factors, to assess the effect of their deviations on the process calculations and to establish control actions during thermal processing to avoid process deviations.

Thermal processing is a complex system, and standard processes are established based on achieving target process lethality (F value) at a critical point, usually the package center. The required PT for a given product depends on the RT, product initial temperature (T_i), cooling water temperature (T_w), and several product related properties such as heating rate index (f_h), heating lag factor (j_h), and cooling lag factor (j_c). It is necessary to understand and estimate the influence of

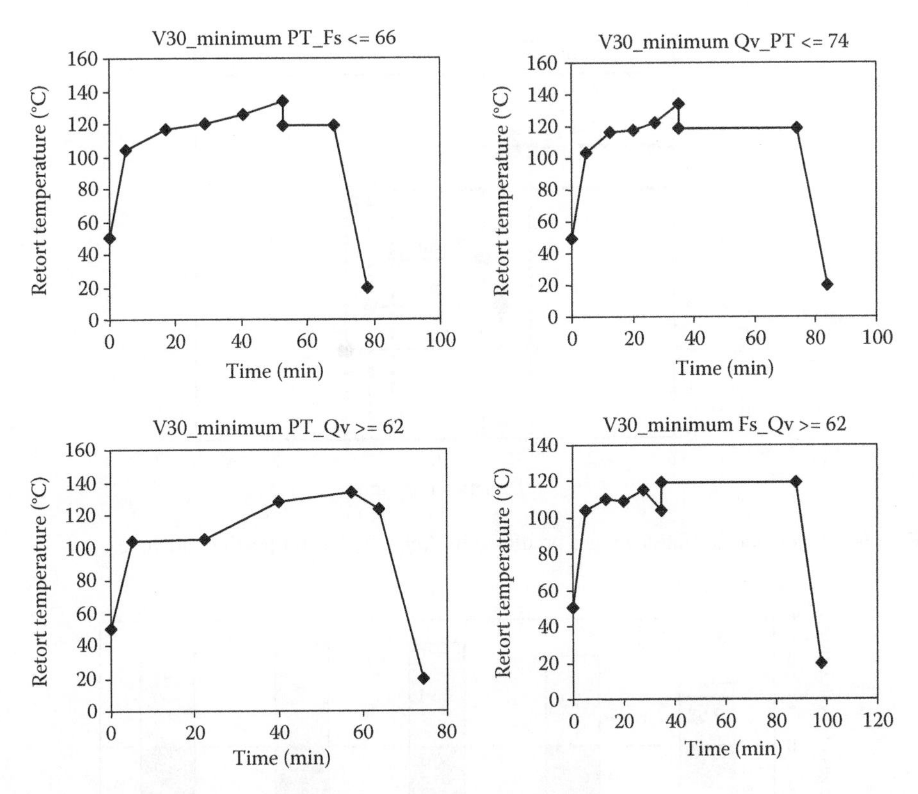

FIGURE 5.8 Optimal MRV profiles obtained by GA–ANN method.

these process parameters and the deviations from their expected values on the required PT. Chen and Ramaswamy [48] developed ANN models for (1) evaluating the relative order of importance of different critical control variables with respect to process calculations and (2) developing predictive models to compensate for their deviations. The critical variables studied were retort temperature, initial temperature, cooling water temperature, heating rate index, heating lag factor, and cooling lag factor. Their ranges of deviation from a set point were selected as −2°C to 2°C for RT, −5°C to 5°C for both T_w and T_i, −2 to 2 min for f_h, and −0.2 to 0.2 for both j_c and j_h. ANN models were developed and used for analysis of different critical variables with respect to their importance on the accumulated lethality, PT, cooling time (CT), and total time (TT) under the given processing conditions. By the use of ANN models, the relative order of importance of critical variables within the deviation ranges was as follows: for F, RT>f_h>j_h>$T_i \times T_w$>T_i>$T_i \times f_h$>RT×T_i>j_c>RT×j_h; for PT, RT>f_h>> j_h>T_i>j_c>$T_i \times T_w$; for CT, j_c>T_w>f_h; for TT, RT> f_h>j_h>j_c>T_w>T_i>$T_i \times j_c$>$T_i \times T_w$. The accepted deviation ranges for various input variables under given control ranges were predicted by NN models, one of which is shown in Figure 5.9. Based on these graphs, it can be easily determined that when the desired F value was set at 6 ± 0.5 min, the maximum acceptable deviation ranges of different variables were ±0.3°C for RT; ±4°C for T_i; ±0.1 for j_h; ±0.8, ±1, ±1.2 min for f_h at f_h=20, 40, 60 min, respectively, and ±0.4 for j_c. NN models were also used for analysis of the combination effect of multiple deviations on F, PT, and CT (shown in Figure 5.10). By the use of this graph, the maximum changes in F and PT for different deviation combinations could be easily determined.

5.6 CONCLUSIONS

As confirmed by a variety of applications reported, the modeling capability of ANNs is not question and that they can be used for complex cases with multiple variables and nonlinear relationships usually difficult for conventional methods. In the food thermal processing, application of ANN is

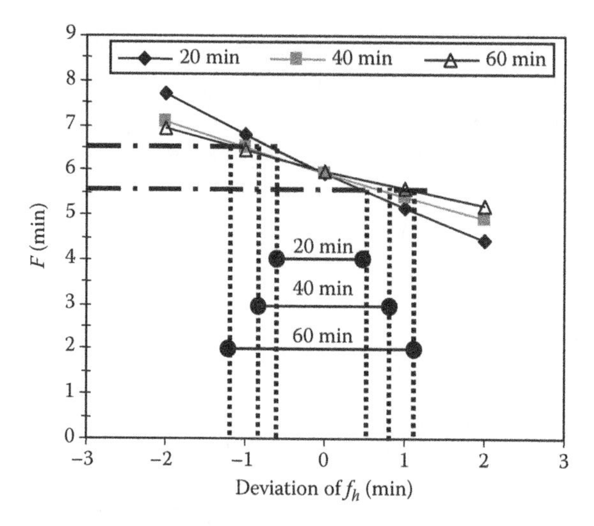

FIGURE 5.9 Acceptable deviation ranges predicted by ANN models for heating rate index, f_h.

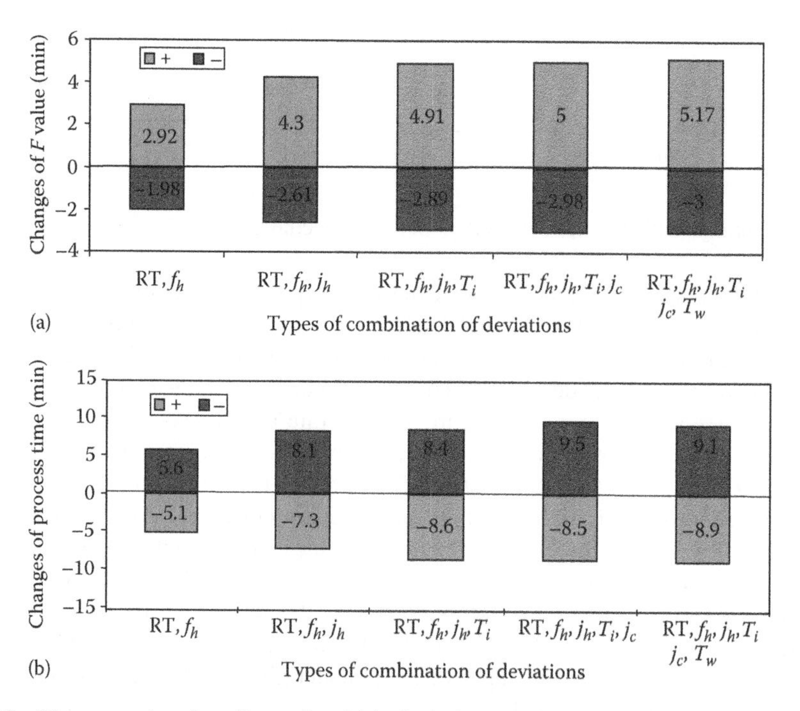

FIGURE 5.10 The comprehensive effects of multiple deviations predicted by ANN models (a) lethality value and (b) heating time.

still relatively new to other academic areas. Although a few studies have been reported as mentioned in this chapter about ANN for modeling, optimization, and critical control points (CCPs) analysis of thermal processing, most of them are still staying on the hypothesis level, meaning that these results have not been used for industrial applications. Furthermore, the online use of NNs in thermal processing area is still blank. Therefore, it leaves more rooms for researchers to make efforts on application of NNs in the thermal processing area.

It should be noted that ANNs are not without limitations. First of all, NNs work like a black box; thus, ANN models cannot give clear internal relationships between input and output variables as provided by other models based on conventional methods. Therefore, NNs should be used as a tool for practical purposes rather than theoretical ones focusing on developing and understanding the intrinsic relationships of various variables. For the practical application, the objective of developing ANN models is to be used for different purposes such as optimization, online control, identification, etc. In order to achieve this goal, NNs must be combined with other techniques, for instance, fuzzy logic, expert system, and GA or other search techniques. Therefore, the future trend for application of NNs should be on developing hybrid methods by using NNs and other techniques which can be more potential to be directly used for industrial purposes, instead of staying at the level only confirming the feasibility of ANN modeling as the most of current works have been done. Second, training ANN model needs enough data which is the most important factor affecting the performance of ANN models. It is impossible to obtain an ANN model with a good performance using a limited or bad distribution data. Thus, NNs are only suitable for problems with a large amount of experimental data, or those that can generate data using a separate computer simulator. In addition, like all other models, trained ANN models can only be used for predictions within the ranges of variable being investigated. Otherwise, the precision of prediction results by ANN models might not be guaranteed.

NOMENCLATURE

Variables

CT Cooling time, min
D Decimal deduction time, min
Dq Decimal destruction time for quality, min
E Time-specific particle concentration function or total square error
En Equivalent unit energy consumption, kJ/kg
E_r Relative average error, %
f Heating or cooling rate index, min
F Accumulated lethality value, min, or Cumulative particle concentration function
Fs Surface cook value, min
g Final temperature difference between can center and retort, °C
h transfer coefficient, W/(m^2 °C)
H Height of the can, mm
j Heat or cooling lag factor
PT Process time or heating time, min
Qv Average quality retention for whole can, %
R Correlation coefficient or ration of diameter to height of can
RT Retort temperature, °C
T Temperature, °C
T1–4 Step temperature for MRV function, °C
U Overall heat transfer, W/(m^2 °C)
V Volume, m^3
w Weight (neural network)
y Output value

Subscripts

c Cooling
di Desired output values
dh Diameter to height
fp Fluid to particle
h Heating
i Index, or initial

j	index
m	Microorganism, or mean value
max	Maximum
min	Minimum
o	Desired value
q	Quality
w	Cooling water

Greek symbols

α	Thermal diffusivity, m^2/s
ε	Error
ρ	Density, kg/m^3 or lethality ratio
θ	Threshold value (neural network)

ABBREVIATIONS

ANN	Artificial neural network
CCP	Critical control point
CRT	Constant retort temperature
GA	Genetic algorithm
MRV	Multiple ramp variable
RTD	Residence time distribution
VRT	Variable retort temperature

REFERENCES

1. WS McCulloch, W Pitts. A logical calculus of the ideas immanent in nervous activity. *Bulletin of Mathematical Biophysics* 5: 115–133, 1943.
2. Anon. *Neural Computing: A Technology Handbook for Professional II/PLUS and NeuralWorks Explorer*. Pittsburgh, PA: NeuralWare, 1993, pp. 1–25.
3. Anon. *Reference Guide: Software Reference for Professional II/PLUS and NeuralWorks Explorer*. Pittsburgh, PA: NeuralWare, 1993, pp. 20–70.
4. S Haykin. *Neural Networks: A Comprehensive Foundation*. New York: Macmillan College Publishing Company, 1994, pp. 65–80.
5. JJ Hopfield. Neural networks and physical systems with emergent collective computational abilities. *Proceedings of the National Academy of Sciences USA*. 79: 2554–2558, 1982.
6. PS Neelakanta, DFD Groff. *Neural Network Modeling: Statistical Mechanics and Cybernetic Perspectives*. Boca Raton, FL: CRC Press, 1994, pp. 33–67.
7. MA Hussain, M Shafiur-Rahman, CW Ng. Prediction of pores formation (porosity) in foods during drying: Generic models by the use of hybrid neural network. *Journal of Food Engineering* 51(3): 239–248, 2002.
8. CR Chen, HS Ramaswamy, I Alli. Prediction of quality changes during Osmo-convective drying of blue berries using neural network models for process optimization. *Drying Technology* 19(3–4): 507–523, 2001.
9. S Sreekanth, HS Ramaswamy, S Sablani. Prediction of psychrometric parameters using neural networks. *Drying Technology* 16(3–5): 825–837, 1998.
10. W Kaminski, P Strumillo, E Tomczak. Genetic algorithms and artificial neural networks for description of thermal processes. *Drying Technology* 14: 2117–2133, 1996.
11. W Kaminski, J Stawczyk, E Tomczak. Presentation of drying kinetics in a fluidized bed by means of radial basis functions. *Drying Technology* 15(6–8): 1753–1762, 1997.
12. W Kaminski, P Strumillo, E Tomczak. Neuro-computing approaches to modeling of drying process dynamics. *Drying Technology* 16(6): 967–992, 1998.
13. JA Hernandez. Optimum operation conditions for heat and mass transfer in foodstuffs drying by means of neural inverse. *Food Control* 20: 435–438.
14. PP Tripathy, S Kumar. Neural network approach for food temperature prediction during solar drying. *International Journal of Thermal Sciences* 48: 1452–1459.

15. A Kosola, P Linko. Neural control of fed-batch baker's yeast fermentation. *Development of Food Science* 36: 321–328, 1994.
16. H Honda, T Hanai, A Katayama, H Tohyama, T Kobayashi. Temperature control of Ginjo sake brewing process by automatic fuzzy modeling using fuzzy neural networks. *Journal of Fermentation and Bioengineering (Japan)* 85(1): 107–112, 1998.
17. O Popescu, DC Popescu, J Wilder, MV Karwe. A new approach to modeling and control of a food extrusion process using artificial neural network and an expert system. *Journal of Food Process Engineering* 24(1): 17–26, 1998.
18. T Eerikainen, YH Zhu, P Linko. Neural networks in extrusion process identification and control. *Food Control* 5(2): 111–119, 1994.
19. G Ganjyal, M Hanna. A review on residence time distribution (RTD) in food extruders and study on the potential of neural networks in RTD modeling. *Journal of Food Science* 67(6): 1996–2002, 2002.
20. TJ Shankar, S Bandyopadhyay. Prediction of extrudate properties using artificial neural networks. *Food and Bioproducts Processing* 89: 29–33, 2007.
21. GS Mittal, JX Zhang. Prediction of freezing time for food products using a neural network. *Food Research International* 33(7): 557–562, 2000.
22. Q Fang, G Bilby, E Haque, MA Hanna, CK Spillman. Neural network modeling of physical properties of ground wheat. *Cereal Chemistry* 75(2): 251–253, 1998.
23. SS Sablani, OD Baik, M Marcotte. Neural network prediction of thermal conductivity of bakery products. *Journal of Food Engineering* 52: 299–304, 2002.
24. T Morimoto, JDe Baerdemaeker, Y Hashimoto. An intelligent approach for optimal control of fruit storage process using neural networks and genetic algorithms. *Computers and Electronics in Agriculture* 18: 205–224, 1997.
25. T Moritmoto, W Purwanto, J Suzuki, Y Hashimoto. Optimization of heat treatment for fruit during storage using neural networks and genetic algorithms. *Computers and Electronics in Agriculture* 19: 87–101, 1997.
26. S Sreekanth, CR Chen, SS Sablani, HS Ramaswamy, SO Prasher. Neural network assisted experimental designs for food research. *Agricultural Sciences, Journal for Scientific Research, SQU (Oman)* 5(2): 97–106, 2000.
27. SS Sablani, HS Ramaswamy, SO Prasher. Neural network applications in thermal processing optimization. *Journal of Food Processing and Preservation* 19(4): 283–301, 1995.
28. SS Sablani, HS Ramaswamy, S Sreekanth, SO Prasher. Neural network modeling of heat transfer to liquid particle mixtures in cans subjected to end-over-end processing. *Food Research International* 30(2): 105–116, 1997.
29. Y Meng, HS Ramaswamy. Neural network modeling of end-over-end thermal processing of particulates in viscous fluids. *Journal of Food Process Engineering* 33: 23–47, 2009.
30. J Manickaraj, N Balasubramanian. Estimation of the heat transfer coefficient in a liquid–solid fluidized bed using an artificial neural network. *Advanced Powder Technology* 19: 119–130, 2008.
31. CR Chen, HS Ramaswamy. Neural computing approach for modeling of residence time distribution (RTD) of carrot cubes in a vertical scraped surface heat exchanger (SSHE). *Food Research International* 33(7): 549–556, 2000.
32. S Kerdpiboon, WL Kerr, S Devahastin. Neural network prediction of physical property changes of dried carrot as a function of fractal dimension and moisture content. *Food Research International* 39: 1110–1118, 2006.
33. H Kurt, M Kayfeci. Prediction of thermal conductivity of ethylene glycol–water solutions by using artificial neural networks. *Applied Energy* 86: 2244–2248, 2009.
34. SS Sablani, MS Rahman. Using neural networks to predict thermal conductivity of food as a function of moisture content, temperature and apparent porosity. *Food Research International* 36: 617–623, 2003.
35. R Eslamloueyan, MH Khademi. Estimation of thermal conductivity of pure gases by using artificial neural networks. *International Journal of Thermal Sciences* 48: 1094–1101, 2009.
36. D Bas, FC Dudak, IH Boyac. Modeling and optimization III: Reaction rate estimation using artificial neural network (ANN) without a kinetic model. *Journal of Food Engineering* 79: 622–628, 2007.
37. W Lou, S Nakai. Application of artificial neural networks for predicting the thermal inactivation of bacteria: A combined effect of temperature, pH and water activity. *Food Research International* 34: 573–579, 2001.
38. A Esnoz, PM Periago, R Conesa, A Palop. Application of artificial neural networks to describe the combined effect of pH and NaCl on the heat resistance of *Bacillus stearothermophilus*. *International Journal of Food Microbiology* 106: 153–158, 2006.

39. M O'Farrell, E Lewis, C Flanagan, WB Lyons, N Jackman. Combining principal component analysis with an artificial neural network to perform online quality assessment of food as it cooks in a large-scale industrial oven. *Sensors and Actuators B: Chemical* 107: 104–112, 2005.
40. M Afaghi, HS Ramaswamy, SO Prasher. Thermal process calculations using artificial neural network models. *Food Research International* 34: 55–65, 2001.
41. SS Sablani, WH Shayya. Computerization of Stumbo's method of thermal process calculations using neural networks. *Journal of Food Engineering* 47: 233–240, 2001.
42. EC Gonçalves, LA Minim, JSR Coimbra, VPR Minim. Modeling sterilization process of canned foods using artificial neural networks. *Chemical Engineering and Processing* 44: 1269–1276, 2005.
43. CR Chen. Application of computer simulation and artificial intelligence technologies for modeling and optimization of food thermal processing, PhD thesis, McGill University, Montreal, Quebec, Canada, 2001.
44. CR Chen, HS Ramaswamy. Prediction and optimization of constant retort temperature (CRT) processing using neural network and genetic algorithms. *Journal of Food Processing Engineering* 25(5): 351–380, 2002.
45. CR Chen, HS Ramaswamy. Prediction and optimization of variable retort temperature (VRT) processing using neural network and genetic algorithms. *Journal of Food Engineering* 53: 209–220, 2002.
46. CR Chen, HS Ramaswamy. Dynamic modeling of retort thermal processing using neural networks. *Journal of Food Processing and Preservation* 26(2): 91–112, 2002.
47. CR Chen, HS Ramaswamy. Multiple Ramp-Variable (MRV) retort temperature control for optimization of thermal processing. *Food and Bioproducts Processing* 82(C1): 1–11, 2004.
48. CR Chen, HS Ramaswamy. Analysis of critical control points for deviant thermal processing using artificial neural networks. *Journal of Food Engineering* 57: 225–235, 2002.
49. S Kupongsak, J Tan. Application of fuzzy set and neural network techniques in determining food process control set points. *Fuzzy Sets and Systems* 157: 1169–1178, 2006.
50. JS Torrecilla, L Otero, PD Sanz. Optimization of an artificial neural network for thermal/pressure food processing: Evaluation of training algorithms. *Computers and Electronics in Agriculture* 56: 101–110, 2004.
51. JS Torrecilla, L Otero, PD Sanz. A neural network approach for thermal/pressure food processing. *Journal of Food Engineering* 62: 89–95, 2005.

6 Modeling Thermal Processing Using Computational Fluid Dynamics (CFD)

Xiao Dong Chen and Da-Wen Sun

CONTENTS

6.1 INTRODUCTION

In the food industry, thermal processing normally refers the processes that heat, hold, and cool a product sequentially, which is required to be free of food-borne illness for a desired period of time. There are many types of thermal processes; among them, pasteurization is a typical one, which reduces the potential of contamination of a special pathogenic microorganism to a predesigned extent. The product will still need to be refrigerated; otherwise, it will not be shelf stable. Sterilization is another example of thermal processes that leads to shelf-stable products in cans, soft containers, or bottles [1]. This process usually employs a much greater temperature than pasteurization.

6.2 BASIC THERMAL PROCESSING PARAMETERS

6.2.1 Decimal Reduction Time D

When a living microorganism population, such as *Escherichia coli*, is subjected to thermal processing at constant temperature (T), its population will reduce. A typical plot of the microbial population over time (N versus t) usually shows an "exponential like" trend. A semilog plot of N versus t may be correlated using a linear fit, yielding a straight line with a negative slope ($-D$):

$$D = \frac{t}{\log N_0 - \log N}, \tag{6.1}$$

where D is called the "decimal reduction time."

In other words, the microbial population reduction may be expressed as

$$\frac{N}{N_0} = 10^{-t/D}. \tag{6.2}$$

Obviously, at different T, D would be different. The higher the temperature, the smaller the D value, and this means that the microorganisms are more vulnerable in a hotter environment. Some typical values of D are given in Table 6.1.

If a liquid is uniformly heated, held, and cooled through three steps as shown in Figure 6.1, the residual live population N_3 may be approximated with the following calculations:

$$N_1 = N_0 \times 10^{-\Delta t_1/\bar{D}_1} \tag{6.3}$$

$$N_2 = N_1 \times 10^{-\Delta t_2/\bar{D}_2} \tag{6.4}$$

$$N_3 = N_2 \times 10^{-\Delta t_1/\bar{D}_3} \tag{6.5}$$

Therefore, combining the previous three equations, one obtains the following:

$$N_3 = N_0 \times 10^{-\left(\frac{\Delta t_1}{\bar{D}_1}+\frac{\Delta t_2}{\bar{D}_2}+\frac{\Delta t_3}{\bar{D}_3}\right)}, \tag{6.6}$$

TABLE 6.1
Typical Z and D_T Values

Microorganism	Z (°C)	D_{121} (min)	Products
Bacillus stearothermophilus	10	4	Vegetables, milk
Bacillus subtilis	4.1–7.2	0.5–0.76	Milk products

Source: Fellows, P.J., *Food Processing Technology (Principles and Practice)*, Woodhead Publishing, Series in Food Science and Technology, Cambridge, U.K., 1996.

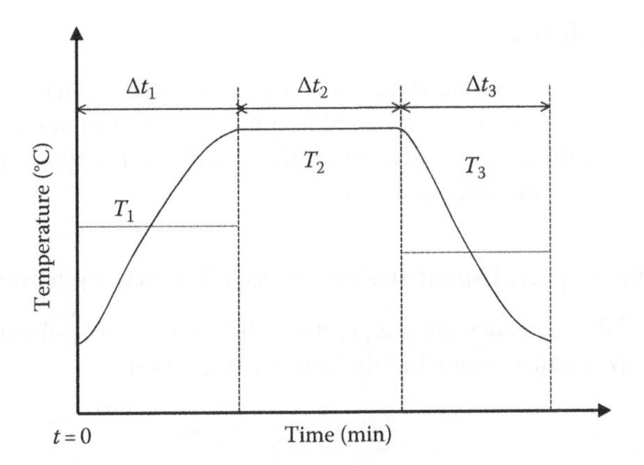

FIGURE 6.1 A typical heating, holding, and cooling curve of a fluid.

where

$\bar{D}_1$ is taken as the value when $T = T_1$
$\bar{D}_2$ is taken as the value when $T = T_2$
$\bar{D}_3$ is taken as the value when $T = T_3$

If the heating region and the cooling period can be approximated with more than one single step taking average T for each period, that is, for the heating period, there are m steps, and for the cooling period, n steps,

$$N_3 = N_0 \times 10^{-\left(\sum_{i=1}^{i=m} \frac{\Delta t_{1,i}}{\bar{D}_{1,i}} + \frac{\Delta t_2}{\bar{D}_2} + \sum_{j=1}^{j=n} \frac{\Delta t_{3,j}}{\bar{D}_{3,j}}\right)}. \tag{6.7}$$

Different parts of the fluid being processed would experience different temperature-time histories, for each part, Equation 6.7 is needed to calculate the individual bacterial deactivation histories.

For example, if the fluid is thermally processed in a pipe (i.e., a tubular heat exchanger), the maximum (center line velocity) is twice the average value at the laminar regime. For the turbulence regime, the maximum velocity is about 1.2 times the average velocity. As such, the turbulence regime provides a more uniform mixing condition for thermal processing. The average residence time is the length of the pipe divided by the average velocity; thus, the microbes that travel at the maximum velocity would not be fully processed.

This is related to "aseptic processing," where food is sterilized or pasteurized in a tubular, helical heat exchanger, scraped surface heat exchanger, microwave, or ohmic heater. The aseptic process has significant quality advantages over classical thermal techniques such as batch (canning) and semibatch operations. In general, flow modeling has to be used to ensure that food products meet the safety requirements. It determines the optimal lengths of the heating, holding, and cooling sections.

6.2.2 Thermal Resistance Constant Z

The thermal resistance constant Z is a parameter representing the microorganism's resistance to temperature rise:

$$Z = \frac{T_2 - T_1}{\log(D_{T_1} / D_{T_2})}. \tag{6.8}$$

6.2.3 Thermal Death Time F

The thermal death time F is the time required to cause a stated reduction in the population of microorganisms or spores. This time is expressed as a multiple of D values. For instance, a 99.99% reduction in microbial population is equivalent to $4D$. Usually, F is expressed as F_T^z for a specific temperature T and a thermal resistance constant Z.

6.2.4 Relationships between Chemical Kinetics and Thermal Processing Parameters

It is generally accepted that at a constant temperature, the microbial population or number concentration (microbe m^{-3}) (N) reduces following the first-order reaction:

$$\frac{\mathrm{d}N}{\mathrm{d}t} = -kN. \tag{6.9}$$

Therefore, the solution for this is

$$\ln\frac{N}{N_0} = -kt. \tag{6.10}$$

Comparing Equation 6.10 with Equation 6.2, it is not difficult to arrive at

$$k = \frac{2.303}{D}. \tag{6.11}$$

During thermal processing, the microbial population is not uniformly distributed within the processing fluid, and also the extent of the deactivation is different at different locations and different timing.

Equation 6.9 may also be directly expressed as

$$\frac{\mathrm{d}c}{\mathrm{d}t} = -kc, \tag{6.12}$$

where c is the mass concentration of the live microbes ($c = m_{microbe}N$, where $m_{microbe}$ is the mass of one microbe).

6.3 FUNDAMENTAL CONSERVATION EQUATIONS FOR COMPUTATIONAL FLUID DYNAMICS

The essential aspect of the computational fluid dynamics (CFD) approach is that regardless of the software being used, all need to solve the governing partial differential equations for continuity, momentum, energy, and mass balances.

For momentum transfer, the governing set of the equations is the Navier–Stokes equations. The fluids of practical interest are usually considered to be noncompressible, except the cases when the extreme high-pressure technology is applied. As such, only the governing equations for incompressible fluids for two typical coordinate systems are described in this section [3].

6.3.1 Cartesian Coordinate System

Continuity

$$\frac{\partial u}{\partial x} + \frac{\partial v}{\partial y} + \frac{\partial w}{\partial z} = 0 \tag{6.13}$$

Momentum

$$\rho\left(\frac{\partial u}{\partial t}+u\frac{\partial u}{\partial x}+v\frac{\partial u}{\partial y}+w\frac{\partial u}{\partial z}\right)=\rho F_x-\frac{\partial p}{\partial x}+\mu\left(\frac{\partial^2 u}{\partial x^2}+\frac{\partial^2 u}{\partial y^2}+\frac{\partial^2 u}{\partial z^2}\right) \tag{6.14}$$

$$\rho\left(\frac{\partial v}{\partial t}+u\frac{\partial v}{\partial x}+v\frac{\partial v}{\partial y}+w\frac{\partial v}{\partial z}\right)=\rho F_y-\frac{\partial p}{\partial y}+\mu\left(\frac{\partial^2 v}{\partial x^2}+\frac{\partial^2 v}{\partial y^2}+\frac{\partial^2 v}{\partial z^2}\right) \tag{6.15}$$

$$\rho\left(\frac{\partial u}{\partial t}+u\frac{\partial w}{\partial x}+v\frac{\partial w}{\partial y}+w\frac{\partial w}{\partial z}\right)=\rho F_z-\frac{\partial p}{\partial z}+\mu\left(\frac{\partial^2 w}{\partial x^2}+\frac{\partial^2 w}{\partial y^2}+\frac{\partial^2 w}{\partial z^2}\right) \tag{6.16}$$

By ignoring the viscous dissipation effect, the internal heat generation or dissipation is due to exothermic or endothermic reactions, the energy balance may be written as

$$\frac{\partial T}{\partial t}+u\frac{\partial T}{\partial x}+v\frac{\partial T}{\partial y}+w\frac{\partial T}{\partial z}=\frac{1}{\rho C_p}\left[\frac{\partial}{\partial x}\left(k\frac{\partial T}{\partial x}\right)+\frac{\partial}{\partial y}\left(k\frac{\partial T}{\partial y}\right)+\frac{\partial}{\partial z}\left(k\frac{\partial T}{\partial z}\right)\right], \tag{6.17}$$

where the thermal conductivity k can be temperature or concentration dependent.

The mass conservation for microbial species and nutrient species may be expressed as

$$\frac{\partial c}{\partial t}+u\frac{\partial c}{\partial x}+v\frac{\partial c}{\partial y}+w\frac{\partial c}{\partial z}=\frac{\partial}{\partial x}\left(D\frac{\partial c}{\partial x}\right)+\frac{\partial}{\partial y}\left(D\frac{\partial c}{\partial y}\right)+\frac{\partial}{\partial z}\left(D\frac{\partial c}{\partial z}\right)+\dot{m}. \tag{6.18}$$

The mass diffusivity D can be a function of temperature and concentration, and $\dot{m}$ is determined by:

$$\dot{m}=-kc \tag{6.19}$$

for either kind of species.

Equation 6.18 is valid only for the small microbes that are assumed to follow nicely the bulk fluid movement (relative velocity is zero).

When food particles are large enough, their movement must be traced using the Lagrange method (this aspect is not included in this chapter for simplicity).

6.3.2 Cylindrical Coordinate System

Continuity

$$\frac{\partial u_r}{\partial r}+\frac{1}{r}\frac{\partial u_\theta}{\partial \theta}+\frac{\partial u_z}{\partial z}+\frac{u_r}{r}=0 \tag{6.20}$$

Momentum

$$\rho\left(\frac{\partial u_r}{\partial t}+u_r\frac{\partial u_r}{\partial r}+\frac{u_\theta}{r}\frac{\partial u_r}{\partial \theta}+u_z\frac{\partial u_r}{\partial z}-\frac{u_\theta^2}{r}\right)$$

$$=\rho F_r-\frac{\partial p}{\partial r}+\mu\left(\frac{\partial^2 u_r}{\partial r^2}+\frac{1}{r}\frac{\partial u_r}{\partial r}+\frac{1}{r^2}\frac{\partial^2 u_r}{\partial \theta^2}+\frac{\partial^2 u_r}{\partial z^2}-\frac{2}{r^2}\frac{\partial u_\theta}{\partial \theta}-\frac{u_r}{r^2}\right) \tag{6.21}$$

$$\rho\left(\frac{\partial u_\theta}{\partial t}+u_r\frac{\partial u_\theta}{\partial r}+\frac{u_\theta}{r}\frac{\partial u_\theta}{\partial \theta}+u_z\frac{\partial u_\theta}{\partial z}-\frac{u_r u_\theta}{r}\right)$$

$$=\rho F_\theta-\frac{1}{r}\frac{\partial p}{\partial \theta}+\mu\left(\frac{\partial^2 u_\theta}{\partial r^2}+\frac{1}{r}\frac{\partial u_\theta}{\partial r}+\frac{1}{r^2}\frac{\partial^2 u_\theta}{\partial \theta^2}+\frac{\partial^2 u_\theta}{\partial z^2}+\frac{2}{r^2}\frac{\partial u_r}{\partial \theta}-\frac{u_\theta}{r^2}\right) \tag{6.22}$$

$$\rho\left(\frac{\partial u_z}{\partial t}+u_r\frac{\partial u_z}{\partial r}+\frac{u_\theta}{r}\frac{\partial u_z}{\partial \theta}+u_z\frac{\partial u_z}{\partial z}\right)=\rho F_z-\frac{\partial p}{\partial z}+\mu\left(\frac{\partial^2 u_z}{\partial r^2}+\frac{1}{r}\frac{\partial u_z}{\partial r}+\frac{1}{r^2}\frac{\partial^2 u_z}{\partial \theta^2}+\frac{\partial^2 u_z}{\partial z^2}\right) \tag{6.23}$$

By ignoring the viscous dissipation effect, the energy balance may be written as

$$\frac{\partial T}{\partial t}+u_r\frac{\partial T}{\partial r}+\frac{u_\theta}{r}\frac{\partial T}{\partial \theta}+u_z\frac{\partial T}{\partial z}=\frac{1}{\rho C_p}\left[\frac{1}{r}\frac{\partial}{\partial r}\left(kr\frac{\partial T}{\partial r}\right)+\frac{1}{r}\frac{\partial}{\partial \theta}\left(k\frac{1}{r}\frac{\partial T}{\partial \theta}\right)+\frac{\partial}{\partial z}\left(k\frac{\partial T}{\partial z}\right)\right] \tag{6.24}$$

The mass conservation is given as

$$\frac{\partial c}{\partial t}+u_r\frac{\partial c}{\partial r}+\frac{u_\theta}{r}\frac{\partial c}{\partial \theta}+u_z\frac{\partial c}{\partial z}=\left[\frac{1}{r}\frac{\partial}{\partial r}\left(Dr\frac{\partial T}{\partial r}\right)+\frac{1}{r}\frac{\partial}{\partial \theta}\left(D\frac{1}{r}\frac{\partial T}{\partial \theta}\right)+\frac{\partial}{\partial z}\left(D\frac{\partial T}{\partial z}\right)\right]+\dot{m} \tag{6.25}$$

Body-fitted coordinate systems are also available in CFD packages but the details are beyond the content of this chapter [3].

6.4 BOUNDARY AND INITIAL CONDITIONS

6.4.1 Velocity Boundary Conditions

The solid wall conditions are that all the velocities at wall (fluid–solid interface) are zero. Inlet and outlet of a system need to be specified with the velocity or mass flow rate values.

6.4.2 Thermal Boundary Conditions

Thermal boundary conditions are in general for one of the following situations: liquid–solid, liquid–liquid (nonimmiscible), or liquid–gas interfaces.

The boundary or interfacial temperature, being constant, may be specified, such as the cases when the convective heat transfer coefficient is very large or when the heat transfer coefficient is naturally large, for example, condensation heat transfer in retort processing, which will be illustrated in detail later.

In general, the interfacial condition would be the conservation of the heat flux from one side (I) to the other (II). For instance, in 1D situation (e.g., the interface is perpendicular to the x direction), one can obtain the following equation:

$$-k_I \left.\frac{\partial T}{\partial x}\right|_I = -k_{II} \left.\frac{\partial T}{\partial x}\right|_{II}. \tag{6.26}$$

To be more general, for the arbitrarily shaped interface, one needs to express the heat flux going through the interface at the direction perpendicular to the local interface.

In cases when one side can be approximated using the *Nusselt* number approach (i.e., $Nu = h \cdot d/k$) to convection heat transfer, Equation 6.26 may be rewritten as [4]

$$-k_I \left.\frac{\partial T}{\partial x}\right|_I = h\left(T_s - T_\infty\right), \tag{6.27}$$

where

T_s is the interface temperature

T_∞ is the bulk temperature in region II

If one side (II) is well insulated (e.g., when an air bubble is considered to exist in a metal can, its insulation effect must be accounted for), Equation 6.26 gives

$$\left.\frac{\partial T}{\partial x}\right|_I = 0 \tag{6.28}$$

6.4.3 Mass Transfer Boundary Conditions

In general, the interfacial condition would be the conservation of the mass flux from one side (I) to the other (II). For instance, in 1D situation (e.g., the interface is perpendicular to the x direction), one can obtain the following equation [4]:

$$-D_I \left.\frac{\partial c}{\partial x}\right|_I = -D_{II} \left.\frac{\partial c}{\partial x}\right|_{II}. \tag{6.29}$$

Note here that the partitioning effect is not considered for simplicity. To be more general, for the arbitrarily shaped interface, one needs to express the mass flux going through the interface at the direction perpendicular to the local interface.

In cases when one side can be approximated using the *Sherwood* number approach (i.e., $Sh = h_m \cdot d/D$) to convention mass transfer, Equation 6.29 may be written as

$$-D_I \left.\frac{\partial c}{\partial x}\right|_I = h_m\left(c_s - c_\infty\right), \tag{6.30}$$

where

c_s is the interface concentration

c_∞ is the bulk concentration in region II

If one side (II) is impermeable, Equation 6.29 gives

$$\left.\frac{\partial c}{\partial x}\right|_I = 0. \tag{6.31}$$

The initial conditions inside a container, in Cartesian coordinate for instance, are usually

$$u = v = w = 0, T = T_0, \text{ and } c = c_0. \tag{6.32}$$

It does however depend on whether the flow in-and-out system is considered or not. If there is flow in and out, then a fully developed flow is usually imposed before the heating or cooling starts.

6.5 SOLUTION METHODS

The methods of grid (or grid block) generation/formation and differential equation discretization for solving the sets of equations mentioned earlier have been described by many authors (e.g., the work by Patankar and Spalding [5]). Though the principles may be similar, the details are not given here as they vary between CFD packages. In food process engineering research and design, existing commercially available CFD packages are commonly employed.

The most notable CFD packages to date are the following:

- PHOENICS (www.cham.co.uk)
- ANSYS CFX (http://www.ansys.com/)
- ANSYS FLUENT (http://www.ansys.com/)

FIDAP was one of the FLUENT series, specializing in optimizing continuous sterilization process, for example, the work by Jung and Fryer [6]. FLUENT series has been used to simulate beer bottle sterilization process, which is similar to that of the sterilization of the can (described in detail later).

The inputs of these computations include the geometrical data (size, shape, etc.) and physical and chemical properties.

6.6 WORKED EXAMPLES

The sterilization of the canned or pouched food liquid is a good example to investigate the interactions of the fluid movement, heat transfer, and species transfer using CFD.

The existence and whereabouts of the so-called slowest heating zone (SHZ) or the coldest zone (CZ), where the temperature of the fluid is the lowest, and its reduction over sterilization time is one of the primary subjects. Due to the fluid movement, stagnation zone (SZ) (usually within a recirculation zone) is also of primary interest as the microbes or nutrient species would not be transported easily out of SZ. Therefore, the movement and size of CZ and the evolution and location of SZ and the interactions of the two are important subjects for CFD analysis to explore, for example, the work by Datta and Teixeira [7,8] and Kumar et al. [9].

If the filled material in a can or a pouch is basically solid, there would be no fluid movement. In this case, heat conduction and species diffusion within the solid matrix can be readily resolved according to classical theories and using relatively simple numerical procedures. Previous studies that used conduction mode as the only heat transfer mechanism pointed to the CZ being the geometrical center. For a vertically placed cylindrical can in a retort device, CZ or the coldest point (CP) is located right at the center of the central axis.

When the fluid flow is involved, CFD is required, under 2D or 3D situations. The process of can sterilization must involve CFD in order to locate the SZs and CZs. Furthermore, the species transfer involves fluid movement effect, which needs to be accounted for in CFD simulations. Though CFD is powerful in the ease of providing the details of the flow, temperature, and concentration fields, the interpretation of the simulation results and laboratory validations are all very important aspects of the CFD modeling exercises.

For either can sterilization or pouch sterilization processes, one must look up the important CFD and experimental studies carried out by Ghani [10] and Ghani et al. [11–26]. In this series of work, the details of the fluid, heat and mass transfer, and their interactions have been demonstrated.

In this chapter, the vertically placed cylindrical can (which is filled completely by a liquid) is considered as an example. In the early studies of Ghani and coworkers [10,11], the main features of the fluid flow pattern and temperature distribution and their evolution during sterilization have been demonstrated. One typical result using sodium carboxy-methyl cellulose (CMC) as a sample viscous fluid is shown in Figure 6.2. There are two circulation regions that can be identified (Figure 6.2a): one is near the side of the cylindrical section, and the other is located at the center region of the bottom surface.

The occurrences of these two regions result in the kind of temperature distribution shown in Figure 6.2b, which can be interpreted as shown in Figure 6.3.

Figure 6.3a shows that when only the bottom plate is heated and the other parts of the container are colder, the circulation starts from the fluid rising from the bottom plate surface. Figure 6.3b shows that if the bottom plate is colder and the wall of the cylindrical section is heated (the top may also be heated), then the circulation should start from the hot sidewall surface and move up to the top due to buoyancy effect and then drop down to the lower region due to the gravity effect. As such, when all the sides are heated including the bottom plate, there are at least two recirculation regions that compete against each other, forming the pattern shown in Figure 6.3c. For different physical properties of the fluid of concern, and the variation in the height-to-diameter ratio, the ratio of the size of the side region circulation to that of the bottom region should vary.

The previous interpretation is schematically shown in Figures 6.3 and 6.4. Figure 6.5 clearly shows how the velocity generated affects the location of the lowest temperature region (SHZ or CZ).

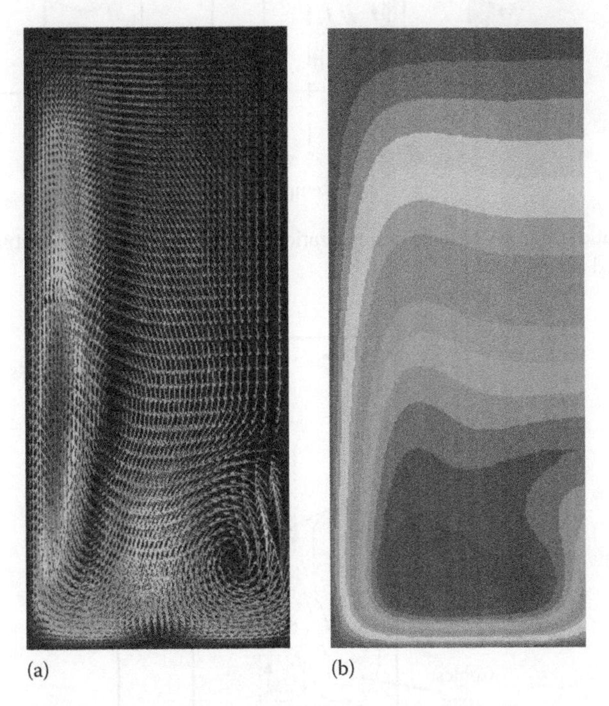

FIGURE 6.2 Velocity contours of a can filled with CMC after being heated by steam condensation for 1157 s (the left side is the can side wall (a); the right side is the center axis of the can (b)). (From Ghani, A.G., Thermal sterilization of canned liquid foods, PhD thesis, Department of Chemical and Materials Engineering, The University of Auckland, New Zealand, 2002.)

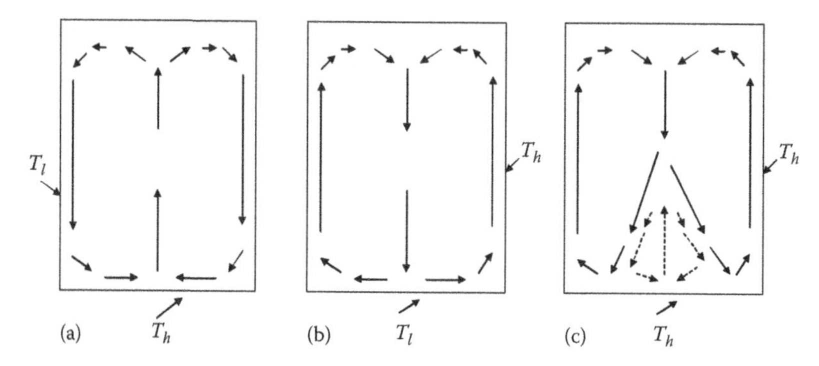

FIGURE 6.3 An interpretation of the flow pattern formation shown in Figure 6.2: (a) only the bottom plate is heated and the other parts of the container are colder, (b) the bottom plate is colder and the wall of the cylindrical section is heated (the top may also be heated), and (c) all the sides are heated including the bottom plate. (From X.D. Chen, unpublished analysis, 2004.)

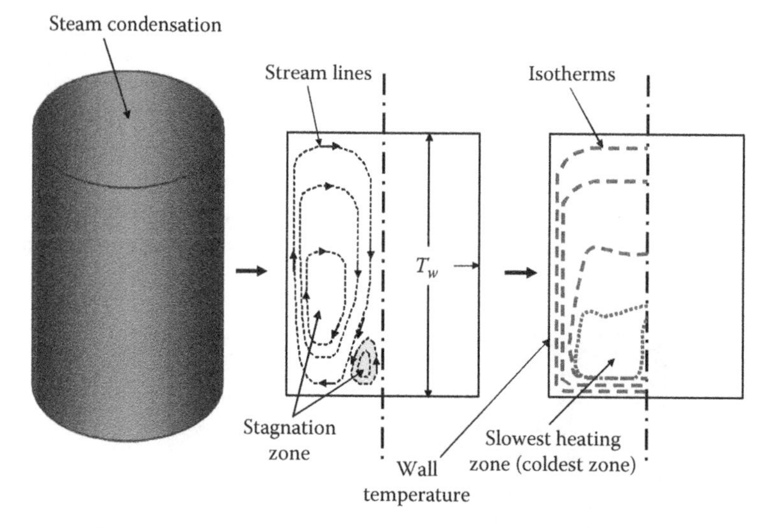

FIGURE 6.4 Schematic summary of the can sterilization process (fluid and temperature interactions). (From X.D. Chen, unpublished analysis, 2004.)

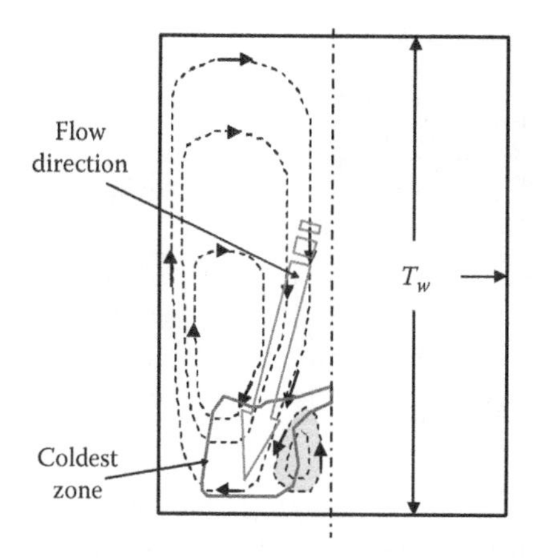

FIGURE 6.5 Illustration of the fluid flow direction and the formation of the coldest zone. (From X.D. Chen, unpublished results, 2004.)

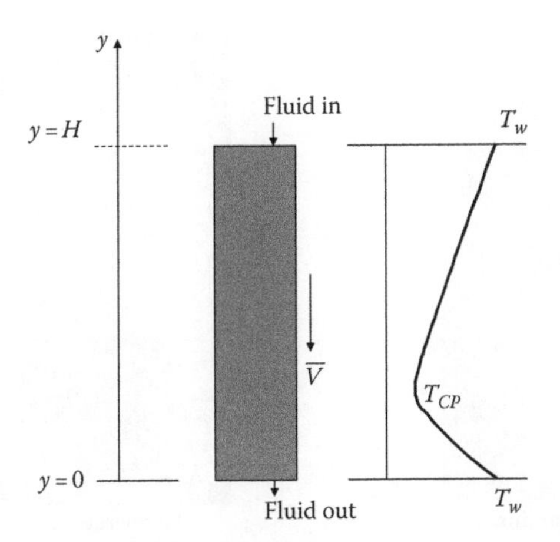

FIGURE 6.6 An equivalent 1D system simulating the effect of natural convection on the development of SHZ or CZ. (From X.D. Chen, unpublished results, 2004.)

As is evident in Figure 6.4, the main fluid flow pathway corresponds nicely the whereabouts of the CZ. Figure 6.5 shows more explicitly such a phenomenon, while the effect of natural convection may be grossly viewed in a 1D manner, that is, a 1D heat transfer affected by an overall downward fluid (at a velocity generated due to natural convection effect; see Figure 6.6). This velocity is determined by the Grashof (*Gr*) number and Prandtl (*Pr*) number [27,28].

A wide range of situations and geometries have been examined in the literature. Figure 6.7 shows an example of vertically placed can sterilization. The can sizes were 40.8 mm in radius and 111 mm in height. The basic equations used are the following. Note that 2D situation is considered due to the symmetrical nature.

Continuity

$$\frac{\partial u_r}{\partial r} + \frac{\partial u_z}{\partial z} + \frac{u_r}{r} = 0 \tag{6.20a}$$

Momentum

$$\rho\left(\frac{\partial u_r}{\partial t} + u_r\frac{\partial u_r}{\partial r} + u_z\frac{\partial u_r}{\partial z}\right) = -\frac{\partial p}{\partial r} + \mu\left(\frac{\partial^2 u_r}{\partial r^2} + \frac{1}{r}\frac{\partial u_r}{\partial r} + \frac{\partial^2 u_r}{\partial z^2} - \frac{u_r}{r^2}\right) \tag{6.21a}$$

$$\rho\left(\frac{\partial u_z}{\partial t} + u_r\frac{\partial u_z}{\partial r} + u_z\frac{\partial u_z}{\partial z}\right) = \rho g - \frac{\partial p}{\partial z} + \mu\left(\frac{\partial^2 u_z}{\partial r^2} + \frac{1}{r}\frac{\partial u_z}{\partial r} + \frac{\partial^2 u_z}{\partial z^2}\right) \tag{6.23a}$$

By ignoring the viscous dissipation effect, the energy balance may be written as

$$\frac{\partial T}{\partial t} + u_r\frac{\partial T}{\partial r} + u_z\frac{\partial T}{\partial z} = \frac{1}{\rho C_p}\left[\frac{1}{r}\frac{\partial}{\partial r}\left(kr\frac{\partial T}{\partial r}\right) + \frac{\partial}{\partial z}\left(k\frac{\partial T}{\partial z}\right)\right]. \tag{6.24a}$$

The mass conservation is given as

$$\frac{\partial c}{\partial t} + u_r\frac{\partial c}{\partial r} + u_z\frac{\partial c}{\partial z} = \left[\frac{1}{r}\frac{\partial}{\partial r}\left(Dr\frac{\partial T}{\partial r}\right) + \frac{\partial}{\partial z}\left(D\frac{\partial T}{\partial z}\right)\right] + \dot{m} \tag{6.25a}$$

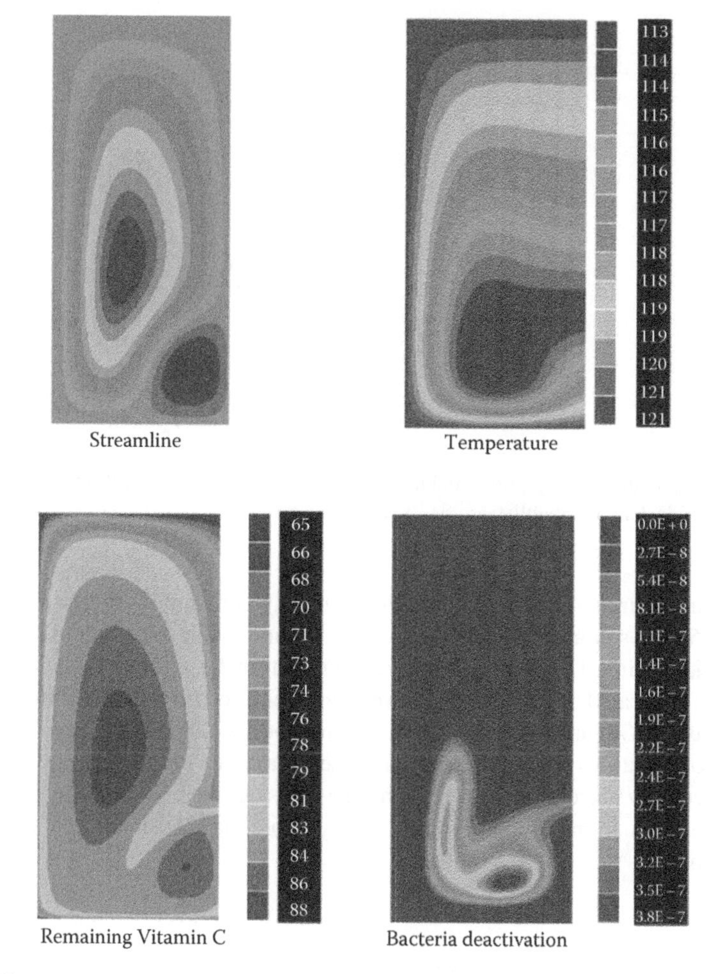

FIGURE 6.7 Temperature, bacteria deactivation, Vitamin C destruction, and streamline profiles in a can filled with concentrated cherry juice (75°Brix) and heated by condensing steam after 2450 s (provided by A. Ghani, 2004).

The Boussinesq approximation has been employed [4,29,30]:

$$\rho = \rho_{ref}\left[1-\beta\left(T-T_{ref}\right)\right], \tag{6.33}$$

where

"*ref*" denotes reference condition

β is the thermal expansion coefficient of the liquid

Equation 6.19 was used to calculate the deactivation rate.

6.7 RECENT CFD APPLICATIONS IN THERMAL PROCESSING

In recent years, CFD has been used to model and simulate many different types of thermal processes [31–41]. These applications include drying, baking, ohmic heating, packaging, etc. The following sections summarize some of these recent applications.

6.7.1 Drying Process

Drying is one of the most common thermal processes for fruits. CFD was used to study the external flow and temperature fields during drying of kiwi fruits [42]. Based on the simulation verified by the experimental results, the local distributions of the surface convective heat transfer coefficients for the fruits were determined to predict the local convective mass transfer coefficients through the analogy between the thermal and concentration boundary layers. Normally, in fruit drying, cabinet dryers are used. However, one disadvantage in using cabinet dryer is that the desired moisture content of end product can be nonuniform. Therefore, CFD was used to investigate seven different geometries of cabinet dryers in order to obtain a uniform distribution of drying air flow and temperature in the dryer [43], resulting in the development of a new cabinet dryer with a side-mounted plenum chamber. The experimental results confirmed that this new dryer had an even distribution of air velocity and temperature throughout the dryer.

In a separate study, a laboratory drying oven with a forced air circulation was designed to store products at a constant, spatially uniform temperature, and an experimentally validated 3D CFD analysis on the flow and thermal processes in the oven was conducted [44]. This CFD model included all heat transfer modes, temperature-dependent air properties, and local heat transfer coefficients on the external walls. With the help of the CFD simulation, the temperature uniformity in the modified prototype was significantly improved.

Spray drying is one of the most common drying methods. During the operation of spray driers, particle deposition is an undesirable phenomenon. A new mathematical model of particle deposition was developed and implemented in CFD software FLUENT [45]. Simulation confirmed a significant re-entrainment of particles into the drier chamber at the chamber walls and a higher deposition intensity near the adjunctions of the ceiling wall, the vertical wall, and the conical wall due to lower wall shear stress.

CFD was also used to model and simulate maltodextrin solutions spray drying to control stickiness [46]. The drying process was simulated at steady state with the consideration of the geometry of the equipment such as drying chamber, air inlet and atomizer, and the relationship between drying air and product properties. The simulation showed that it was possible to determine zones and operating conditions for which particles could be sticky inside the chamber. Such a study provides an approach to evaluate possible stickiness to avoid fouling of the plant or to optimize powder insertion position for agglomeration.

6.7.2 Baking Process

During baking, temperature is the dominating factor that determines the final bread quality as temperatures affect various physiochemical changes, including starch gelatinization, protein denaturation, enzymatic reactions, and browning reactions. CFD modeling is often used as a useful approach to study the unsteady heat transfer in the oven as well as the heating history and temperature distribution in the baking oven chamber in order to facilitate a better understanding of the baking process [47,48]. It can also be used to help the design of the baking process, for example, CFD was used to develop an improved heating system for industrial tunnel baking ovens [49] and to predict radiation heat fluxes incident on the top surface of the food, both across the width of the baking chamber and along its length. A novel gas-fired radiant burner suitable for incorporation into industrial tunnel ovens was designed. The simulations indicated the capability of the new burner in delivering irradiation to a traveling conveyor more uniformly than existing radiant burner designs.

A bakery pilot oven was also modeled by using CFD [50]. The approach used was based on the integration of a heat flux–measuring device into modeled geometry. Heat conduction, convection, and radiation mechanisms are all considered and coupled with turbulent flow, and radiation was shown to be the dominant mode of heat transfer. The model predictions showed a good qualitative agreement with the experimental measurements.

A 2D CFD model was developed to facilitate a better understanding of the baking process of crustless bread [51]. The crustless bread was made by spraying water at prefixed intervals on the surface of the dough and gently baking the dough at controlled temperature in a special type of baking oven. The CFD simultaneously showed that the bread core temperature reached at 95°C at the end of baking, with the moisture of the bread satisfying the normal bread quality.

As shown previously, radiation is the most dominant heat transfer mode in baking ovens. A 3D CFD model was thus developed by including three different radiation models, that is, discrete transfer radiation model, surface to surface, and discrete ordinates [52]. The simulation indicated that all three models predicted almost similar results; thus, the discrete ordinates radiation model was selected for subsequent simulation to study the profiles of temperature and starch gelatinization of crust and crumb of the product. The results indicated that when the temperature of bread center reached 100°C, the baking process was completed at 1500 s.

Finally, a CFD model was employed to investigate the effect of hot air distribution and placement of bread on temperature and starch gelatinization index of bread in a pilot-scale baking oven [53]. The simulation results showed that uneven temperature distribution was caused by the nonuniformity of air flow pattern inside the oven cavity, and compared to baking breads in the bottom tray, bread in upper trays required shorter baking time and lower gelatinization index.

6.7.3 Ohmic Heating

Ohmic heating is an advanced food thermal processing method. Modeling of the ohmic heating process of foods is challenging due to heterogeneous food properties, in particular, variation of the food electrical conductivities. A 3D CFD was used to simulate the ohmic heating of tomato soup in a pouch with user-defined functions for electric field equations [54]. The 3D model was able to identify the potential cold spots over the entire pouch; particularly, it was found that the bottom corners of the pouch were a zone of low current density with a temperature of only 53.3°C even when the peak pouch temperature reached 139°C.

The effects of food electrical conductivities on ohmic heating was studied by CFD with user-defined functions for electric field equations [55], in which a solid–liquid food complex containing three different solid particles with substantially different electrical conductivities and 3% NaCl solution was used. Potato, meat, and carrot were used as the solid food samples, which had lower electric conductivities than carrier medium. The simulation showed the existence of hot spots on the continuous phase in zones perpendicular to the solid cubes and cold spots in between the particles where the current density lacks. In addition, CFD was also used to study the effects of electrical field intensity, electrical conductivity, solid heating, liquid–solid conductivity, and some parameters, such as drying rate, moisture content, and temperature changes during the drying process of potato via ohmic heating [56].

6.7.4 Other Applications

There are many other CFD applications for relevant thermal food processing. CFD was applied to provide insight on the natural convective processes occurring during the sterilization of canned liquid food and to evaluate the contributions to the overall heat transfer rate from different surfaces of the uniformly heated cylindrical can [57]. CFD was also employed to study the heating and cooling cycles during thermal processing of asparaguses in a still can containing 0, 1, 8, and 19 asparaguses, and the results showed that with the increase in the number of asparaguses in the can, the increase in brine velocity was higher at the beginning of the heating and cooling phases [58]. The slowest heating and cooling zones were found to be at the height of about 13.5% of the can height from the bottom and about 13.5% of the can height from the top, respectively. The results also indicated that heating of the asparaguses was not uniform, and the top of the asparaguses received higher heat treatment than the bottom.

Pulsed electric field (PEF) is a novel processing technique at low or moderate temperatures. During PEF processing, the electric field strength and temperature distribution in the treatment chamber are the key processing parameters affecting the treatment efficacy and possibly food sensorial and nutritional quality. In a recent study, a 3D CFD model of a pilot-scale PEF system with colinear configuration of the electrodes was developed [59]. In the model, the fluid dynamics was coupled with the electrical field, and thus, the model could be used to predict the liquid's flow pattern, electric field distribution, temperature increase, and residence time in the treatment chamber. Simulation with a model liquid food and whole milk indicated that the simulated PEF energy dissipated into the liquids was between 4 and 66 kJ kg^{-1} and process temperatures ranged from approximately 25°C to 80°C.

Pasteurization is an important unit operation for inactivation of pathogenic bacteria in eggs, and CFD simulations were performed on thermal pasteurization of egg at 55.6°C [60]. Simulation for a whole egg (with yolk) at two different modes, namely, stationary and rotation (at 2.5 and 5 rpm), revealed that the rotating egg only took 9.5 min to reach the pasteurization temperature, while the stationary egg needed about 30 min. Therefore, rotation of egg made the thermal pasteurization process more efficient and could minimize the thermal damage of the egg's nutrients; it was thus recommended that the food industry adopt rotation as an additional operation during pasteurization of eggs to produce high-quality intact eggs without affecting its functional properties.

6.8 CONCLUSION

In this chapter, fundamental aspects of mathematical modeling of thermal processing of food fluids or suspensions are summarized. In food engineering research and design, a number of commercial software packages have been used. They have been developed in the context of mechanical engineering but have been applied to an ever-increasing list of industrial situations, including food and bioprocessing applications. In thermal processing, CFD packages can be used not only to analyze an existing process but also to optimize the process and to explore scenarios of new, more effective operations. There is no question that food-processing problems are complex and also in most situations involve transient, 3D conditions. Analytical approximation and solutions may no longer be appropriate, and CFD approach combined with chemical or biochemical kinetics equations is necessary. Although many thermal processing problems have been investigated using CFD, there is still a scope of further development which will no doubt be beneficial for the improvement and modernization of the current food industry to make it safer for the consumers. Furthermore, despite powerful in many ways, in using CFD approach, it is always necessary to be able to understand the underlined physics and chemistry in simple terms and indeed to conduct appropriate experiments to validate the calculations for the situations of interest.

NOMENCLATURE

A	Pre-exponential factor (s^{-1})
β	Thermal expansion coefficient (K^{-1})
c	Concentration (mol m^{-3} or kg m^{-3})
C_p	Specific heat capacity (J kg^{-1} K^{-1})
d	Diameter or characteristic dimension (m)
D	Diffusion coefficient (m^2 s^{-1})
D_T	Decimal reduction time at temperature T (min)
E	Activation energy (J mol^{-1})
F	Thermal death time (min)
g	Gravity acceleration (m s^{-2})
Gr	Grashof number $Gr = d^3 \rho^2 g\beta \Delta T/\mu^2$
h	Heat transfer coefficient (W m^{-2} K^{-1})

H	Height of the can (m)
k	Thermal conductivity (W m^{-1} K^{-1})
Nu	Nusselt number $Nu = hd/k$
p	Pressure (N m^{-2} or Pa)
Pr	Prandtl number $Pr = C_p\mu/k$
r	Radius or radial coordinate (m)
R	Radius (m) or the universal gas constant (J mol^{-1} K^{-1})
T	Temperature (°C)
u, v, w	Velocities at x, y, z directions, respectively (m s^{-1})
u	Velocity in general (m s^{-1})

Subscripts

h	High
l	Low
r	Radial coordinate
ref	Reference
θ	Angular coordinate
w	Wall
z	z direction coordinate

ACKNOWLEDGMENT

The author would like to thank Dr. Abdul Ghani of the University of Auckland for his kindness in providing the good simulation diagrams that are used in this chapter and also his invaluable advices in completing the chapter. Also, the author is grateful for the experience in taking part in the collaboration research with Drs. A. Ghani, M.M. Farid, and P. Richards on similar topics covered in this chapter.

REFERENCES

1. RP Singh, DR Heldman. *Introduction to Food Engineering*. 2nd Edn, San Diego, CA: Academic Press, 1993.
2. PJ Fellows. *Food Processing Technology (Principles and Practice)*. Cambridge, U.K.: Woodhead Publishing, Series in Food Science and Technology, 1996.
3. X Zhao, Q Liao. *Mechanics of the Viscous Fluids*. Beijing, China: Mechanical Engineering Press, 1981.
4. FP Incropera, DP DeWitt. *Fundamentals of Heat and Mass Transfer*. 5th Edn, New York: John Wiley & Sons, 2002.
5. SV Patankar, DB Spalding. A calculation procedure for heat, mass and momentum transfer in three dimensional parabolic flows. *International Journal of Heat and Mass Transfer* 15(10): 1787–1806, 1972.
6. A Jung, PJ Fryer. Optimizing the quality of safe food: Computational modeling of a continuous sterilization process. *Chemical Engineering Science* 54: 717–730, 1999.
7. AK Datta, AA Teixeira. Numerical modeling of natural convection heating in canned liquid foods. *Transactions of the ASAE* 30(5): 1542–1551, 1987.
8. AK Datta, AA Teixeira. Numerically predicted transient temperature and velocity profiles during natural convection heating of canned liquid foods. *Journal of Food Science* 53(1): 191–195, 1988.
9. A Kumar, M Bhattacharya, J Blaylock. Numerical simulation of natural convection heating of canned thick viscous liquid food products. *Journal of Food Science* 55(5): 1403–1411, 1990.
10. AG Ghani. Thermal sterilization of canned liquid foods. PhD thesis. Department of Chemical and Materials Engineering, The University of Auckland, New Zealand, 2002.
11. AG Ghani, MM Farid, XD Chen. A CFD simulation of the coldest point during sterilization of canned food. *Proceedings of the 26th Australian Chemical Engineering Conference*, September 28–30, Port Douglas, Queensland, No. 358, 1998.

12. AG Ghani, MM Farid, XD Chen, P Richards. Heat transfer and biochemical changes in liquid food during sterilization using computational fluid dynamics (CFD). *Proceedings of Chemeca'99*, September 26–29, Newcastle, Australia, 1999.
13. AG Ghani, MM Farid, XD Chen, P Richards. Numerical simulation of natural convection heating of canned food by computational fluid dynamics. *Journal of Food Engineering* 41(1): 55–64, 1999.
14. AG Ghani, MM Farid, XD Chen, P Richards. Numerical simulation of biochemical changes in liquid food during sterilization of viscous liquid using computational fluid dynamics (CFD), Part 1. *Proceedings of the 10th World Congress of Food Science and Technology*, October 3–8, Sydney, Australia, 1999 (on CD Rom).
15. AG Ghani, MM Farid, XD Chen, P Richards. An investigation of deactivation bacteria in canned liquid food during sterilization using computational fluid dynamics (CFD). *Journal of Food Engineering* 42(4): 207–214, 1999.
16. AG Ghani, MM Farid, XD Chen, P Richards. Numerical simulation of biochemical changes in liquid food during sterilization of viscous liquid using computational fluid dynamics (CFD), Part 2. *Food Australia* 53(1–2): 48–52, 2000.
17. AG Ghani, MM Farid, XD Chen, C Watson. Numerical simulation of transient two-dimensional profiles of temperature and flow velocity of canned food during sterilization. *Proceedings of the Eighth International Congress on Engineering and Food*, April, Pueblo, Mexico, 2000 (on CD Rom).
18. AG Ghani, MM Farid, XD Chen. Numerical simulation of transient temperature in a 3-D pouch during sterilization using computational fluid dynamics. *Proceedings of Chemeca'2000*, July 9–12, Perth, Australia, 2000 (on CD-ROM).
19. AG Ghani, MM Farid, XD Chen. Analysis of thermal sterilization of liquid food in cans and pouches using CFD. *New Zealand Food Journal* 30(6): 25–30, 2000.
20. AG Ghani, MM Farid, XD Chen, P Richards. A CFD study on the effect of sterilization temperature on bacteria deactivation and vitamin destruction. *Journal of Process Mechanical Engineering* 215(1): 9–17, 2001.
21. AG Ghani, MM Farid, XD Chen. A CFD study on the effect of sterilization temperatures on biochemical changes of liquid food in 3-D pouches. *Proceedings of the Second International Symposium on Applications of Modeling as an Innovative Technology in the Agri-Food Chain*, December 9–13, Palmerston North, New Zealand, 2001 (on CD Rom).
22. AG Ghani, MM Farid, XD Chen. Heat transfer in a 3-D pouch during sterilization using computational fluid dynamics. *Proceedings of the Sixth World Congress of Chemical Engineering*, September 23–27, Melbourne, Australia, 2001 (on CD Rom).
23. AG Ghani, MM Farid, XD Chen. Thermal sterilization of canned food in a 3-D pouch using computational fluid dynamics. *Journal of Food Engineering* 48, 147–156, 2001.
24. AG Ghani, MM Farid, XD Chen. Numerical simulation of transient temperature and velocity profiles in a horizontal can during sterilization using computational fluid dynamics. *Journal of Food Engineering* 51: 77–83, 2002.
25. AG Ghani, MM Farid, XD Chen. Theoretical and experimental investigation of the thermal destruction of Vitamin C in food pouches. *Journal of Computers and Electronics in Agriculture*, 34: 129–143, 2002 (Special Issue on "Applications of CFD in the Agri-Food Industry").
26. AG Ghani, MM Farid, XD Chen. Theoretical and experimental investigation on the thermal inactivation of *Bacillus stearothermophilus* during thermal sterilization in food pouches. *Journal of Food Engineering* 51: 221–228, 2002.
27. AG Ghani, MM Farid, XD Chen. A computational and experimental study of heating and cooling cycles during thermal sterilization of liquid foods in pouches using CFD. *Journal of Process Mechanical Engineering* 217: 1–9, 2003.
28. K Matsuzaki, H Ohba, M Munekata. Numerical analysis of thermal convections in three-dimensional cavity: Influence in approximation models on numerical solutions. *Proceedings of the Sixth International Symposium on Experimental and Computational Aerodynamics of Internal Flows (6-ISAIF)*, Shanghai, China, S Yu, N X Chen, and X Liu (eds.), Vol. 2, 2003, pp. 433–438.
29. DD Gray, A Giorgini. The validity of the Boussinesq approximation for liquids and gases. *International Journal of Heat and Mass Transfer* 19: 545–551, 1976.
30. SM Yang, WQ Tao. *Heat Transfer*. 3rd Edn, Beijing, China: Higher Education Press, 1999.
31. T Norton, D-W Sun. Computational fluid dynamics (CFD)—An effective and efficient design and analysis tool for the food industry: A review. *Trends in Food Science and Technology* 17(11): 600–620, 2006.

32. DW Sun (ed.). *Computational Fluid Dynamics in Food Processing*. Boca Raton, FL: CRC Press/Taylor & Francis, 2007, 740 pp, ISBN 978-0-8493-9286-3.
33. DW Sun. Computational fluid dynamics for the food industry, keynote speech. *Proceedings of 2007 CIGR International Symposium/3rd CIGR Section VI International Symposium on Food and Agricultural Products: Processing and Innovations*, September 24–26, Naples, Italy, 2007.
34. T Norton, D-W Sun. An overview of CFD applications in the food industry, in D-W Sun (ed.), *Computational Fluid Dynamics in Food Processing*. Boca Raton, FL: CRC Press, 2007, pp. 1–41.
35. D-W Sun. CFD: An innovative and effective design tool for the food industry, invited speech. *Proceedings of the 10th International Congress on Engineering and Food (ICEF 10)*, April 20–24, Viña del Mar, Chile, 2008.
36. D-W Sun. Advanced applications of computational fluid dynamics (CFD) in the food industry, keynote speech. *Proceedings of 10th International Congress on Mechanization and Energy in Agriculture*, October 14–17, Antalya, Turkey, 2008.
37. D-W Sun. Recent research advances of computational fluid dynamics (CFD) applications in the food industry, plenary lecture. *Proceedings of IFCON 2008-6th International Food Convention*, December 15–19, Mysore, India, 2008.
38. T Norton, D-W Sun. Computational fluid dynamics in thermal processing, in R Simpson (ed.), *Engineering Aspects of Thermal Processing*. Boca Raton, FL: CRC Press, 2009, pp. 317–363.
39. D-W Sun. Computational fluid dynamics (CFD): An innovative tool for food process design and optimisation, keynote speech. *Proceedings of the Food and Nutrition Simulation Conference—FOODSIM'2010*, June 24–26, CIMO Research Centre, Braganca, Portugal, 2010.
40. D-W Sun. Computational fluid dynamics (CFD): An effective tool for food process design and optimization, plenary lecture. *Proceedings of the 6th Food Science International Symposium*, August 5–11, Beijing, China, 2010.
41. T Norton, D-W Sun. CFD: An innovative and effective design tool for the food industry, in JM Aguilera, GV Barbosa-Cánovas, R Simpson, and J Welti-Chanes (eds.), *Food Engineering Interfaces*, New York: Springer, 2011, pp. 45–68.
42. A Kaya, O Aydin, I Dincer. Experimental and numerical investigation of heat and mass transfer during drying of Hayward kiwi fruits (*Actinidia deliciosa* Planch). *Journal of Food Engineering* 88(3): 323–330, 2008.
43. Y Amanlou, A Zomorodian. Applying CFD for designing a new fruit cabinet dryer. *Journal of Food Engineering* 101(1): 8–15, 2010.
44. J Smolka, AJ Nowak, D Rybarz. Improved 3-D temperature uniformity in a laboratory drying oven based on experimentally validated CFD computations. *Journal of Food Engineering* 97(3): 373–383, 2010.
45. Y Jin, XD Chen. A fundamental model of particle deposition incorporated in CFD simulations of an industrial milk spray dryer. *Drying Technology* 28(8): 960–971, 2010.
46. A Gianfrancesco, C Turchiuli, D Flick, E Dumoulin. CFD modeling and simulation of maltodextrin solutions spray drying to control stickiness. *Food and Bioprocess Technology* 3(6): 946–955, 2010.
47. SY Wong, W Zhou, J Hua. CFD modeling of an industrial continuous bread-baking process involving U-movement. *Journal of Food Engineering* 78(3): 888–896, 2007.
48. SY Wong, W Zhou, J Hua. Designing process controller for a continuous bread baking process based on CFD modelling. *Journal of Food Engineering* 81(3): 523–534, 2007.
49. ME Williamson, DI Wilson. Development of an improved heating system for industrial tunnel baking ovens. *Journal of Food Engineering* 91(1): 64–71, 2009.
50. M Boulet, B Marcos, M Dostie, C Moresoli. CFD modeling of heat transfer and flow field in a bakery pilot oven. *Journal of Food Engineering* 97(3): 393–402, 2010.
51. A Mondal, AK Datta. Two-dimensional CFD modeling and simulation of crustless bread baking process. *Journal of Food Engineering* 99(2): 166–174, 2010.
52. N Chhanwal, A Anishaparvin, D Indrani, KSMS Raghavarao, C Anandharamakrishnan. Computational fluid dynamics (CFD) modeling of an electrical heating oven for bread-baking process. *Journal of Food Engineering* 100(3): 452–460, 2010.
53. A Anishaparvin, N Chhanwal, D Indrani, KSMS Raghavarao, C Anandharamakrishnan. An investigation of bread-baking process in a pilot-scale electrical heating oven using computational fluid dynamics. *Journal of Food Science* 75(9): E605–E611, 2010.
54. S Jun, S Sastry. Reusable pouch development for long term space missions: A 3D ohmic model for verification of sterilization efficacy. *Journal of Food Engineering* 80(4): 1199–1205, 2007.
55. JY Shim, SH Lee, S Jun. Modeling of ohmic heating patterns of multiphase food products using computational fluid dynamics codes. *Journal of Food Engineering* 99(2): 136–141, 2010.

56. MK Moraveji, E Ghaderi, R Davarnejad. Simulation of the transport phenomena during food drying with ohmic heating in a static system. *International Journal of Food Engineering* 6(5): 2010.
57. A Kannan, PC Gourisankar Sandaka. Heat transfer analysis of canned food sterilization in a still retort. *Journal of Food Engineering* 88(2): 213–228, 2008.
58. A Dimou, S Yanniotis. 3D numerical simulation of asparagus sterilization in a still can using computational fluid dynamics. *Journal of Food Engineering* 104(3): 394–403, 2011.
59. R Buckow, S Schroeder, P Berres, P Baumann, K Knoerzer. Simulation and evaluation of pilot-scale pulsed electric field (PEF) processing. *Journal of Food Engineering* 101(1): 67–77, 2010.
60. R Ramachandran, D Malhotra, A Anishaparvin, C Anandharamakrishnan. Computational fluid dynamics simulation studies on pasteurization of egg in stationary and rotation modes. *Innovative Food Science and Emerging Technologies* 12(1): 38–44, 2011.

7 Modeling Thermal Microbial Inactivation Kinetics

Ursula Andrea Gonzales Barron

CONTENTS

7.1 INTRODUCTION

Understanding the way in which populations of microorganisms decrease in response to heat is fundamental to the engineering design of thermal inactivation processes important in the food, pharmaceutical, and bioprocess industries. Thermal inactivation of a microorganism implies a loss of the ability to grow and reproduce, and not a physical destruction or fragmentation. A practical working definition however is "to reduce the number of viable contaminants to a desired level following thermal treatment." This is referred to as the sterility requirement. Thermal sterilization is widely used because of good reliability and economy. Nucleic acids, structural proteins, and enzymes of microorganisms are denaturized and hence inactivated by moist heat treatment. Thermal sterilization can be operated either as a batch or a continuous process. Continuous thermal processes, which operate at steady state, are widely preferred to the batch process because greater control of the process temperature and exposure time is possible. The continuous process provides improved product quality and is less laborious than a batch process. Because of the risks associated with error to public health, heat treatment operations must be conservative.

In order to optimize process conditions and controls to achieve desired results, the effect of heat on the inactivation kinetics of the population needs to be characterized mathematically. Of particular interest are kinetic models that can be reliably used to predict the combined effects of process parameters such as exposure time, temperature, and the food's intrinsic properties. As models are capable of accurately predicting the microbial inactivation, they are highly valued by process engineers responsible for the design of ultrahigh temperature (UHT), high-temperature short-time (HTST) sterilization, and pasteurization processes that operate at temperatures far above those in which experimental data can be obtained. The synthesis and validation of an adequate microbial inactivation model is essential to process optimization and, for longer term, to optimize processing in real time. This chapter commences with the description of the most important primary inactivation models and their applications, from the simplest and oldest log-linear model to more complex models with the capabilities to mathematically represent concavity, shoulder and/or tail of a survival curve. A classification of secondary inactivation models is proposed and models are explained in detail. Tertiary modeling is introduced as a means of model validation. In principle, primary and secondary inactivation models are presented within the framework of static or two-step modeling, and later on, one-step fitting of the most common combinations of primary and secondary models is explained from the point of view of dynamic modeling. Finally, this chapter also features the latest developments in stochastic inactivation modeling as opposed to the regular deterministic modeling.

7.2 MICROBIAL INACTIVATION MODELS

Microbial inactivation by heat has been a key operation in the food industry. In the beginning of the twentieth century, microbial death kinetics received extensive attention in microbiology, with the development of log-linear models by Bigelow [1,2] and Esty and Meyer [3], and constitutes one of the earliest forms of predictive microbiology. This was a time when there was a premium on linear models and linearization methods because of the limited means to handle elaborate algebraic models and solve differential equations. As an approximation, this early log-linear model has had remarkably success in underpinning the thermally processed food industry by predicting time/temperature combinations sufficient to reduce spores of *Clostridium botulinum* type A by 12-log-cycles in nonacid foods. An interesting analysis of thermal inactivation patterns [4] provides insight into the success of the log-linear model, suggesting that specific deviations are not of global significance and identifying situations where these need to be considered.

7.2.1 Primary Inactivation Models

7.2.1.1 First-Order Inactivation Kinetic Model

The first-order kinetic model is a classically thermal death time model, which describes the inactivation of microorganisms with a constant proportion in each successive time period. According to Teixeira [5], the nature of the inactivation transformation can be explained in terms of Eyring's transition-state theory and the Maxwell–Boltzmann distribution of the speed of molecules from molecule thermodynamics. The first-order inactivation kinetics model assumes that all cells have the same heat resistance to a lethal treatment and that their inactivation is stochastic; that is, the death of an individual cell results from a random inactivation of a critical molecule [6]. This results in a linear relationship between the logarithmic inactivation and treatment time isothermal conditions [7]:

$$\frac{dN}{dt} = -k_d(T)N \tag{7.1}$$

$$N = N_0 \exp\left[-k_d(T)t\right] \tag{7.2}$$

$$\ln S(t) = -k_d(T)t \tag{7.3}$$

According to Equation 7.3, a semilogarithmic plot of the isothermal survival curve will always be a straight line with slope $-k_d$. The time to reduce the population by 1 log cycle (base 10), known as the "*D* value," is $\ln(10)/k_d$. An important advantage of this model is the clear physical meaning of the parameter *D*. When log *D* values are plotted against the corresponding temperatures, the reciprocal of the slope is equal to the *z* value, which is the increase in temperature required for a 1-log decrease in *D* value. The objective in most sterilization processes was to reduce the initial spore population by a sufficient number of log cycles so that the final number of surviving spores is one millionth of a spore, interpreted as probability for survival of one in a million (in the case of spoilage causing bacteria), and one in a trillion in the case of pathogenic bacteria such as *C. botulinum*. Given that an exponential decay has no zero, "commercial sterility" is therefore reached when a hypothetical sporal population of *C. botulinum* is reduced by 12 orders of magnitude or 12D. Knowledge of the *D* values of representative strains allows the determination of the *F* value, which is the time required to achieve 12D, assuming a *z* value of 10°C. This is used to integrate the lethal effect of temperature within a given thermal process, and is explained in the following section.

7.2.1.2 Conventional Calculation of Sterility in Thermal Processing

The conventional way of calculating the efficiency of heat treatments in food protection is based on the assumption that survival curves of microbial cells and bacterial spores are governed by a first-order kinetic. The traditional sterilization value or accumulated lethality (*F* value) is defined as the time of a heat treatment at the reference temperature (generally $T_{ref} = 121.1$°C) for a specific *z* value, or as any equivalent heat treatment which would cause the same destruction ratio n_d. The *target F value*, which depends both on the required level of safety and on the heat resistance of the target species of spore or bacteria cell, is

$$F = n_d D_{ref} = \log\frac{N}{N_0} D_{ref} \tag{7.4}$$

where $n = (\log N/N_0)$ being the ratio of decimal reduction represents the *safety level*, while D_{ref} being the time of decimal reduction at the reference temperature represents the *heat resistance*. For low-acid foods ($pH < 4.6$), the minimum microbiologically safe thermal process is generally regarded as an F value of 2.8 min at the slowest heat point of the food (D value of *C. botulinum* at 121.1°C is 0.23 min, and targeted reduction is 12D).

At a constant temperature, the *actual* F value is the product of the heating time and the so-called biological destruction value L, which is a function of temperature, as shown in the following:

$$F = L(T)t = 10^{(T - T_{ref})/z} t \tag{7.5}$$

In standard calculations, the z value is assumed to be 10°C, which corresponds to that of *C. botulinum*. Then the traditional F value is implicitly applied to an ideal strain of *C. botulinum*, the destruction curve of which would be governed by a first-order kinetic and which would be characterized by a z value of 10°C. Because F values are additive, in the case of a variable temperature heat treatment, it can be written as follows:

$$F = \int_0^t 10^{(T - T_{ref})/z} \, dt \tag{7.6}$$

and numerically solved in the following discrete form, with time increments Δt_i of 1 min:

$$F = \sum 10^{(T_i - T_{ref})/z} \Delta t_i \tag{7.7}$$

This approach has served the food industry for over 70 years and, to date, forms the basis of most commercially applied thermal processes.

7.2.1.3 First-Order Inactivation Kinetic Model with Lag Phase

Many survival curves have a shoulder phase before initiation of the exponential decline (Figure 7.1, curve B). A number of reasons have been attributed to the presence of a shoulder. If clumps of microorganisms exist in the suspension, all cells in the clump need to be inactivated before the

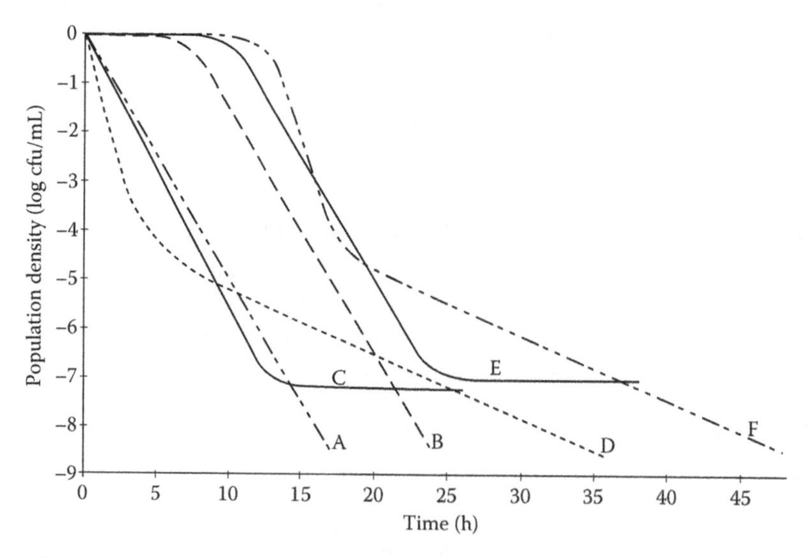

FIGURE 7.1 A representation of six different shapes of survival curves: (A) log-linear, (B) survival with shoulder, (C) survival with tail, (D) biphasic, (E) survival with shoulder and tail, and (F) biphasic with shoulder.

colony-forming ability of the clump is inactivated [8], and the lag phase will be observed. If the lag phase represents a period in which the cells are able to resynthesize a vital component, death ensues only when the rate of destruction exceeds the rate of resynthesis [9]. If thermal inactivation of microorganisms is cumulative rather than instantly lethal, or there are multiple target sites for thermal inactivation [8], the lag-phase may also be observed.

Maintaining the log-linear nature of the exponential decline, Buchanan et al. [10] proposed the following modified model to express the effect of the shoulder phase as a shift lag-phase λ:

$$\log\frac{N}{N_0}=\begin{cases}0 & (t\le\lambda)\\ -\dfrac{t-\lambda}{D} & (t>\lambda)\end{cases} \tag{7.8}$$

The greater advantage of the Buchanan model is its simplicity. It also provides a mathematical means that approximates the way that microbiologists have traditionally estimated growth kinetics graphically [11,12]. Its drawback is that the model can only produce a good fit for growth curves with an abrupt transition from shoulder to decay phase. This model cannot estimate the shoulder time accurately, and its performance varies and depends much on experimental data [13].

Despite the worldwide use of the first-order kinetic model, especially in the canning industry for the so-called 12D process of the proteolytic strains of *C. botulinum* spores, a lot of deviations have been observed particularly at lower temperatures and for vegetative cells, indicating that inactivation kinetics are not always following first-order linear kinetics. Deviations from first-order kinetics such as shoulders and tails (Figure 7.1) are very common, and they require other more complex model types. Such behavior has been observed in a growing number of microorganisms and spores [14–18]. In an extensive survey covering over 120 reported survival curves, van Boekel [18] found that only less than 5% of those examined could be considered log-linear. This led to the conclusion that log-linear survival curves are the exception rather than the rule. Consideration of nonfirst-order inactivation kinetics could be important in the safe application of milder heat processes or those relying on the combined effect of other factors such as pH and water activity [19]. The primary models to be presented in the following sections have been developed to represent survivor curves with tail, shoulders, and tail and sigmoidal.

7.2.1.4 Early Attempts to Model Non-Log Linear Inactivation Curves

Casolari [20] attempted to explain the occurrence of the tail in a survival curve assuming that the death of the organisms is caused by a lethal hit of a water molecule that carries a certain level of energy, higher than the critical level E_d. As such, he regards tailing as a phenomenon produced by the increasingly low probability of collision between water molecules having more than E_d energy and microbial particles. The following pseudomechanistic model was obtained combining probability theory with the Maxwellian distribution of energy:

$$N(t)=N_0^{1/B_C(T)t} \tag{7.9}$$

$$B_C(T)=\left(\frac{N_A}{M_{H_2O}}\right)^2\exp\left(\frac{-2E_d}{RT}\right) \tag{7.10}$$

Geeraerd et al. [21] point out that the main limitation of this model is its inability to model shoulders. Sapru et al. [22,23] proposed a set of differential equations, inspired by the work of Teixeira and his coworkers [24,25]. The Sapru model has been derived specifically to describe the activation of microbial spores during sterilization processes, which implies an initial increase of the activated spore population. They distinguished two types of spores: (1) a dormant, viable population N_D potentially able of producing colonies in an appropriate growth medium after activation, and (2) an active population N_C able of forming colonies in a suitable growth medium, as described in the following:

$$\frac{dN_D}{dt} = -\left(k_{d1} + K_a\right)N_D \tag{7.11}$$

$$\frac{dN_C}{dt} = K_a N_D - k_{d2} N_C \tag{7.12}$$

Structurally, this model is unable to describe a more or less flat initial shoulder as it was designed to describe spore activation and inactivation at high temperatures [21].

7.2.1.5 Logistic or Biphasic Models

Apart from a shoulder, survival curves may also contain a tail. Tails are hypothesized to originate whether from the presence of a subpopulation of more heat-resistant microorganisms or as a consequence of experimental artifacts [21]. The first hypothesis however could obey either a *vitalistic* theory whereby a permanent difference in resistance exists or an *adaptation* theory whereby the difference in resistance is acquired. Nevertheless these theories have in common the assumption of a certain distribution of heat resistance among the microbial population. If some microorganisms are intrinsically more resistant than others, they can survive under testing conditions or the reduction rate of their population will be slowed down. Figure 7.1 (curves C and D) shows two different types of tailing in survivors curves called *biphasic* curves, as they represent a mix of two fractions or subpopulations of different heat resistance.

In earlier studies various logistic equations have been used to model this type of nonlinear survival curves. Cerf [26] proposed the following two-parallel reaction model for describing biphasic curves:

$$\frac{N}{N_0} = f\exp\left(-k_1 t\right) + \left(1-f\right)\exp\left(-k_2 t\right) \tag{7.13}$$

Kamau et al. [27] applied three different forms of logistic equations to fit various shaped survival curves for *Listeria monocytogenes* heated in lactoperoxidase system. For linear curves and biphasic curves, the equations were respectively, as follows:

$$\frac{N}{N_0} = \frac{2}{1+\exp\left(k_d t\right)} \tag{7.14}$$

$$\frac{N}{N_0} = \frac{2f}{1+\exp\left(k_1 t\right)} + \frac{2\left(1-f\right)}{1+\exp\left(k_2 t\right)} \tag{7.15}$$

Based on the logistic models of Kamau et al. [27], Whiting and Buchanan [28] proposed a five-parameter logistic model to take into account significant shoulder and tailing in survival curves. In this model, again two microbial populations are distinguished: a primary, heat-sensitive population and a secondary, more resistant population. This leads to a survival curve with two distinct regions in the linearly decreasing phase, a biphasic curve, as described as follows:

$$\frac{N}{N_0} = \frac{f\left(1+\exp\left(-k_1\lambda\right)\right)}{1+\exp\left(k_1\left(t-\lambda\right)\right)} + \frac{\left(1-f\right)\left(1+\exp\left(-k_2\lambda\right)\right)}{\left(1+\exp\left(k_2\left(t-\lambda\right)\right)\right)} \tag{7.16}$$

Traditional D values may be calculated from Equation 7.4 as $\ln(10)/k$ for each population. The value of λ can be set to zero in case lag phases were not present. It has been successfully applied to survival curves of *L. monocytogenes* [29,30], *Staphylococcus aureus* [31], and *Enterococcus faecium* [32] during HTST pasteurization. Although the model can fit six different shapes of survival curves [33] (Figure 7.1), it is entirely derived from an empirical logistic equation, which again, results in its main weakness. First, the shoulder phase λ does not visually coincide with a time period where the microbial population remains at the inoculum's level. Second, in contrast to what could be intuitively expected, when $\lambda=0$ and $f=1$, the simplified format of the model does not coincide with the classical first-order kinetic model [21], and therefore, it is not capable of accurately fitting the linear phase of curves and those with a shoulder [33]. The Whiting and Buchanan model [28] is also purely empirical.

To overcome these weaknesses, Xiong et al. [33] developed a more robust four-parameter log-logistic model based on both the Whiting and Buchanan [28] inactivation model and the Cerf [26] model, and claimed that it could successfully fit the survival curves of *L. monocytogenes* and *Salmonella enteritidis* PT4. The following equation can model the six different survival curve shapes presented in Figure 7.1:

$$\text{Log}\left(\frac{N}{N_0}\right) = \begin{cases} 0 & \left(0 \le t \le \lambda\right) \\ \log\left(f\cdot\exp\left(-k_1\left(t-\lambda\right)\right)+\left(1-f\right)\cdot\exp\left(-k_2\left(t-\lambda\right)\right)\right) & \left(t>\lambda\right) \end{cases} \tag{7.17}$$

Compared with the Whiting and Buchanan [28] model, it produces a better fit for all types of survival curves [33]. Furthermore, the well-known first-order kinetics model can be derived from Xiong's model but not from the Whiting and Buchanan's model. However, both the Buchanan model and the Cerf's model are special cases of the Xiong's model. For the linear phase of the survival curves, the Whiting and Buchanan model presents a slight curvature while Xiong's model produces a straight line. However, Xiong's model assumes that microbial population in the shoulder phase is constant and still assumes log-linearity in both two microbial population fractions, which may not be the case for all survival curves. The model is not completely dynamic due to its discontinuities from the shoulder phase to the linear death phase (i.e., not differentiable at $t=\lambda$). This suggests that the model cannot cope with realistic temperature fluctuations in a consistent way; that is, if realistic time-varying temperature profiles occur, Xiong's model can only be used if the temperature profile is approximated by a (finite) number of temperature steps [21]. In other words, the model cannot deal with the previous history of the microbial population.

Biphasic modeling has been successfully applied to describe the inactivation kinetics of several microorganisms in several food matrices as a function of temperature and also of pressure. For example, Reyns et al. [34] reported a biphasic inactivation curve for the yeast *Zygosaccharomyces bailii*, with a first part covering four to six decimal reductions and obeying first-order kinetics, followed by a tail corresponding to a small fraction of cells that were inactivated at a much lower rate. Also van Opstal et al. [35] reported biphasic pressure–temperature inactivation curves for *Escherichia coli* in Hepes-buffer and, to a minor extent, in carrot juice for specific pressure/temperature/time combinations. The biphasic models have more flexibility compared to the first-order model and can provide a suitable alternative for fitting inactivation curves with tails. In addition, it preserves the physical meaning of the parameters used in the first-order model.

7.2.1.6 Log-Logistic Inactivation Model

Another variation from the logistic function, the log-logistic model, has been proposed by Cole et al. [36] assuming a distribution of heat sensitivity within the population of heated cells, and has been applied to the thermal destruction of *L. monocytogenes* [36,37], *S.* Typhimurium [38,39], and *E. coli* [19], which is presented as follows:

$$Y = \alpha + \frac{\omega - \alpha}{1 + \exp\left(\left(4\sigma\left(\tau_1 - \log(t)\right)\right)/(\omega - \alpha)\right)} \tag{7.18}$$

The log-logistic modeling approach has also been applied to the survival of *Yersinia enterocolitica* at suboptimal pH and temperature [40]. Modeling survival curves of *L. monocytogenes* Scott A inactivated by high hydrostatic pressure in whole milk, Chen and Hoover [41] found that the log-logistic model produced consistently best fits to all survival curves, while the modified Gompertz model (to be presented next) the poorest. However, the log-logistic model was found to be inferior at predicting inactivation of *L. monocytogenes* at temperature levels other than the experimental temperatures, in contrast to the Weibull model, which provided reasonable predictions [41]. Still, an important disadvantage of this model is the lack of physical meaning of the model parameters.

7.2.1.7 Modified Gompertz Model

The modified Gompertz equation was originally proposed by Gibson et al. [42] to model growth curves and was later adapted to model heat and pressure inactivation kinetics. Bhaduri et al. [43] first demonstrated that the empirical modified Gompertz was effective in modeling nonlinear survival curves for *L. monocytogenes* during heating processes and produced more accurate estimations of thermal inactivation behavior of microbes than the first-order kinetic model. The empirical modified Gompertz is

$$\log N = A - C\exp\left(-\exp\left(-B\left(t - M\right)\right)\right) \tag{7.19}$$

McMeekin et al. [44] derived the kinetic parameters: maximum death rate μ_m, lag-phase duration λ, minimum cell concentration N_0, and tailing ratio q_G from the modified Gompertz model, as follows:

$$\mu_m = \frac{BC}{e} \tag{7.20}$$

$$\lambda = M - \frac{1}{B} + \frac{\log N_0 - A}{BC/e} \tag{7.21}$$

$$\log N_{res} = A - C = \log N_0 + C\exp\left(-\exp\left(BM\right)\right) - C \tag{7.22}$$

$$\log q_G = \log\frac{N_{res}}{N_0} = C\exp\left(-\exp\left(BM\right)\right) - C \tag{7.23}$$

To avoid the direct use of different initial numbers N_0 in the case of multiple experiments, the modified Gompertz was reparameterized as follows and applied to describe nonlinear survival curves of *L. monocytogenes* in infant formula [45,46]:

$$\log\frac{N}{N_0} = C\exp\left(-\exp\left(BM\right)\right) - C\exp\left(-\exp\left(B\left(t-M\right)\right)\right) \tag{7.24}$$

$$\log\frac{N}{N_0} = -C\,\exp\left\{-\exp\left[\frac{\mu_m e}{C}\left(\lambda - t\right)+1\right]\right\} \tag{7.25}$$

This highly flexible Gompertz equation is capable of fitting survival curves, which are linear, display an initial shoulder followed by a linear course, or are sigmoidal.

Other applications of the Gompertz equation include the effect of combined high pressure and mild heat on the inactivation of *E. coli* and *Staphylococcus aureus* in milk and poultry [47] and the inhibition of *Enterobacteriaceae* and clostridia during sausage curing [48]. However, this model is only effective in modeling sigmoidal curves due to its structural limitations [21] and cannot accurately model the linear death phase. This is confirmed in the author's study [21] to model *Bacillus cereus sous vide* heating processes. They found that the Gompertz model is inappropriate to describe the survival curves without a tail phase. The modified Gompertz model has also been found to produce a poorer fit than the log-logistic model and Weibull models [41,49]. As the modified Gompertz is not derived from mechanistic considerations which incorporate all intrinsic and extrinsic variables governing cellular metabolism and interpreting the modeled response in terms of known phenomena and processes, it has not been adequately validated to provide a reasonable biological interpretation of parameters. Consequently, other models were developed to provide a possible interpretation of parameters.

7.2.1.8 Baranyi Model

Baranyi et al. [50] proposed to consider an inactivation curve as the mirror image of the well-known microbial growth model given by Baranyi et al. [51] and outlines a procedure for its transformation. Xiong et al. [33] later adapted their mechanistic microbial growth model to model inactivation curves redefining $\alpha(t)$ as a shoulder adjustment function and $\beta(t)$ as a tailing adjustment function, as given in the following:

$$\frac{\mathrm{d}N}{\mathrm{d}t} = -\mu_m N\alpha\left(t\right)\beta\left(t\right) \tag{7.26}$$

$$\alpha\left(t\right) = 1 - \frac{r^{nC}}{r^{nC} + t^{nC}} \tag{7.27}$$

$$\beta\left(t\right) = 1 - \frac{N_{res}}{N} \tag{7.28}$$

In the integrated logarithm base 10 form, Xiong et al. [33] proposed the use of the following equation:

$$\log\frac{N}{N_0} = \log\left(q_B + \left(1-q_B\right)\exp\left(-\mu_m\left(t - B\left(t\right)\right)\right)\right) \tag{7.29}$$

$$B\left(t\right) = \frac{r}{3}\left(\frac{1}{2}\ln\frac{\left(r+t\right)^2}{r^2 - rt + t^2} + \sqrt{3}\arctan\frac{2t-r}{r\sqrt{3}} + \sqrt{3}\arctan\frac{1}{\sqrt{3}}\right) \tag{7.30}$$

where

n_c is the curvatural parameter

The lag parameter r is the time required for the relative death rate to reach half of the maximum relative death rate μ_m

q_B is the tailing ratio N_{res}/N_0, which can be used to indicate whether or not a tailing region exists

The benefit of using the tailing ratio is to avoid the direct use of the initial numbers N_0, which in the case of multiple experiments, are usually different [45,46]. Comparing this model with the modified Gompertz, Xiong et al. [33] found that for the validation data, the Baranyi equation produced better fit than the Gompertz model, and they highlighted the fact that the first-order kinetics model can be derived from the Baranyi model but not from the Gompertz model, and that the Baranyi model is more robust, as it can use three or less parameters to fit the four commonly observed survival curves: log linear, with tail only, with shoulder only, and with tail and shoulder. Deducted in this way, the modified Baranyi model for microbial inactivation is of empirical nature.

7.2.1.9 Lambert Model

Building upon nonlinear disinfection kinetics, Lambert [52] reformulated a previous model for biocide inactivation [53] using the idea of the Hom time exponent h_d [54] and a maximum achievable log reduction (M_d) to describe the various shapes of the log survivor time plots. The reduction in microbial numbers R_d is defined as follows:

$$\log R_d = M_d\left[1-\exp\left(K_{0,T}t^{h_d}\right)\right] \tag{7.31}$$

The inactivation rate constant $K_{0,T}$ is dependent on temperature and can be substituted by an Arrhenius expression. Likewise, Peleg and Cole [55] suggested that the time exponent h_d itself has temperature dependence. Therefore, Lambert [52] suggested that h_d be substituted with the log of the Arrhenius equation, producing the following five-parameter empirical model:

$$\log R_d = M_d\left[1-\exp\left(-10^{(P_1-(P_2/T))}t^{(P_3-(P_4/T))}\right)\right] \tag{7.32}$$

This model described well data of thermal inactivation of *C. botulinum* (Figure 7.2), *Bacillus stearothermophilus*, *Salmonella anatum*, *Pseudomonas viscose*, and *Streptococcus faecalis*. Furthermore, the Lambert model underwent a purely empirical modification to include the effects of pH and water activity on the thermal inactivation of *Salmonella bedford*. In all cases, the value of M_d reflected the size of the initial inoculum while time exponents higher than one were indicative of lags before the onset of inactivation. This was in agreement with Peleg and Cole [55] who suggested that it is possible for survival curves to change their concavity as a result of changing external conditions.

7.2.1.10 Geeraerd Models

7.2.1.10.1 Log-Linear Decay with Shoulder and Tail

With interest in describing the microbial inactivation during mild heat treatments such as sous vide or cook-chill, Geeraerd et al. [21] developed a non-log linear dynamic inactivation model. They explained that a dynamic model (i.e., differential equations) is needed in order to be able to evaluate the influence of a temperature increase or decrease within a food product, by integrating the model equation over the time course, taking into account the time dependence of the environmental conditions. When trying to identify a static model (i.e., explicit solution of a dynamic model only valid during nonvarying environmental conditions) on dynamic experimental data, this can only

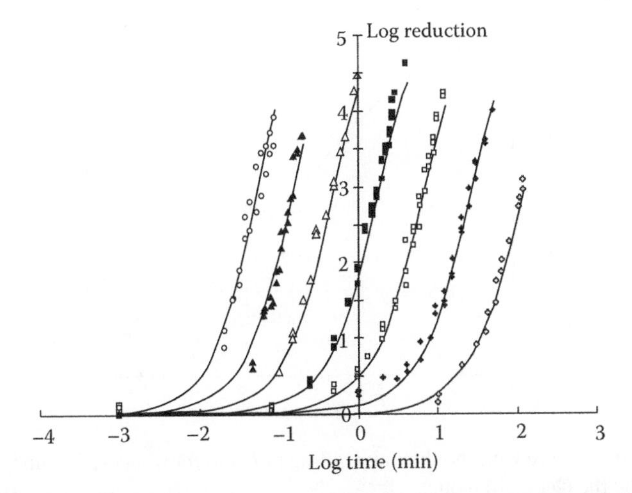

FIGURE 7.2 Use of the Lambert model for the representation of the thermal inactivation of *C. botulinum* between 373.1 and 404.5 K in intervals of 4.8 K. Markers indicate the observed log reductions, solid lines the fitted inactivation curves.

be done by *resetting* certain values at every environmental change, a procedure that becomes very complicated. Geeraerd et al. [21] extended the log-linear model to account for shoulder and tail with transition functions or adjustment factors as two-coupled differential equations:

$$\frac{\mathrm{d}N}{\mathrm{d}t} = -k_d N\left(\frac{1}{1+C_C}\right)\left(1-\frac{N_{res}}{N}\right) \tag{7.33}$$

$$\frac{\mathrm{d}C_C}{\mathrm{d}t} = -k_d C_C \tag{7.34}$$

where the first factor (Equation 7.16) models the log-linear part of the inactivation curve, and the second factor (Equation 7.17) describes the shoulder effect. Equation 7.16 is defined by four parameters: N_0, k_d, N_{res} (residual population density), and $C_C(0)$ (initial physiological state of the cells). It is assumed that in the beginning of the inactivation, a whole pool of components C_C is present around or in each cell; gradually this component is destroyed. Thus, the shoulder phase in the model can be interpreted by Michaelis–Menten kinetics–based adjustment factor $(1/(1+C_C))$. The value of this factor approaches 0 at the beginning of the shoulder region, and approaches 1 toward the end of the shoulder region. The tail phase is reflected by the additional factor $(1-N_{res}/N)$, which implies the existence of a subpopulation N_{res}. In this model, the length of the shoulder, the inactivation rate, and the level of tailing are not dependent upon each other unlike the earlier non-log linear inactivation models of Casolari [20] and Sapru et al. [22]. However, the Geeraerd log linear model with shoulder and tail remains an empirical model and is a reparameterization of the Baranyi model [21,56].

After substituting $C_C(0)$ by "$\exp(k_dS)-1$," with S being a parameter representing the shoulder, the following explicit solution can be obtained:

$$\log N = \log\left\{\left(10^{\log N_0} - 10^{\log N_{res}}\right).\exp(-k_d t)\left(\frac{\exp(k_d S)}{1+\left(\exp(k_d S)-1\right)\exp(-k_d t)}\right)+10^{\log N_{res}}\right\} \tag{7.35}$$

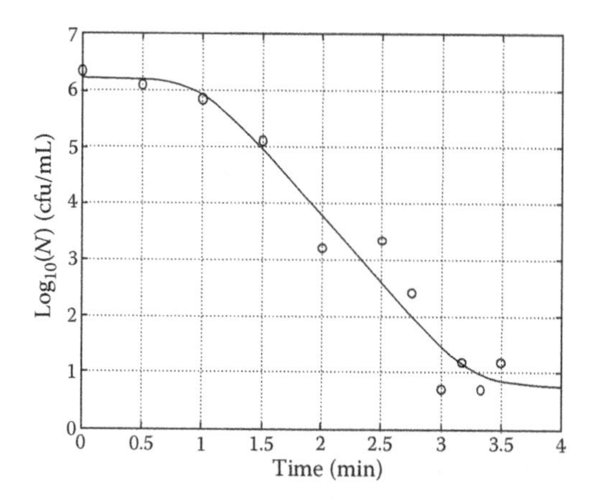

FIGURE 7.3 Inactivation curve with shoulder and tailing of *L. monocytogenes* in minced beef in vacutainers at 50°C as described by the Geeraerd model.

The Geeraerd model [21] is capable of simulating independently a smooth initiation (shoulder phase) and/or saturation (tail phase) during a mild heat treatment (Figure 7.3) and offers a possible interpretation from a mechanistic point of view about microbial survival and the tail phenomenon. This model has been effectively used to represent the survival of *Salmonella enterica* and *L. monocytogenes* [57]; the survival of *Lactobacillus sakei* during a mild thermal inactivation [21], *Monilinia fructigena* and *Botrytis cinerea* during a pulsed white light treatment [58] and the mild temperature inactivation of *E. coli* K12 [59]. However, as the model was developed for mild heat treatment, it seems unable to deal with steep temperature profiles [13].

7.2.1.10.2 Biphasic with Shoulder

In order to describe biphasic curves with shoulder, Geeraerd et al. [60] combined features of the Geeraerd [21] model and the Cerf [26] model. The model is defined by five parameters: log N_0, f, k_1, k_2, and S. If $S=0$ (after identification on experimental data), the model reduces to the Cerf model [60]. In dynamic conditions, the model is characterized by the following equations:

$$\frac{dN_1}{dt} = -k_1 N_1 \left(\frac{1}{1+C_C}\right) \tag{7.36}$$

$$\frac{dN_2}{dt} = -k_2 N_2 \left(\frac{1}{1+C_C}\right) \tag{7.37}$$

$$\frac{dC_C}{dt} = -kC_C \tag{7.38}$$

and in static conditions by the following:

$$Y = \log N_0 + \log\left[\left(f \cdot \exp(-k_1 t) + (1-f) \cdot \exp(-k_2 t) \cdot \frac{\exp(k_1 \cdot S)}{1+\left(\exp(k_1 \cdot S)-1\right) \cdot \exp(-k_1 t)}\right)\right] \tag{7.39}$$

7.2.1.11 Weibull Model

A survival curve, like a dose-response curve, is by definition the cumulative form of the resistances distribution within the microbial population, measured in terms of the time or dose at which an individual cell or spore is inactivated [16,61–63]. This had been proposed several times [14,15] but only later received a more widespread attention in the food microbiology community. In an early research, Kilsby et al. [64] hypothesized that the genotypically homogeneous bacterial population under lethal stress has an underlying distribution of inactivation times. While Peleg [61] hypothesized that in a given population, the resistances to a lethal agent could be described by a Fermi distribution, Kilsby et al. [64] were the first to propose the use of the normal and the Prentice distributions to describe the distribution of the logarithm of inactivation times. This enabled the calculation of a 50% lethal dose for the destruction of bacteria and provided better fit than the models previously developed at that time. Unfortunately, the procedure for fitting these distributions was not proven to be straightforward by their proponents.

However, the Weibull distribution led the way [16,18,61,65,66] within this type of vitalistic [67] or probabilistic models. Since death or inactivation can be considered a failure phenomenon, a Weibullian distribution is expected to be quite common, as it has been observed in many unrelated processes that involve breakup and disintegration.

The model can be expressed as follows, and it becomes the first-order kinetic model when $\beta = 1$:

$$\log \frac{N}{N_0} = -\left(\frac{t}{\chi}\right)^{\beta} \tag{7.40}$$

For $\beta > 1$ (shape parameter), convex curves are obtained, and for $\beta < 1$, concave curves are described. The scale parameter χ can be denoted as the time for the first decimal reduction. This is distinguished from the conventional D value; the significance of the χ value is restricted to the first decimal reduction of surviving spores or cells from N_0 to $N_0/10$ [65].

Although the Weibull model is of empirical nature and the physical meaning of the parameters is less obvious compared to other models, a link with physiological effects has been proposed by van Boekel [18]. $\beta < 1$ was suggested to indicate that the surviving cells at any time point in the inactivation curve have the capacity to adapt to the applied stress, whereas $\beta > 1$ would indicate that the remaining cells became increasingly damaged. The χ scale factor of the distribution is expected to be hardly influenced by temperature [18], and this can indeed be observed in the isothermal inactivation patterns of a variety of microbial cells and spores [17,18,62,68–71]. If the temporal distribution of the lethal events is indeed Weibullian, then the β parameter is expected to be either temperature independent or only weakly affected by temperature [18]. This has been confirmed for several types of microorganisms, which could be described either way with almost the same degree of fit [69,71,72].

Van Boekel [18] and Mafart et al. [65] observed a strong correlation between the parameters β and χ. The dependency of the parameters is due to the model structure; that is, an error in β will be balanced by an error in χ. Such an autocorrelation causes certain instability of parameters estimates. Nevertheless, this drawback can be avoided by fixing the value of β at an average value, characteristic of a strain, so that N_0 and χ can be estimated from a linear regression [65,73].

The main advantages of the model are its simplicity, ease of fit, and capability of modeling linear survival curves as well as those containing shoulder and/or tail regions [74]. It has been effective in modeling survival curves of *C. botulinum*, *B. stearothermophilus* spores, *S.* Typhimurium, *L. monocytogenes*, *B. cereus* spores, *Bacillus pumilus* spores, *Y. enterocolitica*, among others [55,65,66,74,75]. It has also been successfully used to model inactivation of *L. monocytogenes* by simultaneous application of high pressure and mild heat [41] and by high pressure alone [76], high pressure inactivation of a variety of *Vibrio* spp. in pure culture and inoculated oysters [77], high pressure inactivation of *Y. enterocolitica* in phosphate buffer and UHT whole milk [49], among others.

However, the validation shows that the empirical nature of the Weibull model prevents it from building up a strong relationship between the known behavior of microbes and the parameters of the model, which sometimes leads to conflicting results in investigating the effect of environmental factors on its parameters, particularly for the shape parameter β [13].

The application of the Weibull distribution, modeling upward and/or downward concavity of survival curves, does not render the traditional concept of the F value—which was developed under the first-order kinetic assumption. Mafart et al. [65] presented an adaptation of the Bigelow method based on the Weibull distribution for assessing the efficiency of sterilization. When a Weibull distribution is assumed, unlike the first-order kinetic mode, the F values are no longer additive, and the destruction ratio is no longer proportional to this value. Therefore, the F concept loses a great part of its relevance, and instead the decimal reduction ratio n becomes the only convenient indicator of the heat treatment efficiency [65,71]. At variable temperature, the decimal reduction ratio can be computed as follows:

$$n_d = \beta \int_0^t \left[\frac{L(T)}{\chi_{ref}} \right]^{\beta} t^{\beta-1}\, dt \tag{7.41}$$

and in discretized form as follows:

$$n_d = \beta \sum \left[\frac{L(T_i)}{\chi_{ref}} \right]^{\beta} t^{\beta-1} \Delta t_i \tag{7.42}$$

As the standard F value remains an interesting criterion that allows to intrinsically compare several heat treatments, regardless of the target species which is to be destroyed, Mafart et al. [65] also proposed an F value (F_{adj}) adjusted according to the β and z values of the target microorganism as given in the following:

$$F_{adj} = n^{1/\beta} \chi_{ref} \tag{7.43}$$

Albert and Mafart [56] proposed a modified Weibull model to additionally describe the shoulder and tailing phenomena to the concave or convex curves (already described by the simple Weibull model). The model consists of four parameters or three when there is no presence of tailing ($N_{res} = 0$):

$$\log N = \left(N_0 - N_{res} \right) 10^{-(t/\chi)^{\beta}} + N_{res} \tag{7.44}$$

The modified Weibull model possesses the structural requirements as indicated by Geeraerd et al. [21] to describe microbial inactivation. When compared with the Geeraerd model, little difference was found (Figure 7.4), except that the shoulder effect of the modified Weibull is sometimes less marked and that the Geeraerd model does not allow fitting upward concavity curves.

7.2.1.12 Two-Mixed Weibull Distribution Model

The two-mixed Weibull distribution model has been conceived as a new primary model, based on two Weibull distributions of cell resistance, to describe survival curves and change in the pattern with the modifications of resistance of two assumed subpopulations. Despite the number of proposed models, Coroller et al. [78] argued that none of them was flexible enough so that it could reflect all changes of shapes with the intensity of stress or with the physiological state of the cells. In order to describe all shapes of inactivation kinetics, Coroller et al. [78] assumed that the

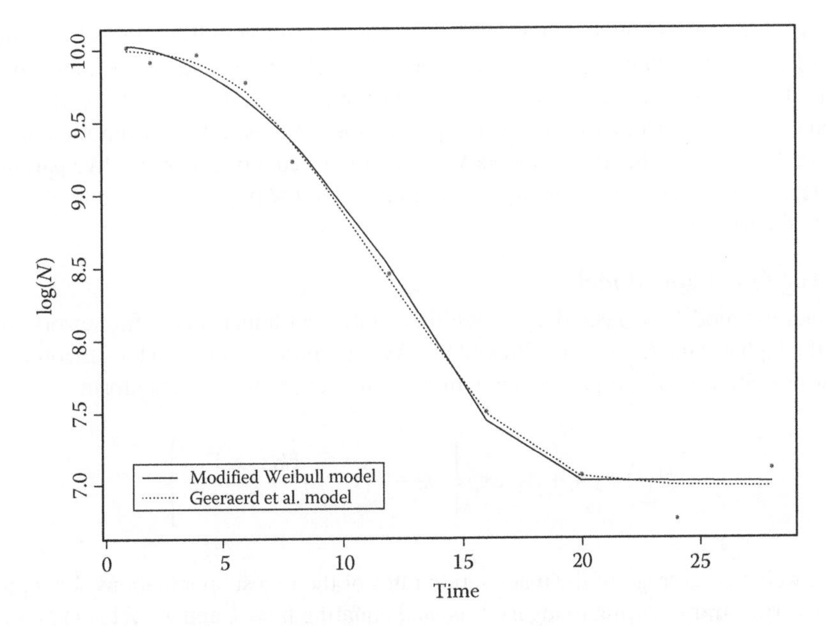

FIGURE 7.4 A comparison between the modified Weibull model and the Geeraerd model describing *L. innocua* kinetics in cooked meat at 55°C.

bacterial population is composed of two groups that differ in their levels of resistance to stress, which are supposed to follow Weibull distributions. As without mathematical transformation, the fraction f of subpopulation 1 in the total population provides insufficient discrimination; a new parameter α_W was introduced based on the logit transformation of f as follows:

$$\alpha_W = \log\left(\frac{f}{1-f}\right) \tag{7.45}$$

Assuming that subpopulation 1 is more sensitive to stress than subpopulation 2 ($\chi_1 < \chi_2$), the general Weibull model becomes

$$N(t) = \frac{N_0}{1+10^{\alpha_W}}\left(10^{-(t/\chi_1)^{\beta_1}+\alpha_W} + 10^{-(t/\chi_2)^{\beta_2}}\right) \tag{7.46}$$

When enumeration at low concentrations is possible, the right part of the curve, corresponding to the most resistant population, subpopulation 2, seems to be convex, like the curve for the most sensitive subpopulation 1. The authors [78] then proposed a simplification of the model by which the shape parameters β_1 and β_2 of the subpopulations be the same. The simplified general Weibull model then becomes

$$N(t) = \frac{N_0}{1+10^{\alpha_W}}\left(10^{-(t/\chi_1)^{\beta}+\alpha_W} + 10^{-(t/\chi_2)^{\beta}}\right) \tag{7.47}$$

When Coroller et al. [78] subjected cells of *L. monocytogenes* and *S.* Typhimurium taken from six stages of growth to stress, the bacterial resistance increased from the end of the exponential phase to the late stationary phase. Moreover, the shapes of the survival curves gradually evolved as the physiological states of the cells changed. The double Weibull simplified model represented the adaptation of cells better than the Whiting and Buchanan [28] model and produced narrowed confidence intervals. Whiting's model presents five parameters, four (k_1, k_2, λ, and f) of which characterize the evolution of the resistance of the overall population with respect to the duration of incubation of subculture. On the other hand, the double Weibull simplified model has also five parameters, but only three (χ_1, χ_2, and α_W) of these parameters are related to the physiological state of the cells and environmental conditions.

7.2.1.13 Log-Quadratic Model

The log-quadratic model has been derived within a statistical linear model framework to approximate both the biphasic model of Cerf [26] and the Weibull model. Using Taylor's theorem and some approximations, Stone et al. [79] arrived to a mixture model of the following form:

$$N(t) = (A_1 + A_2)\exp\left\{-\bar{k}t + f(1-f)\frac{(k_1 - k_2)^2}{2}t^2\right\} \quad (7.48)$$

where $\bar{k}$ is a weighted average of the inactivation rates of the two subpopulations, $\bar{k}=fk_1+(1-f)k_2$. In statistical terms, transforming to logarithms, and equating $\beta' = -\bar{k}$ and $\gamma' = f(1-f)(k_1-k_2)^2/2$, the equation becomes

$$\log N_i = \alpha'_{Ri} + \beta'_{Ti}t_i + \gamma'_{Ti}t_i^2 + \varepsilon_i \quad (7.49)$$

The regression coefficient α' is an intercept used for each replicate, β' is the rate parameter which depends on the temperature; and the random variation ε is assumed to be normally distributed with constant variance across all replicates and temperatures. The regression coefficient γ depends also on the temperature, and its positive value represents concave upward curvature. The merit of this model is its simplicity, the possibility to fit *at once* to a series of isothermal inactivation data with different replicates, alleviating the need for any secondary model on the dependence of β'_T (parameter related to the D values) on temperature. One-step fitting eliminates the asymmetric treatment of parameters and allows joint optimization. Furthermore, confidence intervals about fitted values are readily obtained under the assumed normal distribution of random variation. Data on *C. botulinum* inactivation at different temperatures (isothermal experiments) could be better described by the proposed log-quadratic model than the log-linear model and the two variants of the Weibull model (fixed and variable shape parameter).

Nevertheless, this model has been derived under the assumption that the inactivation rates of the subpopulations k_1 and k_2 are sufficiently closed to each other. Furthermore, Stone et al. [79] warned that the log-quadratic model admits the possibility of a fit that shows increasing microbial counts over time for times greater than some large threshold. Although the situation is very unlikely to arise when data that show decreasing microbial counts over time are interpolated, still it does mean that the log-quadratic model is unsuitable for extrapolation.

7.2.1.14 Power-Law Memory Kernel Model

Kaur et al. [80] developed a survival modeling approach intended to overcome the limitations of the Weibull model. They argued that the Weibull or power-law model was less suited to predict the sigmoidal shape of curves formed due to shoulders and tails and therefore devised a more flexible inactivation model based on fractional differential equations (FDEs) where the effect of prior process history is incorporated by a memory kernel function. They considered a one-term FDE and a two-term FDE. The solution of the one-term FDE leads to the Weibull equation, while two-term FDE is expressed as follows:

$$ {}_0D_t^{\alpha} \log S(t) + a_0 \log S(t) = -a_1 \tag{7.50} $$

where

α and a_0, a_1 are three parameters needed for predicting the log $S(t)$

${}_0D_t^{\alpha}$ is the Riemann–Liouville derivative of fractional order α, defined as follows:

$$ {}_0D_t^{\alpha}\left[\log S(t)\right] \equiv \frac{1}{\Gamma(v)} \int_0^t (t-\varepsilon)^{v-1} \log S(\varepsilon) d\varepsilon \tag{7.51} $$

where

v is a real number

$\Gamma(v)$ is the gamma function

The function $(t-\varepsilon)^{v-1}$ plays the role of a memory kernel

After several transformations, the two-term FDE is solved to produce

$$ \log S(t) = -a_1 t^{\alpha} E_{\alpha,\alpha+1}\left(-a_0 t^{\alpha}\right) \tag{7.52} $$

where $E_{\alpha,\alpha+1}(-a_0 t^{\alpha})$ is the two-parameter (α, $\alpha+1$) Mittag–Leffler function:

$$ E_{\alpha,\alpha+1}\left(-a_0 t^{\alpha}\right) = \sum_{k=0}^{\infty} \frac{\left(-a_0 t^{\alpha}\right)^k}{\Gamma(\alpha k + \alpha + 1)} \tag{7.53} $$

The parameters a_0 and a_1 were found to increase with increasing temperatures for *Salmonella*, *E. coli*, and *L. monocytogenes* in all different meats. The a_1 parameter is directly proportional to the effect of external process evolution on microbial destruction; it acted like a scaling parameter. The function $t^{\alpha} E_{\alpha,\alpha+1}(-a_0 t^{\alpha})$ behaves like a memory kernel and affects the slope or shape of survival curves at different point times. These properties allowed the two-term FDE based model to predict shoulder and tails for all isotherms with greater accuracy as compared to the Weibull and log-linear models. Values of α different from 1 in all cases depicted that bacterial inactivation did not follow integer-order kinetics [80]. Vaidya and Corvalan [81] efficiently predicted microbial isothermal and nonisothermal lethality using a slightly different FDE-based model, and warned that for $\alpha<1$ the model becomes a relaxation model, which tends to revert to its unstressed configuration when the applied thermal stress is reduced. Although the fitting of the power-law memory kernel inactivation model appears somewhat complex, Kaur et al. [80] sustained that the main advantage of this model is its flexibility as it can take into account varied shapes of survival curves. Yet, the FDE model is based on the reasonable assumption that the microbial cell population is a mixture of different subpopulations in a food product, and the total death time can be expressed with the help of memory functions like the Mittag–Leffler functions and fractional order differentiation. The only disadvantage is that the memory kernel model has three parameters in comparison to the two parameters of the Weibull model.

7.2.2 Secondary Inactivation Models

Models that describe the effect of environmental conditions, that is, physical, chemical, and biotic features, on the values of the parameters of a primary model are termed secondary models. Secondary models can be divided into three groups according to their modeling approaches. The first group, encompassed by the Bigelow-type, log-logistic, square-root, and Arrhenius models, was established to provide high quality of fit based on the (at least partial) biological interpretability of the parameter values. The second group of models, that is, polynomial or surface response models and artificial neural network (ANN) models, describes empirically an underlying, unknown relationship between microbial evolution parameters and environmental influences. The third group of models has been deduced on mechanistic grounds based on reaction kinetics and thermodynamic considerations and therefore can be deemed as mechanistic.

7.2.2.1 Bigelow-Type Models

In the classical Bigelow model [2], the decimal reduction time D is only dependent on the temperature as given as follows:

$$D = D_{ref} \times 10^{(T_{ref} - T)/z} \tag{7.54}$$

An example of predicted D values for *B. cereus* and *Clostridium perfringens* spores in pork luncheon meat calculated using the classical Bigelow equation (z values 8.6°C and 8.3°C, respectively) is shown in Figure 7.5 [82]. Mafart and Leguerinel [83] used an extension of Bigelow equation and developed a model with three parameters to describe the effects of temperature and pH on heat resistance. In parallel with bacterial growth kinetics, this appears to be an extension of the gamma concept [84,85]:

$$D = D_{ref} \times 10^{(T_{ref} - T)/z_T} \times 10^{(\mathrm{pH}_{ref} - \mathrm{pH})/z_{\mathrm{pH}}} \tag{7.55}$$

Extending the model of Mafart and Leguerinel [83], Gaillard et al. [86] added to the equation a factor for the effect of water activity to describe the heat resistance of *B. cereus* spores. This model as shown in the following however did not take into account the clear interaction between factors, which limited its range of validity:

$$D = D_{ref} \times 10^{(T_{ref} - T)/z_T} \times 10^{\left((\mathrm{pH}_{ref} - \mathrm{pH})/z_{\mathrm{pH}}\right)^2} \times 10^{(1 - Aw)/z_{Aw}} \tag{7.56}$$

The D_{ref} parameter of the original Bigelow equation can be replaced by the asymptotic D_{ref} to describe the negative inverse of the slope of the log-linear part, after a shoulder region [87,88].

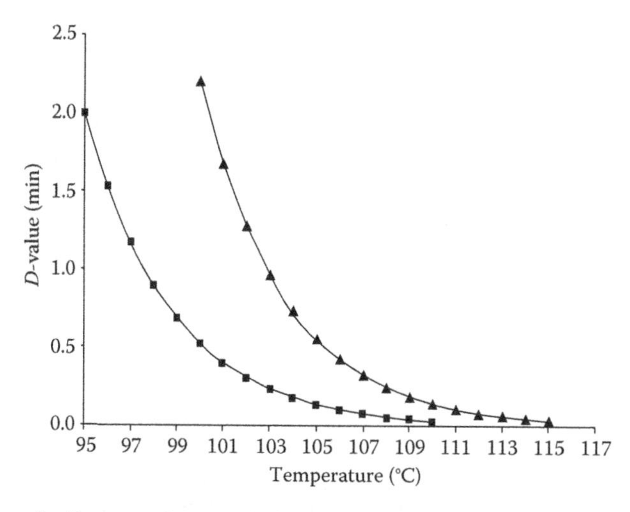

FIGURE 7.5 D values for *B. cereus* (square marker) and *C. perfringens* (triangle marker) spore cocktail in pork luncheon meat predicted by the classical Bigelow model.

Valdramidis et al. [89] compared this Bigelow-type secondary model against a polynomial and an Arrhenius-type model to describe dynamic conditions, and found that the Bigelow-type equation was the most suitable to model the inactivation kinetics of *L. monocytogenes* Scott A in macerated potatoes under surface dry heating conditions at holding temperatures of 90°C and 100°C and lowering A_w values, when coupled with the inactivation primary model of Geeraerd et al. [21].

An equivalent version of the Bigelow equation was proposed to describe the effect of the temperature on the scale parameter χ of the Weibull primary model (Mafart et al., 2002):

$$\chi = \chi_{ref} \times 10^{(T_{ref}-T)/z} \tag{7.57}$$

Later on, Jagannath et al. [70] extended this equation to describe the heating temperature, pH, and water activity influence on the same scale parameter χ of the Weibull model. Using that secondary equation, the thermal inactivation of *B. subtilis* spores heated in milk, rice porridge, and soy sauce were characterized, with $\chi_{ref\text{-}matrix}$ being a constant specific to a food matrix:

$$\chi = \chi_{ref\text{-}matrix} \times 10^{(T_{ref}-T)/z_T} \times 10^{((\mathrm{pH}_{ref}-\mathrm{pH})/z_{\mathrm{pH}})^2} \times 10^{(1-Aw)/z_{Aw}} \tag{7.58}$$

7.2.2.2 Arrhenius-Type Models

The classical Arrhenius equation was derived empirically from thermodynamic considerations [90] to describe the relationship between the bacterial specific death rate and the temperature [44], and is described as follows:

$$k_d = k_0 e^{-E_A/RT} \tag{7.59}$$

In practice, the natural logarithmic transformation format is always used. If k_0 and E_A are constant with temperature, the logarithm of Equation 7.29 describes a linear relationship in a plot of ln(k) against 1/T. Notwithstanding its great popularity, Peleg et al. [71] argued that the Arrhenius equation would not be the best model of temperature effect on microbial inactivation, even if the isothermal rate constant could be meaningful determined. They argument that there is no reason to assume that the destructive process caused by heat must always have a fixed energy of activation E_A, whatever the term means in the context of nonlinear kinetics. Moreover, they point out that the logarithmic transformation of the rate constant is usually unnecessary because its range would rarely, if ever, span several orders of magnitude. The Arrhenius model also implies that the destructive effect of heat is qualitatively the same at low and high temperatures, which is clearly not the case in microbial inactivation.

On the other hand, it has been recognized for several decades that other factors, especially water activity and pH, influence thermal death rates of bacteria. Davey et al. [91] were the first to develop an additive, linear Arrhenius-type model as shown in the following for the combined effect of process temperature and medium pH of thermal resistance of spores:

$$\ln k_d = C_0 + \frac{C_1}{T} + C_2\mathrm{pH} + C_3\mathrm{pH}^2 \tag{7.60}$$

Applicability of these models to thermal inactivation of *C. botulinum*, thiamine denaturation, and aerobic/anaerobic denaturation of ascorbic acid and combined effect of temperature and pH on heat resistance are also described [92,93]. The following alternative equation for the combined effect of temperature and liquid pH was used by Khoo et al. [94], Chiruta et al. [95], Holdsworth [96], and McMeekin et al. [44]:

$$\ln k_d = C_0 + \frac{C_1}{T} + \frac{C_2}{T^2} + C_3\mathrm{pH} + C_4\mathrm{pH}^2 \tag{7.61}$$

Gaillard et al. [86] reparameterized the aforementioned model slightly for predicting the heat resistance of *B. cereus*. Comparing among secondary models, Khoo et al. [94] indicated that this Arrhenius-type model was more suitable than both the square-root model and polynomial model for the representation of the thermal inactivation of *E. coli*, in terms of accuracy of predictions, relative complexity, ease of synthesis and ease of use, and potential for physiological interpretation of model form and coefficients. The major drawback that they observed was the inability of the model to adequately predict the numbers of viable *E. coli* from the start point of tailing when used with a first-order inactivation model.

Cerf et al. [97] proposed another five parameter model to accurately predict the combined effect of inactivation temperature, A_w, and pH on the thermal death of *E. coli*. This model is an extension of Davey's model and is described as follows:

$$\ln k_d = C_0 + \frac{C_1}{T} + C_2\text{pH} + C_3\text{pH}^2 + C_4 A_w^2 \tag{7.62}$$

McMeekin et al. [44] suggested that as some of the coefficients of the Arrhenius-type models present a close correlation, this model could be over-parameterized. Still, in common with many microbiology models, the additive Arrhenius model is of empirical nature.

7.2.2.3 Square-Root Models

The square-root model, a special form of the Ratkowsky–Belehradek model [44] has been used more or less exclusively to model bacterial growth, in both the growth and lag-phase of growth, and more recently in bacterial disinfection using UV light [98]. A square-root model for inactivation was applied by Khoo et al. [94] apparently for the first time to model the combined effects of temperature and pH on the thermal death rate, the model is shown as follows:

$$\sqrt{k_d} = b\left(T - T^*\right)\left(\text{pH} - \text{pH}^*\right) \tag{7.63}$$

where T^* and pH* are the nominal temperature and pH characteristic of the microorganism-medium interaction. Although this square-root model produced a good fit to the *E. coli* inactivation data, Khoo et al. [94] argued that the multiplicative nature of the model results in a more complicated structure than both the Arrhenius-type and the polynomial models.

7.2.2.4 Log-Logistic Model

It is well known that when a microbial population is heated, substantial mortality only occurs when the temperature reaches a certain lethal level. In this lethal range, the inactivation rate increases as the temperature is raised. Peleg et al. [61] and Campanella and Peleg [99] proposed the following log-logistic equation capturing these features to define the reciprocal of the location factor of the Weibull primary model ($b_W = 1/\chi^\beta$) in terms of temperature:

$$b_W\left(T(t)\right) = \ln\left[1 - \exp\left\{k\left(T(t) - T_C\right)\right\}\right] \tag{7.64}$$

This model was applied in modeling the thermal inactivation of *C. botulinum*, *Bacillus sporothermodurans* spores, and *S. enteritidis* cells [71,100]. Still, it is in the authors' opinion [61,99] that the model should only be used for time and temperature conditions covered by the available experimental survival data and not for extrapolation.

7.2.2.5 Polynomial Models

Polynomial or surface response models were first introduced into predictive microbiology by Gibson [42] to describe growth rate and lag time functions of storage temperature, pH, and salt concentration. They had been extensively used during the 1990s [101]. Polynomial models are attractive because they are relatively easy to fit as a multiple linear regression and allow virtually any of the environmental parameters and their interactions to be taken into account [6,11,102]. Despite their complexity, in combination with primary model, the polynomial models can provide reasonable predictions of the behavior of the microbes in food systems [6].

When thermal processes are developed for low-acid foods, understanding the relationship between temperature, pH, and the *D* value of the spore is very important and has been a subject of study in several investigations. A direct linear relationship between pH and the logarithm of the *D* values has been described for *E. coli* [103]. The extent by which the pH reduces the thermal resistance of bacterial spores seems to depend on different factors such as the strain investigated, the substrate [104] and the water activity [105]. The treatment temperature used also influences the effect of the pH on the *D* values. Fernandez et al. [104] studied the interaction between heating temperature and pH on the *D* values of *B. stearothermophilus* spore heated at temperatures between 115°C and 125°C and showed that increasing temperatures reduced the effect of pH. A polynomial model of the following type:

$$k_d = C_0 + C_1T + C_2T^2 + C_3\text{pH} + C_4\text{pH}^2 + C_5T\text{pH} \tag{7.65}$$

was found by Fernandez et al. [106] to closely describe the inactivation rate of *C. sporogenes* and *B. stearothermophilus* spores heated up to 125°C in terms of temperature and pH in a mushroom extract medium.

A second-order response surface model with interaction shown in the following was applied by Valdramidis et al. [89] to predict thermal inactivation rate of *L. monocytogenes* under dry heat conditions in terms of temperature and water activity, and produced results less satisfactory than the Bigelow-type model:

$$k_d = C_0 + C_1T + C_2T^2 + C_3A_w + C_4A_w^2 + C_5TA_w \tag{7.66}$$

They observed that decreasing the A_w of the macerated potato resulted in an increase microbial resistance at temperatures between 55°C and 65°C.

Polynomial models have also been used to assess the influence of other intrinsic properties of food, such as fat content, sugars, citric acid, sodium chloride, etc., on the *D* value of spores [107,108]. Nevertheless, secondary models of the polynomial type have not been employed exclusively to predict not only the death rate k_d or the *D* values but also other parameters from nonlinear primary models. For instance, a polynomial equation of the quadratic type was used by Blackburn et al. [19] to express the parameter τ_1 from the primary log-logistic model in terms of temperature, pH, and salt concentration, to reproduce the observed thermal inactivation of *S. enteritidis* and *E. coli* O157:H7 in broth. Similarly, Mattick et al. [109] used a response surface to predict the shape and location parameters of the Weibull primary inactivation model, in terms of temperature and water activity. They demonstrated that at temperatures higher than 70°C, *Salmonella* cells at low water activity were more heat tolerant than those at a higher water activity, but below 65°C the reverse was true. A polynomial model was used to characterize each of the three parameters of a modified-Gompertz model in terms of milk fat content, pH, and heating temperature to describe the thermal inactivation of *L. monocytogenes* in a formulated milk product [110]. Thus, as a general rule, a polynomial model illustrated in the following can be used to relate *any parameter* of a primary model (y) to the *environmental factors* (x):

$$y = a_0 + \sum a_i x_i + \sum_{i \neq j} a_{ij} x_i x_j + \sum a_i' x_i^2 \tag{7.67}$$

In order to stabilize the variance of the data, logarithmic transformation of the model parameters are generally used [33,42,102]. As disadvantages, polynomial models have properties that limit their usefulness as secondary predictive models. Using polynomial models, the thermal inactivation rate can only be predicted under various conditions within the range of the data set. Additionally, polynomial models are usually quadratic, thus being incapable to apply in complicated nonlinear cases; the collinearity problems between factors may exist, and the sensitivity analysis of input variables is difficult to perform because of the presence of crossed interactions (overparameterization). Finally, they have coefficients without biological interpretation.

7.2.2.6 Artificial Neural Network Models

ANN has been extensively applied in growth kinetics as a robust approach to correlate the primary model parameters with the environmental factors due to two reasons: (1) it can automatically derive mathematical formulae to map the relationships, and (2) it is capable of operating a large number of processing elements (neurons) in parallel to correlate large and complex data sets [111]. Thus, ANN has been successfully applied to predict the behavior of microorganisms, such as *Shigella flexneri* and *L. monocytogenes*, affected by environmental factors [112–120]. Surprisingly, the application of ANN models in microbial death kinetics has been very limited.

Lou and Nakai [121], using a multilayer feed-forward neural network type and a back-propagation learning algorithm, built an ANN model to provide a nonlinear relationship between water activity, pH, and temperature on the thermal inactivation rate of *E. coli* using the data sets of [103]. They selected a network with one hidden layer composed of three neurons while the output layer comprised a single neuron corresponding to the value of the thermal inactivation rate. Comparing with other secondary models, Lou and Nakai [121] concluded that the ANN-based model was far more accurate than both a polynomial model and an Arrhenius-type model [97] in fitting the measured responses, while the predictive accuracy of the Arrhenius-type model was most unsatisfactory. Figure 7.6 shows that the Arrhenius-type model and the polynomial model yielded almost the same predicted inactivation rate values at $A_w=0.928$ and pH=3; however, they did not fit the observed data as well as the ANN model did.

Contrasted with the response surface model using fixed functions, the ANN models are more versatile and less restrictive as they do not impose assumptions pertaining to the form of functions and use flexible basis functions to fit data [115]. The response surface model requires stating the order of the model while the ANN tends to implicitly match the input vector to the output vector for all observations available. Nevertheless, due to the black-box character of the ANN, it is difficult to produce a function expression to interpret the relationships between the environmental

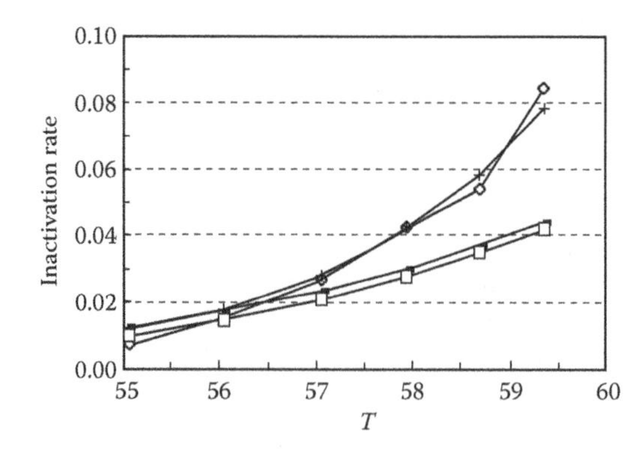

FIGURE 7.6 A comparison among secondary models describing the effect of temperature on the inactivation rate of *E. coli*. Polynomial model (□), Arrhenius-type model (–), Artificial neural network model (+), and observations (◊).

factors and model parameters [121] and impossible to determine how a neural network can reach a particular conclusion [112].

7.2.2.7 Reaction-Type Kinetic Models

The effect of the water activity on the thermal resistance was first taken into account by Reichart [103] who derived a reaction-type kinetic semiempirical model. Considering the stoichiometric equation of the chemical reaction between a "critical structure" of the cell and another reactant molecule, the thermal death was described by an extension of the Eyring's model. Taking into account the effect of the water activity on heat destruction, Reichart [103] modified empirically the extended model and fitted it to experimental data on the heat destruction of *E. coli*, yielding

$$k_d = \frac{k_B T}{h} \exp\left(-\frac{\Delta G^*}{RT}\right) A_w^{n_w} \left(\left|H^+\right|^{n_H} + \left|OH^-\right|^{n_{OH}}\right) \tag{7.68}$$

However, the application of this model is difficult because it requires hydrogen and hydroxyl ion concentrations to be known. Later on, Reichart [122] expanded his reaction-type kinetics model to describe the thermal death rate of microbes as a function of temperature, water activity, pH, and redox potential, as described in the following:

$$\log\frac{k_d}{T} = n_w \log A_w + \log\frac{k_B}{h} + \frac{\Delta S^*}{2.3R} - \frac{\Delta H^*}{2.3RT} - 2n_H \text{pH} - \frac{n_H}{2.3RT}\left[G_{mi} - FaE_0\right] \tag{7.69}$$

The general form of this model was very similar to the empirical model of Reichart and Mohacsi-Farkas [123], which was of the following form:

$$\log k_d = C_0 + C_1 \text{pH} + C_2 E_0 + C_3 \log A_w + C_4 \frac{1}{T} \tag{7.70}$$

Reichart model [122] was successfully validated with data sets of heat destruction of *Lactobacillus brevis* and *Lactobacillus plantarum*, and as a mechanistic model, it allowed the estimation of the thermodynamic parameters of the thermal death of these bacteria.

7.3 MODEL VALIDATION UNDER DYNAMIC CONDITIONS

The validity of thermal inactivation models, a primary model describing the evolution in time of microbial concentration (at static environmental conditions) coupled to a secondary model describing the effect of environment (i.e., temperature, pH, water activity, etc.) on the primary model's thermal death kinetic(s) parameters, has been normally assessed under dynamic conditions. Within the context of thermal inactivation, the most common dynamic condition is the nonisothermal condition of the heat treatment process [23,62], although other changing intrinsic properties may also be recorded if secondary models allow for properties other than temperature.

In actual thermal processes, the product temperature is always a function of the process time because it cannot reach the desired value instantaneously. Therefore, to calculate microbial survival curves under nonisothermal conditions, the product's temperature history must be known and then the momentary lethal effects on the population in question integrated. For the first-order kinetics, there are standard methods for performing the calculation [124]. These methods are however inapplicable when the survival curves are nonlinear. Peleg and Penchina [73], Peleg et al. [62], and Peleg et al. [71] proposed general methods to calculate microbial survival under nonisothermal conditions, which does not require the isothermal curves to be linear. They are similar to the one proposed by

Kormendy and Kormendy [125], except that no inactivation kinetic model is assumed or implied, and there are no restrictions on the mathematical representation of the survival parameters.

The underlying assumptions to compute a thermal inactivation under variable temperature are that (1) the momentary inactivation rate is determined by the momentary temperature and the survival ratio but not by the heat application rate, (2) all the isothermal curves within the pertinent temperature range can be adequately described by the primary model, and (3) the number or organisms are sufficiently large that continuous algebraic expressions can be used to describe the changes in the survival ratio [62].

Basically, Peleg et al. [62] proposed to express the temperature history of the product in algebraic form ($T = f(t)$) with a convenient empirical model, to express the primary model (Weibull)'s parameter in a function of the temperature using a log-logistic model:

$$b_W\left(T(t)\right) = \ln\left[1 - \exp\left\{k\left(T(t) - T_C\right)\right\}\right] \tag{7.71}$$

and to use those equations to integrate the differentiated form of the primary model (i.e., Weibull model) with respect to time, as shown by

$$\frac{\mathrm{d}\log S(t)}{\mathrm{d}t} = -b_W\left(T(t)\right)\beta\left\{\frac{-\log S(t)}{b_W\left(T(t)\right)}\right\}^{(\beta-1)/\beta} \tag{7.72}$$

Incorporating Equations 7.71 through 7.72 produces the nonisothermal inactivation Weibull-log logistic model:

$$\frac{\mathrm{d}\log S(t)}{\mathrm{d}t} = \ln\left\{1 + \exp\left[k_d\left(T(t) - T_c\right)\right]\right\}\cdot\beta\left\{\frac{\log S(t)}{\ln\left\{1 + \exp\left[k_d\left(T(t) - T_c\right)\right]\right\}}\right\}^{(\beta-1)/\beta} \tag{7.73}$$

Notice however that in Peleg et al. [62] and in subsequent work, they used a Weibull parameterization different from the one already presented in Equation 7.49, by replacing $b_W = 1/\chi^{\beta}$ as follows:

$$\log S(t) = -b_W(T)t^{\beta} \tag{7.74}$$

The solution in the form of log(S) versus time can be found for almost any conceivable temperature profile with standard commercial mathematical software. Although "apparently" cumbersome, the differential model has been validated using experimental survival data of *Salmonella* [126], *Listeria* [62], *E. coli* [68], and the spores of *Bacillus sporothermodurans* [69] under dynamic temperature. A schematic of the construction of a nonisothermal curve is shown in Figure 7.7. The survival ratio of any microorganism is calculated at any instant under nonisothermal conditions assuming that the local slope of the nonisothermal curve is that of the isothermal curve at the momentary temperature, but at a time that corresponds to the momentary survival ratio.

However, this integration method requires the knowledge or approximation of mathematical equations describing the process temperature with time. Consequently, this method is less convenient to follow real-time progresses of ongoing thermal processes and to be integrated in the control system of commercial thermal processing equipment. To overcome this limitation, Peleg et al. [71] proposed a way to generate microbial survival curves in real time by converting the Weibull's differential expression (Equation 7.73) into a *difference equation* which can be solved in short incremental intervals. They showed that solving the difference equation for fixed intervals of 1 min

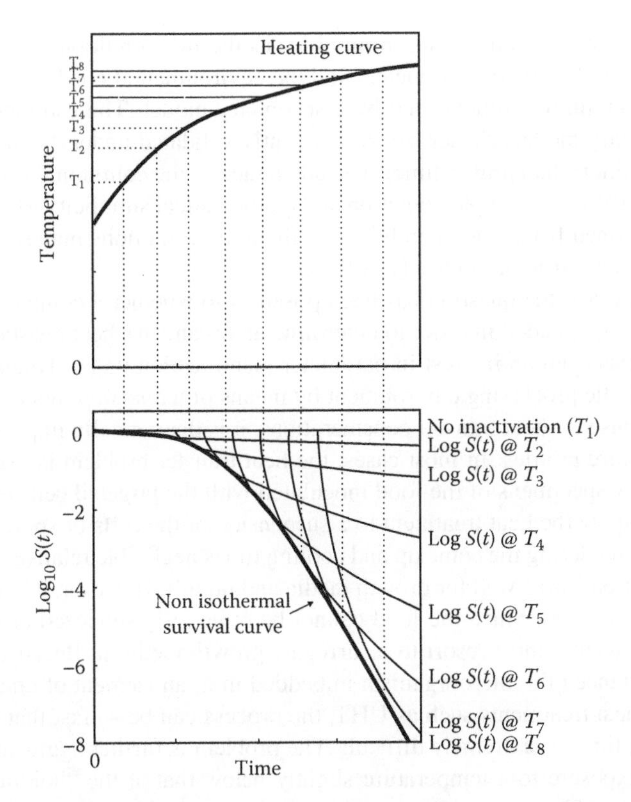

FIGURE 7.7 Schematic of the construction of a nonisothermal inactivation curve from momentary isothermal curves.

$(t_i = t_{i-1} + \Delta t)$ did not produce any noticeable difference from the results due to the integration of the differential equation shown as follows:

$$\frac{\log S(t_i) - \log S(t_{i-1})}{t_i - t_{i-1}} = -\frac{b_W[T(t_i)] + b_W[T(t_{i-1})]}{2}\beta\left\{\frac{-[\log S(t_i) + \log S(t_{i-1})]}{b_W[T(t_i)] + b_W[T(t_{i-1})]}\right\}^{\beta/(\beta-1)} \quad (7.75)$$

Having the actual or good estimates of the microorganism survival parameters β, *k*, and T_C, for any recorded temperature *T*(*t*) at time *t*, the corresponding $b_W[T(t)]$ can be calculated with Equation 7.71, and in the implicit Equation 7.73, what remains unknown is only the log survival ratio *S*(*t*), which can be estimated using a numerical approach, like, for example, the Newton–Raphson's method, at each time interval. Although their method seems flexible enough to be used in different applications, the computational load is increased because an implicit polynomial equation must be solved for each iteration. Chen et al. [127] proposed an alternative numerical algorithm for calculating the survival curves in real-time. Unlike the method of Peleg et al. [71] with the Chen's method [127], all the algebraic operations are explicit, and therefore, calculation speed in real-time is by far reduced.

7.4 DYNAMIC MODELING AND BACTERIAL ADAPTATION

Traditionally, the microbial inactivation (and growth) kinetics have been defined by a *two-step* modeling approach. This is an approach whereby the kinetic parameters of a (primary) model describing the microbial survival with respect to time are estimated at different static environmental conditions, in the first instance and most commonly, at isothermal conditions. In the second instance, a relationship between the kinetic parameters and the environmental condition at which they were obtained (i.e., temperature) is represented by a secondary model. The estimated parameters and primary and secondary models chosen are subsequently validated under dynamic conditions (i.e., environmental condition changing in time). The advantage of the commonly used two-step modeling methodology is the ease of implementation and mathematical simplicity. However, the transposition of results obtained from static conditions to dynamic conditions may require adjustment of the initial mathematical structure in use [128,129].

Thus, in the last years, the question has been posed as to whether a complete set of isothermal survival data are always needed in order to determine an organism's heat resistance parameters. In other words, there has been an interest in parameter identification under dynamic conditions representative of a realistic processing environment by means of a *one-step* modeling procedure. The question arises because for heat transfer considerations, it is theoretically impossible to obtain true isothermal temperature profiles. In most cases, the heat transfer problem is overcome by treating very small or narrow specimens of the food inoculated with the targeted cells or spores. The most effective way is to apply the heat treatment to a suspension of the cells or spores held in or passed through a capillary, rendering the come up and cooling times negligible relative to the holding time. This particular solution works well for growth media and liquids. However, it has serious limitation when solid food is involved. Since the food cannot be effectively squeezed or pumped through a capillary, the experimenter must resort to a surrogate growth medium. Because the medium does affect the heat resistance of a microorganism imbedded in it, an element of uncertainty is inevitably introduced. In heat treatments such as UHT, the process can be so fast that retrieving samples during the "holding time" is extremely difficult. The problem is further aggravated by the fact that even a very short exposure to a temperature slightly below that at the "holding time" can cause significant inactivation. Thus, even if one could find a way to withdraw samples during the short process, accounting for the varying contributions of the come up and cooling times in treatments where different temperatures are reached would be quite difficult [130].

A possible solution is to attempt to determine the survival parameters from nonisothermal treatments in which the inoculated food is heated slowly, albeit not at a rate that will permit adaptation. In principle, it is possible to predict the survival patterns of microorganisms from inactivation data obtained under nonisothermal conditions. By studying the microbial population decline in time, kinetic parameters are estimated at once during the tested dynamic environment. It has been shown that the one-step modeling approach produces more precise estimates than the two-step classical modeling approach, since confidence intervals are narrower due to the increased number of degrees of freedom [131].

Initially, dynamic modeling of microbial inactivation was tested in linearly increasing temperature profiles. For instance, Miles and Mackey [132] described an analytic method for evaluating the inactivation parameters D and z by assuming that microbial kinetics follow first order. Another analytical method that has been evaluated was the equivalent point method in which linear temperature increase and *linear* microbial inactivation kinetics are also assumed [133]. Fernandez et al. [134], Peleg et al. [100], and Conesa et al. [135] suggested the use of nonisothermal methods for the heating phase of the treatment and isothermal methods for the holding phase. Mathematical approaches of using the more general Weibull model for estimating the lethality of nonisothermal dynamic processes have been reported by several researchers [62,65,73,127,136,137]. Recently, Huang [138] compared the performance of the log-linear kinetic model, the Weibull model and the Gompertz model to represent the survival of *L. monocytogenes* in ground beef under dynamic conditions defined by a *linear* temperature profile. To this effect, the differential forms of the inactivation models were used and a numerical solution was found. The modified Gompertz model turned out to be the most accurate kinetic model, and both the log-linear and Weibull models grossly underestimated the survival of the bacteria during constant-rate (linear) heating.

Based on simulation results, Versyck et al. [139] motivated the need of dynamic temperature profiles in microbial inactivation experiments aimed at instantaneous estimation of model parameters from measurements of the microbial population load at dynamic temperature profiles which do not *necessarily* increase linearly. Peleg and Normand [72] demonstrated that a primary model in its differential form (i.e., Weibull model) coupled to a secondary model (log-logistic model) enables the calculation of survival parameters of organisms and spores from data gathered under nonlinear nonisothermal conditions. Nonlinear temperature profiles were characterized by empirical models so that differential equations could be solved. They concluded that, although in principle a single nonisothermal survival curve is sufficient to calculate the survival parameters, it is still recommended that any future database for calculating the survival parameters of a given organism include several heating regimes. This would serve as a validation of the mathematical model and verification of the assumptions on which it is based.

Using the Weibull-log logistic model, Corradini and Peleg [140] adapted Equation 7.69 to derive a set of ordinary differential equations for the characterization of "intact" ($S_{INT}(t)$) and "injured" ($S_{INJ}(t)$) microbial cells subjected to a heat treatment, as given in the following:

$$\frac{d\log S_{TOT}(t)}{dt} = -b_{WTOT}(T(t))\beta_{TOT}\left\{\frac{-\log S_{TOT}(t)}{b_{WTOT}(T(t))}\right\}^{(\beta_{TOT}-1)/\beta_{TOT}}$$

$$\frac{d\log S_{INT}(t)}{dt} = -b_{WINT}(T(t))\beta_{INT}\left\{\frac{-\log S_{INT}(t)}{b_{WINT}(T(t))}\right\}^{(\beta_{INT}-1)/\beta_{INT}} \tag{7.76}$$

$$\frac{d\log S_{INJ}(T)}{dt} = \frac{d\log S_{TOT}(T)}{dt} - \frac{d\log S_{INT}(T)}{dt}$$

Any traditional survival model is explicitly or implicitly based on the notion that a cell exposed to a hostile environment can only be in one of two states: alive and hence countable or dead and hence irrecoverable. Models normally do not take into account the possibility that some of the cells considered dead, or spores assumed inactivated, are actually only injured and might become viable at a later time after the damage repair. There are experimental methods to distinguish the number of "healthy" and "injured" cells. Corradini and Peleg [140] using their model, showed that the number of *Listeria innocua*'s intact cells diminished very rapidly (Figure 7.8) with the result that almost all the survivors were injured to at least some extent. In time, the curves of the total survivors and that of the injured cells become nearly indistinguishable. Only in the beginning of the heat process, during mild conditions, the curves are distinguishable (Figure 7.8). While examples with experimental observations were given for isothermals, the set of differential equations can in the same way be fitted to survival data obtained under dynamic conditions, as shown by the simulations run in [140].

Later on, Valdramidis et al. [128] determined microbial inactivation parameters also under dynamic temperature conditions by solving the differential equations of the primary model of Geeraerd et al. [21] and the Bigelow secondary model. Similarly, they expressed algebraically the temperature evolution as a function of time using a (nonlinear) Dabes-type function. The Levenberg–Marquardt method outperformed the Gauss-Newton optimization method although for any optimization approach initial guesses of the true parameters should be given, which should be based on previous literature information. An illustration of the fitting efficiency of the dynamic modeling for which nonlinearity is exhibited in both the *E. coli* inactivation data and temperature data is shown in Figure 7.9. The one-step determination of parameter estimates obtained under

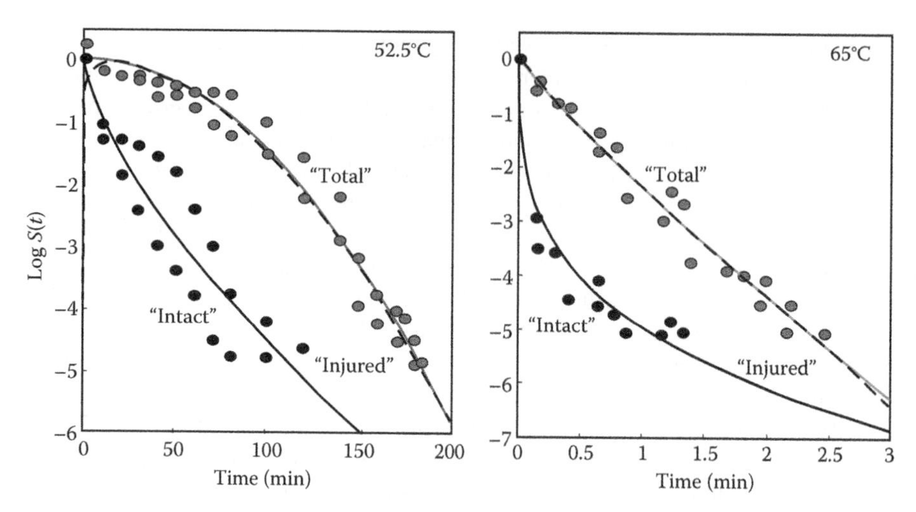

FIGURE 7.8 Survival curves of total and intact *L. innocua* cells at two lethal temperatures as modeled by the Weibull-log logistic model. Notice that with the exception of the very early stage of the treatment, most of the survivors have been injured to at least some extent.

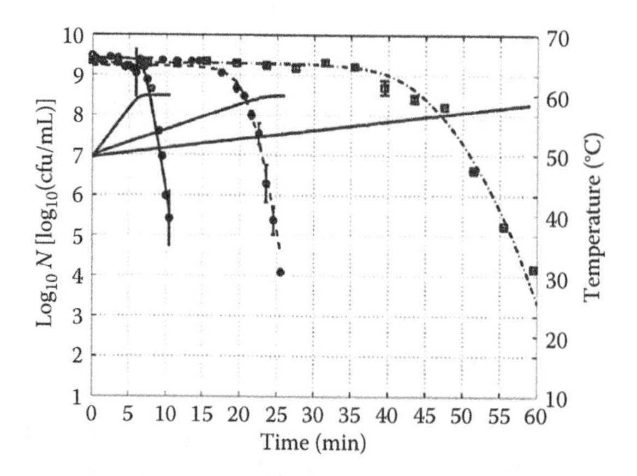

FIGURE 7.9 Fitting efficiency of the dynamic modeling of *E. coli* in brain heart infusion (BHI) broth for which temperature profile data are nonlinear. Data sets obtained at heating rates of 1.64, 0.43, and 0.15°C min^{-1} are shown.

dynamic environments is preferable and is expected to result in reliable predictions within the temperature range under study. Nevertheless, the microbial primary and secondary model chosen have to be considered a priori correct as it cannot be interpreted graphically as in the two-step modeling approach. Furthermore, the scatter as well as the amount of the microbial data can influence the convergence of the optimization algorithm.

Extracting the survival parameters from nonisothermal inactivation data is not a trivial matter, as it requires a special numerical procedure instead of standard nonlinear regression. This is because the dynamic survival curve's model is not an algebraic expression but a numerical solution of a rather complicated differential equation whose coefficients are functions of the temperature profile. While the dynamic or one-step modeling has been demonstrated to be feasible and effective, still this method will not be of use in short heat treatments such as UHT for the same reasons that exclude isothermal experiments as a reliable data source. Because the processing time is very short, the relative weight of the come up and cooling time must be significant, especially at the higher temperatures. For example, if the target temperature is 150°C then one would have to take into account that exposure of even a second at 130°C–145°C might cause appreciable damage to the treated spores. Similarly, because the treatment is very short, a reliable collection of samples at successive time intervals will rarely be an option [141]. Both problems would be eliminated if one could find a way to calculate the survival parameters, not in the conventional manner (i.e., trying to obtain a set of isothermal or nonisothermal survival curves), but directly from the counts of viable spores or cells at the end of the full treatments. Thus, Corradini et al. [130] proposed the so-called three-end points method which basically consists of the extraction of survival parameters from single final survival ratios determined at the end of (at least) three nonisothermal heat regimes. They argued that such possibility would eliminate altogether the need to take samples during the (short) treatment, and with it, all the problems that retrieving intermediate samples causes. Apart from accurate determination of the final survival ratios using as many replications as judged feasible, what would be required is that the temperature profile could be determined accurately and expressed algebraically. Corradini et al. [130] assumed that the microbial inactivation follows the Weibull-log logistic model, and therefore, the three survival curves would be the three numerical solutions of Equation 7.66 for the corresponding three $T(t)$'s but with the same inactivation parameters β, k_d, and T_C. In practice, the inevitable experimental scatter would distort the results of such a calculation. However, averaging the results obtained with more than three final survival ratio triplet combinations, determined in four or more heating regimes, can remove this impediment. Thus, Corradini et al. [142], Peleg et al. [141], and Corradini et al. [130] demonstrated the applicability of the "three-end points method" to estimate survival parameters from the final survival ratios of three or more heat treatments in different microorganisms, such as *E. coli*, *S. enteritidis*, *B. sporothermodurans*, *C. botulinum* spores as well as isobaric data of *E. coli,* and chemical inactivation of *Pseudomonas aeruginosa*. Convergence was guaranteed by the solution of only two equations with two unknowns at each iteration. Nevertheless, the proponents warn that the method may not work effectively for survival models that need four or more parameters for their characterization. This is because the increase in the number of degrees of freedom may result in calculated parameters having totally unrealistic values.

A series of researchers [15,129,143–145] have observed the ability of certain bacterial cells to adjust their metabolism in response to stress in order to increase their survivability. A notable example is *E. coli* whose cells can produce "heat shock" proteins which help them survive mild heat treatments [146]. Other organisms, *Staphylococcus aureus* and *B. cereus*, can develop defensive mechanisms that help them to survive in an acidic environment [147–149]. The cells' ability to augment their resistance is not unlimited however, and it takes time for the cells to activate the protective system and synthesize its chemical elements. Consequently, the effect of heat adaptation on an organism's survival pattern becomes measurable only at or slightly above the "sublethal" temperature range. Under dynamic conditions, therefore, adaptation can be detected only when the heating rate is sufficiently low to allow the cells to respond metabolically to the heat stress prior to their destruction [150]. When adaptation occurs, it is noticed as a gap between survival curves determined at low heating rates and those predicted by kinetic models whose parameters had been determined at high lethal temperatures [21,129,143,145,147]. To account for adaptation at low heating rates, Stasiewicz et al. [151] modified the Weibull inactivation model and assumed that the rate parameter's temperature dependence followed a modified version of the Arrhenius equation.

Corradini and Peleg [150] developed a variant of their Weibull-log logistic model to account for dynamic adaptation. Instead of the Arrhenius equation used by Stasiewicz et al. [151] as secondary model, Corradini and Peleg [150] opted for the log-logistic secondary model as they consider that unlike the Arrhenius, the log-logistic makes a clear distinction between the lethal and nonlethal temperatures without the need of temperature scale compression. First, they analyzed the case of adaptation under constant heating rate $dT(t)/dt = v$, and expressed Weibull rate parameter $b_W(T)$ as a function of the heating rate too. The new Weibull rate parameter $b_W(T,v)$ was formulated by multiplying the original $b_W(T)$ by using the following logistic adaptation factor $f_{adapt}(T,v)$:

$$f_{adapt}(T,v)=\frac{1}{1+\exp\left\{k_{adapt}\left(T_{c\,adapt}(v)-T\right)\right\}} \tag{7.77}$$

Combining the original rate parameter factor (Equation 7.64) with the adaptation factor yields the following heating rate dependent Weibull rate parameter:

$$b_W(T,v)=\frac{\ln\left\{1+\exp\left[k_d(T-T_c)\right]\right\}}{1+\exp\left\{k_{adapt}\left[T_{c\,adapt}(v)-T\right]\right\}} \tag{7.78}$$

As the heating rate v increases, the distinction between $b_W(T,v)$ and $b_W(T)$ progressively disappears. However, even when v approaches infinity, $b_W(T,v)$ cannot be larger than $b_W(T)$ because the denominator can never be less than 1. This newly defined $b_W(T,v)$ was inserted in Equation 7.65 to produce the Weibull inactivation rate model given in the following for stress adaptation:

$$\frac{d\log S(t)}{dt}=\frac{\ln\left\{1+\exp\left[k_d\left(T(t)-T_c\right)\right]\right\}}{1+\exp\left\{k_{adapt}\left[T_{c\,adapt}(v)-T(t)\right]\right\}}\cdot\beta\left\{\frac{\log S(t)}{\dfrac{\ln\left\{1+\exp\left[k_d\left(T(t)-T_c\right)\right]\right\}}{1+\exp\left\{k_{adapt}\left[T_{c\,adapt}(v)-T(t)\right]\right\}}}\right\}^{(\beta-1)/\beta} \tag{7.79}$$

Although this dynamic model has a cumbersome appearance, it is an ordinary differential equation that, for almost any conceivable practical temperature profile $T(t)$, can be solved numerically to produce the survival curves that correspond to a variety of constant heating rates. In cases where the heating regime is not constant, the constant heating rate v of Equation 7.70 should be replaced by the instantaneous heating rate $dT(t)/dt$. This dynamic model could closely represent the heat adaptation of *L. monocytogenes* and *E. coli* under slow heat regimes that could not be accurately represented by the simpler models (without dynamic adaptation) [150].

7.5 STOCHASTIC MODELING

Taub et al. [152], Doona et al. [153], and Ross et al. [154] developed a "quasi-chemical model" on the basis of population dynamics controlled by events at the cellular level. The model's starting point is that the individual cells' growth and division rates, the rate of harmful metabolites production, and that of their mortality, which can occur simultaneously, determine the evolution of a microbial population. Thus, pure linear or nonlinear inactivation, according to the quasi-chemical model, is a special case where mortality is predominant and cells' growth or division is insignificant or nonexistent. Horowitz et al. [155] have recently developed a model based on similar considerations. It differs from the quasi-chemical model in that the cells' division and mortality kinetic rate constants have been replaced by probability rate functions. Both models can account for smooth transitions between growth and inactivation and vice versa. Notice that the primary models presented in the previous sections do not enable these transitions as they have been developed only for inactivation. The probabilistic model described by Horowitz et al. [155] comes in two varieties: stochastic and discrete for individual and small group of cells, and deterministic and continuous for large microbial populations. The latter model is the limit of the former as the number of cells grows. The main rationale for proposing the fully probabilistic approach was to avoid mechanistic and kinetic assumptions concerning processes at the cellular level, whose experimental verification would require considerable efforts. The alternative assumptions were about the general character of the underlying division and mortality probability functions, whose general features can be deduced directly from experimentally observed microbial growth and mortality patterns. Corradini et al. [156] applied the probabilistic modeling approach to explain the emergence of different microbial inactivation patterns in terms of underlying inactivation probabilities at the cellular level. If the underlying mortality probability function remains constant throughout the treatment, the model generates first-order inactivation kinetics. Otherwise, it produces survival patterns that include Weibullian, tailing with a survival residual level, complete elimination, flat shoulder with linear or curvilinear continuation and sigmoid curves. Through simulation, Corradini et al. [156] showed that because of the stochastic nature of the process, the fate of a small group of bacteria can vary considerably from very rapid elimination to survival that seems to defy the odds. In contrast, the fate of a large group or population becomes increasingly predictable because of the "averaging effect." Thus, despite being determined by random events at the cellular level, the survival pattern of a population becomes increasingly deterministic as the population grows. This stochastic model has been also extended for the description of the activation of spores and inactivation of just activated and dormant spores [157]. By linking the survival curve's shape to the underlying probability of events at the individual spore level, the model explained the irregular activation and survival patterns of individual and small group of spores. Because of the way the probabilistic model is constructed, it has the advantage that, unlike most primary models of spore inactivation that are expressed in algebraic form, it has the same form for isothermal and dynamic conditions.

7.6 CONCLUSIONS

Although as the first thermal inactivation model, the first-order kinetics model, was proposed nine decades ago, the body of knowledge on thermal inactivation modeling was enriched considerably only in the past two decades. The suitability of the first-order kinetics for the modeling of heat inactivation, as well as for novel technologies, is being reconsidered. This is because growing experimental evidence of frequent occurrence of deviations has indicated that the log-linear inactivation model is an exception rather than a rule. Several theories have been proposed to explain these deviations which may be different for spores or vegetative cells. Shoulders have been mainly attributed to the occurrence of sublethal injury, multitarget inactivation, cell clumping, or activation phenomena for spores. Tails are considered to reflect the resistance heterogeneity within the populations, either inherent to the bacterial cells or acquired during the treatment. Some authors consider that upward concavity reflects the first-order inactivation of two microbial subpopulations of different heat resistance. An upward concave curve can also be justified by a continuous distribution of resistances within the microbial population. These frequent non-log linear survival curves can be represented effectively by the non-log linear model types presented in this chapter. A number of models have been developed to describe non-log linear curves, with the Geeraerd's, biphasic and Weibull models being the most commonly used primary models. Models based on the Weibull distribution are characterized by its simplicity and versatility. On the other hand, environmental factors, such as composition of the treatment medium, pH, water activity, or addition of preservative substances, strongly affect the resistance of microorganisms to heat. These effects can be modeled using a variety of secondary models, from which Bigelow, Arrhenius, and polynomial models are the most commonly used. Nevertheless, models presented in this chapter are not only used in conventional heat treatment but also have been successfully applied to current alternative methods for microbial inactivation such as ionizing radiation, ultrasound under pressure, high hydrostatic pressure and pulsed electric field. The successful implementation of novel technologies for food preservation relies on the progress in the field of mechanisms of inactivation. An adequate knowledge of the physiological behavior of microorganisms toward inactivation agents is essential for the development of safe foods. It is necessary for understanding the effect of environmental factors on resistance, and it would help interpret kinetics of inactivation and to develop mathematical models based on parameters with a biological meaning and able to predict microbial inactivation in a wider range of conditions. Two types of experimental setups, static and dynamic, are clearly distinguished for the determination of thermal resistance parameters of microbial cells in foods. The traditional two-step modeling procedure is applied to results obtained from static experimental conditions, while the recent one-step modeling procedure is applied to microbial survival estimates obtained from dynamic experimental conditions. Nevertheless, both types of modeling are not without advantages and drawbacks. At present, there is a considerable amount of data on bacterial inactivation accurately described by a variety of models, and although the data published in literature are hampered by the diversity of experimental conditions and model types, effective comparison and summarization of available knowledge on the inactivation rates of microorganisms in foods would be of interest and should be performed through the use of meta-analytical techniques. Resulting meta-analysis models can be used to facilitate the deliberation of expert panels or in microbial risk assessments. Latest research efforts on thermal inactivation modeling point toward bacterial adaptation modeling, inclusion of a memory function to account for the population's history, unification of growth and inactivation models and stochastic modeling at cellular level for small bacterial populations. Characterization of resistant strains may also be useful in mechanistic studies. Finally, the development of mathematical models based on physiological facts to establish treatment conditions is needed. Ideally, models should consider the lethal event, the heterogeneity within the bacterial population, and possible phenomena of adaptation and damage.

NOMENCLATURE

a_0, a_i, a_{ij}, a_i'	Coefficients of a polynomial secondary model
a_0	Parameter of the FDE-based model related to producing complex shapes of survival curves
a_1	Scaling parameter of the FDE-based model
A	Asymptotic log count as t decreases indefinitely
A_1	Initial numbers or concentration of microorganisms or spores of the first population (CFU)
A_2	Initial numbers or concentration of microorganisms or spores of the second population (CFU)
A_w	Water activity
$b_W(T)$	Temperature-dependent rate parameter from the Weibull model
$b_{WTOT}(T)$	Temperature-dependent rate parameter of the total population, intact, and injured cells
$b_{W\,INT}(T)$	Temperature-dependent rate parameter of the intact microbial cells
$b_W(T,v)$	Temperature- and heat rate-dependent rate parameter of the Weibull model modified for bacterial heat adaptation
B	Relative death rate at time M (s^{-1})
$B_C(T)$	Temperature function
C	Difference in value of the upper and lower asymptote
$C_0 - C_4$	Regression coefficients
C_C	Physiological state of cells
$C_C(0)$	Initial physiological state of the cells
D	Time to reduce the population by 1-log cycle
D_{ref}	Time to reduce the population by 1-log cycle at the reference temperature, T_{ref}
${}_0D_t^{\alpha}$	Riemann–Liouville derivative of fractional order α
e	Base of natural logarithm
E_A	Activation energy ($J\ mol^{-1}$)

E_0	Redox potential of the environment
E_d	Energy necessary for a lethal hit (kcal mol^{-1})
$E_{\alpha,\alpha+1}(-a_0t^{\alpha})$	Two-parameter (α, $\alpha+1$) Mittag–Leffler function used in the FDE-based model
f	Proportion of the initial population in the total population
$f_{adapt}(T,v)$	Logistic stress-adaptation factor for the Weibull rate parameter b_W
F	Traditional sterilization value
F_{adj}	Adjusted sterilization value by Weibull model
Fa	Faraday's constant (96485.3°C mol^{-1})
G_{mi}	Free enthalpy in the living cell membrane (J mol^{-1})
$\Delta G^{\neq}$	Free enthalpy of activation of the lethal reaction (J mol^{-1})
h	Planck's constant (6.626×10^{-34} J s)
h_d	Hom's time exponent
$\Delta H^{\neq}$	Enthalpy of activation (J mol^{-1})
i	ith measurement of spore/microbial concentration
k	Microorganism's inactivation parameter or rate at which $b_W(T)$ climbs as temperature increases
$\overline{k}$	Weighted average of the inactivation rates of k_1 and k_2 of the two subpopulations
k_o	Arrhenius equation constant (pre exponential factor)
k_1	Inactivation rate of the primary population (s^{-1})
k_2	Inactivation rate of the secondary population (s^{-1})
k_{adapt}	Constant marking the steepness of the adaptation factor-versus-temperature relationship around the inflection point
k_B	Boltzmann's constant (1.3806×10^{-23} J K^{-1})
k_d	Inactivation rate (s^{-1})
k_{d1}	Inactivation rate constant of the dormant population (s^{-1})
k_{d2}	Inactivation rate constant of the active population (s^{-1})
K_a	Activation constant of the dormant spores (s^{-1})
$K_{0,T}$	Inactivation rate constant at a given temperature (s^{-1})
$L(T)$	Biological destruction value
M	Time at which absolute death rate is maximum (s)
M_d	Maximum log reductions achievable
M_{H_2O}	Mass of 1 mol of water molecules (g mol^{-1})
n_c	Curvatural parameter of the Baranyi model
n_d	Decimal reduction ratio
n_H	Concentration exponent of hydrogen ions in the critical reaction
n_{OH}	Concentration exponent of hydroxyl ions in the critical reaction
n_w	Concentration exponent of water in the critical reaction
N	Number of organisms at time t (CFU)
N_A	Avogadro number (6.022×10^{23} mol^{-1})
N_C	Active population (CFU mL^{-1})
N_D	Dormant and viable population (CFU mL^{-1})
N_o	Initial number of organisms at time $t=0$ (CFU)
N_{res}	Residual number of organisms (CFU)
N_1	Number of organisms in the primary population (CFU)
N_2	Number of organisms in the secondary population (CFU)
pH^*	Nominal pH characteristic of the interaction microorganism-medium
pH_{ref}	Reference pH
P_1,P_2	Arrhenius temperature parameters in the Lambert model
P_3,P_4	Arrhenius time exponent parameters in the Lambert model
q_B	Tailing ratio of the Baranyi model
q_G	Tailing ratio of the Gompertz model

r	Time required for the relative death rate to reach half of the maximum relative death μ_m (s)
R	Universal gas constant (8.314 J mol^{-1} K^{-1})
R_d	Decimal log reduction in microbial numbers
S	Parameter characterizing the shoulder of a survival curve
$S(t)$	Survival ratio defined as $N(t)/N_0$
$S_{INJ}(t)$	Survival ratio of the injured cells
$S_{INT}(t)$	Survival ratio of the intact cells
$S_{TOT}(t)$	Survival ratio of the total population, injured and intact cells
ΔS^*	Entropy of activation
t	Time (s)
T	Temperature (°C, K)
T^*	Nominal temperature characteristic of the interaction microorganism-medium (°C)
T_C	Temperature region where intensive inactivation starts (°C)
T_{Cadapt}	Inflection point of the adaptation factor-versus-temperature relationship (°C)
T_{ref}	Reference temperature (K)
v	Constant rate heating (°C s^{-1})
x_i,x_j	ith and jth environmental factor (i.e., temperature, NaCl, pH)
y	Any parameter from a primary model (i.e., k_d, μ_m, etc.)
Y	Logarithm base 10 of population size at time t
z_T	Increase in temperature required to reduce the D value to one tenth of its value
z_{pH}	pH variation that induces a reduction of one tenth of the D value
z_{Aw}	Difference of A_w from 1, which leads to a one tenth of the D value
α	Upper asymptote in Cole's inactivation model (CFU)
α	Fractional order of the Riemann–Liouville derivative in the memory-kernel model
α'_R	Intercept parameter for each replicate in the log-quadratic model
α_W	Logit transformation of the fraction f of subpopulation 1 in the total population
$\alpha(t)$	Shoulder adjustment function
β	Shape parameter of the Weibull model
β_1	Shape parameter of the general Weibull model for subpopulation 1
β_2	Shape parameter of the general Weibull model for subpopulation 2
β_{INT}	Shape parameter of the Weibull model for the intact population
β'_T	Rate parameter dependent on temperature in the log-quadratic model
β_{TOT}	Shape parameter of the Weibull model for the total population, intact and injured cells
$\beta(t)$	Tailing adjustment function
γ'_T	Quadratic term parameter dependent on temperature in the log-quadratic model
ε	Normally distributed random variation
$\Gamma(v)$	Gamma function of a real number v
λ	Lag phase duration (s)
λ_d	Time for complete inactivation (s)
μ_m	Maximum (exponential) death rate (s^{-1})
ω	Lower asymptote of Cole's inactivation model
σ	Maximum slope of Cole's inactivation model
τ_1	Position of maximum slope of Cole's inactivation model
χ	Scale parameter of the Weibull model (s)
χ_1	Scale parameter of the general Weibull model for subpopulation 1 (s)
χ_2	Scale parameter of the general Weibull model for subpopulation 2 (s)
χ_{ref}	Time of first decimal reduction at the reference temperature T_{ref} (s)
$\chi_{ref-matrix}$	Time of first decimal reduction at the reference temperature T_{ref} specific to a food matrix (s)

REFERENCES

1. WD Bigelow, GS Bohart, AC Richardson, CO Ball. Heat penetration in processing canned foods. *National Canners Association Bulletin* 16-L, 1920.
2. WD Bigelow. The logarithmic nature of thermal death time curves. *Journal of Infectious Diseases* 29: 528–536, 1921.
3. JR Esty, KF Meyer. The heat resistance of the spores of B. botulinus and related anaerobes. *Journal of Infectious Diseases* 31: 650–663, 1922.
4. ED Van Asselt, MH Zwietering. A systematic approach to determine global thermal inactivation parameters for various food pathogens. *International Journal of Food Microbiology* 107: 256–264, 2006.
5. AA Teixeira. Mechanistic models of microbial inactivation behaviour in foods. In: S Brul, S van Gerwen, M Zwietering, eds. *Modelling Microorganisms in Food.* Cambridge, U.K.: Woodhead Publishing Limited, 2007, pp. 198–213.
6. RC Whiting, RL Buchanan. Predictive microbiology. In: MP Doyle, LR Beuchat, TJ Montville, eds. *Food Microbiology, Fundamentals and Frontiers*. Washington, DC: ASM Press, 1997, pp. 728–739.
7. DW Schaffner, TP Labuza. Predictive microbiology: Where are we, and where are we going? *Food Technology* 51: 95–99, 1997.
8. MR Adams, MO Moss. *Food Microbiology*. London, U.K.: The Royal Society of Chemistry, 1997.
9. DAA Mossel, JEL Corry, CB Struijk, RM Baird. *Essentials of the Microbiology of Foods: A Textbook for Advanced Students*. Chichester, U.K.: John Wiley & Sons, 1995.
10. RL Buchanan, MH Golden, RC Whiting. Differentiation of the effects of pH and lactic or acetic concentration on the kinetics of Listeria monocytogenes inactivation. *Journal of Food Protection* 56: 474–478, 1993.
11. RL Buchanan, RC Whiting, WC Damert. When is simple good enough: A comparison for the Gompertz, Baranyi and three-phase linear models for fitting bacterial growth curves. *Food Microbiology* 14: 313–326, 1997.
12. RL Buchanan, MH Golden. Model for the non-thermal inactivation of Listeria monocytogenes in a reduced oxygen environment. *Food Microbiology* 12: 203–212, 1995.
13. H Li, G Xie, A Edmondson. Evolution and limitations of primary mathematical models in predictive microbiology. *British Food Journal* 109: 608–626, 2007.
14. WA Anderson, PJ McClure, AC Baird-Parker, MB Cole. The application of a log-logistic model to describe the thermal inactivation of Clostridium botulinum 213B at temperatures below 121.1 degrees C. *Journal of Applied Bacteriology* 80: 283–290, 1996.
15. JC Augustin, V Carlier, J Rozier. Mathematical modelling of the heat resistance of Listeria monocytogenes. *Journal of Applied Microbiology* 84: 185–191, 1998.
16. M Peleg, MB Cole. Reinterpretation of microbial survival curves. *Critical Reviews in Food Science and Nutrition* 38: 353–380, 1998.
17. A Fernandez, C Salmeron, PS Fernandez, A Martinez. Application of the frequency distribution model to describe the thermal inactivation of two strains of Bacillus cereus. *Trends in Food Science and Technology* 10: 158–162, 1999.
18. MAJS Van Boekel. On the use of the Weibull model to describe thermal inactivation of microbial vegetative cells. *International Journal of Food Microbiology* 74: 139–159, 2002.
19. CD Blackburn, LM Curtis, L Humpheson, C Billon, PJ McClure. Development of thermal inactivation models for Salmonella enteritidis and Escherichia coli O157:H7 with temperature, pH and NaCl as controlling factors. *International Journal of Food Microbiology* 38: 31–44, 1997.
20. A Casolari. Microbial death. In: MJ Bazin, JL Prosser, eds. *Physiological Models in Microbiology* 2. Boca Raton, FL: CRC Press, 1988, pp. 1–44.
21. AH Geeraerd, CH Herremans, JF Van Impe. Structural model requirements to describe microbial inactivation during a mild heat treatment. *International Journal of Food Microbiology* 59: 185–209, 2000.
22. V Sapru, AA Teixeira, GH Smerage, JA Lindsay. Predicting thermophilic spore population dynamics for UHT sterilisation processes. *Journal of Food Science* 57: 1248–1252, 1992.
23. V Sapru, GH Smerage, AA Teixeira, JA Lindsay. Comparison of predictive models for bacterial spore population resources to sterilization temperature. *Journal of Food Science* 58: 223–228, 1993.
24. AC Rodriguez, GH Smerage, AA Teixeira, FF Busta. Kinetic effects of lethal temperatures on population dynamics of bacterial spores. *Transactions of ASAE* 31: 1594–1601, 1988.
25. AA Teixeira, AC Rodriguez. Microbial population dynamics in bioprocess sterilization. *Enzyme Microbiology Technology* 12: 469–473, 1990.
26. O Cerf. Tailing of survival curves of bacterial spores. *Journal of Applied Bacteriology* 42: 1–9, 1977.

27. DN Kamau, S Dores, KM Pruitt. Enhanced thermal destruction of Listeria monocytogenes and Staphylococcus aureus by the lactoperoxidase system. *Applied and Environmental Microbiology* 56: 2711–2716, 1990.
28. RC Whiting, RL Buchanan. Use of predictive microbial modelling in a HACCP program. In: *Proceedings of the Second ASEPT International Conference: Predictive Microbiology and HACCP*. Laval Cedex, France, June 10–11, 1992, pp. 125–141.
29. RL Buchanan, MH Golden, Whiting RC, Phillips JG, Smith JL. Non-thermal inactivation models for Listeria monocytogenes. *Journal of Food Science* 59: 179–188, 1994.
30. RL Buchanan, MH Golden. Interactions between pH and malic acid concentration on the inactivation of Listeria monocytogenes. *Journal of Food Safety* 18: 37–48, 1998.
31. RC Whiting, S Sackitey, S Calderone, K Merely, JG Phillips. Model for the survival of Staphylococcus aureus in non-growth environments. *International Journal of Food Microbiology* 31: 231–243, 1996.
32. WH Ross, H Couture, A Hughes, T Gleeson, RC McKellar. A non-linear mixed effects model for the destruction of Enterococcus faecium in a high-temperature short-time pasteurizer. *Food Microbiology* 15: 567–575, 1998.
33. R Xiong, G Xie, AS Edmondson, RH Linton, MA Sheard. Comparison of the Baranyi model with the modified Gompertz equation for modelling thermal inactivation of Listeria monocytogenes Scott A. *Food Microbiology* 16: 269–279, 1999.
34. KMA Reyns, CCF Soontjes, K Cornelis, CA Weemaes, ME Hendrickx, CW Michiels. Kinetic analysis and modelling of combined high pressure-temperature inactivation of the yeast Zygosaccharomyces bailii. *International Journal of Food Microbiology* 89: 125–138, 2000.
35. I Van Opstal, SCM Vanmuysen, EY Wuytack, B Masschalck, CW Michiels. Inactivation of Escherichia coli by high hydrostatic pressure at different temperatures in buffer and carrot juice. *International Journal of Food Microbiology* 98: 179–191, 2005.
36. MB Cole, KW Davies, G Munro, CD Holyoak, DC Kilsby. A vitalistic model to describe the thermal inactivation of Listeria monocytogenes. *Journal of Industrial Microbiology* 12: 232–239, 1993.
37. PJ Stephens, MB Cole, MV Jones. Effects of heating on the thermal inactivation of Listeria monocytogenes. *Journal of Applied Bacteriology* 77: 702–708, 1994.
38. A Ellison, W Anderson, MB Cole, G Stewart. Modelling the thermal inactivation of Salmonella typhimurium using bioluminescence data. *International Journal of Food Microbiology* 23: 467–477, 1994.
39. G Duffy, A Ellison, W Anderson, MB Cole, GSAB Stewart. Use of bioluminescence to model the thermal inactivation of Salmonella Typhimurium in the presence of a competitive microflora. *Applied Environmental Microbiology* 61: 3463–3465, 1995.
40. CL Little, MR Adams, WA Anderson, MB Cole. Application of a log-logistic model to describe the survival of Yersinia enterocolitica at sub-optimal pH and temperature. *International Journal of Food Microbiology* 22: 63–71, 1994.
41. H Chen, DG Hoover. Modelling the combined effect of high hydrostatic pressure and mild heat on the inactivation kinetics of Listeria monocytogenes Scott A in whole milk. *Innovative Food Science and Emerging Technologies* 4: 25–34, 2003.
42. AM Gibson, N Bratchell, TA Roberts. Predicting microbial growth: Growth response of Salmonellae in a laboratory medium as affected by pH, sodium chloride and storage temperature. *International Journal of Food Microbiology* 6: 155–178, 1988.
43. S Bhaduri, PW Smith, SA Palumbo, CO Turner-Jones, JL Smith, BS Marmer, RL Buchanan, LL Zaika, AC Williams. Thermal destruction of Listeria monocytogenes in liver sausage slurry. *Food Microbiology* 8: 75–78, 1991.
44. TA McMeekin, J Olley, T Ross, DA Ratkowsky. *Predictive Microbiology: Theory and Applications*. New York: Wiley & Sons, 1993.
45. RH Linton, WH Carter, MD Pierson, CR Hackney. Use of a modified Gompertz equation to model nonlinear survival curves for Listeria monocytogenes Scott A. *Journal of Food Protection* 58: 946–954, 1995.
46. RH Linton, WH Carter, MD Pierson, CR Hackney, JD Eifert. Use of a modified Gompertz equation to predict the effects of temperature, pH and NaCl on the inactivation of Listeria monocytogenes Scott A heated in infant formula. *Journal of Food Protection* 59: 16–23, 1996.
47. MF Patterson, DJ Kilpatrick. The combined effect of high hydrostatic pressure and mild heat on inactivation of pathogens in milk and poultry. *Journal of Food Protection* 61: 432–436, 1998.
48. J Bello, MA Sanchez-Fuertes. Application of a mathematical model for the inhibition of Enterobacteriaceae and clostridia during a sausage curing process. *Journal of Food Protection* 58: 1345–1350, 1995.

49. H Chen, DG Hoover. Pressure inactivation kinetics of Yersinia enterocolitica ATCC 35669. *International Journal of Food Microbiology* 87: 161–171, 2003.
50. J Baranyi, A Jones, C Walker, A Kaloti, TP Robinson, BM Mackey. A combined model for growth and subsequent thermal inactivation of Brochothrix thermosphacta. *Applied Environmental Microbiology* 62: 1029–1035, 1996.
51. J Baranyi, TA Roberts, PJ McClure. Some properties of a non-autonomous deterministic growth model describing the adjustment of the bacterial population to a new environment. *Mathematical Medicine and Biology* 10: 293–299, 1993.
52. RJW Lambert. A model for thermal inactivation of microorganisms. *Journal of Applied Microbiology* 95: 500–507, 2003.
53. RJW Lambert. Advances in disinfection testing and modelling. *Journal of Applied Microbiology* 91: 351–363, 2001.
54. LW Hom. Kinetics of chlorine disinfection in an ecosystem. *Journal of the Environmental Division of the American Society of Civil Engineering* 98: 183–194, 1972.
55. M Peleg, MB Cole. Estimating the survival of Clostridium botulinum spores during heat treatment. *Journal of Food Protection* 63: 190–195, 2000.
56. I Albert, P Mafart. A modified Weibull model for bacterial inactivation. *International Journal of Food Microbiology* 100: 197–211, 2005.
57. EJ Greenacre, TF Brocklehurst, CR Waspe, DR Wilson, PDG Wilson. Salmonella enterica serovar Typhimurium and Listeria monocytogenes. Acid tolerance response induced by organic acids at 20°C: Optimization and modeling. *Applied and Environmental Microbiology* 69: 3945–3951, 2003.
58. D Marquenie, AH Geeraerd, J Lammertyn, C Soontjes, JF Van Impe, CW Michiels, BM Nicolai. Combinations of pulsed white light and UV-C or mild heat treatment to inactivate conidia of Botrytis cinerea and Monilinea fructigena. *International Journal of Food Microbiology* 85: 185–196, 2003.
59. VP Valdramidis, N Belaubre, R Zuniga, AM Foster, M Havet, AH Geeraerd, MJ Swain, K Bernaerts, JF Van Impe, A Kondjoyan. Development of predictive modelling approaches for surface temperature and associated microbiological inactivation during hot air decontamination. *International Journal of Food Microbiology* 100: 261–274, 2005.
60. AH Geeraerd, V Vladramidis, JF Van Impe. GInaFiT, a freeware tool to assess non-log-linear microbial survivor curves. *International Journal of Food Microbiology* 102: 95–105, 2005.
61. M Peleg. Microbial survival curves—The reality of flat 'shoulders' and absolute thermal death times. *Food Research International* 33: 531–538, 2000.
62. M Peleg, CM Penchina, MB Cole. Estimation of the survival curve of Listeria monocytogenes during non-isothermal heat treatments. *Food Research International* 34: 383–388, 2001.
63. M Peleg. A model of survival curves having an 'activation shoulder.' *Journal of Food Science* 67: 2438–2443, 2002.
64. DC Kilsby, KW Davies, PJ McClure, C Adair, WA Anderson. Bacterial thermal death kinetics based on probability distributions: The heat destruction of Clostridium botulinum and Salmonella Bedford. *Journal of Food Protection* 63: 1179–1203, 2000.
65. P Mafart, O Couvert, S Gaillard, I Leguerinel. On calculating sterility in thermal preservation methods: Application of the Weibull frequency distribution model. *International Journal of Food Microbiology* 72: 107–113, 2002.
66. R Virto, D Sanz, I Alvarez, CJ Raso. Inactivation kinetics of Yersinia enterocolitica by citric and lactic acid at different temperatures. *International Journal of Food Microbiology* 103: 251–257, 2005.
67. ER Withell. The significance of the variation on shape of time-survivor curves. *Journal of Hygiene* 42: 124–183, 1942.
68. MG Corradini, M Peleg. Demonstration of the Weibull-Log logistic survival model's applicability to non isothermal inactivation of Escherichia coli K12 MG1655. *Journal of Food Protection* 67: 2617–2621, 2004.
69. PM Periago, A van Zuijlen, PS Fernandez, PM Klapwijk, PF ter Steeg, MG Corradini, M Peleg. Estimation of the non-isothermal inactivation patterns of Bacillus sporothermodurans IC4 spores in soups from their isothermal survival data. *International Journal of Food Microbiology* 95: 205–218, 2004.
70. A Jagannath, T Tsuchido, JM Membre. Comparison of the thermal inactivation of Bacillus subtilis spores of foods using the modified Weibull and Bigelow equations. *Food Microbiology* 22: 233–239, 2005.
71. M Peleg, MD Normand, MG Corradini. Generating microbial survival curves during thermal processing in real time. *Journal of Applied Microbiology* 98: 408–417, 2005.
72. M Peleg, MD Normand. Calculating microbial inactivation parameters and predicting survival curves from non-isothermal inactivation data. *Critical Reviews in Food Science and Nutrition* 44: 409–418, 2004.

73. M Peleg, CM Penchina. Modelling microbial survival during exposure to a lethal agent with varying intensity. *Critical Reviews in Food Science and Nutrition* 40: 159–172, 2000.
74. Couvert, S Gaillard, N Savy, P Mafart, I Leguerinel. Survival curves of heated bacterial spores: Effect of environmental factors on Weibull parameters. *International Journal of Food Microbiology* 101: 73–81, 2005.
75. A Fernandez, J Collado, LM Cunha, MJ Ocio, A Martinez. Empirical model building based on Weibull distribution to describe the joint effect of pH and temperature on the thermal resistance of Bacillus cereus in vegetable substance. *International Journal of Food Microbiology* 32: 147–153, 2002.
76. H Chen, DG Hoover. Use of a Weibull model to describe and predict pressure inactivation of Listeria monocytogenes Scott A in whole milk. *Innovative Food Science and Emerging Technologies* 5: 269–276, 2004.
77. X Hu, P Mallikarjunan, J Koo, LS Andrews, ML Jahncke. Comparison of kinetic models to describe high pressure and gamma irradiation used to inactivate Vibrio vulnificus and Vibrio parahaemolyticus prepared in buffer solution and in whole oysters. *Journal of Food Protection* 86: 292–295, 2005.
78. L Coroller, I Leguerinel, E Mettler, N Savy, P Mafart. General model, based on two mixed Weibull distributions of bacterial resistance, for describing various shapes of inactivation curves. *Applied and Environmental Microbiology* 72: 6493–6502, 2006.
79. G Stone, B Chapmen, D Lovell. Development of a log-quadratic model to describe microbial inactivation, illustrated by thermal inactivation of Clostridium botulinum. *Applied and Environmental Microbiology* 75: 6998–7005, 2009.
80. A Kaur, PS Takhar, DM Smith, JE Mann, MM Brashears. Fractional differential equations based modeling of microbial survival and growth curves: Model development and experimental validation. *Journal of Food Science* 73: E403–E414, 2008.
81. N Vaidya, CM Corvalan. An integral model of microbial inactivation taking into account memory effects. *Journal of Food Protection* 72: 837–842, 2009.
82. B Byrne, G Dunne, DJ Bolton. Thermal inactivation of Bacillus cereus and Clostridium perfringens vegetative cells and spores in pork luncheon roll. *Food Microbiology* 23: 803–808, 2006.
83. P Mafart, I Leguerinel. Modelling combined effects of temperature and pH on heat resistance of spores by linear-Bigelow equation. *Journal of Food Science* 63: 6–8, 1998.
84. P Mafart. Taking injuries of surviving bacteria into account for optimising heat treatments. *International Journal of Food Microbiology* 55: 175–179, 2000.
85. L Leguerinel, O Couvert, P Mafart. Relationship between the apparent heat resistance of Bacillus cereus spores and the pH and NaCl concentration of the recovery medium. *International Journal of Food Microbiology* 55: 223–227, 2000.
86. S Gaillard, I Leguerinel, P Mafart. Model for combined effects of temperature, pH and water activity on thermal inactivation of Bacillus cereus spores. *Journal of Food Science* 63: 1–3, 1998.
87. VK Juneja, VS Eblen, HM Marks. Modeling non-linear survival curves to calculate thermal inactivation of salmonella in poultry at different fat levels. *International Journal of Food Microbiology* 70: 37–51, 2001.
88. VK Juneja, HM Marks. Characterizing asymptotic D-values for Salmonella spp. subjected to different heating rates in sous-vide cooked beef. *Innovative Food Science and Emerging Technologies* 4: 395–402, 2003.
89. VP Valdramidis, AH Geeraerd, JE Gaze, A Kondjoyan, AR Boyd, HL Shaw, JF Van Impe. Quantitative description of Listeria monocytogenes inactivation kinetics with temperature and water activity as the influencing factors; model prediction and methodological validation on dynamic data. *Journal of Food Engineering* 76: 79–88, 2006.
90. TP Labuza, D Riboh. Theory and application of Arrhenius kinetics to the prediction of nutrient losses in foods. *Food Technology* 36: 66–74, 1982.
91. KR Davey, SH Lin, DG Wood. The effect of pH on continuous high temperature/short time sterilization of liquid. *American Institute of Chemical Engineering Journal* 24: 537–540, 1978.
92. KR Davey. Linear-Arrhenius models for bacterial growth and death and vitamin denaturations. *Journal of Industrial Microbiology* 12: 172–179, 1993.
93. KR Davey. Extension of the generalised chart for combined temperature and pH. Lebensm–Wiss. *U-Technology* 26: 476–479, 1993.
94. KY Khoo, KR Davey, CJ Thomas. Assessment of four model forms for predicting thermal inactivation kinetics of Escherichia coli in liquid as affected by combined exposure time, liquid temperature and pH. *Transactions of Institution of Chemical Engineers Part C* 81: 129–137, 2003.
95. J Chiruta, KR Davey, CJ Thomas. Thermal inactivation kinetics of three vegetative bacteria as influenced by combined temperature and pH in a liquid medium. *Transactions of the Institution of Chemical Engineers Part C* 75: 174–180, 1997.

96. SD Holdsworth. *Thermal Processing of Packaged Foods*. London, U.K.: Blackie Academic and Professional, 1997.
97. O Cerf, KR Davey, AK Sadoudi. Thermal inactivation of bacteria—A new predictive model for the combined effect of three environmental factors: Temperature, pH and water activity. *Food Research International* 29: 219–226, 1996.
98. SA Amos, KR Davey, CJ Thomas. A comparison of predictive models for the combined effect of UV dose and solids concentration on disinfection kinetics of Escherichia coli for potable water production. *Process Safety and Environmental Protection* 79(3): 174–182, 2001.
99. OH Campanella, M Peleg. Theoretical comparison of a new and the traditional method to calculate Clostridium botulinum survival during thermal inactivation. *Journal of the Science of Food and Agriculture* 81: 1069–1076, 2001.
100. M Peleg, MD Normand, OH Campanella. Estimating microbial inactivation parameters from survival curves obtained under varying conditions—The linear case. *Bulletin of Mathematical Biology* 65: 219–234, 2003.
101. T Ross, P Dalgaard. Secondary models. In: RC McKellar, X Lu, eds. *Modeling Microbial Responses in Food*. Boca Raton, FL: CRC Press, 2004, pp. 63–150.
102. RL Buchanan, J Philips. Response surface model for predicting the effects of temperature, pH, sodium chloride content, sodium nitrite concentration and atmosphere on the growth of Listeria monocytogenes. *Journal of Food Protection* 53: 370–376, 1990.
103. O Reichart. Modelling the destruction of Escherichia coli on the base of reaction kinetics. *International Journal of Food Microbiology* 23: 449–465, 1994.
104. PS Fernandez, MJ Ocio, T Sanchez, A Martinez. Thermal resistance parameters of Bacillus stearothermophilus spores heated in acidified mushroom extract. *Journal of Food Protection* 57: 37–41, 1994.
105. MFG Jermini, W Schmidt-Lorenz. Heat resistance of vegetative cells and asci of two Zygosaccharomyces yeasts in broth at different water activity values. *Journal of Food Protection* 50: 835–841, 1988.
106. PS Fernandez, MJ Ocio, F Rodrigo, M Rodrigo, A Martinez. Mathematical model for the combined effect of temperature and pH on the thermal resistance of Bacillus stearothermophilus and Clostridium sporogenes spores. *International Journal of Food Microbiology* 32: 225–233, 1996.
107. VK Juneja, BS Marmer, JG Phillips, AJ Miller. Influence of the intrinsic properties of food on thermal inactivation of spores of nonproteolytic Clostridium botulinum: Development of a predictive model. *Journal of Food Safety* 15: 349–364, 1995.
108. VK Juneja, BS Eblen. Heat inactivation of Salmonella typhimurium DT104 in beef as affected by fat content. *Letters in Applied Microbiology* 30: 461–467, 2000.
109. KL Mattick, F Jorgensen, P Wang, J Pound, MH Vandeven, LR Ward, JD Legan, HM Lappin-Scott, TJ Humphrey. Effect of challenge temperature and solute type on heat tolerance of Salmonella serovars at low water activity. *Applied and Environmental Microbiology* 67: 4128–4136, 2001.
110. AT Chhabra, WH Carter, RH Linton, MA Cousin. A predictive model to determine the effects of pH, milk fat and temperature on thermal inactivation of Listeria monocytogenes. *Journal of Food Protection* 62: 1143–1149, 1999.
111. Xie G, Xiong R, Church I. Comparison of kinetics, neural network and fuzzy logic in modelling texture changes of dry peas in long time cooking. *Food Science and Technology (Lebensmittel-Wissenschaft und-Technologie)* 31: 639–647, 1998.
112. MN Hajmeer, IA Basheer, YM Najjar. Computational neural networks for predictive microbiology. II. Application to microbial growth. *International Journal of Food Microbiology* 34: 51–66, 1997.
113. YM Najjar, LA Basheer, MN Hajmeer. Computational neural networks for predictive microbiology: I. Methodology. *International Journal of Food Microbiology* 34: 27–49, 1997.
114. ME Ramos-Nino, CA Ramirez, MN Clifford, MR Adams. A comparison of quantitative structure-activity relationships for the effects of benzoic and cinnamic acids on Listeria monocytogenes using multiple linear regression, artificial neural network and fuzzy systems. *Journal of Applied Microbiology* 82: 168–176, 1997.
115. AH Geeraerd, CH Herremans, C Cenes, JF Van Impe. Application of artificial neural networks as a non-linear modular modelling technique to describe bacterial growth in chilled food products. *International Journal of Food Microbiology* 44: 9–68, 1998.
116. AH Geeraerd, CH Herremans, LR Ludikhuyze, ME Hendrickx, JF Van Impe. Modelling the kinetics of isobaric-isothermal inactivation of Bacillus subtilis alpha-amylase with artificial neural networks. *Journal of Food Engineering* 36: 263–279, 1998.
117. M Cheroutre-Vialette, A Lebert. Application of recurrent neural network to predict bacterial growth in dynamic conditions. *International Journal of Food Microbiology* 73: 107–118, 2002.

118. MN Hajmeer, I Basheer. A probabilistic neural network approach for modelling and classification of bacterial growth/no growth data. *Journal of Microbiological Methods* 51: 217–226, 2002.
119. MN Hajmeer, I Basheer. A hybrid Bayesian-neural network approach for probabilistic modelling of bacterial growth/no growth interface. *International Journal of Food Microbiology* 82: 233–243, 2003.
120. MN Hajmeer, I Basheer. Comparison of logistic regression and neural network-based classifiers for bacterial growth. *Food Microbiology* 20: 43–55, 2003.
121. W Lou, S Nakai. Application of artificial neural networks for predicting the thermal inactivation of bacteria: A combined effect of temperature, pH and water activity. *Food Research International* 34: 573–579, 2001.
122. O Reichart. Reaction kinetic interpretation of heat destruction influenced by environmental factors. *International Journal of Food Microbiology* 64: 289–294, 2001.
123. O Reichart, C Mohacsi-Farkas. Mathematical modelling of the combined effect of water activity, pH and redox potential on the heat destruction. *International Journal of Food Microbiology* 24: 103–112, 1994.
124. AA Teixeira. Thermal processing calculations. In: DR Heldman, IB Lund, eds. *Handbook of Food Engineering*. New York: Marcel Dekker, 1992, pp. 563–619.
125. I Kormendy, L Kormendy. Considerations for calculating heat inactivation processes when semilogarithmic thermal inactivation models are non-linear. *Journal of Food Engineering* 34: 33–40, 1997.
126. KL Mattick, JD Legan, TJ Humphrey, M Peleg. Calculating Salmonella inactivation in non-isothermal heat treatments from non-linear isothermal survival curves. *Journal of Food Protection* 64: 606–613, 2001.
127. G Chen, OH Campanella, CM Corvalan. A numerical algorithm for calculating microbial survival curves during thermal processing. *Food Research International* 40: 203–208, 2007.
128. VP Valdramidis, AH Geeraerd, K Bernaerts, JF Van Impe. Identification of non-linear microbial inactivation kinetics under dynamic conditions. *International Journal of Food Microbiology* 128: 146–152, 2008.
129. VP Valdramidis, AH Geeraerd, JF Van Impe. Stress adaptive responses by heat under the microscope by predictive microbiology. *Journal of Applied Microbiology* 103: 1922–1930, 2007.
130. MG Corradini, MD Normand, M Peleg. Prediction of an organism's inactivation patterns from three single survival ratios determined at the end of three non-isothermal heat treatments. *International Journal of Food Microbiology* 126: 98–111, 2008.
131. MAJS Van Boekel. Statistical aspects of kinetic modelling for food science problems. *Journal of Food Science* 61: 477–485, 1996.
132. CA Miles, BM Mackey. A mathematical analysis of microbial inactivation at linearly rising temperatures: Calculation of the temperature rise needed to kill Listeria monocytogenes in different foods and methods for dynamic measurements of D and z values. *Journal of Applied Bacteriology* 77: 14–20, 1994.
133. B Welt, A Teixeira, M Balaban, G Smerage, D Sage. Iterative method for kinetic parameter estimation from dynamic thermal treatments. *Journal of Food Science* 62: 8–14, 1997.
134. A Fernandez, MJ Ocio, PS Fernandez, A Martinez. Effect of heat activation and inactivation conditions on germination and thermal resistance parameters of Bacillus cereus spores. *International Journal of Food Microbiology* 63: 257–264, 2001.
135. R Conesa, P Periago, A Esnoz, A Lopez, A Palop. Prediction of Bacillus subtilis spore survival after a combined non-isothermal-isothermal heat treatment. *European Food Research and Technology* 217: 319–324, 2003.
136. MG Corradini, MD Normand, M Peleg. Calculating the efficacy of heat sterilization processes. *Journal of Food Engineering* 67: 59–69, 2005.
137. A Halder, AK Datta, SSR Geedipalli. Uncertainty in thermal process calculations due to variability in first-order and Weibull kinetics parameters. *Journal of Food Science* 72: E155–E167, 2007.
138. L Huang. Thermal inactivation of Listeria monocytogenes in ground beef under isothermal and dynamic temperature conditions. *Journal of Food Engineering* 90: 380–387, 2009.
139. KJ Versyck, K Bernaerts, AH Geeraerd, JF Van Impe. Introducing optimal experimental design in predictive modelling: A motivating example. *International Journal of Food Microbiology* 51: 39–51, 1999.
140. MG Corradini, M Peleg. A Weibullian model for microbial injury and mortality. *International Journal of Food Microbiology* 119: 319–328, 2007.
141. M Peleg, MD Normand, MG Corradini, AJ van Asselt, P de Jong, PF ter Steeg. Estimating the heat resistance parameters of bacterial spores from their survival ratios at the end of UHT and other heat treatments. *Critical Reviews in Food Science and Nutrition* 48: 634–648, 2008.
142. MG Corradini, MD Normand, C Newcomer, DW Schaffner, M Peleg. Extracting survival parameters from isothermal, isobaric, and iso-concentration inactivation experiments by the "3 end points method." *Journal of Food Science* 74: R1–R11, 2009.

143. M Hassani, P Mañas, J Raso, S Condon, R Pagan. Predicting heat inactivation of Listeria monocytogenes under nonisothermal treatments. *Journal of Food Protection* 68: 736–743, 2005.
144. LM Tamagnini, GB de Sousa, RD Gonzalez, CE Budde. Behaviour of Enterobacter amnigenus and Salmonella Typhimurium in Crottin goat's cheese: Influence of fluctuating storage temperature. *Small Ruminant Research* 76: 177–182, 2008.
145. VP Valdramidis, AH Geeraerd, K Bernaerts, JF Van Impe. Microbial dynamics vs. Mathematical model dynamics: The case of microbial heat resistance induction. *Innovative Food Science and Emerging Technologies* 7: 118–125, 2006.
146. F Arsene, T Tomoyasu, B Bukau. The heat shock response of Escherichia coli. *International Journal of Food Microbiology* 55: 3–9, 2000.
147. M Hassani, G Cebrian, P Mañas, S Condon, R Pagan. Induced thermo-tolerance under nonisothermal treatments of a heat sensitive and a resistant strain of Staphylococcus aureus in media of different pH. *Letters in Applied Microbiology* 43: 619–624, 2006.
148. M Hassani, P Mañas, S Condon, R Pagan. Predicting heat inactivation of Staphylococcus aureus under nonisothermal treatments at different pH. *Molecular Nutrition and Food Research* 50: 572–580, 2006.
149. IS Lee, JL Slonczewski, JW Foster. A low-pH inducible, stationary-phase acid tolerance response in Salmonella Typhimurium. *Journal of Bacteriology* 176: 1422–1426, 1994.
150. MG Corradini, M Peleg. Dynamic model of heat inactivation kinetics for bacterial adaptation. *Applied and Environmental Microbiology* 75: 2590–2597, 2009.
151. M Stasiewicz, BP Marks, A Orta-Ramirez, DM Smith. Modelling the effect of prior sublethal thermal history on the thermal inactivation rate of Salmonella in ground turkey. *Journal of Food Protection* 71: 279–285, 2008.
152. IA Taub, FE Feeherry, EW Ross, K Kustin, CJ Doona. A quasi-chemical kinetics model for the growth and death of Staphylococcus aureus in intermediate moisture bread. *Journal of Food Science* 68: 2530–2537, 2003.
153. CJ Doona, FE Feeherry, EW Ross. A quasi-chemical model for the growth and death of microorganisms in foods by non-thermal and high-pressure processing. *International Journal of Food Microbiology* 100: 21–32, 2005.
154. EW Ross, IA Taub, CJ Doona, FE Feeherry, K Kustin. The mathematical properties of the quasi-chemical model for microorganism growth-death kinetics in foods. *International Journal of Food Microbiology* 99: 157–171, 2005.
155. J Horowitz, MD Normand, MG Corradini, M Peleg. A probabilistic model of growth, division and mortality of microbial cells. *Applied Environmental Microbiology* 76: 230–242, 2010.
156. MG Corradini, MD Normand, M Peleg. Stochastic and deterministic model of microbial heat inactivation. *Journal of Food Science* 75: R59–R70, 2010.
157. MG Corradini, MD Normand, M Eisenberg, M Peleg. Evaluation of a stochastic inactivation model for heated-activated spores of Bacillus spp. *Applied and Environmental Microbiology* 76: 4402–4412, 2010.

Part II

Quality and Safety of Thermally Processed Foods

Part II

Quality and Safety of Thermally Processed Foods

8 Thermal Processing of Meat and Meat Products

Brijesh K. Tiwari and Colm O'Donnell

CONTENTS

8.1 INTRODUCTION

Meat is defined as the edible flesh of animals used as a food, which is composed of bundles of muscle fibers joined by strong connective tissues with fat scattered between them to provide flavor, moisture, and texture. The composition and high moisture content (ca. 60%) of meat creates an ideal environment for the growth and proliferation of pathogenic and spoilage microorganisms [1]. Preservation of meat and meat products requires adequate processing and/or preservation techniques to ensure microbial safety and maintenance of quality over an extended shelf life. Thermal processing of meat and meat products has evolved significantly since a concept of heat treatment to preserve food was introduced by Nicholas Appert [2] and the invention of the pressure cooker or retort. Thermal processing of meat and meat products remains the most important method of preservation, which can be applied by a processor to render a food product commercially sterile, as well as to modify sensory characteristics [1]. Thermal processing of meat can be used alone or in combination with other novel food processing techniques. The primary objective of thermal processing is to enhance food safety and extend product shelf life. Thermal processing also provides an opportunity to develop a wider range of meat and meat products with desired sensory attributes. The range of thermal processing techniques employed for preserving meat and developing new products can be classified as dry, moist, novel thermal (microwave or infrared [IR]), or combination of techniques with minimal changes in meat quality. Thermal processing of meat can broadly be classified as batch type using a cooker or retort to provide heating, holding, and cooling phase or continuous with product on conveyors into a tunnel with heating, holding, and cooling operations taking place in series on a continuous basis [3,4].

Thermal processing can be semicontinuous where either heating and/or holding phase is continuous, and cooling is batch process depending on the application. The type of thermal processing employed for meat and meat product largely depends on the nature of the raw meat and characteristics of the final product. This chapter discusses recent developments in thermal processing of meat products.

8.2 COOKING OF MEAT

Cooking is the most common term used for the preparation of food using heat. Cooking of meat involves moist heat application either in a retort or pressure cooker. The amount of heat transferred to the meat depends on the heat transfer coefficient of the heating medium, cooking time, and temperature. Cooking time and temperature combinations play an important role in ensuring the stability of meat products. A range of products can be obtained by altering cooking time and temperature. Meat cooking methods include (1) dry heat method, (2) moist heat method, and (3) novel heat method (Figure 8.1).

These methods employ varying levels of temperature and time to achieve numerous objectives, which include the following:

1. Pathogenic and spoilage microbial inactivation
2. Extension of shelf life
3. Increased palatability of meat by improving flavor and texture

8.2.1 Dry Heating Methods

8.2.1.1 Roasting

Roasting or oven cooking of meat employs dry heat either from an open flame, oven, or other heat source suitable for slow cooking of meat. Roasting of meat involves heat transfer from the surroundings to the meat surface and consequently induces a temperature gradient inside the product, leading to an increase in internal temperature resulting in a significant loss in weight [5]. This loss is attributed to moisture loss and induces changes in proteins and lipids leading to internal transport of moisture due to protein denaturation. Such protein denaturation causes the shrinkage of the meat fiber network, resulting in a mechanical force that expels the excess interstitial moisture toward the surface [6]. Roasting generally causes nonenzymatic or Millard browning, which imparts a characteristic flavor to the meat product.

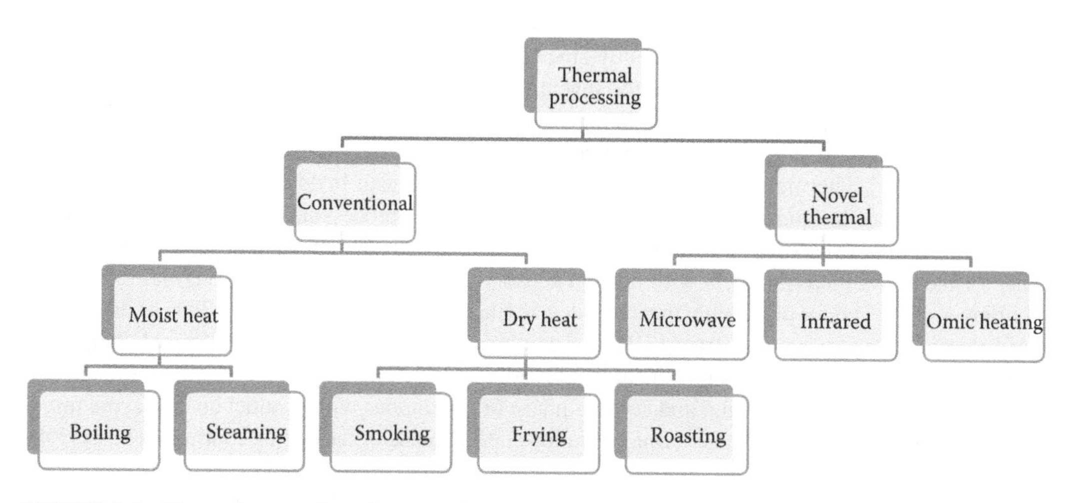

FIGURE 8.1 Thermal processing of meat and meat products.

Goñi and Salvadori [5] developed an oven cooking model for beef to predict evaporation loss, dripping loss (DL), cooking time, and the internal moisture content variation, which depends on the water-holding capacity of beef. Both convection and radiation heat transfer from the surrounding to the meat surface occur during roasting of meat, whereas heat transfer through conduction occurs from meat surface to the core of the meat subjected to roasting.

To model heat transfer in meat products during roasting, meat can be considered as a solid matrix. Heat transfer can be expressed as

$$\rho C_p \frac{\partial T}{\partial t} = \nabla(k\nabla T) \tag{8.1}$$

The surface temperature of meat during roasting is usually about 100°C due to high oven or roaster temperature (>100°C). High surface temperature causes evaporation of surface moisture, which can be quantified as water vapor flux (j_{evap}). The amount of heat associated with the evaporative flux is added to the boundary conditions for the heat balance as given in the following equation [5]. This mass flux depends on the water activity of the meat surface and the relative humidity of the oven or roaster:

$$-nk\nabla T = h(T_s - T_o) + \varepsilon\sigma\left(T_s^4 - T_o^4\right) + \lambda j_{evap} \tag{8.2}$$

where

$$j_{evap} = k_g(a_w P_{sat}(T_s) - RHP_{sat}(T_o)) \tag{8.3}$$

Protein that comprises ~20% of meat muscle (w/w) acts as a sponge, and a substantial amount of moisture is lost as a result of dripping. Hence, the weight loss occurring during roasting is attributed to both evaporative losses (EL) from the surface and drip loss in the form of exudates. Exudate is a solution of sarcoplasmic proteins [7]. Goñi and Salvadori [5] predicted EL by surface integration of evaporative mass flux and DL as the difference of moisture content at any given process time (t):

$$EL(t) = \int_0^t \left[\int_\Gamma j_{evap}(\Gamma, t)\,d\Gamma\right] dt \tag{8.4}$$

$$DL(t) = m_s[c_i - \bar{C}(t)] \tag{8.5}$$

where

$$\bar{C}(t) = \frac{1}{V}\int_\Omega c(\Omega, t)\,d\Omega \tag{8.6}$$

Hence, total losses can be written as the sum of EL and drip losses as given in the following:

$$TL(t) = EL(t) + DL(t) \tag{8.7}$$

Heat and mass transfer models for roasting of meat are based on many assumptions. Obuz et al. [8] modeled cylindrical beef samples cooked in a forced-air convection oven using the following assumptions to reduce the level of complexity in the model:

1. Meat samples were homogeneous and cylindrical in geometry.
2. Initial moisture distributions in samples were uniform.
3. Ambient temperature and moisture were step functions of time.
4. Vaporization of water was restricted to the roast surface only.
5. Roast shrinkage was negligible.
6. The effect of crust formation on physical properties was negligible.
7. Fat transport was negligible.
8. The problem is axisymmetric.

For constant properties and heat generation within the oven, the heat transfer equation for cylindrical coordinates is given as follows:

$$\frac{\partial T}{\partial t} = \alpha\left[\frac{\partial^2 T}{\partial r^2} + \frac{1}{r}\frac{\partial T}{\partial r} + \frac{1}{r^2}\frac{\partial^2 T}{\partial \theta^2} + \frac{\partial^2 T}{\partial z^2}\right] \tag{8.8}$$

The initial temperature when the cylindrical meat is placed in the oven is T_0, and as time progresses, heat is transferred to the meat surface resulting in EL and conductive heat transfer into the meat. Surface heat transfer to the meat and convection gain or loss from the surface can be expressed as follows [8]:

$$h(T - T_\infty) + q = -k\alpha\left[\frac{\partial T}{\partial r} + \frac{\partial T}{\partial \theta} + \frac{\partial T}{\partial z}\right] \tag{8.9}$$

Mass transfer occurring during roasting by diffusion of moisture through the meat and evaporation from the roast surface and moisture loss occurring as a drip loss should be considered in predicting losses. Obuz et al. [8] determined the mass flow rate (D_m) at the surface of the roast by the difference in moisture contents of the air and roast surface as outlined by Singh et al. [9]:

$$q = hA(T_s - T_\infty) - L_v D_m \tag{8.10}$$

The amount of energy required to evaporate moisture from the meat surface can be determined by multiplying the mass flow rate by the latent heat of vaporization. Hence, the heat and moisture transfer can be determined from the following:

$$D_m = k_m A(P_s - P_\infty) \tag{8.11}$$

8.2.1.2 Smoking

Smoking is a dry heat treatment method often employed for flavoring or cooking of meat by exposing it to smoke. Smoking of meat can be either cold smoking or hot smoking. Cold smoking is used to enhance meat flavor, and the process temperature is usually <40°C. Generally, cold smoked meat and meat products are roasted prior to consumption. Hot smoking involves the exposure of meat to smoke and heat in a controlled environment in a temperature range of 75°C–85°C. During hot smoking, cooking of meat and flavor enhancement occur simultaneously. In addition to cooking, brown color also develops on the meat surface as a result of the Maillard reaction on the surface of smoked products.

8.2.1.2.1 Smoking Method

Smoking of meat is one of the oldest methods to cook meat. Several types of smoking chambers are available commercially. The major components of an ideal smoking unit consist of an enclosed

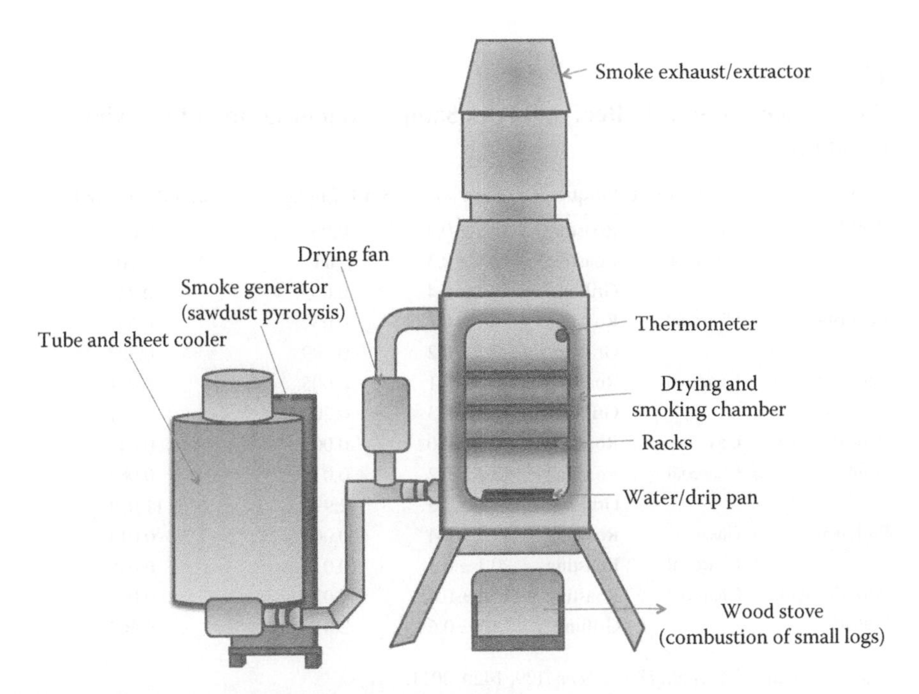

FIGURE 8.2 Typical smoke chamber and its components.

fire box or heat source and smokebox or smoking chamber. The source of heat used for smoking of meat and meat products is generally wood, sawdust, or charcoal. Smoke from the firebox flows through a connecting pipe or opening into the smoking chamber (Figure 8.2) [12]. Smoke generated from the wood is an antimicrobial agent and an antioxidant, which prevents or slows rancidification of animal fats. Wood smoke also contains a large number of polycyclic aromatic hydrocarbons (PAHs) and their alkylated derivatives [10], which are considered carcinogenic. PAHs are formed during incomplete combustion processes, which occur in varying degree whenever wood, coal, or oil is burnt [11]. PAHs are also formed in foods during smoke curing, broiling, roasting, and grilling over open fires or charcoal. Combustion temperature during the generation of smoke is critical. The presence of PAH in smoked meat and meat products is a public health concern and the European Union established a maximum level of 5 ng g^{-1} for benzo[α]pyrene in smoked meat (Commission Regulation (EC) 208/2005). Stumpe-Vīksna et al. [13] investigated the effect of wood (Apple, Alder, Alder + juniper, Spruce, Maple, Hazel, Plum, AspenBird-cherry, and Rowantree) used for the smoking of meat on the formation of PAH. They observed that the wood species used has a significant influence on the amount of PAH in smoked meat with samples smoked with Apple-tree and Alder containing the smallest PAH concentrations, while the samples smoked with spruce had the highest concentrations of PAH. This indicates that the selection of the wood species as a heating source has a significant influence on the amount of PAH in smoked meat. Table 8.1 indicates that charcoal-roasted, gas-roasted, and charcoal-grilled beef and pork samples contained PAHs with charcoal grilling producing the highest PAH levels [14]. Due to the carcinogenic potential of PAHs, researchers are currently investigating alternative roasting and grilling techniques with minimal or no PAHs without influencing sensorial acceptability of meat and meat product.

8.2.1.2.2 Heat and Mass Transfer during Smoking

Heat and mass transfer inside the smoking chamber depends on several extrinsic factors such as smoke temperature, smoking time, heating source, and intrinsic factors such as moisture, fat content, and product characteristics. Sebastian et al. [14] developed a model to investigate the relative influence of several physical phenomena on the process performances and on the

TABLE 8.1
PAH Concentrations in Beef and Pork Samples under Different Cooking Conditions

Meats	Cooking Technique		Fat (%)	B(a)P (µg kg^{-1})[a]	Total PAHs (µg kg^{-1})
Beef loin	Gas	Roasting	18.7 ± 0.3	0.003	0.025
	Charcoal	Roasting	18.4 ± 0.3	nd	0.007
		Grilling	19.2 ± 0.4	0.055	0.753
Beef ribs	Charcoal	Roasting	6.7 ± 0.1	0.032	0.05
		Grilling	7.5 ± 0.2	0.199	0.787
Beef ribs with sauces	Charcoal	Roasting	12.3 ± 0.1	0.005	0.021
		Grilling	11.5 ± 0.3	0.202	0.797
Pork shoulder loin	Gas	Roasting	13.4 ± 0.0	0.002	0.047
	Charcoal	Roasting	13.2 ± 0.2	0.018	0.063
		Grilling	14.9 ± 0.3	2.994	11.033
Pork belly	Gas	Roasting	17.7 ± 0.4	0.005	0.013
	Charcoal	Roasting	20.1 ± 0.1	0.014	0.048
Pork ribs with sauces	Charcoal	Roasting	16.6 ± 0.6	0.026	0.069
		Grilling	14.0 ± 0.6	2.819	9.507

Source: Chung, S.Y. et al., *Food Chem.*, 129, 1420, 2011.
[a] Mean value of 10 samples.

evolution of pork meat characteristics. They investigated the effect of (1) radiative heat transfer between the radiant plates and the product or the smoke, (2) convective heat transfer between the plates and the pyrolysis or combustion smoke, (3) convective heat transfer between the product and the pyrolysis smoke, (4) conductive heat transfer between the combustion smoke and the radiant plates, (5) fat and moisture phase changes at the surface or inside the product, (6) mass transfer of vapor between the product and the pyrolysis smoke, and (7) mass transfer of water and melted fat flowing out of the product and falling down to the bottom of the cooking chamber. The variables in the heat and mass balance equations are time and space dependent along the z-axis as shown in Figure 8.3 based on temperatures and moisture or fat contents [14] and are given as follows:

1. Energy balance in pyrolysis smoke

$$\rho_{ps}\cdot\tau_{ps}\cdot C_{P_{ps}}\cdot\frac{\partial(T_{ps})}{\partial t}+\frac{q_{ps}}{l}\cdot C_{P_{ps}}\cdot\frac{\partial(T_{ps})}{\partial x}=h_{s.ps}\cdot(T_s-T_{ps})+h_{p.ps}\cdot(T_p-T_{ps}) \quad (8.12)$$

2. Vapor mass balance in pyrolysis smoke

$$\rho_{ps}\cdot\tau_{ps}\cdot\frac{\partial(W_{ps})}{\partial t}+\frac{q_{ps}}{l}\cdot\frac{\partial(W_{ps})}{\partial x}=Fm_v \quad (8.13)$$

3. The combustion smoke

$$\rho_{cs}\cdot\tau_{cs}\cdot C_{P_{cs}}\cdot\frac{\partial(T_{cs})}{\partial t}+\frac{q_{cs}}{l}\cdot C_{P_{cs}}\cdot\frac{\partial(T_{cs})}{\partial x}=h_{s.cs}\cdot(T_s-T_{cs}) \quad (8.14)$$

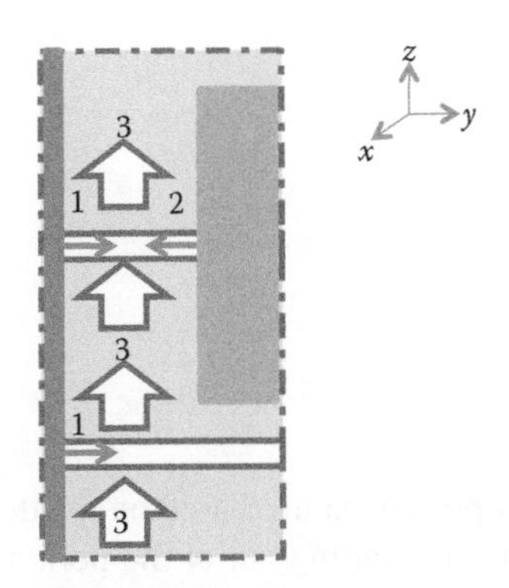

Pyrolysis smoke
1: Heat convection with the plate
2: Heat convection and vapor transfer with the product
3: Pyrolysis smoke circulation

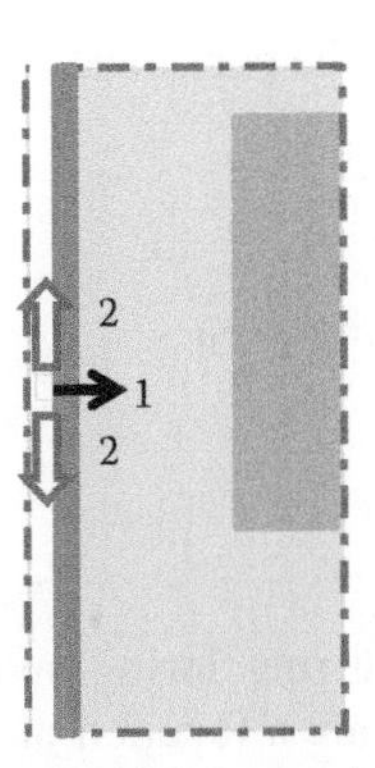

Combustion smoke
1: Heat convection and radiation with the plate
2: Combustion smoke circulation

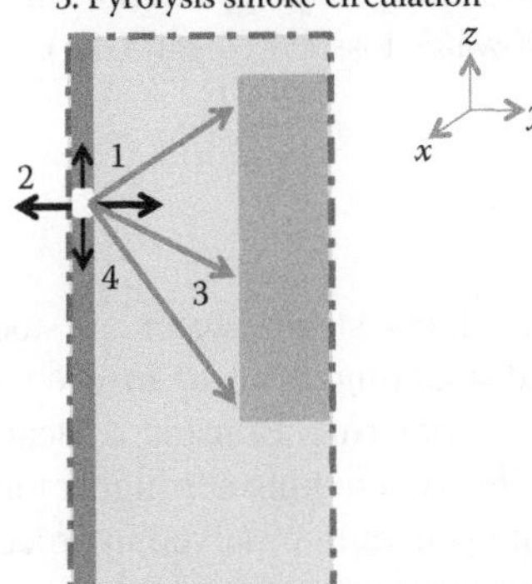

Radiant plate
1: Heat convection with the pyrolysis smoke
2: Heat convection and radiation with the combustion smoke
3: Radiation transfer with the product
4: Heat conduction through the plate

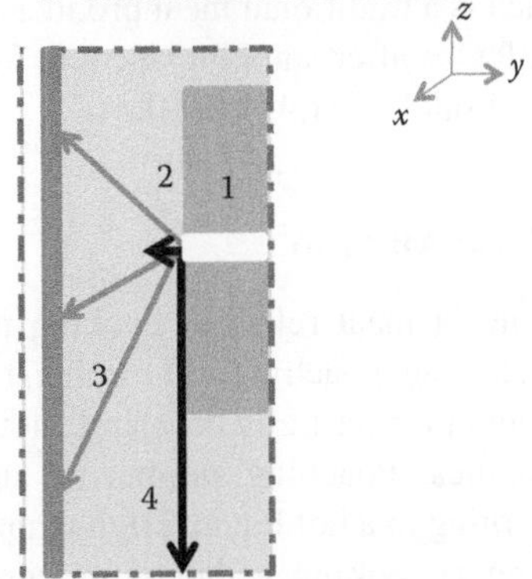

Product
1: Phase changes: - fat fusion and water evaporation
2: Heat convection with the pyrolysis smoke
3: Radiation transfer with the plate
4: Water and fat drip

FIGURE 8.3 Heat and mass transfers between solid matters and gases inside a representative elemantary volume. (Adapted from Sebastian, P. et al., *J. Food Eng.*, 70(2), 227, 2005.)

4. Energy balance at the radiant plate

$$\rho_s \cdot \tau_s \cdot C_{P_s} \cdot \frac{\partial(T_s)}{\partial t} = h_{s.cs} \cdot \left(T_{cs} - T_s\right) + k_s \cdot \tau_s \cdot \frac{\partial^2(T_s)}{\partial z^2} + \frac{\sigma}{L} \cdot \int_{L-L_1-L_p}^{L-L_1} \varphi_s(\zeta_1) \cdot \left(T_p(\zeta_1)^4 - T_s^4\right) \mathrm{d}(\zeta_1) \tag{8.15}$$

5. Energy balance in the pork meat

$$\rho_p \cdot \tau_p \cdot C_{P_p} \cdot \frac{\partial(T_p)}{\partial t} = h_{p.ps} \cdot \left(T_{ps} - T_p\right) + \frac{\sigma}{L} \cdot \int_0^L \varphi_p(\zeta_2) \cdot \left(T_s(\zeta_2)^4 - \mathrm{T}_p^4\right)$$

$$\times \mathrm{d}\,(\zeta_2) - \Delta h_v \cdot Fm_v - \Delta h_f \cdot \dot{\varnothing}_{mf} \tag{8.16}$$

6. Moisture mass balance in the pork meat

$$\rho_p \cdot \tau_p \cdot \frac{\partial(W_p)}{\partial t} = -Fm_v - Fm_g \tag{8.17}$$

7. Fat mass balance in the pork meat

$$\frac{dX_p}{dt} = \Gamma \cdot \frac{dW_p}{dt} \tag{8.18}$$

The heat and mass transfer occurring during smoking of pork meat are based on the drying kinetics of the meat product, which are a function of (1) moisture coming out of the pork and lost due to evaporation or drops to the bottom of the smoking chamber, (2) fat fusion kinetics occurring inside the product as a result of melting and fat loss from the product, and (3) convective and radiative heat transfers within the smoking chamber. A total weight loss up to 38.3% is reported for Boucané, which is a traditional meat product obtained by salting, drying, and hot smoking pork belly [15]. These losses after smoking account for the high water loss (31.0 kg/100 kg), fat melting (6.2 kg/100 kg), and salt loss (1.4 kg/100 kg).

8.2.2 Moist Heat Method

Moist heat cooking of meat refers to cooking methods involving steam, water, or stock. These include braising, stewing, poaching, simmering, boiling, and steaming. Braising involves scorching the meat surface and then partially covering with liquid and simmering. Braising is ideal for cooking tougher cuts of meat. Poaching, simmering, and boiling involve cooking at a higher temperature (100°C) by submerging in a hot liquid. High temperatures are generally observed in processes such as meat canning, retort-cooking, and pressure-cooking. Steaming involves steam cooking of meat at a temperature of about 100°C (may increase if pressurized, e.g., pressure cooker). Steam cooking is ideal compared to hot water at comparable temperature because the heat transfer is better in case of steam than hot water, since the latent heat from condensing steam aids in the heating of the product. However, if the moist heat cooking time is prolonged, meat with high connective tissue content will become tenderer as the collagen dissolves [16].

8.2.3 Novel Thermal Processing of Meat

The major disadvantages of conventional thermal cooking of meat and meat products are longer cooking times and nonuniform heating [17]. Novel thermal methods such as dielectric heating, radio frequency (RF), or microwave reduce cooking time without compromising meat safety. Heat transfer in conventional meat cooking is mainly by conduction, convection, or radiation, whereas in case of RF or microwave heating, the product is heated internally and not through the meat surfaces.

IR cooking is of particular interest to the processed meat sector because conventional cooking ovens, using high-velocity hot air convection, can cause surface deterioration, overheating, oxidation, charring, impingement damage, low yields, difficult emissions, and high-energy costs. IR radiation has intrinsic advantages in achieving these goals as there is no direct intention or necessity to heat the air, thus oven temperatures and humidity are reduced. This results in savings in energy and conservation of cooked out material, which will not be lost to evaporation by high oven temperatures. A further advantage of this method is the ease at which heat can be applied evenly over a broad surface area.

Laycock et al. [17] studied the effect of RF cooking on the color, water-holding capacity, texture of ground or comminuted meat and muscle. They observed a significant reduction in cooking time for ground beef, comminuted meat, and muscle compared to cooking in a water bath (Figure 8.4) to a core temperature of 72°C with lower juice losses and acceptable color, water-holding capacity,

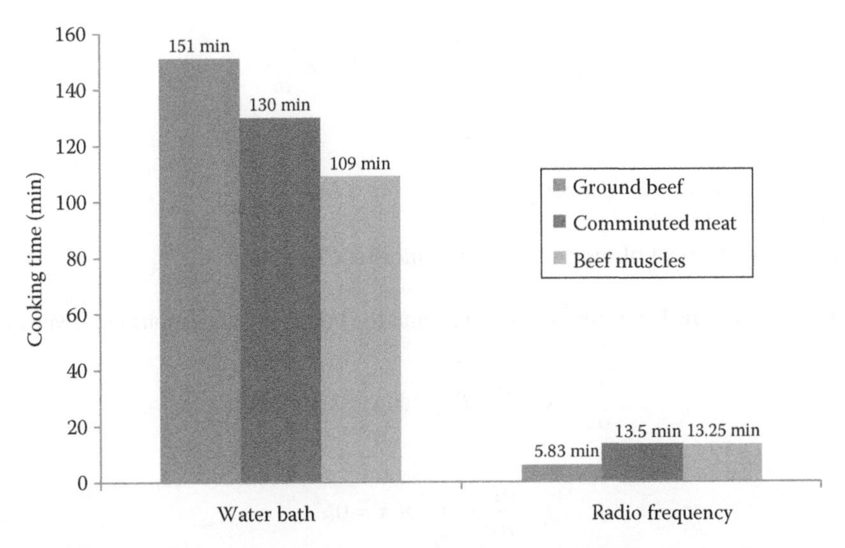

FIGURE 8.4 Effect of radio frequency and water bath cooking on cooking time (min) of beef meat. (Adapted from Laycock, L. et al., *Meat Sci.*, 65, 959, 2003.)

and texture. Microwave cooking of meat is achieved by the absorption of microwave energy by rotation of the dipolar water molecules and by translation of the ionic components of the meat [17]. The propagation of microwaves into meat depends on (1) the dielectric and thermal properties, (2) the composition and geometry of product, and (3) the power level [17–23]. Jeong et al. [18] studied the effect of fat level with and without salt on the cooking pattern and physicochemical properties of ground pork patties cooked by microwave energy (900 W). They observed that the temperatures at the edges of the pork patties increased faster than those at the center or the midway positions within the microwave cooker. However, for patties with NaCl added, the temperature of the center position was higher than that of the midway position. Total cooking loss, drip loss, reduction in diameter, and thickness were higher in patties with 20% fat compared to those with 10% fat, indicating the influence of fat content and salt on cooked meat using microwave energy. Yarmand et al. [24] studied the effect of domestic microwave (700 W), industrial microwave (12,000 W), and conventional oven cooking of goat and lamb meat to internal temperature of 70°C. They observed higher cooking losses in microwave heating compared to conventional oven. They observed a cooking loss for goat meat of 26.6% for conventional cooking and 29.3% for domestic microwave cooking.

Sheridan and Shilton [25] studied the efficacy of cooking hamburger patties by mid-IR sources (λ_{max} = 2.7 μm) and far-IR (λ_{max} = 4.0 μm). They observed that the change in core temperature followed closely the change in surface temperature with short cooking time and is independent of the fat content of the samples when subjected to the higher energy source. However, with the lower energy source, the rate of core temperature rise is dependent on the fat content, and target core temperature can be achieved quickly for the samples containing high fat.

Heat transfer during IR heating depends on IR heating source and meat characteristics. These include temperature difference between the heating source and the meat surface, emissivity of heating source, meat surface absorptivity, reflectivity, and radiation penetration depth [25]. Shorter wavelength radiation is more penetrative [26]. The presence of protein and fat in meat samples have principal absorption bands at wavelengths of 3.0–4.0 μm and at 5.5–6.5 μm and on up to 10.0 μm [25,27]. Heat transfer during IR heating within the meat is by conduction and by diffusion of moisture, fat, liquids, and vapors. Heat and mass transfer during the cooking of beef patties by long wavelength, far-IR radiation was investigated by Shilton et al. [28]. For modeling purpose, Shilton et al. [28] considered beef patties as infinite slab and unsteady state 1D heat transfer in a slab can be described by

$$\rho_s C_p \frac{\partial T}{\partial t} = \frac{\partial}{\partial x}\left(k \frac{\partial T}{\partial x}\right) + \rho_d \lambda \frac{\partial C}{\partial t} \tag{8.19}$$

where
- T is the temperature of the beef patty
- x is the thickness of the beef patty
- C is the moisture content of the beef patty (kg moisture/kg solid)

The initial conditions are $T = T_i$ for all locations, and the boundary conditions are given as follows:

$$k \frac{\partial T}{\partial x} = \sigma\left(T_\infty^4 - T_s^4\right) \quad \text{at } x = X, 0 < t \leq t_f \tag{8.20}$$

$$\frac{\partial T}{\partial x} = 0 \quad \text{at } x = 0 \tag{8.21}$$

where
- k (W m^{-1} K^{-1}) is the thermal conductivity of the beef patty
- σ is the Stefan–Boltzmann constant (5.669×10^{-8} W m^{-2} K^{-4})

This allows for the absorption of the far-IR radiation by the beef patty and enables the temperature of the IR source to be used as an initial boundary condition.

Since physical and thermal properties for the beef patties vary with temperature and fat content, Shilton et al. [25] calculated density (ρ) and thermal conductivity (k) using the following polynomial expressions proposed by Choi and Okos [29]:

$$\rho = r_1 - r_2 T - r_3 T^2 \tag{8.22}$$

$$k = k_1 + k_2 T - k_3 T^2 \tag{8.23}$$

where k_1, k_2, k_3, r_1, r_2, and r_3 are constants. Heat transfer from the meat surface to the core during IR heating occurs partly by internal convection and the movement of moisture and fat and partly by conduction. To model heat transfer in beef patties, Equation 8.19 can be modified to accommodate internal convection designated an effective heat transfer coefficient [28] as given by

$$k = \left(k_1 + k_2 T - k_3 T^2\right) + (h_{eff_1} + h_{eff_2} \cdot T) \tag{8.24}$$

where h_{eff1} (W m^{-1} K^{-1}) and h_{eff2} (W m^{-1} K^{-2}) are constants.

EL during IR heating can be estimated by using the following equation for unsteady state 1D mass transfer if beef patties are considered as infinite slab [28]:

$$\frac{\partial C}{\partial x} = D \frac{\partial^2 C}{\partial x^2} \tag{8.25}$$

where
- C is the moisture content in the beef patty (kg moisture/kg solid)
- D is the mass diffusion coefficient of vapor through the beef patty (m^2 s^{-1})

The initial conditions are $C = C_i$ for all locations, and the boundary conditions are given by [28]

$$D\frac{dT}{dx} = K_m(C_\infty - C_s) \quad \text{at } x = X, 0 < t \leq t_f \tag{8.26}$$

$$\frac{\partial C}{\partial x} = 0 \quad \text{at } x = 0$$

where

K_m is the mass transfer coefficient (m s^{-1})

D is mass diffusion coefficient that can be calculated from the following equation, which was originally derived to calculate mass diffusivities during air drying processes of food [29]:

$$D = D_x \cdot \exp\left(\frac{D_T}{T}\right) \cdot \exp\left(\frac{D_c}{C}\right) \tag{8.27}$$

where D_x, D_T, and D_c are constants.

Heat and mass transfer models described earlier showed good prediction with experimental data for beef patties up to fat content of 30% by including internal fat and moisture convection terms in the model. However, if internal fat and moisture convection terms are not used, the models show good prediction only for beef patties with 0% fat content. This indicates that the heat transfer in beef patties is strongly influenced by internal movement of melted fat by convection during the cooking process [28].

8.2.4 *Sous-Vide* Processing

Sous-vide processing and products were originated in France in the 1970s when a French chef experimented with vacuum sealing and slow cooking as a way to extend shelf life and prevent shrinkage of foie gras (fat liver made of the liver of a duck or goose) [30]. *Sous-vide* means "under vacuum" in French, and the term describes both the process and end product. It is characterized as a combination of mild thermal processing and vacuum packaging to preserve meat products. *Sous-vide* processing involves fresh or raw meat placed in a pouch or a semirigid tray, vacuum sealed, cooked with mild heat, and stored at low temperature of 4°C until they are to be served. The obvious advantage of *sous-vide* processing is that the hermetic seal prevents loss of moisture, and shrinkage is minimized [30]. Concerns associated with *sous-vide* processing involve the microbiological safety of the product because psychrotrophic food-borne pathogens and spore-forming organisms including *Clostridium botulinum* may survive the mild heat treatment and be capable of harming the consumer when improper storage or recooking methods are applied [31]. Table 8.2 shows some recommended heat treatments for *sous-vide* products intended for extending shelf life. Table 8.2 shows that time–temperature combinations vary with microorganism and with regulatory agencies. For example, the U.K. Department of Health [32] recommended that 70°C for 2 min is sufficient to achieve microbial safety; however, this was later changed to 90°C for 4.5 min [33].

Díaz et al. [34] studied spoilage during the refrigerated storage of cooked pork loin processed by the *sous-vide* method (cooked at an oven temperature/time of 70°C/12 h with an internal temperature/time of 70°C/11 h). They observed that the sensory spoilage preceded microbiological spoilage for *sous-vide* pork loin. They further observed that the pork loin was unacceptable after 10 weeks of chilled storage at 4°C. This loss of acceptance was mainly due to the deterioration of meaty flavor and odor along with the loss of appearance, juiciness, and firmness. Literature reveals that the shelf life of *sous-vide* cooked meat-based products has been established at 4–7 weeks, depending on the product, processing, and storage conditions for fresh oxblade roast, sirloin roast, sirloin steaks, beef

TABLE 8.2
Recommended Processing Conditions for *Sous-Vide* Processing

Source	Temperature (Internal Temperature) and Time Conditions	Target Organism
French regulations (Ministère de l'Agriculture 1974, 1988)	70°C for 40 min can be stored for 6 days under chilled conditions (4°C)	*Enterococcus faecalis*
	70°C for 100 min stored for 21 days under chilled conditions (4°C)	*Enterococcus faecalis*
	70°C for 1000 min stored for 42 days under chilled conditions (4°C)	*Enterococcus faecalis*
Department of Health Guidelines (1989), United Kingdom	70°C for 2 min stored for 5 days under chilled conditions (4°C)	*L. monocytogenes*
Sous Vide Advisory Committee (SVAC)	80°C for 26 min or 90°C for 4.5 min Stored for 8 days under chilled conditions (4°C)	*C. botulinum* type E
Advisory Committee on the Microbiological Safety of Food (ACMSF)	90°C for 10 min stored for >10 days under chilled conditions (4°C)	*C. botulinum*
European Chilled Food Federation Botulinum Working Party	70°C for 2 min can be stored for short duration whereas products processed	*L. monocytogenes*
	90°C for 10 min can be stored for longer duration	*C. botulinum*

Source: Adapted from Ghazala, S. and Trenholm, R., Hurdle and HACCP concepts in *sous vide* and cook-chill products, in: S Ghazala, ed., *Sous vide and Cook-Chill Processing for the Food Industry*, Aspen Publishers, Inc., Gaithersburg, MD, 1998.

tournados, veal, skinless chicken breasts, racks of lamb, leg of lamb, pork loins, turkey breasts, and blue-eye trevalla fillets [35].

8.3 CHANGES IN MEAT DURING COOKING

Physicochemical changes occur during thermal processing of meat leading to weight loss, modifications of water-holding capacity, texture, muscle fiber shrinkage, color, and flavor [36]. Physicochemical changes occurring during thermal processing strongly depend on the denaturation of meat protein, moisture loss composition, and characteristics of the muscles, the heating method, as well as the time/temperature evolution during cooking [37,38]. Mora et al. [38] investigated the effect of various cooking methods for turkey breast involving dry air, low-steam and high-steam heating conditions on cooking parameters (temperature, cook value, and yield), and textural and sensory properties. They observed that the low-steam cooking conditions (RH = 35%) significantly increased cooking yield. Low-steam-cooked samples were significantly different from high-steam samples, producing a higher cooking yield (7% higher than the high steam cooking), probably due to less meat longitudinal shrinkage, leading to lower water loss upon cooking. Dry air cooking requires higher cooking time compared to the low steam and high steam with a uniform crust formation and some nonenzymatic browning on turkey breast. A study shows that the cooking of beef rib-eye and brisket, pork neck steak and belly, veal chop, and rolled breast meat cut can reduce absolute fat content by ~17.9%–44.4% [39]. Table 8.3 indicates that the cooking influences the content of several nutrients in meat in different ways depending on the cut and the cooking process parameters such as time, medium, and temperature.

TABLE 8.3
Effect of Cooking Process on the Losses (%) in Nutritional Composition of Various Meat Cuts

Type of Meat Cut	Cooking Process	Cooking Loss	SFA	MUFA	PUFA	Retinol	α-Tocopherol	Thiamine	Riboflavin	Nicotinic Acid
Beef rib-eye	Grilled to a core temperature of 72°C	29.1	13.3	26.1	15.5	34.0	14.1	74.3	49.3	39.5
Beef brisket	Boiled for 1 h in stock-pot	35.9	22.8	17.9	7.1	39.3	19.3	100.0	83.4	65.2
Pork neck steak	Grilled to a core temperature of 72°C	32.4	25.2	24.7	16.6	15.4	11.0	73.9	53.4	51.3
Pork belly	Grilled to a core temperature of 72°C	24.9	16.2	17.9	13.0	24.9	16.6	75.7	48.8	50.0
Veal chop	Grilled to a core temperature of 72°C	25.9	20.6	20.3	18.1	28.0	21.8	73.1	48.6	50.5
Veal rolled breast	Steamed for 45 min in antiadhesive pan stock-pot with multisteamer	39.5	44.8	38.6	48.4	41.6	16.7	89.6	67.6	64.0

Source: Adapted from Gerber, N. et al., *Meat Sci.*, 81(1), 148, 2009.
SFA, saturated fatty acids; MUFA, mono unsaturated fatty acids; PUFA, poly unsaturated fatty acids.

Cooking time and temperature also have an effect on the tenderness of meat. Combes et al. [40] observed a two-phase increase in rabbit meat toughness with increase in cooking temperature up to 50°C and from 65°C to 80°C with a decrease in toughness at temperatures between 50°C and 60°C. Literature indicates that the first increase in meat toughness is attributed to the denaturation and thermal shrinkage of the connective tissue and the second increase in toughness to the denaturation of myofibrillar proteins [41,42]. Similarly, an increase and decrease in the cooking losses are reported in the temperature range of 50°C–100°C for bovine *M. semitendinosus* [43]. Palka and Daun [43] observed an increase in cooking losses with an increasing internal temperature of meat with greatest increments in cooking losses observed between 50°C–60°C and 60°C–70°C, probably due to thermal denaturation of myofibrillar proteins and collagen shrinkage. Denaturation of proteins during thermal processing can cause a loss of up to 20%–40%, mainly in the form of moisture and fat [43–46]. Cooking temperature also decreases the volume of meat due to denaturation of proteins, causing shrinkages and losses occurring in moisture and fat. A volume reduction up to 30.5% is reported during cooking of ground beef patties at 75°C for 20 min [44]. The rates of moisture and fat losses increase with increase in temperature up to 75°C with >30% moisture and >40% fat loss occurring at 75°C. The rates of moisture and fat losses reduce with increase in temperature above 70°C. Moisture losses occur mainly by evaporation from the meat surface or as a drip in a form of exudates whereas, fat is lost as a drip [45]. Some studies show an increase in fat and protein content as a result of moisture loss. The loss of moisture is primarily due to denaturation. Denaturation of protein causes less water to be entrapped within the protein structures held by capillary forces [47]. The level of protein denaturation depends on cooking conditions such as holding time and core temperature [48]. The chemical changes occurring in muscle proteins during thermal cooking principally depend on the cooking temperature and type of muscle protein. The loss of solubility and denaturation of sarcoplasmic proteins occur in the temperature range of 40°C–60°C, whereas denaturation of myofibrillar proteins occurs at a much lower temperature range of 40°C–50°C [48]. Braeckman et al. [46] studied the influence of different IR-grilling/hot air cooking conditions on moisture and fat content and on texture and color attributes of meat patties (hamburgers). It was observed that higher air velocity and air temperature during thermal processing led to shorter cooking times. It resulted in less significant differences in moisture and total weight loss of the processed meat patties.

Flavor development in thermally processed meat depends mainly on fat. For example, strong flavor observed in old animal meat is due to changes in the oxidation levels of fats. The heating of fatty acids in the presence of air enhances oxidation, thus modifying flavor profiles of cooked products. Thermal processing triggers the development of pleasant flavors and organoleptic enhancements in meat products depending on the cooking method. For example, pan-fried meat samples (*M. longissimus dorsi* and *M. biceps femoris*) are reported to have more intensity of roasted meat flavor, less intensity of both boiled meat flavor and piggy flavor compared to oven-cooked meat [16]. Core temperature and holding time also dictate flavor development in meat and meat products. In general, an increase in core temperature results in an increase in flavor and simultaneous decrease in juiciness and tenderness of meat and meat products [47,49]. Sensorial analysis of minced pork prepared as patties on a pan, steaks prepared on a pan, whole roasts prepared in a pot, and whole roasts prepared in an oven at 140°C or 90°C indicated that the sensation of the sensory attribute differs according to core temperature and cooking technique [16]. Pan-fried samples had the highest score as far as roasted pork odor and roasted pork flavor were concerned at a core temperature of 80°C and the lowest score of other odor and flavor attributes (boiled odor/flavor and piggy odor/flavor) compared to samples with core temperatures of 75°C and 65°C [16].

8.4 EFFECTS OF THERMAL PROCESSING ON MICROORGANISMS

The pathogens of human health concern in meat and meat products include *Escherichia coli* O157:H7, *Salmonella* spp., *Listeria monocytogenes*, *Staphylococcus aureus*, and *Clostridium*

TABLE 8.4
Safe Internal Temperature for Meat and Meat Products

Meat and Meat Products	Safe Internal Temperature (°C)
Beef, veal, and lamb (pieces and whole cuts)—medium rare	63
Beef, veal, and lamb (pieces and whole cuts)—medium	71
Beef, veal, and lamb (pieces and whole cuts)—well done	77
Pork (pieces and whole cuts)	71
Ground meat and meat mixtures (e.g., burgers, sausages, meatballs, meatloaf, casseroles)—beef, veal, lamb, and pork	71
Others (hot dogs, stuffing, and leftovers)	74

Source: Adapted from CFIA, 2011 Canadian Food Investigation Agency. Food, thermometer food safety tips, available at: www.inspection.gc.ca/english/fssa/concen/tipcon/thermo.pdf, accessed date: November 4, 2011, 2011.

perfringens and *C. botulinum*. While progress is being made in the control of these pathogens, some of them will continue to be of concern well into the future [50–52]. Cooking of meat causes the inactivation of both spoilage and pathogenic organisms depending on the roasting temperature, holding time, type of microorganism, and raw material characteristics. Table 8.4 shows some recommended safe internal temperature for meat and meat products to achieve desired microbial load reduction [53]. Studies show that broiling was more effective than grilling or frying when meat samples were cooked to internal temperatures of 60°C or 65°C for inactivation of *E. coli* O157:H7 in beef [54–58]. Shen et al. [55] investigated the inactivation of *E. coli* O157:H7 in moisture-enhanced restructured nonintact beef cooked to 65°C using different cooking appliances set at different temperatures. They observed that the initial set temperatures of 204°C–260°C, regardless of appliance, resulted in 3.3- to 5.5-log reductions compared to 1.5- to 2.4-log reductions obtained at 149°C (CFU g^{-1}) in another study. Shen et al. [58] compared the thermal inactivation of *E. coli* O157:H7 in nonintact beefsteaks of different thicknesses by different cooking methods. They observed that the extent of pathogen inactivation decreased in order of roasting to a core temperature of 65°C (2.0–4.2 log CFU g^{-1}) > pan broiling (1.6–2.8 log CFU g^{-1}) ≥ double pan broiling (1.1–2.3 log CFU g^{-1}). Cooking of thick steaks (4.0 cm) causes greater reductions in *E. coli* counts (2.3–4.2 log CFU g^{-1}) compared to thinner samples (1.1–2.9 log CFU g^{-1}). This indicates that increased steak thickness allowed greater inactivation of *E. coli* O157:H7 due to increased holding time. Generally, under normal roasting conditions, meat products are not sterilized; however, microbial load is reduced to enhance shelf life of meat and meat products provided roasts are stored under refrigerated and hygienic conditions to avoid recontamination. Olds et al. [59] investigated the effect of cooling methods on growth of *C. perfringens* ATCC 10388 in cooked turkey roasts. They observed that the rapid cooling of roasts to 5°C is essential to avoid proliferation of *C. perfringens*.

E. coli O157:H7 is considered a pathogen of significance in beef and ground beef products. *Listeria* spp. is ubiquitous in nature and widely reported in ready-to-eat meat products. One of the most popular meat products is fermented sausage, which originated from Mediterranean countries with a wide variety of fermented sausages produced around the world. Dry/semidry fermented sausages are characterized as low risk for listeriosis [60]. However, due to several number of foodborne outbreaks involving fermented sausages, some countries have imposed regulations on sausage production and the production process for fermented products to provide a 5-log reduction of *E. coli* [70]. Table 8.5 shows the time and temperature combinations, which can achieve >5-log reductions.

If thermal process deviations (heating or cooling) occur, spores of *C. perfringens* if present in raw meats utilized for processing of cooked products may be heat-activated, germinate, and grow to

TABLE 8.5
Heat Treatments of Fermented Sausages Giving Reductions of *E. coli* O157:H7 > 5 log[a]

Products	Fermentation Temperature (°C)	Heat Treatment[b] Temperature (°C)/Time (h)	Comments	References
Pepperoni	36	63/<1 or 53/1	No visible changes of product	[61]
Summer sausage	29–41	54/1 (pH 4.6)	Effect dependent on pH. At pH 5.0 $\log_{10}$ reduction= 4.6	[62]
Lebanon bologna	26.7–37.8	43.3/20 or 46.1/10 or 48.9/3		[63]
Salami	—	50 or 55 or 60/0–6	*E. coli* inoculated in minced salami. High recovery of injured cells	[64]
Lebanon bologna	26.7–37.8	37.8/24 + 43.3/24		[65]
Pepperoni	—	58.3/1 or 61/0.3	Sausages treated at 58.3°C gave changes in sensory properties	[66]
Salami	25	54/NI	Heating after fermentation or after maturation	[67]
Lebanon bologna	26.7–37.7	48.9/10.5	Effect of heating to 46.1°C reduced by NaCl and nitrite to 2 $\log_{10}$ reduction	[68]
Snack sticks	Acidified with acid	69.8/~0.08		[69]

Source: Olds, D.A. et al., *J. Food Protect.*, 69(1), 112, 2006. With permission.
[a] For some processes fermentation together with heat treatment give >5 $\log_{10}$ reduction.
[b] Core temperature; +, both treatments included; NI, no information.

hazardous levels during cooling or improper storage. The time/temperature guidelines for cooling cooked products specifies that the maximum internal temperature should neither remain between 54.4°C and 26.7°C for more than 1.5 h nor between 26.7°C and 4.4°C for more than 5 h [71]. The U.S. Food and Drug Administration (FDA) Division of Retail Food Protection recognized that inadequate cooling was a major food safety problem and established a recommendation that all food should be cooled from 60°C to 21°C in 2 h and from 21°C to 5°C in 4 h [72]. Byrne et al. [73] reported that the cooking protocol for meat of 70°C for 2 min is sufficient to achieve at least a 6-log reduction in *Bacillus cereus* and/or *C. perfringens* vegetative cell, but spores will survive during thermal treatment of pork luncheon roll. They also suggested that the generation time for *C. perfringens* may be as short as 8 min [74]; hence, it is essential that cooked meat is cooled as rapidly as possible to prevent bacterial growth and multiplication [75,76].

IR heating is also reported to inactivate bacteria, spores, yeast, and mold in meat and meat products. Resistance of bacteria, yeasts, and molds to IR heating may be different due to their structural and compositional differences. In general, spores are more resistant than vegetative cells. Inactivation of *E. coli* due to both IR heating and thermal conductive heating follows first-order kinetics. The apparent death rate constant is comparatively higher for pasteurization by IR heating compared to thermal conductive heating. At any given temperature, the death rate constant is higher for far-IR heating than conductive heating, which demonstrates that far-IR heating is potentially more efficient than conductive heating for pasteurization [77]. IR radiation has a poor penetration capacity. However, the surface temperature of meat materials increases rapidly, and heat is transferred inside food materials by thermal conduction. Huang [78] indicated the suitability of IR for the surface pasteurization of turkey frankfurters. IR heating to 80°C, 75°C, and 70°C reduced the counts of *L. monocytogenes* by 4.5, 4.3, and 3.5 log units, respectively. IR-heated turkey samples were slightly darker than the controls after treatments, but refrigerated storage resulted in no significant difference in color parameters. Huang and Sites [79] described an IR pasteurization process for the inactivation of *L. monocytogenes* on ready-to-eat meats such as hotdogs. The increase in surface temperature to 70°C, 75°C, 80°C, or 85°C resulted in a 1.0-, 2.1-, 3.0-, or 5.3-log reduction, respectively, during the initial come-up period. Holding the product at a higher temperature of 80°C for 3 min or 85°C or 2 min led to an additional bacterial inactivation of 6.4 or 6.7 logs, respectively.

8.5 KINETICS OF THERMAL PROCESSING

The use of kinetic models that describe the microbial responses during thermal processing is a quantitative approach to design and optimize thermal treatment focusing on the production of safe meat products. These models can be further exploited to quantitatively describe the influence of processing conditions on food safety. Consequently, the effects of intrinsic, extrinsic, and/or processing factors on the resulting microbial proliferation in food products or food model systems can be evaluated. Ideally these models should be parsimonious, flexible, and built upon parameters based on the physiological mechanism of inactivation [80]. There is significant evidence to suggest that the microbial inactivation of microorganisms follows an exponential profile when subjected to heat. The procedure for estimating kinetic parameters of high-moisture foods including meat and meat products is well established. This involves a two-step procedure including plotting the concentration versus time at varying constant temperatures. The slope obtained from the first step is the rate constant. Rate constants are plotted against the inverse of temperature and fitted to the Arrhenius equation, which provides kinetic model parameters such as the activation energy from the slope, half-life, and D_{value}.

D_{value} can be calculated by using the following equation and is the time required at temperature T to reduce a microbial population by 90%:

$$D_{value} = \frac{2.303}{K} \tag{8.28}$$

z value can be calculated from D value as follows and is defined as the number of degrees of temperature change necessary to cause the D value to change by a factor of 10, measured in °C:

$$z = \frac{T_2 - T_1}{\log D_2 - \log D_1} \tag{8.29}$$

where

C_t is microbial load or concentration at time t (min)
C_0 is initial microbial load or concentration ($t = 0$)
K (min^{-1}) is first-order degradation rate constant

Thermal treatments are based on the commonly used sterilization value (F value) and pasteurization value (P value). Where microbial inactivation processes are described as log-linear, a linear relationship exists between the logarithm of the microbial population level and the treatment time [81]. Valdramidis et al. [82] proposed the use of the mild heat value (MH value), which is defined as the time needed to achieve a predefined microbial reduction at a reference temperature and a known thermal resistant constant, z, for log-linear or nonlog-linear microbial inactivation kinetics. Traditionally, the F value [82] is defined as the time required in order to achieve a specific reduction in microbial numbers at a given temperature, and it thus represents the total time–temperature combination received by a food. Similarly, the P value is the corresponding thermal death value under pasteurization conditions. The F value is given as follows:

$$F_{Tref} = \int_0^t 10^{(T(t)-T_{ref})/z} \mathrm{d}t \tag{8.30}$$

Equation 8.30 is valid if the survival curve obeys first-order kinetics. Despite the worldwide use of this approach, especially in the meat canning industry for the so-called 12D process of the proteolytic strains (Group I) of *C. botulinum* spores [83,84], a lot of deviations from log-linearity have been observed [85]. As other authors acknowledge [87–90], the success of the canning industry in using the F value as a measure of the heat processes efficacy could be attributed to overprocessing and not to the calculation method's correctness [83]. These deviations are evident particularly at lower temperatures than sterilization and for vegetative cells [86,87]. A concept of MH value is also applicable to the mild heat treatment of meat and meat products. Over the last 30 years, a number of microbial inactivation models have been developed for describing nonlog-linear microbial inactivation kinetics [88] (Table 8.6).

In case of nonlinearity in microbial inactivation curves, the thermal death time needs to include the possibility of nonlog-linear microbial survival curves [81]. For example, an alternative approach for evaluating the efficacy of a thermal process assumes that the microbial heat resistances follow a Weibulian frequency distribution model [85–90]. Nevertheless, this approach takes into account only two types of nonlog-linearity (i.e., concave, convex) and is not retaining classical parameters (like the z value) for evaluating the achieved microbial reduction.

Thermal microbial inactivation kinetics can be described based on the actual (micro) environmental conditions (*<env>*) such as temperature, pressure, salt concentration, water activity, etc., and the physiological state (*<phys>*) of the species, for instance, as influenced by temperature history [81] as given in the following, this expression can then be coupled with differential equations that describe the dynamics of the physiological state [91]:

$$\frac{\mathrm{d}N}{\mathrm{d}t} = -k(N, <env>, <phys>) \cdot N \tag{8.31}$$

where N is the cell density of the microbial species (CFU mL^{-1}).

TABLE 8.6
Some Models for Microbial Inactivation Kinetics

Inactivation Model	Model Equation and Parameters
Cerf model	$\log N_t = \log(N_0) + \log(f \exp^{-k_{\max 1} t} + (1-f)\exp^{-k_{\max 2} t}$
Gompertz model	$N_t = \alpha + \gamma \cdot \exp[-\exp(-\beta(t-\mu))]$ N_t = log CFU mL^{-1} t = elapsed time, μ = the inflection point, β = slope parameter, γ = range, and α = final CFU mL^{-1}.
Weibull model	$\ln\left[\frac{N_t}{N_0}\right] = -\frac{1}{2.303}\left(\frac{t}{\alpha}\right)^{\beta}$ β = shape factor and α = inactivation rate constant
Geeraerd model	$\log_{10}(N) = \log_{10}(N_0) - \frac{k_{max}(t)}{\ln(10)} + \frac{\log_{10} e^{(k_{max} S_l)}}{(1 + e^{(k_{max} S_l)} - 1)} \cdot e^{(-k_{max} t)}$

A sound set of differential equations, which is a subcase of Equation 8.31 and describes the microbial inactivation kinetics by incorporating physiological adjustments during the microbial inactivation experiments, is the dynamic, nonlog-linear model of Geeraerd et al. [91]. This model is constructed for microbial inactivation by mild heating [81]. The microbial inactivation parameters under the isothermal conditions can be estimated by taking into account the shoulder and tailing effects [81]:

$$\frac{\mathrm{d}N}{\mathrm{d}t} = -k_{max} \cdot \left[\frac{1}{1+C_c}\right] \cdot \left[1 - \frac{N_{res}}{N}\right] \cdot N \tag{8.32}$$

$$\frac{\mathrm{d}C_c}{\mathrm{d}t} = -k_{max} \cdot C_c \tag{8.33}$$

where

N represents the microbial cell density (CFU mL^{-1})
C_c is related to the physiological state of cells (—)
k_{max} denotes the specific inactivation rate (min^{-1})
N_{res} is the residual population density (CFU mL^{-1})

The microbial inactivation parameters can be estimated by assuming first-order inactivation kinetics for all the inactivation data when omitting the second and third factor of the right-hand side of Equation 8.32 or by assuming first-order inactivation kinetics only for the log-linear portion of the inactivation kinetics [81]. This is a model for describing nonlinearities that incorporate shoulder and/or tailing effects, and it automatically reduces to log-linear inactivation kinetics if the data do not include these effects.

8.6 CONCLUSIONS

Thermal processing is one of the most widely used methods of meat preservation and has contributed significantly to the development of the processed meat industry. Thermal processing of meat and meat products not only ensures microbial safety but also improves palatability. Some of the traditional cooking methods such as smoking raise human health concerns because of PAHs (especially benzo (α)pyrene). PAHs can be formed when fat and juices from meat grilled directly over an open fire falls onto the fire, and these flames containing PAHs adhere to the meat surface. However, the use of microwave or IR cooking prior to exposure to high temperatures can substantially reduce PAHs. Recent advances in thermal processing of meat technology including the use of IR, microwave, ohmic heating, and nonthermal techniques such as high-pressure processing, ultrasound, pulsed electric fields, in addition to the use of natural antimicrobial agents, and their combinations with conventional techniques, will ensure food safety with minimal impact on the nutritional composition and sensorial quality of meat and meat products.

NOMENCLATURE

Equations 8.1 through 8.7

a_w Water activity
c Moisture content, dry basis
c_{wb} Moisture content, wet basis
$\bar{C}$ Average moisture content, dry basis

C_P	Specific heat of meat (J kg^{-1} K^{-1})
DL	Dripping weight loss (kg)
EL	Evaporative weight loss (kg)
ε	Beef emissivity
Γ	Surface of geometric model
h	Convective heat transfer coefficient (W m^{-2} K^{-1})
i	Initial
k	Thermal conductivity of meat (W m^{-1} K^{-1})
k_g	Mass transfer coefficient (kg Pa^{-1} m^{-2} s^{-1})
λ	Latent heat of evaporation (J kg^{-1})
m_s	Dry solid mass (kg)
o	Oven
Ω	Domain of geometric model
P_{sat}	Water vapor pressure (Pa)
RH	Oven relative humidity
ρ	Meat density (kg m^{-3})
s	Surface
σ	Stefan–Boltzmann constant (5.67×10^{-8} W m^{-2} K^{-4})
T	Temperature (°C)
Φ_w	Volumetric fraction of water

Equations 8.8 through 8.11

A	Area (m^2)
α	Thermal diffusivity ($\alpha = k/\rho C_p$)
D_m	Mass flow rate (kg s^{-1})
h	Convective heat transfer coefficient (W m^{-2} K)
k_m	Mass transfer coefficient (kg m^{-2}, Pa s)
L_v	Latent heat of vaporization of water (J kg^{-1})
P	Partial vapor pressure of water in the oven (Pa)
P_s	Vapor pressure of water at the roast surface (Pa)
q	Heat flow rate (W)
r	Distance in radial direction (m)
t	time (s)
T	Temperature (K)
T_s	Temperature at the surface of the roast (K)
T_∞	Temperature of the air in the oven (K)
θ	Distance in circumferential direction (m)
z	Distance in axial direction (m)

Equations 8.12 through 8.18

cr	critical
cs	Combustion smoke
C	Moisture content (kg moisture/kg solid)
Cp	Heat capacity (J kg^{-1} K^{-1})
dp	Dry product
D	Mass diffusivity (m^2 s^{-1})
D_c	Constant in Equation 8.27 (kg moisture/kg solid)
D_T	Constant in Equation 8.27 (K)
D_x	Constant in Equation 8.27 (m^2 s^{-1})
ε	Emissivity (dimensionless)

f	Final
Fm	Evaporation mass flux (kg s^{-1}m^{-2})
g	Gravity
Γ	Fat drip factor (dimensionless)
h	Convective heat transfer coefficient (W m^{-2} K^{-1})
h_{eff1}	Constant in Equation 8.24 (W m^{-1} K^{-1})
h_{eff2}	Constant in Equation 8.24 (W m^{-1} K^{-2})
hm	Lewis mass transfer coefficient (m s^{-1})
is	Isenthalpic
k	Thermal conductivity (W m^{-1} K^{-1})
k_1	Constant in Equation 8.23 (W m^{-1} K^{-1})
k_2	Constant in Equation 8.23 (W m^{-1} K^{-2})
k_3	Constant in Equation 8.23 (W m^{-1} K^{-3})
K_m	Mass transfer coefficient (m s^{-1})
l	Width (m)
L	Length (m)
λ	Latent heat of vaporization (kJ kg^{-1})
md	Mixed drying
mf	Melted fat
M	Molar mass (kg mol^{-1})
μ	Density (kg m^{-1} s^{-1})
ω	Drip and evaporation mass flow ratio (dimensionless)
p	product
ps	Pyrolysis smoke
P	Pressure (Pa)
Φ	gray view factor (dimensionless)
$\dot{\Phi}$	Melting mass flux (kg s^{-1} m^{-2})
q	Mass flow (kg s^{-1})
r_1	Constant in Equation 8.22 (kg m^{-3})
r_2	Constant in Equation 8.22 (kg m^{-3} K^{-1})
r_3	Constant in Equation 8.22 (kg m^{-3} K^{-2})
ρ	Density (kg m^{-3})
s	Radiant plate ("s" as sheet)
st	Steady state
σ	Stephan–Boltzmann constant (W m^{-2} K^{-4})
t	Time (s)
T	Temperature (K)
T_∞	Infinite temperature (K)
τ	Thickness (m)
Θ	Surface mass (kg m^{-2})
U	Velocity (m s^{-1})
v	Vapor
v_{sat}	Saturated vapor
w	Water
W	Dry-based moisture content (kg kg^{-1} d.b.)
x	Thickness of beef patty (m)
X	Dry-based fat content (kg kg^{-1} d.b.)
ξ	Coordinate (m)
z	height (m)
ζ	Height (m)
∞	Infinite temperature

Equations 8.28 through 8.37

c	Moisture content, dry basis
c_{wb}	Moisture content, wet basis
$\bar{C}$	Average moisture content, dry basis
Ct	Water activity
C_P	Specific heat of meat (J kg^{-1} K^{-1})
DL	Dripping weight loss (kg)
EL	Evaporative weight loss (kg)
ε	Beef emissivity
Γ	Surface of geometric model
h	Convective heat transfer coefficient (W m^{-2} K^{-1})
i	Initial
k	Thermal conductivity of meat (W m^{-1} K^{-1})
k_g	Mass transfer coefficient (kg Pa^{-1} m^{-2} s^{-1})
λ	Latent heat of evaporation (J kg^{-1})
m_s	Dry solid mass (kg)
o	Oven
Ω	Domain of geometric model
P_{sat}	Water vapor pressure (Pa)
Φ_w	Volumetric fraction of water
RH	Oven relative humidity
ρ	Meat density (kg m^{-3})
s	Surface
σ	Stefan–Boltzmann constant (5.67×10^{-8} W m^{-2} K^{-4})
T	Temperature (°C)

REFERENCES

1. GH Zhou, XL Xu, Y Liu. Preservation technologies for fresh meat—A review. *Meat Science* 86: 119–128, 2010.
2. S Featherstone. A review of developments in and challenges of thermal processing over the past 200 years—A tribute to Nicolas Appert. *Food Research International*, doi:10.1016/j.foodres.2011.04.034, 2011.
3. P Richardson, ed. *Thermal Technologies in Food Processing Philip Richardson*, 14 Chapters. Cambridge, England: Woodhead Publishing Ltd., 2001, pp. 1–3.
4. P Fellows. *Food Processing Technology: Principle and Practice*. Cambridge, England: Woodhead Publishing Ltd., 1997.
5. SM Goñi, VO Salvadori. Prediction of cooking times and weight losses during meat roasting. *Journal of Food Engineering* 100:1–11, 2010.
6. EW Godsalve, EA Davis, J Gordon, HT Davis. Water loss rates and temperature profiles of dry cooked bovine muscle. *Journal of Food Science* 42(4):1038–1045, 1977.
7. E Tornberg. Effects of heat on meat proteins—Implications on structure and quality of meat products. *Meat Science* 70(3):493–508, 2005.
8. E Obuz, TH Powell, ME Dikeman. Simulation of cooking cylindrical beef roasts. *Likensmittel-Wissenshaft und-Technology* 35:637–644, 2002.
9. N Singh, RG Akins, LE Erickson. Modeling heat and mass transfer during the oven roasting of meat. *Journal of Food Process Engineering* 7:205–220, 1984.
10. M Obiedzinski, A Borys. Identification of polynuclear aromatic hydrocarbons in wood smoke. *Acta Alimentaria Polonica* 27(3):169–173, 1977.
11. S Wretling, A Eriksson, GA Eskhult, B Larsson. Polycyclic aromatic hydrocarbons (PAHs) in Swedish smoked meat and fish. *Journal of Food Composition and Analysis* 23:264–272, 2010.
12. SY Chung, RR Yettella, JS Kim, K Kwon, MC Kim, DB Min. Effects of grilling and roasting on the levels of polycyclic aromatic hydrocarbons in beef and pork. *Food Chemistry* 129:1420–1426, 2011.
13. I Stumpe-Vīksna, V Bartkevičs, A Kukāre, A Morozovs. Polycyclic aromatic hydrocarbons in meat smoked with different types of wood. *Food Chemistry* 110:794–797, 2008.

14. P Sebastian, D Bruneau, A Collignan, M Rivier. Drying and smoking of meat: Heat and mass transfer modeling and experimental analysis. *Journal of Food Engineering* 70(2):227–243, 2005.
15. I Poligné, A Collignana, G Trystramc. Characterization of traditional processing of pork meat into boucané. *Meat Science* 59:377–389, 2001.
16. C Bejerholm, MD Aaslyng. The influence of cooking technique and core temperature on results of a sensory analysis of pork-depending on the raw meat quality. *Food Quality and Preference* 15:19–30, 2003.
17. L Laycock, P Piyasena, GS Mittal. Radio frequency cooking of ground, comminuted and muscle meat products. *Meat Science* 65:959–965, 2003.
18. JY Jeong, ES Lee, JH Choi, JY Lee, JM Kim, SG Min, YC Chae, CJ Kim. Variability in temperature distribution and cooking properties of ground pork patties containing different fat level and with/without salt cooked in microwave energy. *Meat Science* 75:415–422, 2007.
19. S Ryyänen, PJO Risman, T Ohlsson. Hamburger composition and microwave heating uniformity. *Journal of Food Science* 69(7):187–196, 2004.
20. TP Shuka, RC Anantheswaran. Ingredient interactions and product development for microwave heating. In: AK Datta, RC Anantheswaran, eds. *Handbook of Microwave Technology for Food Applications*. New York: Marcel Dekker, 2001, pp. 355–395.
21. O Sipahioglu, SA Barringer, C Bircan. The dielectric properties of meat as a function of temperature and composition. *Journal of Microwave Power and Electromagnetic Energy* 38(3):161–170, 2003.
22. RF Shiffman. Food product development for microwave processing. *Food Technology* 40(6):94–98, 1986.
23. L Zhang, JG Lyng, N Brunton, D Morgan, B McKenna. Dielectric and thermophysical properties of meat batters over a temperature range of 5–85°C. *Meat Science* 68(2):173–184, 2004.
24. MS Yarmand, A Homayouni. Effect of microwave cooking on the microstructure and quality of meat in goat and lamb. *Food Chemistry* 112:782–785, 2009.
25. PS Sheridan, NC Shilton. Analysis of yield while cooking beefburger patties using far infra red radiation. *Journal of Food Engineering* 51(1):3–11, 2002.
26. A Hashimoto, Y Yamazaki, M Shimizu, O Sei-Ichi. Drying characteristics of gelatinous materials irradiated by infrared radiation. *Drying Technology* 12(5):1029–1052, 1994.
27. M Dagerskog, L Ostertrom. Infra-red radiation for food processing I, fundamental properties. *Lebensmittel Wissenschaft und Technologie* 12(4):237–242, 1979.
28. N Shilton, P Mallikarjunan, P Sheridan. Modeling of heat transfer and evaporative mass losses during the cooking of beef patties using far-infrared radiation. *Journal of Food Engineering* 55:217–222, 2002.
29. Y Choi, M Okos. Effects of temperature and composition on the thermal properties of foods. In: M Le Maguer, P Jelen, eds. *Food Engineering and Process Applications, Transport Phenomena*, Vol. 1. London, U.K.: Elsevier Science, 1985, pp. 93–101.
30. S Ghazala, HS Ramaswamy, JP Smith, MV Simpson. Thermal process simulations for *sous vide* processing of fish and meat foods. *Food Research International* 28(2):117–122, 1995.
31. S Ghazala, R Trenholm. Hurdle and HACCP concepts in *sous vide* and cook-chill products. In: S Ghazala, ed. *Sous vide and Cook-Chill Processing for the Food Industry*. Gaithersburg, MD: Aspen Publishers, Inc., 1998, pp. 294–310.
32. E Hyytia-Trees, E Skytta, M Mokkila, A Kinnunen, M Lindstrom, L Lahteenmaki, R Ahvenainen, H Korkeala. Safety evaluation of *sous-vide*-processed products with respect to nonproteolytic *Clostridium botulinum* by use of challenge studies and predictive microbiological models. *Applied and Environmental Microbiology* 66(1):223–229, 2000.
33. Department of Health. *Chilled and frozen foods. Guidelines on Cook-Chill and Cook-Freeze Catering Systems*. London, U.K.: HMSO, 1989.
34. P Díaz, G Nieto, MD Garrido, S Bañón. Microbial, physical–chemical and sensory spoilage during the refrigerated storage of cooked pork loin processed by the *sous vide* method. *Meat Science* 80(2):287–292, 2008.
35. H Nyati. An evaluation of the effect of storage and processing temperature on the microbiological status of sousvide extended shelf-life products. *Food Control* 11:471–476, 2000.
36. H Walsh, S Martins, EE O' Neill, JP Kerry, T Kenny, P Ward. The effects of different cooking regimes on the cook yield and tenderness of non-injected and injection enhanced forequarter beef muscles. *Meat Science* 84(3):444–448, 2010.
37. M Christensen, PP Purslow, LM Larsen. The effect of cooking temperature on mechanical properties of whole meat, single muscle fibres and perimysial connective tissue. *Meat Science* 55:301–307, 2000.
38. B Mora, E Curti, E Vittadini, D Barbanti. Effect of different air/steam convection cooking methods on turkey breast meat: Physical characterization, water status and sensory properties. *Meat Science* 88(3):489–497, 2011.

39. N Gerber, MRL Scheeder, C Wenk. The influence of cooking and fat trimming on the actual nutrient intake from meat. *Meat Science* 81(1):148–154, 2009.
40. S Combes, J Lepetit, B Darche, F Lebas. Effect of cooking temperature and cooking time on Warner–Bratzler tenderness measurement and collagen content in rabbit meat. *Meat Science* 66:91–96, 2003.
41. PE Bouton, PV Harris. Changes in the tenderness of meat cooked at 50–65°C. *Journal of Food Science* 46:475–478, 1981.
42. PV Harris, WR Shorthose. Meat texture. In: R Lawrie, ed. *Developments in Meat Science*, Vol. 4, 1988, pp. 245–296.
43. K Palka, H Daun. Changes in texture, cooking loss, and myofibrillar structure of bovine M Semitendinosus during heating. *Meat Science* 51:237–243, 1999.
44. Z Pan, RP Singh. Physical and thermal properties of ground beef during cooking. *Lebensmittel-Wissenschaft und Technologie* 34:437–444, 2001.
45. BK Oroszvári, CS Rocha, I Sjöholm, E Tornberg. Permeability and mass transfer as a function of the cooking temperature during the frying of beefburgers. *Journal of Food Engineering* 74:1–12, 2006.
46. L Braeckman, F Ronsse, P Cueva Hidalgo, J Pieters. Influence of combined IR-grilling and hot air cooking conditions on moisture and fat content, texture and colour attributes of meat patties. *Journal of Food Engineering* 93:437–443, 2009.
47. MD Aaslyng, C Bejerholm, P Ertbjerg, HC Bertram, HJ Andersen. Cooking loss and juiciness of pork in relation to raw meat quality and cooking procedure. *Food Quality and Preference* 14(4):277–288, 2003.
48. H Martens, E Stabursvik, M Martens. Texture and colour changes in meat during cooking related to thermal denaturation of muscle proteins. *Journal Texture Studies* 13:291–309, 1982.
49. HR Cross, PR Durland, SC Seideman. Sensory qualities of meat. In: PJ Bechtel, ed. *Muscle of Food.* Orlando, FL: Academic Press, 1986, pp. 279–320.
50. RT Bacon, JN Sofos. Food hazards: biological food; characteristics of biological hazards in foods. In: RH Schmidt, GE Rodrick, eds. *Food Safety Handbook.* New York: Wiley Interscience, 2003, pp. 157–195.
51. JN Sofos, I Geornaras. Overview of current meat hygiene and safety risks and summary of recent studies on biofilms, and control of *Escherichia coli* O157:H7 in nonintact, and *Listeria monocytogenes* in ready-to-eat, meat products. *Meat Science* 86:2–14, 2010.
52. JN Sofos. Challenges to meat safety in the 21st century. *Meat Science* 78:3–13, 2008.
53. CFIA. 2011 Canadian Food Investigation Agency. Food, thermometer food safety tips. Available at: www.inspection.gc.ca/english/fssa/concen/tipcon/thermo.pdf. Accessed date: November 4, 2011, 2011.
54. K McDonald, D-W Sun. Predictive food microbiology for the meat industry: A review. *International Journal of Food Microbiology* 52:1–27, 1999.
55. C Shen, I Geornaras, KE Belk, GC Smith, JN Sofos. Inactivation of *Escherichia coli* O157:H7 in moisture-enhanced nonintact beef by pan-broiling or roasting with various cooking appliances set at different temperatures. *Journal of Food Science* 76:64–71, 2011.
56. SB Sporing. *Escherichia coli* O157:H7 risk assessment for production and cooking of blade tenderized beef steaks. MSc thesis. Kansas State University, Manhattan, NY, 1999.
57. A Mukherjee, Y Yoon, JN Sofos, GC Smith, KE Belk, JA Scanga. Evaluate survival/growth during frozen, refrigerated, or retail type storage, and thermal resistance, following storage of *Escherichia coli* O157:H7 contamination on or in marinated, tenderized or restructured beef steaks and roasts which will minimize survival or enhance destruction of the pathogen. Final report submitted to the Natl Cattlemen's Beef Assn by the Center for Red Meat Safety, Department of Animal Science, Colorado State University, Fort Collins, CO, 2007.
58. C Shen, JM Adler, I Geornaras, KE Belk, GC Smith, JN Sofos. Inactivation of *Escherichia coli* O157:H7 in nonintact beef steaks of different thickness by pan-broiling, double panbroiling or roasting using five types of cooking equipment. *Journal of Food Protection* 73:461–469, 2010.
59. DA Olds, AF Mendonca, J Sneed, B Bisha. Influence of four retail food service cooling methods on the behavior of *Clostridium perfringens* ATCC 10388 in turkey roasts following heating to an internal temperature of 74 degrees C. *Journal of Food Protection* 69(1):112–117, 2006.
60. FDA/FSIS (Food and Drug Administration/Food Safety and Inspection Service). Quantitative assessment of the relative risk to public health from foodborne *Listeria monocytogenes* among selected categories of ready-to-eat foods. Available at: http://www.fda.gov/Food/ScienceResearch/ResearchAreas/RiskAssessmentSafetyAssessment/defaulthtm. Accessed November 3, 2011, 2003.
61. JC Hinkens, NG Faith, TD Lorang, P Bailey, D Buege, CW Kaspar, JB Luchansky. Validation of pepperoni processes for control of *Escherichia coli* O157:H7. *Journal of Food Protection* 59(12):1260, 1996.

62. M Calicioglu, NG Faith, DR Buege, JB Luchansky. Viability of *Escherichia coli* O157:H7 in fermented semidry low-temperature-cooked beef summer sausage. *Journal of Food Protection* 60(10):1158–1162, 1997.
63. KR Ellajosyula, S Doores, EW Mills, RA Wilson, RC Anantheswaran, SJ Knabel. Destruction of *Escherichia coli* O157:H7 and Salmonella typhimurium in Lebanon bologna by interaction of fermentation pH, heating temperature, and time. *Journal of Food Protection* 61(2):152–157, 1998.
64. G Duffy, DCR Riordan, JJ Sheridan, BS Eblen, RC Whiting, IS Blair, D McDowell. Differences in thermotolerance of various *Escherichia coli* O157:H7 strains in a salami matrix. *Food Microbiology* 16(1):83, 1999.
65. KJK Getty, RK Phebus, JL Marsden, JR Schwenke, CL Kastner. Control of *Escherichia coli* O157:H7 in large (115 mm) and intermediate (90 mm) diameter Lebanon-style bologna. *Journal of Food Science* 64(6):1100, 1999.
66. DCR Riordan, G Duffy, JJ Sheridan, RC Whiting, IS Blair, DA McDowell. Effects of acid adaptation, product pH, and heating on survival of *Escherichia coli* O157:H7 in pepperoni. *Applied and Environmental Microbiology* 66(4):1726, 2000.
67. LL Duffy, PB Vanderlinde. *E. coli* and salami manufacture—Meeting the challenge of the ANZFA requirements. *Food Australia* 52(7):269–270, 2000.
68. N Chikthimmah, RC Anantheswaran, RF Roberts, EW Mills, SJ Knabel. Influence of sodium chloride on growth of lactic acid bacteria and subsequent destruction of *Escherichia coli* O157:H7 during processing of Lebanon bologna. *Journal of Food Protection* 64(8):1145–1150, 2001.
69. SK Stoltenberg, KJK Getty, H Thippareddi, RK Phebus, TM Loughin. Fate of *Escherichia coli* O157:H7 during production of snack sticks made from beef or a venison/beef fat blend and directly acidified with citric or lactic acid. *Journal of Food Science* 71(6):228–235, 2006.
70. AL Holck, L Axelsson, TM Rode, M Høy, I Måge, O Alvseike, TM L'Abée-Lund, MK Omer, PE Granum, E Heir. Reduction of verotoxigenic *Escherichia coli* in production of fermented sausages. *Meat Science* 89(3):286–295, 2011.
71. USDA-FSIS. Performance Standards for the production of processed meat and poultry products, Proposed Rule. FSIS Directive 7111.1, US Department of Agriculture, Food Safety and Inspection Service, Washington, DC Federal Register, 2001, 66:12589–12636.
72. FDA Division of Retail Food Protection. Food Code. U.S. Department of Health and Human Services, Public Health Service. Food and Drug Administration, Pub. No. PB97 141204, 2001, Washington, DC.
73. B Byrne, G Dunne, DJ Bolton. Thermal inactivation of *Bacillus cereus* and *Clostridium perfringens* vegetative cells and spores in pork luncheon roll. *Food Microbiology* 23:803–808, 2006.
74. A Andersson, U Ronner, PE Granum. What problems does the food industry have with the spore-forming pathogens *Bacillus cereus* and *Clostridium perfringens*? *International Journal of Food Microbiology* 28:145–155, 2006.
75. RM Kalinowsk, RB Tompkin, PW Bodnaruk, WP Pruett. Impact of cooking, cooling, and subsequent refrigeration on the growth or survival of *Clostridium perfringens* in cooked meat and poultry products. *Journal of Food Protection* 66:1227–1232, 2003.
76. RJ Danler, EA Boyle, CL Kastner, H Thippareeddi, DY Fung, RK Phebus. Effects of chilling rate on outgrowth of *Clostridium perfringens* spores in vacuum-packaged cooked beef and pork. *Journal of Food Protection* 66:501–503, 2003.
77. J Sawai, Y Isomura, T Honma, H Kenmochi. Characteristics of the inactivation of *Escherichia coli* by infrared irradiative heating. *Biocontrol Science* 11:85–90, 2006.
78. L Huang. Infrared surface pasteurization of turkey frankfurters. *Innovative Food Science Emerging Technology* 5:345–351, 2004.
79. L Huang, J Sites. Elimination of *Listeria monocytogenes* on hotdogs by infrared surface treatment. *Journal Food Science* 73:27–31, 2008.
80. JF Van Impe, F Poschet, AH Geeraerd, KM Vereecken. Towards a novel class of predictive microbial growth models. *International Journal of Food Microbiology* 100:97–105, 2005.
81. VP Valdramidis, AH Geeraerd, BK Tiwari, PJ Cullen, A Kondjoyan, JF Van Impe. Estimating the efficacy of mild heating processes taking into account microbial non-linearities: A case study on the thermisation of a food stimulant. *Food Control* 22(1):137–142, 2011.
82. CO Ball. Thermal process time for canned food. *National Research Council Bulletin* 7(37), 1923.
83. CR Stumbo. *Thermobacteriology in Food Processing*. New York: Academic Press, 1965.
84. ICMSF. *ICMSF: Microorganisms in Food 5, Characteristics of Microbial Pathogens*. London, U.K.: Blackie Academic and Professional, 1996.

85. MG Corradini, MD Normand, M Peleg. Expressing the equivalence of non-isothermal and isothermal heat sterilization process. *Journal of the Science of Food and Agriculture* 86(5):785–792, 2006.
86. LH Huang. Thermal inactivation of *Listeria monocytogenes* in ground beef under isothermal and dynamic temperature conditions. *Journal of Food Engineering* 90(3):380–387, 2009.
87. VP Valdramidis, AH Geeraerd, K Bernaerts, JF Van Impe. Microbial dynamics versus mathematical model dynamics: the case of microbial heat resistance induction. *Innovative Food Science & Emerging Technologies* 7(1–2):80–87, 2006.
88. AH Geeraerd, V Valdramidis, JF Van Impe. GInaFiT, a freeware tool to assess non-log-linear microbial survivor curves. *International Journal of Food Microbiology* 102(1):95–105, 2005.
89. P Mafart, O Couvert, S Gaillard, I Leguerinel. On calculating sterility in thermal preservation methods: Application of the Weibull frequency distribution model. *International Journal of Food Microbiology* 72(1–2):107–113, 2002.
90. AS Sant'Ana, A Rosenthal, PR Massaguer. Heat resistance and the effects of continuous pasteurization on the inactivation of Byssochlamys fulva ascospores in clarified apple juice. *Journal of Applied Microbiology* 107(1):197–209, 2009.
91. AH Geeraerd, CH Herremans, JF Van Impe. Structural model requirements to describe microbial inactivation during a mild heat treatment. *International Journal of Food Microbiology* 59(3):185–209, 2000.

9 Thermal Processing of Poultry Products

Paul L. Dawson, Sunil Mangalassary, and Brian W. Sheldon

CONTENTS

9.1 INTRODUCTION

Thermal processing of poultry meat and egg products serves several purposes including elimination of pathogenic microorganisms, improvement of sensory properties (flavor, texture, appearance, etc.), stabilization of color in cured products, extraction of high-value components (fat rendering, lysozyme), or removal of undesirable components. Poultry products that are subjected to thermal processing include raw meat that is cooked, ready-to-eat (RTE) meat, liquid egg, shell eggs, and rendered meat. On January 1, 2003, the Food Safety Inspection Service (FSIS) of the U.S. Department of Agriculture revised the requirements for handling meat products under Title 9 of the U.S. Code of Federal Regulations. These regulations are specifically described in this chapter under the headings of cooked meat, canned meat, rendered products, and egg products. Comments on thermal processing of canned or retort pouch-type products that were made by representatives of the National Food Processors Association (NFPA) on May 10, 2001, at an FSIS public meeting on RTE Performance Standards are as follows:

> Following a food poisoning incident in 1971 in which the failure to properly apply a thermal process to a commercially canned product led to fatal consequences, the National Canners Association (now NFPA) petitioned the Food and Drug Administration (FDA) to promulgate new regulations to address the problem. Elements of this new program were designed to control the primary food safety hazards associated with canning operations - the survival of spores of *Clostridium botulinum* which could subsequently germinate and produce the deadly botulism toxin within the anaerobic environment of the sealed can. Consumption of even small amounts of this potent toxin, in the absence of prompt administration of antitoxin, can quickly lead to paralysis and death of any consumer, not just those who are immunocompromised or fall in some other special risk category.

Thermal processing is still the favored method to ensure microbiological safety of poultry products even though newer technologies such as irradiation, high-intensity electric fields, ultrahigh pressure, and high-intensity light are finding niche applications in the food industry. Thermal processing will remain the dominant method used to impart safety, flavor, and value to poultry products. Consumer demand for convenience and food service needs has prompted the development of a myriad of cooked poultry products requiring a minimum amount of on-site preparation by the consumer or food service workers. This marketing shift has been accompanied by a shift in the regulatory approach to ensuring product safety. Inspection by governmental agencies, while still required, has given way to standards requiring the producer of the product to ensure and document the safety of their products. Regulations now require the processor to "prove" their product safety. Since challenge studies (inoculation with a pathogen) in a commercial setting are not possible, research involving bacterial inactivation models has become prevalent. This chapter will summarize the applicable regulations and discuss the research pertaining to thermal processing of poultry meat and egg products.

9.2 MEAT

There are many factors that affect the thermal process used for treating food products. From a regulatory standpoint, the primary concern is product safety that focuses on assuring that all microbial pathogens are reduced to a level that renders the product safe. This translates into either a minimum log reduction in population or to a zero tolerance for some target microorganisms. In general, bacteria survive better in meat products than in test solutions, buffered media or growth media. Numerous studies have evaluated the relative heat resistance of *target microorganisms* as influenced by the medium [1–3]. These studies are widespread since *thermal inactivation* is affected by many factors. Bacterial heat resistance is generally higher in meat than in buffered media or peptone water [4–6]. Some variables that influence the effectiveness of a *heat inactivation* process include pH [2,3,7–9], fat content [10], and water activity [11,12].

Pasteurization using hot water or steam is employed as an effective method for reducing levels of pathogenic bacteria on the surface of poultry meat. Goksoy et al. [13] performed a series of experiments to determine the relationship between hot water immersion temperature and residence time on the appearance of chicken skin and/or meat. The results showed that since the changes caused to samples by heat treatment were initially textural rather than colorimetric and thus could be identified visually but not instrumentally with a chroma meter. The visual examination revealed that there is no immersion heat treatment (below 90°C) capable of reducing thermotolerant microorganisms, on poultry without causing adverse changes in the product. A device to surface pasteurize meat without producing a cooked appearance was built and tested [14]. The researchers [14] achieved a rapid heating by condensing pure, thermally saturated steam onto the meat surface in the absence of noncondensible gases, which could keep up with poultry slaughter line speeds.

9.2.1 Cooking of Poultry Meat

9.2.1.1 Current Research on Cooking Poultry Meat

Adequate thermal processing or *cooking* of meat products is one of the most important critical control points for eliminating pathogenic bacteria and viruses and preventing *food-borne diseases*. Cooking is a commonly employed critical control point in the hazard analysis and critical control point (*HACCP*) food safety program of meat processors and is one of the last control points applied to a food product before consumption [15]. The thermal inactivation of *pathogenic organisms* is expressed in terms of D and z values. The D value or thermal reduction time is defined as the time at a specified temperature required to reduce a population of organisms by 1 log cycle or 90%. The relationship between D value and temperature is defined in terms of the z value and is defined as the temperature change in degree Celsius needed to bring about a 1 log cycle change in D value. The determination of D and z values for a particular microorganism is useful in designing a thermal process that targets that specific microorganism [16].

Due to the variety of factors affecting bacterial lethality, many researchers have attempted to model the effect of these multiple factors on process lethality (i.e., deriving D and z values for target microorganisms). *Salmonella* spp., *Listeria monocytogenes*, *Campylobacter jejuni*, and *Escherichia coli* O157:H7 are the four most common bacteria reported in the literature for which heat inactivation studies have been conducted. For example, Fain et al. [10] reported D values for *L. monocytogenes* in lean ground turkey of 81.3, 2.6, and 0.6 at 51.7°C, 57.2°C, and 62.8°C, respectively. Veeramuthu et al. [17] published D values for *Salmonella* Senftenberg (211.35, 13.24, and 3.43 min) and *E. coli* O157:H7 (79.4, 5.47, and 1.69 min) in ground turkey processed at 55°C, 60°C, and 65°C, respectively. Moreover, Blankenship and Craven [4] recorded D values for *C. jejuni* in cooked chicken of 15.2–0.25 min when processed between 49°C and 57°C. Murphy et al. [18,19] reported on the thermal inactivation kinetics for *Salmonella* spp. and *L. monocytogenes* in ground chicken. They found that the D value of a six serotype mixture of *Salmonell*a equaled the sum of the D values for each individual serotype.

Juneja et al. [20] discovered that the D values (at 55°C, 57.5°C, 60°C, 62.5°C, and 65°C) for a four-strain mixture of *E. coli* O157:H7 in lean ground chicken were lower than the D values obtained for a 90% lean ground beef product. In a subsequent study, Juneja et al. [21] generated survival curves (58°C–65°C) for an eight-serotype cocktail of *Salmonella* mixed in ground poultry samples containing varying levels of fat (1%–12%). They found that for a given increase in fat level, ground chicken had higher D values than ground turkey. At the higher processing temperatures, the D values were longer for ground turkey than ground chicken. A general trend of increasing D value with fat level was found with chicken, which was not found in turkey meat (Table 9.1).

Cooking methods can also affect bacterial inactivation as was demonstrated by Schnepf and Barbeau [22] with *S.* Typhimurium-inoculated chicken cooked by microwave, convection, or conventional ovens. Cooking method can also influence the degree of bacterial inactivation on the meat surface. For example, an air impingement oven (a popular commercial oven) having a high air velocity will dehydrate the meat surface faster than other methods such as microwave and steam cooking. Thus, the inactivation of bacterial cells on poultry meat surfaces may in some instances be more impacted by oven relative humidity than by the water activity of the product. This observation was supported by Murphy et al. [23] who discovered that the higher humidity environment

TABLE 9.1
***D* and *z* Values for Selected Poultry Meat Products**

Organisms	Products	Temperature (°C)	*D* Value (min)	*z* Value (°C)	Reference
S. aureus	Chicken a la king	60	5.37		[140]
L. monocytogenes	Minced chicken	70	—	6.7	[120]
	Ground turkey	51.7	81.3		[10]
		57.2	2.6		
		62.8	0.6		
	Turkey bologna	61	2.1	4.44	[39]
		65	0.27		
S. Senftenberg	Ground turkey	55	211.35		[17]
		60	13.24		
		65	3.43		
E. coli O157:H7	Ground turkey	55	79.4		
		60	5.47		
		65	1.69		
	Chicken	70	0.38–0.55		[141]
	Turkey bologna	55	4.83	13.9	[41]
		60	0.76		
		65	0.20		
		70	0.15		
S. Typhimurium	Chicken broth	55	4.16		[37]
	Turkey bologna	57	4.6	5.56	[39]
		60	1.4		
S. 8 serotype cocktail: *S.* Thompson *S.* Enteritidis phage type 13A *S.* Enteritidis phage type 4 *S.* Typhimurium *S.* Hadar *S.* Copenhagen *S.* Montevideo *S.* Heidelberg	Chicken 2% fat	58	7.38		[21]
	Chicken 6.3% fat	58	7.33		
	Chicken 9% fat	58	8.54		
	Chicken 12% fat	58	9.04		
	Chicken 2% fat	60	4.83		
	Chicken 6.3% fat	60	4.68		
	Chicken 9% fat	60	5.40		
	Chicken 12% fat	60	5.50		
	Chicken 2% fat	62.5	1.14		
	Chicken 6.3% fat	62.5	1.16		
	Chicken 9% fat	62.5	1.16		
	Chicken 12% fat	62.5	1.30		
	Chicken 2% fat	65	0.41		
	Chicken 6.3% fat	65	0.51		
	Chicken 9% fat	65	0.53		
	Chicken 12% fat	65	0.50		
	Turkey 2% fat	58	7.50		
	Turkey 6.3% fat	58	7.71		
	Turkey 9% fat	58	6.91		
	Turkey 12% fat	58	7.41		
	Turkey 2% fat	60	4.56		
	Turkey 6.3% fat	60	4.94		
	Turkey 9% fat	60	5.13		
	Turkey 12% fat	60	5.43		

(*continued*)

TABLE 9.1 (continued)
***D* and *z* Values for Selected Poultry Meat Products**

Organisms	Products	Temperature (°C)	*D* Value (min)	*z* Value (°C)	Reference
	Turkey 2% fat	62.5	1.53		
	Turkey 6.3% fat	62.5	1.85		
	Turkey 9% fat	62.5	1.45		
	Turkey 12% fat	62.5	1.78		
	Turkey 2% fat	65	0.59		
	Turkey 6.3% fat	65	0.55		
	Turkey 9% fat	65	0.57		
	Turkey 12% fat	65	0.59		
C. jejuni	Turkey bologna	53	4.53	8.3	[41]
		55	3.20		
		60	0.64		
		62	0.42		

in steam-injected ovens increased *Salmonella* spp. inactivation by >2 $\log_{10}$ cycles compared to a dry-air cooking method attaining the same endpoint temperature (EPT). Marks et al. [24] and Murphy et al. [25] demonstrated that water bath–based heating studies over predicted *Salmonella* and *Listeria* spp. inactivation by as much as 5 $\log_{10}$ cycles compared to chicken breast patties heated in a dry-air convection oven. Millio and Ricke [26] combined organic acids at 2.5% (sodium acetate [SA], sodium butyrate [SB], sodium lactate [SL], or sodium propionate [SP] adjusted to pH 4) with low-temperature treatments to reduce *Salmonella* spp. on raw chicken model media. Exposure to pH 4 water at 55°C or the organic acid salt solutions at room temperature had no effect on *Salmonella* populations while all organic acid salt treatments yielded significant *S.* Typhimurium reductions, ranging from 1 log (SA) to almost 4 logs (SB).

Different approaches are used for the verification of thermal processing adequacy in meat products. Orta-Ramirez and Smith [15] categorized these into the following headings: thermocouples and thermometers, color determination, EPT indicators, enzymatic methods, immunoassays, physical methods, and time–temperature integrators. Thermal energy is an important factor causing color changes in heat-processed meat and poultry products. Determination of color changes in meat or meat juice exudates has been suggested as a means of estimating EPT. Ang and Huang [27] found that color of precooked chicken leg meat may provide a rapid estimation of EPT to which the product has been heated but is limited by packaging method and storage time, which significantly influenced the color values.

The catalytic activity, defined as the decomposition of H_2O_2 into H_2O and O_2, has been evaluated as a potential indicator of heat treatment endpoint for chicken patties. The catalytic activity test offered a simple and rapid procedure for estimating the EPT of processed chicken patties cooked to 73°C [28] and a 3°C–4°C error in EPT prediction was observed. Bogin et al. [29] developed a biochemical method, which could be performed under field conditions for verifying heat treatments for turkey breast meat. The activities of 12 enzymes and residual soluble proteins were examined following various heat treatments under different conditions. The enzymes aspartate aminotransferase, creatine kinase, malic dehydrogenase, isocitric dehydrogenase, and aldolase were shown to be valuable markers for the evaluation of heat treatment.

To predict heat penetration during thermal processing, heat transfer models have been widely used for meat products. Chen et al. [30] developed a 2D axisymmetric finite element (FE) model to simulate coupled heat and mass transfer during convection cooking of regularly shaped chicken patties, under actual oven transient conditions. The predicted transient center temperature under various cooking condition had an error of 3.8°C–5.7°C, as compared to experimental data. The best prediction was obtained when both thermal conductivity and specific heat were modeled as state-dependent functions in the simulations.

From a culinary point of view, there are generally two ways to cook poultry meat: dry heat and moist heat. For the most part with cooking moist heat includes braising, stewing, and steaming, while dry heat methods include roasting, broiling, pan-broiling/frying, and grilling. Due to the toughness of meat from older birds as a result of greater connective tissue cross-linking slow, moist heat is typically used, while dry heat cooking is preferred for meat from younger animals. Ghite et al. [31] reported that cooking time (5, 10, and 15 min), cooking temperature (120°C, 150°C, and 180°C), and cooking method (hot-air or hot-air steam) significantly affected cooking losses and tenderness of chicken breasts. Higher temperatures and longer cooking times increased cooking losses and shear force as might be expected; however, cooking temperature had a greater effect than cooking time on losses and shear force. Cooking temperature of 120°C had cooking losses and shear forces from 14.75 to 18.46 g/100 g and from 2865 to 4341 g force/100 g meat, respectively, while cooking at 180°C resulted in losses and shear forces from 24.12 to 36.99 g/100 g and from 3659 to 7128 g force/100 g, respectively. Moist heat only slightly improved the tenderness of chicken breast meat compared to dry heat cooking.

Chicken flavor was produced from chicken fat first oxidized using lipoxygenase from soybean meal then thermally reacted with amino acids and reducing sugar via the Maillard reaction to form 3-methyl-2-thiophenecarboxaldehyde, hexanal, 1-(2-furanyl)-ethanone, dihydro-2-methyl-3(2H)-furanone, anethole, 2-(methylthio)-thiophene, tetrahydro-2,5-dimethylfuran, 4-methyl-5-thiazole-ethanol, and nonanol [32].

9.2.1.2 Regulations for Cooking Poultry Meat

The United States Department of Agriculture-Food Safety and Inspection Service (USDA-FSIS) implemented a 7 $\log_{10}$ reduction in population of *Salmonella* for fully and partially cooked poultry products [33]. The *USDA-FSIS* provided additional regulations in their "Guidelines for Cooked Poultry Rolls and Other Cooked Poultry Products" as published as part of the Draft Compliance Guidelines for RTE Meat and Poultry Products.

> Cooked poultry rolls and other cooked poultry products should reach an internal temperature of at least 71.1°C (160°F) prior to being removed from the cooking medium. However, cured and smoked poultry rolls and other cured and smoked poultry should reach an internal temperature of at least 68.3°C (155°F) prior to being removed from cooking medium. In cooked RTE products where heat will be applied incidental to a subsequent processing procedure, the product may be removed from the media for subsequent processing provided that it is immediately fully cooked to a 71.1°C (160°F) internal temperature.
>
> Establishments producing cooked poultry rolls and other cooked poultry products should have sufficient monitoring equipment, including recording devices, to assure that the temperature (accuracy assured within 0.56°C (1°F)) limits of these processes are being met. Data from the recording devices must be made available to USDA-FSIS program employees upon request.

These FSIS cooking guidelines also suggested new time–temperature combinations for cooking RTE poultry products having different fat levels.

Canadian Food Inspection Agency, Bureau of Food Safety and Consumer Protection Retail Food states the following in their information bulletins:

> Section B.22.026. No person shall sell poultry, poultry meat or poultry meat by-product that has been barbecued, roasted or broiled and is ready for consumption unless the cooked poultry, poultry meat or poultry meat by-product
>
> (a) at all times
>
> (i) has a temperature of 4.4°C (40°F) or lower, or 60°C (140°F) or higher, or
>
> (ii) has been stored at an ambient temperature of 4.4°C (40°F) or lower, or 60°C (140°F) or higher, and

(b) carries on the principal display panel of the label a statement to the effect that the food must be stored at a temperature of 4.4°C (40°F) or lower, or 60°C (140°F) or higher.

The Illinois Department of Public Health requires that:

- Other foods, such as poultry; stuffed fish, meat or pasta; or stuffing containing fish, meat or poultry must be cooked to 73.9°C (165°F) or above for at least 15 seconds. Not cooking to the established temperatures could result in salmonella poisoning.

9.2.2 Ready-to-Eat Meat

9.2.2.1 Current Research on Ready-to-Eat Meat

The majority of recent recalls of meat and poultry products involve RTE products. *Surface contamination* of RTE meat with pathogenic organisms generally occurs during postprocess handling. To control this contamination, surface pasteurization thermal processes involving steam or hot water may be effectively applied either prior to packaging or as a postpackage pasteurization application.

A series of three papers from the same research group described the impact of *in-package pasteurization* of Vienna sausages on controlling spoilage microorganisms, primarily *lactic acid bacteria* (LAB) [34–36]. Depending on the severity of the applied thermal treatment, LAB were reduced from 84.4% of the total bacterial population to between 52.9% and 74.6% of the total population for Vienna sausages stored for 128 days at 8°C [34]. However, in-package pasteurization did not delay the rate of spoilage because of an increase in *Bacillus* spp. populations detected in the heat-treated sausages. Decimal reduction values (*D* values) of LAB (*Lactobacillus sake*, *Leuconostoc mesenteroides*, and *Lactobacillus curvatus*) in packaged Vienna sausages ranged from 14.4 to 52.9 s at 57°C, 60°C, and 63°C [35]. The combination of organic acid addition and in-package pasteurization or pasteurization alone extended the microbiological shelf life of Vienna sausages (i.e., reaching a 5×10^6 cfu/g total aerobic plate count) fourfold compared to nontreated samples [36].

Several cooking thermal inactivation studies have been conducted for *Listeria* spp. and/or *Salmonella* spp. in chicken patties [25,37,38], ground beef [39], packaged ground chicken [19], turkey, chicken, and chicken broth [40] sous vide beef mince and solid beef pieces [41], and packaged low-fat (mixed species including poultry) bologna [42,43]. A wide variety of heating temperatures were tested in these studies. Juneja et al. [40] reported a *D* value (55°C) for *S.* Typhimurium of 4.16 min in chicken broth, whereas McCormick et al. [42] reported *D* values of 278 (4.6 min) and 81 s (1.4 min) for *S.* Typhimurium inoculated on the surface of individually packaged bologna slices cooked at 57°C and 60°C, respectively. McCormick et al. [42] also reported *D* values for *L. monocytogenes* on packaged mixed species bologna of 124 (2.1 min) and 16.2 s (0.27 min) at 61°C and 65°C, respectively. When in-package pasteurization was combined with nisin impregnated packaging films, *L. monocytogenes* inhibition during storage was enhanced [40]. Hughes et al. [44] found *Campylobacter jejuni D* values for in-package pasteurized turkey bologna were 272.0, 192.1, 38.4, and 25.2 s at 53°C, 55°C, 60°C, and 62°C, respectively. These researchers also reported *E. coli* O157:H7 *D* values for in-package pasteurized turkey bologna were 289.5, 45.8, 15.8, and 9.1 s at 55°C, 60°C, 65°C, and 70°C, respectively. Selected *D* and *z* values for poultry meat products are shown in Table 9.1.

Murphy and Berrang [45] discovered that postprocess pasteurization of fully cooked vacuum packaged chicken breast strips with hot water or steam at 88°C for 10–35 min significantly lowered the *Listeria innocua* populations. Muriana et al. [46] reported a 2–4 log reduction of *L. monocytogenes* populations in RTE deli style vacuum packaged whole or formed turkey, ham and roast beef following a postpackage pasteurization process at 90.6°C–96.1°C (195°F–205°F) for 10 min. Murphy et al. [47] examined the effect of packaging film thickness on thermal inactivation of *Salmonella* sp. and *L. innocua* in cooked chicken breast meat. Their findings indicated that increasing film thicknesses reduced heating rates and subsequently reduced the level of inactivation of the test organisms. In a subsequent publication, Murphy et al. [48] suggested a model to use in predicting the amount of heating time required to achieve a 7 $\log_{10}$ reduction in *Salmonella* spp. or *L. innocua* populations for different thicknesses of fully cooked, vacuum-packaged chicken breast meat products pasteurized in a hot water cooker at 90°C.

The same research group [48] also evaluated a postcook in-package pasteurization process against *L. monocytogenes* inoculated on a fully cooked turkey breast meat product. The effectiveness of the heat treatment was affected by product surface roughness. About 50 min of heating time was needed to achieve a 7 $\log_{10}$ CFU/cm^2 reduction on products having a convoluted surface roughness of up to 15 mm in depth.

Gande and Muriana [49] recorded a 1.25–3.5 log reduction in *L. monocytogenes* populations following a prepackaging surface pasteurization treatment (60–120 s at 246.1°C–398.9°C) of a fully cooked meat product removed from its primary packaging wrap. When the turkey bologna was subjected to both a pre- (60 s) and postpackaging (45 or 60 s) pasteurization process, they detected a 2.7–4.3 log reduction in *L. monocytogenes* populations. These relatively low log reductions for the time used for surface pasteurization can be explained by the relatively large meat mass compared to individual slices. The effect of thickness on heat penetration is well documented, but the effect of sample thickness on surface heating rate is less well documented. Mangalassary et al. [50] found that thicker and higher fat turkey bologna samples had a slower surface heating rates than thinner and lower fat samples. For example, at 80°C the in-package pasteurization time for a 5 log reduction of *L. monocytogenes* on the surface of turkey bologna was 0.72, 4.56, and 7.12 min for 4, 12, and 20 mm thick samples [50]. The rates of surface heating for 4, 12, and 20 mm think bologna in the package are shown in Figure 9.1.

9.2.2.2 Regulations for Packaged Ready-to-Eat Poultry Meat

The U.S. regulations governing the thermal processing of RTE meat has evolved from a specific EPT requirement to a lethality requirement [51]. This evolution roughly follows the move from an USDA inspection-based meat and poultry processing system to a food processing-derived HACCP-based program. The responsibility for documenting the safety of food products shifts from the government to the processor. The 1999 USDA-FSIS ruling (CFR 381.150) specified a lethality of

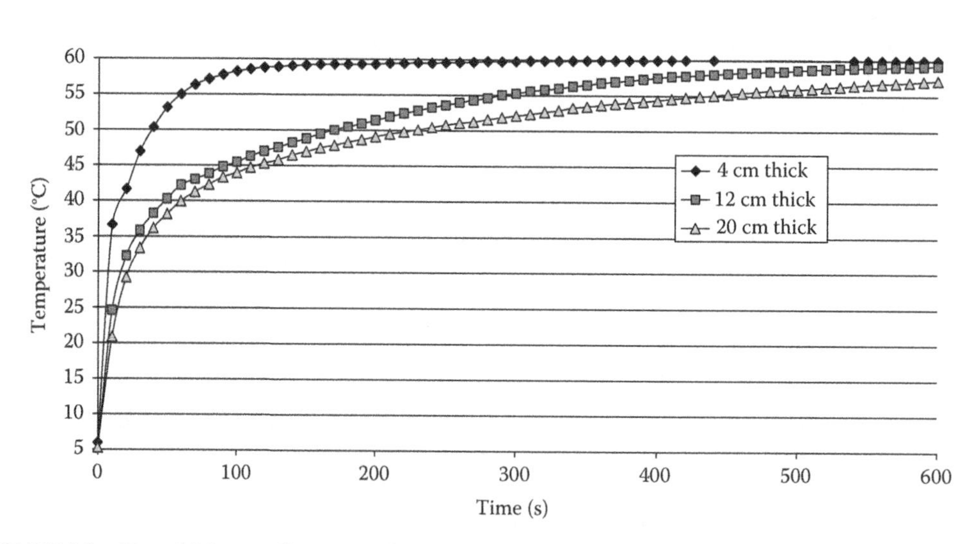

FIGURE 9.1 Meat thickness effect on surface heating rate in a 60°C water bath for in-package pasteurized bologna.

heat treatment for RTE whole muscle poultry meat of 7.0 $\log_{10}$ cycles for *Salmonella* spp. Processors are therefore required to validate their thermal process for each product using a "scientifically supportable means."

RTE meat and poultry products are products that have been processed, so that they may be safely consumed without further preparation by the consumer (i.e., without cooking or application of some other lethality treatment to destroy pathogens; 66FR 39:12590). On February 27, 2001, FSIS published a proposed rule titled "Performance Standards for the Production of Processed Meat and Poultry Products" (66FR 12590). The proposed regulations include lethality and stabilization performance standards and Listeria testing requirements.

Except for thermally processed, commercially sterile products, the performance standards for lethality for all RTE poultry products require a 7.0 $\log_{10}$ reduction of *Salmonella* throughout the finished product. Moreover, except for thermally processed, commercially sterile products, the performance standards require no growth of *Clostridium botulinum* and no more than 1 $\log_{10}$ growth of *Cl. perfringens* throughout the RTE meat and poultry product. For poultry products, an endpoint cooking temperature necessary to achieve a 7.0 $\log_{10}$ reduction of *Salmonella* is recommended.

Postprocessing lethality treatments such as steam pasteurization, hot water pasteurization, and radiant heating have been developed to prevent or eliminate postprocessing contamination by *L. monocytogenes*. The postprocessing lethality treatment that reduces or eliminates the pathogen must be included in the establishment's HACCP plan. The interim final rule for the control of *L. monocytogenes* in RTE meat and poultry products includes three alternative approaches that establishments can choose in applying a postprocess lethality treatment.

1. Employ both a postprocess lethality treatment and a growth inhibitor to control *Listeria* on RTE products. Establishments opting for this alternative will be subject to FSIS verification activity that focuses on the efficacy of the postprocess lethality treatment. Sanitation programs are important but are generally designed into the degree of lethality required for safety as delivered by the postprocess lethality treatment.
2. Employ either a postprocess lethality treatment or a growth inhibitor to control *Listeria* on RTE products. Establishments opting for this alternative will be subject to more frequent FSIS verification activity than for alternative (1).
3. Employ sanitation measures only. Establishments opting for this alternative will be targeted with the most frequent level of FSIS verification activity. Within this alternative, FSIS will place increased scrutiny on operations that produce hotdogs and deli meat. In a 2001 risk ranking, FSIS and the FDA identified these products as posing a relatively high risk for illness and death.

Currently, there are no regulations governing the surface pasteurization of cooked meat products. A process would need to be established for each product based on the desired level of pathogen destruction at the product surface [48].

9.2.3 Canned and Retorted Poultry Meat

9.2.3.1 Current Research on Canned and Retorted Poultry Meat

Thermally processed, commercially sterile meat and poultry products are commonly referred to as canned products although the containers can be flexible, such as pouches, or semirigid, as in lunch bowls [52]. The incoming quality of raw materials (refrigerated chicken legs and breasts) used for the manufacture of canned chicken products and the outgoing quality of the finished products were evaluated by Turek et al. [53]. Analysis of the canned products showed complete sterility and good physicochemical properties.

Lyon et al. [54] studied the texture profiles of canned, boned chicken meat from birds treated under different processing regimes. They concluded that textural fibrousness and cohesiveness of mass associated with shortening, aging chilling times, and subsequent cooking of chicken meat may be an asset for products that undergo further heat treatment such as retorting.

Vacuum-canned, commercial, mechanically deboned chicken meat was initially challenged by Thayer et al. [55] with either *Cl. botulinum* spores or *Salmonella* Enteritidis (SE) irradiated at 0, 1.5, and 3.0 KGy and then stored at 5°C for 0, 2, and 4 weeks. None of the samples stored at 5°C developed botulinum toxin. However, when the samples were temperature abused at 28°C, they became toxic within 18 h and demonstrated obvious signs of spoilage such as can swelling and putrid odors.

Studies were carried out by Rywotycki [56] to determine the effect of seasonal variations in the temperature of the cooling water used, water consumption and the cooling time of pasteurized turkey meat cuts canned in containers of different size and shapes, ranging from 0.455 to 1.365 kg. Season markedly influenced cooling water temperature and hence the amount of water required for cooling; temperatures were the highest in the summer and the lowest in the winter. Season also had a marked effect on the shelf life of the canned product; bacterial counts were lower in cans produced in winter months than those produced in the summer with no significant differences noted between products manufactured in the autumn or spring.

Sauvaget and Auffret [57] described the benefits of adding citric acid to canned chicken, turkey, and rabbit for maintaining the integrity and firmness of muscles, bones, and cartilage during the canning process. Zhang et al. [58] compared three different types of thermocouples and receptacle designs with regard to their effect on the observed heat penetration rate of a conduction-based heat treatment of chicken gravy packaged in small diameter containers (202 × 204 and 211 × 300 cans). Their results indicated that stainless steel receptacles and needle-type thermocouples inserted into 202 × 204 cans significantly increased the apparent heat penetration rates.

9.2.3.2 Regulations for Commercially Sterile Products

A canned meat/poultry product is defined by the USDA as a "product having a water activity of 0.85 or higher and receives a thermal process either before or after being packed in a hermetically sealed container." The USDA further defines commercially sterile as a "condition achieved by application of heat, sufficient alone or in combination with other treatments and/or treatments to render the product free of microorganisms capable of growing in the product at non-refrigerated conditions (over 10°C) at which the product is intended to be held during distribution and storage." Subpart X section 381.300–381.311 of the Code of Federal Regulations stipulates the thermal processing requirements for canned poultry products. For a low-acid food receiving a thermal or other bacteriocidal process, that process must be validated to achieve a probability of 10^{-9} that there are *Cl. botulinum* spores in the container capable of growing or a 12 $\log_{10}$ reduction of *Cl. botulinum*, assuming an initial load of ≤1000 spores per container.

9.2.4 Rendering

9.2.4.1 Current Research on Rendered Poultry

The demand for poultry meat has increased considerably in recent years. The change in consumer preference for portioned or further processed poultry has resulted in increasing amount of *by-products* and underutilized products. Skin and subcutaneous fat account for about 19% of carcass weight [59,60].

The basic purpose of *rendering* is to produce stable products of commercial value, free from disease bearing organisms, from raw materials that are often unsuitable or unfit for human consumption. Rendering involves two basic processes of first separating the fat and then drying the residue. The most common method used to extract fat from the tissue is by heating [61]. Different types of thermal processing are involved in rendering of poultry fat.

Piette et al. [62] inoculated one set of finely homogenized chicken skin samples with *Acinetobacter* sp., *Bronchothrix thermosphacta, Candida tropicalis, Debaromyces hansenii, Enterobacter agglomerans, Enterococcus faecalis*, a *Lactobacillus* spp., or *Pseudomonas fluorescens* while the control samples were not inoculated. Both samples were heated to 80°C or 50°C to extract fat. Extraction of fat at 80°C resulted in nearly complete inactivation of the indigenous and inoculated micro flora resulting in microbiological counts in the rendered fat of generally below detection level. Conversely, a large number of organisms (3.69–7.28 $\log_{10}$ cfu/g) survived the 50°C extraction process.

A second study was undertaken by the same researchers [63] to determine the influence of extraction temperature on the recovery of fat from chicken skin. A maximum amount of fat (89.6% of the initial fat) was recovered from homogenized skin heated to 80°C, whereas the 50°C rendering temperature resulted in lowest fat recovery (51.5% of skin fat content).

Sheu and Chen [64] studied the yield and quality characteristics of edible broiler skin fat obtained from five rendering methods: microwave rendering, conventional oven rendering, water cooking, griddle rendering, and deep fat frying. Microwave rendering produced the highest fat yields (47.5%) followed by deep-fat frying (33.4%), conventional oven baking (31.6%), griddle rendering (25.8%), and water cooking (24.8%). The moisture content of the rendered skin fat was the highest for the water cooking process (1.43%) and the lowest for the conventional oven baked process (0.19%). No significant differences in free fatty acid contents were observed among the rendering methods. More recently, Kinley et al. [65] tested 150 samples of different types of meals (feather, meat, meat/bone, poultry, and blood) from various U.S. rendering facilities. They found an average pH of from 6.16 to 7.36 and the moisture content ranging from 1.9% to 11.5%. The total bacterial counts ranged from 1.7 to 6.68 $\log_{10}$ CFU/g, with the highest in blood meal and the lowest in meat meal.

Bioactive peptides can be isolated from poultry meat tissues using various thermal and enzymatic treatments.

9.2.4.2 Regulations for Rendered Poultry Meat

Chapter III, Part 315 of the U.S. Food Safety and Inspection Service, Department of Agriculture describes rendering or other disposal of carcasses and parts passed for cooking. Section 315.1 addresses carcasses and parts passed for cooking; rendering into lard or tallow.

> Carcasses and parts passed for cooking may be rendered into lard in accordance with Sec. 319.702 of this subchapter or rendered into tallow, provided such rendering is done in the following manner:
>
> (a) When closed rendering equipment is used, the lower opening, except when permanently connected with a blowline, shall first be sealed securely by a Program employee; then the carcasses or parts shall be placed in such equipment in his presence, after which the upper opening shall be securely sealed by such employee. When the product passed for cooking in the tank does not consist of a carcass or whole primal part, the requirements for sealing shall be at the discretion of the circuit supervisor. Such carcasses and parts shall be cooked for a time sufficient to render them effectually into lard or tallow, provided all parts of the products are heated to a temperature not lower than 76.7°C (170°F) for a period of not less than 30 minutes.
>
> (b) At establishments not equipped with closed rendering equipment for rendering carcasses and parts passed for cooking into lard and tallow, such carcasses or parts may be rendered in open kettles under the direct supervision of a Program employee. Such rendering shall be done during regular hours of work and in compliance with the requirements as to temperature and time specified in paragraph (a) of this section.

PART 315, Rendering or other disposal of carcasses and parts passed for cooking describes carcasses and parts passed for cooking and their utilization for food purposes after cooking.

> Carcasses and parts passed for cooking may be used for the preparation of meat food products, provided all such carcasses or parts are heated to a temperature not lower than 76.7°C (170°F) fpr a period of not less than 30 minutes either before being used in or during the preparation of finished product.

9.3 EGG PRODUCTS

As foods of animal origin, eggs may be contaminated with a variety of potentially pathogenic microorganisms. Understanding and controlling the microbiological composition of eggs are issues of importance both in terms of ensuring product safety and providing acceptable refrigerated shelf life. Because many *egg products* provide nearly ideal growth media for a wide variety of bacterial pathogens, tolerances for viable microorganisms in finished products are quite small.

Over the last century, the growth of the egg products industry has been linked to advances in egg processing, packaging, and distribution technologies. At the expense of Grade A *shell eggs*, a variety of processed egg products have emerged as growing segments of the total egg market [66]. Refrigerated *liquid egg products* include *liquid whole egg* (LWE), scrambled egg mixes (i.e., whole egg plus nonfat-dried milk solids and vegetable oil), *egg white* (albumen), *egg yolk* (EY), and various blends of yolk and white [66]. Some nonegg ingredients that are sometimes added to preserve or improve the physical characteristics and functional performance of egg products include carbohydrates (e.g., corn syrup, sucrose), hydrocolloids (e.g., gums, starches), salt, and citric acid [62]. In a conventional egg processing facility, the sequence of operations for producing liquid egg involves sorting to remove leakers and inedibles, movement of eggs through an egg washer, rinsing in a solution containing 100–200ppm available chlorine, mechanical egg breaking, collection (either as whole egg or separated albumen and yolk), homogenization (either before or after heat treatment), pasteurization, cooling, packaging, quality assurance testing, and distribution (either frozen or refrigerated). The details of these procedures are provided by Stadelman and Cotterill [67].

Among the advantages cited for the use of refrigerated liquid egg over shell eggs or frozen egg blends are consistent quality, convenience, storage space savings, elimination of freezing costs, and labor savings at the point of use. Because such products are thermally processed (pasteurized) before packaging and distribution, they offer improved shelf stability and microbiological safety relative to fresh shell eggs [66].

9.3.1 Development and Purpose of Egg Pasteurization

9.3.1.1 Historical Background

The storage stability and safety of refrigerated egg products are closely linked to the microbiological quality of the final, packaged product at the time when it enters the distribution chain. The primary means of ensuring the microbiological safety of liquid egg products is the use of appropriate egg pasteurization processes. The term pasteurization refers to the heat treatment of foods at temperatures below those needed for complete sterilization [68]. As a result of product pasteurization, some microorganisms are inactivated, some may be attenuated (i.e., sublethally injured) while bacterial spores may be stimulated to germinate [68]. In the current regulatory context, minimum pasteurization processes have been designed to destroy certain target microbial pathogens in specific foods. Enhanced microbial shelf stability is an additional benefit of pasteurization, provided that the spoilage organisms present in the raw product are relatively heat sensitive or are prevented from multiplying through the use of controlled refrigeration, freezing, or other preservation technologies.

Pasteurization of liquid egg products was first used commercially in the United States in 1938, primarily as a means of extending the shelf life of frozen liquid egg products [69]. A small, gradual increase in the numbers of companies using thermal treatments for liquid egg occurred in the 1940s and 1950s. Brant et al. [70] reviewed the early literature on batch pasteurization of liquid egg. At present, most egg processors in the United States utilize continuous egg pasteurization processes. Regulations promulgated by the USDA and the FDA regarding the incidence of *Salmonella* in processed eggs products made liquid pasteurization virtually mandatory in 1966 [70]. The Egg Products Inspection Act of 1970 brought egg pasteurization requirements in the United States under more uniform regulatory control [71].

9.3.1.2 Egg Pasteurization and *Salmonella* Safety

Current USDA regulations stipulate that liquid, frozen, and dried whole egg, yolk, and white be pasteurized or otherwise treated to inactivate all viable salmonellae [68]. It is important to note that *Salmonella* is the only bacterial pathogen specifically addressed within the context of these regulations [72]. A limiting factor in the development of egg pasteurization processes is the fact that the time and temperatures that inactivate salmonellae in egg products are at or near those that adversely affect the physical and functional properties of the egg proteins [71]. USDA-mandated egg pasteurization specifications, listed in the Code of Federal Regulations (Title 7, Section 59.570), require that every particle of egg be held for at least a specified time and temperature to "assure complete pasteurization" (subsection b) and to produce "a *Salmonella*-negative product" (subsection c) [73].

The pasteurization requirements prescribed by the USDA vary according to the egg product due to the susceptibility of certain egg fractions (e.g., unsupplemented egg white) to heat-induced protein denaturation. Current minimum time–temperature combinations for the average particle range from 6.2 min at 55.6°C (for unsupplemented albumen) to 3.5 min at 63.3°C (for salt whole egg, sugar yolk, or salt yolk) [73]. The liquid egg pasteurization requirements of a number of Asian, European, and United Kingdom countries were summarized by the International Commission on Microbiological Specifications for Foods [71]; for LWE, these processes range from 2.8 min at 61.7°C (Northern Ireland) to 9 min at 65.2°C (Germany). The predicted microbial lethality of such processes is prescribed only for *Salmonella* spp. In the United States, a 3.5 min holding time at 60°C is said to yield a nine-log-cycle (9D) reduction in the viable *Salmonella* population (where 1D results in a 90% reduction in the target population) in the average particle of LWE. Under laminar flow conditions, a 9D process for inactivation of salmonellae would be equivalent to only a 4.5D process within the fastest-moving particle of egg. The time–temperature combinations prescribed in 1966 for other liquid egg products (Table 9.2) were reportedly designed to provide "approximately equal pasteurization effectiveness" [72].

TABLE 9.2
Requirements for the Conventional Pasteurization of Liquid Egg Products in the United States (FDA CFR 9:590, 570)

Liquid Egg Products	Minimum Temperature (°C)	Minimum Time (min)
Albumen (without use of chemicals)	56.7	3.5
	55.6	6.2
Whole egg	60.0	3.5
Whole-egg blends (less than 2% added nonegg	61.1	3.5
ingredients)	60.0	6.2
Fortified whole egg and blends (24%–38% egg	62.2	3.5
solids, 2%–12% added nonegg ingredients)	61.1	6.2
Salted whole egg (2% or more salt added)	63.3	3.5
	62.2	6.2
Sugared whole egg (2%–12% sugar added)	61.1	3.5
	60.0	6.2
Plain yolk	61.1	3.5
	60.0	6.2
Sugared yolk (2% or more sugar added)	63.3	3.5
	62.2	6.2
Salted yolk (2%–12% salt added)	63.3	3.5
	62.2	6.2

Source: Nisbikawa, Y. et al., *Int. J. Food Microbiol.*, 18, 271, 1993.

In recent years, a renewed interest has emerged to more accurately define the microbiological safety of current USDA-mandated pasteurization processes. Although the implementation of uniform egg pasteurization requirements for liquid egg products has clearly improved consumer food safety, experimental results included in the findings of Schuman and Sheldon [74] and Palumbo et al. [75] indicate that the relative microbial lethality of various liquid egg pasteurization processes does differ relative to that predicted for *Salmonella* spp. in LWE. This issue represents an active area of microbiological egg safety research and an ongoing issue of regulatory policy debate.

9.3.1.3 Egg Pasteurization and Shelf-Life Extension

In addition to providing a margin of *Salmonella* safety, a secondary benefit of liquid egg pasteurization is the extension of the refrigerated shelf life of the product. As noted previously, such conventional pasteurization processes do not render the product sterile. Vegetative and spore-forming microorganisms that survive pasteurization may cause liquid egg spoilage during storage, handling, and distribution. Furthermore, microbial contaminants capable of spoiling liquid egg may be introduced into the product during postpasteurization handling [76]. Pasteurized LWE and whole egg blends currently constitute the majority of the liquid egg products produced in the United States [73]. The shelf life of conventionally pasteurized LWE is relatively short, ranging from 12 days at 2°C to 5 days at 9°C [77]. Because of such shelf-stability limitations, liquid egg processors have relied either upon rapid (<14 days) transport and use requirements for refrigerated LWE or frozen storage and distribution of the pasteurized, packaged product [76]. While extending shelf life, the process of freezing and thawing whole egg products yields several undesirable physicochemical changes, including gelation of the yolk proteins, phase separation, color changes, and increased viscosity [78]. The functional properties of whole egg products may also be diminished as a result of frozen storage [79,80].

9.3.1.4 Ultrapasteurized, Aseptically Packaged Liquid Whole Egg

In 1987, Ball et al. [76] documented the development of *ultrapasteurization* processes (i.e., heating at >60°C for <3.5 min) which, when coupled with *aseptic processing* and packaging, yielded LWE with a shelf life of 3–6 months at 4°C [76]. At present, commercially ultrapasteurized LWE has a code-dated shelf life of 10 weeks at 4°C [81]. The extended shelf life of this nonsterile product is dependent on the use of (1) raw LWE of excellent initial microbiological quality, (2) unique *high-temperature short-time* thermal treatments, (3) packaging of the product using an aseptic filler within a sterile zone, and (4) maintenance of proper refrigeration temperatures (1°C–4°C) throughout distribution and retail storage. In 1994, 175–200 million pounds of extended shelf-life LWE were produced using this ultrapasteurization and aseptic packaging technology [81]. Ultrapasteurized LWE products offer a number of advantages over both frozen liquid egg and shell eggs, including consistent product quality, convenience and portion control, storage space savings, and economy (in terms of time and labor savings and the elimination of the need for frozen distribution) [81].

Relative to conventionally pasteurized LWE, the ultrapasteurized product offers enhanced microbiological shelf stability (at 1°C–4°C) and a greater assurance of *Salmonella* safety. In addition, achieving a predicted 9D inactivation of the bacterial pathogen *L. monocytogenes* was one of the design criteria used by Ball et al. in the development of egg ultrapasteurization processes [82]. While conventional LWE pasteurization represents only a 2.1D to 2.7D process with respect to *L. monocytogenes* inactivation [83], the egg ultrapasteurization processes of Ball et al. were demonstrated to provide process lethalities against *Listeria* of at least 6.7 to >7.3 log cycles [84].

9.3.1.5 Extended Shelf-Life Liquid Egg Substitutes

Liquid egg substitutes represent another category of *extended shelf-life* liquid egg products. These products were developed nearly 20 years ago to simulate the features of LWE, yet with little or none of the cholesterol or fat normally present in yolk or whole egg. Egg substitutes consist primarily of liquid egg white plus a variety of nonegg ingredients (e.g., nonfat dried milk, vegetable oils, emulsifiers, stabilizers, gums, artificial color, vitamins, and minerals) that mimic to an extent the appearance, flavor, and texture of LWE [81]. These products are pasteurized and filled into a variety of retail packages, generally in a nonaseptic manner. The finished product has a pH of 6.4–8.0 and a code-dated refrigerated shelf life of 8–12 weeks at 1°C–4°C [85].

9.3.1.6 In-Shell Pasteurized Eggs

Because of the increased concern over the egg-associated food-borne pathogen SE, research efforts have focused on developing an *in-shell pasteurization* process for producing pathogen-free shell eggs that can be used as it is or in products receiving little or no additional heating prior to consumption. During the 1940s and 1950s, the thermal treatment of eggs to prevent embryonic growth in fertile eggs, to reduce the incidence of spoilage during long-term storage and to maintain internal quality received considerable research attention. Stadelman [86] presents a concise review of that research with particular attention directed toward the practice of "*thermostabilization*," a patented process [87] whereby shell eggs are placed in heated water or oil to extend storage life and prevent spoilage and deterioration of egg shell quality. Hou et al. [88] documented the feasibility of combining water bath and hot air oven processing to reduce viable SE populations within inoculated intact shell eggs. In a related study, Schuman et al. [89] evaluated the effects of water-immersion heat treatments on the inactivation of SE within intact shell eggs. Six pooled strains of SE (ca. 3×10^8 cfu, inoculated near the center of the yolk) were completely inactivated within 50–57.5 min at a water bath temperature of 58°C and within 65–75 min at 57°C (an 8.4D–8.5D process per egg). Haugh unit values and albumen whip times increased during heating, although yolk index and albumen pH values were unaffected. The authors [89] concluded that broken-out whole egg or yolk from immersion-heated shell eggs could provide *Salmonella*-free ingredients for the preparation of a variety of minimally cooked foods of interest to consumers and foodservice operators.

9.3.1.7 Dehydrated Eggs

Dehydrated egg is usually pasteurized by holding at relatively low pasteurization temperatures for a long time. For instance, dried egg white with glucose removed is held for 7 days in a hot room at a minimum temperature of 130°F (54°C). Glucose must first be removed (usually enzymatically) to prevent the Maillard browning reaction resulting in both browning and loss of essential amino acids such as lysine. More recently, a 7 log reduction of avian influenza virus in dried egg white would take 2.6 days at 54.4°C [90].

9.3.1.8 Egg Waste

Chiu and Wei [91] determined the optimal thermal process for eggs from hatchery waste for use as animal feed using response surface methodology. Eggs processed for 23 h at a fixed temperature of 65°C, resulted in higher protein digestibility in vitro (89.6%) and dry matter (88.5%) content of waste egg meal with lower electricity consumption (82.4 kWh/60 kg).

9.3.2 General Egg Safety Considerations

Foods of animal origin including dairy products, meat, poultry, and eggs have long been recognized as the primary source of many of the bacteria responsible for food-borne infections and intoxications [92,93]. Based on the published literature, a list of currently recognized bacterial pathogens with a reasonable likelihood of occurrence in blended raw egg products could include *Salmonella* spp., *Staphylococcus aureus, Bacillus cereus, L. monocytogenes, Aeromonas hydrophila, Yersinia enterocolitica*, and *Campylobacter* spp. Between 1973 and 1987, eggs (unspecified product types) were linked to only 1% of all outbreaks and 1% of all cases with known vehicles [94]. The proportion of salmonellosis outbreaks associated with eggs increased from none (in 1973–1975), to 1% (1976–1978), 4% (1979–1981), 3% (1982–1984), and 8% (1985–1987). In a more recent review of international food-borne disease surveillance statistics, Todd [95] reported that contaminated meat products were linked to more outbreaks of food-borne disease (i.e., 16%) in the United States than all other food product categories. Of the 22 countries surveyed, France and Spain were the only two nations to report that eggs and egg products were the vehicles most frequently involved in outbreaks of food-borne illness.

9.3.2.1 *Salmonella*

Of the previously listed pathogens, *Salmonella* species have the greatest historical association with contaminated eggs and egg products. The bacterial pathogens *L. monocytogenes* and *A. hydrophila* pose particular concerns if present in chilled liquid egg products due to their ability to multiply in foods at proper refrigerated storage temperatures (i.e., 5°C). Over the last decade, the egg-associated serotype *S.* Enteritidis has emerged as one of the leading causes of food-borne salmonellosis in the United States, Canada, the United Kingdom, France, Spain, and other nations [95]. In the United States, the number of cases of salmonellosis caused by *S.* Enteritidis increased slowly but steadily between 1976 and 1987. Since 1988, the isolation rate for *S.* Enteritidis from clinical cases of salmonellosis has risen sharply, possibly as a result of this serotype becoming more invasive in egg-laying flocks [96]. Since 1990, the number of salmonellosis caused by *S.* Enteritidis have exceeded those attributed to all other serovars, including *S.* Typhimurium [97]. Epidemiological studies indicate that Grade A shell eggs are a major source of *S.* Enteritidis infections in humans [98]. Between January 1985 and May 1987, 35 food-borne *S.* Enteritidis outbreaks of known cause were reported to the Centers for Disease Control (CDC); of these, 27 outbreaks (77%) were epidemiologically linked to the consumption of foods that contained eggs or eggs alone, either inadequately cooked or eaten raw in such foods as hollandaise sauce, homemade eggnog, or Caesar salad dressing [98]. Between 1985 and 1991, a total of 380 *S.* Enteritidis outbreaks were reported in the United States, involving 13,056 cases and 50 deaths. Grade A shell eggs were implicated in 82% of the outbreaks [99].

Without question, the implementation of the Egg Products Inspection Act of 1970 and its accompanying liquid egg pasteurization requirements has dramatically reduced the incidence of egg-associated salmonellosis in the United States [100]. The use of pasteurized liquid egg products (rather than shell eggs) for the preparation of a variety of egg dishes is viewed as the major means for preventing salmonellosis outbreaks, especially in foodservice settings and nursing homes [101]. When pasteurized liquid egg products are properly packaged, stored, and handled by food preparers, egg-associated risks of salmonellosis may be significantly reduced. To date, no salmonellosis outbreaks in the United States, England, or Wales have been associated with foods containing pasteurized liquid egg [101].

9.3.2.1.1 *Thermal Resistance of Salmonella in Eggs and Egg Products*

Because liquid egg pasteurization processes were designed to ensure inactivation of *Salmonella,* numerous investigators have sought to define the thermal inactivation kinetics of this important pathogen [71,101]. The conventional pasteurization process for LWE (i.e., 3.5 min at 60°C for the average particle) was based on the assumption that typical strains of *Salmonella* have a *D* value at 60°C of ~0.4 min in blended whole egg [68]. Despite differences in *Salmonella* serovars/strains and in the experimental procedures used, this assumption has proven to be reasonably accurate, based on kinetic data from a variety of research laboratories.

Representative decimal reduction time data for *S.* Oranienburg and *S.* Typhimurium in various whole egg, yolk, and egg white products are presented in Table 9.2 [71]. The predicted *D* value for *S.* Oranienburg in LWE (pH 7.0) at 60°C is 0.3 min when extrapolated from the decimal reduction time curve of Cotterill et al. [102]. Humphrey et al. [103] reported *D* values of 0.20 and 0.26 min for two *S.* Typhimurium strains heated in LWE at 60°C. As presented in Table 9.3, *S.* Oranienburg and *S.* Typhimurium were more heat resistant in plain yolk than in LWE (pH 7.0), in contrast, these strains were more heat sensitive in unsupplemented albumen (pH 9.2–9.5) than in whole egg. Similar heat resistance trends were documented by Humphrey et al. [103], who evaluated the heat

TABLE 9.3
Thermal Resistance of *Salmonella* spp. in Various Liquid Egg Products

Test Organisms	Products	*D* Value (min) at °C							*z* Value (°C)	References
		50	52	54	56	58	60	62		
S. Oranienburg	Whole egg pH 7.0	—	—	4	1.7	—	—	—	4.5	[96]
	Whole egg + 10% sucrose	—	—	25	10	3.5	—	—	4.3	
	Whole egg + 10% salt	—	—	30	12	6	3	—	7.0	
S. Typhimurium	Whole egg pH 5.5	—	—	17	5.5	—	—	—	4.2	[142]
	Whole egg + sorbic acid 100 ppm	—	—	14	4.5	—	—	—	4.1	
	Whole egg + β-propiolactone 100 ppm	—	—	11	4.5	—	—	—	4.5	
	Whole egg + benzoic acid 500 ppm	—	—	6.5	2.2	—	—	—	4.1	
S. Typhimurium	Egg yolk	—	—	9.5	3	1.3	0.4	—	4.4	[103]
	Egg yolk + 10% sucrose	—	—	75	25	10	4	—	4.8	
	Egg yolk + 10% salt	—	—	110	45	17	6	2	4.6	
S. Typhimurium	Egg white pH 9.2	10	3.5	1	—	—	—	—	4.2	[103]
	Egg white pH 9.2 + 10% sucrose	20	7	2.5	0.8	—	—	—	4.3	
	Egg white pH 7.3 + aluminum salt	—	22	6.5	2.5	—	—	—	4.2	
Mixture of three strains	Egg white pH 9.5	9	3.5	1.1	—	—	—	—	4.8	[143]
	Egg white pH 9.5 + 0.5% polyphosphate	1.1	0.6	0.3	—	—	—	—	4.6	

resistance of three *S*. Enteritidis isolates and two *S*. Typhimurium isolates. *D* values for the five isolates ranged from 0.20 to 0.44 min in LWE heated at 60°C [103]. In a similar study of 17 strains of *S*. Enteritidis (primarily phage type 8), Shah et al. [100] reported that *D* values at 60°C ranged from 0.20 to 0.52 min in LWE (mean = 0.32 min). Even with a *D* value of 0.52 min (the highest value reported in the study), conventional pasteurization of LWE at 60°C for 3.5 min would provide a predicted lethality of 6.7 log cycles. Baker [104] used a continuously stirred three-neck flask to assess the thermal resistance of nine strains of *S*. Enteritidis in raw LWE. *D* values at 60°C ranged from 0.31 to 0.69 min (mean = 0.42 min). When using a plate pasteurizer, Michalski et al. [105] reported a greater than 9D reduction in *S*. Enteritidis in whole egg at 60.0°C for 3.5 min. They attributed the greater kill to better mixing due to turbulent flow in the right-angle turns in the holding tube of the plate pasteurizer. Based on the previous decimal reduction time data, a 3.5 min process at 60°C would yield a 5.1–11.3 log reduction in the viable *S*. Enteritidis population in LWE.

The published thermal resistance data for *Salmonella* spp. heated in liquid egg white or liquid yolk have varied more widely according to the strains tested, the investigator, and the laboratory methods used. For example, Humphrey et al. [103] reported a *D* value of 1.0 min for *S*. Typhimurium heated at 55°C in albumen. Using a different strain, Garibaldi et al. [106] obtained data that yielded a predicted *D* value of only 0.58 min at 55°C. Palumbo et al. [107] determined survival of a six-strain mixture of *S*. Enteritidis, Typhimurium, and Senftenberg (not 775 W) in egg white using a submerged vial technique. As reported in previous studies [108], these researchers also reported that *Salmonella* is much more heat resistant at low albumen pH. Log reductions using a 3.5 min holding time at 56.6°C were 0.97 at pH 7.8, 1.64 at pH 8.2, 2.20 at pH 8.8, and 4.24 at pH 9.3. Michalski et al. [105] reported a 7.5 *D* value for a five-strain cocktail of *S*. Enteritidis in egg white (pH 9.0, 100 μL capillary tubes). In liquid egg yolk, the *D* values at 60°C reported by Humphrey et al. [103] (*D* = 0.8 min) and Garibaldi et al. [109] (*D* = 0.4 min) also differed substantially for different strains of *S.* Typhimurium. Palumbo et al. [75] reported *D* values at 60°C of 0.55–0.75 min for six *Salmonella* spp. heated in unsupplemented liquid yolk. The *D* value for 10% sugared at 63.3°C was 0.72 min while 10% salted yolk had an observed *D* value of 11.50 min at 63.3°C. Generally *Salmonella* are more heat resistant in yolk, but yolk is less sensitive to higher temperatures as it affects functional properties. However, the addition of salt and sugar apparently further increases the heat resistance of these organisms in yolk products, especially salted yolk at 10% and 20% [75,109]. In contrast, Michalski et al. [105] observed an 8D reduction in 10% salted yolk at 63.3°C and a 1D reduction in egg yolk containing both 5% salt and 5% sugar using capillary tubes. These noted differences in heat resistance may be partially explained by the use of capillary tubes that largely eliminate the come-up time and the effect of viscosity [110].

In summary, the thermal resistance data discussed earlier indicate that conventional pasteurization processes for liquid yolk, liquid albumen (pH 9.2), and LWE are predicted to provide minimum process lethalities for *Salmonella* spp. of 7.75, 6.2, and 5.1 log cycles, respectively. It should be noted that differences in the experimental methodologies used in assessing the thermal resistance of salmonellae (and other pathogens) in liquid egg may impact the accuracy of the *D* values reported. In addition to the major factors affecting the wet-heat inactivation of microorganisms (e.g., proximate composition of the heating menstruum, moisture content, pH, inoculum concentration, growth conditions, etc.), the geometry of the heating vessel (e.g., test tubes, reaction vials, capillary tubes, or flasks) and its headspace volume and orientation within the heating bath may bias the resultant batch kinetic data [111–113]. The analysis of subsamples drawn from a single large heated vessel appears to be particularly unreliable. For example, Dabbah et al. [114] reported exceptionally high *D* values (i.e., >6.0 min at 60°C) when *Salmonella* spp. were heated in LWE in a 300 mL stirred glass flask. While there remains no general consensus on the best experimental methodology for bacterial thermal resistance testing, the advantages of using small (i.e., very low volume capillary tubes), sealed, and fully immersed heating vessels have been documented [74,112,115].

Several investigators have evaluated the efficacy of simulated domestic or foodservice cooking methods as means to inactivate *S.* Enteritidis in Grade A shell eggs. Saeed and Koons [116] asked persons who identified themselves as regular egg consumers to cook sets of 11 eggs by each of three methods. As expected, the individuals varied considerably in the times used to cook the product to their preferred degree of doneness. When the contents of shell eggs contained 10^1–10^3 CFU/mL, viable salmonellae were detected in 24% of fried eggs, 15% of scrambled eggs, and 10% of omelets tested. Baker [104] reported that endpoint internal temperatures of 74°C for scrambled eggs and 61°C–70°C for fried eggs were necessary to ensure the inactivation of S. *Enteritidis*. With respect to hard-cooked egg preparation, Licciardello et al. [117] and Baker et al. [118] demonstrated that when shell eggs were internally inoculated with 10^8–10^9 *Salmonella* cells, the pathogen was effectively inactivated when the eggs were placed directly into boiling water for a minimum of 7–8 min. While the aforementioned procedures provide valuable guidance for the safe preparation of fully cooked egg dishes, there is a need for the egg industry to provide refrigerated shell eggs with a greater assurance of microbiological safety to retail consumers and the foodservice industry. Shell egg safety improvements of this kind would be particularly valuable if they would permit the safe preparation of a variety of foods (e.g., soft-boiled eggs, soft-poached eggs, "sunny-side" fried eggs, gourmet sauces, salad dressings, custards, etc.) that call for the use of fresh shell eggs as ingredients.

9.3.2.2 Listeria and Aeromonas

Over the last 10–15 years, both *L. monocytogenes* and *A. hydrophila* have been identified as "emerging" food-associated pathogens. Common features of concern to public health agencies and food processors include the psychrotrophic nature of these pathogens, their wide distribution in nature and raw agricultural commodities, their ability to persist in moist low-nutrient niches in processing plants, the special techniques required to identify low numbers of each pathogen amid a diverse background microflora, and the fact that the infectious dose levels for *L. monocytogenes* and *A. hydrophila* are currently unknown [119,120]. *L. monocytogenes* is relatively resistant to environmental extremes (e.g., low pH, high sodium chloride levels, reduced water activity, low nutrient density, and freeze–thaw cycling) and is among the most heat-resistant vegetative bacterial pathogens of concern in foods [119]. Although there have been no documented cases of listeriosis linked to the consumption of eggs or egg products, *L. monocytogenes* has been associated with laying hens, raw and cooked poultry meat products, and commercially blended raw LWE. Of the 23 avian species in which listeriosis has been documented, chickens have remained the most common avian host for nearly 60 years [121]. Interest in risks associated with poultry-borne listeriae has increased as a result of several cases of human listeriosis in the United States and the United Kingdom that were linked to consumption of turkey frankfurters and RTE, cook-chill chicken [121].

In the first published survey of its kind, Leasor and Foegeding [122] reported that *Listeria* spp. were isolated from 15 of 42 (36%) previously frozen samples of raw, commercially broken LWE obtained from 6 of 11 (54%) commercial egg processors located throughout the United States. Of the 15 *Listeria-positive* samples, 15 contained *L. innocua*, and 2 samples contained *L. monocytogenes*. In a survey of the largest egg processing plant in Northern Ireland, Moore and Madden [123] isolated *Listeria* spp. from 125 of 173 (72%) in-line filters used to remove shell debris from raw blended LWE. The only species isolated were again *L. innocua* (62%) and *L. monocytogenes* (37.8%). Additional LWE samples (~140 mL each) were collected just prior to the pasteurizer on nine consecutive days, and the highest concentration of *Listeria* spp. obtained was ~40 listeriae per milliliter of liquid egg.

The results of a large, unpublished U.S. national survey conducted by the USDA (Agricultural Marketing Service, Poultry Division, Egg Products Inspection Branch) in 1992 demonstrated that 32.8% of all raw liquid egg products ($n = 1555$) and 2.9% of all conventionally pasteurized liquid egg products ($n = 1292$) tested positive for the presence of *Listeria* spp. In the same survey, LWE products containing citric acid showed *Listeria* spp. incidence rates of 40.0% (for 90 samples of raw LWE) and 3.2% (for 95 samples of conventionally pasteurized LWE). *Listeria* spp. were also isolated from 42 of 98 (42%) of all environmental processing plant swab samples collected during the USDA national survey. In a survey of two egg washing facilities in Southeastern Ontario, Laird et al. [124] isolated *L. innocua* in egg wash water (pH 10.2–11.8) and certain environmental samples (i.e., preloaders, sewer drains, and floors). In summary, the aforementioned surveys indicate that *L. monocytogenes* and *L. innocua* occur in commercially processed raw liquid egg products at relatively high incidence rates, yet at fairly low concentrations (estimated at <100 CFU/mL). Nonetheless, the refrigerated storage life of extended shelf-life liquid egg products permits ample time for this pathogen to multiply to high cell populations if present (as a thermal process survivor or a postprocess contaminant) in the finished, packaged product.

9.3.2.2.1 Thermal Resistance of Listeria in Eggs and Egg Products

The thermal resistance of *L. monocytogenes* in a variety of foods and microbiological media has been reviewed by Mackey and Bratchell [125] and Farber and Peterkin [119]. The literature to date supports the contention that *L. monocytogenes* is among the most heat-resistant nonspore-forming pathogens associated with foods, with a calculated z_D-value range (in most foods and broths) of 6.7°C–6.9°C [125]. In the aftermath of the 1983 milk-associated listeriosis outbreak in Massachusetts, the adequacy of fluid milk pasteurization processes to inactivate *L. monocytogenes* was the subject of numerous publications. In a summary of this research, Mackey and Bratchell [125] predicted that high-temperature, short-time pasteurization conditions (71.7°C for 15 s) and vat pasteurization conditions (63°C for 30 min) would achieve 5.2D and 39D reductions in the viable *Listeria* populations, respectively. These results are noteworthy in that a similar "5D" lethality standard for *L. monocytogenes* could be considered in the context of evaluating desirable or achievable pathogen reducing requirements for continuously pasteurized (conventional) liquid egg products.

In 1990, Foegeding and Leasor [83] published a study in which *D* values were determined for five strains of *L. monocytogenes* in sterile raw LWE. The experimental units consisted of 0.05 mL samples of inoculated LWE within sealed glass capillary tubes that were heated by immersion in a preheated water or oil bath. *D* values for strain F5069, the most heat-resistant isolate tested, averaged 22.6, 7.1, 1.4, and 0.20 min at 51°C, 55.5°C, 60°C, and 66°C, respectively [83]. For the five strains evaluated, minimum conventional pasteurization conditions for LWE (60°C for 3.5 min) were predicted to represent 2.1D–2.7D processes with respect to *L. monocytogenes*. In a subsequent study, Foegeding and Stanley [84] determined thermal death times (*F* values) for *L. monocytogenes* F5069 in raw LWE using the immersed sealed capillary tube procedure (0.05 mL of liquid egg per tube). This *Listeria* strain was eliminated from inoculated samples (5.0×10^6 to 2.0×10^7 CFU/tube) after 16, 8, 4.5, 1.6, and 0.6 min at 62°C, 64°C, 66°C, 69°C, and 72°C.

Bartlett and Hawke [126] evaluated the heat resistance of *L. monocytogenes* strains Scott A (a clinical isolate) and HAL 975E1 (an egg isolate) in the following five liquid egg products: LWE, 10% NaCl whole egg (LWEN), 10% sucrose whole egg (LWES), 10% NaCl egg yolk (EYN), and 10% sucrose egg yolk (EYS). The presence of salt decreased the water activity (a_w) of the products to a greater extent than sucrose; all water activity values were >0.91 except for salt yolk ($a_w = 0.867$). Survivor curves were constructed and *D* values calculated based on data from a 0.2 mL submerged sealed test tube procedure. The relative heat resistance of *L. monocytogenes* in the five products was as follows: heat resistance in salted EY > salted LWE >> EYS ≥ sucrose LWE ≥ LWE. These thermal resistance trends were very similar to those reported by Cotterill et al. [102] for *Salmonella* Oranienburg in liquid egg products. Based on current U.S. conventional egg pasteurization requirements, Bartlett and Hawke [126] predicted that process lethalities against *L. monocytogenes* Scott A would range from 0.2 log cycle (for salt EY) to 1.8 log cycles (for sucrose LWE). Similar thermal resistance trends were reported by Palumbo et al. [75], who determined *D* values for six pooled strains of *Salmonella* spp. and for five pooled strains of *L. monocytogenes* inoculated in plain liquid EY and in various EY products containing added salt and/or added sucrose. Both pathogens were more heat resistant in EY + 10% NaCl than in EY + 10% sucrose or in plain EY. Based on *D* values derived by using a submerged capped test tube procedure, the lethality of USDA-mandated conventional egg pasteurization processes was estimated to range from 0.3 log cycles (in EY + 10% NaCl) to 6.1 log cycles (in plain EY) for *Salmonella* spp. and from 0.2 log cycles (in EY + 10% NaCl) to 3.3 log cycles (in EY + 10% sucrose) for *L. monocytogenes* [75].

In summary, the aforementioned four publications indicate that the margin of safety provided by conventional pasteurization requirements for LWE, plain yolk, and products containing added sucrose is not large, especially if *L. monocytogenes* is present in the raw bulk tank at levels >100 CFU/mL. In NaCl-supplemented LWE or yolk, conventional pasteurization would be inadequate to inactivate even 10 CFU/mL of *L. monocytogenes*. These findings take on additional significance in light of a recent USDA document that details the criteria for approval to produce and market liquid egg products with an "extended shelf life" (i.e., >4 weeks at 4.4°C) [127]. In order to gain regulatory approval to produce such products, companies must pasteurize the product at 60°C for 3.5 min. If alternative thermal processes are used, the company must provide laboratory data demonstrating that the thermal process yields a 7 log reduction in the viable *L. monocytogenes* population. As noted previously, the data of Foegeding and Leasor [83] and Bartlett and Hawke [126] clearly demonstrate that a 3.5 min process at 60°C represents only a 1.7D–2.7D *Listeria* inactivation process. Thus, the proposed 7D lethality requirement appears to be unduly harsh and may not permit production of liquid egg products of acceptable organoleptic and functional quality. At the time of this review, no studies documenting the thermal resistance of *Listeria* spp. in liquid egg white or in liquid egg substitutes were available in the published scientific literature.

Two research groups have evaluated the survival potential of *L. monocytogenes* in eggs heated under simulated foodservice or domestic conditions. Using inoculated shell eggs (containing $>5\times10^5$ CFU/g of egg contents), Urbach and Schabinski [128] reported that viable listeriae were isolated from fried eggs cooked in a manner which coagulated the white yet left a "soft yolk." In a detailed study, Brackett and Beuchat [129] reported that frying inoculated whole eggs "sunny-side up" until the albumen was partially coagulated reduced both low (10^2 CFU/g) and high (10^5 CFU/g) *L. innocua* by only 0.4 log CFU/g. In contrast, cooking one or three scrambled eggs on a skillet to an endpoint internal temperature of 70°C–73°C reduced the lower inoculum by >1.8 log cycles and the higher inoculum by 3.0 log cycles. The authors [129] concluded that although it is unlikely that large numbers of listeriae would survive such cooking procedures unless present at >10 CFU/g, it would be prudent for persons who might be especially susceptible to listeriosis to consume only "thoroughly cooked" eggs.

The significance of *A. hydrophila* as a potential food-borne pathogen has been the subject of publications by Buchanan and Palumbo [130], Morgan and Wood [131] and Abeyta et al. [132]. Aeromonads have been isolated from a wide variety of refrigerated foods of animal origin, including finfish, shellfish, raw milk, whipped cream, butter, veal, beef, lamb, pork, and poultry products [133]. Poultry products from which *A. hydrophila* have been isolated include raw chicken (unspecified cut) and chicken liver [134], chicken thigh meat [135], and fresh broiler carcasses [136]. *Aeromonas* spp. have also long been recognized as potential spoilage organisms in checked or cracked shell eggs [137].

Although published data on the incidence of *Aeromonas* spp. in raw or pasteurized LWE produced in the United States are currently unavailable, the pathogen has been isolated from commercially broken raw LWE in Japan [138] and Australia [139]. In the Australian study, MacKenzie and Skerman [139] reported that *A. hydrophila* was the fourth most predominant strain present in samples of spoiled LWE, and aeromonads were isolated with higher frequency when liquid egg was stored at 4°C or 10°C as compared to 20°C or 25°C. Kraft et al. [140] characterized the microflora present in fresh raw LWE obtained from two Iowa egg breaking plants. The five most prevalent bacterial genera (of 148 total isolates) were *Pseudomonas* (47.3%), *Arthrobacter* (18.2%), *Aeromonas* (17.6%), *Escherichia* (8.8%), and *Micrococcus* (2.7%).

9.3.2.2.2 Thermal Resistance of Aeromonas in Eggs and Egg Products

The documented growth potential of *A. hydrophila* in LWE at 6.7°C (slight temperature abuse) underscores the importance of ensuring the adequacy of egg pasteurization/ultrapasteurization processes with respect to *Aeromonas* inactivation. To date, only two published studies have addressed the heat resistance of *A. hydrophila* (human clinical isolate) in an egg product. Nishikawa et al. [141] reported that a 4 min process at 55°C was sufficient to inactivate the pathogen (initially inoculated at 10^8 CFU/mL) in raw liquid EY heated in capped test tubes dipped in a preheated water bath. The authors [141] also demonstrated that *A. hydrophila* was substantially less heat resistant than a clinical isolate of *S.* Typhimurium when heated in liquid yolk. Because the study was conducted at only one water bath temperature (55°C, without correction for come-up times), neither D nor z_D values were reported. In the second study, Schuman et al. [110] sought to kinetically characterize the heat resistance of *A. hydrophila* in raw LWE using an immersed sealed capillary tube procedure. Decimal reduction times of four individual strains of *A. hydrophila* at 48°C, 51°C, 54°C, 57°C, and 60°C were found to range from 3.62–9.43 min (at 48°C) to 0.026–0.040 min (at 60°C). Both egg processing plant isolates were more heat resistant than the American Type Culture Collection (ATCC) strains. The z_D values were 5.02°C–5.59°C, similar to those for other nonspore-forming bacteria. Although this study indicated that *A. hydrophila* is substantially less heat resistant than *Salmonella* LWE, it is important for egg processors to take measures to prevent postpasteurization contamination of liquid egg products by *Aeromonas* spp. and other psychrotrophic pathogens, including *L. monocytogenes*. At the time of this review, no other reports on the thermal resistance of *Aeromonas* spp. in other liquid egg products or in shell eggs cooked under simulated foodservice conditions were available in the published scientific literature.

Although several investigators have attempted to kinetically characterize the thermal inactivation of *A. hydrophila* in model buffer systems and in skim milk [142,143], each study yielded nonlinear (tailing) survivor curves, which complicated the analysis of the thermal destruction data. The previous studies yielded diphasic (tailing) inactivation curves in which surviving populations of 10^2–10^5 CFU/mL of solution were detected long after the initial linear phase of inactivation. For purposes of calculating *D* values, these authors [143] disregarded the tailing portions of the survivor curves, while noting the potential significance of microbial subpopulations of apparently greater heat resistance than the rest of the inoculum. Using a buffered peptone system, Stecchini et al. [144] reported that the inactivation of *A. hydrophila* in 9 mL capped test tubes was a nonlinear process, best described mathematically by a complex hyperbolic function.

9.4 CONCLUSIONS

Application of heat continues to be the most efficient method for preserving food. Current issues facing producers of poultry products food processors in general include the persistence of *L. monocytogenes* in RTE products, antibiotic resistant bacteria, and handling of animal coproducts. The USDA has issued regulations to address the Listeria problem in RTE meat as described in this chapter. The use of antibiotics in animal production is also being addressed by both poultry companies and governing bodies worldwide by reducing or eliminating their use. Between 40% and 50% of a carcass is not destined for human consumption thus must be handled and recycled if possible. Poultry meal from this nonedible portion is often used as a animal feed component. Concern over the bacterial levels in these products reentering the food cycle through the feed has been debated. The survival of *Salmonella* in poultry meat and bone meal has been shown to be well below the levels found in foods for human consumption. The high temperatures and times used in the rendering process virtually eliminate most vegetative bacterial cells. Other issues such as mad cow disease that have not yet affected the poultry industry may affect the handling of these coproducts in the future.

Thermal processing remains the best and most reliable process used to render products safe for consumption by humans and use as animal feed. The fact that the microorganisms can only develop limited resistance to heat and this resistance is lost without continued exposure along with the relatively low cost have kept thermal processing the chosen method.

REFERENCES

1. UM Abdul-Raouf, LR Beuchat, MS Ammar. Survival and growth of *Escherichia coli* O157:H7 in ground, roasted beef as affected by pH, acidulates, and temperature. *Applied and Environmental Microbiology* 59:2364–2368, 1993.
2. J Chiruta, KR Davey, CJ Thomas. Thermal inactivation kinetics of three vegetative bacteria as influenced by combined temperature and pH in a liquid medium. *Food and Bioproducts Processing* 75:174–180, 1997.
3. VK Juneja, BS Marmer, JG Philips, AJ Miller. Influence of intrinsic properties of food on thermal inactivation of spores of nonproteolytic *Clostridium botulinum*: Development of a predictive model. *Journal of Food Safety* 15:349–364, 1995.
4. LC Blankenship, SE Craven. *Campylobacter jejuni* survival in chicken meat as a function of temperature. *Applied and Environmental Microbiology* 4:88–92, 1982.
5. RG Bell, KM DeLacy. Heat injury and recovery of *Streptococcus faecium* associated with the souring of club-packed luncheon meat. *Journal of Applied Bacteriology* 57:229–236, 1984.
6. S Ghazala, D Coxworhty, T Alkanani. Thermal kinetics of *Streptococcus faecium* in nutrient broth/sous vide products under pasteurization conditions. *Journal of Food Processing and Preservation* 19:243–257, 1995.
7. O Reichart. Modeling the destruction of *Escherichia coli* on the basis of reaction kinetics. *International Journal of Food Microbiology* 23:449–465, 1994.

8. S Gaillard, J Leguerinel, P Mafort. Model for combined effects of temperature, pH, and water activity on thermal inactivation of *Bacillus cereus* spores. *Journal of Food Science* 63:887–889, 1998.
9. VK Juneja, BS Eblen. Predicative thermal inactivation model for *Listeria monocytogenes* with temperature, pH, NaCl, and sodium pyrophosphate as controlling factors. *Journal of Food Protection* 62:986–993, 1999.
10. AR Fain, JE Line, AB Moran, LM Martin, RV Lechowich, JM Carosella, WL Brown. Lethality of heat to *Listeria monocytogenes* Scott A: *D*-value and *Z*-value determinations in ground beef and turkey. *Journal of Food Protection* 54:756–761, 1991.
11. LC Blankenship. Survival of *Salmonella typhymurium* experimental contaminant during cooking of beef roasts. *Applied and Environmental Microbiology* 35:1160–1165, 1978.
12. SJ Goodfellow, WL Brown. Fate of *Salmonella* inoculated into beef for cooking. *Journal of Food Protection* 51:598–605, 1978.
13. EO Goksoy, C James, JEL Corry, SJ James. The effect of hot-water immersions on the appearance and microbiological quality of skin–on chicken–breast pieces. *International Journal of Food Science and Technology* 36:61–69, 2001.
14. AI Morgan, N Goldberg, ER Radewonuk, OJ Scullen. Surface pasteurization of raw poultry meat by steam. *Lebensmittel-Wissenschaft und-Technologie* 29:447–451, 1996.
15. A Orta-Ramirez, DM Smith. Thermal inactivation of pathogens and verification of adequate cooking in meat and poultry products. *Advances in Food Nutrition Research* 44:147–194, 2002.
16. IJ Pflug. Evaluating a ground beef patty cooking process using the general method of process calculation. *Journal of Food Protection* 60:1215–1223, 1997.
17. GJ Veeramuthu, JF Price, CE Davis, AM Booner, DM Smith. Thermal inactivation of *Escherichia coli* O157:H7, *Salmonella Senftenberg*, and enzymes with potential as time-temperature indicators in ground turkey thigh meat. *Journal of Food Protection* 61:171–175, 1998.
18. RY Murphy, BP Marks, ER Johnson, MG Johnson. Inactivation of *Salmonella* and *Listeria* in ground chicken breast meat during thermal processing. *Journal of Food Protection* 62:980–985, 1999.
19. RY Murphy, BP Marks, ER Johnson, MG Johnson. Thermal inactivation kinetics of *Salmonella* and *Listeria* in ground chicken breast meat and liquid medium. *Journal of Food Science* 65:706–710, 2000.
20. VK Juneja, OP Snyder, BS Marmer. Thermal destruction of *Escherichia coli* 0157:H7 in beef and chicken: Determination of D and z values. *International Journal of Food Microbiology* 35:231–237, 1997.
21. VK Juneja, BS Eblen, HM Marks. Modeling non-linear survival curves to calculate thermal inactivation of *Salmonella* in poultry of different fat levels. *International Journal of Food Microbiology* 70:37–51, 2001a.
22. M Schnepf, WE Barbeau. Survival of *Salmonella typhimurium* in roasting chickens in a microwave, convection and a conventional electric oven. *Journal of Food Safety* 9:245–252, 1989.
23. RY Murphy, ER Johnson, JA Marcy, MG Johnson. Survival and growth of *Salmonella* and *Listeria* in the chicken breast patties subjected to time and temperature abuse under varying conditions. *Journal of Food Protection* 64:23–29, 2001.
24. BP Marks, H Chen, RY Murphy, ER Johnson. Incorporating pathogen lethality kinetics into a coupled heat and mass transfer model for convection cooking of chicken patties. IFT Abstract 79B-19. Presented July 27, 1999, Chicago, IL.
25. RY Murphy, LK Duncan, ER Johnson, MD Davis. Effect of overlapping chicken patties during air/steam impingement cooking on thermal inactivation of *Salmonella* Senftenberg and *Listeria innocua*. *Journal of Applied Poultry Research* 10:404–411, 2001.
26. SR Millio, SC Ricke. Synergistic reduction of Salmonella in a model raw chicken media using combined thermal and acidified organic salt intervention treatment. *Journal of Food Science* 75:M121–M125, 2010.
27. CYW Ang, YW Huang. Color changes of chicken leg patties due to end point temperature, packaging and refrigerated storage. *Journal of Food Science* 59:26–29, 1994.
28. CYW Ang, F Liuand, YW Huang. Catalatic activity as an indicator of end-point temperatures in breaded, heat processed chicken patties. *Journal of Muscle Foods* 7:291–302, 1996.
29. E Bogin, BA Israeli, I Klinger. Evaluation of heat treatment of turkey breast meat by biochemical methods. *Journal of Food Protection* 55:787–791, 1992.
30. H Chen, BP Marks, RY Murphy. Modeling coupled heat and mass transfer for convection cooking of chicken patties. *Journal of Food Engineering* 42:139–146, 1999.
31. M Ghita, V Stanescu, L Tudor, LI Ilie, M Gonciarov, R Popa. Research concerning the influence of processing temperatures for tenderness of chicken meat. *Lucrari Stintifice Medicina Veterinara* 43(2):216–219, 2010.

32. Z Qui, X Jianchun, S Baoguo, Z Fuiping. Studies on the preparation of chicken flavor from chicken fat oxidized by enzymatical catalysis and thermal reaction. *Journal of Chinese Institute of Food Science and Technology* 10(4):124–129, 2010.
33. FSIS. Performance standards for production of certain meat and poultry products. 9 CFR Parts 3318, 320, and 382. *Federal Register* 64(3):723–749, 1999.
34. CM Franz, A von Holy. Bacterial populations associated with pasteurized vacuum-packed Vienna sausages. *Food Microbiology* 13:165–174, 1996.
35. CM Franz, A von Holy. Thermotolerance of meat spoilage lactic acid bacteria and their inactivation in vacuum-packaged Vienna sausages. *International Journal of Food Microbiology* 29:59–73, 1996.
36. GH Dykes, LA Marshall, D Meissner, A von Holy. Acid treatment and pasteurization affect the shelf life and spoilage ecology of vacuum-packaged Vienna sausages. *Food Microbiology* 13:69–74, 1996.
37. RY Murphy, ER Johnson, LK Duncan, MD Davis, MG Johnson, JA Marcy. Thermal inactivation of *Salmonella* spp. and *Listeria innocua* in the chicken breast patties processed in a pilot scale air convection oven. *Journal of Food Science* 66:734–741, 2001c.
38. RY Murphy, ER Johnson, BP Marks, MG Johnson, JA Marcy. Thermal inactivation of *Salmonella* spp. and *Listeria innocua* in ground chicken breast patties processed in an air convection oven. *Poultry Science* 80:515–521, 2001.
39. SE Smith, JL Maurer, A Orta-Ramirez, ET Ryser, DM Smith. Thermal inactivation of Salmonella Typhimurium DT 104, and *Escherichia coli* O157:H7 in ground beef. *Journal of Food Protection* 66:1164–1168, 2001.
40. VK Juneja, BS Eblen, GM Ransom. Thermal inactivation of *Salmonella* spp. in chicken broth, beef, pork, turkey, and chicken: Determination of *D*- and Z-values. *Journal of Food Science* 66:146–152, 2001.
41. DJ Bolton, CM McMahon, AM Doherty, JJ Sheridan, DA McDowell, IS Blair, and D Harrington. Thermal inactivation of *Listeria monocytogenes* and *Yersinia enterocolitica* in minced beef under laboratory conditions of a sous-vide prepared mince and solid beef cooked in a commercial retort. *Journal of Applied Microbiology* 88:626–632, 2000.
42. K McCormick, IY Han, BW Sheldon, JC Acton, PL Dawson. *D* and Z values for *Listeria monocytogenes* and Salmonella *Typhimurium* in packaged low fat RTE turkey bologna subjected to a surface pasteurization treatment. *Poultry Science* 82:1337–1342, 2003.
43. K McCormick, IY Han, BW Sheldon, JC Acton, PL Dawson. In package pasteurization combined with nisin-impregnated films to inhibit Listeria monocytogenes and Salmonella Typhimurium in turkey bologna. *J. Food Science* 70(11):M52–57, 2005.
44. LA Hughes, BW Sheldon, PL Dawson. Decimal reduction values and *z*-values of *Escherichia coli* O157:H7 and *Campylobacter jejuni* on surfaces of bologna. *Poultry Science* 81(Suppl. 1):15, 2002.
45. RY Murphy, ME Berrang. Effect of steam and hot water post-process pasteurization on microbial and physical property measures of fully cooked vacuum packaged chicken breast strips. *Journal of Food Science* 67:2325–2329, 2002.
46. PM Muriana, W Quimby, CA Davidson, J Grooms. Post-package pasteurization of RTE deli meats by submersion heating for reduction of *Listeria monocytogenes. Journal of Food Protection* 65:963–969, 2002.
47. RY Murphy, LK Duncan, JA Marcy, ME Berrang, KH Driscoll. Effect of packaging film thickness on thermal inactivation of Salmonella and *Listeria innocua* in fully cooked chicken breast meat. *Journal of Food Science* 67:3435–3440, 2002.
48. RY Murphy, LK Duncan, KH Driscoll, JA Marcy. Lethality of *Salmonella* and *Listeria innocua* in fully cooked chicken breast meat products during post-cook in-package pasteurization. *Journal of Food Protection* 66:242–248, 2003.
49. N Gande, PM Muriana. Pre-package surface pasteurization of Ready-To-Eat meats with radiant heat oven for reduction *Listeria monocytogenes. Journal of Food Protection* 66:1623–1630, 2002.
50. S Mangalassary, PL Dawson, J Rieck, IY Han. Thickness and compositional effects on surface heating rate of bologna during in-package pasteurization. *Poultry Science* 83:1456–1461, 2004.
51. T Betley. Surface pasteurization of ready-to-eat meat products. *Workshop on Thermal Processing of Ready-To-Eat Meat Products* conducted at Ohio State University. March 4–6, 2003.
52. P Uhler. Thermally processed, commercially sterile products. Available at: http://www.fsis.usda.gov/OPPDE/rdad/FRPubs/97-013N/PUhler-canning. Accessed March 10, 2004.
53. P Turek, J Murcko, J Nagej. Effect of raw materials used on the quality of canned poultry products. *Hydina* 32:128–135, 1990.
54. BG Lyon, CE Lyon, JP Hudspeth. Texture profiles of canned boned chicken as affected by chilling-aging time. *Poultry Science* 3:1475–1478, 1994.

55. DW Thayer, G Boyd, CN Hubtanen. Effects of ionizing radiation and anaerobic refrigerated storage on indigenous microflora, Salmonella and *Clostridium botulinum* types A and B in vacuum-canned, mechanically deboned chicken meat. *Journal of Food Protection* 58:752–757, 1995.
56. R Rywotycki. Comparative estimation of cooling and bacteriological state of pasteurized canned poultry meat. *Chlodnictwo* 35:48–54, 2000.
57. D Sauvaget, P Auffret. Additives for use in preparation of poultry or rabbits, whole or in pieces, by canning (Patent). French Patent Application. FR 2,815,517 A1, 2002.
58. Z Zhang, L Weddig, S Economides. The effect of thermocouple and receptacle type on observed heating characteristics of conduction-heating foods packaged in small metal containers. *Journal of Food Process Engineering* 25:323–335, 2002.
59. PL Hayse, WW Marion. Eviscerated yield, component parts, and meat, skin and bone ratios in the chicken broiler. *Poultry Science* 52:718–722, 1973.
60. S Lesson, JD Summers. Production and carcass characteristics of the broiler chicken. *Poultry Science* 42:1207–1209, 1980.
61. JE Swan. Animal by-product processing. In: YH Hui, ed. *Encyclopedia of Food Science and Technology*. New York: John Wiley & Sons, Inc., 1992, pp. 39–43.
62. G Piette, M Hundt, L Jacques, M Lapointe. Effect of low extraction temperature on microbiological quality of rendered chicken fat recovered from skin. *Poultry Science* 79:1499–1502, 2000.
63. G Piette, M Hundt, L Jacques, M Lapointe. Influence of extraction temperature on amounts and quality of rendered chicken fat recovered from ground or homogenized skin. *Poultry Science* 80:496–500, 2001.
64. B Kinley, J Rieck, P Dawson, X Jiang. Analysis of *Salmonella* and enterococci isolated from rendered animal products. *Canadian Journal of Microbiology* 56(1):65–73, 2010.
65. KS Sheu, TC Chen. Yield and quality characteristics of edible broiler skin fat as obtained from five rendering methods. *Journal of Food Engineering* 55:263–269, 2002.
66. AEB. *Egg Products*, Publication. K-0001. Park Ridge, IL: American Egg Board, 1994.
67. WJ Stadelman, OJ Cotterill. *Egg Science and Technology*, 3rd edn. Westport, CT: AVI Publishing Co., 1986.
68. GJ Banwart. Control of microorganisms by destruction. In: GJ Banwart, eds. *Basic Food Microbiology*, abridged textbook edn. Westport, CT: AVI Publishing Co., 1979, pp. 642–701.
69. FE Cunningham. Egg-product pasteurization. In: WI Stadelman, OJ Cotterill, eds. *Egg Science and Technology*. Westport, CT: AVI Publishing Co., 1986, pp. 243–268.
70. AW Brant, GW Patterson, RW Walters. Batch pasteurization of liquid whole egg. I. Bacteriological and functional property evaluations. *Poultry Science* 47:878–884, 1968.
71. RP Elliott, BC Hobbs. Eggs and egg products. In: International Commission on Microbiological Specifications for Foods. *Microbial Ecology of Foods, Vol II: Food Commodities*. New York: Academic Press, 1980, pp. 521–566.
72. USDA. *Egg Pasteurization Manual*. ARS 74–48. Poultry Laboratory, Agricultural Research Service, U.S. Department of Agriculture, Albany, CA, 1969.
73. Anonymous. CDC recommends institutions use pasteurized eggs. *Food Chemistry News,* October 25, 1993.
74. JD Schuman, BW Sheldon. Thermal resistance of *Salmonella* spp. and *Listeria monocytogenes* in liquid egg yolk and egg white. *Journal of Food Protection* 60:634–638, 1997.
75. MS Palumbo, SM Beers, S Bhaduri, SA Palumbo. Thermal resistance of *Salmonella* spp. and *Listeria monocytogenes* in liquid egg yolk and egg yolk products. *Journal of Food Protection* 58:960–966, 1995.
76. HR Ball, M Hamid-Samimi, PM Foegeding, KR Swartzel. Functionality and microbial stability of ultrapasteurized, aseptically-packaged, refrigerated whole egg. *Journal of Food Science* 52:1212–1218, 1987.
77. LR York, LE Dawson. Shelf-life of pasteurized liquid whole egg. *Poultry Science* 52:1657–1658, 1973.
78. OJ Cotterill. Freezing egg products. In: WI Stadelman, OJ Cotterill, eds. *Egg Science and Technology*. Westport, CT: AVI Publishing Co., 1986, pp. 217–239.
79. R Jordan, BN Luginbill, LE Dawson, CJ Echterling. The effect of selected pretreatments upon the culinary qualities of eggs frozen and stored in a home-type freezer. *Food Research* 17:1–7, 1952.
80. K Ijichi, HH Palmer, H Lineweaver. Frozen whole egg for scrambling. *Journal of Food Science* 35:695–698, 1970.
81. J Giese. Ultrapasteurized liquid whole eggs earn 1994 IFT Food Technology Industrial Achievement Award. *Food Technology* 48(9):94–96, 1994.
82. RJ Ronk. Liquid eggs deviating from the standard of identity; temporary permit for market testing. *Federal Register* 54:1794–1795, 1989.

83. PM Foegeding, SB Leasor. Heat resistance and growth of *Listeria monocytogenes* in liquid whole egg. *Journal of Food Protection* 52:9–14, 1990.
84. PM Foegeding, NW Stanley. *Listeria monocytogenes* F5069 thermal death times in liquid whole egg. *Journal of Food Protection* 53:6–8, 1990.
85. HR Ball, M.G. Waldbaum, Co., Gaylord, MN, Personal communication, 1995.
86. WJ Stadelman. The preservation of quality in shell eggs. In: WJ Stadelman, OJ Cotterill, eds. *Egg Science and Technology*. Binghamton, NY: The Haworth Press, 1991, pp. 63–72.
87. EM Funk. Process for preserving eggs for edible consumption. U.S. Patent 2,423,233, 1947.
88. H Hou, RK Singh, PM Muriana, WJ Stadelman. Pasteurization of intact eggs. *Food Microbiology* 13:93–101, 1996.
89. B Schuman, BW Sheldon, JM Vandepopuliere, HR Ball, Jr. Immersion heat treatments for inactivation of *Salmonella enteritidis* with intact eggs. *Journal of Applied Microbiology* 83:438–444, 1997.
90. C Thomas, DE Swayne. Thermal inactivation of H5N2 high-pathogenicity avian influenza virus in dried egg white with 7.5% moisture. *Journal of Food Protection* 72:1997–2000, 2009.
91. WZ Chiu, HW Wei. Optimization of the thermal condition for processing hatchery waste eggs as meal for feed. *Poultry Science* 90:1080–1087, 2011. doi:10.3382/ps.2010–01127.
92. D Roberts. Sources of infection: *Food Lancet* 336:859–861, 1990.
93. JM Jay. Microbiological food safety. *Critical Reviews in Food Science and Nutrition* 31:177–190, 1992.
94. NH Bean, PM Griffin, JS Goulding, CB Ivey. Foodborne disease outbreaks, a 5-year summary, 1983–1987. *Journal of Food Protection* 53:711–728, 1990.
95. ECD Todd. Worldwide surveillance of foodborne disease: The need to improve. *Journal of Food Protection* 59:82–92, 1996.
96. C Thornton. *Salmonella* Enteritidis: The undefined threat. *Egg Industry* 96:14–22, 1991.
97. Anonymous. *Salmonella:* A chicken and egg phenomenon. *American Society of Microbiology News* 59:386–387, 1993.
98. ME St. Louis, DL Morse, ME Potter, TM DeMelfi, JJ Cuzewich, RV Tauxe, PA Blake. The emergence of Grade A eggs as a major source of *Salmonella* Enteritidis infections. *Journal of the American Medical Association* 259:2103–2107, 1988.
99. B Mishu, J Koehler, LA Lee, D Rodriguez, FH Brenner, P Blake, RV Tauxe. Outbreaks of *Salmonella Enteritidis* infections in the United States, 1985–1991. *Journal of Infectious Diseases* 169:547–552, 1994.
100. DB Shah, JC Bradshaw, JT Peeler. Thermal resistance of egg associated epidemic strains of *Salmonella* Enteritidis. *Journal of Food Science* 56:391–393, 1991.
101. TJ Humphrey. Contamination of eggs with potential human pathogens. In: RG Board, R Fuller, eds. *Microbiology of the Avian Egg*. London, United Kingdom: Chapman and Hall, 1994, pp. 93–112.
102. OJ Cotterill, J Glauert, GF Krause. Thermal destruction curves for *Salmonella oranienburg* in egg products. *Poultry Science* 52:568–577, 1973.
103. TJ Humphrey, PA Chapman, B Rowe, RJ Gilbert. A comparative study of the heat resistance of salmonellas in homogenized whole egg, egg yolk or albumen. *Epidemiology and Infection* 104:237–241, 1990.
104. RC Baker. Survival of *Salmonella enteritidis* on and in shelled eggs, liquid eggs, and cooked egg products. *Dairy, Food and Environmental Sanitarians* 10:273–275, 1990.
105. CB Michalski, RE Brackett, YC Hung, GOI Ezeike. Use of capillary tubes and plate heat exchanger to validate USDA pasteurization protocols for elimination of *Salmonella* Enteritidis from liquid egg products. *Journal of Food Protection* 62:112–117, 1999.
106. JA Garibaldi, K Ijichi, HG Bayne. Effect of pH and chelating agents on the heat resistance and viability of *Salmonella typhimurium* TM-1 and *Salmonella senftenberg* 775W in egg white. *Applied Microbiology* 18:318–322, 1969.
107. MS Palumbo, SM Beers, S Bhaduri, SA Palumbo. Thermal resistance of *Listeria monocytogenes* and *Salmonella* spp. in liquid egg white. *Journal of Food Protection* 59:1182–1186, 1996.
108. OJ Cotterill. Equivalent pasteurization temperatures to kill *Salmonella* in liquid egg white at various pH levels. *Poultry Science* 47:352–365, 1968.
109. JA Garibaldi, RP Straka, K Ijichi. Heat resistance of *Salmonella* in various egg products. *Applied Microbiology* 17:491–496, 1969.
110. JD Schuman, BW Sheldon, PM Foegeding. Thermal resistance of *Aeromonas hydrophila* in liquid whole egg. *Journal of Food Protection* 60:231–236, 1997.
111. HJ Beckers, PSS Soentoro, EHM Delfgou-van Asch. The occurrence of *Listeria monocytogenes* in soft cheeses and raw milk and its resistance to heat. *International Journal of Food Microbiology* 4:249–256, 1987.

112. CW Donnelly, EH Briggs, LS Donnelly. Comparison of heat resistance of *Listeria monocytogenes* in milk as determined by two methods. *Journal of Food Protection* 50:14–17, 1987.
113. JM Jay. High-temperature food preservation and characteristics of thermophilic microorganisms. In: JM Jay, ed. *Modern Food Microbiology*. New York: Van Nostrand Reinhold, 1992, pp. 335–351.
114. R Dabbah, WA Moats, VM Edwards. Survivor curves of selected *Salmonella Enteritidis* serotypes in liquid whole egg homogenates at 60°C. *Applied Microbiology* 50:1772–1776, 1971.
115. IJ Pflug, RG Holcomb. Principles of the thermal destruction of microorganisms. In: SS Block, ed. *Disinfection, Sterilization, and Preservation*. Philadelphia, PA: Lea and Febiger, 1983, pp. 751–803.
116. AM Saeed, CW Koons. Growth and heat resistance of *Salmonella Enteritidis* in refrigerated and abused eggs. *Journal of Food Protection* 56:927–931, 1993.
117. JJ Licciardello, JTR Nickerson, SA Goldblith. Destruction of salmonellae in hard-boiled egg. *American Journal of Public Health* 55:1622–1628, 1965.
118. RC Baker, S Hogarty, W Poon, DV Vadehra. Survival of *Salmonella typhimurium* and *Staphylococcus aureus* in eggs cooked by different methods. *Poultry Science* 62:1211–1216, 1983.
119. JM Farber, PI Peterkin. *Listeria monocytogenes,* a food-borne pathogen. *Microbiology Review* 55:476–511, 1991.
120. CAST. Foodborne pathogens: Risks and consequences. Task Force Report No. 122. Council for Agricultural Science and Technology, Ames, IA, 1994.
121. NA Cox, JS Bailey, ET Ryser. Incidence and behavior of *Listeria monocytogenes* in poultry and egg products. In: ET Ryser, EH Marth, eds. *Listeria, Listeriosis, and Food Safety*. New York: Marcel Dekker, 1999, pp. 565–595.
122. SB Leasor, PM Foegeding. *Listeria* species in commercially broken raw liquid whole egg. *Journal of Food Protection* 52:777–780, 1989.
123. J Moore, RH Madden. Detection and incidence of *Listeria* species in blended raw egg. *Journal of Food Protection* 56:652–654, 660, 1993.
124. JM Laird, FM Bartlett, RC McKellar. Survival of *Listeria monocytogenes* in egg washwater. *International Journal of Food Microbiology* 12:115–122, 1991.
125. BM Mackey, N Bratchell. The heat resistance of *Listeria monocytogenes*: A review. *Letters of Applied Microbiology* 9:89–94, 1989.
126. FM Bartlett, AE Hawke. Heat resistance of *Listeria monocytogenes* Scott A and HAL 957E1 in various liquid egg products. *Journal of Food Protection* 58:1211–1214, 1995.
127. USDA-AMS. Criteria for approval to use an expiration date on containers of liquid egg products packaged for extended shelf-life under refrigerated conditions (Revision 1, February 3, 1992). U.S. Department of Agriculture, Agricultural Marketing Service, Washington, DC, 1992.
128. H Urbach, CI Shabinski. Zur listeriose des Menschenal. *Zur Hygiene* 141:239–248, 1955.
129. RE Brackett, LR Beuchat. Survival of *Listeria monocytogenes* on the surface of egg shells and during frying of whole and scrambled eggs. *Journal of Food Protection* 55:862–865, 1992.
130. RL Buchanan, SA Palumbo. *Aeromonas hydrophila* and *Aeromonas sobria* as potential food poisoning species: A review. *Journal of Food Safety* 7:15–29, 1985.
131. DR Morgan, LW Wood. Is *Aeromonas* sp. a foodborne pathogen? Review of clinical data. *Journal of Food Safety* 9:59–72, 1988.
132. C Abeyta, SA Palumbo, GN Stelma. *Aeromonas hydrophila* group. In: YH Hui, JR Gorham, KD Murrell, DO Cliver, eds. *Foodborne Disease Handbook: Diseases Caused by Bacteria*, Vol. 1. New York: Marcel Dekker, 1994, pp. 35–59.
133. LR Beuchat. Behavior of *Aeromonas* species at refrigeration temperatures. *International Journal of Food Microbiology* 13:217–224, 1991.
134. SA Palumbo, F Maxino, AC Williams, RL Buchanan, DW Thayer. Starch-ampicillin agar for the quantitative detection of *Aeromonas hydrophila. Applied and Environmental Microbiology* 50:1027–1030, 1985.
135. AJC Ockrend, BE Rose, B Bennett. Incidence and toxigenicity of *Aeromonas* species in retail poultry, beef and pork. *Journal of Food Protection* 50:509–513, 1987.
136. HM Barnhart, OC Pancorbo, DW Dreesen, EB Shotts, Jr. Recovery of *Aeromonas hydrophila* from carcasses and processing water in a broiler processing operation. *Journal of Food Protection* 52:646–649, 1989.
137. RG Board. The properties and classification of the predominant bacteria in rotten eggs. *Journal of Applied Bacteriology* 28:437–453, 1965.
138. N Sashihara, H Mizutani, S Takayama, H Konuma, A Suzuki, C Imai. Bacterial contamination of liquid (frozen) whole egg. *Journal of the Food Hygienic Society of Japan* 20(2):127–136, 1979.
139. KA MacKenzie,VBD Skerman. Microbial spoilage in unpasteurized liquid whole egg. *Food Technology in Australia* 34:524–528, 1982.

140. AA Kraft, JC Ayres, CS Torrey, RH Salzer, CAN DaSilva. Coryneform bacteria in poultry, eggs and meat. *Journal of Applied Bacteriology* 29:161–166, 1966.
141. Y Nisbikawa, J Ogasawara, T Kimura. Heat and acid sensitivity of motile *Aeromonas:* A comparison with other food-poisoning bacteria. *International Journal of Food Microbiology* 18:271–278, 1993.
142. SA Palumbo, AC Williams, RL Buchanan, JC Phillips. Thermal resistance of *Aeromonas hydrophila. Journal of Food Protection* 50:761–764, 1987.
143. S Condón, ML Garcia, A Otero, FJ Sala. Effect of culture age, pre-incubation at low temperature and pH on the thermal resistance of *Aeromonas hydrophila. Journal of Applied Bacteriology* 72:322–326, 1992.
144. ML Stecchini, I Sarais, A Ciomo. Thermal inactivation of *Aeromonas hydrophila* as affected by sodium chloride and ascorbic acid. *Applied and Environmental Microbiology* 59:4166–4170, 1993.

10 Thermal Processing of Fishery Products

María Isabel Medina Méndez and
José Manuel Gallardo Abuín

CONTENTS

10.1 INTRODUCTION

Heating is the treatment mostly applied in fish canning industry and is also used in a wide variety of processes such as cooking, pasteurization, and sterilization of marine products. Each type of process has specific objectives and provokes different chemical and sensorial changes. Sterilization of canned fish is aimed to avoid the bacteriological and microbiological contamination, but the chemical effects affecting color and flavor can also be significant. Industrial treatments usually employed in foodstuffs can mainly modify proteins and lipids and induce interactions among components. Marine lipids with high amounts of polyunsaturated fatty acids (PUFA) are very prone to suffer modifications by heat. The degree, rate, and nature of these reactions depend on the type of product, quality of raw material, and the thermal processing employed.

The objective of this chapter is to review the effects that thermal processes have on quality of seafood. The information of the general principles governing the heat transfer, evaluation, and determination of heat penetration data are not emphasized here.

10.2 MAIN OPERATIONS IN FISH THERMAL PROCESSING

Precooking is relatively severe heat treatment before sterilization. Fish is usually cooked with steam at 100°C–102°C. The main objectives of fish precooking are

1. To prevent its loss during retorting due to partial dehydration of fish muscle
2. To remove lipids to avoid off-flavor
3. To coagulate the protein
4. To produce more palatable food
5. To improve the digestibility and to eliminate the meat from the shell in shellfish

The time of precooking depends on the fish species, size, the quality of the raw material and the temperature along the backbone. A consequence of thorough cooking is the destruction or reduction of vegetative cells of pathogens that may have been introduced in the process before cooking.

Pasteurization is a mild or moderate treatment, usually performed on fishery products placed into a hermetically sealed container. This process is used to extend the refrigerated shelf life of different seafood.

The purposes of pasteurization are

- To improve the safety of the product during an extended refrigerated shelf life, which involves eliminating the spores of *Clostridium botulinum* type E and nonproteolytic B and F that are the types of *C. botulinum* most commonly found in fish
- To eliminate or reduce other target pathogens (e.g., *Listeria monocytogenes, Vibrio vulnificus*)

For pasteurization processes that target nonproteolytic *C. botulinum*, generally a reduction of 6 orders of magnitude in the level of pathogens is suitable. This is called a "6D" process. Examples of properly pasteurized products are blue crabmeat pasteurized to a cumulative lethality of $F_{85°C}$ ($F_{185°F}$) = 31 min, z = 9°C (16°F); *surimi*-based products pasteurized at an internal temperature of 90°C (194°F) for at least 10 min. The control of later refrigeration is critical for the safety of these products.

Sous vide is a French term, which means under vacuum. The *sous vide* involves the cooking of fish inside a hermetically sealed vacuum package. The *sous-vide* technology or cooking under vacuum defined those foods that are cooked in stable containers and stored in refrigeration. Because these products are processed at low temperatures (65°C–95°C), the sensorial and nutritional characteristics are maximized in comparison with the sterilized products. The final product is not sterile, and its shelf life depends on the applied thermal treatment and the storage temperature. As the pasteurization does not produce the commercial sterility, the final fish product requires refrigerated storage. There are not many studies on the effect of the *sous-vide* pasteurization on quality of seafood and more research is necessary.

Thermal processing for canned fishery products: The heat treatment applied in sterilization of canned fish manufacture is aimed to eliminate pathogenic microorganisms, together with others that cause spoilage during the storage. The bacteria vary in their resistance to moist heat and some can form resistant spores. Among the spore-forming microorganisms, the *C. botulinum* of types A and B is the most heat resistant, constituting a potential health hazard. For low-acid products with pH greater than 4.5, such as canned fish, the anaerobic conditions are ideal for growth, and toxin production by *C. botulinum*. Therefore, its destruction is the critical parameter used in the heat

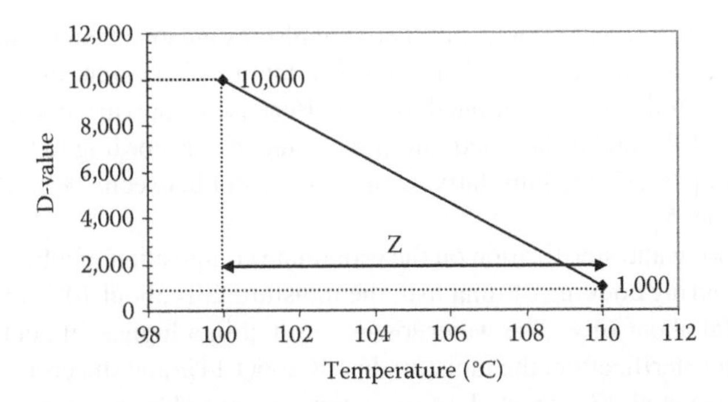

FIGURE 10.1 Curve of decimal reduction time, D value, and z value.

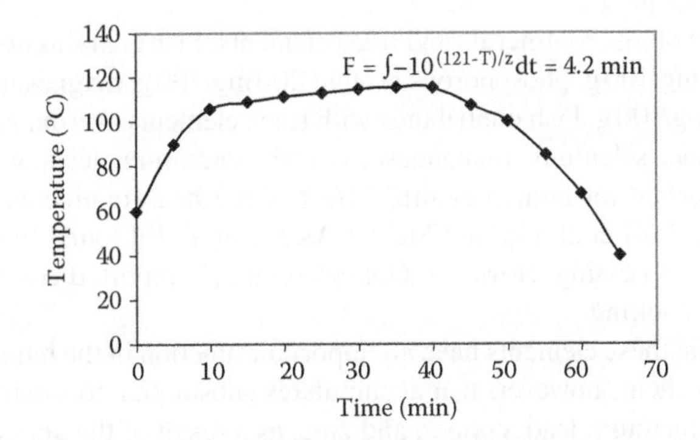

FIGURE 10.2 Curve time/temperature during sterilization of canned mussels.

processing. It has been established that the minimum thermal process sufficient for safety should achieve 12 decimal reductions in the population of *C. botulinum* spores; this is known as a 12D process. For such heat process the probability of *C. botulinum* spore survival is 10–12 or one in a million. This probability of survival is commercially acceptable and does not represent a significant health risk. When the thermal process is sufficient to fulfill the criteria of safety and prevention of nonpathogenic spoilage under normal conditions of storage, the product is "commercially sterile." Figures 10.1 and 10.2 show the calculation of z-values and F for canning of mussel performed at our pilot plant.

There are infinite numbers of time–temperature combinations that can produce commercial sterility but is important to establish a combination, which provides a final product that retains quality attributes. The effect of rotation to reduce the process time and to maximize the quality retention in the thermal processing of mackerel in oil has been reported by Ansar et al. [1]. The results obtained for these authors show that the process time at a given F_0 and temperature decreased with increasing rotation compared to the stationary retort sterilization for the same lethality. For this reason, the optimization of the thermal processes for nutrition and quality retention is necessary. In this chapter, we have summarized the main aspects affecting seafood components subjected to thermal stress.

10.3 PROXIMATE COMPOSITION OF FISH MUSCLE

The muscle of fish contains a series of chemical constituents such as proteins, lipids, minerals, and vitamins. Around 60 chemical elements are present in fish muscle, 75% is oxygen, 10% hydrogen, 9.5% carbon, 3% nitrogen, 1.5% calcium, and other traces of elements.

The main obstacle to establish the chemical composition in marine species is the high seasonal variability, especially in their lipid concentration. Intrinsic factors affecting these variations are genetic, morphological, and physiologic, but there are also exogenous factors such as those derived of life conditions, feeding, and geographical area. The fat or lipid content in fish is highly variable in fish species due to all the aforementioned aspects. Because of fat variations, and their influence on quality and flavor, fish can be classified into three categories according to their fat content: lean, with a fat content of up to 2%; medium fatty, with a fat content between 2% and 5%; and fatty, with a fat content exceeding 5%.

The effect of cooking and sterilization on the proximate composition of albacore was reported by Gallardo et al. [2]. During cooking of tuna fish, the moisture falls about 10%, the protein increased about 9%, and the fat about 61%. The water losses resulted in a higher fat content in cooked tuna than in raw fish. After sterilization, the moisture loss is about 14% and the protein content increased about 24%. Ruiz-Roso et al. [3] reported an important moisture loss after precooking of sardines. Other authors have also reported the effects of different cooking methods on the *proximate composition* of different species [4–7].

Fish flesh is a source of macrominerals and trace elements. Fish contains an important amount of potassium around 300 mg/100 g, phosphorous around 200 mg/100 g, magnesium about 25 mg/100 g, and calcium about 15 mg/100 g. Fish contributes with trace elements as iron, zinc, copper, and other elements such as iodine, selenium, manganese, copper, cadmium, lead, vanadium, cobalt, and boron, which are important for human health. Effect of the heat treatments on the elements was reported by Ackurt [8], Gall et al. [5], and Steiner-Asiedu et al. [9] found little or no effect on the minerals after thermal processing. However, Gokoglu et al. [7] reported considerable effect on the *mineral* content due to cooking.

There is no doubt that these elements have an important function in the human body, and that fish is a valuable source of them; however, fish accumulates substantial concentrations of heavy metals, such as cadmium, mercury, lead, copper, and zinc, as a result of the agricultural and industrial activities. As a consequence of their toxicity, the concentration of heavy metals in fish flesh can be essential. During cooking by baking and steam blanching, a reduction in the heavy metal concentrations related to the loss of water and uncoagulated proteins has been reported by Atta et al. [10]. The reduction depends on the medium of cooking, temperature, and time.

The *struvite* is a crystal of magnesium ammonium phosphate ($MgNH_4PO_4{\cdot}6H_2O$) that is found in canned products such as shrimp, crab, tuna, salmon, and other seafood. The struvite crystals are formed by the union of the mineral constituents to fish meat during sterilization. The formation of struvite in canned tuna is related to pH and is more likely to occur in fish having pH values > 6.0. The struvite is not harmful, and it dissolves in dilute acid such as vinegar or lemon.

10.4 VITAMINS

Fish is a source of *vitamins*, in which concentration depends on species, anatomical part of the fish, and age. Most of the vitamins are degraded to some extension during heating. The fat-soluble vitamins (vitamins A, D, E, and K) are more stable than the water-soluble vitamins (B1, B2, B6, niacin, folic acid, pantothenic acid, and vitamin C). During canning, *vitamin B1* (Thiamine) and *vitamin C* are destroyed substantially [11]. *Thiamine* is the most thermolabile B vitamins, and its losses during precooking and after conventional sterilization were about 42% and 90%, respectively. The losses after the sterilization at high vacuum flame are about 68%. *Niacin* and *riboflavin* showed higher heat stability than thiamine, and the retention in canned albacore was about 60%–70%, respectively.

Many kinetic data are available for thiamine in vegetable, milk, and meat, but there are few kinetic studies in fish. The kinetics of thermal degradation of thiamine was determined by Banga et al. [12] in canned tuna using an unsteady state experimental procedure. The values calculated by these authors for D and z were 90.5 min (at 121.1°C) and 31.4°C, respectively (Table 10.1). The D found is in agreement with that reported by Suparno et al. [13] for the destruction of thiamine in

TABLE 10.1
Kinetic Parameters for Thiamine and Surface L Value Thermal Degradation

Parameter	Thiamine	Surface L Value
D_{ref} (min)	90.5 ± 2.1	1,468 ± 112
z (°C)	31.4 ± 0.6	44.0 ± 3.6
$s(y)^2$	$3{,}683 \times 10^{-4}$	$12{,}417 \times 10^{-4}$
Predicted versus observed correlation	y = 0.968x + 1.71	y = 0.962x + 2.7
R^2	0.988	0.908

Source: From Springer Science+Business Media: *Zeitschrift für Lebensmittel-Untersuchung und-Forschung*, Kinetics of thermal degradation of thiamine and surface colour in canned tuna, 197, 1993, 127–131, Banga, J.R. et al. With permission.

L value (Luminosity, Universal Chromatic Scale), Dref, z, kinetic parameters; $s(y)^2$, variance of the regression; R^2, correlation coefficient between observed and predicted values.

trout, but the z value was higher. The kinetic model permits the simulation and optimization of the process and provides the maximum retention of nutrient.

10.5 SENSORY CHARACTERISTICS

The effect of thermal processing on the *sensory* quality in marine products is difficult to predict because of intra- and interspecific variability of fish species and factors such as appearance, odor, color, flavor, and texture.

The *color* of canned tuna is an important quality attribute. *Myoglobin* is a monomeric globular hemoprotein with a molecular weight of 18 kDa, and it is localized in the red muscle fiber. Three forms of myoglobin contribute to the color of muscle: the reduced or deoxymyoglobin, which is purple-red; the red oxymyoglobin associated with a high-quality product; and the metmyoglobin, which is brown and is responsible for discoloration of meat during storage. Cooking favors the formation of metmyoglobin because the oxidation is accelerated by the action of temperature. In general, fish myoglobins, especially in tuna fish, are sensitive to autooxidation, and this autooxidation increases during thermal processing.

The thermal denaturation midpoint of tuna myoglobin is 85°C. This denaturation can produce a color change denominated "*greening*," associated with cooking of albacore, yellowfin, skipjacks, and other tuna species. The formation of the green pigment was studied in model systems by Grosjean et al. [14]. This pigment was found in a system composed of tuna myoglobin, *trimethylamine oxide* (TMAO), and cysteine heated in phosphate buffer. The presence of a cysteine side chain in the fish protein allows the reaction of the sulfhydryl group with an oxidant as TMAO and the free cysteine. Some works have reported the relationships between trimethylamine (TMA) and TMAO and the myoglobin concentration of tuna muscle for predicting tuna greening [15–18].

Other important discoloration is the *bluing*, which has been detected in king crab, blue crab, and Dungeness crab [19–21]. The blue discoloration is generally associated with the heat processing, and an important review was reported by Boon [22]. The blueing occurs in pasteurized and canned crabmeat. It is developed during the heat treatment, and its effect increases during storage. It has been related to the presence of copper and iron. A significant correlation between copper concentration and discoloration intensity was found in Dungeness canned crab by Elliot and Harvey [21] and Babbit et al. [23].

The oxidation of phenolic compounds, such as tyrosine, to produce *melanin* is the other cause of discoloration [24]. Tyrosinases and phenooxidases present in the live crab initiate the oxidation of phenols to melanins. The posterior nonenzymatically oxidation and polymerization of these intermediate phenolics in canned crabmeat may proceed to form colored chromophores, mainly in the presence of metals. According to Babbit et al. [23] the holding time of live crab before cooking and the times of cooking have a significant effect on the extent of the blueing of crab. The cooking at 100°C for 30 min of fresh crab was ideal for inactivating the polyphenoloxidases and preventing blueing but not the cooking for 15 min. However, cooking at 100°C for 30 min does not prevent the blueing when the raw material is not fresh.

Sreenath et al. [25] describe the effect of three different retort temperatures (115°C, 121.1°C, and 130°C) with a common F_0 value of 8 min on the sensory properties of mackerel canned in brine. The results of this study show that thermal processing at 130°C yielded a product that was superior with respect to instrumental color and texture to those processed at 115°C and 121.1°C and was rated high by the sensory panelists. According to Lund [26], the changes in the *sensory quality* by the thermal processing are closely related to chemical reactions, which usually have a temperature dependence that can be described by a z value of 33°C. The temperature dependence of sensory quality changes during the thermal processing was determined by Ohlsson [27] in fish, meat, and vegetable products. The experimentally determined z values ranged from 13°C to 34°C, with an average value of 23°C. The average of z values obtained by Ohlsson is in fair agreement with the z value for the thiamine degradation (26.7°C).

Kinetics of thermal degradation of surface color in canned tuna was reported by Banga et al. [12]. The z value found by these authors was $z=44$°C that is much higher than that found by Ohlsson [27] in fish ($z=25$°C). The minimal sugar content of tuna can explain this difference.

10.6 PROTEINS

The muscle of fish mainly consists of *myofibrillar* proteins, 65%–75% of the total muscle proteins. They are contractile proteins as *myosin* and *actin* and regulatory proteins as tropomyosin and troponin, and other minor proteins. Other proteins of fish muscle are sarcoplasmic proteins or water-soluble proteins, which account for 20%–30% of the total proteins. They are mostly enzymes and are involved in postmortem biochemical changes. Fish flesh also contains connective tissue proteins, about 3%, which are related with the texture of fish fillets.

According to Johnston [28], the resulting variation in muscle cellularity and the associated changes in the connective tissue matrix are thought to be important determinants of fish flesh characteristics. Muscle cellularity is the major determinant of texture [29,30]. It has been suggested that variations in the quality of fillet along fish are due to variations in the size distributions of the muscle fibers and to the collagen present. Differences in the texture of fish muscle have been also related to the lipid and water content [31]. The lipid and water content reduced the structural factors of the muscle, lowering its mechanical strength. Kanoh et al. [32] found that the dark muscle of yellowfin tuna, whose fiber diameter is inferior than the ordinary muscle, has firmer texture than the ordinary muscle, and proposed that the fiber size is important for the tenderness of fish meat. The differences in toughness changes during thermal processing have a high correlation with fiber sizes and ultrastructures of the two types of fish muscle.

The contribution of muscle fibers to firmness was studied in cooked muscle of five fish species using optical and scanning electron micrography by Hatae et al. [29]. These authors propose that when muscle is cooked, the sarcoplasmic protein is released from the contracting muscle fiber and is coagulated in the interstitial spaces.

In general during fish heating, *sarcoplasmic* and *myofibrillar proteins* are coagulated and denaturated. The extent of these changes depends on the temperature and time and affects the yields and final quality of the fishery product. In heat denaturation of flesh proteins, H bonds, which are involved in the secondary and tertiary structures of proteins, are broken resulting in an unfolding of the native configuration. During the early stages of cooking (30°C–50°C), there is an unfolding of peptide chains, partial denaturation of sarcoplasmic proteins, resulting in toughness and decreased water-holding capacity. When the temperature increases (50°C–70°C), stable cross-linkages are formed with denaturation and coagulation of proteins.

The thermal processing is important to reduce heat labile pathogens and to obtain a palatable product; by this reason, it is necessary to establish accurately the conditions of fish cooking. Several studies have been carried out to determine the *end-point* temperature in meat products, but the studies on marine products are limited. For the control of heating temperature in fishery products, two methods based on the thermal denaturation of fish proteins have been proposed. Doesburg and Papendorf [33] used the *coagulation test* to determine the degree of the heating of muscle and Chan et al. [34] measured the amounts of sarcoplasmic and myofibrillar proteins extractable from fish muscle heated to temperatures of up to 100°C. Isoelectric focusing (IEF) of sarcoplasmic proteins was used for determining temperature in cooked fish and smoked fish because of the presence of parvalbumins, which are heat-stable proteins [35]. The changes in the protein patterns of parvalbumins may reflect the degree of the heating. The suitability of these methods was investigated in fish-heated extracts, heated minced fish muscle, and smoked herring and mackerel by Rehbein [36]. According to this author, herring heated at 55°C or 60°C could be differentiated by IEF from herring that was smoked at 65°C–70°C. Different protein patterns were also obtained by IEF for mackerel smoked at 55°C, 60°C, 65°C, or 70°C. From these results, it can be inferred that IEF or coagulation test can be used for determining the temperatures of fish heating of up to 65°C.

The coagulation method was used for determining the *end-point* temperature of heated blue marlin meat in the range of 1°C–67°C [37]. From the analysis of proteins by sodium dodecyl-sulfate-polyacrylamide gel electrophoresis (SDS-PAGE) and by studying the enzymatic activity, lactate dehydrogenase was found to be responsible for coagulation at 67°C. The assessing of the end-point temperature of heated fish and shellfish using the coagulation method was also reported by Uddin et al. [38]. The coagulation test could determine the end-point temperature of shellfish meats between 60°C and 67°C, and the results were confirmed by SDS-PAGE and enzyme activity determination. A thermostable protein with a molecular weight of 35 kDa was detected in heated shellfish meats up to 108°C. This protein was found to be tropomyosin subunit in scallop adductor muscle. According to the results obtained from these authors, the tropomyosin could be used as an indicator of the end-point temperature. However, the applicability of the enzyme activities can have great difficulties because they depend on the physiological condition of the fish.

Differential scanning calorimetry (DSC) can also be used to investigate the thermal stability of proteins and to estimate the cooking temperature of the seafood products. The thermal stability of fish myosin of different species in the range of 20°C–80°C by DSC was reported by Ogawa et al. [39]. The denaturation process was different depending on the species. The DSC curves of myosin of sardine, stone flounder, sea bream, and carp had two peaks; trout, bigeye tuna, and yellow tail showed three peaks; and horse mackerel only one peak.

The thermal denaturation of hake myofibrillar proteins using the DSC was studied by Beas et al. [40]. Two endothermic transitions with T_{max} values at 46.5°C and 75.3°C and a shoulder at 51°C were obtained by these authors. Martens and Vold [41] reported similar transitions for cod muscle. The DSC of the exudative sarcoplasmic fraction of the whole muscle showed three peaks at 45.2°C, 59°C, and 75.5°C and contributed to denaturation peaks.

The protein denaturation during fish precooking implies texture and binding changes, which allow for a better differentiation between red and white muscle and an easier separation of muscle from bone. The most labile proteins are the sarcoplasmic proteins and myosin, and their denaturation is sufficient to ensure textural changes needed. The effects of the thermal protein denaturation and moisture loss in skipjack tuna during steam cooking were reported by Bell et al. [42]. The DSC thermogram showed that the first peak at 52°C corresponds to myosin denaturation, the second peak at 59°C corresponds to collagen, and the third peak at 68°C to actin. During the precooking of

TABLE 10.2
Effect of Heating on the Content[a] of –SH Groups and S–S Bonds in Pollock and Mackerel Proteins

	Temperature	Pollock				Mackerel			
	(°C)	N	–SH+S–S/2	–SH	S–S/2	N	–SH+S–S/2	–SH	S–S/2
Untreated		12	11.5 ± 0.4	7.0 ± 0.5	4.4. ± 0.3	6	10.8 ± 0.3	7.2 ± 0.3	3.7 ±0.4
heated	40	3	11.3 ± 0.4	7.1 ± 0.3	4.3 ± 0.4				
for	50	3	11.3 ± 0.7	7.5 ± 0.1	3.9 ± 0.6	3	10.9 ± 0.4	8.1 ± 0.3	2.8 ± 0.2
20 min	60	3	11.0 ± 0.2	6.6 ± 0.4	4.5 ± 0.5	3	11.4 ± 0.4	8.1 ± 0.1	3.3 ± 0.4
	70	3	11.6 ± 0.5	6.1 ± 0.2	5.5 ± 0.4	3	10.8 ± 0.5	7.2 ± 0.3	3.5 ± 0.7
	80	3	11.6 ± 0.5	6.3 ± 0.1	5.3 ± 0.5	3	10.8 ± 0.2	6.6 ± 0.3	4.1 ± 0.5
	95	3	11.6 ± 0.6	5.2 ± 0.5	6.5 ± 0.3	4	11.2 ± 0.4	6.1 ± 0.5	5.2 ± 0.4
	115	4	10.4 ± 0.3	3.9 ± 0.5	6.6 ± 0.2	2	9.9	9.9	6.3

Source: From Opstevedt, J. et al., *J. Agric. Food Chem.*, 32, 929. Copyright 1984 American Chemical Society. With permission.

N, number of determinations.

[a] SH/16 g of N; average ± standard deviation.

skipjack, the structural muscle proteins decrease in dimension upon reaching their thermal denaturation temperatures and the shrinkage of the muscle fibrils and tissue occurs.

Sulfhydryl (SH) groups and disulfide bonds are important in maintaining structure and functions of native proteins. At temperatures above 90°C, the oxidation of SH occurs and leads to the formation of covalent S–S bonds. The effect of heating on the –SH groups and S–S bonds was studied by Opstevedt et al. [43] in Alaska Pollock and Pacific mackerel (Table 10.2). The heating for 20 min at temperatures ranging from 40°C to 115°C showed a linear decrease in the –SH groups and an increase in the content of S–S disulfide bonds at temperatures from 50°C to 115°C.

The heating at 115°C causes a loss in cystine plus cysteine [44,45]. The loss of cysteine plus cystine produces H_2S. The release of H_2S during seafood sterilization is a problem because the corrosion of tins and discoloration of cans, and it provokes an offensive off-flavor when the can is opened. The production of H_2S indicates the destruction of essential sulfur-containing amino acids.

The effect of heat processing on bigeye tuna and halibut at 115°C and 124°C varying in F_0 values of 8, 21, and 32 min was reported by Tanaka and Kimura [46]. At 115°C, the electrophoretic patterns of proteins were similar in both fish and the bands of myosin heavy chain disappeared at any value of F_0. The actin and tropomyosin bands decreased when the F_0 value increased. At 124°C, the protein degradation was smaller as compared with that at 115°, because the thermal processing time necessary to attain the equal lethality was far shorter than that at 115°C.

Taking advantage of denaturation, heat induces capability of fish muscle proteins for forming gels [47]. Thermal gelation of fish protein, for example, in making kamaboko was due to myosin; light meromyosins combined with each other, after which heavy meromyosins combined to form the gel. Actin was degraded on heating and the fragments aggregated to form a curd-like substance, so that actin had no gel-forming ability [48]. Heating and cooling cycles enhance the setting process for gelation of fish paste and improve textural quality [49].

10.7 LIPIDS

Fish *lipids* are highly concentrated in PUFA, especially eicosapentaenoic acid (20:5 n-3, EPA) and docosahexaenoic acid (22:6 n-3, DHA). This composition leads to a high nutritional value of fish products associated with the well-known beneficial effects of n-3 PUFA on human health.

The n-3 PUFA decrease the serum cholesterol and triacylglycerol levels and prevent the coronary heart disease [50]. *Hydrolysis* and main *oxidation* are the major alterations of fish lipids during thermal processing. Lipid degradation is often focused on the reactivity of PUFA since they can produce a significant number of polar compounds including volatile derivatives by action of heat via the process of lipid oxidation [51]. In this way, the highly unsaturated fatty acid composition renders fish flesh extremely susceptible to oxidation and rapid degradation during processes that involve thermal treatments. In addition, oxidation of n-3 fatty acids and the effect of oxidized lipids on proteins and amino acids in fish muscle cause loss of nutritional value [52].

Cooking and sterilization provokes *lipolysis* and *oxidation* during fish canning, but there are also a significant effect of filling medium. Thermal treatments associated with fish cooking and sterilization decrease the initial lipid content of flesh mainly by loss of *triacylglycerols* [3,53]. *Phospholipids* are also affected, being choline phospholipids affected most during cooking, in particular lyso forms. Lipolytic enzymes in fish are mostly active at ordinary cold storage temperatures. Heat treatment resulting in inactivation of enzymes is of considerable practical importance. A significant increase of the amount of free fatty acids as consequence of triacylglycerols hydrolysis occurred in fish muscle subjected to thermal processing [53]. The application of ^{13}C-Nuclear Magnetic Resonance (^{13}C-NMR) Spectroscopy has shown a preferential stereospecificity of lipid *hydrolysis* in the internal position of the glycerol moiety after thermal stress, resulting in a larger increase of free DHA [54]. The ratio of 1,3-diacylglyceride and 1,2-diacylglyceride quantities confirms this finding. These results contrast with the enzymatic hydrolysis during fish frozen storage, occurring preferentially in the sn-1 and sn-3 position of triacylglycerols, with a consequent cleavage of saturated and monounsaturated fatty acids. After thermal processing, an increase in free PUFA can be of special relevance for suffering oxidation. The most susceptible compounds at thermal stress are *plasmalogens*, 1-O-alk-1-enyl-glycerophospholipids [55]. At sterilization temperatures (115°C), the weak ether bonds of these compounds are broken, and as a result, the amount of plasmalogens in canned brine and canned oily tuna decrease by 50%. Plasmalogens have been more susceptible to damage during heat processing than their corresponding acyl derivatives. This behavior may be explained by the aldehydogenic chemical structure easily open to acidic attacks.

Overall quality and flavor of the thermally processed product can be highly influenced by lipid oxidation rending to *rancidity*. Initial raw composition, process conditions, and packing substrate have a special relevance. Canning process provokes increases of primary, secondary, and tertiary lipid oxidation products in canned tuna [56]. The highest thiobarbituric acid index (TBA-i) values were found in tuna canned muscle using brine as dipping medium, thus indicating lower protection in the muscle kept in a highly aqueous environment than using oily packing media.

^{13}C-NMR spectroscopy has been applied to elucidate the mechanism of lipid oxidation occurring during thermal treatment of fish [57]. Effects of temperature and time of processing have been studied by means of a model system of lipids extracted from salmon (*Salmo salar*) muscle to simulate industrial conditions of canning. Unsaturated fatty acids located at sn-2 position of the glycerol moiety are most prone to suffering from oxidative damages. Regarding to the mechanism of the reaction, the results inferred from olefinic and methylenic resonances indicated a higher susceptibility of the closest allylic sites to the carbonyl group, followed by those placed near to the methyl terminal group. Unsaturations located in the middle of the carbon chain did not show notorious damages. The glyceryl region provided an unusual resonance at 53.4 ppm that could be assigned to a hydroxylic compound formed during oxidation.

Oxidation in fish during thermal processing has also been studied by determining volatile production with a static headspace gas chromatographic system [58]. The major *volatiles* formed included acetaldehyde, propanal, heptane, 2-ethylfuran, pentanal, and hexanal. The formation of *2-ethylfuran* had the highest independent contribution for the prediction of oxidative stability of fish muscle at 4 days at 40°C and 150 min at 100°C. The formation of 2-ethylfuran in oxidized n-3 PUFA in fish can be explained by the 14-hydroperoxide of eicosapentaenoate (20:5 n-3), and the 16-hydroperoxide of docosahexaenoate (22:6 n-3), which can undergo b-cleavage to produce a conjugated diene radical, which can react with oxygen to produce a vinyl hydroperoxide. Cleavage of this vinyl hydroperoxide by loss of a hydroxyl radical forms an alkoxyl radical that undergoes cyclization to produce 2-ethylfuran.

Associated with *hydrolysis* and *oxidation*, overprocessed canned samples such as those subjected to sterilization treatments involving high temperatures and long times lead a significant decrease of n-3 fatty acids (110°C and 130°C during 120 min). NMR techniques have been proposed as useful tools to select the optimal conditions of thermal fish processes based on minimum lipid hydrolysis and oxidation and maximum n-3 PUFA retention [59]. Dahl and Malcata [60] have employed a thermal oxidation model of sardine oil to predict oxidation stability of PUFA. PUFA with less than four double bonds were relatively stable to oxidation for up to 10 h at 50°C–70°C, and the qualitatively richest pattern of volatiles was obtained when the reaction was performed at 80°C.

Canned fish products in oily media showed the dilution of the natural lipids of fish muscle by triacylglycerols from packing oils [61]. As a result, a significant increase of the fatty acids present in vegetable oils used as dipping media, such as oleic and linoleic acids, is found in the final canned flesh. There is also an exchange of n-3 PUFA from fish flesh to filling oil, resulting in the oil enriched with n-3 PUFA. Regarding to hydrolysis, the extent and mechanism of the reaction seem to be independent on the filling medium used. Oxidation can be minimized by employing packing media containing natural antioxidants such as extra virgin olive oil rich in polyphenols or certain vegetable oils rich in tocopherol isomers [56,62]. In tuna oxidized at 40°C and 100°C, 400 ppm of the extra virgin olive oil polyphenols was an effective antioxidant as compared with 100 ppm of a 1:1 mixture of the synthetic antioxidants butylated hydroxytoluene and butylated hydroxyanisole. The extra virgin olive oil polyphenols were effective antioxidants when added to heated tuna muscle in the presence of either brine or refined olive oil. The extra virgin olive oil polyphenols had higher antioxidant activity in the brine samples than in the refined olive oil. Their hydrophobic character and their low solubility into an aqueous media may explain the higher antioxidant activity of extra virgin olive oil polyphenols in tuna packed in brine since they are totally adsorbed into fish flesh.

10.7.1 Volatiles

The *flavor* profiles of cooked fish differ in the aroma notes and in the intensity of these notes. 1-Octen-3-one, (Z)-1,5-octadien-3-one, (E)-2-nonenal and (E;Z)-2,6-nonadienal play an important role in fresh fish-like odors due to their low odor threshold values [63]. During boiling and cooking, important thermal degradation of volatiles can occur in addition to the detection of *off-flavor* substances resulting from longer periods of frozen storage. Volatiles of canned tuna, canned salmon, dried and smoked fish, and boiled fish have been widely described [64–67].

10.7.2 Effects of Heating on Formation of Cholesterol Oxides

Cholesterol is stable under normal conditions in air; however, it is oxidized by thermal stress and by exposure to light, to produce cholesterol oxides. *Cholesterol oxides* including 7β-hydroxycholesterol, 7-ketocholesterol, α-epoxide, β-epoxide, cholestane triol, and 25-hydroxycholesterol were identified and quantified. These oxides are atherogenic, carcinogenic, and mutagenic compounds [68–71]. The oxidation of cholesterol during heating in an air oven at high temperature was studied by Osada et al. [72]; the cholesterol was stable for 24 h at 100°C but was unstable at temperatures up to 120°C. The *cholesterol oxides* in fish products were determined by gas–liquid chromatography and mass spectrometry by Ohshima et al. [73]. These authors suggested that cholesterol oxidation in fish products proceeds in conjunction with oxidation of PUFA during the storage in air. The radicals produced from PUFA oxidation act as accelerator of the oxidation of cholesterol. The effects of grilling on formation of cholesterol oxides in seafood products were reported by Ohshima et al. [74].

10.8 LOW-MOLECULAR-WEIGHT COMPONENTS

10.8.1 Sugars

In general, the content of free *sugars* and sugar phosphates in fish muscle and shellfish is low. The phosphorylated sugars are unstable during cooking [75]. Free ribose can be involved in the browning of canned fish because of the reaction of carbonyl groups such as reducing sugars or lipid oxidation products with amino compound as lysine. These reactions are significant in reducing the nutritive value, mainly when the fish is heated severely. Free ribose can be eliminated during cooking and the possible browning of the canned fish can be avoided.

10.8.2 Free Amino Acids

Free amino acids and *creatine* are the major components of the total nonprotein nitrogen (NPN) compounds in teleost muscle, and they are recognized as contributors to the flavor [76]. Free amino acids of fish are associated with autolysis and bacterial actions in the earlier stages of alteration [77]. Mackie and Ritchie [78] demonstrated that there was considerable variation in the concentration of amino acids along the length of fish, and variations were also found from the head to tail portions of cod fillets.

Free amino acids can be significantly affected by heating. Considerable loss of histidine, about 86%, after thermal processing of herring at 116°C for periods of up to 5 h has been reported [75].

The effect of the cooking and sterilization on the free amino acids content in albacore was reported by Pérez-Martin et al. [79]. After steam cooking, the losses were not significant. According to these authors, during the sterilization under standardized conditions (F = 6), the changes were not significant (Table 10.3); however, when F = 13, significant differences were found, confirming the negative effect of the sterilization time. These authors suggest that the analysis of the free amino acids would be useful to know the extent of thermal processing.

10.8.3 Amines

10.8.3.1 Total Volatile Bases

The *total volatile basic amines* (TVB-N) include *TMA*, *dimethylamine* (DMA), ammonia, and other volatile nitrogenous compounds associated with the spoilage of marine products. *TMAO* is present in a large number of fish and shellfish and it is generally accepted that TMAO is the main source of TMA and DMA. Endogenous enzymes in fish and exogenous enzymes produced by bacteria during frozen or ice preservation are responsible for the reduction of TMAO to TMA and DMA. TMAO can be reduced by cysteine in the presence of hemoglobin or iron as catalyst. TMA is associated with the fishy odor; however, this fishy odor is produced when TMA reacts with lipids. In gadoid frozen species, TMAO is reduced to DMA and formaldehyde (FA), which induces textural toughening of fish flesh.

The rate of thermal decomposition of TMAO varied with the fish species and was low in white fish and high in red fleshed fish [80]. This difference was due to the different concentration of hemoproteins, free amino acids, and –SH basis. The thermal decomposition of TMAO takes place at 55°C–60°C in dark muscle and above 80°C in ordinary muscle.

Precooking produces an increase in TMA, which is proportional to cooking time. Large concentrations of ammonia, DMA, and TMA were found in herring subjected to overprocessed conditions: heated for up to 5 h at 120°C [81].

Tokunaga [82] studied the relationship between the freshness of raw albacore and the quality of canned albacore. The ratio of TMAO remaining undecomposed to the total amines content (TMAO + TMA + DMA) might be an index of freshness of the raw albacore. In the thermal decomposition of TMAO, TMA and DMA are produced in the ratio of 2:1, respectively. This

TABLE 10.3
Effect of Sterilization Time on the Free Amino Acid Content in Albacore Muscle (mg/100 Dry Weight)

			Canned			
Amino Acid	Fresh	SD	115°C, 60 min	SD	115°C, 100 min	SD
Asp	36.9	4.7	32.7	4.2	32.1	4.4
Glu	55.7	5.2	49.3	4.8	37.3	5.1
Ser	34.5	5.8	30.5	5.0	31.0	5.4
His	2320.0	39.7	2050.1	37.6	1461.6	40.5
Gly	56.7	6.1	49.9	5.6	27.8	5.1
Thr	71.5	9.7	63.7	10.1	50.0	11.2
Arg	273.0	21.7	240.5	18.9	163.8	22.3
Tau	199.1	11.1	173.2	9.6	122.4	9.5
Tyr	49.3	3.6	42.4	3.9	31.0	3.4
Ala	80.6	3.2	70.2	3.8	58.9	4.0
Trp	97.1	4.1	83.6	4.4	76.7	3.8
Met	62.5	3.3	53.1	3.0	41.2	3.5
Val	26.6	3.1	23.6	2.9	16.0	2.8
Phe	150.5	3.9	134.0	3.3	117.5	3.0
Ile	48.1	5.1	40.9	4.3	38.4	3.6
Leu	68.8	5.7	60.2	5.1	48.1	4.9
Lys	109.3	5.6	95.2	4.8	81.1	5.1
Total	3740.2 a		3293.1 b		2434.9 c	

Source: From Springer Science+Business Media: *Zeitschrift für Lebensmittel-Untersuchung und- Forschung*, Changes in free amino acids content in albacore (*Thunnus alalunga*) muscle during thermal processing, 187, 1988, 432–435, Pérez-Martin, R.I. et al. With permission.

Values followed by the same letter are not significantly different (standard deviation [SD], four samples). Albacore caught in 1985.

author [82] suggested the use of the ratio of DMA over TMA (DMA-N/TMA-N) × 100 to evaluate the quality of canned albacore. This ratio was approximately 50% if the fish employed for canning was highly fresh, 40% when the fish was stored for 70 h before canning, and 30% or lower when the albacore was stored about for 116 h before canning. However, TMAO concentration in fish varies widely with season, size, and age of fish, and TMA is formed during thermal processing. Therefore, it is difficult to use TMAO as a reliable measure of the freshness of the raw fish that is used for canning.

Changes in volatile bases and TMAO during canning of albacore were reported by Gallardo et al. [83] (Table 10.4). The *TVB-N* levels increased during cooking and sterilization. The effect of the process, in the order of increasing TVB-N, was R < C < E118 < E115 < E110. The results suggest that, under given conditions of thermal processing, the levels of TVB content may be of practical value in assessing the initial quality of albacore. If the thermal processing conditions are not known, high levels of TVB-N in canned product indicate either a poor quality of raw fish or overprocessing.

Determinations of the TVB-N in several fish species, prior to its cooking and after sterilization were carried out by Yeannes et al. [84]. Increases in the contents of the bases were confirmed. Using different specimens of bonito with different degrees of spoilage, the contents of the volatile bases in the cooked bonito were significant correlated with the values of the raw material when the

TABLE 10.4
Changes in TVB, Individual Bases, and TMAO during Processing of Canned Tuna in 120 mL Rectangular Cans

			Canned at		
Compounds	**Raw Fish**	**Precooked Fish**	**110°C 90 min**	**115°C 55 min**	**118°C 40 min**
TVB	280	340	450	430	410
	(20)	(14)	(21)	(19)	(22)
Ammonia	250	310	420	410	380
	(34)	(16)	(25)	(17)	(24)
DMA	2.4	5.2	7.0	7.4	7.7
	(0.6)	(1.9)	(1.9)	(1.7)	(1.5)
TMA	4.4	13	16	17	19
	(1.3)	(2.2)	(2.9)	(3.3)	(4.1)
TMAO	19	8.4	4.2	3.3	1.9
	(6.3)	(4.9)	(2.5)	(1.9)	(2.1)

Source: From Gallardo, J.M. et al., *Int. J. Food Sci. Technol.*, 25, 78, 1990. With permission.

Values are concentration (mg N kg^{-1} fish/canned contents) (standard deviation [SD]).

processing conditions are held constant. A correlation between the bases content of sterilized bonito and the cooked product was detected too.

10.8.3.2 Biogenic Amines

Biogenic amines are aliphatic, alicyclic, or heterocyclic organic bases of low molecular weight, which have been found in different foods, such as cheese, beer, wine, fermented meat, and marine products. They are formed by action of bacterial enzymes on free amino acids and the most important are histamine, *putrescine*, *cadaverine*, spermidine, spermine, tyramine, agmatine, tryptamine, and 2-phenylethylamine, which are produced by decarboxylation of free amino acids.

Histamine, produced by decarboxylation of histidine during microbial decomposition of scombroid fish, has received most of the attention. It is the main compound responsible for the intoxication named scombroid poisoning due to the ingestion of fish species such as mackerel, tuna, and mahi-mahi or pelagic species as sardines and herring [85]. Putrescine and cadaverine have been reported to enhance the toxicity of histamine [85].

The determination of the biogenic amines is important not only from the point of view of their toxicity but also because they can be used as indicators of the degree of freshness or spoilage of food [86].

Most of the biogenic amines are stable to thermal processing, and their presence in canned fish can be an index of the quality of raw material. The content of histamine, cadaverine, and tyramine was determined by Schulze and Zimmermann [87] in canned tuna and sardines. Concentrations of 7–13 mg/kg of histamine were detected in canned albacore, 14–36 mg/kg in canned skipjack, 20–40 mg/kg in canned mackerel, 25–48 mg/kg in canned sardines, and 24–55 mg/kg in saury pike.

Luten et al. [88] reported that histamine is stable during cooking. However, Sims et al. [89] reported that histamine formation could take place during the sterilization, and they also found that thermal processing lowered histamine, putrescine, and cadaverine concentrations in skipjack tuna. Mietz and Karmas [90] proposed a chemical quality index (QI) based on biogenic amines, which reflected the quality loss in canned tuna:

$$QI = (HI + PU + CA)/(1 + SP + SM)$$

where
  QI is the quality index
  HI is histamine
  PU is putrescine
  CA is cadaverine
  SP is spermidine
  SM is spermine

The increase in the QI was correlated with the decrease of sensory scores on canned tuna.

Veciana-Nogués et al. [91] reported that no significant differences in thermal processing of tuna were found for histamine, putrescine, cadaverine, tyramine, and agmatine (Table 10.5). Only changes were detected for spermine and spermidine, which decreased after heat processing. For this reason, histamine, cadaverine, putrescine, and tyramine seem to be appropriated as quality indices to assess the raw material used in canned tuna.

10.8.3.3 Nucleotides

Fish spoilage has been determined usually by the measurement of the concentration of total basic nitrogen (TVB-N) and TMA. However, the loss of freshness, which precedes microbiological spoilage, is associated with autolytic reactions. Freshness of fish muscle and the nucleotide degradation has been studied by Gill [92]. One of the most important biochemical changes is the hydrolysis of adenosine-5′-triphosphate (ATP). After the death of fish, the degradation of ATP proceeds according the following sequence:

TABLE 10.5
Contents of Biogenic Amines (μg/g) through Canning Process (n = 12)

Amines (μg/g)	Raw Fish	Before Cooking	After Cooking	After Packing	End Product
HI	0.32 (O.62)[a]*	0.55 (1.44)*	0.040 (0.65)*	0.54 (0.90)*	0.63 (0.83)*
TY	0.32 (0.67)*	0.08 (0.18)*	0.24 (0.63)*	0.17 (0.35)*	0.15 (024)*
SE	1.80 (1.44)*	2.08 (2.06)*	2.32 (2.35)*	2.05 (2.42)*	1.80 (1.41)*
AG	0.94 (0.77)*	0.17 (0.44)*	0.15 (0.36)*	0.32 (0.53)*	0.40 (0.29)*
CA	0.25 (0.45)*	0.17 (0.31)*	0.33 (0.64)*	0.19 (0.32)*	0.21 (0.29)*
PU	0.29 (0.51)*	0.22 (0.28)*	0.27 (0.13)*	0.13 (0.20)*	0.32 (0.57)*
PHE	nd[b]	nd	0.10 (0.30)	0.27 (0.90)*	nd
TR	nd	0.11 (0.38)	0.12 (0.80)	0.01 (0.20)	0.12 (0.90)
SD	5.10 (1.74)*	4.31 (1.34)*	3.46 (1.21)*	3.51 (1.72)*	2.82 (1.29)**
SM	14.25 (5.91)*	19.91 (5.55)*	11.89 (3.34)*	12.89 (3.65)*	8.32 (2.67)**

Source: From Veciana-Nogués, M.T. et al., *J. Agric. Food Chem.*, 45, 4324–4328. Copyright 1997 American Chemical Society. With permission.

[a] Mean value and (standard deviation); values in the same column bearing the same number of asterisks are not different ($p > 0.05$); statistical comparisons were not performed for PHE and TR because these amines are found in only a low percentage of samples.

[b] nd, not detected (<0.25 μg/g for HI, PU, CA, AG, and PHE and 1 μg/g for SD, SM, TY, SE, and TR).

$$ATP \rightarrow ADP \rightarrow AMP \rightarrow IMP \rightarrow HxR \rightarrow Hx$$

where
- ADP is *adenosine diphosphate*
- AMP is *adenosine monophosphate*
- IMP is *inosine monophosphate*
- HxR is *inosine*
- Hx is *hypoxanthine*

Most of the enzymes involved in the breakdown of ATP, after the death, to IMP are autolytic. ATP is rapidly degraded to AMP and to IMP by dephosphorylation. However, the degradation of IMP to HxR and then to Hx is mainly due to spoilage bacteria. The concentration of HxR and Hx increases with the length of the storage period. The dephosphorylation of IMP via HxR and Hx varies with the species and occurs within the period of edibility during the ice storage. These changes are implicated in the loss of fresh flavors; IMP has been shown to impart a fresh flavor and Hx produces flavors described as bitter. The concentrations of ATP and its degradation products are usually employed as indices of freshness of refrigerated seafood [92–94].

The rate of formation of HxR and Hx varies in different species of fish; in some HxR accumulates and in others the HxR is broken down rapidly and Hx accumulates [95,96]. The rate of breakdown of ATP and derivatives is dependent on different factors, such as intraspecies differences [97], the capture [98], and seasonal variations. On the other hand, the breakdown of ATP and derivatives depends on the time and storage temperature.

Hughes and Jones [99] reported that the hypoxanthine is stable under the conditions of canning of herring and suggested that the hypoxanthine analysis is a practical value in assessing the freshness of raw material. Crawford [100] reported a comparison of the hypoxanthine content of raw tuna and the canned product, and a significant relation was found. Gallardo [101] reported the heat stability of hypoxanthine during the canning of sardines.

Tokunaga et al. [102] reported that IMP and HxR were relatively stable during the sterilization of fish. For tuna, which is an inosine-forming species, the IMP ratio (IMP/IMP + HxR + Hx) seems to be the most appropriate for determining quality. Veciana-Nogués et al. [91] showed that the IMP ratio in canned tuna was useful as quality indicator of the raw material used in canned tuna. No significant differences were found between samples in contents of IMP, HxR, and Hx. These authors [91] confirmed that the IMP ratio can be used as a freshness indicator of the raw tuna used for canning. The IMP ratio obtained for these authors was much higher than 0.114, which had been proposed by Fujii et al. [103] as the minimum value for fish acceptance. The effects of two retort conditions, common retort at 115°C for 90 min and high-temperature short time (HTST) at 125°C for 9 min, on the ATP-related compounds in pouched fish muscle were reported by Kuda et al. [104]. The IMP ratio obtained for chub mackerel, yellowfin tuna, pink salmon, and pink shrimp indicated that the HTST was the best process.

10.8.4 Ethanol

Ethanol has been used for many years as an objective indicator of seafood quality. Ethanol is produced from carbohydrates via glycolysis and by decarboxylation of amino acids. It is a common metabolite of a variety of bacteria and has been used as an index of quality for fish including canned fish [105–109]. Although ethanol is volatile, it is heat stable and may be used to assess the quality of canned fish products.

10.9 EFFECT ON THE NUTRITIVE VALUE

In general, the *nutritional value* of the fish and shellfish is not seriously damaged during thermal processing [75,110,111]. Changes in nutritional quality of raw, precooked, and canned tuna were studied by Seet et al. [112]. The cans were sterilized at 115°C and 121°C for 120 and 95 min, respectively. The amino acid concentration remained constant for raw, precooked, and canned tuna, except for histidine and sulfur amino acids. The protein digestibility decreased by 2.2% during thermal processing at 115°C and by 1.8% at 121°C.

The effect of heat processing on *amino acids* content in bigeye tuna and halibut processed at 115°C and 124°C, varying in F values of 8, 21, and 32 min, was reported by Tanaka and Kimura [46]. No significant changes were detected in the overall composition, and degradation of amino acids during thermal processing was not detected. For the study of the content of nutrients and availability of lysine in the manufacturing of canned tuna, two different sterilization times (60 and 90 min) were studied by Castrillon et al. [110]. The net protein utilization was 83.6% for fresh bonito and 84.1% for canned tuna during sterilization for 60 min; however, when the time of sterilization was increased to 90 min, the effect was negative and the net protein utilization was 82.9%.

A more detailed study was carried out by Navarro et al. [113] who demonstrated that overprocessing (115°C during 90 min) caused negative changes in the protein. It did not seem to affect lysine, since its *availability* does not decrease. However, it acted on some of the other less abundant amino acids, which upon deteriorating led to a decrease in the global nutritional utilization and damage to the protein quality.

Changes in the nutritional value of mussels during steam cooking for 15 min and sterilization at 115°C for 70 min were reported by Lema et al. [114]. No significant differences for *digestibility* between mussels cooked by steam and sterilized mussels were found; however, the sterilization affected the biological value.

Lysine is the most sensitive of the amino acids, and its availability serves as index of damage even when it is not the limiting amino acid. Lysine has been widely investigated although others such as histidine, cysteine, tryptophan, and arginine are affected also by heating. Hurrel and Carpenter [115] demonstrated that in fish, because the small concentration of reducing sugars, the loss of lysine were slight; Seet et al. [112] reported that the amount of available lysine during precooking of albacore decreased about 2%, and there were no changes in available lysine in further sterilization in high vacuum flame sterilized. However, when the conventionally sterilization was used, the canned tuna lost 10% of the total available lysine since the sterilization time to ensure commercial sterility was longer than that of the high vacuum flame sterilization. However, none of these results isolated allows kinetic parameters to be calculated. Mathematical models for the prediction of the effects of heating on the availability of lysine and protein digestibility in fish products require knowledge of kinetic parameters, to quantify the effects and to predict the quality of the canned fish.

Banga et al. [116] evaluate the changes in availability of lysine and protein digestibility during different thermal processing of albacore muscle and construct kinetic models, which allow these changes to be predicted for a given treatment. According to these authors [116], the values for the kinetic parameters indicate that the classical methods of sterilization of tuna, temperatures of retort under 125°C and F values under 12 min with $z = 10°C$, have only a slight effect on availability of lysine and protein digestibility. The computer simulation of the process with the software described in Banga et al. [117] also indicated no significant changes.

10.10 OPTIMIZATION OF THERMAL PROCESSING

The maximal retention of organoleptic property, quality, and nutritional value is essential for designing conditions of thermal fish processing.

Due to the kinetic character of chemical and physical reactions causing changes during the thermal processing of fish, research on mathematical modeling for the *prediction* of these changes and their *optimization* require knowledge of kinetic parameters. These models allow for quantifying the effects of the thermal processing and predicting the final quality, and they may be used in automatic control.

The prediction of *precooking times* for albacore by computer simulation was reported by Pérez-Martin et al. [118], who developed a semiempirical model of precooking. A numerical simulation to predict the internal temperature profile of skipjack tuna during precooking and cooling was also performed by Zhang et al. [119].

The application of z-transfer functions to canned tuna fish thermal processing was reported by Ansorena et al. [120]. This methodology was validated with canned seafood and different size containers and retort temperature. The results obtained indicate that this methodology can to be used in the design of the thermal processing leading not only to more efficient processes but also to product quality.

The *inactivation kinetics* of microorganisms and quality factors show different temperature sensitivities. According to Lund [26], thermal inactivation of microorganisms is much more dependent of the temperature than the losses of quality factors. In fish canneries, the thermal processing at constant temperature, which is effective, is applied at solid food as tuna. Optimal sterilization conditions and different number of methods have been developed for the prediction of nutrient degradation during the thermal processing at constant temperature [121–125].

The use of variable sterilization temperature could improve the quality and nutrient retention. Teixeira et al. [126] were the first to compare the advantages of variable-temperature process over the traditional constant-temperature process. Banga et al. [117] developed a method for the optimization of the thermal processing of conduction-heated canned foods. These authors [117] obtained a significant increase of quality factor retention at the surface with a variable retort temperature profiles as against the optimum constant-temperature profile.

Silva et al. [127] optimized overall quality and the nutrient retention in foods preserved by heat by the use of two objective functions: volume average retention and volume average cook value.

10.11 CONCLUSIONS

In this review, an attempt has been made to include the more relevant research on the effects that the fish thermal processing have on the main fish constituents such as proteins, amino acids, and lipids and on the sensory characteristics and nutritive value. The main literature about the use of mathematical models useful in the optimization of the thermal processing of fish products and in the enhancing of quality was reviewed. It can be concluded from the available data in relation to the thermal degradation of nutrients and sensory characteristics that the sensory characteristics and nutritive value of heat preserved fish products remained satisfactory. In future, the new advances in the basic sciences, mainly by the application of proteomics, metabolomics, nutrigenomics, and the mass spectrometry, will contribute to characterizing and evaluating the chemical modifications of the proteins and others constituents in marine products during thermal processing. An improvement of the ability to predict the effect of the thermal processing on the chemical degradation of constituents and loss of nutritional value can be expected.

NOMENCLATURE

The D value or decimal reduction time is the heat treatment time, expressed in minutes, required to reduce the number or concentration of spores, microorganisms, or quality factors to 10% of the original value.

z value is the temperature increment required to reduce the D value to 10% of the original one.

F, or sterilizing value, is the equivalent heat treatment time expressed in minutes required to achieve a reduction of the number of spores or vegetative cells of a particular microorganism.

F_0 is the sterilizing value at reference temperature 121.1°C and z = 10°C.

REFERENCES

1. A Ansar Ali, B Sudhir, AK Mallick, T Srinivasa Gopal. Effect of rotation on process time of thermally processed mackerel in oil in aluminum cans. *Journal of Food Process Engineering* 31: 139–154, 2008.
2. JM Gallardo, RI Pérez-Martin, JM Franco, S Aubourg. Chemical composition and evolution of nitrogen compounds during the processing and storage of canned albacore (*Thunnus alalunga*). In: *Proceedings of the IUFoST International Symposium on Chemical Changes during Food Processing*, November, Valencia, Spain, 1984, pp. 51–58.
3. B Ruiz-Roso, I Cuesta, M Pérez, E Borrego, B Pérez-Olleros, G Varela. Lipid composition and palatability of canned sardines. Influence of the canning process and storage in olive oil for five years. *Journal of the Science of Food and Agriculture* 77: 244–250, 1998.
4. J Mai, J Shimp, J Weilhrauch, JE Kinsella. Lipids of fish fillets: Changes following cooking by different methods. *Journal of Food Science* 43: 1669–1674, 1978.
5. KL Gall, WS Otwell, JA Koburger, H Appledorf. Effects of four cooking methods on proximate, minerals, and fatty acid composition of fish fillets. *Journal of Food Science* 48: 1068–1074, 1983.
6. P Puwastien, K Judprasong, E Kettwan, K Vasanachitt, Y Nakngamanong, L Bhattacharjee. Proximate composition of raw and cooked Thai freshwater and marine fish. *Journal of Food Composition and Analysis* 12: 9–16, 1998.
7. N Gokoglu, P Yerlikaya, E Cengiz. Effects of cooking methods on the proximate composition and mineral contents of rainbow trout (*Oncorhynchus mykiss*). *Food Chemistry* 84: 19–22, 2003.
8. F Ackurt. Nutrient retention during preparation and cooking of meat and fish by traditional methods. *Gida Sanayii* 20: 58–66, 1991.
9. M Steiner-Assiedu, K Julshamn, Ø Lie. Effect of local processing methods (cooking, frying, and smoking) on three species from Ghana: Part I—Proximate composition, fatty acids, minerals, trace elements, and vitamins. *Food Chemistry* 40: 309–321, 1991.
10. MB Atta, LA El-Sebaie, MA Noaman, HE Kassab. The effect of cooking on the heavy metals in fish (*Tilapia milotica*). *Food Chemistry* 58: 1–4, 1997.
11. ST Seet, WD Brown. Nutritional quality of raw, precooked and canned albacore tuna (*Thunnus alalunga*). *Journal of Food Science* 48: 288–289, 1983.
12. JR Banga, AA Alonso, JM Gallardo, RI Pérez-Martin. Kinetics of thermal degradation of thiamine and surface colour in canned tuna. *Zeitschrift für Lebensmittel- Untersuchung und-Forschung* 197: 127–131, 1993.
13. J Suparno, AJ Rosenthal, SW Hanson. Kinetics of the thermal destruction of thiamine in the white flesh of rainbow trout (*Salmo gairdneri*). *Journal of the Science of Food and Agriculture* 53: 101–106, 1990.
14. O Grosjean, BF III Cobb, B Mebine, WD Brown. Formation of a green pigment from tuna myoglobins. *Journal of Food Science* 34: 404–409, 1969.
15. C Nagaoka, N Suzuki. Detection of green tuna before cooking. *Food Technology* 18: 777–781, 1964.
16. M Yamagata, K Horimoto, C Nagaoka. Assessment of green tuna: Determining trimethylamine oxide and its distribution in tuna muscles. *Journal of Food Science* 34: 156–159, 1969.
17. M Yamagata, K Horimoto, C Nagaoka. Factors concerning green tuna. *Food Technology* 24: 198–199, 1970.
18. M Yamagata, K Horimoto, C Nagaoka. Accuracy of predicting occurrence of greening in tuna based on content of trimethylamine oxide. *Journal of Food Science* 36: 55–57, 1971.
19. HS Groninger, JA Dassow. Observations of the blueing of king crab (*Paralithodes camtschatia*). *Fishery Industrial Research* 2: 47–52, 1964.
20. ME Waters. Blueing of processed crab meat. II. Identification of some factors involved in the blue discoloration of canned crab meat (*Callinectes sapidus*). National Marine Fisheries Services Special Sciences Report Fisheries. U.S. Department of Commerce. National Oceanic and Atmospheric Administration (NOAA). Washington, DC: NMFS, 1971.
21. HH Elliot, EW Harvey. Biological methods of blood removal and their effectiveness in reducing discoloration in canned Dungeness crab meat. *Food Technology* 5: 163–166, 1951.
22. DD Boon. Discoloration in processed crabmeat: A review. *Journal of Food Science* 40: 756–761, 1975.
23. JK Babbit, DK Law, DL Crawford. Blueing discoloration in canned crab meat (*Cancer magister*). *Journal of Food Science* 38: 1101–1103, 1973.
24. JK Babbit. Blueing discoloration of Dungeness crab. In: CR Martin, J Flick, ChE Hobard, DR Ward, eds. *Chemistry and Biochemistry of Marine Food Products*. Westport, CT: Avi Publishing Company, 1982, pp. 423–428.

25. PG Sreenath, S Abhilash, CN Ravishankar, R Anandan, TK Srinivasa. Heat penetration characteristics and quality changes of Indian mackerel (*Rastrelliger kanagurta*) canned in brine at different retort temperatures. *Journal of Food Process Engineering* 32: 893–915, 2009.
26. DB Lund. Maximizing nutrient retention. *Food Technology* 31: 71–78, 1977.
27. T Ohlsson. Optimal sterilization temperatures for sensory quality in cylindrical containers. *Journal of Food Science* 45: 1517–1521, 1980.
28. I Johnston. Muscle development and growth: Potential implications for flesh quality in fish. *Aquaculture* 177: 99–115, 1999.
29. K Hatae, A Tobimatsu, M Takeyama, JJ Matsumoto. Contribution of the connective tissues on the texture differences of various fish species. *Bulletin of the Japanese Society of Scientific Fisheries* 52: 2001–2008, 1986.
30. K Hatae, F Yoshimatsu, JJ Matsumoto. Role of muscle fibres in contributing firmness of cooking fish. *Journal of Food Science* 55: 693–696, 1990.
31. E Dunajski. Texture of fish muscle. *Journal of Texture Studies* 10: 301–318, 1979.
32. S Kanoh, JM Aleman Polo, Y Kariya, T Kaneko, S Watabe, K Hashimoto. Heat-induced textural and histological changes of ordinary and dark muscle of yellowfin tuna. *Journal of Food Science* 53: 673–678, 1988.
33. JJ Doesburg, D Papendrof. Determination of degree of heating of fish muscle. *Journal of Food Technology* 4: 17–26, 1969.
34. JK Chan, TA Gill, AT Paulson. The dynamics of thermal denaturation of fish myosins. *Food Research International* 25: 117–123, 1992.
35. J Yamuda. Isoelectric focusing patterns of heated-denatured water-soluble proteins of fish and shellfish muscle. *Bulletin of Tokai Regional Fisheries Research Laboratory* 111: 37–41, 1983.
36. H Rehbein. Determination of the heating temperature of fishery products. *Zeitschrift für Lebensmittel-Untersuchung und-Forschung* 195: 417–422, 1992.
37. M Uddin, S Ishizaki, M Tanaka. Coagulation test for determining end-point temperature of heated blue marlin meat. *Fisheries Science* 66: 153–160, 2000.
38. M Uddin, S Ishizaki, M Ishida, M Tanaka. Assessing the end-point temperature of heated fish and shellfish meat. *Fisheries Science* 68: 768–775, 2002.
39. M Ogawa, T Ehara, T Tamiya, T Tsuchiya. Thermal stability of fish myosin. *Comparative Biochemistry and Physiology* 106B: 517–521, 1993.
40. VE Beas, JR Wagner, M Crupkin, MC Añon. Thermal denaturation of hake (*Merluccius hubbsi*) myofibrillar proteins: A differential scanning calorimetric and electrophoretic study. *Journal of Food Science* 55: 683–687, 1990.
41. H Martens, E Vold. Studies of muscle protein denaturation. In: *Proceedings of the 22nd European Meeting Meat Research Work*, Malmo Sweden, 1976. pp. J9.3–9.4.
42. JW Bell, BE Farkas, SA Hale, TC Lanier. Effect of thermal treatment on moisture transport during steam cooking of skipjack tuna (*Katsuwonus pelamis*). *Journal of Food Science* 66: 307–313, 2001.
43. J Opstevedt, R Miller, RW Hardy, J Spinelli. Heat induced changes in sulfhydryl groups and disulfide bonds in fish protein and their effect on protein and amino acid digestibility in rainbow trout. *Journal of Agricultural and Food Chemistry* 32: 929–935, 1984.
44. EL Miller, AW Hartley, DC Thomas. Availability of sulphur amino acids in protein foods. 4. Effect of heat treatment upon the total amino acid content of cod muscle. *British Journal of Nutrition* 19: 565–573, 1965.
45. J Bjarnason, KJ Carpenter. Mechanisms of heat damage in proteins. *British Journal of Nutrition* 24: 313–329, 1970.
46. M Tanaka, S Kimura. Effect of heating condition on protein quality of retort pouched fish meat. *Nippon Suisan Gakkaishi* 54: 265–270, 1988.
47. H Ichikawas, T Dobashi, S Kondo. Muscle proteins as foods: Heat induced gel formation capability of fish muscle proteins from research on myosin. *New Food Industry* 43(8): 11–16, 2001.
48. T Tsuchiya. Thermal gelation mechanism of muscle proteins. *Kenkyu Hokoku-Asahi Garasy Zaidan* 56: 263–70, 1990.
49. I Shoichiro, K Toshiyuki, T Munehiko, T Munehiko, T Takeshi, A Keishi. Thermal gelation of fish muscle proteins treated with electric pulses and air-blowing. *Nippon Shokuhin Kagaku Kogaku Kaishi* 42(2): 131–136, 1995.
50. G Pigott, B Tucker. Science opens new horizons for marine lipids in human nutrition. *Food Reviews International* 3: 105–138, 1987.

51. R Hsieh, J Kinsella. Oxidation of polyunsaturated fatty acids: Mechanisms, products and inhibition with emphasis on fish. *Advances in Food Research and Nutrition Research* 33: 233–341, 1989.
52. HO Hultin. Lipid oxidation in fish muscle. In: GJ Flick, RE Martin, eds. *Advances in Seafood Biochemistry*. Lancaster, PA: Technomic Publishing Company, 1992, pp. 99–122.
53. S Aubourg, C Sotelo, J Gallardo. Changes in flesh lipids and fill oils of albacore (*Thunnus alalunga*) during canning and storage. *Journal of Agricultural and Food Chemistry* 38: 809–812, 1990.
54. I Medina, R Sacchi, S Aubourg. Carbon-13 nuclear magnetic resonance monitoring of free fatty release during fish thermal processing. *Journal of the American Oil Chemists' Society* 71 (5): 479–482, 1994.
55. I Medina, S Aubourg, R Pérez Martín Analysis of 1-alk-O-enyl glycerophospholipids of albacore tuna *(Thunnus alalunga)* and their alterations during thermal processing. *Journal of Agricultural and Food Chemistry* 41: 2395–2399, 1993.
56. I Medina, R Sacchi, L Biondi, S Aubourg, L Paolillo. Effect of packing media on the oxidation of canned tuna lipids. Antioxidant effectiveness of extra virgin olive oil. *Journal of Agricultural and Food Chemistry* 46(3): 1150–1157, 1998.
57. I Medina, R Sacchi, I Giudicianni, S Aubourg. Oxidation in fish lipids during thermal stress as studied by 13C-NMR. *Journal of the American Oil Chemists' Society* 75(2): 147–154, 1998.
58. I Medina, MT Satué, E Frankel. Static headspace-gas chromatography analysis to determinate oxidation of fish muscle during thermal processing. *Journal of the American Oil Chemists' Society* 76(1): 231–236, 1999.
59. I Medina, R Sacchi, S Aubourg. Application of NMR to design industrial conditions in the thermal processing of canned fish. *Zeitschrift für Lebensmittel-Untersuchung und- Forschung* 210(3): 176–178, 1999.
60. S Dahl, FX Malcata. Characterization of forced oxidation of sardine oil: Physicochemical data and mathematical modeling. *Journal of the American Oil Chemists' Society* 76(5): 633–641, 1999.
61. MB Hale, T Brown. Fatty acids and lipid classes of three underutilized species and changes due to canning. *Marine Fisheries Review* 45: 4–6, 1983.
62. I Medina, MT Satué, JB German, E Frankel. Comparison of natural polyphenols antioxidant from extra virgin olive oil with synthetic antioxidants in tuna lipids during thermal oxidation. *Journal of Agricultural and Food Chemistry* 47: 4873–4879, 1999.
63. DB Josephson, RC Lindsay. Enzymatic generation of volatile aroma compounds from fresh fish. In: TH Parliament, R Croteau, eds. *Biogeneration of Aromas*, ACS Symposium Series 317. Washington, DC: American Chemical Society, 1986, pp. 201–219.
64. H Sakakibara, J Ide, T Yanai, I Yajima, K Hayashi. Volatile flavor compounds of some kinds of dried and smoked fish. *Agricultural and Biological Chemistry* 54: 9–16, 1990.
65. R Przybylski, NAM Eskin, LJ Malcolmson. Development of a gas chromatographic system trapping and analyzing volatiles from canned tuna. *Canadian Institute of Science and Technology Journal* 24(3–4): 129–135, 1991.
66. B Girard, S Nakai. Static headspace gas chromatographic method for volatiles in canned salmon. *Journal of Food Science* 56: 1271–1278, 1991.
67. C Milo, W Grosch. Changes in the odorants of boiled salmon and cod as affected by the storage of the raw material. *Journal of Agricultural and Food Chemistry* 44: 2366–2371, 1996.
68. JR Guyton, BL Black, CL Seidel. Focal toxicity of oxysterols in vascular smooth muscle cell culture. *American Journal of Pathology* 137: 425–434, 1990.
69. MR Wrensch, NL Petrakis, LD Gruenka, R Miike, VL Ernster, EB King, WW Hauck, JC Craig, WH Goodson. Breast fluid cholesterol and cholesterol β-epoxide concentrations in women with benign breast disease. *Cancer Research* 49: 2168–2174, 1989.
70. CW Kendall, M Koo, E Sokoloff, AV Rao. Effect of dietary oxidized cholesterol on azoxymethane-induced colonic preneoplasia in mice. *Cancer Letters* 66: 241–248, 1992.
71. LL Smith, VB Smart, GAS Ansari. Mutagenic cholesterol preparations. *Mutation Research* 68: 23–30, 1979.
72. K Osada, T Kodama, K Yamada, M Sugano. Oxidation of cholesterol by heating. *Journal of Agricultural and Food Chemistry* 41: 1198–1202, 1993.
73. T Ohshima, N Li, C Koizumi. Oxidative decomposition of cholesterol in fish products. *Journal of American Oil Chemists' Society* 70: 596–600, 1993.
74. T Ohshima, K Shozen, H Ushio, C Koizumi. Effects of grilling on formation of cholesterol oxides in seafood products rich in polyunsaturated fatty acids. *Zeitschrift für Lebensmittel- Untersuchung und-Forschung* 29: 94–99, 1996.

75. A Aitken, JJ Connell. Fish. In: RJ Priestley, ed. *Effects of Heating on Foodstuffs*. London, U.K.: Applied Science Publishers, 1979, pp. 219–254.
76. NR Jones. The free amino acids of fish II. Fresh skeletal muscle from lemon sole (*Pleuronectes microcephalus*). *Journal of the Science of Food and Agriculture* 10: 282–286, 1959.
77. J Liston. Microbiology in fishery sciences. In: JJ Connell, ed. *Advances in Fish Science and Technology*. Surrey, England: Fishing News Book, 1980, pp. 138–157.
78. IM Mackie, AH Ritchie. Free amino acids of fish flesh. In: *Proceedings of the 2nd International Congress of Food Science and Technology*, Madrid, Spain, August 22–27, 1974, Vol. I, pp. 29–38.
79. RI Pérez-Martin, JM Franco, S Aubourg, JM Gallardo. Changes in free amino acids content in albacore (*Thunnus alalunga*) muscle during thermal processing. *Zeitschrift für Lebensmittel- Untersuchung und-Forschung* 187: 432–435, 1988.
80. T Tokunaga. On the thermal decomposition of trimethylamine oxide in muscle of some marine animals. *Bulletin of the Japanese Society of Scientific Fisheries* 41: 535–546, 1975.
81. RB Hughes. Chemical studies on the herring (*Clupea harengus*) I. Trimethylamine oxide and volatile amines in fresh, spoiling, and cooked herring flesh. *Journal of the Science of Food and Agriculture* 10: 431–436, 1959.
82. T Tokunaga. Studies on the quality evaluation of canned marine products. I. Determination of the ratio of dimethylamine to trimethylamine in canned albacore. *Bulletin of the Japanese Society of Scientific Fisheries* 41: 547–553, 1975.
83. JM Gallardo, RI Pérez-Martin, JM Franco, S Aubourg, CG Sotelo. Changes in volatile bases and trimethylamine oxide during the canning of albacore (*Thunnus alalunga*). *International Journal of Food Science & Technology* 25: 78–81, 1990.
84. MI Yeannes, CE del Valle, HM Lupin. Generación de bases nitrogenadas volátiles durante el proceso de elaboración de conservas de pescado. *Revista de Agroquímica y Tecnología de Alimentos* 23: 585–590, 1983.
85. SL Taylor. Histamine food poisoning: Toxicology and clinical aspects. *Critical Reviews of Toxicology* 17: 91–128, 1986.
86. A Halász, A Barath, L Simon-Sakardi, W Holzapfel. Biogenic amines and their production by microorganisms in food. *Trends in Food Science and Technology* 5: 42–49, 1984.
87. K Schulze, K Zimmermann. Influence of various storage conditions on the development of biogenic amines in tuna and mackerel. *Fleischwirtschaft* 62: 906–910, 1982.
88. JB Luten, W Bouquet, LAY Seuren, MM Burggraaf, G Riekwel-Booy, P Durand, M Etienne, JP Gouyou, A Landrein, A Ritchie, M Leclerq, R Guinet. Biogenic amines in fishery products: Standardization methods within E.C. In: HH Huss, M Jakobsen, J Liston, eds. *Quality Assurance in the Fish Industry*. Amsterdam, the Netherlands: Elsevier Science Publishers, 1992, pp. 427–439.
89. GG Sims, G Farn, RK York. Quality indices for canned skipjack tuna: Correlation of sensory attributes with chemical indices. *Journal of Food Science* 57: 1112–1115, 1992.
90. JL Mietz, E Karmas. Chemical quality index of canned tuna as determined by high-pressure liquid chromatography. *Journal of Food Science* 42: 155–158, 1977.
91. MT Veciana-Nogués, A Mariné-Font, MC Vidal-Carou. Biogenic amines in fresh and canned tuna. Effects of canning on biogenic amine contents. *Journal of Agricultural and Food Chemistry* 45: 4324–4328, 1997.
92. TA Gill. Objective analysis of seafood quality. *Food Reviews International* 6: 681–714, 1990.
93. NR Jones, J Murray. Degradation of adenine and hypoxanthine nucleotide in the muscle of chilled stored trawled cod (*Gadus callarias*). *Journal of the Science of Food and Agriculture* 13: 475–480, 1962.
94. JM Ryder. Determination of adenosine triphosphate and its breakdown products in fish muscle by high performance liquid chromatography. *Journal of Agricultural and Food Chemistry* 33: 678–680, 1985.
95. DJ Fraser, D Pitts, WJ Dyer. Nucleotide degradation and organoleptic quality in fresh and thawed mackerel held at and above ice temperature. *Journal of Fisheries Research Board of Canada* 25: 239–253, 1968.
96. DH Greene, JK Babbit, KD Keppond. Patterns of nucleotide catabolism as a freshness indicator in flatfish from the Gulf of Alaska. *Journal of Food Science* 55: 1236–1238, 1990.
97. RE Martin, RJH Gray, MD Pierson. Quality assessment of fresh fish and the role of the naturally occurring microflora. *Food Technology* 32: 188–198, 1978.
98. T Lowe, JM Ryder, JF Carrager, RMG Wells. Flesh quality in snapper, *Pagrus aurata*, affected by capture stress. *Journal of Food Science* 58: 770–773, 796, 1993.

99. RB Hughes, NR Jones. Measurement of hypoxanthine concentration in canned herring as an index of the freshness of the raw material with a comment on flavour relation. *Journal of the Science of Food and Agriculture* 17: 434–436, 1966.
100. L Crawford. A comparison of the hypoxanthine content of raw tuna muscle to the canned products. *Bulletin of the Japanese Society of Scientific Fisheries* 36: 1136–1139, 1970.
101. JM Gallardo. El contenido en hipoxantina como índice del grado de frescura del pescado y de los productos pesqueros. *Informes Técnicos del Instituto de Investigaciones Pesqueras* 58: 1–8, 1978.
102. T Tokunaga, H Iida, K Nakamura, S Terano, T Furukawa, K Sato, Ch Hoshino. Study on the several chemical tests for estimating the quality of canned product. I. On the canned mackerel. *Bulletin of Tokai Regional Fisheries Research Laboratory* 107: 1–10, 1982.
103. Y Fujii, K Shudo, K Nakamura, S Ishikawa, M Okada. Relationship between the quality of canned fish and its contents of ATP break-down III ATP- breakdowns in canned albacore and skipjack in relation to the organoleptic inspection. *Bulletin of the Japanese Society of Scientific Fisheries* 39, 39–69, 1973.
104. T Kuda, M Fujita, H Goto, T Yano. Effects of retort conditions on ATP-related compounds in pouched fish muscle. *LWT-Food Science and Technology* 41: 469–473, 2008.
105. F Hillig. Determination of alcohol in fish and eggs products. *Journal of the Association of Official Analytical Chemists* 41: 776–781, 1958.
106. PA Lerke, RW Huck. Objective determination of canned tuna quality: Identification of ethanol as a potential useful index. *Journal of Food Science* 42: 755–758, 1977.
107. DM Cosgrove. A rapid method for estimating ethanol in canned salmon. *Journal of Food Science* 43: 641–642, 1978.
108. H Iida, T Tokunaga, K Nakamura. The relationship between the sensory judgment of canned albacore and its ethanol content. *Bulletin of Tokai Regional Fisheries Research Laboratory* 104: 77–82, 1981.
109. TA Hollingworth, HR Throm. Correlation of ethanol concentration with sensory classification of decomposition in canned salmon. *Journal of Food Science* 47: 1315–1317, 1982.
110. AM Castrillón, MP Navarro, JM Gallardo, RM Ortega, RI Pérez-Martin, G Varela. Effect of processing on the quality and nutritional value of fish preserves. In: *Proceedings of the IUFoST International Symposium on Chemical Changes during Food Processing*, November, Valencia, Spain, 1984, pp. 33–37.
111. E Yañez, D Balleter, G Donoso. Effects of drying temperature on the quality of fish protein. *Journal of the Science of Food and Agriculture* 21: 426–428, 1970.
112. ST Seet, JR Heil, SJ Leonard, WD Brown. High vacuum flame sterilization of canned diced tuna: Preliminary process development and quality evaluation. *Journal of Food Science* 48: 364–369, 1983.
113. MP Navarro, AM Castrillóm, RM Ortega, JM Gallardo, G Varela. Changes in the nutritional utilization of fish protein as a function of the sterilization time of its preserve. In: *Proceedings of the IUFoST International Symposium on Chemical Changes during Food Processing*, November, Valencia, Spain, 1984, pp. 38–42.
114. L Lema, FJ Mataix, G Varela. Effect of the processing in the preparation of preserves on the nutritional value of mussel protein (*Mytilus edulis*). In: *Proceedings of the IUFoST International Symposium on Chemical Changes during Food Processing*, November, Valencia, Spain, 1984, pp. 343–346.
115. RI Hurrel, KJ Carpenter. Maillard reactions in foods. In: T Hojem, O Kvale, eds. *Physical, Chemical and Biological Changes in Food Caused by Thermal Processing*. London, U.K.: Applied Science Publishers, 1977, pp. 168–184.
116. JR Banga, AA Alonso, JM Gallardo, RI Pérez Martín. Degradation kinetics of protein digestibility and available lysine during thermal processing of tuna. *Journal of Food Science* 57: 913–915, 1992.
117. JR Banga, RI Pérez-Martin, JM Gallardo, JJ Casares. Optimization of the thermal processing of conduction heated canned foods: Study of several objective functions. *Journal of Food Engineering* 14: 25–51, 1991.
118. RI Pérez-Martin, JR Banga, MG Sotelo, SP Aubourg, JM Gallardo. Prediction of precooking times for albacore (*Thunnus alalunga*) by computer simulation. *Journal of Food Engineering* 10: 83–95, 1989.
119. J Zhang, BE Farkas, SA Hale. Precooking and cooling of skipjack tuna (*Katsuwonus pelamis*): A numerical simulation. *Lebensmittel- Wissenschaft und-Technologie* 35: 607–616, 2002.
120. M R Ansorena, C del Valle, VO Salvadori. Application of transfer functions to canned tuna fish thermal processing. *Food Science Technology International* 16: 43–51, 2010.
121. AA Teixeira, JR Dixon, JW Zahradnik, GE Zinsmeister. Computer optimization of nutrient retention in the thermal processing of conduction-heated foods. *Food Technology* 23: 137–147, 1969.
122. MK Lenz, DB Lund. The lethality-Fourier number method: Experimental verification of a model for calculating average quality factor retention in conduction-heated canned foods. *Journal of Food Science* 42: 998–1001, 1977.

123. HAC Thijssen, PJA Kerkhof, AAA Liefkens. Short-cut method for the calculation of sterilization conditions yielding optimum quality of packaged foods. *Journal of Food Science* 43: 1096–1103, 1978.
124. HAC Thijssen, LHP Kechen. Calculation of optimum sterilization conditions for packed conduction-type foods. *Journal of Food Science* 45: 1267–1272, 1292, 1980.
125. MM Nadkarni, TA Hatton. Optimal nutrient retention during the thermal processing of conduction-heated canned foods: Application of the distributed minimum principle. *Journal of Food Science* 50: 1312–1321, 1985.
126. AA Teixeira, GE Zinsmeister, JW Zahradnik. Computer simulation of variable report control and container geometry as a possible means of improving thiamine retention in thermally processed foods. *Journal of Food Science* 40: 656–659, 1975.
127. C Silva, M Hendrickx, F Oliveira, P Tobback. Critical evaluation of commonly used objective functions to optimize overall quality and nutrient retention of heat-preserved foods. *Journal of Food Engineering* 17: 241–258, 1992.

11 Thermal Processing of Dairy Products

Alan L. Kelly, Nivedita Datta, and Hilton C. Deeth

CONTENTS

11.1 INTRODUCTION: WHY HEAT MILK?

Milk is a fluid product produced by female mammals that provides the nutritional requirements of the neonate. As such, it is remarkably rich in many essential food constituents, such as amino acids (supplied for ingestion in the form of protein), lipids, sugar (lactose), minerals, and vitamins. However, this richness also makes milk a fertile medium for the growth of microorganisms, which either spoil the product or present a risk of food poisoning or food-borne disease. In addition, there are a number of enzymes of different types (e.g., proteases, lipases) in milk that can contribute to undesirable changes during storage (e.g., lipolysis can induce rancidity), and treatment or intervention to inactivate or control such activities is often essential to the prolonging of the acceptable shelf life of dairy products.

Thus, the majority of milk consumed today is processed to ensure safety and prolong shelf life. The same applies to the huge range of products that can be made from milk, including cheese, fermented milks (e.g., yogurt), and food ingredients. A range of preservation strategies have evolved over the millennia that milk has been a part of the human diet; for example, preservation by fermentation was known in biblical times. However, by far the most common strategy used to preserve milk for the last 100 years or more has been the application of heat (i.e., thermal processing), which will be the focus of this chapter. Milk may also be heated to establish specific product properties, for example, inhibiting oxidation during storage of whole milk powder (WMP) or generating stronger gel structures in yogurt; product-specific thermal processes will also be discussed here.

11.2 HISTORICAL DEVELOPMENT OF THERMAL PROCESSING OF MILK

The historical development of the thermal processing of milk was reviewed by Westhoff [1] and Holsinger et al. [2].

In the period 1864–1866, the famous French scientist Louis Pasteur discovered that the spoilage of wine and beer could be prevented by heating to around 60°C for several minutes; in recognition of this profoundly important work, mild thermal processes applied to milk are now referred to as *pasteurization*. Although pasteurization was originally developed to improve the stability and quality of food products, it soon became apparent that it offered consumers protection against hazards associated with the consumption of raw milk. Early in the twentieth century, a particular benefit was a reduction in the risk of transmission of tuberculosis from infected cows, through their milk, to humans. Developments in technology and widespread implementation of pasteurization occurred early in the twentieth century. The first commercial pasteurizers, in which milk was heated at 74°C–77°C, were produced in Germany in 1882 [1], the first commercially operated milk pasteurizer was installed in Bloomville, New York, in 1893 [2], and the first law requiring pasteurization of liquid (beverage) milk was passed by the authorities in the city of Chicago in 1908.

Many early pasteurization processes used conditions not very different from those proposed by Pasteur; generally milk was heated to 62°C–65°C for at least 30 min, followed by rapid cooling to less than 10°C (such processes are now referred to as *low-temperature long-time*, LTLT, pasteurization). However, *high-temperature short-time* (HTST) pasteurization, in which milk is heated to 72°C–74°C and held for 15–30 s, using a continuous-flow plate heat exchanger, gradually became the standard industrial procedure for the heat treatment of liquid milk. Compared to the LTLT process, the HTST process has the advantages of reduced heat damage and flavor changes, and increased throughput [3].

Later developments in the thermal processing of milk led to the introduction (in the 1940s) of *ultrahigh-temperature* (UHT) processes, in which milk is heated to temperatures of 135°C–140°C for 2–5 s [4]. UHT treatments can be applied using a range of heat exchanger technologies (e.g., indirect processes using plate or tubular heat exchangers or direct processes where steam is injected into milk), and essentially result in a sterile product that is shelf stable for at least 6 months at room temperature. A key-enabling technology to ensure long shelf life was the parallel development of aseptic packaging systems, where the sterile milk leaving the heat exchanger is packaged into a sterile hermetically sealed containers under conditions that prevent recontamination [5]. Eventual deterioration of UHT milk quality generally results from physicochemical phenomena (e.g., age gelation) rather than microbiological or enzymatic processes.

Sterilized milk products may also be produced using in-container retort systems. In fact, such processes were developed before the work of Pasteur; in 1809, Nicholas Appert developed an in-container sterilization process for the preservation of a range of food products, including milk. *In-container sterilization* is generally used today for concentrated (condensed) milks; typical processes involve heat treatment at 110°C–120°C for 10–20 min.

In recent years, there has been considerable interest in extending the shelf life of pasteurized milk [7]. This has given rise to the term *extended shelf-life* (ESL) milk. While such milk can be produced by processes other than heat (e.g., microfiltration [6], the most common type of ESL treatment is high-temperature treatment, mostly in the range 120°C–130°C for a short time (<1–4 s). ESL milk has a refrigerated shelf life of ≥40 days. It cannot be stored at room temperature, as the heat treatment conditions applied are not sufficient to ensure inactivation of the spores of bacteria capable of growing at room temperature.

As stated earlier, the primary function of the thermal processing of milk is to kill undesirable microorganisms. Modern pasteurization is a very effective means of ensuring that liquid milk is free of the most heat-resistant pathogenic bacteria likely to be present in raw milk; today, most health risks linked to the consumption of pasteurized milk are probably due to postpasteurization contamination of the product. UHT or in-container sterilized milks, on the other hand, are virtually free of even spore-forming thermophilic bacterial species.

11.3 OVERVIEW OF HEAT TREATMENTS APPLIED TO MILK

As with any food product, the choice of heat treatment applied to milk depends on a trade-off between, first, the degree of microbial inactivation required to ensure safety and extend the shelf life by an acceptable factor and, second, changes in quality of the product; these two consequences of heating are usually, in effect, inversely correlated.

Overall factors that should be considered in determining the heat treatment required for a particular product include the following:

- Postheating growth potential of spore-forming bacteria. For example, low pH conditions or storage at low temperatures suppress outgrowth of spore-forming bacteria; thus, products that include these additional hurdles, for example, pasteurized or fermented milks, may not require as extensive heat treatment as higher pH products or products to be stored at room temperature.
- Consumer preference. For example, in some countries, such as Ireland and Australia [8], consumer preference is clearly aligned toward pasteurized milk, and the more strongly cooked flavors associated with UHT milk are undesirable.
- Target market. For example, the requirements for processing of infant formulae are among the most stringent in the dairy industry, due to the vulnerability of that group of consumers to food-borne illness.

Heat treatments applied to milk are generally defined by the maximum temperature of heating and the holding time at that temperature. The processes most commonly applied to milk and dairy products are summarized in Table 11.1.

TABLE 11.1
Types of Heat Treatment Applied to Milk and Their Microbiological Effects

Process	Typical Conditions	Heat Exchanger	Microbiological Effect	Typical Microbial Quality
Thermization	57°C–68°C for 15 s	PHE	Destroys psychrotrophs	3- to 4-log reduction in psychrotrophs[a]
Pasteurization	63°C for 30 min	Batch	Destroys vegetative pathogens	Total plate count <100,000 cfu/mL[b]
Pasteurization	72°C–74°C for 15–30 s	PHE	Destroys vegetative pathogens	Total plate count <100,000 cfu/mL[b]
Yogurt mix heating	85°C–95°C for 2–30 min	PHE/THE	Destroys all pathogens and most spores	—
Preheating for producing milk powder	70°C–135°C for 1 s–5 min	PHE/THE/Other	Depends on treatment	—
High temperature pasteurization	120–130 for <1–4 s	PHE/THE	Destroys vegetative bacteria and most spores	—
UHT (indirect)	135°C–140°C for 3–5 s	PHE/THE/SSHE	Destroys all bacteria, including spores	Total plate count <100 cfu/mL after 15 days at 30°C[c]
UHT (direct)	135°C–140°C for 3–5 s	Steam infusion or injection system	As UHT (indirect)	Total plate count <100 cfu/mL after 15 days at 30°C[c]
In container sterilization	110°C–120°C for 10–20 min	Batch or continuous (hydrostatic) retort	As UHT	Total plate count <100 cfu/mL after 15 days at 30°C[c]

PHE, plate heat exchanger; THE, tubular heat exchanger; SSHE, scraped-surface heat exchanger.

[a] Data from Pearce, L., Kinetic studies on the heat inactivation of MAP in UHT milk during turbulent flow pasteurization, in *Proceedings of Workshop on Revisiting Heat Resistance of Microorganisms in Milk*, Kiel, Germany, 2003, pp. 1–41.

[b] Data from Gallmann, P.U., Possible quality improvements/benefits of extended shelf life (ESL) milk, in *Proceedings of the IDF World Dairy Summit Meeting on Extended Shelf Life (ESL) Milk*, Dresden, Germany, 2000, pp. 1–9.

[c] Data from Lewis, M.J., *Int. J. Dairy Technol.*, 52, 121, 1999.

11.4 EFFECTS OF HEATING ON MILK CONSTITUENTS

The primary constituents of milk include fat (as an emulsion), protein (a heterogeneous population, some of which are in colloidal form, while others are dissolved in the aqueous phase, including enzymes), lactose, salts, and vitamins. Milk is an exceedingly complex raw material for processing; it has a range of constituents whose nature, stability, and properties are changed by the types of heat treatments applied to it. Many properties of milk (sensory, nutritional, and physicochemical) consequently change as a result of thermal processing.

Perhaps the most significant changes on heating milk are those that involve proteins. Most (70%–80%) of the protein in milk is casein, a family of four relatively hydrophobic proteins (α_{s1}-, α_{s2}-, β-, and κ-caseins). As is the case for many hydrophobic substances found in an aqueous environment, the caseins are found in milk in colloidal micelles [ranging from 50 to 500 nm in size (average ~120 nm) with an average mass of ~10^8 Da; there are 10^{14}–10^{16} micelles mL^{-1} milk]. One of the caseins, κ-casein, is a glycoprotein with a hydrophilic sugar-rich C-terminal (often called the glycomacropeptide, GMP); κ-casein is found at the surface of the micelle, where the GMP protrudes into the serum (forming the so-called hairy layer) and stabilizes the micelle by acting in a role analogous to the amphiphilic nature of a classical emulsifying agent. The hairy layer is negatively charged and is largely responsible for the steric repulsion which prevents casein micelles from aggregating.

The remainder (20%–30%) of the proteins in milk is the so-called whey proteins, which are globular proteins found in the serum (aqueous) phase of milk; the major whey proteins are β-lactoglobulin (β-lg) and α-lactalbumin (α-la). In terms of thermal processing, β-lg is of particular significance; due to an uneven number of cysteine residues, β-lg possesses a free thiol group which, when the protein is in the native state, is located within the interior of the structure of the molecule. Heating to temperatures >75°C, however, results in unfolding of the molecule, and exposure of the thiol group, which can then undergo thiol-disulfide interchange reactions with disulfide bond-containing proteins (e.g., β-lg itself, α-la, κ-, and α_{s2}-caseins, milk fat globule membrane proteins and the indigenous milk alkaline proteinase, plasmin). Such interactions have profound implications for the quality of heated milk and dairy products such as yogurt. For example, binding of denatured β-lg to the casein micelle, *via* disulfide bond formation with κ-casein, fundamentally and negatively affects the rennet coagulation properties of milk, as will be discussed later. Denatured β-lg may also associate with κ-casein via hydrophobic interactions. Exposure of free thiol groups on denaturation of β-lg results in a cooked flavor of the milk, while providing antioxidant activity.

Interactions of denatured β-lg with plasmin greatly reduce the proteolytic activity in milk. The inactivation of milk enzymes is an intensively researched consequence of heating milk, due to the use of such enzymes as indices of the severity of heat treatment applied as will be discussed in detail later in this chapter.

Severe heating has a number of specific effects on the caseins, including dephosphorylation, deamidation of glutamine and arginine residues, cleavage, and formation of covalent cross-links; these changes may result in protein–protein interactions that contribute to thermal instability. There have been very few studies of the heat-induced hydrolysis of casein, perhaps surprisingly as quite a number of peptides can be released from all the principal caseins on heating at 140°C for prolonged time (up to 120 min), depending on pH at heating [12].

Heat causes some alterations to the mineral balance in milk; for example, heating milk causes precipitation of calcium phosphate and reduces calcium ion activity [13]. Heating milk also affects lactose, the nature and extent of such effects depending on environmental conditions and the severity of heating; heat-induced changes in lactose include degradation to acids (with a concomitant decrease in pH), isomerization (e.g., to lactulose), production of compounds such as furfural, and interactions with amino groups of proteins [14]. In the Maillard reaction, lactose or lactulose reacts with an amino group, such as the ε-amino group of lysine residues, in a complex (and not yet fully understood) series of reactions with a variety of end products [14,15]. The early stages of the Maillard reaction (condensation reactions between the carbonyl group of lactose and the ε-amino group of lysine of milk proteins) result in the formation of the protein-bound Amadori product, lactulosyllysine, which then degrades to a range of advanced Maillard products, including hydroxymethylfurfural (HMF), furfurals, and formic acid. The most obvious consequence of the Maillard reaction is a change in the color of milk (browning), due to the formation of pigments called melanoidins, or advanced-stage Maillard products. Extensive Maillard reactions can result in polymerization of proteins [15], and also alter the flavor and nutritive quality of dairy products, in the latter case through reduced digestibility of the caseins and loss of available lysine.

Moderately intense heating processes applied to milk primarily cause the isomerization of lactose to lactulose. Lactulose, a disaccharide of galactose and fructose, is of interest as a bifidogenic factor and also as a laxative. More severe treatments (e.g., sterilization) will result preferentially in Maillard reactions. Products of heat-induced changes in lactose, such as lactulose may be used as indices of heat treatment of milk [16,17].

Feinberg et al. [17] compared a number of indicators for heat treatment of milk (including furosine, lactulose, various parameters relating to whey protein denaturation, and lactoperoxidase activity) in terms of their possible use as tracers for discrimination between treatments from thermization to sterilization; it was found that no single parameter could universally discriminate the treatments, but that a multivariate analysis was the only promising solution in this regard.

11.5 OVERVIEW OF THE MICROBIOLOGY OF MILK

Milk from a healthy cow is secreted free of microorganisms, but it becomes contaminated with many microorganisms during direct contact with various sources (Figure 11.1). Three main types of microorganisms are the key factors for assessing quality and safety of heat-treated milk, for example,

- Thermoduric/heat-resistant bacteria
- Bacteria that enter through postprocessing contamination (PPC)
- Bacteria that can grow at refrigeration temperature

Raw milk has a short shelf life even under highly hygienic refrigeration conditions [18] and requires processing as soon as possible after arrival in the dairy plant to ensure the production of high-quality processed milk and dairy products. Heat treatment is the most common and single most effective industrial processing procedure for reducing bacterial numbers in milk, thereby enhancing its shelf life and ensuring the absence of pathogenic organisms. Milk preservation by pasteurization is based on three unit operations: heat treatment, rapid cooling, and refrigerated storage of the cooled milk. Thermal treatment and refrigerated storage are two important processes which are

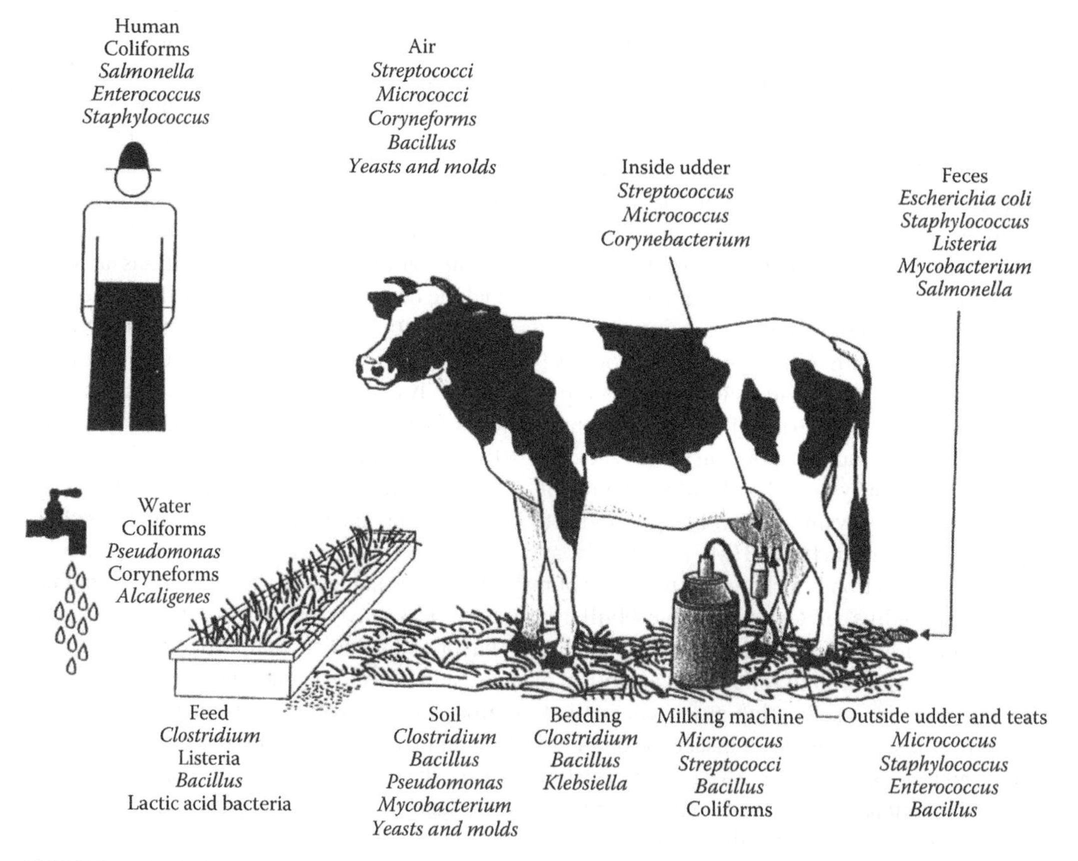

FIGURE 11.1 Sources of contamination of raw milk. (From Frank, J.F. and Hassan, A.N., Microorganisms in milk, in *Encyclopedia of Dairy Science*, Roginski, H., Fuquay, J.W., and Fox, P.F., eds., Academic Press, London, U.K., 2002, pp. 1786–1796.)

fundamental for the safety of heat-treated milk. The thermal process parameters of pasteurization are sufficient to kill pathogens and the majority of vegetative microorganisms but not sufficient to inactivate microbial endospores. Refrigerated storage prevents, or severely retards, the growth of organisms that are capable of surviving pasteurization, and germination of bacterial spores during the shelf life of pasteurized milk.

The quality and safety of heat-treated milk depend on raw milk quality, processing conditions, PPC, and storage temperature [19]. The former will be discussed here, while the latter factors will be considered later in this chapter.

Even if produced under highly hygienic conditions, raw milk contains a heterogeneous population of bacteria, some of which are pathogenic and others of which can cause spoilage during milk storage. Fortunately, pathogens are generally not determinants of the shelf life of thermally treated milk and dairy products [11,20]. This is due to their low number in raw milk, their inability to grow at refrigeration temperatures and their destruction by pasteurization.

In exception to the aforementioned generalization, one potentially pathogenic bacterium in raw milk, *Bacillus cereus*, can be a significant spoilage agent due to the ability of its spores to survive pasteurization and the ability of some strains to proliferate in pasteurized milk at refrigeration temperature. This bacterium can produce enzymes such as a proteinase, which hydrolyses casein to produce an intensely bitter flavor, and phospholipase, which hydrolyses the phospholipids of the milk fat globule membrane and causes instability of the fat emulsion. In unhomogenized milk, this can cause the defect known as bitty cream, but this is not observed in homogenized milk. *B. cereus* in milk seldom causes food poisoning as the contaminated product develops a strong, unacceptable flavor [19] and also because toxigenic strains are seldom psychrotrophic. In addition, large numbers of this bacterium are required to produce food poisoning; Ryser [21] reported that most outbreaks of *B. cereus* poisoning were caused by contamination levels of at least 10^6 cfu/g. Innovative sanitary measures at the farm and the processing plant, together with modernization of dairy processing plants, have reduced the incidence of disease outbreaks associated with pasteurized milks (to <1% of total reported outbreaks). The persisting outbreaks highlight the requirement for strict vigilance at every stage of the manufacturing process and cold chain distribution of thermally processed milks. Additionally, the use of cutting-edge technologies in dairy processing provide new processes, new products (e.g., ESL milk), new packaging materials, and new marketing patterns; each of these needs to be evaluated in order to assure the safety of heat-treated milk.

In addition to *B. cereus*, other common types of spore-forming bacteria frequently present in raw milk are *Bacillus licheniformis* and *Geobacillus stearothermophilus,* although the types and numbers of organisms vary considerably between milks. For example, some husbandry practices such as stall-feeding of silage are known to significantly increase the numbers of certain species. *G. stearothermophilus,* which is thermophilic and produces extremely heat-resistant spores, causes a defect known as "flat-sour" spoilage in UHT milk stored at elevated temperature. In recent studies, a very heat-resistant mesophilic species of *Bacillus* has been isolated from UHT milk products in several countries and has been named *Bacillus sporothermodurans.* Although it can multiply to a maximum of about 10^5 cfu/mL in milk without causing a change in its composition or sensory properties, its extreme heat stability and its ability to grow at room temperature make it a most unwelcome organism in dairy products [22–24].

Psychrotrophic (i.e., able to tolerate cold) bacteria such as *Pseudomonas* spp. are very common in raw milk. These bacteria can produce extracellular enzymes such as proteinase, lipase, and phospholipase, when the counts of such bacteria exceed ~10^6 cfu/mL in milk. These enzymes can significantly and negatively affect the quality of milk and dairy products. A factor of particular significance is their high heat stability, which is much higher than that of the parent bacteria, with the enzymes frequently surviving processes that result in total elimination of the bacteria. For example, *Pseudomonas* proteinases can survive UHT treatment (135°C–150°C for 2–8 s) and cause bitterness and premature gelation in UHT milk [25]. *Pseudomonas* lipases can cause rancid off-flavors in cheese [26] and UHT milk [27] and have been reported to cause a soapy off-flavor in flavored UHT milk recombined from milk powder containing residual enzyme [10]. These heat-stable enzymes do not present a problem in pasteurized milk and cream because these products have a short shelf life and are stored under refrigeration conditions. As will be discussed in the next section of this chapter, thermization of raw milk when it arrives at a factory is used to extend the storage life of milk by controlling the growth of psychrotrophs, thereby reducing the likelihood of heat-resistant enzymes being present in subsequently processed milk and dairy products, for example, milk powder [28,29].

The microbiology of milk is obviously profoundly affected by the magnitude of heat treatment to which it is subjected. Milk heated at temperatures even slightly higher than those used for conventional HTST pasteurization, for example, 80°C compared to 72°C, perhaps unexpectedly, has a lower keeping quality than pasteurized milk [30,31]. This has been attributed to activation of dormant spores present in raw milk [32] and destruction of the antibacterial enzyme lactoperoxidase [33].

When the temperature of heat treatment is increased to 115°C–120°C for holding times of 1–5 s, the number of spore-forming bacteria in milk and cream is reduced, and the shelf life of the products is extended [6,34–36]. However, to eliminate the most heat-resistant spores, much more severe heating is required. This is the basis of UHT treatments, which heat milk at 135°C–140°C for 3–5 s. Westhoff and Dougherty [37] studied the heat resistance of *Bacillus* spores in UHT milk and concluded that heating milk at 139°C–154°C for 9 s was required to eliminate the most heat-resistant aerobic mesophilic bacilli. This is similar to the conditions required for the inactivation of *B. sporothermodurans*: 148°C for 10 s or 150°C for 6 s [38].

Several other bacteria that may be present in raw milk are of significance for specific products and heating processes; these will be discussed later in this chapter.

11.6 THERMIZATION OF MILK: TECHNOLOGY AND FUNCTIONS

Thermization is a subpasteurization operation sometimes applied to raw milk that is intended to be held after intake for extended periods of refrigerated storage before manufacture of certain products (e.g., Dutch-type cheese). The primary issue of concern in such cases is growth of psychrotrophic bacteria during cold storage, and consequent secretion of heat-stable extracellular proteinases and lipases by such bacteria into the milk, as discussed earlier.

Thermization, which typically involves heating milk to 57°C–68°C for 15 s, will inactivate psychrotrophic bacteria such as pseudomonads and thereby prevent enzyme production [39]. Thermized milk has markedly better microbial quality during refrigerated storage than raw milk, but the net keeping quality depends on the precise heating regime applied and the raw milk quality; the natural antimicrobial system in milk (i.e., the lactoperoxidase-thiocyanate-hydrogen peroxide system) is unaffected by thermization. Samelis et al. [40] studied in detail the impact of thermization at 60°C or 67°C for 30 s on the microflora of caprine or ovine milk for cheese-making, and in particular characterized inactivation of nonstarter lactic acid bacteria, such as lactococci and bacilli, which may subsequently contribute to cheese ripening.

Other thermization (i.e., subpasteurization) treatments may be used to prolong the shelf life of cultured milk products (either before or after packaging), by inactivating starter and other microflora in the product, preventing defects such as postacidification. Thermization also reduces the stability of extracellular proteinases and lipases produced by *Pseudomonas* bacteria to subsequent heat treatment, for example, in UHT processes, by inducing so-called low-temperature inactivation [41,42]. However, in some cases, such low-temperature inactivation may be due to the formation of enzyme–casein aggregates, rather than autoproteolysis, which can occur in the absence of milk proteins [43].

It is possible that thermization also stimulates the germination of *B. cereus* spores, facilitating their inactivation by subsequent pasteurization. A process known as tyndallization involves heating milk in several successive heat treatments to successively germinate and inactivate spores; however, the efficiency of such processes has never been satisfactorily demonstrated, and they are rarely used [31].

Thermization has little effect on the cheese-making properties of milk and may in fact increase cheese yield by preventing premanufacture hydrolysis of milk protein by bacterial proteinases and by partially reversing low temperature dissociation of β-casein and calcium from casein micelles [44].

Potentially useful indicators for the efficiency of thermization in the ranges 50°C–60°C or 60°C–70°C are the lysosomal enzyme α-L-fucosidase and the enzyme phospho-hexose-isomerase, respectively [45,46]. The potential use of α-L-fucosidase as an index of thermization is influenced by the variability in its activity in raw milk, which was studied in detail by Di Noni [47]; this author concluded that this enzyme may not be an ideal indicator for such processes, as raw milk samples with high levels of the enzyme could potentially yield false-negative results.

Inosine may also be used as an index of heating at 62°C, at which temperature its concentration increases with holding time, although measurement of this compound may not differentiate thermization from pasteurization processes [39]. It is normally accepted that thermized milk should retain active alkaline phosphatase, which may allow it to be distinguished from pasteurized milk; Levieux et al. [48] studied the inactivation kinetics of alkaline phosphatase on heating milk at 50°C–60°C for 5–60 min using a monoclonal-antibody-based immunoassay and concluded that this method could be used to determine mild heat treatments applied to milk.

11.7 PASTEURIZATION OF MILK

11.7.1 Technology of Pasteurization

Pasteurization of milk is probably the largest volume liquid processing operation in the modern food industry and is consequently a highly optimized and controlled procedure. While a few plants (e.g., small cheese factories) may still employ the older LTLT process (heating milk to 62°C–65°C for 30 min), most processes today are of the HTST type (e.g., heating milk to 72°C–74°C for 15–30 s).

HTST pasteurization is performed in plate heat exchangers, which are highly efficient due to extensive use of heat regeneration, whereby the majority of heat required to heat the cold incoming raw milk is supplied by the outgoing hot pasteurized milk, which is thereby cooled in the process. Regeneration is never completely efficient and pasteurizers generally include a heating section where the milk is brought to the final pasteurization set temperature by a hot water circuit, and a final cooling section where the pasteurized milk is cooled to 5°C by chilled water. In a liquid milk processing plant, other unit operations such as separation (and standardization of fat content) and homogenization are often included in line with the pasteurizer, the milk being routed out midway through the regeneration section, when it is at the optimal temperature for these processes to be applied, without requiring separate heating and cooling of milk for that purpose.

Modern pasteurization plants include a number of safety features designed to ensure the safety of the final product and suitability for consumption. These include

- Temperature sensors, either at the end of the heating or holding sections, controlling a flow diversion valve that rejects milk which has not attained the required set temperature
- External insulated holding tubes to ensure correct holding times at the pasteurization temperature
- Booster pumps to increase the pressure of the pasteurized milk flowing through the regenerative heating sections, so that, if leaks occur, pasteurized milk will flow into raw milk and not vice versa

The technology of pasteurization was described in detail by Kelly and O'Shea [3]; a diagram of a typical modern HTST pasteurization plant is presented in Figure 11.2.

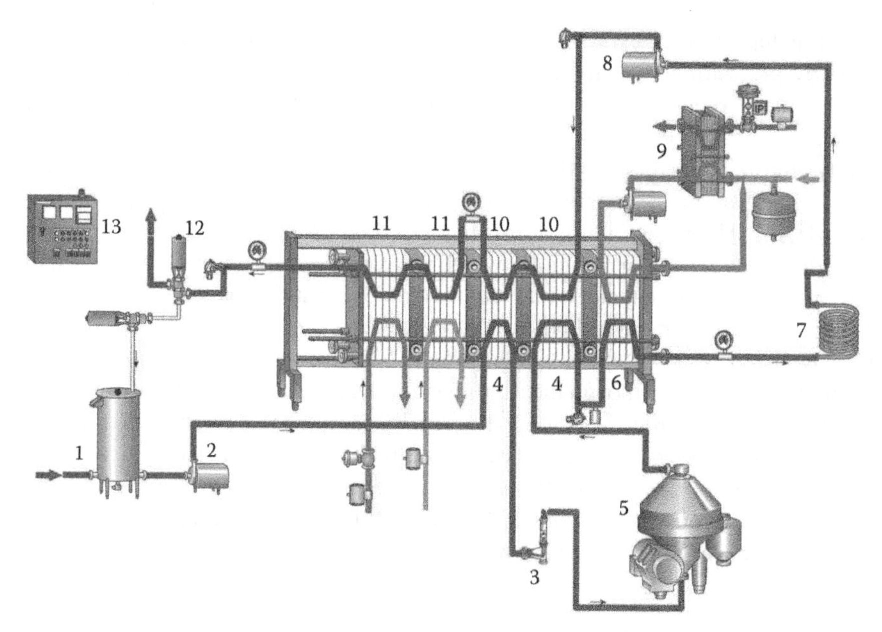

FIGURE 11.2 A complete milk pasteurization plant, including (1) balance tank, (2) feed pump, (3) flow controller, (4) regenerative preheating sections, (5) centrifugal clarifier, (6) heating section, (7) holding tube, (8) booster pump, (9) hot-water heating system, (10) regenerative cooling sections, (11) cooling sections, (12) flow diversion valve, and (13) control panel. (Reproduced by courtesy of Tetra Pak A/B, Lund, Sweden.)

The time–temperature combination traditionally used for HTST (72°C–74°C for 15 s) was based on the thermal inactivation kinetics of two bacteria (*Mycobacterium tuberculosis* and *Coxiella burnettii*) then considered to be the most heat-resistant vegetative pathogenic bacteria likely to be present in raw milk; the similarity of their death kinetics to the inactivation kinetics of the indigenous enzyme, alkaline phosphatase, made that a particularly suitable indicator enzyme, or time–temperature integrator (TTI), for determination of the efficacy of pasteurization.

In recent years, evidence for the survival of *Listeria monocytogenes* and, in particular, *Mycobacterium avium* subsp. *paratuberculosis* (see later) in pasteurized milk has raised industry concern, and this has become the subject of ongoing research [49,50].

11.7.2 Effects of Pasteurization on Milk Constituents

Overall, HTST pasteurization has little effect on the constituents of milk; whey protein denaturation is minimal (<5%); and the texture, flavor, and color of pasteurized milk should compare well to that of raw milk [51]. The extents of denaturation of immunoglobulins and agglutinins, which affect the antimicrobial and creaming properties of milk, respectively, are minor at 72°C but increase at higher temperatures.

As mentioned earlier, the milk constituent of perhaps greatest significance in terms of pasteurization is the enzyme, alkaline phosphatase [52]. A negative result (residual activity below a set of maximum value) in a phosphatase test (assay) is regarded as an indication that milk has been pasteurized correctly [53]. Detailed kinetic studies on the thermal inactivation of alkaline phosphatase under conditions similar to pasteurization have been published [54,55].

The antimicrobial lactoperoxidase system is affected little by pasteurization at 72°C, but higher temperature treatments have progressively greater effect [57]. To test for overpasteurization (excessive heating) of milk, lactoperoxidase, being more heat-stable than alkaline phosphatase and being inactivated at ~80°C, is often used as an indicator enzyme (Storch test [57]); lactoperoxidase may also be used as an index of the efficacy of pasteurization of cream, which must be heated more severely than milk to ensure the killing of target bacteria, due to the protective effect of the fat therein. Detailed models for the inactivation kinetics of both alkaline phosphatase and lactoperoxidase in milk samples of different fat contents were recently developed by Claeys et al. [56]. Recently, other enzymes have been studied as TTIs for heat treatments (particularly at temperatures above 72°C) that may be applied to milk; these include catalase, lipoprotein lipase, acid phosphatase, *N*-acetyl-β-glucosaminidase, and γ-glutamyl-transferase [53,57].

11.7.3 Effects of Pasteurization on Milk Safety

The most important pathogens that can grow in milk at refrigeration temperature, together with their resistance to pasteurization, are summarized in Table 11.2. As stated already, pasteurization was designed to destroy vegetative pathogenic microorganisms, and the twin hurdles of pasteurization and refrigeration are generally sufficient to ensure the safety of pasteurized milk. For example, *Staphylococcus aureus*, a mastitis pathogen, produces a heat-stable enterotoxin that causes food poisoning after consumption of toxin-containing food. However, under normal circumstances, this pathogen does not pose a threat in pasteurized milk since it cannot grow or produce toxin at refrigeration temperatures and is readily killed by pasteurization. Similarly, while many strains of *Escherichia coli*, a commensal microbe found in the gut of cows and also in raw milk, can grow rapidly at refrigerated temperatures, one particular strain, *E. coli* O157:H7, a food-borne pathogen of international importance, is unable to grow in pasteurized milk at temperatures below 10°C [58]. In addition, it is readily destroyed by pasteurization, and hence, consumption of pasteurized milk does not pose a significant risk from this pathogen [59].

In terms of spore-forming bacteria that can be present in raw milk, *Clostridium perfringens*, a food poisoning bacterium, produces very heat-resistant spores that survive pasteurization, but *Cl. perfringens* does not pose a health hazard in pasteurized milk due to the inability of the spores to germinate and grow at refrigeration temperature. In contrast, *B. cereus* in pasteurized milk has caused food poisoning outbreaks as its spores can survive heat treatment and some strains are capable of growing at low temperatures [60]. However, as was discussed earlier, this bacterium is rarely detected in milk stored below 5°C that has been processed in a properly maintained dairy plant [61].

TABLE 11.2
Heat and Cold Stability of Pathogenic Bacteria in Milk

Organism	Survives Pasteurization	Growth at 6°C
S. aureus	No	No
Campylobacter jejuni	No	No
Salmonella spp.	No	No
E. coli	No	?
L. monocytogenes	No	Yes
Y. enterocolitica	No	Yes
B. cereus	Yes (spores)	Yes[a]
Clostridium spp.	Yes (spores)	No[b]

Source: Adapted from Muir, D.D., *J. Soc. Dairy Technol.*, 49, 24, 1996.
[a] Certain strain only.
[b] Some nonproteolytic spp. can grow.

TABLE 11.3
Major Disease Outbreaks Involving Pasteurized Milk

Causative Organism	Location	Year	Number of Cases	Cause	Reference
S. aureus	United States	1914–1942	29	PPC	[21]
Campylobacter jejuni	United Kingdom	1978–1984	27 3	Improper pasteurization PPC	[69]
Salmonella spp.	United States	1985	1600	PPC	[21,63]
Salmonella Typhimurium	United States	2000	93	PPC	
E. coli	United States	1994	18	PPC	[21]
L. monocytogenes	United States	1983	49	Unknown	[21,64,181]
	United States	1994	45	PPC	
L. monocytogenes	United States	2007	5 (3 deaths)	PPC	[181]
L. monocytogenes					[74]
Y. enterocolitica	United States	1975	217	Poor raw material handling	[21]
	United States	1982	172		[21]
	United States	1995	10	PPC	[21]
	Europe	1985	36	PPC PPC	[187]
B. cereus	Europe	1980	280	Unknown	[60]

Thus, while hygienic milk production and handling, pasteurization and refrigeration are effective measures to ensure the safety of milk, significant problems can arise if pathogenic organisms that can proliferate at low temperatures contaminate pasteurized milk (Table 11.3).

In 1985, an unusually large outbreak of salmonellosis in Chicago, involving more than 16,000 cases, was associated with the consumption of low-fat (2%) pasteurized milk [62]; a subsequent investigation concluded that the pathogen entered the milk as a postpasteurization contaminant. Similarly, *Campylobacter jejuni* in raw or inadequately pasteurized milk was implicated in several outbreaks of gastroenteritis in Great Britain from 1978 to 1984 [67]; some cases were traced to pasteurized milk in glass bottles whose lids had been pecked by magpies and jackdaws, which are probable carriers of this bacterium [69]. *Yersinia enterocolitica*, which causes yersiniosis in children younger than 7 years of age [69], is heat-labile and completely inactivated by pasteurization [70]; however, several outbreaks of yersiniosis have been reported in the United States and Europe [68]. This bacterium can grow in milk during refrigeration and thus poses a potential health hazard in underpasteurized milk and milk contaminated with the organism after heat treatment.

Although *L. monocytogenes* is more heat tolerant than most other nonspore-forming pathogens, pasteurization should result in its total destruction [71]; however, some evidence exists of it being linked to a serious food poisoning incident involving pasteurized milk in the United States during 1983 [21]. However, Rebagialti et al. [72] showed that this pathogen can survive pasteurization by invading phagocytes in the milk. In this way, it can remain viable in pasteurized milk and replicate during subsequent refrigerated storage. Thus, removal of this form of *Listeria* is required to ensure the safety of heat-treated milk. Wen et al. [73] demonstrated that, after being grown at 35°C to its long-term survival phase, *L. monocytogenes* increased its heat resistance, $D_{62.8°C}$ value, by 19-fold. Furthermore, during the transition from log, through stationary and death phases to long-term survival phase, *Listeria* changed its cellular morphology from rods to cocci. The authors hypothesized that the increase in heat resistance during this change was due to cytoplasmic condensation.

In 2007, an outbreak of listeriosis in Boston was attributed to flavored milk. The milk was found to be pasteurized adequately but the *L. monocytogenes* was a postpasteurization contamination [64]. The processing plant had no environmental monitoring program for *Listeria*. A major environmental source of *Listeria* is enclosed areas such as milk meters and within hollow rollers in a conveyer, which are ideal sites for biofilm formation [74].

As indicated in Table 11.3, most of the food poisoning outbreaks due to consumption of milk are believed to have been either due to inadequate heating or to postheating contamination. One means by which pathogenic organisms can contaminate pasteurized milk is by leakage of raw milk into the pasteurized milk in the regeneration section of pasteurizers. This should not occur in properly installed pasteurizers where a booster pump prevents flow of milk from the raw into the pasteurized streams.

Maintaining proper hygiene in raw milk handling and ensuring adequate pasteurization can thus prevent contamination of pasteurized milk with pathogenic microorganisms. However, control of spoilage bacteria in pasteurized milk is more difficult. The main causative agents, thermoduric and psychrotrophic bacteria, are discussed in the following sections.

11.7.4 Effects of Pasteurization on Milk Shelf Life

Thermoduric organisms are those that will survive pasteurization. The main types of these organisms in milk and cream are shown in Table 11.4.

Common examples that can be isolated from pasteurized milk and cream include coryneforms, micrococci, and some streptococci, which all grow very slowly at refrigeration temperatures [76], and also spore-forming organisms such as *Bacillus* spp. whose spores can survive much more intense heating conditions, such as 80°C for 10 min. The spores can be activated, germinated, and grow in heat-treated products under certain conditions, subsequently causing spoilage. As mentioned earlier, some may even compromise the safety of the heat-treated product.

Another approach to extending the shelf life of HTST-pasteurized milk is to use pulsed electric field (PEF) processing in addition to heat treatment. Sepulveda et al. [77] applied PEF treatment (35 kV/cm, total time 11.5 μs) at 65°C directly after thermal pasteurization and, in a separate trial, 8 days after pasteurization. The shelf life was extended from 44 days (HTST) to 60 days (HTST + PEF) or 78 days (HTST + 8 days + PEF). The main effect of PEF was an extension of the lag phase of bacterial growth in the milk.

TABLE 11.4
Heat Resistant Organisms Recovered from Milk and Cream after Heat Treatment

	Isolates Recovered after Heating (%)			
	Milk		Cream	
Bacterial Type	63°C/30 min	80°C/10 min	63°C/30 min	80°C/10 min
Bacillus spp.	54	61	55	65
Coryneform group	46	37	31	35
Other Gram-positive	0	2	0	0
Gram-negative	0	0	14	0

Source: Adapted from Phillips, J.D. et al., *J. Soc. Dairy Technol.*, 34, 109, 1981.

PPC with Gram-negative, heat-sensitive psychrotrophic bacteria is the most important cause of spoilage in pasteurized milk and cream. These bacteria, which produce off-flavors due to enzymatic action when the bacterial count exceeds about 10^7 cfu/mL [30,78], enter the product from nonsterile surfaces of processing and packaging equipment, the air and packaging material. Constant monitoring of the PPC level with sensitive test methods such as the *Psychrofast* test [79] and strict attention to the cleanliness of surfaces with which the product comes in contact after heat treatment will ensure low contamination levels and maximize the shelf life of the product. PPC can be minimized by heating plant surfaces at 95°C for 30 min and packaging under ultra-clean conditions and completely avoided by applying aseptic techniques downstream of the holding tube; however, such facilities are expensive and not normally used for pasteurized milk. The presence of high counts of coliform bacteria in pasteurized milk is classically symptomatic of PPC [80].

As mentioned earlier, psychrotrophic bacteria such as *Pseudomonas* spp. and their heat-stable enzymes can present particular problems for pasteurized milk products, either if growth and enzyme production has occurred before pasteurization, or if PPC occurs.

11.7.5 Significance of Storage Temperature

Storage temperature has a vital role in determining the shelf life of heat-treated products. Ideally, pasteurized milk should be kept at or below 4°C; if the storage temperature is raised to 10°C, the shelf life is reduced by at least 50% [4]. Zadow [81] reported that shelf life is reduced to 50% by raising the temperature of products by only 3°C.

Recently, Bedeltavana et al. [82] investigated PPC caused by *Pseudomonas fluorescens* and, using molecular typing methods, established the route of contamination in a pasteurization plant. Based on a combination of 16S–23S PCR ribotyping with proteolytic and lipolytic profiles of *P. fluorescens*, they found 87% of the isolates were common to those in the condensed water on the filling machine. The condensed water was contaminated with *P. fluorescens* from the raw milk and the contamination was spread *via* aerosols in the open space of the filing areas, with resultant contamination of the filling machine and packaging materials.

Milk from cows with mastitis has an elevated somatic cell count (SCC) and high-SCC milk contains plasmin and lipase that are able to withstand pasteurization [83]. Plasmin hydrolyses the casein to peptides, some of which can cause bitterness [84] and astringency [85] while lipase breaks down milk lipids to free fatty acids (FFA) that cause a rancid flavor [83,86,87]. Ma et al. [83] demonstrated that the bitter flavor appeared at 46 days while the rancid flavor was generated at 68 days in pasteurized milk from high-SCC raw milk, indicating that milk plasmin activity is the key factor limiting the shelf life of pasteurized milk during refrigerated storage. Barbano et al. [88] observed that the bitter off-flavor was detected at 28 days in 2% fat pasteurized milk made from raw milk with a SCC of 340,000 cells/mL (milk bacterial count was <20,000) while the processed milk made from raw milk with a SCC of 25,000 cells/mL and of the same bacterial quality did not produce any bitterness until 55 days. Similarly, Santos et al. [87] demonstrated that the number of days for detection of bitter off-flavor in 2% fat pasteurized milk decreased from 56 to 18 days when the SCC of the raw milk increased from 25,000 to 1,000,000 cells/mL.

When microfiltration was used to remove virtually all of the somatic cells from skim milk, the degree of proteolysis caused by plasmin did not decrease. This indicates that removal of somatic cells does not reduce the plasmin activity in the pasteurized skim milk. This phenomenon is attributable to the fact that the inactive zymogen of the enzyme, plasminogen, is activated by an enzyme produced by the somatic cells, and this activation starts in the early stages of milk production while the milk is still in the udder. Therefore, the use of raw milk with low SCC provides an important option for extending the shelf life of pasteurized milk during refrigerated storage [65].

11.7.6 Emerging Issues

M. avium ssp. *paratuberculosis* (MAP) (*M. paratuberculosis*) was discovered 100 years ago and is the causative bacteria of Johne's disease in cattle. Milk can be contaminated by MAP through the mammary gland of infected cows but also, and more importantly, through contamination of milk during the milking process by feces from infected cows [89].

Since MAP has also been implicated in Crohn's disease in humans [90], reports of its possible survival during HTST pasteurization have caused some concern and sparked research activity in several laboratories [91]. Progress has been slow because of difficulties in growing the organism; it is extremely slow growing (taking up to 4 months to culture for counting purposes [92]) and hence can be overgrown by any organisms whose growth is not inhibited during incubation. Trials to determine the resistance of this bacterium to pasteurization have yielded mixed results with some laboratories reporting complete destruction under simulated commercial conditions while others have reported survival of some organisms in some trials [49,93–95]. Inactivation during pasteurization appears to depend on the strain of the organism and the number of organisms present. Some strains can be inactivated completely by classical HTST conditions (72°C for 15 s) at levels of 100 cfu/mL but small numbers (4–16 cfu/mL) have been shown to survive higher temperature (90°C for 15 s) pasteurization of milk with an initial high (10^6 cfu/mL) MAP count [49].

Increasing the holding time during pasteurization processes has been found to be more effective in inactivating MAP in milk than increasing the pasteurization temperature. Homogenization before pasteurization has been shown to increase the lethality of the heat treatment on MAP. These findings suggest that heat penetration is more important than the intensity of heat applied; this may be related to the tendency of the organism to form clumps [95].

Cronobacter sakazakii, previously known as *Enterobacter sakazakii* [96]), in powdered infant formula has been implicated in disease outbreaks. It is considered as an emerging pathogen and the causative agent of life-threatening illnesses in infants, including meningitis [97]. The mortality rate of infants associated with *C. sakazakii* meningitis has been reported to be 33%–80% [98].

Since the thermal resistance of *C. sakazakii* is insufficient to survive a standard pasteurization process (its $D_{72°C}$ value is only 1.3 s), contamination of powdered infant formula must occur post-pasteurization during drying or packaging [99]. Two possible PPC routes have been suggested; the organism may survive near the drying equipment and enter the product at the filling stage, or dry ingredients may be used in the product without proper pasteurization [100]. In dry conditions, this opportunistic bacterium can survive at temperatures of up to 47°C and has been frequently identified in warm and dry environments, such as in the vicinity of drying equipment in milk powder factories [101]. In an attempt to determine the cause of survival of *C. sakazakii* in powdered infant formulae, Breeuwer et al. [102] observed that it can tolerate higher osmotic and dehydration stresses than other Enterobacteriaceae. Its high tolerance to desiccation is associated with rapid accumulation of trehalose in the cell wall. Osaili et al. [100] reported that extended dry storage of infant milk formulae containing *C. sakazakii* increased its heat tolerance during reconstitution of the milk powder with hot water. This pathogen is also able to form biofilms, which confers protection against environmental stresses and increases its survival in the processing environment, as well as in the milk powder.

Paenibacillus is a genus of Gram-positive, endospore-forming bacteria, originally included within the genus *Bacillus*, which is now recognized as a significant organism in the dairy industry. *Paenibacillus polymyxa*, formerly known as *Bacillus polymyxa* [103], has been reported to be an important spoilage bacterium in pasteurized milk. More recently, Scheldeman et al. [104] isolated *Paenibacillus lactis* from raw and UHT milk, and Fromm and Boor [105] and Ranieri and Boor [106] reported that *Paenibacillus* spp. were the predominant microorganisms in pasteurized milk in New York after refrigerated storage for 17 days.

Paenibacillus spp. are psychrotolerant and capable of forming endospores that can survive thermal processing. The spores, which are present in dairy cattle feed, including silage and feed concentrate [24], can contaminate raw milk and thereby create a route for entering pasteurized milk in a viable form. The spores can germinate in the pasteurization process, and the resulting vegetative cells are able to grow at refrigeration temperature and cause spoilage in pasteurized milk. *Paenibacillus* spp. produce proteases that can cause bitterness in pasteurized milk [107]. Thus, removal of spores such as those of *Paenibacillus* spp. from raw milk using microfiltration in combination with heat treatment would extend the shelf life of the processed milk [88].

11.8 ULTRAPASTEURIZATION OR ESL PROCESSING

The most recent form of heat treatment introduced for milk and other liquid dairy products, such as custard, is ultrapasteurization or ESL processing, which provides additional shelf life for pasteurized products, which is beneficial to both consumers and retailers. It is a continuous heat treatment of intensity between HTST pasteurization and UHT sterilization, and hence, the properties of ESL milk differ from those of both pasteurized and UHT milk. In particular, ESL milk does not have a strong cooked flavor characteristic of UHT milk. ESL milk can also be produced by microfiltration or centrifugation (bactofugation), with or without a pasteurization step [108], but this is not discussed in this chapter.

The conditions of ESL processing vary between processors, since there is no single standard definition for ESL milk; however, a reasonable definition is the following given by Rysstad and Kolstad [109]: *ESL products are products that have been treated in a manner to reduce the microbial count beyond normal pasteurisation, packaged under extreme hygienic conditions, and which have a defined prolonged shelf life under refrigeration conditions.*

A range of different temperature–time combinations for the high-temperature holding section have been suggested for producing ESL milk. Generally, temperatures >100°C for very short times are used, with the most common conditions being 120°C–130°C for ~1–4 s. This results in a reasonable extension of shelf life of milk, with little flavor impairment, provided of course that the final heating and initial cooling stages are not excessively long [110]. Blake et al. [111] concluded that flavor impairment commenced at temperatures above 134°C. Unfortunately, higher pasteurization temperatures, in the range of 72°C–~90°C, are detrimental to the shelf life of milk compared with HTST pasteurization at 72°C and are therefore unsuitable for ESL processing [112]. Two possible reasons for the shorter shelf life of milk heated in this temperature range are the inactivation of the antibacterial enzyme lactoperoxidase and the activation of bacterial spores, which can subsequently germinate and grow with little competition. Ranjith [6] showed that milk treated at temperatures ≤117.5°C had total counts of >10^6 cfu/mL after 13 days, whereas milks treated at temperatures ≥120°C showed counts of <10^2 cfu/mL after storage at 7°C for >40 days. Heating at ≥120°C destroys all nonspore-forming organisms, as well as most psychrotrophic spore-forming organisms such as *B. cereus* and *Bacillus circulans.*

A key aspect of the heating conditions used to produce ESL milk with a flavor similar to that of pasteurized milks is short heating, holding, and cooling times. This is most easily achieved by direct steam heating, either steam injection or steam infusion, for example, in the APV Pure-Lac™, Instant Infusion, and High-Heat Infusion systems [113]. However, many commercial ESL products are produced on plate and tubular indirect heating plants.

ESL milk is not "commercially sterile," which means that some (spore-forming) bacteria that can grow at room temperature are not destroyed by the processing conditions. Consequently, all ESL products have to be stored under refrigeration. The shelf life of refrigerated ESL milk is variously cited from 14 to 90 days. However, with the heating conditions indicated earlier (i.e., 120°C–130°C for 1–4 s), the shelf life is largely determined by the method of packaging. The lowest suitable packaging level is ultraclean, while aseptic packaging is the highest [114]; however, aseptic packaging is a more costly option than ultraclean packaging. As a rule of thumb, a shelf life of up to ~40 days can be achieved with ultraclean packaging, while longer times are possible with aseptic packaging. Since there is always the possibility of postheat-treatment contamination in nonaseptic packaging, long shelf lives can be more consistently achieved with aseptic packaging.

In ESL milk produced by heating at 127°C for 5 s, Mayr et al. [115] found Gram-positive nonspore-forming bacteria to be more common spoilage organisms than Gram-negatives or spore formers. A range of Gram-positive genera, including *Rhodococcus, Anquinibacter, Arthrobacter, Microbacterium,* other coryneforms, *Enterococcus, Staphylococcus,* and *Micrococcus,* were identified. Possible sources of contamination were considered to be the air and packaging materials. In addition, most of the spore formers were considered to be postprocessing contaminants. The most common spore formers in the ESL milk (127°C for 5 s) were *B. licheniformis* (73%), followed by *Bacillus subtilis, B. cereus, Brevibacillus brevis,* and *Bacillus pumilus* [182]. In milks directly heated at 120°C–132°C for 4 s, *B. licheniformis, Bacillus coagulans,* and *B. cereus* were the spore-forming species [111]. Since *B. cereus* can grow at 7°C and is a potential pathogen [6], processors must ensure that there is minimal risk of this organism contaminating ESL milk and proper temperature control must be maintained in the cold chain.

11.9 UHT PROCESSING OF MILK

11.9.1 Technology of UHT and Effects on Milk Constituents and Stability

UHT processing of milk involves heating milk in a continuous process to temperatures higher than 135°C for a few seconds, cooling rapidly and aseptically packaging the milk into sterile containers [116,122]. The sequence of steps in UHT processing is shown in Figure 11.3.

UHT treatment must be sufficient to produce a commercially sterile product (one in which bacterial growth will not occur under normal storage conditions) but not severe enough to cause chemical changes which result in an unacceptable flavor, color, and/or nutritive loss. In general, the heating should be equivalent to a minimum of 9-log reduction of thermophilic spores (sometimes referred to as a B^* value > 1) and a maximum reduction of 3% in the level of thiamine (sometimes referred to as a C^* value < 1). The temperature and holding time commonly used for UHT processing are 135°C–140°C and 2–5 s, respectively. Since the nominal holding time is the average time a particle spends in the holding tube, it is important to ensure that the time taken for the fastest particle is sufficient to inactivate the target microorganisms, that is, *Clostridium* and *Bacillus* spores.

The microbiological failure rate for PPC in UHT products is quite low (0.02%) [11]. Higher failure rates indicate significant inadequacies in the aseptic packaging step. Interestingly, Cerf and Davey [75] concluded that most of the bacteriological failures in UHT milk packages (which they reported as 1–4 per 100,000 packs) could be explained statistically on the basis of residence time distribution whereby a very small percentage of spores pass through the holding tube too fast to be destroyed. This suggests that many of the failures may not be due to PPC.

The spores of some species such as *G. stearothermophilus* and *B. sporothermodurans* are extremely heat resistant and can resist UHT treatment conditions. The latter species is of particular concern to the dairy industry as it is mesophilic and hence can grow at the temperature at which UHT milk is normally stored. It has already caused the closure of some UHT-processing plants [10].

11.9.1.1 Heating Mode

Two main types of UHT process are used commercially: *direct* heating in which milk comes into direct contact with the heating medium, steam; and *indirect* heating in which the heating medium, steam or superheated water, is separated from the milk by a stainless steel plate or wall of a tube. A less common type of indirect heating involves electrical heating of the stainless steel tubes carrying the milk. The electrical resistance of the stainless steel causes what is known as joule heating when an electrical current is passed through the walls of the tubes. In direct systems, heating to, and cooling from, the high temperature is very fast due to transfer of the latent heats of condensation and evaporation, respectively, between the steam and liquid milk; cooling is achieved in a vacuum

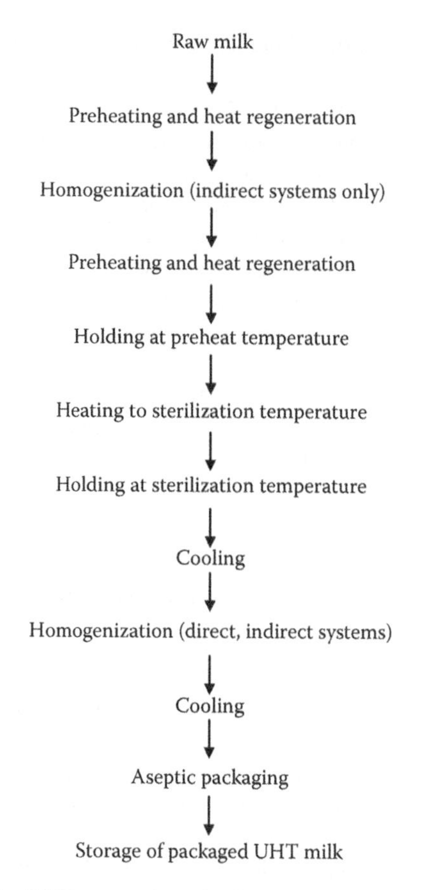

FIGURE 11.3 Steps involved in UHT processing of milk. (From Datta, N. et al., *Aust. J. Dairy Technol.*, 57, 211, 2002.)

chamber. Direct heating requires sterilization temperatures 3°C–4°C higher than indirect heating to achieve an equal sterilization effect because of the greater heat input during the heat-up phase of indirect heating [66].

Systems combining direct and indirect heating are available commercially (e.g. the *High-Heat Infusion* system of APV [9] and the *Tetra Therm® Aseptic Plus Two* system of Tetra Pak [117]. These achieve better energy economy than conventional direct systems (due to greater heat regeneration) and cause greater inactivation of heat-resistant spores and less chemical damage, especially flavor change and burn-on, than conventional indirect UHT systems [118,183]. A variation of direct heating, Innovative Steam Injection (ISI), in which milk is heated by steam injection to temperatures up to 180°C for a fraction of a second, was developed by Huijs et al. [119]. This process is able to inactivate highly heat-resistant spores without causing excessive chemical change in milk.

11.9.1.2 Time–Temperature Profiles

The time–temperature curve or profile of a UHT process, which depends on the configuration of the plant, is a significant factor determining the quality of the final product. The greater the area under the curves at temperatures greater than about 80°C, the greater the chemical, that is, quality, changes induced in the milk. The time–temperature profiles for some UHT systems are shown in Figure 11.4. However, each UHT installation presents a unique profile [110].

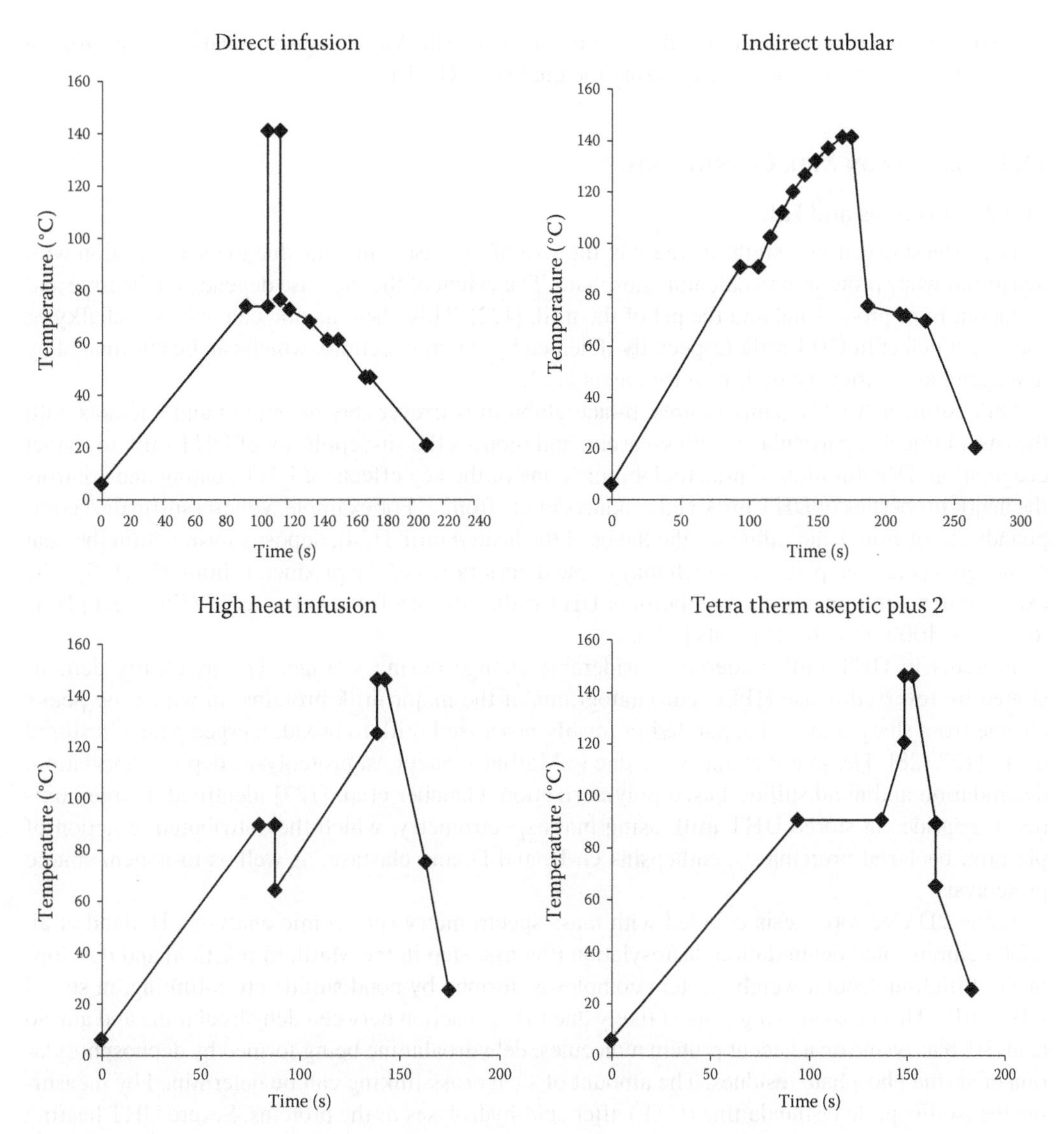

FIGURE 11.4 Time–temperature profiles of various heating modes in UHT processing. (From Datta, N. et al., *Aust. J. Dairy Technol.*, 57, 211, 2002.)

Time–temperature profiles are much more useful for describing the behavior of a particular UHT processing plant than the nominal holding times and temperatures which are usually the only conditions cited. Two plants can have the same holding times and temperatures but differ greatly in their overall effect on milk. This is particularly relevant when direct and indirect plants are compared, and when pilot plants are used to simulate commercial plants [110]. This deficiency can be overcome by considering the overall B^* and C^* values of the plants. These can be calculated from the detailed time–temperature profile of the whole plant and the formulae for the B^* and C^* parameters. The calculations are facilitated by computer programs such as Excel [110,120]) and commercially available programs such as *NIZO Premia* [121]. In addition to B^* and C^* values, the changes occurring in milk constituents, such as denaturation of whey proteins and formation of lactulose, as well as changes such as browning and fouling, can be calculated based on the kinetics of the changes. Therefore, if the time–temperature profile of a plant is known, these programs can be used to predict the changes that will occur in the plant. This knowledge is valuable for optimizing the processing conditions and maximizing the quality of UHT products.

11.9.2 Effects on Milk Constituents

11.9.2.1 Proteins and Fat

UHT processing causes a slight increase in the size of the casein micelle due to its association with denatured whey proteins and calcium phosphate. The extent of the increase depends on the time and temperature of processing, and the pH of the milk [122]. This phenomenon can lead to a chalky or astringent defect in UHT milk (especially if heated by steam injection), which can be eliminated by homogenization after the high-heat treatment [123].

On treatment at UHT temperatures, β-lactoglobulin is irreversibly denatured and interacts with the casein micelle, particularly with κ-casein, and reduces the susceptibility of UHT milk to rennet coagulation. Denaturation of β-lactoglobulin is one of the key effects of UHT heating and controls the major properties of UHT milk and products made from it. For example, volatile sulfhydryl compounds are liberated and influence the flavor of the heated milk [124], deposits form within the heat exchangers [122] and plasmin, which may cause deterioration of the product, is inhibited [125]. The extent of denaturation of β-lactoglobulin in UHT milk can vary from as low as 35% in direct plants to close to 100% in indirect plants [126].

Proteins in UHT milk undergo considerable change during storage. This is clearly demonstrated by reversed-phase HPLC chromatograms of the major milk proteins, in which the peaks change from sharp and well separated in freshly processed milk to broad, merged peaks in stored milks [127,128]. The major changes are due to Maillard reactions, proteolysis, dephosphorylation, deamidation and nondisulfide-based polymerization. Gaucher et al. [127] identified a large number of peptides in stored UHT milk using mass spectrometry, which they attributed to action of plasmin, bacterial proteinases, cathepsins G, B, and D, and elastase, as well as to nonenzymatic proteolysis.

Using 2D electrophoresis coupled with mass spectrometry (proteomic analysis), Holland et al. [129] demonstrated deamidation, lactosylation (the first step in the Maillard reaction) and development of high-molecular-weight protein complexes, formed by nondisulfide cross-linking, in stored UHT milk. This cross-linking is most likely due to the reaction between dehydroalanine and amino acids such as lysine on adjacent protein molecules, dehydroalanine being formed by dephosphorylation of serine phosphate residues. The amount of such cross-linking can be determined by measuring the isodipeptide lysinoalanine (LAL) after acid-hydrolyses of the proteins. Severe UHT heating [130] and storage of UHT milk, especially at high ambient temperatures such as 45°C [128], have been shown to result in the formation of LAL.

UHT processing causes virtually no physical or chemical changes in the structure, properties or nutritional value of milk fat. However, during storage, oxidation can take place, resulting in the development of stale flavor, due to the formation of volatile aliphatic aldehydes and methyl ketones [131]. Active packaging has been proposed to reduce these off-flavors in UHT milk by either absorbing such flavor compounds [132] or reducing oxygen levels in the milk [133].

11.9.2.2 Enzyme Inactivation

The milk alkaline proteinase, plasmin, and its inactive precursor, plasminogen, have considerable heat resistance. They are more susceptible to inactivation by indirect UHT processing than by direct heating; 19% of the original plasmin and 37% plasminogen activity remained in directly processed milk, while no residual plasmin activity and 19% of the plasminogen remained in indirectly processed milk [134,135].

Appropriate preheating conditions have been shown to inactivate plasmin sufficiently to prevent proteolysis, and hence development of bitterness, in UHT milk during storage. Newstead et al. [136]

found that treatment of at least 90°C for 30 s was required to inactivate milk plasmin and prevent proteolysis of directly processed UHT milk during storage, while van Asselt et al. [137] concluded that pretreatment at 80°C for 300 s inactivated plasmin and prevented proteolysis in ISI-treated milk (150°C–180°C for 0.2 s). UHT milk made from poor-quality raw milk, that is, with high bacterial and SCC, is much more susceptible to development of bitterness during storage than was that made from good quality milks [138]. This can be attributed to the action of plasmin and bacterial proteinases. Heat-stable proteinases from pseudomonads have been reported to retain 20%–40% of their activity after exposure to UHT conditions of 140°C for 5 s [139].

11.9.2.3 Protein–Sugar Interactions

UHT treatment causes extensive Maillard reactions in milk; products of such are useful as indicators of the thermal history of the UHT milk and for predicting the sensory and nutritional quality of UHT-processed milk [120,140]. The Maillard reaction also progresses during storage, particularly at elevated temperatures; this can be observed by the appearance of lactosylated proteins, particularly whey proteins, on 2D electrophoresis gels [129].

11.9.2.4 Lactose Isomerization

Lactose isomerizes to lactulose (galactose-fructose) during UHT processing. The extent of isomerization depends on the severity of heat treatment. Lactulose is a reliable index of heat treatment of UHT milk as it is not present in unheated milk and it changes little during storage of the milk [140,141]. Levels of lactulose have been reported to range from 90 to 250 mg/L in directly heated milk and from 310 to 570 mg/L in indirectly heated milk [142]. According to EU Directive 92/46, 1995, the lactulose content of UHT milk should be <600 mg/L, as higher levels are indicative of sterilized milk.

11.9.2.5 Minerals and Vitamins

UHT processing transfers minerals from the aqueous phase to the casein micelle and reduces ionic calcium levels by 10%–20%. This, in addition to the interaction of whey proteins with the casein micelle, reduces the susceptibility of UHT milk to coagulation by rennet. Some calcium phosphate is rendered insoluble at the high temperatures used in UHT heating and deposits on the surfaces of the heat exchanger (fouling).

The amount of calcium in the ionic form in milk has a major influence on the stability of milk during UHT processing [143]. This is particularly significant for fouling [145] and sedimentation [144] of caprine milk, which has a naturally high ionic calcium concentration. These problems can be overcome by reduction of ionic calcium by addition of stabilizing salts, such as sodium citrate or phosphates, or by ion exchange [144,145].

The fat-soluble vitamins (A, D, and E) and some of the water-soluble vitamins (pantothenic acid, nicotinic acid, riboflavin, and biotin) are largely unaffected by UHT treatment, but losses of 20% and 30%, respectively, in thiamine and vitamin B_{12} can occur during UHT treatment [122]. The levels of ascorbic acid and folic acid are markedly reduced in UHT milk containing a significant level of oxygen during UHT processing and storage [146].

11.9.2.6 Flavor Compounds

At least two different types of flavors develop in UHT milk: the "cooked" or "heated" flavor, which develops during processing, and the "stale" flavor, which develops during storage. Volatile sulfur compounds, such as methanethiol, dimethyl sulfide, and dimethyl trisulfide, produced mainly from denatured β-lactoglobulin, are responsible for the "cooked" flavor [147], which disappears rapidly in the presence of oxygen or oxidizing agents. Aliphatic aldehydes are major contributors to the stale flavor [131]. UHT milk kept at 25°C has maximum flavor acceptability between 3 and 5 weeks, that is, after the cooked flavor disappears and before the stale flavor appears [148].

11.9.3 Physical Stability of UHT Milk

In contrast to pasteurized products, storage temperature has a less significant effect on the shelf life of UHT products, as bacterial contamination is rare. However, the temperature of storage has a significant effect on the shelf life of these products through its effect on their physicochemical properties [149]. Several changes that can occur in UHT milk during storage will be discussed in this section.

11.9.3.1 Gelation

Gelation during storage is a common problem of UHT milk, which ultimately limits its shelf life [66]. A major initiating factor is partial proteolysis of the caseins, by either plasmin or residual heat-resistant bacterial proteinases. Milk sterilized by indirect heating gels more slowly during storage than milk treated by direct methods [134]. Similarly, higher processing temperatures delay gelation; Topçu et al. [138] reported that UHT milk processed at 150°C rather than 140°C showed less proteolysis and took longer to gel during storage. These effects appear to be due to the greater inactivation of the proteinases and greater stabilization of the casein micelle by complexation with denatured whey proteins during the more severe indirect heating. The addition of sodium hexametaphosphate (0.1%) to raw milk before processing delays the onset of gelation of UHT milk during storage [149].

11.9.3.2 Fat Separation

Despite homogenization of milk in the UHT process, a layer of fat occasionally develops on the surface of the milk during storage. Less fat separation occurs in milk processed with direct steam injection than with indirect heating [150], due to the additional homogenization effect of the steam injection. The efficiency of homogenization has a major effect on fat separation. Poor homogenization, which can occur if the homogenizer valve seats are damaged, can result in some large fat globules escaping proper homogenization. Only a small percentage of these large globules is sufficient to cause significant fat separation [151].

11.9.3.3 Sedimentation

Sedimentation in UHT milk is due to destabilization of casein micelles. The amount of sediment increases with the time and temperature of heating and with the time and temperature of storage [152]. The pH of the milk is very important as severe sedimentation occurs at $pH<6.5$. Sedimentation occurs more readily in concentrated milk than in normal strength milk but less in reconstituted milk [122]. Goat's milk is particularly susceptible to sedimentation, which has been attributed to its high ionic calcium content. Sediment volume decreases with increasing homogenization pressure and addition of chemicals such as trisodium citrate or disodium hydrogen phosphate (0.025%–0.1%) to raw milk [116,144].

11.10 IN-CONTAINER STERILIZATION OF MILK AND CONCENTRATED MILK

In-container sterilization (e.g., in bottles, cans, or jars), as mentioned earlier, is one of the oldest thermal processing strategies applied to milk, predating even the work of Pasteur. Today, most processors producing long-life milk use UHT systems, due to the advantages of high throughput and continuous operation. In addition, the long times for which products are exposed to high temperatures in sterilization processes result in extensive deterioration in the quality of the products (e.g., extensive development of brown colors due to Maillard reactions and creation of strong cooked flavors, due to production of volatile sulfhydryl compounds from β-lg). Typical treatment conditions used for in-container sterilization processes for dairy products are 110°C–120°C for 10–20 min [153]. Garrote et al. [153] studied in detail the impact of processing variables on the convective heat transfer coefficients during sterilization of canned evaporated whole milk and modeled the impact of process variables such as temperature and rotation speed on heat transfer.

The products most commonly treated in-container today are sterilized concentrated milk and sweetened condensed milk; in the former case, long shelf life is achieved by sterilization applied to the packaged product while, in the latter, thermal treatment reduces the microbial load of the product, but microbial growth is ultimately prevented by the effect of added sucrose on the osmotic pressure in the medium.

Sterilization has traditionally been applied in autoclaves or batch retorts; however, in most plants today continuous sterilization in hydrostatic retorts is used. A diagram of a hydrostatic tower sterilizer is shown in Figure 11.5; the principle of such systems is as follows. A central chamber, in which sterilization is achieved, is maintained at sterilization temperature by steam under pressure, which is counterbalanced on the inlet and discharge sides by columns of water that generate a hydrostatic pressure sufficient to maintain the conditions of temperature and pressure required for sterilization. The water in the inlet "leg" is warmed to achieve preheating of the product, while that in the discharge "leg" is cooled and serves to reduce the temperature of the containers and their contents poststerilization. Presterilization of product in a UHT-type process may be applied before packaging, to reduce the heat load that needs to be applied in the hydrostatic retort.

Other designs of continuous retort for sterilization of milk products include horizontal sealed rotary-valve retorts, in which product is passed through a steam zone in which sterilization is achieved, and horizontal continuous rotating autoclaves ("cooker-coolers"). In the latter system, the design incorporates three cylindrical vessels at different pressures and temperatures through which the product containers pass in a helical manner, thereby progressively moving through the retort, while simultaneously being rotated, to increase rates of heat transfer and reduce thermal damage.

While, as discussed already, milk is a relatively heat-stable system, under certain conditions it can be coagulated during the process of sterilization [154]; the mechanism of heat-induced coagulation of milk (i.e., the heat stability of milk) has been the subject of considerable amounts of research [155,156].

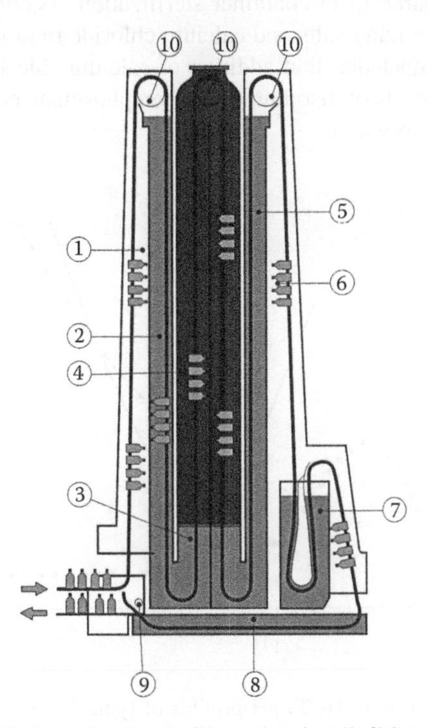

FIGURE 11.5 Hydrostatic vertical continuous sterilizer, showing (1–3) heating stages, (4) steam-filled sterilization section, (5–9) cooling stages, and (10) upper shafts and wheels. (Reproduced by courtesy of Tetra Pak A/B, Lund, Sweden.)

The heat stability of milk (expressed as the number of minutes at a particular temperature, for example, 140°C, before visible coagulation of the milk occurs—the heat coagulation time, HCT) is highly dependent on pH, with large differences in heat stability being apparent over quite a narrow range of pH values (Figure 11.6). Indeed, pH is probably the most significant factor influencing the heat stability of milk. Most milk samples show what is referred to as a type A heat stability-pH profile, with a pronounced minimum and maximum, typically around pH 6.7 and 7.0, respectively; heat stability decreases on the acidic side of the maximum and increases on the alkaline side of the minimum. The rare type B profile does not exhibit a minimum, and heat stability increases progressively throughout the pH range 6.4–7.4.

As well as pH, a number of other factors affect the heat stability of milk [155–157]. These include the following:

- Levels of Ca^{2+} or Mg^{2+}, reduction of which increases stability in the pH range 6.5–7.5
- Hydrolysis of lactose, which increases heat stability throughout the pH range
- Addition of κ-casein, which eliminates the minimum in the type A profile
- Addition of β-lactoglobulin to type B milk, which converts it to a type A profile
- Addition of phosphates, which increase heat stability
- Addition of oxidizing agents, which convert a type A to a type B profile, or reducing agents, which decrease heat stability
- Presence of alcohols and sulfhydryl-blocking agents, which reduce the heat stability of milk

Omoarukhe et al. [157] reported that while addition of insoluble calcium salts to milk had little effect on heat stability, adding soluble calcium salts (particularly calcium chloride) accelerated gelation and also showed that milk became unstable under conditions of UHT processing at lower levels of addition of calcium compared to in-container sterilization. Tsioulpas et al. [154] also studied the impact of addition of stabilizing salts and calcium chloride to raw milk on stability during in-container sterilization, and concluded that addition of calcium chloride promoted coagulation on heating, but addition of low levels of disodium hydrogen phosphate could increase stability during storage, in terms of reducing browning.

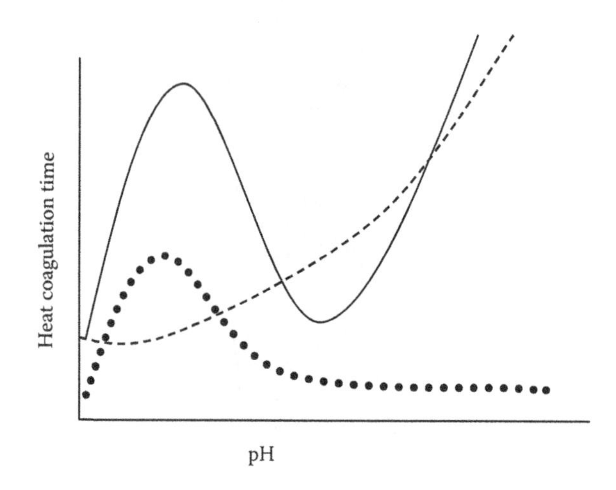

FIGURE 11.6 Heat coagulation time (HCT)-pH profiles of typical type A bovine milk (whole line) and type B or serum protein-free milk (dashed line), as determined at 140°C, or concentrated milk (dotted line), as determined at 120°C. (From Fox, P.F. and Kelly, A.L., in *Proteins in Food Processing*, Yada, R., ed., Woodhead Publishing, New York, 2004.)

Concentrated (e.g., evaporated) milk is much less thermally stable than unconcentrated milk and its heat stability profile, normally assayed at 120°C for that reason, is quite different. It shows a maximum at ~pH 6.4, with decreasing stability at higher and lower pH values (Figure 11.6). The possibility that concentrated milk will coagulate on sterilization may be reduced by preheating the milk, adjustment of the pH of the milk, or addition of stabilizing salts (e.g., polyphosphates).

11.11 THERMAL PROCESSING DURING MANUFACTURE OF OTHER DAIRY PRODUCTS

11.11.1 Evaporation and Spray Drying

The two largest bulk commodity milk powders produced today are WMP and skim milk powder (SMP); a number of different heat treatments may be applied to milk before evaporation prior to manufacture of milk powder. While the principal objective of such heating is to ensure the safety of the final powder, the functionality and applications of the powder are also largely determined at this point. For example, SMP is frequently classified on the basis of its content of remaining native whey protein (expressed as a whey protein nitrogen index, WPNI); milk used for manufacture of so-called low-heat powders has been mildly heated, and the powders contain a high proportion of native whey protein, while high-heat powders have low levels of native whey protein due to more severe heating during processing [158–161]. Faka et al. [162] showed that reducing the concentration of free calcium ions could significantly improve the heat stability of reconstituted low-heat SMP.

The denaturation and aggregation of proteins in milk powders of different heat classification was studied in detail by Patel et al. [158]; the correlation between WPNI and qualitative studies of proteins using sodium dodecyl sulfate (SDS) and capillary electrophoresis was examined in terms of exact levels of denaturation and interaction of individual proteins. Williams et al. [160] concluded that good heat stability for milk powders could be achieved by balancing the level of whey protein denaturation with the formation of aggregates and the proportion of small casein micelles in the milk and thus was influenced both by heating regime applied and perhaps seasonal variability in milk quality.

The subsequent applications of SMP are largely determined by its heat classification. For example, for cheese-making, low-heat powders are recommended, as denatured whey protein would interfere with the rennet coagulation process (see the following). On the other hand, for yogurt, where denatured whey proteins can become a structural element of the acid gel, high-heat powders are favored.

Evaporation plants are typically equipped with flexible thermal processing capabilities, with heat exchangers capable of treating milk at different temperatures, with different holding times, to allow production of powders of different heat classification on a single plant. Preheating may be performed using any of a range of heat exchangers, including plate heat exchangers, spiral heat exchangers wrapped around the tubes in the evaporator itself or using very-short-time steam injection heating systems. Direct heat exchangers are preferred over indirect systems, as biofilms of thermophilic bacteria may develop within indirect heat exchangers [163,164]. *B. cereus* spores in milk powder are a particular health hazard because both reconstitution and pasteurization can induce their germination and outgrowth.

Denaturation at later stages of the powder-manufacturing process (e.g., during evaporation or spray drying) is relatively minor compared to that caused at the preheating stage [165]. The rate of the increase in temperature during preheating may affect whey protein interactions. Slower indirect heating favors whey protein–whey protein interactions and higher overall denaturation of whey proteins than more rapid direct heating methods, which favor extensive casein–whey protein interactions [163,166].

The properties and applications of the major heat classification groups of SMP are outlined in Table 11.5. In manufacture of WMP, heat classification is not used; the milk is generally heated to 85°C to inactivate indigenous lipase activity and expose antioxidant free sulfhydryl groups [167].

TABLE 11.5
Food Applications of Skim Milk Powder (SMP) of Different Heat Classes

Heat Classification	Heat Treatments Applied	WPNI	Functional Properties	Food Applications
Low heat	70°C for 15 s	>6.0	Solubility, lack of cooked flavor	Recombined milk, milk Standardization, cheese
Medium heat	85°C for 1 min 90°C for 30 s 105°C for 30 s	1.5–6.0	Emulsification, foaming, water absorption, viscosity, color, flavor	Ice cream, chocolate, confectionery
High heat	90°C for 5 min 120°C for 1 min 135°C for 30 s	<1.5	Heat stability, gelation, water absorption	Recombined evaporated milk
High–high heat	>120°C for >40 s	<1.5	Flavor, water binding, color	Bakery, recombined evaporated milk

Source: Adapted from Singh, H. and Creamer, L.K., *J. Dairy Res.*, 58, 269, 1991.

11.11.2 Yogurt and Cheese Manufacture

Most cheese is made today from pasteurized milk for public health reasons, although some varieties continue to be made from raw milk, and even a relatively mild treatment such as pasteurization is believed to result in perceptible changes in the quality of such varieties [168].

Severe heat treatment is undesirable for many cheese varieties, especially hard varieties such as Cheddar, principally due to denaturation of whey proteins. Denatured whey protein can interfere with both the primary stage of rennet coagulation (i.e., hydrolysis of the Phe_{105}–Met_{106} bond of κ-casein) and later gel assembly and structure development, in both cases due to steric hindrance at the surface of the micelle due to binding of denatured β-lactoglobulin to κ–casein [169]. Cheese made from milk that has been heated under conditions more severe than conventional pasteurization is generally regarded as yielding an inferior quality product, with defects in flavor and texture.

Nonetheless, there has been significant interest in this area, due to the economic advantages of incorporating whey protein in the curd and thereby increasing cheese yield [170]. Process modifications that can be used to compensate in part for the negative effects of severely heating cheesemilk include adjustment of the pH of the heated milk to ~6.2, to favor the action of chymosin and solubilize some heat-precipitated calcium phosphate. Alternatively, heat and pH adjustment may be applied to whey to yield a microparticulated whey protein precipitate, which can be added to cheese milk and thereby incorporated into the curd (the Centri-whey process [171].

For soft cheese, or acid-coagulated cheese varieties (e.g., Cottage cheese or Quarg), severe heat treatment and high levels of whey protein denaturation are favored and applied quite commonly. Similarly, in the case of fermented milks (e.g., yogurt), high-heat treatment (e.g., 85°C–95°C for 2–30 min) is commonly used to induce extensive whey protein denaturation and association with caseins, in addition to destroying pathogenic and spoilage organisms; in such cases, the denatured whey protein acts as a structural element in the gel, and gives a firmer texture which is less likely to synerese (i.e., expel whey) during storage [172–174]. The mechanism of formation of structure on acidification of heated milk has been studied intensively in recent years [166,175–177]. The denatured whey proteins also become susceptible to aggregation during acidification, as their isoelectric point is approached [178]. Attempts have been made to use UHT heating (which also denatures whey proteins) for yogurt manufacture but the resulting yogurt has lower viscosity and gel strength than yogurt made by the conventional process [179,180].

11.12 CONCLUSIONS

Milk is a complex product from chemical, physicochemical, enzymatic, and microbiological perspectives. Its potential to act as a vector for food pathogens and its inherent tendency to spoil rapidly, even at refrigeration temperatures, necessitate rapid processing to preserve and stabilize the product and protect the consumer. Heat treatment is one of the oldest means for achieving these goals, and a suite of different thermal processes, applied using a range of heat exchanger and process technologies, now exists in the modern dairy industry. However, heating milk has a range of other effects on its constituents and characteristics, which must be considered as part of the net implications of applying any thermal process to milk.

REFERENCES

1. DC Westhoff. Heating milk for microbial destruction: A historical outline and update. *Journal of Food Protection* 41: 122–130, 1978.
2. VH Holsinger, KT Rajkowski, JR Stabel. Milk pasteurization and safety: A brief history and update. *Revue Scientifique et Technique De L'Office International des Epizooties* 16: 441–451, 1997.
3. AL Kelly, N O'Shea. Pasteurizers: Design and operation. In: H Roginski, JW Fuquay, PF Fox, eds. *Encyclopedia of Dairy Science*. London, U.K.: Academic Press, 2002, pp. 2237–2244.
4. M Lewis, N Heppell. *Continuous Thermal Processing of Foods: Pasteurization and UHT Sterilisation.* Gaithersburg, MD: Aspen Publishers, Inc., 2000.
5. G Robertson. The paper beverage carton: Past and present. *Food Technology* 56(7): 46, 48, 50–52, 2002.
6. HMP Ranjith. High temperature pasteurization. *Proceedings of the IDF World Dairy Summit Meeting on Extended Shelf Life (ESL) Milk*, September 16–20, 2000, Dresden, Germany, pp. 1–9.
7. PH Larsen. Microfiltration for pasteurized milk. In: *Heat treatments and alternative methods. International Dairy Federation* 9602: 315–325, 1996.
8. ML Perkins, HC Deeth. A survey of Australian consumers' attitude towards UHT milk. *Australian Journal of Dairy Technology* 56: 28–34, 2001.
9. LB Fredsted. APV Unit Systems—The innovative steps ahead. *Scandinavian Dairy Information* 10: 50–51, 1996.
10. International Dairy Federation. *Bacillus sporothermodurans*—A *Bacillus* forming highly heat-resistant spores. *IDF Bulletin* 357: 1–27, 2000.
11. DD Muir. The microbiology of heat-treated fluid milk products. In: RK Robinson, ed. *Dairy Microbiology*. New York: Elsevier Applied Science, 1990, pp. 209–243.
12. JCA Hustinx, TK Singh, PF Fox. Heat-induced hydrolysis of sodium caseinate. *International Dairy Journal* 7: 207–212, 1997.
13. DJ Oldfield. Heat induced whey protein reactions in milk: Kinetics of denaturation and aggregation as related to milk powder manufacture. PhD thesis, Massey University, Palmerston North, New Zealand, 1998.
14. J O'Brien. Heat-induced changes in lactose: Isomerisation, degradation, Maillard browning. In: PF Fox, ed. *Heat-Induced Changes in Milk.* Special Issue 9501. Brussels, Belgium: International Dairy Federation, 1995, pp. 134–170.
15. MAJS Van Boekel. Effect of heating on Maillard reactions in milk. *Food Chemistry* 62: 403–414, 1998.
16. I Birlouez-Aragon, P Sabat, N Gouti. A new method of discriminating milk heat treatment. *International Dairy Journal* 12: 59–67, 2002.
17. M Feinberg, D Dupont, T Efstathiou, V Louapre, JP Guyonnet. Evaluation of tracers for the authentication of thermal treatments of milks. *Food Chemistry* 98: 188–194, 2006.
18. DD Muir, JD Phillips. Prediction of the shelf life of raw milk during refrigerated storage. *Milchwissenschaft* 39: 7–11, 1984.
19. MJ Lewis. Microbiological issues associated with heat-treated milks. *International Journal of Dairy Technology* 52: 121–125, 1999.
20. DD Muir. The shelf-life of dairy products: 1. Factors influencing raw milk and fresh products. *Journal of the Society of Dairy Technology* 49: 24–32, 1996.
21. ET Ryser. Public health concerns. In: EH Marth, JL Steele, eds. *Applied Dairy Microbiology*. New York: Marcel Dekker, 1998, pp. 263–404.

22. B Pettersson, F Lembke, P Hammer, E Stackebrant, G Priest. *Bacillus sporothermodurans*, a new species producing highly heat-resistant endospores. *International Journal of Systematic Bacteriology* 46: 759–764, 1996.
23. N Klijn, L Herman, L Langeveld, M Vaerewijck, A Wagendrop, I Huemer, A Weerkamp. Genotypical and phenotypical characterization of *Bacillus sporothermodurans* strains, surviving UHT sterilisation. *International Dairy Journal* 7: 421–428, 1997.
24. P Scheldeman, L Herman, S Foster, M Heyndrickx. *Bacillus sporothermodurans* and other highly heat-resistant spore formers in milk. *Journal of Applied Microbiology* 101: 542–555, 2006.
25. BA Law, AT Andrews, ME Sharpe. Gelation of ultra-high-temperature-sterilized milk by proteases from a strain of *Pseudomonas fluorescens* isolated from raw milk. *Journal of Dairy Research* 44: 145–148, 1977.
26. BA Law, ML Sharpe, HR Chapman. The effect of lipolytic Gram-negative psychrotrophs in stored milk on the development of rancidity in Cheddar cheese. *Journal of Dairy Research* 43: 459–468, 1976.
27. DM Adams, TG Brawley. Heat resistant bacterial lipases and ultra-high temperature sterilisation of dairy products. *Journal of Dairy Science* 64: 1951–1957, 1981.
28. GF Senyk, RR Zall, WF Shipe. Subpasteurization heat treatment to inactivate lipase and control bacterial growth in raw milk. *Journal of Food Protection* 45: 513–515, 1982.
29. IG West, MW Griffiths, JD Phillips, AWM Sweetsur, DD Muir. Production of dried skim milk from thermised milk. *Dairy Industries International* 51: 33–34, 1986.
30. MA Schröder, MA Bland. Effect of pasteurization temperature on keeping quality of whole milk. *Journal of Dairy Research* 50: 569–578, 1984.
31. MJ Lewis. Improvements in the pasteurization and sterilisation of milk. In: G Smit, ed. *Dairy Processing: Improving Quality*. Cambridge, U.K.: Woodhead Publishing, Ltd., 2003, pp. 81–103.
32. JV Brown, R Wiles, GA Prentice. The effect of different time-temperature pasteurization conditions upon the shelf life of single cream. *Journal of the Society of Dairy Technology* 32: 109–112, 1980.
33. N Barrett, AS Grandison, MJ Lewis. Contribution of lactoperoxidase to the keeping quality of pasteurized milk. *Journal of Dairy Research* 66: 73–80, 1999.
34. MW Griffiths, Y Hurvois, JD Phillips, DD Muir. Elimination of spore forming bacteria from double cream using sub-UHT temperatures: Processing conditions. *Milchwissenschaft* 41: 403–405, 1986.
35. MW Griffiths, Y Hurvois, JD Phillips, DD Muir. Elimination of spore forming bacteria from double cream using sub-UHT temperatures: Effect of processing conditions on spores. *Milchwissenschaft* 41: 474–477, 1986.
36. B Borde-Kekona, MJ Lewis, WF Harrigan. Keeping quality of pasteurized and high pasteurized milk. In: AT Andrews, J Varley, eds. *Biochemistry of Milk and Milk Products*. London, U.K.: Royal Society of Chemistry, 1994, pp. 157–161.
37. DC Westhoff, SL Dougherty. Characterization of *Bacillus* species isolated from spoiled ultra-high-temperature processed milk. *Journal of Dairy Science* 64: 572–580, 1981.
38. P Hammer, F Lembke, G Suhren, W Heesschen. Characterization of a heat resistant mesophilic *Bacillus* species affecting the quality of UHT-milk. In: *Heat Treatments and Alternative Methods*. Brussels, Belgium: International Dairy Federation; 9602, pp 9–16, 1996.
39. L Stepaniak, EO Rukke. Thermisation of milk. In: H Roginski, JW Fuquay, PF Fox, eds. *Encyclopedia of Dairy Sciences*. London, U.K.: Academic Press, 2002, pp. 2619–2623.
40. J Samellis, A Lianou, A Kakouri, C Delbes, J Rojeli, B Bogovic-Matijasic, MC Montel. Changes in the microbial composition of raw milk induced by thermization treatments applied prior to traditional Greek hard cheese processing. *Journal of Food Protection* 72: 783–790, 2009.
41. RK Owusu, A Makhzoum, P Knapp. Heat inactivation of lipase from psychrotrophic *Pseudomonas fluorescens* P38—Activation parameters and enzyme stability and low or ultra-high temperatures. *Food Chemistry* 44: 261–268, 1992.
42. EP Shokker, MAJS Van Boekel. Mechanism and kinetics of inactivation at 40°–70°C of the extracellular proteinase from *Pseudomonas fluorescens* 22F. *Journal of Dairy Research* 65: 261–272, 1998.
43. L Chen, RM Daniel, T Coolbear. Detection and impact of protease and lipase activities in milk and milk powders. *International Dairy Journal* 13: 255–275, 2003.
44. J Dziuba, B Muzinska. An effect of the low-temperature induced processes of dissociation-association on the molecular state of casein. *Milchwissenschaft* 53: 551–555, 1998.
45. G Zehetner, C Bareuther, T Henle, H Klostermeyer. Inactivation of endogenous enzymes during heat treatment of milk. *Netherlands Milk and Dairy Journal* 50: 215–226, 1996.
46. RC McKellar, P Piyasena. Predictive modelling of inactivation of bovine milk alpha-L-fucosidase in a high-temperature short-time pasteurizer. *International Dairy Journal* 10: 1–6, 2000.

47. I De Noni. Study on the variability of fucosidase activity in bovine milk by means of HPLC. *International Dairy Journal* 16: 9–17, 2006.
48. D Levieux, N Genieux, A Levieux. Inactivation-denaturation kinetics of bovine milk alkaline phosphatase during mild heating as determined by using a monoclonal antibody-based immunoassay. *Journal of Dairy Research* 74: 296–301, 2007.
49. IR Grant, MT Rowe, L Dundee, E Hitchings. *Mycobacterium avium* ssp. *paratuberculosis*: Its incidence, heat resistance and detection in milk and dairy products. *International Journal of Dairy Technology* 54: 2–13, 2001.
50. ET Ryser. Pasteurization of liquid milk products. Principles, public health aspects. In: H Roginski, JW Fuquay, PF Fox, eds. *Encyclopedia of Dairy Sciences*. London, U.K.: Academic Press, 2002, pp. 2232–2237.
51. H Nursten. Heat induced changes in the flavour of milk. In: PF Fox, ed. *Heat-Induced Changes in Milk*. Brussels, Belgium: International Dairy Federation, 1995, pp. 308–317.
52. S Ur-Rehman, NY Farkye. Enzymes indigenous to milk: Phosphatases. In: H Roginski, JW Fuquay, PF Fox, eds. *Encyclopedia of Dairy Sciences*. London, U.K.: Academic Press, 2002, pp. 934–938.
53. RA Wilbey. Estimating the degree of heat treatment given to milk. *Journal of the Society of Dairy Technology* 49: 109–112, 1996.
54. RC McKellar, HW Modler, H Couture, A Hughes, P Mayers, T Gleeson, WH Ross. Predictive modelling of alkaline phosphatase inactivation in a high-temperature short-time pasteurizer. *Journal of Food Protection* 57: 424–430, 1994.
55. W Lu, P Piyasena, GS Mittal. Modelling alkaline phosphatase inactivation in bovine milk during high-temperature short-time pasteurization. *Food Science and Technology International* 7: 479–485, 2001.
56. WL Claeys, AM Van Loey, ME Hendrickx. Kinetics of alkaline phosphatase inactivation and of β-lactoglobulin denaturation in milk with different fat content. *Journal of Dairy Research* 69: 541–553, 2002.
57. RC McKellar, S Liou, HW Modler. Predictive modelling of lactoperoxidase and γ-glutamyl transpeptidase inactivation in a high-temperature short-time pasteurizer. *International Dairy Journal* 6: 295–301, 1996.
58. M Kasrazadeh, C Genigeorgis. Potential growth and control of *Escherichia coli* O157:H7 in soft Hispanic type cheese. *International Journal of Food Microbiology* 22: 289–300, 1994.
59. JY D'Aoust, CE Park, RA Szabo, ECD Todd, DB Emmons, RC McKellar. Thermal inactivation of *Campylobacter* species, *Yersinia entercolitica*, and hemorrhagic *Escherichia coli* O157:H7 in fluid milk. *Journal of Dairy Science* 71: 3230–3236, 1988.
60. P van Netten, A van de Moosdijk, P van Hoensel. Psychrotrophic strains of *Bacillus cereus* producing enterotoxin. *Journal of Applied Bacteriology* 69: 73–79, 1990.
61. A Christiansson. *Bacillus cereus*. In: H Roginski, JW Fuquay, PF Fox, eds. *Encyclopedia of Dairy Sciences*. London, U.K.: Academic Press, 2002, 123–128.
62. CA Ryan, MK Nickles, NT Hargrett-Bean, ME Potter, T Endo, L Mayer, CW Langkop, C Gibson, RC McDonald, RT Kenney, ND Puhr, PJ McDonnell, RJ Martin, ML Cohen, PA Blake. Massive outbreak of antimicrobial resistant salmonellosis traced to pasteurized milk. *Journal of the American Medical Association* 258: 3269–3274, 1987.
63. SJ Olsen, M Ying, MF Davis. Multidrug-resistant *Salmonella* Typhimurium infection from milk contaminated after pasteurization. *Emerging Infectious Diseases* 10: 932–935, 2004.
64. M Cumming, P Kludt, B Matyas, A DeMaria, T Stiles, L Han, M Gilchrist, P Neves, E Fitzgibbons, S Condon. Outbreak of *Listeria monocytogenes* infections associated with pasteurized milk from a local dairy, Massachusetts, 2007. *Morbidity and Mortality Weekly Report* 57: 1097–1100, 2008.
65. D Barbano. The role of milk quality and mastitis control in addressing future dairy food marketing opportunities in a global economy. In: *National Mastitis Council Regional Meeting Proceedings*, New York, 2004, pp. 1–5.
66. N Datta, HC Deeth. Age gelation of UHT milk. *Transactions IChemE Part C-Food and Bioproducts Processing* 79: 197–210, 2001.
67. DN Hutchinson, FJ Bolton, PM Hinchliffe, HC Dawkins, SD Horsley, EG Jessop, PA Robertshaw, DE Counter. Evidence of udder excretion of *Campylobacter jejuni* as the cause of a milk-borne *Campylobacter* outbreak. *Journal of Hygiene–Cambridge* 94: 205–210, 1985.
68. JY D'Aoust. Manufacture of dairy products from unpasteurized milk: A safety assessment. *Journal of Food Protection* 52: 906–914, 1989.

69. T Riordan, TJ Humphrey, A Fowles. A point source outbreak of campylobacter infection related to bird-pecked milk. *Epidemiology and Infection* 110: 261–265, 1993.
70. S Toora, E BuduAmoako, RF Ablett, J Smith. Effect of high-temperature short-time pasteurization, freezing and thawing and constant freezing, on the survival of *Yersinia enterocolitica* in milk. *Journal of Food Protection* 55: 803–805, 1992.
71. MP Doyle, KA Glass, JT Beery, GA Garcia, DJ Pollard, RD Schultz. Survival of *Listeria monocytogenes* in milk during high-temperature, short-time pasteurization. *Applied and Environmental Microbiology* 53: 1433–1438, 1987.
72. V Rebagliati, R Philippi, M Rossi, A Troncoso. Prevention of foodborne listeriosis. *Indian Journal of Pathology and Microbiology* 52: 145–149, 2009.
73. J Wen, R Anantheswaran, S Knabel. Changes in barotolerance, thermotolerance, and cellular morphology throughout the life cycle of *Listeria monocytogenes*. *Applied Environmental Microbiology* 75: 1581–1588, 2009.
74. RB Tompkin. Control of *Listeria monocytogenes* in the food-processing environment. *Journal of Food Protection* 65: 709–725, 2002.
75. O Cerf, KR Davey. An explanation of non-sterile (leaky) milk packs in well-operated UHT plants. *Transactions IChemE Part C-Food and Bioproducts Processing* 79: 197–210, 2001.
76. H Seiler, S Stoer, M Busse. Identification of coryneform bacteria isolated from milk immediately after heating and following refrigerated storage. *Milchwissenschaft* 39: 346–348, 1984.
77. DR Sepulveda, MM Góngora-Nieto, JA Guerrero, GV Barbosa-Cánovas. Production of extended-shelf life milk by processing pasteurized milk with pulsed electric fields. *Journal of Food Engineering* 67: 81–86 2005.
78. HC Deeth, T Khusniati, N Datta, RB Wallace. Spoilage patterns of skim and whole milks. *Journal of Dairy Research* 68: 228–242, 2002.
79. HM Craven, SR Forsyth, PG Drew, BJ Macauley. A new technique for early detection of gram-negative bacteria in milk. *Australian Journal of Dairy Technology* 49: 54–56, 1994.
80. International Dairy Federation. Catalogue of tests for the detection of post-pasteurization contamination in milk. *Bulletin* 281, 1993.
81. JG Zadow, Extending the shelf-life of dairy products. *Food Australia* 41: 935–937, 1989.
82. A Bedeltavana, M Haghkhah, A Nazer. Phenotypic characterization and PCR-ribotyping of *Pseudomonas fluorescens* isolates, in tracking contamination routes in the production line of pasteurized milk. *Iranian Journal of Veterinary Research* 11: 222–232, 2010.
83. Y Ma, C Ryan, DM Barbano, DM Galton, M Rudan, K Boor. Effects of somatic cell count on quality and shelf-life of pasteurized fluid milk. *Journal of Dairy Science* 83: 264–274, 2000.
84. RL Rouoseff. Bitterness in food products: An overview. In: RL Rouoseff, ed. *Bitterness in Foods and Beverages. Developments in Food Science*, Vol. 25. Amsterdam, the Netherlands: Elsevier, 1990, pp. 1–14.
85. L Lemieux, RE Simard. Astringency, a textural defect in dairy products. *Lait* 74: 217–240, 1994.
86. MV Santos, Y Ma, Z Caplan, DM Barbano. Sensory threshold of off-flavors caused by proteolysis and lipolysis in milk. *Journal of Dairy Science* 86: 1601–1607, 2003.
87. MV Santos, Y Ma, DM Barbano. Effect of somatic cell count on proteolysis and lipolysis in pasteurized fluid milk during shelf-life storage. *Journal of Dairy Science* 86: 2491–2503, 2003.
88. DM Barbano, Y Ma, MV Santos. Influence of raw milk quality on fluid milk shelf life. *Journal of Dairy Science* 89(E. Suppl.): E15–E19, 2006.
89. MJ Nauta, JWB van der Giessen. Human exposure to *Mycobacterium paratuberculosis* via pasteurized milk. A modelling approach. *The Veterinary Record* 143: 293–296, 1998.
90. RJ Chiodini. Crohn's disease and the mycobacterioses: A review and comparison of two disease entities. *Clinical Microbiology Reviews* 2: 90–117, 1989.
91. International Dairy Federation. *Mycobacterium paratuberculosis International Dairy Federation Bulletin* 362, 2001.
92. SS Nielsen, KK Neilsen, A Huda, R Condron, MT Collins. Diagnostic techniques for *paratuberculosis*. *International Dairy Federation Bulletin* 362: 5–17, 2001.
93. L Pearce, T Truong, RA Crawford, GF Yates, S Cavaignac, GW de Lisle. Effect of turbulent-flow pasteurization on survival of *Mycobacterium avium subsp paratuberculosis* added to raw milk. *Applied and Environmental Microbiology* 67: 3964–3969, 2001.
94. J Stabel, L Pearce, R Chandler, P Hammer, N Klijn, O Cerf, MT Collins, C Heggum, P Murphy. Destruction by heat of *Mycobacterium paratuberculosis* in milk and milk products. *International Dairy Federation Bulletin* 362: 53–61, 2001.

95. L Pearce, T Truong, RA Crawford, GW de Lisle. Kinetic studies on the heat inactivation of MAP in UHT milk during turbulent flow pasteurization. *Proceedings of Workshop on Revisiting Heat Resistance of Microorganisms in Milk*, May 5–8, 2003, Kiel, Germany, pp. 1–36.
96. C Iversen, N Mullane, BD McCardell, B Tall, A Lehner, S Fanning, R Stephan, H Joosten. *Cronobacter* gen. nov., a new genus to accommodate the biogroups of *Enterobacter sakazakii*, and proposal of *Cronobacter sakazakii* gen. nov., comb. nov., *Cronobacter malonaticus* sp. nov., *Cronobacter turicensis* sp. nov., *Cronobacter muytjensii* sp. nov., *Cronobacter dublinensis* sp. nov., *Cronobacter genomospecies* 1, and of three subspecies, *Cronobacter dublinensis* subsp. *Dublinensis* subsp. nov., *Cronobacter dublinensis* subsp. *lausannensis* subsp. nov. and *Cronobacter dublinensis* subsp. *lactaridi* subsp. nov. *International Journal of Systematic and Evolutionary Microbiology* 58: 1442–1447, 2008.
97. C Block, O Peleg, N Minster, B Bar-Oz, A Simhon, I Arad, M Shapiro. Cluster of neonatal infections in Jerusalem due to unusual biochemical variant of *Enterobacter sakazakii. European Journal of Clinical Microbiology Infectious Diseases* 21: 613–616, 2002.
98. KK Lai. *Enterobacter sakazakii* infections among neonates, infants, children and adults. *Medicine* 80: 113–122, 2001.
99. M Nazaro-White, J Farber. Thermal resistance of *Enterobacter sakazakii* in reconstituted dried infant formula. *Letters in Applied Microbiology* 24: 9–13, 1997.
100. T Osaili, A Al-Nabulsi, R Shaker, M Ayyash, A Olaimat, A Abu Al-Hasan, K Kadora, R Holley. Effects of extended dry storage of powdered infant milk formula on susceptibility of *E. sakazakii* to hot water and ionizing radiation. *Journal of Food Protection* 71: 934–939, 2008.
101. R Shaker, T Osaili, W Al-Omary, Z Jaradat, M Al-Zuby. Isolation of *Enterobacter sakazakii* and other *Enterobacter* sp. from food and food production environments. *Food Control* 18: 1241–1245, 2007.
102. P Breeuwer, A Lardeau, M Peterz, H Joosten. Desiccation and heat tolerance of *Enterobacter sakazakii. Journal of Applied Microbiology* 95: 967–973, 2003.
103. JM Jay. *Modern Food Microbiology*, 6th ed. Gaithersburg, MD: Aspen Publishers, Inc., 2000, pp. 1–625.
104. P Scheldeman, K Goossens, M Rodrı´guez-Dı´az, A Pil, J Goris, L Herman, P De Vos, NA Logan, M Heyndrickx. *Paenibacillus lactis* sp. nov., isolated from raw and heat-treated milk. *International Journal of Systematic and Evolutionary Microbiology* 54: 885–891, 2004.
105. HI Fromm, K Boor. Characterization of pasteurized fluid milk shelf-life attributes. *Journal of Food Science* 69: 207–214, 2004.
106. ML Ranieri, KJ Boor. Short communication: Bacterial ecology of high-temperature, shot-time pasteurized milk processed in the United States. *Journal of Dairy Science* 92: 4833–4840, 2009.
107. NH Martin, ML Ranieri, SC Murphy, RD Ralyea, M Wiedmann, KJ Boor. Results from raw milk microbiological tests do not predict shelf-life performance of commercially pasteurized fluid milk. *Journal of Dairy Science* 94: 1211–1222, 2011.
108. W Hoffmann, C Kiesner, I Clawin-Rädecker, D Martin, K Einhoff, PC Lorenzen, H Meisel, P Hammer, G Suhren, P Teufel. Processing of extended shelf life milk using microfiltration. *International Journal of Dairy Technology* 59: 229–235, 2006.
109. G Rysstad, J Kolstad. Extended shelf life milk—Advances in technology. *International Dairy Journal* 59: 85–96, 2006.
110. H Tran, N Datta, MJ Lewis, HC Deeth. Processing parameters and predicted product properties of industrial UHT milk processing plants in Australia. *International Dairy Journal* 18: 939–944, 2008.
111. M Blake, BC Weimer, DJ McMahon, PA Savello. Sensory and microbial quality of milk processed for extended shelf life by direct steam injection. *Journal of Food Protection* 58: 1007–1013, 1995.
112. D Schmidt, SJ Cromie, TW Dommett. Effects of pasteurisation and storage conditions on the shelf life and sensory quality of aseptically packaged milk. *Australian Journal of Dairy Technology* 44: 19–24, 1989.
113. J Anderson. Instant and high heat infusion. In: P Richardson, ed. *Thermal Technologies on Food Processing*. Cambridge, U.K.: Woodhead Publishing, Ltd., 2001, pp. 229–240.
114. PU Gallmann. Possible quality improvements/benefits of extended shelf life (ESL) milk. *Proceedings of the IDF World Dairy Summit Meeting on Extended Shelf Life (ESL) Milk*, September 16–20, 2000, Dresden, Germany, pp. 1–9.
115. R Mayr, K Gutser, M.Busse, H Seiler. Gram positive non-sporeforming recontaminants are frequent spoilage organisms of German retail ESL (extended shelf life) milk. *Milchwissenschaft* 59: 262–266, 2004.
116. RS Mehta. Milk processed at ultra-high-temperature—A review. *Journal of Food Protection* 43: 212–225, 1980.
117. K Bake, KB Theis. Tetra Therm Aseptic Plus. New perspectives for aseptic processing technology. *Deutsche-Milchwirtschaft* 46: 422–423, 1995.

118. N Datta, AJ Elliott, ML Perkins, HC Deeth. Ultra-high temperature (UHT) treatment of milk: Comparison of direct and indirect modes of heating. *Australian Journal of Dairy Technology* 57: 211–227, 2002.
119. G Huijs, A van Asselt, R Verdurmen, P de Jong. High speed milk. *Dairy Industries International* 6911: 30–32, 2004.
120. E Browning, MJ Lewis, D MacDougall. Predicting safety and quality parameters for UHT-processed milks. *International Dairy Journal* 54: 111–119, 2001.
121. F Smit, P de Jong, J Straatsma, M Verschueren. NIZO Premia as knowledge management tool for industry. *Voedingsmiddelentechnologie* 34: 23–26, 2001.
122. H Burton. *Ultra-High-Temperature Processing of Milk and Milk Products*. London, U.K.: Elsevier Applied Science Publishers Ltd., 1988, pp. 1–354.
123. H Hostettler, K Imhof. Studies on sedimentation and formation of a mealy-chalky texture in UHT steam-injected milk. *Milchwissenschaft* 18: 2–6, 1963.
124. Z Al-Attabi, BR D'Arcy, HC Deeth. Volatile sulphur compounds in UHT milk. *Critical Reviews in Food Science and Nutrition* 49: 28–47, 2009.
125. E Enright, AL Kelly. The influence of heat treatment of milk on susceptibility of casein to proteolytic attack by plasmin. *Milchwissenschaft* 54: 491–493, 1999.
126. RLJ Lyster, TC Wyeth, AG Perkin, H Burton. Comparison of milks processed by the direct and indirect methods of ultra-high-temperature sterilisation. V. Denaturation of the whey proteins. *Journal of Dairy Research* 38: 403–408, 1971.
127. I Gaucher, D Molle, V Gagnaire, F Gaucheron. Effects of storage temperature on physico-chemical characteristics of semi-skimmed UHT milk. *Food Hydrocolloids* 22: 130–143, 2008.
128. JS Al-Saadi, HC Deeth. Cross-linking of proteins and other changes in UHT milk during storage at different temperatures. *Australian Journal of Dairy Technology* 63: 93–99, 2008.
129. J Holland, R Gupta, HC Deeth, PF Alewood. Proteomic analysis of temperature-dependent changes in UHT milk. *Journal of Agricultural and Food Chemistry* 59: 1837–1846, 2011.
130. S Cattaneo, F Masotti, L Pellegrino. Effects of overprocessing on heat damage of UHT milk. *European Food Research and Technology* 226: 1099–1106, 2008.
131. ML Perkins, AJ Elliott, BR D'Arcy, HC Deeth. Stale flavour volatiles in Australian commercial UHT milk during storage. *Australian Journal of Dairy Technology* 60: 231–237, 2005.
132. M Czerny, P Schieberle. Influence of the polyethylene packaging on the adsorption of odour-active compounds from UHT-milk. *European Food Research and Technology* 225(2): 215–223, 2007.
133. ML Perkins, K Zerdin, ML Rooney, BR D'Arcy, HC Deeth. Active packaging of UHT milk to prevent the development of stale flavour during storage. *Packaging Technology and Science* 20: 137–146, 2007.
134. B Manji, Y Kakuda, DR Arnott. Effect of storage temperature on age gelation of ultra-high temperature milk processed by direct and indirect heating systems. *Journal of Dairy Science* 69: 2994–3001, 1986.
135. E. Cauvin, P Sacchi, R Rasero, RM Turi. Proteolytic activity during storage of UHT milk. *Industrie Alimentari* 38: 825–829, 1999.
136. DF Newstead, G Paterson, SG Anema, CJ Coker, AR Wewala. Plasmin activity in direct-steam-injection UHT-processed reconstituted milk: Effects of preheat treatment. *International Dairy Journal* 16: 573–579, 2006.
137. AJ Van Asselt, APJ Sweere, HS Rollema, P de Jong. Extreme high-temperature treatment of milk with respect to plasmin inactivation. *International Dairy Journal* 18: 531–538, 2008.
138. A Topçu, E Numanoglu, I Saldamli. Proteolysis and storage stability of UHT milk produced in Turkey. *International Dairy Journal* 16: 633–638, 2006.
139. MW Griffiths, JD Phillips, DD Muir. Thermostability of proteinases and lipases from a number of species of psychrotrophic bacteria of dairy origin. *Journal of Applied Bacteriology* 50: 289–303, 1981.
140. AJ Elliott, A Dhakal, N Datta, HC Deeth. Heat-induced changes in UHT milks-Part 1. *Australian Journal of Dairy Technology* 58: 3–10, 2003.
141. AJ Elliott, N Datta, HC Deeth. Heat-induced and other chemical changes in commercial UHT milks. *Journal of Dairy Research* 72: 442–446, 2005.
142. MM Calvo, JM Klett, MP Santos, A Olano. Comparative study of lactulose determination in milk by means of gas chromatography and liquid chromatography. *Rev Esp Lech* 16: 11–13, 1987.
143. MJ Lewis. The measurement and significance of ionic calcium in milk—A review. *International Journal of Dairy Technology* 64: 1–13, 2011.
144. T Boumpa, A Tsioulpas, AS Grandison, MJ Lewis. Effects of phosphates and citrates on sediment formation in UHT goats' milk. *Journal of Dairy Research* 75: 160–166 2008.

145. S Prakash, N Datta, MJ Lewis, HC Deeth. Reducing fouling during UHT treatment of goat's milk. *Milchwissenschaft* 62: 16–19, 2007.
146. I Andersson, R Oste. Loss of ascorbic acid, folacin and vitamin B_{12} and changes in oxygen content of UHT milk. II. Results and discussion. *Milchwissenschaft* 47: 299–302, 1992.
147. PA Vazquez-Landaverde, JA Torres, MC Qian. Quantification of trace volatile sulfur compounds in milk by solid-phase microextraction and gas chromatography-pulsed flame photometric detection. *Journal of Dairy Science* 89: 2919–2927, 2006.
148. B Blanc, G Odet. Appearance, flavour and texture aspects: Recent developments. *New monograph on UHT milk. International Dairy Federation Bulletin* 133: 25–49, 1981.
149. HR Kocak, JG Zadow. Controlling age gelation of UHT milk with additives. *Australian Journal of Dairy Technology* 40: 58–64, 1985.
150. JA Ramsey, KR Swartzel. Effect of ultra high temperature processing and storage conditions on rates of sedimentation and fat separation of aseptically packaged milk. *Journal of Food Science* 49: 257–262, 1984.
151. GC Hillbrick, DJ McMahon, HC Deeth. Electrical impedance particle size method (Coulter counter) detects the large fat globules in poorly homogenised UHT processed milk. *Australian Journal of Dairy Technology* 53: 17–21, 1998.
152. KR Swartzel. The role of heat exchanger fouling in the formation of sediment in aseptically processed and packaged milk. *Journal of Food Processing and Preservation* 7: 247–257, 1983.
153. RL Garrote, ER Silva, RD Roa, M Ayala. Determining convective heat transfer coefficient during sterilisation of canned evaporative whole milk. *LWT – Food Science and Technology* 43: 724–728, 2010.
154. A Tsioulpas, A Koliandris, AS Grandison, MJ Lewis. Effects of stabiliser addition and in-container sterilisation on selected properties of milk related to casein micelle stability. *Food Chemistry* 122: 1027–1034, 2010.
155. JE O'Connell, PF Fox. Heat-induced coagulation of milk. In: P Fox, PLH McSweeney, eds. *Advanced Dairy Chemistry. Volume 1. Proteins*. New York: Kluwer Academic–Plenum Publishers, 2003, pp. 879–945.
156. JA Nieuwenhuijse, MAJS van Boekel. Protein stability in sterilised milk and milk products. In: PF Fox, PLH McSweeney, eds. *Advanced Dairy Chemistry. Volume 1. Proteins*. New York: Kluwer Academic–Plenum Publishers, 2003, pp. 947–974.
157. ED Omoarukhe, N On-Nom, AS Grandison, MJ Lewis. Effects of different calcium salts on properties of milk related to head stability. *International Journal of Dairy Technology* 63: 504–511, 2010.
158. HA Patel, SG Anema, SE Holroyd, H Singh, LK Creamer. Methods to determine denaturation and aggregation of proteins in low-, medium- and high-heat skim milk powders. *Lait* 87: 251–268, 2007.
159. H Singh. Interactions of milk proteins during the manufacture of milk powders. *Lait* 87: 413–423, 2007.
160. RPW Williams, L D'Ath, B Zisu. Role of protein aggregation in heat-induced heat stability during milk powder manufacture. *Dairy Science and Technology* 88: 121–147, 2008.
161. Y Miyamoto, K Matsumiya, H Kubouchi, M Noda, K Nishimura, Y Matsumura. Effects of heating conditions on physicochemical properties of skim milk powder during production process. *Food Science and Technology Research* 15: 631–638, 2009.
162. M Faka, MJ Lewis, AS Grandison, H Deeth. The effect of free Ca^{2+} on the heat stability and other characteristics of low-heat skim milk powder. *International Dairy Journal* 19: 386–392, 2009.
163. R Early. Milk concentrates and milk powders. In: R Early, ed. *The Technology of Dairy Products,* 2nd edn. London, U.K.: Blackie Academic and Professional, 1998, pp. 228–300.
164. SA Scott, JD Brookes, J Rakonjac, KMR Walker, SH Flint. The formation of thermophilic spores during the manufacture of whole milk powder. *International Journal of Dairy Technology* 60: 109–117, 2007.
165. H Singh, LK Creamer. Denaturation, aggregation and heat stability of milk protein during the manufacture of skim milk powder. *Journal of Dairy Research* 58: 269–283, 1991.
166. L Donato, F Guyomarc'h, F. Formation and properties of the whey protein/kappa-casein complexes in heated skim milk—A review. *Dairy Science and Technology* 89: 3–29, 2009.
167. G Hols, PJJM van Mil. An alternative process for the manufacture of whole milk powder. *Journal of the Society of Dairy Technology* 44: 49–52, 1991.
168. DH Chambers, E Estever, A Retiveau. Effect of milk pasteurization on flavour properties of seven commercially available French cheese types. *Journal of Sensory Studies* 25: 494–511, 2010.
169. P Kethireddipalli, AR Hill, DG Dalgleish. Protein interactions in heat-treated milk and effect on rennet coagulation. *International Dairy Journal* 20: 838–843, 2010.

170. AL Kelly, T Huppertz, JJ Sheehan. Pre-treatment of cheese milk: Principles and developments. *Dairy Science and Technology* 88: 549–572, 2008.
171. P Piyasena, J Chambers. Influence on whey protein on syneresis of raw milk curds. *International Journal of Food Science and Technology* 38: 669–675, 2003.
172. AY Tamime, HC Deeth. Yoghurt, technology and biochemistry. *Journal of Food Protection* 43: 939–977, 1980.
173. J Mottar, A Bassier, M Joniau, J Baert. Effect of heat-induced association of whey proteins and casein micelles on yoghurt texture. *Journal of Dairy Science* 72: 2247–2256, 1989.
174. JA Lucey, H Singh. Formation and physical properties of acid milk gels: A review. *Food Research International* 30: 529–542, 1998.
175. AJ Vasbinder, AC Alting, RW Vischers, CG De Kruif. Texture of acid milk gels: Formation of disulphide cross-links during acidification. *International Dairy Journal* 13: 29–38, 2003.
176. SG Anema. Heat and/or high-pressure treatment of skim milk; changes to the casein micelle size, whey proteins and the acid gelation properties of the milk. *International Journal of Dairy Technology* 61: 245–252, 2008.
177. M Morand, F Guyomarc'h, MH Famelart. How to tailor heat-induced whey protein/kappa-casein complexes as a means to investigate the acid gelation of milk–A review. *Dairy Science and Technology* 91: 97–126, 2011.
178. JA Lucey, T Van Vliet, K Grolle, T Guerts, P Walstra. Properties of acid casein gels made by acidification with glucono-D-lactone. 2. Syneresis, permeability and microstructural properties. *International Dairy Journal* 7: 389–397, 1997.
179. A Kucukcetin, K Weidendorfer, J Hinrichs. Graininess and roughness of stirred yoghurt as influenced by processing. *International Dairy Journal* 19: 50–55, 2009.
180. A Kucukcetin, K Weidendorfer, J Hinrichs. Effect of heat treatment dry matter on graininess and roughness of stirred skim milk yoghurt. *Milchwissenschaft-Milk Science International* 63: 269–272, 2008.
181. C Dalton, C Austin, J Sobel. An outbreak of gastroenteritis and fever due to *Listeria monocytogenes* in milk. *New England Journal of Medicine* 336: 100–105, 1997.
182. R Mayr, K Gutser, M.Busse, H Seiler. Indigenous aerobic sporeformers in high heat treated (127°C, 5 s) German ESL (extended shelf life) milk. *Milchwissenschaft* 59: 143–146, 2004.
183. HC Deeth, N Datta. Ultra-high temperature treatment (UHT) (B). Heating systems. In: H Roginski, JW Fuquay, PF Fox, eds. *Encyclopedia of Dairy Sciences*. London, U.K.: Academic Press, 2002, pp. 2642–2652.
184. JF Frank, AN Hassan. Microorganisms in milk. In: *Encyclopedia of Dairy Science*, H Roginski, JW Fuquay, PF Fox, eds., Academic Press, London, U.K., 2002, pp. 1786–1796.
185. PF Fox and AL Kelly. In: *Proteins in Food Processing*, Yada, R., ed., Woodhead Publishing, New York, 2004.
186. JD Phillips, MW Griffiths, DD Muir. Factors affecting the shelf-life of pasteurized double cream. *Journal of the Society of Dairy Technology* 34:109–113, 1981.
187. MH Greenwood, WL Hooper, JC Rodhouse. The source of yersinia spp. in pasteurized milk: An investigation at a diary. *Epidemiology and infection* 104: 351–360, 1990.

12 Ultrahigh Temperature Thermal Processing of Milk

Pamela Manzi and Laura Pizzoferrato

CONTENTS

12.1 INTRODUCTION

Ultrahigh-temperature (UHT) milk is the most common form of milk in many parts of the world. The main purpose of heating is to make milk safe for human consumption (by killing pathogenic bacteria) and to extend its shelf life (by killing microorganisms and inactivating enzymes).

In the European Union, according to Council Directive [1], industrial heat treatments of milk can be classified, depending on temperature and time of heating, into the following:

- Pasteurized milk obtained by means of a treatment involving a high temperature for a short time (HTST, at least 71.7°C for 15 s) or a pasteurization process using a different time and temperature combination to obtain an equivalent effect.
- UHT milk obtained by applying a continuous flow of heat at high temperature (not less than 135°C) for a short time (but at least 1 s). This treatment has the aim to destroy all residual spoilage microorganisms and their spores, thus considerably extending milk shelf life, permitting milk to be transported or held for a long period before using. According to the European law, no deterioration should be observed in the UHT milk product after 15 days in a closed container at a temperature of +30°C; where necessary, provision can also be made for a period of 7 days in a closed container at a temperature of +55°C. Two main types of UHT thermal processing are distinguished: direct and indirect heating. If the UHT milk treatment process is performed by direct contact of milk and steam (direct UHT), the steam should be obtained from potable water. Moreover, the use of this process must not cause any dilution of the treated milk. In the indirect UHT thermal processing, tubular, plate, or scraped-surface heat exchangers are employed to transfer heat to milk.
- Sterilized milk heated and sterilized in hermetically sealed wrappings or containers, the seal of which must remain intact.

Obviously, during the heating process, numerous changes in the chemical composition and in the physical and organoleptic characteristics of milk occur, even if heating conditions are now optimized to maintain as much as possible original milk quality. Some of these heat-induced changes are reversible, such as lactose mutarotation, alterations in ionic equilibrium and protein conformation, casein association, agglutination of fat globules, and fat crystallization, while others are irreversible, such as denaturation of whey proteins and their interaction with casein micelles, the Maillard reaction, and changes in fat globule composition and vitamin content [2–4].

Several heat-induced markers [e.g., bioavailable lysine, furosine, hydroxymethylfurfural (HMF), lactulose] related to these modifications have been studied to determine the quality of milk. These so-called heat-damage indicators can be used to control and check the heat treatments given to milk.

In this chapter, we focus on milk subjected to the UHT treatment process, which is today the most common thermal treatment technique for milk. In particular, we discuss heat-damage indicators related to alterations in UHT milk nutrient composition (carbohydrates, protein, lipids, vitamins, and minerals) and to the Maillard reaction, as well as the combined use of some of these indicators. We further discuss the safety and organoleptic aspects of UHT milk in relation to its thermal treatment.

12.2 EFFECT OF UHT TREATMENT ON MILK NUTRIENTS

12.2.1 CARBOHYDRATES

Lactose is the main carbohydrate in milk (about 4.8% in cow's milk); it is a disaccharide composed of D-glucose and D-galactose linked by 1–4 β-glycosidic bonds. Other carbohydrates present in milk include free monosaccharides (glucose, fructose, and galactose), aminosugars (glucosamine and galactosamine), and oligosaccharides. Levels of milk sugars can change as a consequence of heat treatment and storage [5].

12.2.1.1 Lactose

Lactose can isomerize to lactulose (4-O-β-galactopyranosyl-D-fructose) in heat-treated milk via the Lobrey de Bruyn–Alberda van Ekenstein transformation. Lactulose isomerization is catalyzed by free amino groups of casein and is strictly dependent on time of heating, heating temperature, and pH.

Lactulose content in milk can be determined enzymatically (detection limit = 100 mg/L) as well as by gas chromatographic analysis (detection limit = 10 mg/L). An enzymatic-spectrophotometric method with higher detection sensitivity has been developed [6]; this method, with a detection limit of 2.5 mg/L, is particularly useful in samples with low lactulose content. Using this method, Marconi and coworkers [6] analyzed 90 milk samples from different heat treatments: lactulose content ranged from 2 to 1146 mg/L (Table 12.1).

Normally, lactulose does not occur in fresh milk and highly pasteurized milk but only in UHT and sterilized milk. Lactulose content is therefore considered a suitable indicator of heat treatment [7]. It has been proposed to stipulate a lactulose content of less than 600 mg/L for UHT milk and more than 600 mg/L for sterilized milk [8,9].

A study on market milks of Czech origin has shown that lactulose content differs significantly ($P < 0.01$) among milk types (liquid and powder) [10]. The authors further found that lactulose levels in all the UHT milk samples analyzed were below the proposed limit (600 mg/L) necessary for differentiation of UHT from sterilized milk [10]. Average contents were found to be very low (35.5 mg/L) in high-temperature-treated milk (Table 12.1), low in dried milk (229.7 mg/kg), and very high in condensed milk (2451.3 mg/L) [10].

This research confirms the work of Mosso and coworkers [11], who analyzed lactulose in 14 samples of UHT milk and 2 samples of sterilized milk. For the UHT milk samples (whole, skimmed, and partially skimmed), lactulose content ranged from 18.6 to 62.0 mg/100 mL. The two sterilized milk samples had lactulose contents of 115 mg/100 mL (showing a brown color) and 64 mg/100 mL [11].

During storage, lactulose concentration in UHT milk can either decrease (via β-elimination to galactose, tagatose, and formic acid) or increase (via lactose isomerization), the former process

TABLE 12.1
Content of Lactulose (mg/100 mL) in Different Heat-Treated Milk

Minimum	Maximum	Mean	Type of Treatment	References
5	10		UHT direct	[3]
15	75		UHT indirect	[3]
49.3	114.7	74.4	Sterilized	[6]
27.5	42.1	34.8	UHT indirect	[6]
12.6	23.3	16.5	UHT direct with injection system	[6]
8.3	12.0	10.7	UHT direct with infusion system	[6]
3.2	7.9	5.8	High temperature pasteurized	[6]
0.2	0.6	0.4	Pasteurized	[6]
		35.5	High temperature pasteurized	[10]
19.7	62.0		UHT	[11]
		64.0	Sterilized—whole milk	[11]
		115.0	Sterilized—low-fat milk	[11]
11.7	85.3		UHT direct	[120]
54.3	65.0		UHT indirect	[120]

being more temperature dependent than the latter [12]. Andrews [12] found that light did not affect lactulose concentration during milk storage at 18°C for 115 days. Lactulose formation increased as a direct function of OH concentration and lactose concentration but was unaffected by protein concentration or dissolved oxygen.

Recently, lactulose has also gained interest in the field of clinical nutrition. Being an indigestible sugar, lactulose has been found to promote the growth of *Bifibacterium bifidus* and hence could be useful as a prebiotic compound [13].

12.2.1.2 Free Monosaccharides and Aminosugars

To verify whether galactose might be used as an indicator to distinguish between different thermal processes applied to milk, Romero and coworkers [14] examined the evolution of galactose in four commercial whole UHT milk samples (two directly and two indirectly heat-treated products) during storage at temperatures of 6°C, 20°C, 30°C, 40°C, and 50°C until milk expiration date (90 days). Galactose concentration (determined by an enzymatic method) remained constant only in samples stored at 6°C, while the concentration progressively increased at all other storage temperatures. No significant differences in galactose formation were observed between samples with respect to direct or indirect heat treatment or initial galactose content [14].

Troyano and coworkers [15] found that the content of monosaccharides (glucose and galactose) and aminosugars (*N*-acetylglucosamine, *N*-acetylgalactosamine, and myo-inositol) ranged from 2 to 14 mg/100 mL in just-processed UHT milk. Some authors have observed an increase in monosaccharide and aminosugar content depending on time and temperature of storage [15], the increase being particularly clear in UHT milk derived from raw milk with a high psychrotrophic activity.

Storage effects on the levels of galactose and *N*-acetylglucosamine in UHT milk were investigated in two commercial and two experimental batches of UHT milk (whole and skim) stored at 10°C–30°C for 120 days [16].

In agreement with the aforementioned research, an increase in galactose and *N*-acetylglucosamine contents with time was observed, the increase being more pronounced at higher storage temperatures. Increase in monosaccharide content is probably due to dephosphorylation of the respective sugar phosphates (levels of which decrease at 10°C–20°C), although the authors hypothesized that during storage at 30°C, changes in galactose levels may also be due to other processes, possibly involving glycosidases [16].

With the aim to understand the biochemical processes taking place during the shelf life of UHT milk, Recio and coworkers [17] studied changes in the free monosaccharide fraction and noncasein N contents of five batches of commercial UHT milk during 3 months of storage. They observed that during storage, milk batches with a high residual proteolytic activity showed a considerable increase in galactose, *N*-acetylglucosamine, and *N*-acetylgalactosamine with time, whereas glucose and myo-inositol contents remained unaltered. No changes occurred in batches with a slight or negligible proteolytic activity.

12.2.2 Proteins

Milk proteins can be classified into two main fractions: casein (about 80% of total milk protein) and whey proteins (about 20%). The casein fraction, composed of α-s1-, α-s2-, β-, and κ-caseins, precipitates at pH 4.6. In particular, κ-casein can be hydrolyzed by rennin to give para-κ-caseinate that is insoluble in the presence of calcium ions and precipitates as a curd. Usually, casein exists in colloidal particles, so-called casein micelles, containing calcium and phosphorus, some enzymes, citrate, and milk serum.

Whey proteins do not precipitate at pH 4.6; they are water soluble and sensible to temperature. The main components of whey protein are β-lactoglobulin, α-lactalbumin, bovine serum albumin, and immunoglobulins. β-Lactoglobulin represents about 50% of the total whey protein and is the main whey protein in bovine, ovine, caprine, and buffalo milks, while α-lactalbumin is the principal protein in human milk.

12.2.2.1 Denaturation of Whey Protein

Several physical and chemical changes occur in whey protein during thermal processing of milk, and these changes (denaturation) affect the functional and sensory properties of milk. During the heating process, whey proteins containing sulfhydryl residues undergo various changes, resulting in the formation of (1) a protein complex between β-lactoglobulin and κ-casein, with consequent modification of rennet coagulation behavior and heat stability, (2) typical off-flavors, and (3) unusual amino acids (lysinoalanine) [18]. In particular, under certain conditions of concentration and pH, β-lactoglobulin is able to form gels [19,20].

Whey proteins show different thermal stabilities: α-lactalbumin > β-lactoglobulin > bovine serum albumin > immunoglobulins. While denaturation of α-lactalbumin can be a good parameter to describe high-temperature treatments, such as sterilization, denaturation of β-lactoglobulin is useful to describe thermal treatments from pasteurization to UHT processing [4,21].

α-Lactalbumin is frequently chosen as an indicator of heat treatment. To characterize its native and heat-denatured forms, monoclonal antibodies specific in inhibition enzyme-linked immunosorbent assay (ELISA) have been proposed [22]. Using this method, it is possible to differentiate among raw, pasteurized, UHT, and sterilized milk even if the α-lactalbumin concentration of the original raw milk is unknown. However, as mentioned earlier, this technique is mainly suited to detect UHT treatment and sterilization because of the heat stability of α-lactalbumin [22].

Fukal and coworkers [23] investigated the use of immunochemical probes to discriminate among different heat treatments of milk. They raised polyclonal antibodies against 10 immunogens: five native milk proteins, α + β-casein, κ-casein, whole casein, α-lactalbumin, and β-lactoglobulin, and the corresponding five pasteurized milk proteins. Their results showed no significant differences in the immunoreactivity of anticasein antibodies in heat-treated milk compared to raw milk. In contrast, using a combination of antibodies against native α-lactalbumin and β-lactoglobulin, immunoreactivity decreased in heated samples and allowed ($P < 0.001$) a categorization of milk as raw, pasteurized, UHT treated, or bath sterilized [23].

Irreversible denaturation of whey proteins during heating of milk under commercial processing conditions has been widely studied. Oldfield and coworkers [24] observed that native protein concentration decreased with increased heating time at all temperatures studied. Kinetics and thermodynamics of denaturation varied among the whey proteins: temperature dependence was revealed for α-lactalbumin and β-lactoglobulin. Aggregation of lactoglobulin was due to disulfide linkages, while aggregation of α-lactalbumin was apparently more complex, involving the formation of disulfide-linked aggregates and hydrophobic interactions [24].

12.2.2.2 Proteolysis

During UHT milk storage, one of the main problems limiting shelf life is gelation. Gelation is caused by modifications of casein micelle aggregation, producing a 3D network. The mechanism is not completely known, but it is probably due to proteolysis induced by either heat-resistant native proteinases (plasmin) or microbial proteinases. While plasmin, a native alkaline serine proteinase, is almost completely inactivated by the UHT process, its precursor, plasminogen, is particularly resistant to heat. Plasminogen is converted to plasmin by an activator whose activity increases after heat treatment [25].

Usually, plasmin attacks β-casein and α-2-casein, while the microbial proteinases prefer to attack κ-casein. Note that as little as 1 ng of bacterial proteinase per milliliter is sufficient to cause gelation of UHT milk during storage [26].

It is possible to discriminate between the types of proteolysis by reversed phase high-pressure liquid chromatography (RP-HPLC) and the fluorescamine method: peptides produced by bacterial proteinases are less hydrophobic and elute earlier on RP-HPLC than those produced by a plasmin activity. Acid precipitation allows fractionation of peptides; in the pH 4.6-soluble fraction are total proteolysis peptides, whereas the 12% trichloroacetic acid-soluble fraction contains those generated by bacterial proteinases only. If proteolysis is caused by plasmin, this may be a sign of the UHT processing conditions (notably direct UHT processing) being too mild; if proteolysis is caused by bacterial proteinases, this may be a sign of low quality of raw milk [25].

Effects of heat treatment on the plasmin system in milk, with particular reference to plasminogen activators derived from somatic cells, were studied in depth by Kennedy and Kelly [27]. Milk samples were heated at various temperatures (63°C–138°C) for set times (10–90 s), simulating industrial procedures. Results showed that thermal plasmin inactivation is a first-order reaction, under UHT treatment, and that plasmin is less stable in high (800,000 cells/mL) than in low (<150,000 cells/mL) somatic cell count milk. On the contrary, the results also showed that at lower temperatures (63°C–90°C), plasmin is slightly more stable in high than in low somatic cell count milk. Moreover, plasminogen activators derived from somatic cells were confirmed to be stable over the entire range of temperatures studied [27]. The correlation between para-κ-casein and related peptides, and proteolysis was recently investigated in stored raw, pasteurized, and UHT milk samples [28]. Para-κ-casein-derived compounds were observed in raw and pasteurized milk stored at 6°C for more than 5 days, and in UHT milk stored at 20°C for 30, 60, or 90 days. The presence of these compounds was attributed to the action of proteinases from psychrotrophic bacteria on κ-casein [28].

Garcia-Risco and coworkers [29] investigated the effect of storage on proteolysis. They studied protein degradation and distribution of caseins and whey proteins in the soluble (supernatant fraction) and colloidal phases (pellet fraction) during storage for 5 months at room temperature in six commercial UHT milks using capillary electrophoresis and sodium dodecyl sulfate-polyacrylamide gel electrophoresis (SDS-PAGE). During storage, a decrease of all caseins and, in particular, κ-casein was observed due to proteolytic activity.

Cauvin and coworkers [30] found that storage temperature did not significantly affect proteolysis of casein during storage of UHT milk, although casein concentration decreased and proteose–peptone concentration increased in UHT milk samples from two factories, stored at 4°C and room temperature for 3 months. This was attributed to a plasminogen-derived activity.

The effect of homogenization of milk on proteolysis varies depending on whether it precedes or follows heating. Homogenization of milk before UHT treatment has been found to decrease activities of proteinase, glycosidase, and phosphatase during storage. During homogenization, casein and whey proteins interact with the fat globule membrane, the morphology of casein micelles changes, giving rise to small micellar particles, and a complex formation between κ-casein and β-lactoglobulin occurs. All of these aspects may be responsible for the reduced proteolytic degradation observed in homogenized UHT milk samples [31]. In particular, heat-induced binding of the whey proteins α-lactalbumin and β-lactoglobulin to milk fat globule membranes has been studied: α-lactalbumin behavior did not change with temperature, and the total amount of protein bound was ~0.2 mg/g fat. On the other hand, β-lactoglobulin interaction with milk fat globules did increase with temperature between 65°C and 85°C, from 0.2 to 0.7 mg/g fat [32].

Homogenization after heating facilitates proteolysis in skim milk that contains high levels of soluble β-lactoglobulin and κ-casein, with mainly κ-casein susceptible to degradation. In contrast, whole milk samples show the lowest levels of proteolysis, probably due to an increased complex formation between κ-casein and β-lactoglobulin [33].

12.2.2.3 Free Amino Acids

Heat treatments could potentially affect the content of free amino acids in milk. However, concentration of free amino acids in milk heated by thermization/pasteurization, UHT, and sterilization processes is reported to be not markedly influenced by temperature or holding time during processing [34].

Gandolfi and coworkers [35] investigated the occurrence in milk of free and bound D-amino acids. Significant amounts of free D-alanine, D-aspartic acid, and D-glutamic acid were found in raw cow's milk. The amount of D-amino acids in milk did not increase by pasteurization, UHT treatment, or sterilization. In contrast, D-alanine content in raw milk samples increased during cold storage at 4°C, and hence, it was suggested that D-alanine might be considered an indicator of bacterial milk contamination [35]. Indeed, the presence of D-amino acids in milk, and in food, in general, can be ascribed not only to thermal racemization but also to microbial activity [36].

In processed foods, some unusual amino acids can be detected that do not naturally occur in raw foods. Their presence is due to processing conditions, such as heat and alkali treatments of foods. In particular, the production of lysinoalanine, an unnatural amino acid, is a two-step process consisting of (1) production of dehydroalanine from the degradation of cystine, *O*-phosphoserine, or *O*-glycoserine and (2) reaction of dehydroalanine with the free amino group of lysine to form lysinoalanine [37]. The molecule of lysinoalanine seems to reduce protein digestibility due to inhibition of proteolytic metalloenzymes [37]. Moreover, lysinoalanine formation causes a decrease in lysine bioavailability for nutritionally important, high-lysine proteins such as casein. The presence of lysinoalanine in milk-based infant formulas [38] and in different processed milks [39] has been investigated in order to verify whether lysinoalanine, determined by HPLC method, can be a good indicator of heat treatment of milk. Results confirmed that lysinoalanine content is lower in raw and pasteurized milk than in UHT milk (Table 12.2). Lysinoalanine can also be utilized as a sensitive indicator in dairy products in which caseinate has been added [40].

12.2.2.4 Protein Quality and Digestibility

Heat treatments might also negatively affect milk nutritional quality and, in particular, protein digestibility. Efigenia and coworkers [41] evaluated the effect of heat treatment on milk nutritional quality for samples of raw, pasteurized, domestic boiled, and UHT milk: fat, protein, and N contents were not significantly modified by any of these treatments. Protein efficiency ratio, digestibility, and in vitro liver protein synthesis, studied by feeding trials in rats, were unaffected by heat treatments, with the exception of a lower rate of protein synthesis in the liver of rats fed with boiled milk [41].

Alkanhal and coworkers [42] studied the modification of protein digestibility and quality in UHT milk during storage by comparing apparent digestibility, true digestibility, biological value, and net protein utilization in Sprague-Dawley rats fed with pasteurized, fresh UHT, and reconstituted (obtained from skim milk powder) UHT milk, all stored for 0, 3, or 6 months at 37°C, as a source of protein in their diet. Results of this study are reported in Figure 12.1. Interestingly, biological value and net protein utilization of pasteurized milk did not change with storage time, while biological value and net protein utilization of UHT milk were lower for stored milk than directly used

TABLE 12.2
Content of Lysinoalanine (mg/1000 g Protein) in Different Heat-Treated Milk

Minimum	Maximum	Type of Treatment	References
<0.4		Pasteurized	[40]
<15		Pasteurized	[37]
4	24	Raw	[39]
17	47	Pasteurized	[39]
13	69	High pasteurized	[39]
49	186	UHT	[39]
<400		UHT	[37]
224	653	Sterilized	[39]
6	90	UHT direct	[120]
48	61	UHT indirect	[120]

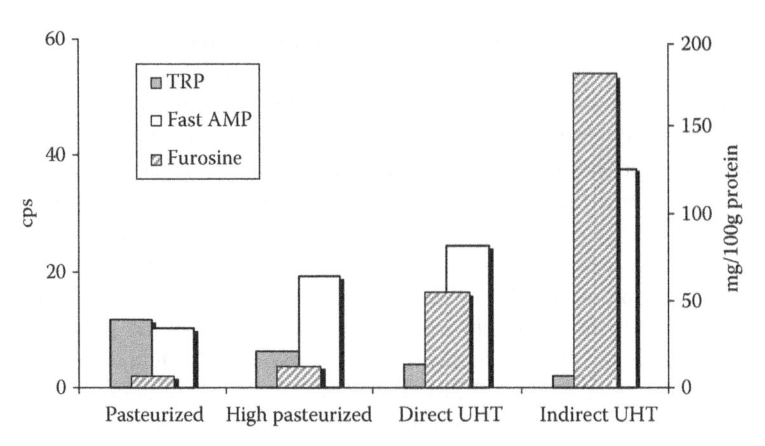

FIGURE 12.1 Fluorescence of advanced Maillard products (fast AMP), soluble tryptophan (TRP), and furosine content in different heat-treated milk. (Modified by Birlouez-Aragon, M. et al., *Int. Dairy J.*, 8, 771, 1998.)

(0-month) milk. Protein quality of reconstituted UHT milk was poorest at all storage times and further declined after 6 months of storage (Figure 12.1).

Lacroix and coworkers [43] investigated in rats the fate of dietary nitrogen after the ingestion of industrially heat-treated dairy proteins. For this purpose, male Wistar rats were fed with mixed meal containing [^{15}N]-labeled milk products such as nonheat-treated products (micellar casein [CAS], milk soluble protein isolate [MSPI], and microfiltered milk [MF]) or heat-treated products (pasteurized at 72°C for 20 s, pasteurized at 96°C for 5 s, UHT, or spray-dried milk). The results of this study showed that digestibility was around 96% except for spray-dried milk (94%) and MSPI milk soluble protein isolated (98%). Ingested nitrogen recovered in some tissues (liver, intestinal mucosa) was 19.3% for spray, 16.7% for MF, and around 14%–15% for other products. Moreover, the biological value of purified protein fractions appeared to be lower than what appeared in products containing total milk protein. No correlation was detected between the nutritional quality of milk proteins and heat treatments ranging from low pasteurization to UHT, despite the presence of a large quantity of heat treatment markers for UHT. Probably to evaluate a loss of nutritional quality, other markers besides those for the heat treatment were required.

Lacroix and coworkers followed also the fate of dietary nitrogen in humans [44]. The authors studied in humans the nutritional consequences of the chemical and physical modifications of milk proteins induced by technological treatments. With this aim, the authors compared microfiltered, pasteurized, and UHT milks in which proteins were labeled with ^{15}N. The authors studied the dietary nitrogen (N) metabolism following a single meal, corresponding to 500 mL of either microfiltered, pasteurized, or UHT defatted milk.

The kinetics of dietary N transfer to serum amino acids (AA), proteins, and urea was significantly higher in UHT than in either the pasteurization or the microfiltration group, but there was no significant difference between them. Pasteurized milk, compared to microfiltered milk, exhibited similar postprandial utilization values of dietary N. The authors also reported an enhancement of the anabolic use of dietary N in plasma proteins after the ingestion of UHT milk, probably due to structural modifications of proteins and especially the desegregation of casein micelles under UHT treatment. According to the authors [44], further trials are necessary to confirm whether the higher deamination rate of UHT proteins is due to more rapid gastric emptying or if it is also caused by the blockade of certain AA residues.

12.2.2.5 Inactivation of Endogenous Milk Enzymes

Milk contains enzymes that have the ability to catalyze specific chemical reactions. In milk, both endogenous and exogenous enzymes are present: endogenous enzymes are mainly hydrolases (lipase, the heat-stable plasmin, and alkaline phosphatase, destroyed by minimum pasteurization temperatures), and exogenous enzymes, generally heat stable, are produced by psychrotrophic bacteria. As expected, enzyme activities can be considerably affected by thermal processing.

Girotti and coworkers [45] analyzed raw, pasteurized, and UHT milk samples with different fat contents and cream samples for xanthine oxidase (EC 1.1.3.22). In pasteurized milk, xanthine oxidase activity depended on fat content, while in UHT milk, enzyme activity completely disappeared due to heat treatment.

Martin and coworkers [46] studied adenosine deaminase (EC 3.5.4.4), which oxidatively deaminates adenosine to inosine, in model and practical experiments. Activity of adenosine deaminase in milk increased up to about 75°C but was no longer evident at 85°C. Activity was present in short-time heated milk samples but not in high-temperature and UHT milk samples. Adenosine deaminase showed a higher thermal stability than alkaline phosphatase, and its activity was not influenced by homogenization [46].

Zehetner and coworkers [47] studied the stability of four endogenous milk enzymes, α-L-fucosidase (EC 3.2.1.51), phosphohexose isomerase (EC 5.3.1.9; glucose-6-phosphate isomerase), phosphodiesterase I (EC 3.1.4.1), and α-mannosidase (EC 3.2.1.24) during heating (temperature range of 52°C–86°C) for heating times up to 1600 s. At a heating time of up to 800 s, α-fucosidase was inactivated at temperatures between 52°C and 62°C and phosphohexose isomerase at temperatures between 56°C and 66°C. Phosphodiesterase I was inactivated under pasteurization conditions, and α-mannosidase was inactivated within the lower range of high-temperature pasteurization. Measurement of the activities of the four endogenous milk enzymes in commercial heat-treated milk samples showed no enzyme activity in 21 UHT milk samples, and in 18 pasteurized milk samples only phosphodiesterase activity (30%) could be detected [47].

12.2.3 Lipids

Milk can be described as an oil-in-water emulsion with the fat globules dispersed in the continuous serum phase. The total fat content of milk, which reflects the energy requirements of the newborn, ranges from 3.5% (cow and woman) to 50% (aquatic mammals) according to the species and varies with breed, age of the animal, stage of lactation, and type of feed [19]. The lipids associated with the milk fat globule are mostly (98%) triglycerides; the other lipids amount to less than 2% of the total, and of this, around 0.5% is diacylglycerols and monoacylglycerols, the balance being made up of phospholipids, sterols, free fatty acids, and traces of other substances such as the fat-soluble vitamin.

The heat load influence on milk will depend on the total fat content because fat affects the viscosity of milk and therefore hinders heat transfer. Pellegrino [48] evaluated the effect of fat content on heat-induced changes for bulk milk (>20 t) with different fat contents, either pasteurized or UHT processed under direct or indirect industrial conditions. Results indicated that fat did indeed protect milk components against tested heat-induced changes (lactulose formation, decreased furosine levels, and whey protein denaturation).

12.2.3.1 Fatty Acids

The saturated fatty acids are about 63% of total milk fatty acids, and in particular, the short-chain fatty acids, butyric, caproic, caprylic, and capric acid (C4, C6, C8, and C10, respectively), are abundant (8%–9%) in ruminant milk fat. Short-chain fatty acids are responsible for the characteristic dairy flavor and aroma, and they can contribute either to the desirable aromas of some cheeses or to rancid flavor.

About 40% of total fatty acids consist of the saturated C16 palmitic and C18 stearic acids, the remainder of the saturates being mainly lauric (C12) and myristic (C14) acid. Among the unsaturated fatty acids, C18:1 oleic acid is abundant (~30% of total fat), but the percentage can vary with the feed of the animal (higher in summer with pasture) [18]. The level of polyunsaturated fatty acids is lower in ruminant milk than in monogastric milk. In particular, cow's milk fat contains low levels of the polyunsaturates C18:2 linoleic (2.4%) and C18:3 linolenic (0.8%) acid [18].

In milk, a small amount of *trans* fatty acids (unsaturated fatty acids with at least a double bond in *trans* configuration) occurs; their presence is due to the hydrogenation action of the microorganisms of the rumen [49]. Generally, *trans* fatty acids have been found to increase low-density lipoprotein (LDL) cholesterol and decrease high-density lipoprotein (HDL) cholesterol, and a relationship between risk of coronary heart disease and high intake of *trans* fatty acids has been hypothesized [50,51].

In contrast, *Butyrivirus fibrisolvens*, a microorganism of the rumen, is able to convert linoleic acid into 24 isomers, most of which, such as 9-*cis*, 11-*trans* octadecadienoic acid (rumenic acid)—the main isomer in ruminant milk fat, are responsible for beneficial effects (conjugated linoleic acid [CLA]) [52]. Actually, in the last decade, several favorable biological activities of CLA have been hypothesized [53]: CLA is thought to possess anticarcinogenetic properties [54] and the capacity to modify lipid metabolism [55] and prostaglandin production in specific tissues [56]. Milk fat is a good source of CLA (0.6%); according to Creamer and MacGibbon [19], New Zealand milk contains the highest level, probably due to breeding cattle mainly on pasture.

UHT treatment can increase the levels of free fatty acids in milk, with a consequent increase in acid degree value (ADV). Choi and Jeon [57] studied free fatty acids and ADV in direct and indirect UHT milk samples during 12 weeks of storage at 23°C. The total amount of free fatty acids increased slightly during storage from 77.9 to 86.2 ppm and from 127.1 to 141.2 ppm in directly and indirectly processed UHT milk, respectively. The higher concentration of free fatty acids observed for indirectly processed UHT milk was ascribed to the more severe heat treatment and the resulting higher lipolysis. ADV increased from 0.62 to 0.77 meq KOH in 100 g fat in directly processed UHT milk and from 1.39 to 1.50 meq KOH in 100 g fat in indirectly processed UHT milk [57].

Milk fat, even if it is composed mainly of saturated fatty acids, can undergo oxidative reactions. UHT milk samples processed at 140°C, stored at 9°C, 15°C, 25°C, 35°C, 45°C, and 55°C, were analyzed for changes brought about by lipid oxidation and proteolysis [58]. The thiobarbituric acid (TBA) value of these samples increased from 0.012 to 0.021, 0.023, 0.032, and 0.065 after 16 weeks of storage at 15°C, 25°C, 35°C, and 45°C, respectively, reaching 0.22 after 12 weeks of storage at 55°C. The level of free amino groups, expressed as glycine, was 18.14 μmol/mL in the fresh sample, increasing to 21.92, 23.08, 28.40, 33.56, and 36.06 μmol/mL of milk after 16 weeks of storage at 9°C, 15°C, 25°C, 35°C, and 45°C, respectively, reaching 32.90 μmol eq of glycine/mL of milk after 12 weeks of storage at 55°C. Both proteolysis and lipid oxidation were strongly related to storage temperature and could be described by zero-order reaction kinetics. A storage temperature of 35°C was found to be critical for proteolytic changes [58].

12.2.3.2 Stability of Fat Globules

Another class of lipids in milk is phospholipids, which, even if present in small amounts (<1%), are essential in the formation of fat globules. These fat globules (ranging from 0.1 to 15 μm in diameter) are surrounded by a membrane that stabilizes the fat globules in an emulsion within the aqueous environment of milk. Mechanical agitation can disrupt the membrane, thus allowing the formation of larger fat globules, while homogenization of milk is able to decrease the size of fat globules (below 1 μm).

The fat membrane is made up of a lipid fraction, a protein fraction, trace metals, and enzymes. Phosphatidylcholine, phosphatidylethanolamine, sphingomyelin, triacylglycerols, diacylglycerols and free fatty acid, cerebrosides, and sterols are the basis of the lipid fraction. The protein fraction of the membrane consists of up to 40 polypeptides (15,000–240,000 Da), the enzyme xanthine oxidase (20%), butyrophilin, and several minor proteins. The relative amount of the protein fraction can vary according to the season [59]. As the heat treatment of milk can induce interaction between the fat globule membrane and the milk protein, some authors [59] have investigated the behavior of fat membranes during heat treatment.

Their results show that polymerization occurs via intramolecular disulfite bonds of xanthine oxidase and butyrophilin during heat treatment.

Separation of the fat layer adversely affects shelf life of UHT milk. In order to investigate the effect of homogenization on the formation of the cream layer in UHT milk, Hillbrick and coworkers [60] examined milk samples obtained by indirect UHT treatment, using plate heat exchangers, and direct UHT treatment, using steam injection. All milk samples, stored upside down at 20°C for 2 weeks, showed a cream layer, but the layer thickness was negatively correlated to pressure of homogenization (0, 17.3, 34.5 MPa; $P<0.05$). When homogenization pressure remained the same, particle size distribution and thickness of fat layers were similar, regardless of UHT plant configuration used. The authors concluded that homogenizing milk after or before heating produces milk of similar microstructure that forms cream layers of similar thicknesses during storage. Nevertheless, it was recommended that UHT milk be homogenized before treatment with high temperature, to avoid the use of aseptic homogenizers [60].

According to Ye and coworkers [61], heat treatment of whole milk caused the association of whey proteins (β-lactoglobulin and α-lactalbumin) with the milk fat globule membrane. These associations started at relatively low temperatures (60°C–65°C) and increase with increasing temperature and heating time. The changes in the composition of milk fat globule membrane seem to play an important role in determining the functional properties of many dairy products (protein stability during renneting, acidification, heat treatments, and storage).

Changes occurring in soluble phosphoglycerides in UHT milk during storage (at 10°C, 20°C, and 30°C for up to 4 months) were evaluated in comparison with pasteurized milk samples [62]. Glycerophosphocholine and glycerophosphoethanolamine disappeared during storage in UHT milk manufactured from raw milk of poor microbial quality, while α-glycerophosphate increased. UHT milk samples manufactured from raw milk of better microbial quality and submitted to severe heat processes did not display changes in phosphoglycerides during storage [62].

12.2.4 Vitamin

During processing, milk, as other foods, can suffer vitamin losses not only due to heat but also due to exposure to oxygen, light, and pH variations or combinations of these factors. Vitamins A, D, and E, β-carotene, and vitamins of the B complex, present in raw milk, are generally stable to heat, and UHT thermal processing does not modify their content.

12.2.4.1 Ascorbic Acid (Vitamin C)

Ascorbic acid (vitamin C) content of cow's milk is about 10–20 mg/L. Ascorbic acid content is somewhat higher for sheep milk, but all the same, 25% of this vitamin is lost during pasteurization. In UHT milk, the content of vitamin C depends on the treatment, the presence of oxygen, and storage conditions.

While milk is not regarded as a reliable or important source of vitamin C in the human diet, vitamin C is of interest here because of its potential as a heat damage indicator. Indeed, vitamin C is heat sensitive, and in the presence of oxygen or enzymes, it is oxidized to dehydroascorbic acid, with a lower vitamin C activity, and other inactive degradation products. Note, however, that heat treatments may cause enzyme inactivation, microorganism destruction, and oxygen removal, thus indirectly improving vitamin stability. This double action of heat treatment has been hypothesized to explain unexpected results of a study analyzing ascorbic acid in milk samples in Italy. In this study, processed milk samples were obtained from Italian central dairies operating under standardized conditions: pasteurization HTST at 72°C, 75°C, 80°C, 85°C, and 90°C for 15 s, and sterilization UHT at 140°C for 3 s [63]. After milk collection, heat treatment, transportation, and sampling in the laboratory, the ascorbic acid retention was low at low pasteurization temperatures and increased with increasing pasteurization temperature. The maximum value was 80°C–85°C; at higher temperature (90°C and UHT), ascorbic acid content decreased.

High pasteurization temperatures apparently stabilize ascorbic acid, especially close to 80°C, when by exclusion of oxygen from milk, a protective action takes place. Indeed, at 80°C–85°C, peroxidase is inactivated and other enzymes do not survive. Heat processing can also reduce the microbial population of the product, and a good correlation ($r=0.99$) has been observed between ascorbic acid content and residual bacterial population [63].

12.2.4.2 Vitamins of the B-Complex

Heat treatment causes slight or no variation in riboflavin content (B2), and no significant difference in riboflavin content can be detected between raw and UHT milk samples. Tagliaferri and coworkers [64] studied the behavior of vitamin B2 in milk subjected to different heat treatments (pasteurization at 75°C, 82°C, or 89°C, direct or indirect UHT). No significant vitamin losses were observed, and the authors concluded that vitamin B2 is not a reliable heat index. Indeed, vitamin B2 losses in milk are mainly caused by other factors, especially exposure to sunlight or fluorescent light [64].

Study of vitamin B1 (thiamine) in milk subjected to different heat treatments (pasteurization at 75°C, 82°C, and 89°C, direct and indirect UHT) showed similar results: significant loss of vitamin B1 due to the heat treatment was not observed. Therefore, vitamin B1 is also not a suitable indicator of thermal processing induced changes in milk. More intense heat treatments (e.g., sterilization, boiling, autoclaving) are necessary to induce vitamin B1 losses in milk [65].

Several folate fractions have been identified in milk samples with HPLC analysis, including tetrahydrofolic acid, 5-methyltetrahydrofolic acid (5-MTHFA), the main form, and 5-formyltetrahydrofolic acid [66]. There is significant seasonal variation in the total folate concentration of milk, with a peak occurring in summer.

Oxygen is known to induce degradation of 5-MTHFA. Effects of the presence of different oxygen levels on the thermal degradation of 5-MTHFA in the UHT region at 110°C, 120°C, 140°C, and 150°C were investigated in a buffer model food system [67]. Results showed that in the presence of oxygen, the overall folate degradation is a second-order reaction; Arrhenius activation energy values for aerobic and anaerobic degradation were 106 and 62 kJ/mol, respectively. Results confirmed the importance of degassing milk before heat treatment to maximize recovery of folate [67].

The concentration of folate-binding protein (FBP), which might have an impact on folate absorption, is significantly lower in UHT milk and fermented milk, both of which are processed at temperatures of >90°C, than in pasteurized milk [68]. Wigertz and coworkers [69] investigated the relation between retention of 5-methyltetrahydrofolate and FBP concentration. Raw and pasteurized milk samples contained similar amounts of FBP: 211 and 168 nmol/L, respectively, while UHT-processed milk and yogurt, both processed at high temperatures, contained only 5.2 and 0.2 nmol/L FBP, respectively. All folates in raw and pasteurized milk were found to be protein bound, while folates in UHT processed milk and yogurt occurred freely. Raw milk, pasteurized milk, UHT milk, and yogurt contained 44.8, 41.1, 36.1, and 35.6 µg/L 5-methyltetrahydrofolates, respectively, after deconjugation. It was concluded that significant losses of 5-methyltetrahydrofolates take place as a result of processing. Results supported the equimolar ratio of FBP and folates in raw and pasteurized milk.

Heat processing of milk reduces the amount of 5-methyltetrahydrofolate and the concentration and folate-binding capacity of FBP [70]. What can be the implications of FBP denaturation on folate bioavailability in milk and to what extent this denaturation can affect folate retention during processing and storage are questions that need further investigation [70].

12.2.4.3 Fat-Soluble Vitamin: β-Carotene, Retinol, and Vitamin E

The β-carotene content of milk varies markedly, depending on the animal species, feed, and season (e.g., fresh pasture is richer in carotenoids than hay or silage). A naturally yellowish color of cow's milk is associated with the presence of carotenoids. Milk of goat and buffalo does not contain carotenoids, probably because of a different animal metabolism, and usually dairy products from these animals are whiter than cow products. The β-carotene content of milk is unaffected by UHT processing.

The main form of vitamin A in milk, all-*trans* retinol, comprises a β-ionone ring with a side chain composed of two isoprene units and four double bonds in the *trans* configuration. Among retinol *cis* isomers, 13-*cis* retinol has the largest biopotency (75%) relative to the all-*trans* retinol, while 9-*cis* retinol has a biological activity of only 19% [71]. The other isomers have even lower vitamin A activities. As a consequence, these compounds should be quantified separately in order to obtain a more accurate vitamin A assessment in nutritional studies.

Even if vitamin A is generally considered a heat-resistant compound [72,73], data about retinol stability sometimes are contradictory, depending on whether the analytic method employed involved separation of the isomers. Indeed, some authors report no significant vitamin A losses from UHT processing [74], while others, using chromatographic methods that allow separation of vitamin A isomers, do report significant losses [75].

Samples of raw milk, not subjected to thermal processing, from various species show no conversion of the predominant all-*trans* isomers to *cis* isomers [76]. On the other hand, the most ubiquitous 9-*cis* and 13-*cis* isomers, resulting from all-*trans* retinol isomerization reactions, are observed in many types of dairy products, including cheeses, UHT milk, margarine, and butter, in varying concentration, depending on processing and storage conditions [75,77,78]. Given the natural variability of the all-*trans* retinol content in raw unprocessed milk, it has been proposed to express the degree of isomerization due to milk processing by the ratio 13-*cis*/all-*trans* retinol [79].

Panfili and coworkers [79] found an average 13-*cis*/all-*trans* retinol ratio of 2.6% for pasteurized milk undergoing a mild heat treatment (high-quality milk), while pasteurized milk treated with a temperature ranging from 72°C and 76°C for 15 s showed an average 13-*cis*/all-*trans* retinol ratio of 6.4%. Milk subjected to more severe heat treatments showed a higher degree of 13-*cis* isomerization (a 13-*cis*/all-*trans* retinol ratio of 15.7% for UHT milk and 33.5% for sterilized milk), which is consistent with increased thermal interconversion of the retinol isomers. Also, in pasteurized and UHT creams, the 13-*cis* isomer level appeared to be related to heat treatment (13-*cis*/all-*trans* retinol ratio of 3.0% for pasteurized cream and 14.4% for UHT cream) [76].

Also, Murphy and coworkers [80] found significant differences ($P<0.01$) in *cis*/*trans* isomerization between pasteurized and UHT milk samples. On average, UHT milk showed a higher degree of isomerization than pasteurized milk, while sterilized milk samples showed the highest degree of isomerization.

Cis/*trans* isomerization could also be directly promoted by light, with the relative amounts of the different *cis* isomers depending on the wavelength and the light permeability of the container [80]. A comparison among glass, plastic, and paperboard containers showed no significant loss of all-*trans* retinol in milk contained in paperboard boxes, while the rate of loss was significantly lower in plastic than in glass containers [80].

Actually, sterilized milk in plastic bottles had lower 13-*cis*/all-*trans* ratios than sterilized milk in glass bottles, probably because of different light permeability of the packaging, confirming the results obtained by other authors [80].

In general, values of all-*trans* retinol and 13-*cis* retinol are useful for a more complete evaluation of the nutritional quality of milk and dairy products, while the degree of isomerization, expressed by the ratio 13-*cis*/all-*trans* retinol, can represent an indicator of the treatment the product has undergone, relative to the lipid fraction and in particular to the unsaponifiable fraction. This indicator can support the already proposed treatment indices regarding protein and carbohydrate modifications [3,4].

Lim [81] studied the effects of storage (10 days at 4°C–5°C or 180 days at −20°C) and season on the vitamin A (retinol and β-carotene) content of low-temperature long-time (LTLT) pasteurized and UHT (aseptically or nonaseptically filled) milk. Vitamin A contents of LTLT, UHT nonaseptically, and UHT aseptically filled milks were, respectively, 40.9, 41.7, and 47.0 μg/100 mL in summer, and 23.2, 27.2, and 26.6 μg/100 mL in winter. No significant difference in vitamin A contents between LTLT and UHT milks was observed, either before or after storage. Vitamin A losses were greater in all milk types when stored at –20°C for 180 days than at 4°C–5°C for 10 days [81].

Tocopherols (vitamin E) are sensitive to oxidation, heat, and light, but usually small or no losses of α-tocopherol (the most important E vitamins) are observed after UHT treatments [82,83] α-tocopherol losses from 3% to 14% were observed in UHT milk after 1 month of storage at 30°C, losses increasing with longer storage time [82].

12.2.5 Minerals

The most common cations (positively charged ions) in milk are K, Na, Ca, and Mg. The most common anions are chloride and inorganic phosphate; citrate, sulfate, carbonate, and bicarbonate are also present but in lesser amounts. Of these minerals, K, Na, and Cl are almost entirely diffusible, while Ca, Mg, inorganic phosphate, and citrate are nondiffusible and associated with the casein micelles.

Milk salts thus exist in an equilibrium between the liquid and colloidal phase: this equilibrium can be affected by processing. In particular, the distribution of diffusible salts (low-molecular-weight ions and complexes) and nondiffusible salts (bound to proteins) can be modified by acidification, addition of salts, heat treatment, high pressure, ultrafiltration, and cold storage [84]. Heat treatments cause a decrease in soluble salt content, and conversion of a mineral from the liquid to the colloidal form is proportional to heat treatment severity [84].

De La Fuente and coworkers [85] analyzed samples of commercial UHT whole milk and skim milk for contents of total and soluble minerals (Ca, P, and Mg). Wide intervals of variation in mineral distribution were found, which the authors attributed to the nature of the samples and the processing conditions.

When acid whey is obtained from pasteurized, UHT, or sterilized milk, calcium recovery is at least 80% of total milk calcium. Ionic calcium represents about 50% of total calcium in all whey extracts. Total calcium in whey from UHT-treated samples is lower than that from pasteurized samples, probably due to modifications of calcium chemical form induced by heating. The severity of heat treatment does not significantly affect the ionic calcium percentage in whey [86].

Levels of Se, Fe, Cu, Zn, Na, K, Ca, and Mg were investigated in three samples of six different kinds of sterilized (UHT) whole milk and in 151 samples of fresh milk collected in the Canary Islands [87]. In raw milk, average mineral levels were Fe (0.515 mg/L), Cu (0.0769 mg/L), Zn (4.41 mg/L), Na (534.1 mg/L), K (1424 mg/L), Ca (1653 mg/L), Mg (113.9 mg/L), and Se (16.44 μg/L). Mean concentrations of Fe and Zn were significantly lower for UHT milk than for raw milk. The Se content was similar for UHT and raw milk, but the Cu level was significantly higher in UHT than in raw milk. Mean concentrations of Na, K, and Mg in raw and sterilized milk were not significantly different [87].

An interesting study [88] evaluated chemical elements in Brazilian commercial milks, by Instrumental Neutron Activation Analysis. In particular, the authors studied Br, Ca, Co, Cs, Fe, K, Na, Rb, and Zn in UHT commercial milk (skim or whole milk). The results were in agreement with the literature: Ca, K, and Zn presented very low variability (<10%) among the studied samples, while Br, Cs, and Rb showed the higher variation. Moreover considering the results, expressed on dry basis, the data showed a significant difference between skim milk and whole milk for the elements Br, Ca, K, Na, Rb, and Zn: the authors assumed that the removal of fat caused a concentration effect in the dry matter of skim milks [88].

In order to investigate the effect of heat treatments, and the resulting protein denaturation, on mineral bioavailability, Hagemeister and coworkers [89] conducted a trial with Goettingen minipigs feeding with different heat-treated milk samples. Trials with raw, pasteurized, or UHT-sterilized milks showed no significant effect of either pasteurization or UHT sterilization on bioavailability of Fe, Cu, or Zn from milk.

12.3 MAILLARD REACTION IN UHT MILK

The most studied chemical reaction in heat-treated milk is, undoubtedly, the Maillard reaction, in which amino groups (mainly casein-bound lysine residues) and reducing sugars (mainly lactose) are the main reactants [90]. The Maillard reaction consists of several steps, strictly dependent on temperature, pH, water activity, and type of sugar and amino group involved [91].

In the early stage of the Maillard reaction, condensation of the reactive free amino group of lysine with the free carbonyl group of lactose occurs, via formation of a Schiff's base, which is subsequently transformed via the Amadori rearrangement into the Amadori product (ε-deoxyketosyl compounds). In milk, this Amadori product is lactulosyl-lysine (bound to protein).

The initial stage of the Maillard reaction has important nutritional consequences: formation of lactulosyl-lysine represents a considerable loss of biologically available lysine. The amount of lysine loss in the Maillard reaction has been proposed as a marker of the severity of heat treatment [90].

The Maillard reaction, on severe industrial treatments, can carry on with the Amadori product undergoing a cleavage, leading, through different pathways, to many reactive compounds. In the final stages of the Maillard reaction, polymeric reactions lead to brown pigments (melanoidins), responsible for food odors and flavors.

12.3.1 Furosine: An Early Maillard Reaction Index

The early stage of the Maillard reaction can be monitored through the amount of furosine (ε-*N*-2-furoylmethyl-L-lysine) in milk. Furosine arises from acid hydrolysis (6 N HCl at 110°C for 24h) of lactulosyl-lysine [92]. Furosine detection can be performed by several methods: reversed-phase HPLC [93–95], microbore (1.0mm i.d.) and narrow-bore (2.0mm i.d.) reversed-phase HPLC [96,97], or capillary electrophoresis [98,99]. Furosine can be considered a good indicator of the earliest stage of the Maillard reaction because it gives an assessment of unavailable lysine produced during heat treatment of milk [100].

In Italy, the furosine level in pasteurized milk is regulated by law and should not exceed 8.6mg/100g protein [101]. No legal limit has been set for the furosine level in UHT milk. This is probably due to the concurrent use of different types, direct and indirect, of UHT processes that lead to large variations in furosine levels in UHT milk available on the Italian market [48]. Indeed, Panfili and coworkers [79] found that furosine content of Italian UHT milk ranged from 66 to 263mg/100g protein. These data are comparable to those found in other countries (Table 12.3).

Ferrer and coworkers [102] studied the furosine content in store-brand and name brand UHT milk from the Spanish market; furosine levels ranged from 65.5 to 310.6mg/100g protein for store-brand and from 40.3 to 50.7mg/100g protein for name-brand UHT milk (Table 12.3). In Belgium, Van Renterghem and De Block [103] determined furosine in direct UHT milk and indirect UHT milk, using either a plate heat exchange system or tubular heat exchange system. Furosine content in direct UHT milk ranged from 35 to 109mg/100g protein. In the tubular system, the furosine level ranged from 118 to 193mg/100g protein, while in the plate system, data were scattered within the range of 108–240mg/100g protein (Table 12.3).

It is noted that standardization of milk protein will affect furosine as index of heat processing. While fat standardization of milk is already a common practice in the dairy industry, steps are being undertaken to also standardize the protein content. There are several methods to alter protein concentration of milk: addition of ultrafiltrate milk retentate or permeate and addition of fractions of whey protein [104]. Rattray and coworkers [105,106] have found that milk protein standardization through addition of acid whey permeate results in lower levels of furosine than protein standardization obtained with skim milk permeate. This difference is probably due to a decrease in both lactose and protein concentrations when using acid whey permeate, while skim milk permeate only causes a decrease in protein.

TABLE 12.3
Content of Furosine (mg/100 g Protein) in Different UHT Milk

Minimum	Maximum	Countries	Type of Treatment	References
35	109	Belgium	UHT direct	[103]
118	193	Belgium	UHT indirect—tubular	[103]
108	240	Belgium	UHT indirect—plate	[103]
220	372	Belgium	Sterilized	[103]
66	263	Italy	UHT	[79]
3.0	3.2	Europe	Raw milk	[93]
3.5	4.8	Europe	Pasteurized	[93]
56	200	Europe	UHT	[93]
297	415	Europe	Sterilized in bottle	[93]
65	311	Spain	UHT, store brand	[102]
40	51	Spain	UHT, name brand	[102]
42	206	Italy	UHT direct	[120]
150	182	Italy	UHT indirect	[120]

12.3.2 Furfural Compounds: Intermediate Maillard Reaction Indices

Intermediate compounds of the Maillard reaction may be formed in heated milk, but only in very small amounts compared to the Amadori product. Among these compounds, the furfurals must be included. In particular, HMF and other furfural compounds (furfural, furylmethylcetonem, and methylfurfural) can be generated during heat treatment of milk or storage at inadequate temperature.

From a chemical point of view, HMF is the product of the dehydration of hexoses (free or linked to protein) due to the presence of concentrated acids. HMF is formed not only from Amadori compounds but also from sugars (lactose isomerization). It is considered a good index of the severity of milk heat treatment [107] and is suitable as a marker of the most severe heat treatments (UHT and sterilized milk).

Traditionally, HMF is determined by a colorimetric method [108], but this method has low specificity. At present, capillary electrophoresis [109] and RP-HPLC [110] appear to be the most powerful techniques for HMF determination. The analytical separation of HMF can be achieved by micellar electrokinetic chromatography (MEEC) employing sodium dodecyl sulfate as the anionic surfactant or by a simple isocratic HPLC method using reversed-phase microbore columns [111].

Using HMF as a marker, Ferrer and coworkers [102] observed differences between store-brand and name-brand UHT milk. Free HMF was found to be present only in store-brand UHT milk, levels ranging from 8.24 to 50.9 μg/100 mL of milk, probably due to lower milk quality.

Several factors possibly affecting the use of HMF as a marker have been investigated. In particular, the total HMF level in milk appears to be affected by milk fat concentration [107]. The HMF level in commercial UHT milk (stored below 50°C) is also related to temperature and time of storage, and increases with higher temperature [112].

12.3.3 Fluorescent Products: Advanced Maillard Reaction Indices

According to Van Boekel [91], several Maillard products are fluorescent. Fluorescence is typical of intermediate products; they show an initial phase of fluorescence development followed by a decrease. On the basis of this concept, an alternative method to evaluate the severity of milk heat treatment is based on the fluorimetric measurement of denaturated proteins (290 nm excitation and 340 nm emission) and of the fluorescent advanced Maillard reaction products (330 nm excitation and 420 nm emission). It has been suggested that fluorimetric measurement of tryptophan can be

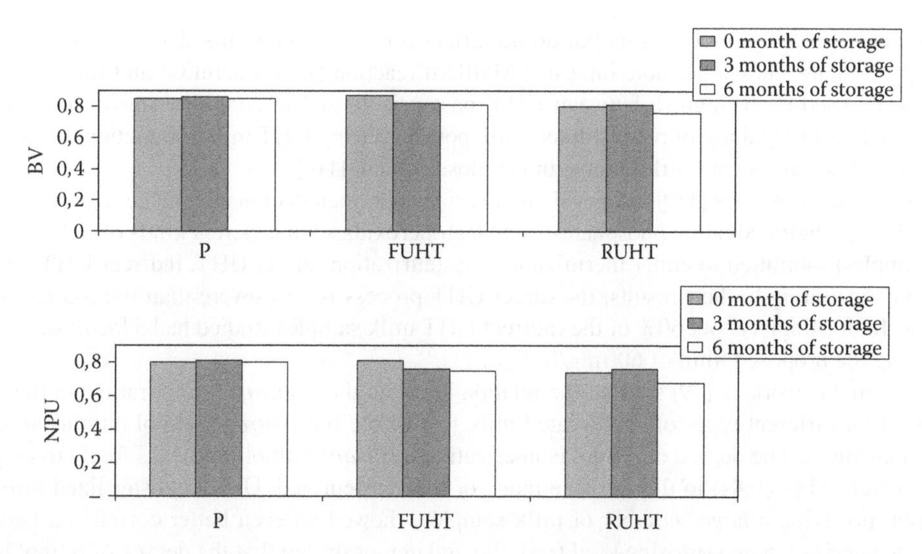

FIGURE 12.2 Biological value and net protein utilization in pasteurized and UHT (fresh and recombinant) during storage. (Modified by Alkanhal, H.A. et al., *Int. J. Food Sci. Nutr.*, 52, 509, 2001.)

considered a marker of protein denaturation during heat treatment, because it is well correlated with β-lactoglobulin, while fluorescent Maillard products can reflect the loss of nutritional quality during processing [113–115].

The fluorimetric measure of tryptophan is more helpful for milk subjected to mild heat treatment (pasteurization), while the fluorescent Maillard products (pyrrole and imidazole derivates) are more significant for indirectly heated UHT milk (Figure 12.2). Using the data represented in Figure 12.2, we find a correlation of $r = 0.94$ between furosine content and fluorescence of Maillard advanced products in the direct and indirect UHT milk samples analyzed.

12.4 HYPHENATED MARKERS OF UHT TREATMENT

Milk heat treatments can produce very complex effects among milk constituents. Simultaneous study of several heat-induced parameters improves the classification of industrial processed milks and provides deeper knowledge of what actually happened in heat-treated milk.

Both furosine and lactulose are good thermal indicators of heat damage [116]. In a detailed study, Pellegrino and coworkers [117] compared the extent of Maillard reaction and lactose isomerization in 46 samples of direct and indirect UHT milk and in-bottle sterilized milk processed under defined conditions. The authors found that evaluation of the ratio furosine/lactulose gives good results only in milk subjected to high temperature, because in pasteurized milk, the Maillard reaction is predominant. With values of furosine rising up to 50 mg/100 g protein, the relation between furosine and lactulose was well described by a linear regression line ($r = 0.994$). High correlations between lactulose and furosine levels were also observed in 160 samples collected from the European market [117].

Pellegrino and coworkers [117] also evaluated the effect of preheating (80°C–90°C) in the UHT process using the furosine/lactulose ratio. The preheating procedure can negatively affect milk quality because during this process the Maillard reaction can carry on, while lactulose is not produced. Indeed, the consequence of preheating was a shift of the regression line toward high levels of furosine; this could also explain why the ratio furosine/lactulose of some commercial samples fell out of the confidence limits of the regression line.

Thus, the combined measure of furosine (describing early Maillard reaction) and lactulose (describing the extent of lactose isomerization) values for sterilized milk is useful to provide additional information about the thermal history of the product. For instance, high pH values of raw milk enhance lactose isomerization but do not affect furosine, while milk drying, prolonged preheating, and long storage promote intensive Maillard reaction [117]. Lactulose and furosine levels can also be used to distinguish between UHT, pasteurized, and in-container sterilized milks, as well as to detect illegal use of reconstituted milk powder during UHT milk production, resulting in higher furosine values with little change in lactulose content [118].

Mortier and coworkers [119] have evaluated various parameters of heat damage (furosine, lactulose, β-lactoglobulin, alkaline phosphatase, and lactoperoxidase) in different kinds of Belgium milk (154 samples) submitted to either thermization, pasteurization, direct UHT, indirect UHT, or sterilization. According to their results, the direct UHT process is less severe than the indirect UHT process. Interestingly, about 60% of the indirect UHT milk samples studied had a lactulose content exceeding the proposed limit of 600 mg/L [8,9].

Panfili and coworkers [79] studied the relation between the *cis/trans* retinol ratio and the furosine level for different types of heat-treated milk to validate the *cis/trans* retinol ratio as an index of thermal stress. The degree of retinol isomerization (*cis/trans* retinol ratio) was found to be positively correlated ($r=0.84$) to the furosine index of raw, pasteurized, UHT, and sterilized milks. A research, involving a larger number of milk samples, showed an even better correlation between retinol isomerization and furosine level ($r=0.94$) and demonstrated that the degree of retinol isomerization can be a useful index of thermal stress, especially for mild heat treatments (P. Manzi, G. Panfili, and L. Pizzoferrato, unpublished data).

Cattaneo and coworkers [120] faced up to the problem of the lack of an upper limit to processing UHT conditions: according to the authors in the EU legislation [1] the designation of UHT milk appears incomplete. In their work, the effects of severe sterilization conditions were investigated on various parameters of heat damage (lactulose, furosine, galactosyl-β-pyranone, and lysinoalanine) in UHT milk processed under the control of the industrial conditions.

When a severe heat process was applied, an extensive increase of the values of all the indicators occurred, but no direct relation was found between their levels and the nominal time–temperature conditions. According to these results [120], lactulose and furosine are strictly correlated in UHT milk but in case of overheating, the advanced Maillard reaction took place extensively and the galactosyl-β-pyranone formation was strongly enhanced. To simulate milk overprocessing, UHT milk was submitted to numerous treatments: the galactosyl-β-pyranone increased from 7.4 in the indirect UHT condition to 28.7 μmol/L during a second treatment (148°C for 20 s) and to 74.3 during a third treatment (148°C for 20 s). The formation of lysinoalanine in UHT drinking milk on storage has been studied, and it seems to be the most sensitive index of storage conditions among the four parameters studied. According to the authors [120], besides the time/temperature conditions, other parameters such as the type of process, the plant characteristics, and the plant management conditions could also contribute to the final heat damage of UHT milk. Only the addition of more details (e.g., an upper limit to the heat treatment) could guarantee the consumer the final product quality.

The problem of overheating milk has also been studied by Seiquer and coworkers [121] who evaluated, in particular, the effect on calcium availability of some Maillard reaction parameters. They studied calcium solubility after the in vitro digestion, and calcium balance assays in rats using UHT milk and overheated milk (three cycles of sterilization at 116°C for 16 min). Their main results showed that the Maillard reaction proceeded extensively in the overheating milk with a significant increase in furosine content (from 202 mg/100 g of protein in UHT milk to 348 mg/100 g of protein in overheating milk); moreover, the Maillard reaction rate was also confirmed by the progression of the color parameters to red and yellow zones in overheating milk.

The thermal damage to milk affects calcium availability, according to their in vitro and in vivo assays: after the in vitro digestion, a negative effect on calcium solubility was observed in overheating milk, and in the in vivo experiment were detected lower values of calcium absorption and retention in rats fed with overheating milk. The authors concluded [121] that special attention should be focused on milk-processing conditions, especially when milk and dairy products are the main dietary components, such as in early infancy.

As mentioned earlier, the UHT processing is able to induce different changes in milk depending on the time-temperature parameters. Generally, all the studies have considered only one or few physicochemical characteristics and focused on one part of the life of the product. Nowadays, new multidimensional statistical approaches can be used in order to consider the possible effects of different parameters. In particular, in the study of Gaucher and coworkers [122], the effects of season, milking region, sterilization process, and storage conditions were evaluated on UHT milk physicochemical characteristics. According to the results of this study, spring milks had higher values of pH, lactose, soluble phosphate, and micellar hydration than milks collected in autumn, while the region effect modified numerous parameters related to the global composition and the casein micelles. The UHT process seemed to decrease in noncasein nitrogen content and in micellar hydration and increased in casein micelle size. Concerning storage conditions, at 40°C, there were a decrease of pH and an increase of noncasein and nonprotein nitrogen contents of milks and were observed low values of stability, evaluated by the ethanol and phosphate tests of UHT milk.

A multivariate approach has also been utilized by Feinberg and coworkers [123]. The authors proved to discriminate commercial milk samples (pasteurized, high pasteurized, UHT, indirect UHT, and sterilized) using some tracers such as furosine; lactulose; native α-lactalbumin; denaturized α-lactalbumin; percent of denaturized α-lactalbumin; FAST index, tryptophan fluorescence, β-lactoglobulin, and lactoperoxidase). Analysis of variance clearly demonstrated that no tracer could be selected to universally discriminate milks from all these technologies, and the authentication technique must be based on a multivariate approach. According to the authors [123], statistical technique is very sensitive to the structure of the data set used for building the discriminant functions and the number of samples could be critical. Moreover, nearly all milk samples could be correctly classified by combining two tracers only, but in storage conditions, in order to reach an acceptable level of discrimination, it was necessary to combine at least four tracers into a multivariate model.

12.5 SAFETY

12.5.1 BACTERIA

As regards microbiological safety, control points for milk products are raw material quality, processing conditions (temperature and time), postprocessing contamination, and storage temperature.

In UHT methods, performing the effective heating process in closed arrangements provides a very high bactericide effect, and using aseptic package systems, it decreases the contamination risk.

Nevertheless, for ambient stable UHT products, the main concern is with heat-resistant spoilage bacteria, such as *Bacillus stearothermophilus* and *Bacillus sporothermodurans* [124,125].

Paenibacillus strains have been isolated from a variety of sources: one distinctive characteristic of *Paenibacillus* strains is their ability to degrade some macromolecules (proteins polysaccharides and polyaromatic hydrocarbons) and to fix nitrogen. Their endospores cause specific problems to the food industry, and although *Paenibacillus* spores in raw and pasteurized milk are not predominant, their presence is due to the contamination of feeds (silage and concentrate) for dairy cattle.

Scheldeman and coworkers [126] isolated endospore-forming bacteria from contaminated industrial UHT milk: based on their phenotypic, phylogenetic, and genomic distinctiveness. These strains, isolated from both raw and heat-treated milk, were placed in the genus *Paenibacillus* as *Paenibacillus lactis* sp. Nov.

Gallardo and coworkers [127] examined the microbiological quality of 10 samples each of pasteurized, UHT, and sterilized milks from Spain. Total aerobic counts, total Enterobacteria, total coliforms and *Escherichia coli*, sulfite-reducing Clostridia, *Staphylococcus aureus*, Salmonella spp., Shigella spp., *Clostridium perfringens*, and yeasts and fungi were determined. All samples complied with Spanish microbiological standards.

In Turkey [128], 75 commercial UHT milk samples coming from five different companies were analyzed for a microbiological control (total aerobic mesophilic microorganism). The aerobic mesophilic microorganisms were not detected in any samples of four trades, whereas, the maximum counts of 2.04 log10 CFU/g were detected in samples of one trade. Furthermore, 6.67% of samples belonging to this trade did not comply with the Turkish Food Codex. These results are a significant contribution on determining microbial quality of the UHT milk consumed in Turkey. The authors [128] concluded that it is important both to choose a good quality of milk used in the preparation of UHT milk and to monitor regularly the microbiological quality of UHT milk.

Guida and coworkers [129] analyzed 72 samples of direct UHT milk (135°C for 1 s), classified into great, middle, and small commercial delivery, for the presence of mesophylic bacteria according to European law [1,130]. About 14% of total milk samples analyzed were out of range, in particular, 25% from great distribution, 17% from middle distribution, and none from small distribution. The authors suggested that contamination could be due to heat-resistant spores or after-packaging contaminations.

Recently, the contamination of UHT milk occurred also in other countries such as Brazil, where the presence of mesophylic microorganisms were identified in UHT milk [131]. In particular, the presence of the *B. sporothermodurans* was studied in 511 samples commercialized in the state of Rio Grande Do Sul. All the samples were in accordance with the legal standard for viable aerobic mesophilic microorganisms (maximum 102 CFU/mL), but in 10.8% of the samples were detected *B. sporothermodurans* colonies. Probably in some circumstances, these highly resistant spores are able to survive to UHT treatments.

12.5.2 Aflatoxins

The contamination of food with aflatoxins is a notorious problem. From epidemiological studies, it is known that aflatoxins are toxic and carcinogenic compounds. Human exposure to aflatoxins can result from ingestion of contaminated food. Aflatoxin M1 (AFM1), a hydroxylated metabolite of aflatoxin B1 (AFB1), can be present in milk via animals consuming feed contaminated with AFB1. AFB1 has been classified as class 1 or human carcinogen and AFM1 as class 2 or probable carcinogen [132].

It is important that a regular monitoring of aflatoxins in animal feed be done, to reduce AFB1 contamination in dairy products, because milk represents a main constituent of the diet of children and growing young, that is, the population most at risk.

The Food and Agriculture Organization [133] established in 1997 permissible levels for AFM1 in milk ranging from 0.05 to 0.5 ppm. The European Community and Codex Alimentarius in 2001 accepted a maximum level of AFM1 in liquid milk not higher than 50 ng/L [134], while according to U.S. regulations, the level of AFM1 in milk should not be higher than 500 ng/L.

Heat processing of milk should not affect the AFM1 concentration because AFM1 is heat stable (decomposition temperature ranges from 237°C to 306°C). Nevertheless, published data on the effect of heat treatment on milk AFM1 levels seem to be contradictory [135]. This can be due to the fact that AFM1 is not stable in milk (contamination disappeared after 6 days storage at 0°C), and not all methods used to determine AFM1 have the same accuracy, precision, and sensitivity.

Many analytical and immunological methods such as TLC, HPLC, and ELISA are available for estimation of AFM1 in milk. HPLC methods for aflatoxin analysis were developed using both normal and reverse phase systems with UV and fluorescence detectors. These methods are very precise, selective, and sensitive. Recently, thanks to the availability of monoclonal and polyclonal antibodies against aflatoxins, various simple sensitive and specific ELISA tests have been developed for aflatoxin analysis.

Martins and Martins [136] investigated the occurrence of AFM1 in raw and UHT commercial milk in Portugal. Only two UHT-treated milk samples exceeded the limit of 0.05 ppm, while the other 68 UHT-treated milk samples had low AFM1 levels (from <0.005 to 0.05 ppm). According to these authors, the findings are similar in other European countries, and data are not worrisome for public health, even though they recommend that more samples should be analyzed.

In Greece, the levels of AFM1 were recently investigated in samples of raw, pasteurized, UHT, and concentrated milks collected in supermarkets [137]. Of the 297 samples analyzed, only two samples of cow's milk, one sample of sheep's raw milk, and two samples of concentrated milk exceeded the limit of 50 ng/L. Results suggest that current regulation of AFM1 levels in milk in Greece is effective.

In Italy, the presence of AFM1 has been assessed in 127 milk samples (UHT, pasteurized, raw) from the Campania and Calabria regions. Several milk samples (14%) contained AFM1 levels that exceeded the EU limit of 50 ng/L [138].

A study on milk produced in Argentina showed very low AFM1 contamination of milk [139]. The authors ascribed this finding to cow-feeding practices; in Argentina, pasture is the ordinary practice, while in most other countries, cows are stabled.

In Brazil, Garrido and coworkers [140] investigated the presence of M1 and M2 aflatoxins in 60 UHT and 79 pasteurized commercial milk samples collected from supermarkets. None of the milk samples were found to be contaminated with aflatoxin M2, but AFM1 was detected in 29 samples (20.9%) at levels of 50–240 ng/L. Occurrence of AFM1 was high in both commercial pasteurized and UHT milk. The contamination level, according to MERCOSUR technical regulations, should not be considered a serious public health problem, even if in some cases, it exceeded the limit (50 ng/L) permitted by the European Union.

During the summer of 2003, the climatic conditions encouraged the development of AFB1 in animal feed in Italy. Nachtmann and coworkers [141] from November 2003 to July 2005 studied the presence of AFM1 in commercial milk in a specific North West Italian region. Three hundred sixteen samples of fresh pasteurized and UHT milk were studied using immunoaffinity columns and the use of HPLC combined with fluorimetric measurement. Only two samples proved to be not in conformity with the current regulations (50 ng/L). The authors concluded that in normal situations, aflatoxins problem is not one of the main causes of health risk in this Italian region, and however, it is necessary to monitor continuously the animal feed in order to assure the public health.

Recently, several papers have been written upon the contamination level with AFM1 of UHT milk commercialized in Turkey and Iran.

In Turkey, in 2006, the levels of AFM1 in UHT milk were examined by the ELISA technique and 129 samples of commercial UHT whole milk were analyzed. A total of 58.1% of Milk samples were contaminated, 53% were below the limit, the remaining 47% were well above the limit permitted by the European Union, while four samples exceeded the prescribed limit of U.S. regulations [142].

A few years later, in 2008, always in Turkey, Tekinsen and Eken [143] investigated 100 UHT milk and 132 kashar cheese samples for AFM1 by ELISA method. In this study, 31% of UHT milk and 27.3% of kashar cheese samples exceeded the maximum tolerable limit of the European Community (EC). According to the authors, these data constitute a human health risk in Turkey: therefore, milk and dairy products have to be controlled continuously by the Turkish public health authorities [143]. In a similar way, numerous studies have been carried on this subject in Iran too. Between 2009 and 2010, Rahimi and coworkers [144] studied 40 samples of raw milk, 48 samples of pasteurized milk, and 48 samples of UHT milk, by ELISA technique. All the samples were lower than 500 ng/L but 119 samples had higher concentrations than the maximum tolerance accepted by some European countries (50 ng/L).

A recent work of Heshmati and Milani [145] examined AFM1 in 210 UHT milk commercialized in Iran by the ELISA method, and the main results revealed the presence of AFM1 in 116 UHT milk samples, and in 70 samples, the levels of AFM1 were higher than the maximum tolerance limit accepted by European countries.

Finally, a recent screening survey [146] was carried out to determine AFM1 in 225 commercial liquid milk samples (116 pasteurized milk and 109 UHT milk samples) in central part of Iran.

AFM1 was detected in 71.5% pasteurized milk samples and in 62.3% UHT milk sample. The author [146] stresses that AFM1 contamination levels of milk is different from one study to another, but these alterations might be ascribed to different factors such as geographical area, analytical method used, and seasons variability. In any case, the results of this study indicated that the contamination of the samples with AFM1 could be a serious public health problem.

12.5.3 Chemical Contamination

Environmental contamination caused by human activities (industrial processing, packaging procedures, accidents during transport, and storage before consumption) can affect milk safety. Chemical pollution, in particular by organic substances (e.g., polychlorinated biphenyls [PCBs]), is one of the major problems. This is a well-known concern, and worldwide, quality control structures focus on identifying hazards and periodically surveying milk all along the supply chain. While data on chemical contamination of breast milk are abundant in the international scientific literature, food chain monitoring results are seldom published. This is unfortunate, as availability of these results would be useful for the whole scientific community.

Milk and cheeses, collected from Washington, DC, area retail stores, have been analyzed for polychlorinated dibenzo-*p*-dioxin and dibenzofuran contamination [147]. Contaminants were identified and quantified and expressed in international toxic equivalent (I-TEQ). Milk contamination ranged from 0.05 to 0.12 I-TEQ. Similar values have been observed in milk collected in rural areas with no apparent sources of contamination [148]. In particular, the I-TEQ values found for the Washington milk samples are higher than the I-TEQ values of German milk [149].

In Italy, Fabietti and coworkers [150] examined 35 commercial milk samples for the presence of the principal aromatic hydrocarbons (benzene, toluene, ethylbenzene, and xylenes). Milk samples were collected from different dairies and included pasteurized whole, semiskimmed and skim milk, and UHT whole and semiskimmed milk. The levels of the two most common aromatic hydrocarbons (benzene and toluene) were relatively low (0.12–100 μg/kg), while the levels of ethylbenzene and total xylenes (meta + ortho + para) were below the detection limit of 0.05 μg/kg. Benzene and toluene levels were higher in whole and semiskimmed milk than in skim milk. No differences were observed between pasteurized and UHT milk with similar fat contents. Results indicate that the levels of aromatic hydrocarbons in commercial milk in Italy are similar to those found in other foods and are sufficiently low to consider milk safe from the point of view of chemical contamination [150].

Recently, chemometric methodologies such as the principal component analysis (PCA) and hierarchical cluster analysis (HCA) have been used for monitoring the authenticity of UHT milk: these methodologies are an interdisciplinary field involving multivariate statistics, mathematical modeling and computing, and it is very useful to obtain information from chemical data. Souza and coworkers [151] used these chemometric methodologies to 100 samples analyzed for adulterants (starch, chlorine, formol, hydrogen peroxide, and urine) to verify milk adulteration in certain geographical regions of Brazil: the substance used for the reconstitution of the milk was an isolated case, and samples positive for the hydrogen peroxide were frequently positive for formol as well [151].

12.6 SENSORY QUALITY

UHT milk, either freshly prepared or stored, has a sensorial characteristic called cooked note. Stale and oxidized flavors occur due to the formation of aldehydes and ketones.

An interesting recent review [152] took several types of volatile sulfur compounds, present in UHT milk (hydrogen sulfide, methanethiol, carbonyl sulfide, carbon disulfide, dimethyl sulfide, dimethyl disulfide, dimethyl sulfone, dimethyl sulfoxide, and dimethyl trisulfide) into consideration. These compounds were associated with cooked flavor, more intense in indirectly processed UHT milk than the directly processed one. These compounds are difficult to isolate, detect, and quantify because of their volatility and reactivity: according to the author, the methods of detection and quantification have improved in recent years. In particular, the preconcentration methods such as solid-phase microextraction and the gas chromatography equipment with selective detectors techniques could be very useful in the detection of these compounds.

Valero and coworkers [153] studied the evolution of volatile components of commercial whole and skim UHT milk during 4 months of storage. Sensory analyses showed a decrease in quality during storage in either whole or skim milk, and after 3 months of storage, there was a slight stale flavor.

The concentration of volatile compounds, such as aldehydes and ketones, provides discriminant information about storage time and storage temperature of whole and partially skimmed UHT milk. Some volatile compounds, such as acetone, 2-butanone, 2-pentanone, 2-heptanone, 3-methylbutanal, pentanal, hexanal, heptanal, dimethyl disulfide, toluene, and limonene, are reportedly useful to differentiate milk subjected to different heat treatments [154].

In whole milk, 2-pentanone and 2-heptanone occurred during the heat treatment, and dimethyl sulfide and dimethyl disulfide were responsible for the cooked note. Skim milk, instead, showed a different pattern of volatile compounds with hydrocarbons representing the main group [154,155].

During storage, UHT milk develops off-flavors due to the release of lower chain-length fatty acids by psychrotrophic microorganisms. Indeed, in UHT milk samples, stored at 22°C and 37°C and monitored more than 33 days, volatile and free fatty acids showed a sharp increase after 13–14 and 20–21 days, respectively, due to lipases from psychrotrophic spores [156]. Hence, good-quality raw milk should be used for UHT milk to prevent rancid flavor development [156]. Actually, it is known that milk with a high ADV has an acceptable lipolyzed flavor [57].

Iwatsuki and coworkers [157] examined the effect of pasteurization temperature on sensory properties of UHT milk (treated at 120°C, 130°C, or 140°C for 2 s); the sensory analysis of the milk was conducted by an expert panel. Using multiple regression analysis, it was found that the palatability of UHT milk is judged mainly on aroma and aftertaste, freshness, and intensity of milk flavor. Results of sensory and principal component analyses showed that milk body, related to milk flavor, viscosity, and fattiness, becomes stronger with increasing UHT temperature [157].

Sensory properties and palatability of UHT milk (heated at 130°C for 2 s) were also evaluated as a function of homogenization pressure [158]. Evaluation of sensory properties by an expert panel and PCA indicated that milk homogenized at lower pressure has greater fattiness, body, milk flavor, sweetness, and thickness. Palatability, as judged by a panel of housewives, improved when milk was homogenized at low pressure. In terms of physicochemical properties, the authors found that with increased homogenization pressure, fractionation of fat globules occurs, resulting in an increase in viscosity, degree of whiteness, and refractive index for the UHT milk [158].

UHT milk is very popular in Europe, Asia, and South America, while consumption in the United States is very low because of the cooked flavor and the high cost of UHT milk.

For this reason in 2009, Oupadissakoon and coworkers [159] studied the sensory properties of a wide range of commercial UHT milk from various countries (France, Italy, Japan, Korea, Peru, Thailand, and the United States); the authors also compared UHT to control pasteurized and sterilized milk samples. The authors used five highly trained panelists: this group had a minimum of 2000 h of testing experience on fresh milk, UHT milk, soy milk, yogurt, ice cream, and cheese; and they used flavor and texture profiling to describe the sensory properties of each milk sample. The main results [159] showed that UHT milk samples varied widely in flavor and texture characters but UHT milk samples did not vary within a country. Fat content of samples did not correlate with dairy fat flavor or with viscosity.

In the study of the sensory quality of UHT milk, the packaging material has to be considered: it is fundamental to reduce the losses of aroma compounds in milk, but it is also possible an interaction of aroma compounds with the packaging material itself.

Generally, glass is considered to be the optimum packaging material for liquid foods; however, most of the UHT milk produced today is stored in packaging materials consisting of three layers (polymers as outer layer, thin films of silica or alumina to inhibit the permeation of oxygen as the second layer and polyethylene be able to adsorb volatile as the third layer).

To analyze this phenomenon, aroma compounds in UHT milk packed in glass or polyethylene bottles were compared by applying the aroma extract dilution analysis (AEDA) [160]. According to the results of this study, lactones, aldehydes, and free fatty acids are considered important odor-active compounds in UHT milk: these compounds were weakly adsorbed to a glass bottle, whereas the adsorption to a polyethylene packaging was much stronger. A strong influence of the chemical structure on the extent of adsorption was detected: aldehydes had a much higher affinity to polyethylene than lactones and acids.

Moreover, this study [160] indicated that, during storage, the overall aroma of UHT milk packed in polyethylene could change due to adsorption or to permeation through packaging, but all the results suggested that the aroma of UHT milk may be changed in a more pronounced way in a polyethylene bottle as compared to a glass bottle.

12.7 CONCLUSIONS

Alterations in the chemical composition of milk due to thermal processing are extremely complex. They depend on the composition of the raw milk, time, and temperature of thermal processing, other processing conditions (e.g., homogenization), and time and temperature of storage. Some heat-induced chemical alterations affect milk nutritional quality (e.g., lysine loss, isomerization of all *trans* retinol), some the physical appearance (e.g., casein proteolysis), and others the organoleptic characteristics.

As regards safety of UHT milk, contamination with heat-resistant bacteria, aflatoxin, and chemical compounds remains of some concern. The problem here is not so much with thermal processing, but with the quality of the raw milk, provided processing and domestic handling procedures are properly performed. Continuous monitoring seems warranted.

The growing scientific interest in heat-induced modifications of chemical, biological, and sensorial characteristics of UHT milk is a response to the growing consumer demand for information on food quality and food safety. This aspect will be more and more important in the future, when development of milk products, as other foods, will be specifically planned on consumer concerns.

A comprehensive description of the heat-induced changes will be necessary to allow more complete milk labeling, extending to nutritional and functional components. This will be a new, important challenge for the food chemist.

ACKNOWLEDGMENTS

The authors thank Dr. Sandra Bukkens for editing of the manuscript and her valuable suggestions and Dr. Emma Malinconico for the English revision.

REFERENCES

1. Council Directive 92/46/EEC, June 16, 1992. *Official Journal of the European Communities* L 268.
2. MM Hewedy, C Kiesner, K Meisser, J Hartkopf, HF Erbersdobler. Effect of UHT heating of milk in an experimental plant on several indicators of heat treatment. *Journal of Dairy Research* 61: 305–309, 1994.
3. J O'Brien. Heat-induced changes in lactose: Isomerization, degradation, Maillard browning. In: PF Fox, ed. *Heat Induced Changes in Milk*, Special Issue 9501. Brussels, Belgium: International Dairy Federation, 1995, pp. 134–170.
4. L Pellegrino, I Resmini, W Luf. Assessment (indices) of heat treatment of milk. In: PF Fox, ed. *Heat Induced Changes in Milk*, Special Issue 9501. Brussels, Belgium: International Dairy Federation, 1995, pp. 409–453.

5. E Troyano, I Martinez-Castro, A Olano. Kinetics of galactose and tagatose formation during heat treatment of milk. *Food Chemistry* 45: 41–43, 1992.
6. E Marconi, MC Messia, A Amine, D Moscone, F Vernazza, F Stocchi, G Palleschi. Heat treated milk differentiation by a sensitive lactulose assay. *Food Chemistry* 84: 447–450, 2004.
7. J De Block, M Merchiers, R Van Renterghem, R Moermans. Evaluation of two methods for the determination of lactulose in milk. *International Dairy Journal* 6: 217–222, 1996.
8. International IDF Standard no. 147. Heat-treated milk determination of lactulose content. Method using high-performance liquid chromatography. *International Dairy Federation*, Brussels, Belgium, 2007.
9. European Commission (EC). Projet de décision de la commission fixant le limites et les méthodes permettant de distinguer les différents types de lait de consommations traits thermiquement, Dairy Chemist's Group. Document VI/5726, 1992.
10. L Vorlova, M Smutna, A Halamickova. Lactulose in market milks. *Bulletin-Potravinarskeho Vyskumu* 41: 51–58, 2002.
11. C Mosso, R Del Grosso, R Ferro. Lactulose determination in UHT milk. *Industrie Alimentari* 33: 648–650, 1994.
12. G Andrews. Lactulose in heated milk. *Bulletin of the International Dairy Federation* 238: 45–52, 1989.
13. C Schumann. Medical, nutritional and technological properties of lactulose. An update. *European Journal of Nutrition* 41(S1): 17–25, 2002.
14. C Romero, FJ Morales, S Jimenez-Perez. Effect of storage temperature on galactose formation in UHT milk. *Food Research International* 34: 389–392, 2001.
15. E Troyano, M Villamiel, A Olano, J Sanz, I Martinez-Castro. Monosaccharides in myo-inositol in commercial milk. *Journal of Agricultural and Food Chemistry* 44: 815–817, 1996.
16. J Belloque, M Villamiel, R Lopez-Fandino, A Olano. Release of galactose and N-acetylglucosamine during storage of UHT milk. *Food Chemistry* 72: 407–412, 2001.
17. MI Recio, M Villamiel, I Martinez-Castro, A Olano. Changes in free monosaccharides during storage of some UHT milks: A preliminary study. *Zeitschrift für Lebensmittel Untersuchung und Forschung* 207: 180–181, 1998.
18. PF Fox, PLH McSweeney. *Dairy Chemistry and Biochemistry*. London, U.K.: Thompson Science, 1998.
19. LK Creamer, AKH MacGibbon. Some recent advances in the basic chemistry of milk proteins and lipids. *International Dairy Journal* 6: 539–568, 1996.
20. GM Kavanagh, AH Clark, SB Ross-Murphy. Heat-induced gelation of globular proteins. Part 3. Molecular studies on low pH β-lactoglobulin gels. *International Journal of Biological Macromolecule* 28: 41–50, 2000.
21. K Kondal-Reddy, MH Nguyen, K Kailasapathy, JG Zadow, JF Hardham. Kinetic study of whey protein denaturation to assess the degree of heat treatment in UHT milk. *Journal of Food Science and Technology* 36: 305–309, 1999.
22. S Jeanson, D Dupont, N Grattard, O Rolet-Repecaud. Characterization of the heat treatment undergone by milk using two inhibition ELISAs for quantification of native and heat denatured alpha-lactalbumin. *Journal of Agricultural and Food Chemistry* 47: 2249–2254, 1999.
23. L Fukal, L Karamonova, M Vitkova, P Rauch, GM Wyatt. Distinguishing raw, pasteurized, UHT-treated and bath-sterilized commercial milks by their interaction with immunoprobes against caseins and whey proteins. *Italian Journal of Food Science* 14: 207–215, 2002.
24. DJ Oldfield, H Singh, MW Taylor, KN Pearce. Kinetics of denaturation and aggregation of whey proteins in skim milk heated in an ultra-high temperature (UHT) pilot plant. *International Dairy Journal* 8: 311–318, 1998.
25. N Datta, HC Deeth. Diagnosing the cause of proteolysis in UHT milk. *Lebensmittel Wissenschaft und Technologie* 36: 173–182, 2003.
26. BC Richardson, DF Newstead. Effect of heat-stable proteinases on the storage life of UHT milk. *New Zealand Journal of Dairy Science* 14: 273–279, 1979.
27. A Kennedy, AL Kelly. The influence of somatic cell count on the heat stability of bovine milk plasmin activity. *International Dairy Journal* 7: 717–721, 1997.
28. B Miralles, L Amigo, M Ramos, I Recio. Analysing para-kappa-casein and related peptides as indicators of milk proteolysis. *Milchwissenschaft* 58: 412–415, 2003.
29. MR Garcia-Risco, M Ramos, R Lopez-Fandino. Proteolysis, protein distribution and stability of UHT milk during storage at room temperature. *Journal of Science of Food and Agriculture* 79: 1171–1178, 1999.
30. E Cauvin, P Sacchi, R Rasero, RM Turi. Proteolytic activity during storage of UHT milk. *Industrie Alimentari* 38: 825–829, 1999.

31. MR Garcia-Risco, M Villamiel, R Lopez-Fandino. Effect of homogenisation on protein distribution and proteolysis during storage of indirectly heated UHT milk. *Lait* 82: 589–599, 2002.
32. M Corredig, DG Dalgleish. Effect of different heat treatments on the strong binding interactions between whey proteins and milk fat globules in whole milk. *Journal of Dairy Research* 63: 441–449, 1996.
33. MR Garcia-Risco, M Ramos, R Lopez-Fandino. Modifications in milk proteins induced by heat treatment and homogenization and their influence on susceptibility to proteolysis. *International Dairy Journal* 12: 679–688, 2002.
34. H Meisel, E Schlimme. Casein-bound phosphorus and contents of free amino acids in milk subjected to different heat treatments. *Kieler Milchwirtschaftliche Forschungsberichte* 47: 289–295, 1995.
35. I Gandolfi, G Palla, L Delprato, F De Nisco, R Marchelli, C Salvadori. D-Amino acids in milk as related to heat treatments and bacterial activity. *Journal of Food Science* 57: 377–379, 1992.
36. G Palla, R Marchelli, A Dossena, G Casnati. Detection by capillary gas chromatography and by reversed-phase high-performance liquid chromatography with L-phenylalaninamides as chiral selectors. *Journal of Chromatography* 475: 45–53, 1989.
37. M Friedmann. Chemistry, biochemistry, nutrition and microbiology of lysinoalanine, lanthionine and histidinoalanine in food and other proteins. *Journal of Agricultural and Food Chemistry* 47: 1295–1319, 1999.
38. A D'Agostina, G Boshin, A Rinaldi, A Arnoldi. Updating on the lysinoalanine content of commercial infant formulae and beicost products. *Food Chemistry* 80: 483–488, 2003.
39. V Faist, S Drush, C Kiesner, I Elmadfa, HF Erbersdobler. Determination of lysinoalanine in foods containing milk protein by high-performance chromatography after derivatisation with dansyl chloride. *International Dairy Journal* 10: 339–349, 2000.
40. L Pellegrino, P Resmini, I De Noni, F Masotti. Sensitive determination of lysinoalanine for distinguishing natural from imitation mozzarella cheese. *Journal of Dairy Science* 79: 725–734, 1996.
41. M Efigenia, B Povoa, T Moraes-Santos. Effect of heat treatment on the nutritional quality of milk proteins. *International Dairy Journal* 7: 609–612, 1997.
42. HA Alkanhal, AA Al-Othman, FM Hewedi. Changes in protein nutritional quality in fresh and recombined ultra high temperature treated milk during storage. *International Journal of Food Science and Nutrition* 52: 509–514, 2001.
43. M Lacroix, J Leonil, C Bos, G Henry, G Airinei, J Fauquant, D Tome, C Gaudichon. Heat markers and quality indexes of industrially heat-treated [15N] milk protein measured in rats. *Journal of Agriculture and Food Chemistry* 54: 1508–1517, 2006.
44. M Lacroix, C Bon, C Bos, J Leonil, R Benamouzig, C Luengo, J Fauquant, D Toé, C Gaudichon. Ultra high temperature treatment, but not pasteurization, affects the postprandial kinetics of milk proteins in humans. *Journal of Nutrition* 138: 2342–2347, 2008.
45. S Girotti, S Lodi, E Ferri, G Lasi, F Fini, S Ghini, R Budini. Chemiluminescent determination of xanthine oxidase activity in milk. *Journal of Dairy Research* 66: 441–448, 1999.
46. D Martin, C Kiesner, PC Lorenzen, E Schlimme. Adenosine deaminase (EC 3.5.4.4): A potential indicator of heat treatment for the distinction of short-time and high-temperature pasteurized milk from the market. *Kieler Milchwirtschaftliche Forschungsberichte* 50: 225–233, 1998.
47. G Zehetner, C Bareuther, T Henle, H Klostermeyer. Inactivation of endogenous enzymes during heat treatment of milk. *Netherlands Milk and Dairy Journal* 50: 215–226, 1996.
48. L Pellegrino. Influence of fat content on some heat-induced changes in milk and cream. *Netherlands Milk and Dairy Journal* 48: 71–80, 1994.
49. G Jahreis, P Möckel, F Schöne, U Möller, H Steinhart. The potential anticarcinogenic conjugated linoleic acid, *cis*-9, *trans*-11 C18:2, in milk of different species: Cow, goat, ewe, sow, mare, woman. *Nutrition Research* 19: 1541–1549, 1999.
50. RP Mensink, MB Katan. *Trans* monounsaturated fatty acids in nutrition and their impact on serum lipoprotein levels in men. *Progress in Lipid Research* 32: 111–122, 1993.
51. LPL Van de Vijver, AFM Kardinaal, C Couet, A Aro, A Kafatos, L Steingrimsdottir, JA Amorim Cruz, O Moreiras, W Becker, JMM Van Amelsvoort, S Vidal-Jessel, I Salminen, J Moschandreas, N Sigfusson, I Martins, A Carbajal, A Ytterfors, G Van Poppel. Association between *trans* fatty acid intake and cardiovascular risk factors in Europe: The TRANSFAIR study. *European Journal of Clinical Nutrition* 54: 126–135, 2000.
52. JL Sébédio, S Gnaedig, JM Chardigny. Recent advances in conjugated linoleic acid research. *Current Opinion on Clinical Nutrition and Metabolism Care* 2: 499–506, 1999.
53. D Kritchevsky. Conjugated linoleic acid. *Nutrition Bulletin* 25: 25–27, 2000.

54. MK McGuire, MS Yongsoon Park, RA Behre, LY Harrison, TD Shultz. Conjugated linoleic acid concentrations of human milk and infant formula. *Nutrition Research* 17: 1277–1283, 1997.
55. MW Pariza, Y Park, ME Cook. The biologically active isomers of conjugated linoleic acid. *Progress in Lipid Research* 40: 283–298, 2001.
56. T Nakanishi, T Koutoku, S Kawahara, A Murai, M Furuse. Dietary conjugated linoleic acid reduces cerebral prostaglandin E2 in mice. *Neuroscience Letters* 341: 135–138, 2003.
57. IW Choi, IJ Jeon. Composition of free fatty acids in ultra-high-temperature processed milk and their contribution to acid degree value. *Foods and Biotechnology* 4: 11–13, 1995.
58. RRB Singh, GR Patil, R Balachandran. Kinetics of lipid oxidation and proteolysis in stored UHT milk. *Milchwissenschaft* 56: 250–253, 2001.
59. A Ye, H Singh, MW Taylor, S Anema. Characterization of protein components of natural and heat-treated milk fat globule membranes. *International Dairy Journal* 12: 393–402, 2002.
60. GC Hillbrick, DJ McMahon, WR McManus. Microstructure of indirectly and directly heated ultra-high-temperature (UHT) processed milk examined using transmission electron microscopy and immunogold labelling. *Lebensmittel Wissenschaft und Technologie* 32: 486–494, 1999.
61. A Ye, H Singh, Mw Taylor, S Anema. Interactions of whey proteins with milk fat globule membrane proteins during heat treatment of whole milk. *Lait* 84: 269–283, 2004.
62. J Belloque, AV Carrascosa, R Lopez-Fandino. Changes in phosphoglyceride composition during storage of ultra high-temperature milk, as assessed by 31P-nuclear magnetic resonance: Possible involvement of thermoresistant microbial enzymes. *Journal of Food Protection* 64: 850–855, 2001.
63. L Pizzoferrato. Examples of direct and indirect effects of technological treatments on ascorbic acid, folate and thiamine. *Food Chemistry* 44: 49–52, 1992.
64. E Tagliaferri, R Sieber, U Buetikofer, P Eberhard, JO Bosset. Evaluation of milk deterioration after various heat and mechanical treatments and light exposure of different durations. III. Determination of vitamin B2 using a new reversed phase-HPLC method. *Mitteilungen aus dem Gebiete der Lebensmitteluntersuchung und Hygiene* 83: 467–491, 1992.
65. E Tagliaferri, JO Bosset, P Eberhard, U Buetikofer, R Sieber. Evaluation of milk deterioration after various heat and mechanical treatments and light exposure of different durations. II. Determination of vitamin B1 using a newly developed RPHPLC method. *Mitteilungen aus dem Gebiete der Lebensmitteluntersuchung und Hygiene* 83: 435–441, 1992.
66. K Wigertz, M Jaegerstad. Analysis and characterization of milk folates from raw, pasteurized, UHT-treated and fermented milk related to availability in vivo. In: *Proceedings of Bioavailability'93: Nutritional, Chemical and Food Processing Implications of Nutrient Availability*, Part II, May 9–12, 1993, Ettlingen, Germany, pp. 421–431.
67. U Viberg, M Jagerstad, R Osta, I Sjoholm. Thermal processing of 5-methyltetrahydrofolic acid in the UHT region in the presence of oxygen. *Food Chemistry* 59: 381–386, 1997.
68. K Wigertz, UK Svensson, M Jagerstad. Folate and folate-binding protein content, in dairy products. *Journal of Dairy Research* 64: 239–252, 1997.
69. K Wigertz, I Hansen, M Hoier-Madsen, J Holm, M Jagerstad. Effect of milk processing on the concentration of folate-binding protein (FBP), folate-binding capacity and retention of 5-methyltetrahydrofolate. *International Journal of Food Sciences and Nutrition* 47: 315–322, 1996.
70. KM Forssen, MI Jagerstad, K Wigertz, CM Witthoft. Folates and dairy products: A critical update. *Journal of the American College of Nutrition* 19: 100S–110S, 2000.
71. H Weiser, G Somorjai. Bioactivity of *cis* and *dicis* isomers of vitamin A esters. *International Journal for Vitamin and Nutrition Research* 62: 201–208, 1992.
72. H Burton. *Ultra-High Temperature Processing of Milk and Milk Products*. New York: Elsevier Applied Science Publishers, 1988.
73. I Andersson, R Öste. Nutritional quality of heat processed liquid milk. In: PF Fox, ed. *Heat-Induced Changes in Milk*, Special Issue 9501. Brussels, Belgium: International Dairy Federation, 1995, pp. 279–307.
74. I Le Marguer, H Jackson. Stability of vitamin A and ultra-high temperature processed milks. *Journal of Dairy Science* 66: 2452–2458, 1983.
75. DC Woollard, H Indyk. The HPLC analysis of vitamin A isomers in dairy products and their significance in biopotency estimations. *Journal of Micronutrient Analysis* 2: 125–146, 1986.
76. G Panfili, P Manzi, L Pizzoferrato. Influence of thermal and other manufacturing stresses on retinol isomerization in milk and dairy products. *Journal of Dairy Research* 65: 253–260, 1998.
77. JW Erdam, CL Poor, JM Dietz. Factors affecting the bioavailability of vitamin A, carotenoids and vitamin E. *Food Technology* 42: 214–221, 1988.

78. RL Fellman, PS Dimick, R Hollender. Photooxidative stability of vitamin A fortified 2% low fat milk and skim milk. *Journal of Food Protection* 54: 113–116, 1991.
79. G Panfili, E Chimisso, P Manzi, L Pizzoferrato. Il grado di isomerizzazione del retinolo può rappresentare un indice di processo in prodotti lattiero-caseari? *Scienza e Tecnica Lattiero-Casearia* 50: 138–146, 1999.
80. PA Murphy, R Engelhardt, SE Smith. Isomerization of retinyl palmitate in fortified skim milk under retail fluorescent light. *Journal of Agricultural and Food Chemistry* 36: 592–595, 1988.
81. JW Lim. Influence of storage and season on the variation of vitamin A in LTLT pasteurized and UHT processed commercial market milk. *Korean Journal of Dairy Science* 14: 193–201, 1992.
82. C Vidal-Valverde, R Ruiz, A Mediano. Effects of frozen and other storage conditions on alfa tocopherol content of cow milk. *Journal of Dairy Science* 76: 1520–1525, 1993.
83. E Quattrucci, S Adorisio, G Panfili, L Pizzoferrato. Effect of technological processes on vitamin content of milk. In: *Proceedings of the 6th European Conference on Food Chemistry, Strategies for Food Quality Control and Analytical Methods in Europe*, September 22–26, 1991, Hamburg, Germany, pp. 315–320.
84. MA De La Fuente. Changes in the mineral balance of milk submitted to technological treatments. *Trends in Food Science and Technology* 9: 281–288, 1998.
85. MA De La Fuente, M Juarez, A Olano. Distribution of minerals between the soluble and colloidal phases in commercial UHT milks. *Milchwissenschaft* 56: 194–198, 2001.
86. M Carbonaro, M Lucarini, G Di Lullo. Composition and calcium status of acid whey from pasteurized, UHT treated and in bottle sterilized milks. *Nahrung* 44: 422–425, 2000.
87. EM Rodriguez-Rodriguez, M Sanz-Alaejos, C Diaz-Romero. Mineral concentrations in cow's milk from the Canary Islands. *Journal of Food Composition and Analysis* 14: 419–430, 2001.
88. LGC Santos, EA De Nadai Fernandes, FS Tagliaferro, MA Bacchi. Characterization of Brazilian commercial milks by instrumental neutron activation analysis. *Journal of Radioanalytical and Nuclear Chemistry* 276: 107–112, 2008.
89. H Hagemeister, J Scholtissek, CA Barth. Influence of heat treatment of milk on bioavailability of iron, copper and zinc. *Archiv für Tierzucht* 41: 261–268, 1998.
90. MAJS Van Boekel. Effect of heating on Maillard reactions in milk. *Food Chemistry* 62: 403–412, 1998.
91. MAJS Van Boekel. Kinetic aspects of the Maillard reaction: A critical review. *Nahrung* 45: 150–159, 2001.
92. PA Finot, R Deutsch, E Bujard. The extent of the Maillard reaction during the processing of milk. *Progress in Food & Nutrition Science* 5: 345–355, 1981.
93. P Resmini, L Pellegrino, G Battelli. Accurate quantification of furosine in milk and dairy products by a direct HPLC method. *Italian Journal of Food Science* 3: 173–183, 1990.
94. L Pizzoferrato, P Manzi, V Vivanti, I Nicoletti, C Corradini, E Cogliandro. Indirect determination of ε-deoxyketosyl lysine in processed milk. In: *Proceedings of Chemical Reactions in Foods III*, September 5–27, 1996, Prague, pp. 154–158.
95. L Pizzoferrato, P Manzi, V Vivanti, I Nicoletti, C Corradini, E Cogliandro. Maillard reaction in milk-based foods: Nutritional consequences. *Journal of Food Protection* 61: 235–239, 1998.
96. I Nicoletti, E Cogliandro, C Corradini, D Corradini. Analysis of e-*N*-2-furosylmethyl-L-lysine (furosine) in concentrated milk by reversed phase chromatography with microbore column. *Journal of Liquid Chromatography & Related Technologies* 20: 791–729, 1997.
97. I Nicoletti, E Cogliandro, C Corradini, D Corradini, L Pizzoferrato. Determination of furosine in hydrolyzate of processed milk by HPLC using a narrow bore column and diode-array detector. *Journal of Liquid Chromatography & Related Technologies* 23: 717–726, 2000.
98. A Tirelli, L Pellegrino. Determination of furosine in dairy products by capillary zone electrophoresis: A comparison with the HPLC method. *Italian Journal of Food Science* 7: 379–385, 1995.
99. D Corradini, G Cannarsa, C Corradini, I Nicoletti, L Pizzoferrato, V Vivanti. Analysis of e-*N*-2-furosylmethyl-L-lysine (furosine) in dried milk by capillary electrophoresis with controlled electroosmotic flow using *N,N,N′,N′*-tetramethyl1,3-butanediamine in the running electrolyte solution. *Electrophoresis* 17: 1–5, 1996.
100. I Gandolfi, P Manghi, E Bassi, C Salvadori, P Cagnasso. Indici di trattamento termico nel latte. *Laboratorio* 2000: 90–94, 1999.
101. Ministero delle Politiche Agricole e Forestali, Decreto Ministeriale DM 15.12.2000. Fissazione dei valori massimi di furosina nei formaggi freschi a pasta filata e nel latte (crudo e pastorizzato perossidasi-positivo). Gazzetta Ufficiale della Repubblica Italiana, n.31, February 7, 2001.

102. E Ferrer, A Alegria, G Courtois, R Farre. High performance liquid chromatographic determination of Maillard compounds in store-brand and name-brand ultra-high temperature treated cow's milk. *Journal of Chromatography A* 881: 599–606, 2000.
103. R Van Renterghem, J De Block. Furosine in consumption milk and milk powders. *International Dairy Journal* 6: 371–382, 1996.
104. W Rattray, P Jelen. Protein standardization of milk and dairy products. *Trends in Food Science and Technology* 7: 227–234, 1996.
105. W Rattray, P Gallmann, P Jelen. Influence of protein standardization and UHT heating on the furosine value and freezing point of milk. *Lait* 77: 296–305, 1997.
106. W Rattray, P Gallmann, P Jelen. Nutritional, sensory and physico-chemical characterization of protein-standardized UHT milk. *Lait* 77: 279–296, 1997.
107. FJ Morales, S Jimenez-Perez. HMF formation during heat-treatment of milk-type products as related to milkfat content. *Food Chemistry and Toxicology* 64: 855–859, 1999.
108. M Keeney, R Bassette. Detection of intermediate compounds in the early stages of browning reaction in milk products. *Journal of Dairy Science* 42: 954–960, 1959.
109. FJ Morales, S Jimenez-Perez. Hydroxymethylfurfural determination in infant milk-based formulas by micellar electrokinetic capillary chromatography. *Food Chemistry* 72: 525–531, 2001.
110. FJ Morales, C Romero, S Jimenez-Perez. Evaluation of heat-induced changes in Spanish commercial milk: Hydroxymethylfurfural and available lysine content. *International Journal of Food Science and Technology* 31: 411–418, 1996.
111. C Corradini, I Nicoletti, G Cannarsa, D Corradini, L Pizzoferrato, V Vivanti. Microbore liquid chromatography and capillary electrophoresis in food analysis. In: G Sontag, W Pfannhauser, eds. *Current Status and Future Trends in Analytical Food Chemistry*, Vol. 2. Vienna, Austria: Austrian Chemical Society, 1995, pp. 299–302.
112. FJ Morales, C Romero, S Jimenez-Perez. A kinetic study of 5-hydroxymethylfurfural formation in Spanish UHT milk stored at different temperatures. *Journal of Food Science and Technology* 34: 28–32, 1997.
113. I Birlouez-Aragon, M Nicolas, A Metais, N Marchond, J Grenier, D Calvo. A rapid fluorimetric method to estimate the heat treatment of liquid milk. *International Dairy Journal* 8: 771–777, 1998.
114. I Birlouez-Aragon, J Leclere, CL Quedraogo, E Birlouez, JF Grongnet. The FAST -method, a rapid approach of the nutritional quality of heat-treated foods. *Nahrung* 3: 201–205, 2001.
115. I Birlouez-Aragon, P Sabat, N Gouti. A new method for discriminating milk heat treatment. *International Dairy Journal* 12: 59–67, 2002.
116. D De Rafael, M Villamiel, A Olano. Formation of lactulose and furosine during heat treatment of milk at temperatures of 100–120°C. *Milchwissenschaft* 52: 76–78, 1997.
117. L Pellegrino, I De Noni, P Resmini. Coupling lactulose and furosine indices for quality evaluation of sterilized milk. *International Dairy Journal* 5: 647–659, 1995.
118. A Montilla, M Calvo, G Santa-Maria, N Corzo, A Olano. Correlation between lactulose and furosine in UHT-heated milk. *Journal of Food Protection* 59: 1061–1064, 1996.
119. L Mortier, A Braekman, D Cartuyvels, R van Renterghem, J De Block. Intrinsic indicators for monitoring heat damage of consumption milk. *Biotechnologie, Agronomie, Société et Environnement* 4: 221–225, 2000.
120. S Cattaneo, F Masotti, L Pellegrino. Effects of overprocessing on heat damage of UHT milk. *European Food Research and Technology* 226: 1099–1106, 2008.
121. I Seiquer, C Delgado-Andrade, A Haro, MP Navarro. Assessing the effects of severe heat treatment of milk on calcium bioavailability: In vitro and in vivo studies. *Journal of Dairy Science* 93: 5635–5643, 2010.
122. I Gaucher, T Boubellouta, E Beaucher, M Piot, F Gaucheron, E Dufour. Investigation of the effects of season, milking region, sterilisation process and storage conditions on milk and UHT milk physico-chemical characteristics: A multidimensional statistical approach. *Dairy Science and Technology* 88: 291–312, 2008.
123. M Feinberg, D Dupont, T Efstathiou, V Louapre, JP Guyonnet. Evaluation of tracers for the authentication of thermal treatments of milks. *Food Chemistry* 98: 188–194, 2006.
124. MJ Lewis. Microbiological issues associated with heat treated milks. *International Journal of Dairy Technology* 52: 121–125, 1999.
125. O Guillaume-Gentil, P Scheldeman, J Marugg, L Herman, H Joosten, M Heyndrickx. Genetic heterogeneity in *Bacillus sporothermodurans* as demonstrated by ribotyping and repetitive extragenic palindromic-PCR fingerprinting. *Applied and Environmental Microbiology* 68: 4216–4224, 2002.

126. P Scheldeman, K Goossens, M Rodriguez-Diaz, A Pil, J Goris, L Herman, P De Vos, NA Logan, M Heyndrickx. *Paenibacillus lactis* sp. nov., isolated from raw and heat-treated milk. *International Journal of Systematic and Evolutionary Microbiology* 54: 885–891, 2004.
127. CS Gallardo, E Sinde, JA Gonzalez, A Castillo, LA Rodriguez. Effectiveness of heat treatment on different types of milks. *Alimentaria* 296: 53–57, 1998.
128. KK Tekinsen, M Elmali, Z Ulukanli. Microbiological Quality of UHT Milk Consumed in Turkey. *Internet Journal of Food Safety* 7: 45–48, 2007.
129. M Guida, G Marino, G Melluso. Problematiche igenistiche relative alla shelf-life del latte UHT. *Industrie Alimentari* 52: 712–716, 2003.
130. Council Directive 92/47/EEC, June 16, 1992. *Official Journal of the European Communities* L.268.
131. FD Neumann, RU Salvatori, C Majolo, H Froder. Occurrence of *Bacillus sporothermodurans* in UHT milk commercialized in the state of Rio Grande Do Sul, Brazil. *Global Veterinaria* 4(2): 156–159, 2010.
132. LJ Rothschild. IARC classes AFB1 as class 1 human carcinogen. *Food Chemistry* 34: 62–66, 1992.
133. FAO. *Worldwide Regulation Micotoxine 1995. A Compendium, Food and Nutrition.* Paper 64. Rome, Italy: FAO, 1997.
134. Codex Alimentarius Commissions. *Comments submitted on the draft maximum level for aflatoxin M1 in milk.* Codex committee on food additives and contaminants, 33rd sessions, Hauge, The Netherlands, 2001.
135. IYS Rustom. Aflatoxin in food and feed: Occurrence, legislation and inactivation by physical methods. *Food Chemistry* 59: 57–67, 1997.
136. ML Martins, HM Martins. Aflatoxin M1 in raw and ultra high temperature-treated milk commercialised in Portugal. *Food Additives and Contaminants* 17: 871–874, 2000.
137. V Roussi, A Govaris, A Varagouli, NA Botsoglou. Occurrence of aflatoxin M1 in raw and market milk commercialized in Greece. *Food Additives and Contaminants* 19: 863–868, 2002.
138. L Grasso, G Scarano, A Salzillo, L Serpe. Presence of aflatoxin M1 in milk and milk derivatives from the Campania and Calabria regions. *Rivista di Scienza dell'Alimentazione* 30: 29–34, 2001.
139. CE Lopez, LL Ramos, SS Ramadan, LC Bulacio. Presence of aflatoxin M1 in milk for human consumption in Argentina. *Food Control* 14: 31–34, 2003.
140. NS Garrido, MH Iha, MR Santos-Ortolani, RM Duarte-Favaro. Occurrence of aflatoxins M1 and M2 in milk commercialized in Ribeirao Preto-SP, Brazil. *Food Additives and Contaminants* 20: 70–73, 2003.
141. C Nachtmann, S Gallina, M Rastelli, GL Ferro, L Decastelli. Regional monitoring plan regarding the presence of aflatoxin M1 in pasteurized and UHT milk in Italy. *Food Control* 18: 623–629, 2007.
142. N Unusan. Occurrence of aflatoxin M1 in UHT milk in Turkey. *Food and Chemical Toxicology* 44: 1897–1900, 2006.
143. KK Tekinsen, HS Eken. Aflatoxin M1 levels in UHT milk and kashar cheese consumed in Turkey. *Food and Chemical Toxicology* 46: 3287–3289, 2008.
144. E Rahimi, A Shakerian, M Jafariyan, M Ebrahimi, M Riahi. Occurrence of aflatoxin M1 in raw, pasteurized and UHT milk commercialized in Esfahan and Shahr-e Kord, Iran. *Food Security* 1: 317–320, 2009.
145. A Heshmati, JM Milani Contamination of UHT milk by aflatoxin M1 in Iran. *Food Control* 21: 19–22, 2010.
146. AA Fallah. Assessment of aflatoxin M1 contamination in pasteurized and UHT milk marketed in central part of Iran. *Food and Chemical Toxicology* 48: 988–991, 2010.
147. DG Hayward. Determination of polychlorinated dibenzo-p-dioxin and dibenzofuran background in milk and cheese by quadrupole ion storage collision induced dissociation MS/MS. *Chemosphere* 34: 929–939, 1997.
148. JR Startin, M Rose, C Wright, I Parker, J Gilbert. Surveillance of British foods for PCDDs and PCDFs. *Chemosphere* 20: 793–798, 1990.
149. P Furst, C Furst, K Wilmers. Survey of dairy products for PCDDs, PCDFs, PCBs and HBB. *Chemosphere* 25: 1039–1048, 1992.
150. F Fabietti, M Delise, A Piccioli-Bocca. Aromatic hydrocarbon residues in milk: Preliminary investigation. *Food Control* 11: 313–317, 2000.
151. SS Souza, AG Cruz, EHM Walter, JAF Faria, RMS Celeghini, MMC Ferreira, D Granato, A de S. Sant'Ana. Monitoring the authenticity of Brazilian UHT milk: A chemometric approach. *Food Chemistry* 124: 692–695, 2011.
152. Z Al-Attabi, BR D'Arcy, HC Deeth. Volatile sulphur compounds in UHT milk. *Critical Reviews in Food Science and Nutrition* 49(1): 28–47, 2009.
153. E Valero, M Villamiel, B Miralles, J Sanz, J Martinez-Castro. Changes in flavour and volatile components during storage of whole and skimmed UHT milk. *Food Chemistry* 72: 51–58, 2001.

154. G Contarini, M Povolo, R Leardi, PM Toppino. Influence of heat treatment on the volatile compounds of milk. *Journal of Agricultural and Food Chemistry* 45: 3171–3177, 1997.
155. G Contarini, M Povolo. Volatile fraction of milk: Comparison between purge and trap and solid phase microextraction techniques. *Journal of Agricultural and Food Chemistry* 50: 7350–7355, 2002.
156. AK Adhikari, OP Singhal. Effect of heat resistant micro-organisms on the fatty acid profile and the organoleptic quality of UHT milk during storage. *Indian Journal of Dairy Science* 45: 272–277, 1992.
157. K Iwatsuki, Y Mizota, M Sumi, K Sotoyama, M Tomita. Effect of pasteurization temperature on the sensory characteristics of UHT processed milk. III. Studies on the sensory characteristics and physicochemical properties of UHT processed milk. *Journal of the Japanese Society for Food Science and Technology* 47: 538–543, 2000.
158. K Iwatsuki, H Matsui, Y Mizota, K Sotoyama, M Sumi, M Tomita. Effect of homogenizing pressure on physicochemical properties and sensory characteristics of UHT processed milk. *Journal of the Japanese Society for Food Science and Technology* 48: 126–133, 2001.
159. G Oupadissakoon, DH Chambers, E Chambers. Comparison of the sensory properties of UHT milk from different countries. *Journal of Sensory Studies* 24(3): 427–440, 2009.
160. M Czerny, P Schieberle. Influence of the polyethylene packaging on the adsorption of odour-active compounds from UHT-milk. *European Food Research and Technology* 225: 215–223, 2007.

13 Thermal Processing of Canned Foods

Z. Jun Weng

CONTENTS

13.1 INTRODUCTION

Thermal processing of *canned foods* can be divided into two major process methods, *in-container sterilization* and in-flow sterilization. The in-flow process refers to aseptic processing and aseptic packaging, which are covered in the previous chapter. The in-container process generally refers to the canning process, in which the prepared food is filled into a package before sealing and sterilization. The food could be packaged in the metal cans, glass bottles, plastic bottles, retortable pouches, retortable cartons, rigid and semirigid plastic containers, etc. Although the *canning* process started from Nicholas Appert's time in 1810 [1], food processing via canning process still provides a universal and economic method for preserving and processing foods. It is one of the most convenient and safe food products in today's life.

The customer eating trends and the food retail trends (such as Wal-Mart) are the key driver on impacting the canning industry. Today's consumer requires foods to be safe, fresh-like quality, healthy, nutritional, novel, ethnic, and convenience. The most pervasive consumer trend in the food industry is the increasing demand for convenience foods. The general trend of the average meal preparation time in the United States has fallen from 30 to 15 min [2]. Convenience and quality is the primary driver that strongly affects every aspect of food canning process, including food formulation, packaging, and processing. In this section, we are going to discuss the recent development on the *sterilizer* systems for thermal processing of canned foods, the latest thermal process calculation methods, and computer-based thermal process real-time control methods for ensuring food safety and quality.

13.2 CURRENT CANNING STERILIZATION SYSTEMS

The basic flowchart of canning operation can be described in Figure 13.1. The food is formulated first and filled into a container. The container could be metal containers, glass jars, plastic bottles, retort pouch, etc. After filling, the container is hermetically sealed. The sealed product then undergoes a *sterilization* (or *pasteurization*) process to obtain the scheduled food safety (*commercial sterility* and/or the desired lethality $F_{T_{ref}}^{z}$ value) and the desired quality. The sterilization process is the critical control point for ensuring food safety. The sterilizer should be able to consistently deliver the commercial sterility to every canned product from batch to batch.

There are many types of sterilizers or retorts used in the canning industry [3,4]. The most common sterilizers used in the canning industry are batch retorts, continuous hydrostatic sterilizers, and continuous rotary (reel and spiral) sterilizers. Modern batch and continuous sterilizers continue to evolve to maximize the process efficiency and product quality, and to adapt various packaging containers such as nitrogen-containing retortable pouch, easy-open cans, and retortable cartons. Based on the product heat transfer characteristics and the packaging materials, the application of various sterilizers can be grossly summarized in Table 13.1. System and system life cycle cost, production throughput, product quality and uniformity, packaging material and geometry, variation of container sizes, and future package trends are among the most important factors when choosing a proper sterilization system. High-temperature short-time (HTST) process generally aided by rotational process usually is considered for maximizing the product quality and throughput. However some products such as canned seafood are sensitive to heat and shear force, a proper sterilization temperature needs to be properly optimized.

13.2.1 BATCH RETORT

The batch retort represents the most versatile and flexible sterilization systems to meet the increasing market demand on the product innovation. It is used for both conduction-heating and convention-heating products. Almost every kind of packaging formats used in the canning industry

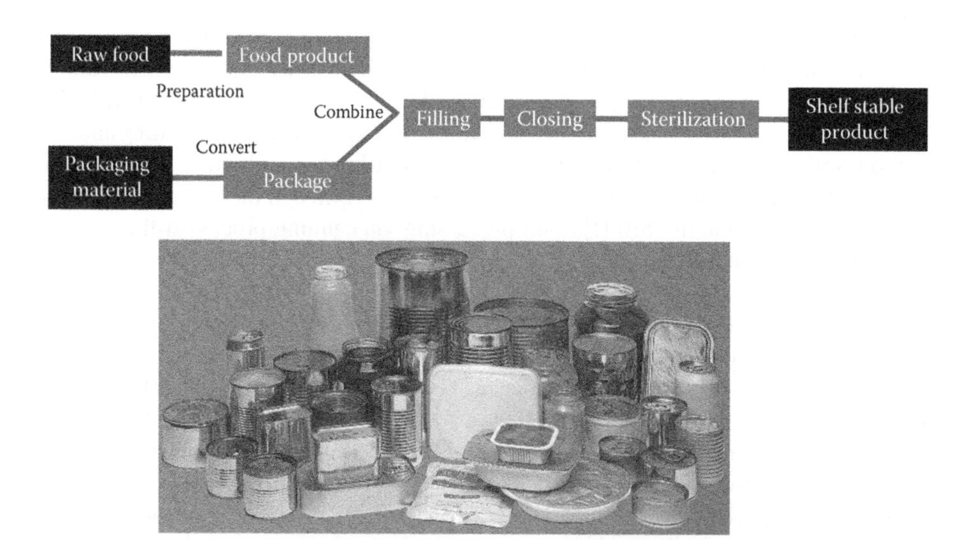

FIGURE 13.1 The general flowchart of canning process and canned foods in various package formats.

TABLE 13.1
Sterilization Applications for Different Heat Transfer Modes and Packages of Canned Foods

Operation	Equipment Types	Product Flexibility	Container Flexibility
Continuous sterilizers	Rotary reel and spiral	Convection + semiconvection	3-piece cans and some 2-piece cans
	Hydrostatic	Conduction	All cans, jars, and bottles
Batch retorts	Still	Convection + conduction	All cans
	Semiautomatic		Flexible packages
	Automated batch		Glass jars and bottles

can be properly processed in a batch retort. The batch retort is usually classified into still retort and rotary retort based on the agitation mode. Generally the rotational batch retort uses end-over-end rotation mode to enhance the product heat transfer in the container and thus increase the product throughput and quality. Different heating media and heating methods, such as steam, steam/air mixture, water cascading, water spray, or water full immersion, are used in various batch retort systems [3].

The major development in the batch retort is perhaps on handling various packaging materials and the system automation. Modern batch retort provides independent control of temperature and pressure for protecting the package integrity during heating and cooling phases. One of these batch retorts is the water spray batch retort, which is increasingly used in the modern canning industry. In these retorts, water is usually stored in the bottom of the retort vessel and circulated by a pump with a high flow rate and sprayed (or cascaded) on the packages. Usually, the water is indirectly heated by steam through a heat exchanger in the circulation line. The primary application of this system is for the food packaged in a container that requires overpressure for balancing the internal pressure such as pouch, bottles, pressurized cans, etc. Figure 13.2a shows a typical Steam Water Spray™ (SWS) *automated batch retort system.* Both superheated water spray and steam are used as heating medium, and air is used as *overpressure* for balancing the product internal pressure. The system utilizes direct steam injection for ensuring fast and uniform come-up. The high-velocity water sprays from various angles intensely mix steam, water, and air and create a homogenous temperature distribution inside the vessel. The system uses indirect cooling method and the sterile water is cooled over a plate heat exchanger, which saves water, water treatment chemicals, and energy. The system is controlled by the LOG-TEC™ Process Management System, which can automatically correct process deviations using its embedded mathematical models such as NumeriCAL™ model without intervention of an authorized person.

There is an increasing demand toward fully automatic retort loading and unloading for reducing the labor cost. It is not uncommon that the entire cook room with a dozen of batch retorts is operated

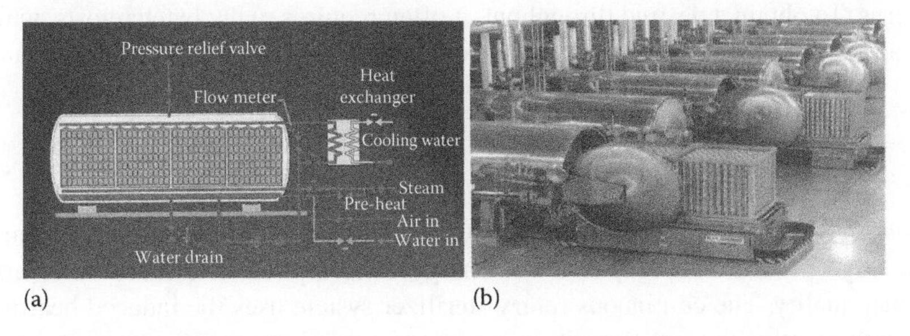

(a) (b)

FIGURE 13.2 Steam Water Spray™ (SWS™) automated batch retort system. (a) System concept view and (b) fully automated SWS system with automatic loading/unloading system. (From John Bean Technologies Corporation (formerly part of FMC Corporation), Chicago, IL.)

(a) (b)

FIGURE 13.3 SuperAgi™ agitating batch retort system. (a) Water spray inside rotating drum and (b) industry installation for canned beverages. (From John Bean Technologies Corporation (formerly part of FMC Corporation), Chicago, IL.)

by one supervising person (Figure 13.2b). Wider ranges of process times, temperatures, and pressures can often be selected for economically processing products in a fully automated batch retorts.

Food industry continues to advance batch retort systems in order to reduce the carbon foot print and improve production efficiency and product quality. *SuperAgi™* was introduced to the food industry in 2006 [5]. The process water sprays were mounted inside the rotating drum (Figure 13.3a), which eliminates obstructions between process water and food packages. This design provides capability for the SWS process to employ high-speed product agitation technology and allows deeper and more process water penetration for a better temperature distribution. The sterilizer was designed with multifunctional process modes to meet industry's increasing innovation on packages and demand on delivering optimum process conditions. It can use either water spray, full or partial water immersion technology with various agitation modes for every package and product combination.

Shaka® batch retort system has been studied for some time and has recently shown interesting progress in commercial applications [6,7]. The major principle of the *Shaka* retort is the back and forth container agitation technology [7]. It was found that the heat transfer inside the container was improved dramatically for a variety of food products such as liquid- or semiliquid-based foods in cans, glass, and pouches when shaking the product container back and forth in a retort. It uses vigorous horizontal agitation to mix the contents of the container and transfer heat rapidly throughout the contents.

The major concern on the batch retorts perhaps is the product quality uniformity within a single batch and from batch to batch. It is common in a single batch that the outer layer products in a basket are usually processed more severely than those inside the basket, which may result in a nonuniform quality distribution of the processed products. Product throughput and energy consumption are also major concerns. To obtain a desired throughput, it often requires many batch retorts, which require plant floor space and increase the energy consumption, especially during venting and come-up phases.

13.2.2 Continuous Rotary Sterilizer

The *continuous rotary sterilizer* is a fully automated continuous processing system, which is designed for high product throughput, lower energy consumption, and more uniformity of processed product quality. The continuous rotary sterilizer system uses the induced heating principle by intermittently rotating the can along its axis (side) and thus increases the rate of heat penetration (Figure 13.4). The system enables using the HTST process principle for maximizing the product quality and throughput.

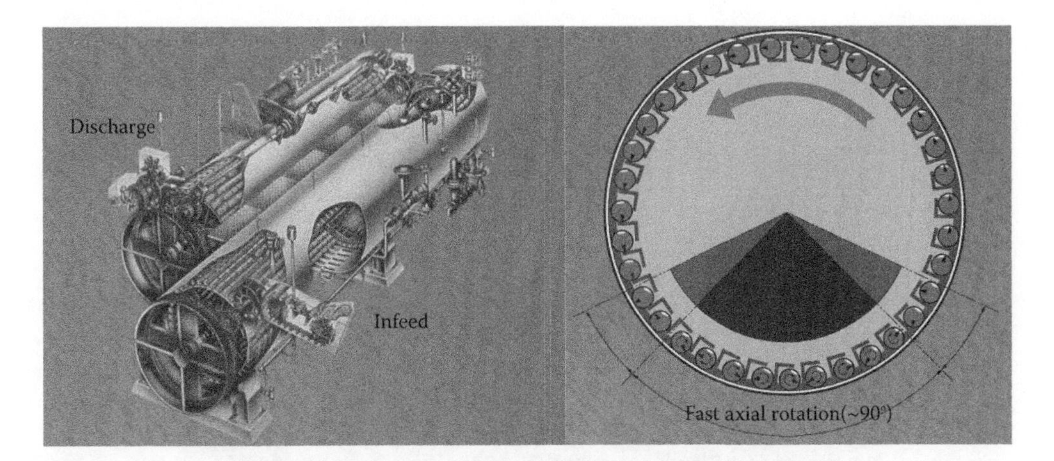

FIGURE 13.4 The process principle of JBT continuous rotary (reel and spiral) sterilizer system (a system cut-away diagram and b induced convection-heating). (From John Bean Technologies Corporation (formerly part of FMC Corporation), Chicago, IL.)

The seamed cans enter the line directly from the closing machine. A feed device delivers the cans via the infeed valve to the rotating reel of the sterilizer. The reel, working in conjunction with the stationary spirals in the shell, carries the cans through the cooking and cooling system [4]. Since each container travels exactly the same trajectory, its process time and temperature profile and cooling profile are exactly identical. A uniform cooking/cooling and a uniform quality for every container thus can be ensured. The continuous rotary sterilizer is widely used in the food industry. More than 7500 JBT FoodTech (formerly FMC) continuous rotary shells have been placed in production around the world since 1921 for processing more than half of the world's canned foods. The major applications are for fruit, vegetables, ready meals, soups, infant formula, and dairy products. Depending on the process time and can size, the system can deliver up to 1500 cans per minute.

To meet the industry demands on reducing the overall cost and increasing the consumer convenience, the continuous rotary sterilizer is consistently evolving to meet these challenges. Figure 13.5 shows JBT continuous rotary sterilizer with overpressure capability for handling different packaging formats such as plastic containers, reduced metal plate thickness, peelable lids, and easy-open-ends. These packaging formats require overpressure to balance the can internal pressure during processing. Superheated water with circulation and the steam-air mixture (top portion) at the same temperature are used as the heating media. The air serves as the overpressure to balance the product internal pressure during processing. Steam and air are mixed in an overhead mixing tube and introduced into the cooker shell over the whole length (Figure 13.5b). Pure *steam-air mixture* has also been successfully used to process the pressure-sensitive product such as rigid plastic containers. In this case, the steam and air are mixed by the reel rotation and can movement.

JBT introduced a large diameter, high-volume continuous rotary pressure sterilizer line with 2.84 m (112″) diameter (Figure 13.6). Comparing to the traditional continuous rotary pressure sterilizer with diameter of 1.47 m (58″), the can holding capacity is nearly doubled. This 100% larger capacity vessel allows continuous rotary pressure sterilizer to operate at higher processing volumes without doubling up on machinery and thus increase the production efficiency and simplicity.

The continuous rotary sterilizer is an ideal system for processing products that require large throughput with limited can size variation. The system generally requires a cylindrical container with limited can diameter and can height variation for each application. Since the product content is intensely mixed by rotation, it may also not be suitable for those products sensitive to the shear force.

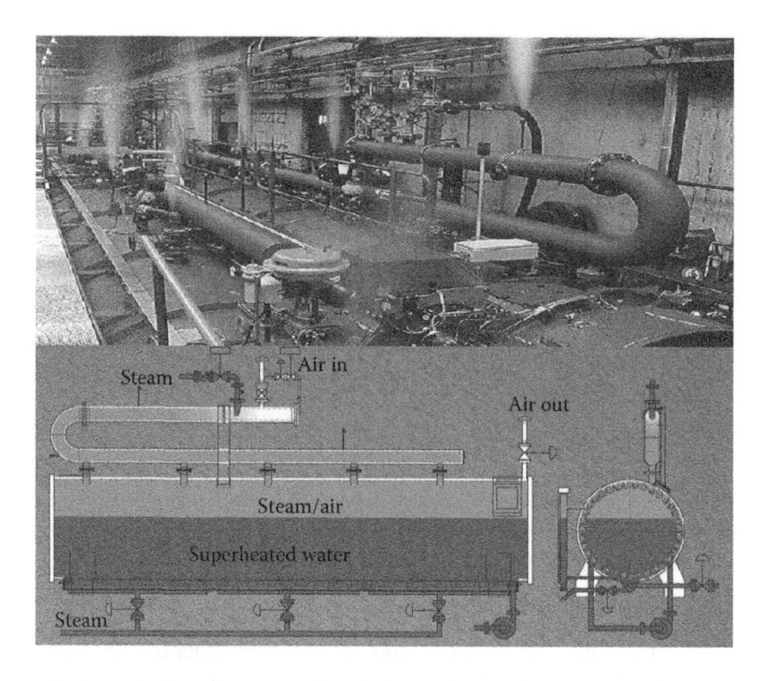

FIGURE 13.5 JBT FoodTech continuous rotary sterilizer with overpressure capability. (a) System in operation and (b) system processing principle. (From John Bean Technologies Corporation (formerly part of FMC Corporation), Chicago, IL.)

FIGURE 13.6 Large diameter (2.84 m) continuous reel and spiral rotary pressure sterilizer system. (From John Bean Technologies Corporation (formerly part of FMC Corporation), Chicago, IL.)

13.2.3 Hydrostatic Sterilizer

The *hydrostatic pressure sterilizer* is another kind of continuous container handling system used in the food canning industry (Figure 13.7). This type of sterilizer is particularly well suited for processing products that require long cook and cool times and high container throughput and for those that derive little or no benefit from container agitation. *Hydrostatic sterilizer* operates on the hydrostatic pressure principle, with the pressure of saturated steam in the chamber (steam dome) balancing the weight of two water columns (feed and discharge legs) adjacent to the steam dome (Figure 13.7a). Typical products presently being sterilized in the hydrostatic sterilizer include baby foods in jars, pet foods, soup, pumpkin, canned meat and stews, and beverages. Typically, a stick of containers is mechanically transferred onto a carrier, which is attached to a moving chain (conveyor). The product is conveyed through a temperature-controlled hot water column (feed leg), a controlled steam chamber (steam dome) set to a processing temperature, a precool

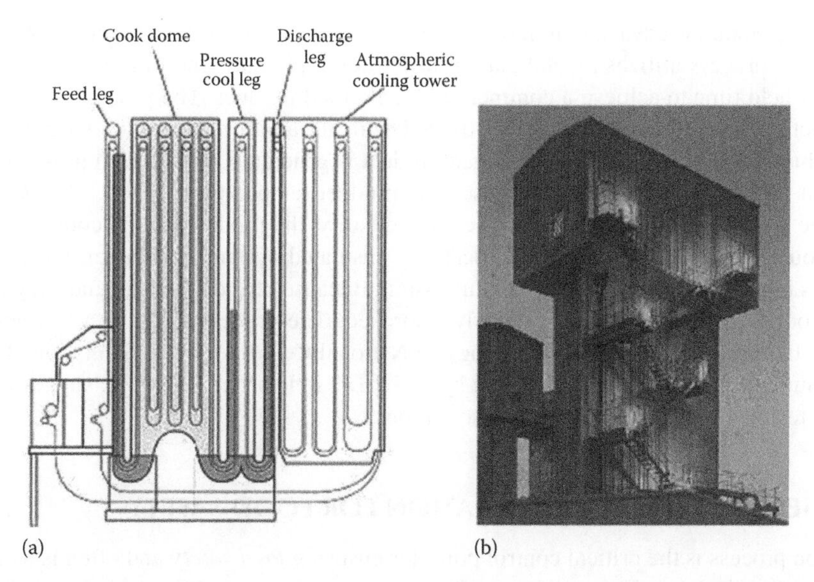

FIGURE 13.7 Continuous hydrostatic sterilizer: (a) standard steam chamber with pressure cool and (b) a hydrostatic sterilizer in operation. (From John Bean Technologies Corporation (formerly part of FMC Corporation), Chicago, IL.)

discharge water column (discharge leg) and a water cascading cooling system, and finally exits the system at the container unloading station. The speed of the conveyor is controlled such that the residence time of each carrier in the steam dome meets the scheduled process time (and thus scheduled lethality) at the minimum required temperature parameters assigned by a competent process authority. Generally, in an ideal and well-controlled process, all products processed in a hydrostatic sterilizer receive exactly the same time–temperature process treatment and a uniform sterility.

The hydrostatic sterilizer can be used to sterilize the product packaged in metal cans, glass jars, plastic bottles, and lately the retortable pouch with the capability of agitating the product for faster heat transfer. However, the initial investment for such a system is usually higher than other systems, and the system is usually limited to certain range of can sizes (diameter), which depend on the designed carrier profile.

13.2.4 Nontraditional Sterilizers

The latest development for processing shelf-stable packaged foods in canning industry perhaps is the development of nontraditional sterilization technologies. In this section, we will briefly introduce two new technologies, *microwave-assisted* and *pressure-assisted thermal processes* [8]. Microwave-assisted water full immersion thermal process was developed in Washington State University in a pilot scale [9]. The packaged food is immersed in the hot process water, that is, above 121.1°C under pressure to maintain the package integrity. The product in the package is then rapidly heated up to the desired process temperature by applying the patented microwave technology at 915 MHz [10]. The product then is held in the processing water for a predetermined time to achieve adequate lethality at the "slowest heating zone" ("cold" spot) inside the package. This process is designed in a continuous mode. Because of rapid heating assisted by microwave technology, the packaged product can be commercially sterilized in a short time, resulting in a significantly improved product quality. The process filing of a low-acid food (mashed potato) packaged in microwavable single-serve poly tray was accepted by U.S. Food and Drug Administration (FDA) in 2009 [8,9], suggesting a significant advance of applying microwave technology in food preservation.

Another technology advance in food preservation is the pressure-assisted thermal process [8,11,12]. This process utilizes a combination of ultrahigh pressure and mild process temperature over a short hold time to achieve a commercial sterile food product. The process is a batch mode. The product packaged in retort pouch is treated with ultrahigh pressure (about 500–800 MPa). The microbial lethal heat inside the packaged product is generated by ultrahigh pressure compression, which rapidly and uniformly increases the product temperature to 80°C–130°C regardless the package size [8]. Upon achieving the desired lethality, the product can be cooled down quasi-instantaneously by decompression. In a perfectly insulated system (adiabatic), the product will return to its initial temperature upon decompression. Because of short heating and very rapid cooling, the product quality can be significantly improved. The process filing of a low-acid product (mashed potato) processed by this technology in National Center for Food Safety and Technology (NCFST, Summit-Argo, ILL) was accepted by U.S. FDA in February 2009 [8], suggesting another significant technology advance in food preservation.

13.3 THERMAL PROCESS CALCULATION FOR FOOD SAFETY

Sterilization process is the critical control point for ensuring *food safety* and often is a critical unit operation on maximizing the production efficiency or throughput. An establishment of a proper thermal process is based on the knowledge of the heat resistance of microorganisms and the heat transfer rate of a specific product. Process modeling and calculation software plays an important role in optimizing sterilization processes while ensuring the food safety. There is an increasing interest in the canning industry on using new *process modeling* and *process calculation method* for food safety and for optimizing the product quality and process efficiency [1,13].

13.3.1 Principles of Thermal Process Calculation

The calculation of the sterilization value or the $F_{T_{ref}}^{z}$ value of the canned foods is based on the *microbial death kinetics* and the product temperature at the critical point or the slowest heating zone ("cold" spot) in the slowest heating container. Equation 13.1 is widely used for the canned food thermal process evaluation and calculation [1,14,15], in which the first-order thermal inactivation kinetics of the microorganism and traditional *z-value model* are used:

$$F_{T_{ref}}^{z}\ \text{value} = \int_{0}^{t} 10^{(T-T_{ref})/z}\,dt \tag{13.1}$$

where

T is the slowest heating zone product temperature in a container
t is the heating and cooling time
T_{ref} is the reference temperature for the microorganism of concern
z is the thermal characteristic of the target microorganism in the lethal temperature range

For the low-acid canned foods (pH greater than 4.6 and a_w greater than 8.5) [16], the commercial sterility is achieved by destruction of all pathogenic organisms together with nonpathogenic spoilage microorganisms, such as *Clostridium sporogenes* (PA 3679), so that the canned foods can be shelf stable under normal storage temperature [4,16]. In the canning industry, the Fo value is often used for the low-acid canned foods and refers to the sterilization value ($F_{T_{ref}}^{z}$ value) with z value of 10°C (18°F) and the reference temperature (T_{ref}) of 121.1°C (250°F). The z value of 10°C is used as the thermal characteristics for the pathogenic microorganism, *Clostridium botulinum* spores. The *C. botulinum* spore is the most heat-resistant pathogen and produces the most deadly toxin if it is allowed to grow. It is the target pathogenic microorganism to be killed during thermal processing for low-acid shelf-stable canned foods.

Equation 13.1 is also used for the thermal process calculation of the low-acid canned foods with an extended refrigerated shelf life and for shelf-stable canned acid (such as fruits) and acidified foods under normal storage conditions. In this case, the food industry sometimes uses the pasteurization value ($P_{T_{ref}}^{z}$ value) instead to the sterilization value ($F_{T_{ref}}^{z}$ value), since foods generally are processed below or around 100°C. For the acid or acidified foods, the process is used to kill organisms affecting the shelf stability of the products stored at the ambient temperature. For the low-acid foods packaged in the reduced oxygen atmosphere and stored/distributed under refrigerated conditions, the process is used to kill the vegetative *pathogenic bacteria* together with *C. botulinum* spores of the nonproteolytic types B for an extended shelf life. The thermal characteristics of the most thermal-resistant pathogens or the target pathogen need to be carefully studied for each application. In some cases, the z value may not be constant in the lethal temperature range. For example, the *C. botulinum* spores of the nonproteolytic type B in the Surimi product show a broken thermal-death-time (TDT) curve or two z values with z = 7.0°C for temperature less than 90°C and z = 10°C for temperature above 90°C [17].

Recently there are some challenges on the validity of the first-order thermal inactivation kinetics of the concerned microorganisms [18]. The major concern was the nonlog-linear survivor curve, or the "shoulder and tail" phenomenon under constant heat treatment. Instead of the traditional first-order thermal inactivation kinetics model, various alternative thermal inactivation kinetic models for microbial survivor curves have been proposed [18,19,20]. These alternative kinetic models could fundamentally change the traditional thermal process calculation methods, such as Ball Formula method [21], and need to be scientifically challenged. Instead of using traditional $F_{T_{ref}}^{z}$ value for the food safety evaluation, the new concept, Food Safety Objective (FSO) has been introduced [18]. The FSO defines the maximum frequency or concentration of a microorganism hazard in a food at the time of consumption that will provide an appropriate level of protection. Obviously these trends challenge the traditional mathematical methods in thermal processing calculation.

13.3.2 Thermal Process Calculation Models and Software

There are a number of mathematical methods proposed for thermal process calculations [22]. In calculating the sterilization value at the slowest heating zone in the container, all proposed mathematical methods are based on Equation 13.1. The difference though is on how the product temperature (T) at the slowest heating zone inside the packaged product is obtained. The general method uses the physically measured product temperatures only (or called physical simulation), whereas the mathematical models use mathematical models for predicting the product temperatures. The obtained product temperatures are then utilized in Equation 13.1 for integrating the sterilization value. Thus, the process calculation methods could be divided into following four categories based on product temperatures obtained:

- The general method using physically measured time–temperature data [1,14]
- The semiempirical mathematical methods using a constant heating or cooling retort temperatures, such as Ball's formula method [21,23], Hayakawa's method [22], Stumbo's method [1,22], and Gillespy method [22]
- The theoretical models such as pure conduction or convection heating models [23,24] and computational fluid dynamic (CFD) models based on Navier–Stoke's equations [25,26]
- The *innovative models* for handling variable retort temperatures

The *general method* purely utilizes the physically measured product temperatures in lethal rate integration (Equation 13.1). Numerical integration procedures such as the Simpson's rule or the trapezoidal method are often used for the integration purpose. Since the actual least lethal zone product temperatures are used in the calculation, it is the most accurate method for evaluating a given set of heat-penetration data. However, the general method is generally not recommended for thermal process development (or design), since it represents the process instance that is under evaluation. The daily and seasonal process conditions, such as the initial product temperature variation, the retort process temperature stability, the come-up profile, and the cooling profile, may vary from cycle to cycle. The retort could be only partially loaded and the process might experience sterilization temperature deviations. These variations could lead to the uncertainty of the food safety of the delivered process.

Except for the general method, almost all published methods used mathematical models to correlate the retort temperatures to the product temperatures. Except some simple heat transfer modes such as pure convection or pure conduction heating products, the majority of canned food products exhibit complicated heat transfer modes. Therefore, the majority of mathematical methods were based on semiempirical equations such as *Ball formula method* [21,22,23] by using heating and cooling factors (jh, fh, jc, fc). The Ball Formula method has served in the food industry for more than half a century and is still widely used in the canning industry. It is flexible in terms of ranging initial product temperatures and retort temperatures in process design. It can be used for any kind of foods packaged in any kind of containers, since a dimensionless factor, j value (jh or jc) is used to correlate the initial product temperature and is independent on the retort temperatures within reasonable range. However, because of time limitations when Ball developed his method half a century ago, there were some assumptions made in his mathematical model [21,23] such as jc = 1.41 and fc = fh. For conduction heating products, the Ball formula method often results in underestimating the actual lethality value (conservative). However, in many cases when the product jc value is substantially less than 1.41 such as products packed in the thin pouch, the Ball formula method could lead to overestimating the actual product lethality value [27]. Moreover, the Ball formula method only handles the constant process temperature. It is of food industry's interests to have a properly optimized thermal process for the product quality and efficiency [13,28].

13.3.3 Innovative Model

With rapid development in the information technologies, the more sophisticated mathematical modeling methods, which can handle variable process retort temperatures (classified as *innovative models*) for canned food thermal process calculations have been discussed and published in the literature [29–34]. NumeriCAL™ [35] and CTemp [36] are two typical examples of software package used in the food industry. The major advantages of these published methods are that they can handle various complex heat transfer modes and variable process temperatures, thus the process temperature deviations.

Since it was introduced in 1988, *NumeriCAL* software is adopted by almost all major food processors for thermal process evaluation and thermal process design of canned foods in North America and its benefit on increasing process efficiency, and product quality has been recognized in the food industry [28]. Its modeling results are accepted by both U.S. FDA and U.S. Department of Agriculture (USDA) regulatory agencies. NumeriCAL core algorithm (J. Manson, personal communication, 1991) is based on the finite difference method for handling the variable retort process temperatures. Since NumeriCAL utilizes heating and cooling factors, it is capable of simulating product temperatures at a single point inside a container for any transient heat transfer modes and for any types of food packaged in all styles of containers. Figure 13.8 shows Windows®-based NumeriCAL v5.0 software [35]. It consists of two major modules: analyze and calculate modules. The analyze module is used to analyze the heat penetration data and to develop NumeriCAL heating and cooling factors for the worst case container. There are a number of nonlinear optimization methods embedded in the model, which can be used to automatically fit the heat penetration curve. The calculate module is used for the thermal process calculation (or thermal process design) and thermal process deviation analysis.

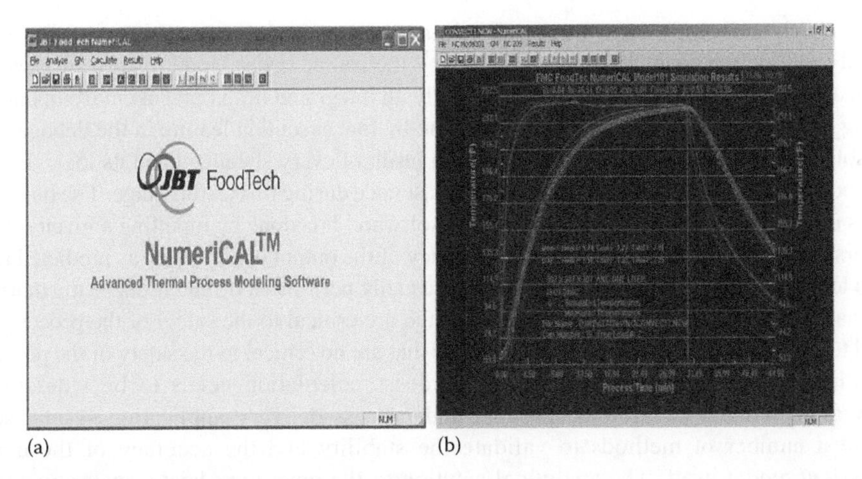

FIGURE 13.8 Microsoft Windows®–based NumeriCAL™ software (version 5.0) designed for thermal process modeling and calculation (a: front page and b: process modeling).

It is not the purpose of this chapter to describe each method. It is important however that software used for thermal process calculations is properly validated and that a validation program based on the software life cycle approach is used (John W. Larkin and Steve Spanik, personal communication, 2004). The software key components, especially the core algorithm, that are critical to the safety of the process should be validated to a high degree [37]. The software validation for the automated controller system has been discussed in the literature [37–41].

13.3.4 Challenge Thermal Process Calculation Software

Thermal process development of canned foods in the food industry generally involves two steps, modeling and calculation. *Process modeling* is to fit or model a set of heat penetration data for developing heating and cooling factors, such as Ball and NumeriCAL methods or other physical parameters. The second step is to use these obtained parameters for the process calculation or process design. Basically the second step is the model prediction step, since it predicts the new thermal process time and temperature for other process conditions, such as different initial product temperature and process retort temperature, and for the process temperature deviation analysis. It is important that the parameters (jh, fh…) used in the new process development are properly validated for each application.

Figure 13.9 shows an overview of verification through the *software life cycle activities* [38]. The left-hand side of the flowchart describes software design activities and the right-hand side of the flowchart describes various stages of testing activities. Testing activities consist of two methods, black box (the functional test) and the white box (the structural test) tests [38,39]. The white box testing method

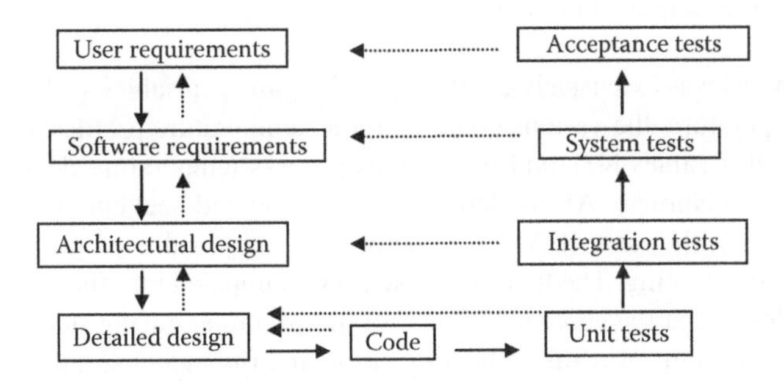

FIGURE 13.9 Overview of verification through the software life cycle activities. (Verify: dashed arrow line and sequence: solid arrow line.)

is used to check and verify the correct paths through the program code. This activity is generally performed by those who can access and understand the source code. Modern software development tool, such as Visual C++ from Microsoft, is typically an integrated development environment for coding, debugging, and execution. In particular, the line-by-line execution feature in the debugging mode enables software developers to check the execution result of every statement and its logic. Each statement of code needs to be successfully executed at least once during this testing stage. The black box test method is used to test the results generated by the software. It is done by inputting a given set of input parameters to the software and checking the accuracy of the output results, such as predicted temperatures and lethality value. The black box test can be usually performed by end users or the thermal process specialists. Those components that influence and are critical to the safety of the process must be validated to a higher degree than those components that are no critical to the safety of the process [37].

A mathematical model used for thermal process calculation needs to be validated in its accuracy and predictability for the product and process delivery application system selected. There are a number of methods to validate the stability and the accuracy of the developed *mathematical model* itself. The analytical solution to the governing heat transfer equation in a special condition (e.g., constant processing and cooling temperatures) is a method often employed to challenge the theoretical model. The microbial challenge is another method used to challenge the model where the physical measurement is difficult to be performed such as in the multiphase aseptic processing system. However, the majority of mathematical models used for thermal processing of canned foods were tested against the physical measurement data such as temperatures.

Weng presented a case study on challenging NumeriCAL model for thermal process calculations [42]. He provided, but not limited to, a list of user requirements, which are critical in thermal process calculations for the food safety and the production efficiency:

1. The software/model should be able to model the entire heat penetration curve including heating and cooling portions for various heat-transfer modes packaged in various containers. The software should show the general method accuracy on lethality value calculations and be able to model the products exhibiting either simple or broken semilogarithmic heating or cooling curves with jh value less than 1.0 up to (but not limited to) around 50.0.
2. The software/model should be able to determine or predict the new processes with the following variations under reasonable range:
   a. Different process temperatures (RT) and initial product temperatures (IT).
   b. Variable come-up temperature profiles.
   c. Variable retort temperatures.
   d. Large g value (g is the temperature difference between retort temperature and the slowest heating zone product temperature at the end of heating).
   e. Variable cooling profiles.
3. The software/model should be able to handle various process temperature deviations.
4. The software/model should be able to incorporate various thermal inactivation kinetic models of the microorganism of concern.

The NumeriCAL model was extensively challenged [42] against variables such as the initial temperature, the retort temperature, the various come-up retort temperature profiles, the cooling temperature variations, larger g values [43], and the various process temperature deviations. Figure 13.10 shows one of the typical NumeriCAL modeling results for canned beef chunk in gravy processed in a hydrostatic sterilizer with NumeriCAL jh value of 4.70 and jc value of 1.00 due to rapid depressurization during initial cooling. The thermocouple tip was impaled into the geometric center of the beef chunk, which located at the geometric center of the container during sterilization process. The product shows a broken semilogarithmic heating curve and the model simulated product temperatures (including cooling phase) agree well with the measured product temperatures. It is noted that from Figures 13.10 through 13.14, the product temperature of the heating portion is plotted in the

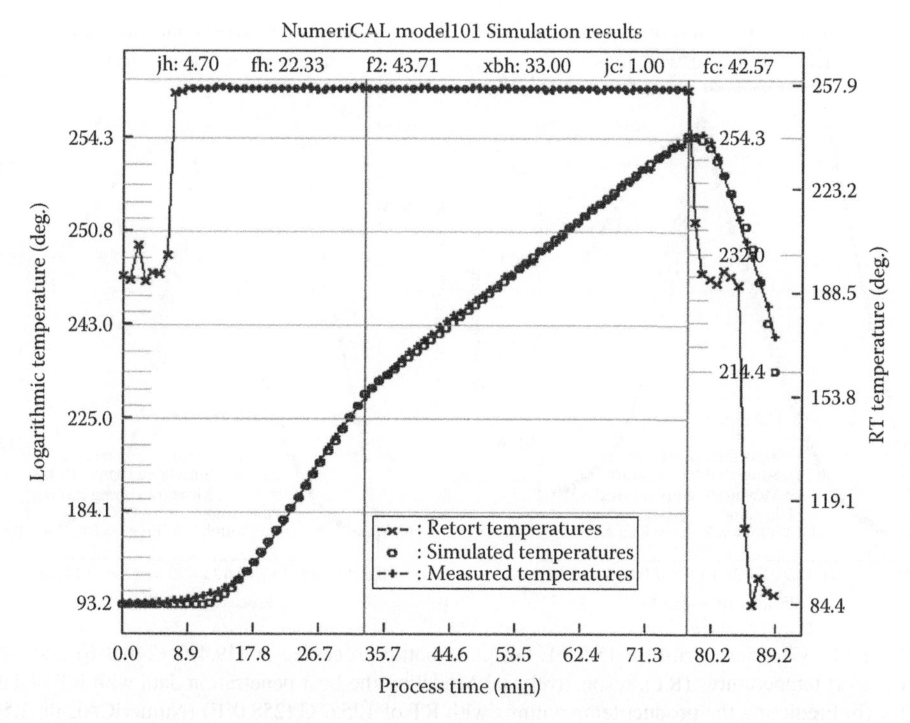

FIGURE 13.10 A typical NumeriCAL modeled product temperature profile (°F) for canned beef chunk in gravy (300 × 405 can size) processed in a hydrostatic sterilizer (NumeriCAL jh: 4.70, fh: 22.33, f2: 43.71, xbh: 33.00, jc: 1.00, fc: 42.57).

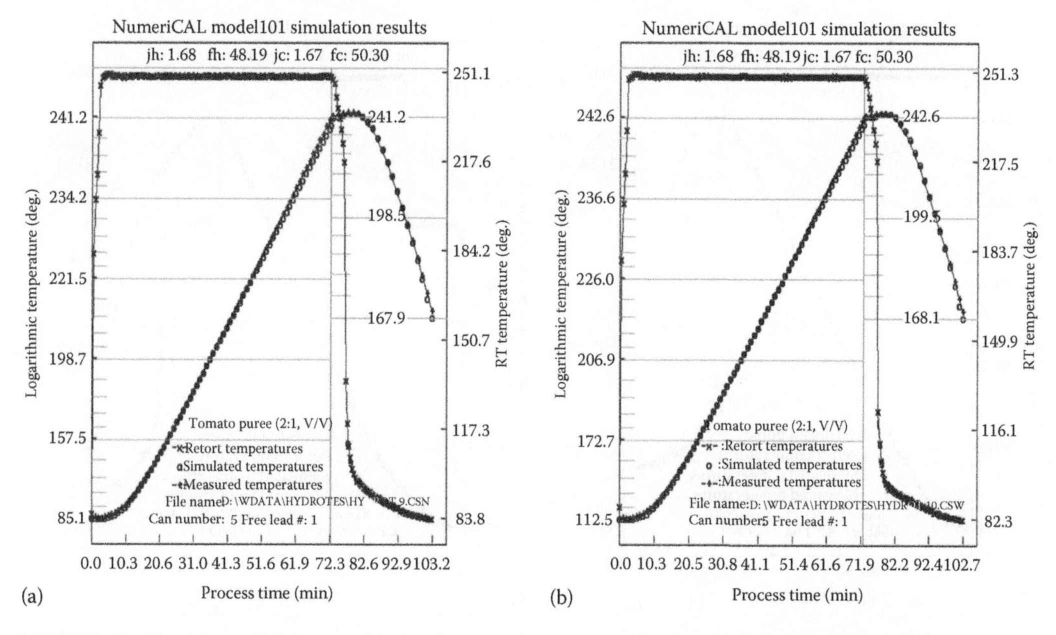

FIGURE 13.11 NumeriCAL simulated product temperatures (°F) under different initial product temperatures (IT) with 15.2°C difference (tomato puree, 2:1 v/v in 300 × 405 cans). (a) Modeling the heat penetration data with IT of 29.5°C (85.1°F). (b) Predicting product temperatures with IT of 44.72°C (112.5°F) (NumeriCAL jh:1.68, fh:48.19, jc:1.67, fc:50.30).

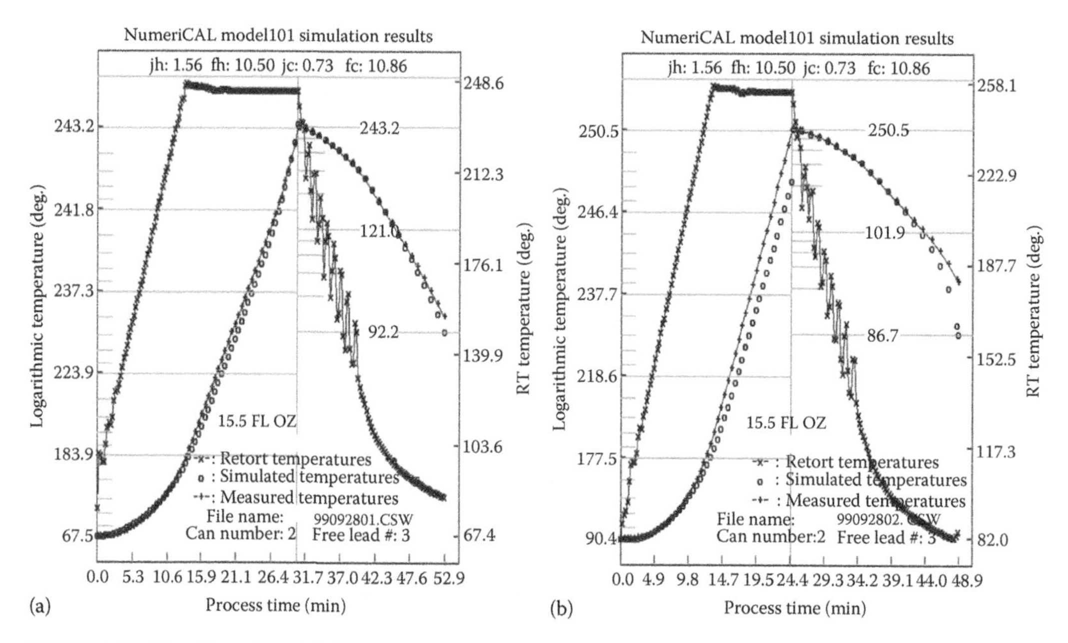

FIGURE 13.12 Chocolate drink in 15.5 FL OZ glass bottle processed at 119.4°C (246.9°F) and 125.6°C (258.0°F) retort temperatures (RT), respectively. (a) Modeling the heat penetration data with RT of 119.4°C (246.9°F). (b) Predicting the product temperatures with RT of 125.6°C (258.0°F) (NumeriCAL jh: 1.56, fh: 10.50, jc: 0.73, fc: 10.86).

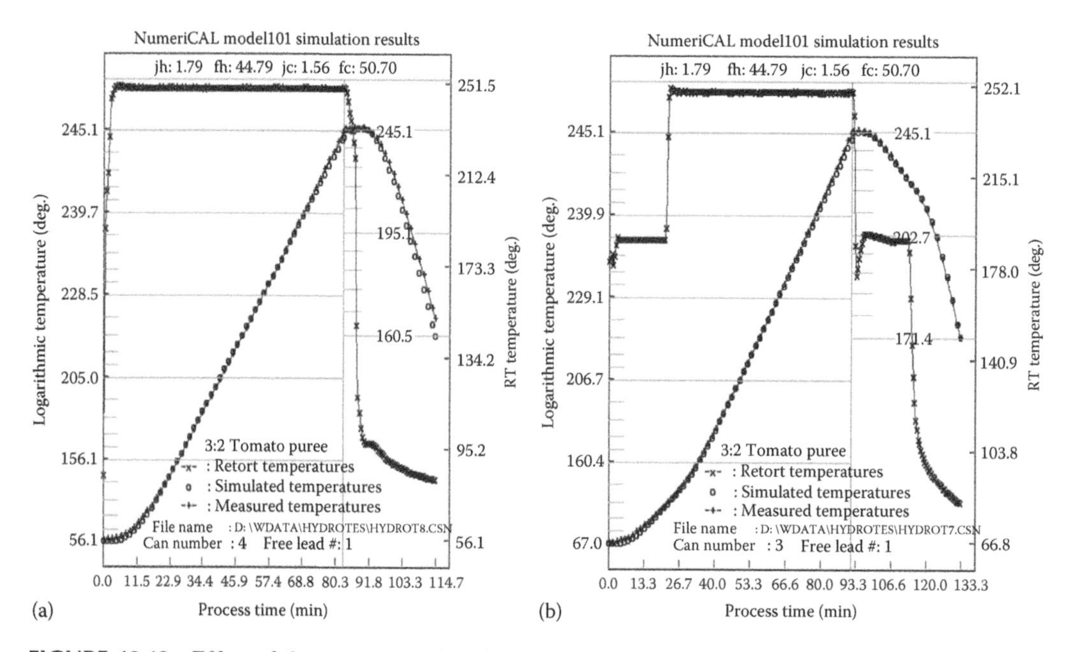

FIGURE 13.13 Effect of the come-up and cooling retort temperature profiles on the NumeriCAL modeling results (°F). Product: 300 × 405 canned condensed mushroom soup (2:1, v/v) processed in a hydrostatic sterilizer simulator. (a) Modeling the heat penetration data. (b) Predicting product temperatures with longer come-up and cooling profiles (NumeriCAL jh: 1.79, fh: 44.79, jc: 1.56, fc: 50.70).

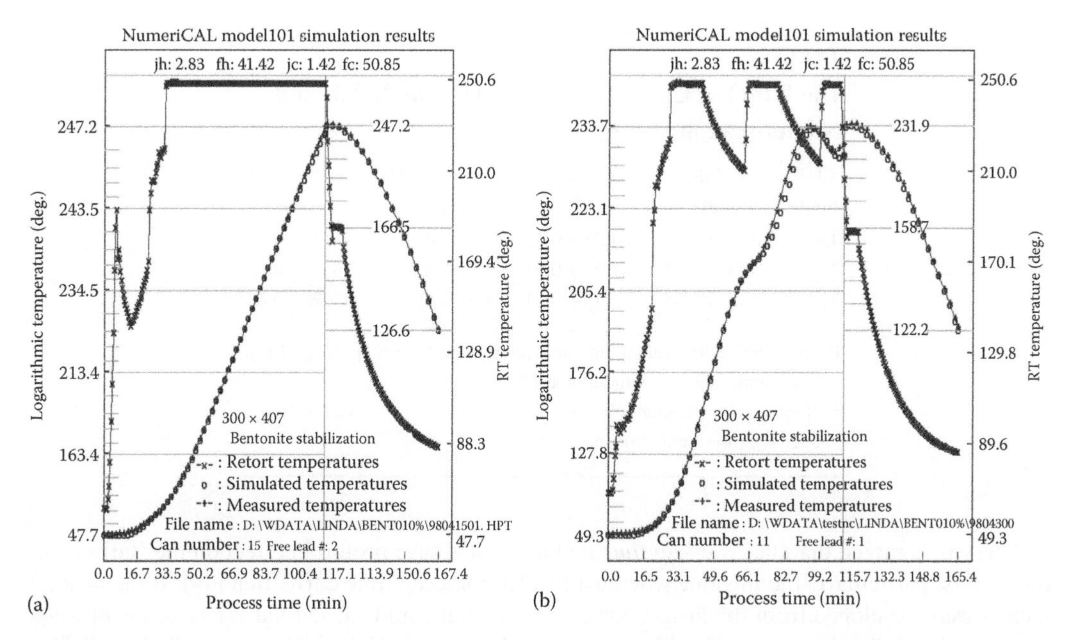

FIGURE 13.14 NumeriCAL simulation results (°F) under process deviation conditions. Product: 10% bentonite in 300 × 407 cans. (a) Normal heat penetration data modeling results. (b) Model predicted time–temperature under multiple process temperature deviation conditions (NumeriCAL jh: 2.83, fh: 41.42, jc:1.42, fc:50.85).

well-known inversed logarithmic scale [1,15,21], which is labeled in the left y axis; the cooling portion is plotted in the logarithmic scale, which is labeled directly on the cooling curve; and the retort temperature is plotted in the linear scale, which is labeled in the right y axis for comparison purpose.

Figure 13.11 shows a case study on the model simulation results for different initial product temperatures. The left curve is the modeling results for developing heating (jh, fh) and cooling (jc, fc) factors with an initial product temperature of 29.5°C. The right curve is the predicted product temperatures under different initial product temperature (15.2°C higher) by using the same input heating and cooling factors as the left curve. The agreement between the predicted and measured product temperatures (Figure 13.11, right curve) suggests that the model predictability is not affected by the initial product temperatures within the range studied.

Figure 13.12 shows the chocolate drink packaged in 15.5 FL OZ glass bottles and processed in 119.4°C and 125.6°C retort temperatures, respectively. The left curve shows modeling results on developing NumeriCAL heating and cooling factors (jh: 1.56, fh: 10.50, jc: 0.73, fc: 10.86). This set of heating/cooling factors was then used to predict the production temperature for a higher retort sterilization temperature (125.6°C) as shown in the right curve. It is clearly shown that the model predicted product temperatures are on the conservative side. The result suggests that the commercial sterility value might not be obtained if the input heating/cooling factors were developed at a higher temperature in this case.

Figure 13.13 shows the effect of the come-up and cooling profiles on the NumeriCAL model predicted results. The left curve shows a normal sterilization process with 4.5 min come-up time for developing NumeriCAL heating and cooling factors (jh: 1.79, fh: 44.79, jc: 1.56, fc: 50.70). This set of input heating and cooling factors was then used to predict the product temperatures for the same product with longer come-up and cooling downtimes (21.5 min for simulating hydrostatic sterilizer feed and discharge leg residence time). The results were plotted in the right curve and show that the model predicted product temperatures agree well with the measurement data. The result suggests that the NumeriCAL heating and cooling factors used for the process design in this case are not affected by either come-up profile or cooling-down profile.

TABLE 13.2
The Effect of Come-Up Time (CUT) on Ball jh and NumeriCAL jh Values

CUT (min)	Ball jh	Ball fh	NumeriCAL jh	NumeriCAL fh
4.5[a]	1.77 ± 0.05	44.85 ± 0.42	1.74 ± 0.03	46.77 ± 0.39
7.0[b]	1.90 ± 0.05	41.99 ± 0.56	1.71 ± 0.07	46.01 ± 0.15
21.5[b]	1.49 ± 0.05	45.30 ± 0.13	1.70 ± 0.02	47.09 ± 0.40

Product: 300 × 405 condensed mushroom soup (2:1, v/v) processed in a hydrostatic sterilizer simulator at 121.1°C.

[a] Normal come-up time (close to linear).

[b] Feed leg time simulation at 87.8°C.

Berry recognized that the *come-up time* and its profile have profound effect on the Ball's heating factors, particularly the jh value when using 42% come-up time correction [44]. If the heating factors were developed from the fastest come-up profile, it could cause food safety problem when these heating factors were used for the process development for the sterilizer actually having longer come-up time. He recommended using a true jh value for the process development.

Table 13.2 shows various come-up time effects in a hydrostatic sterilizer simulator on both Ball's and NumeriCAL jh values. From feed leg time 4.5 to 21.5 min, we found that the Ball jh values ranged from 1.49 to 1.90, whereas NumeriCAL obtained jh value was consistent, which was around 1.71. Since NumeriCAL uses actual come-up and cooling down retort temperatures, the developed jh values reflect the true jh value, which has less effect on the come-up time profile.

Figure 13.14 shows the NumeriCAL simulation results under process deviation conditions (10% Bentonite in 300 × 407 cans) with jh value larger than 2.0. Left curve shows the normal heat penetration data modeling results with NumeriCAL jh of 2.83, fh of 41.42 min, jc of 1.42 and fc of 50.85 min. This set of input heating and cooling factors was then used to predict the product temperatures for the same product with two process temperature deviations. The results were plotted in the right curve and demonstrated that the model predicted product temperatures were in a good agreement with physically measured product temperatures under process temperature deviation conditions. The results suggest that the model could be used in the computer-based real-time control for correcting the process temperature deviations.

13.4 INTELLIGENT THERMAL PROCESS CONTROL

To ensure the optimized process to be adequately delivered to the product, the critical control points used in the process calculation need to be satisfied and properly controlled. However, a sterilizer temperature deviation can often occur during a process due to steam supply interruption or failure. Figure 13.15 shows a typical industrial example of process temperature deviations (two times), which were recorded in a hydrostatic sterilizer. When the *process temperature deviation* occurs, the product might not be safe and could cause potential public health problems. For the low-acid or acidified canned food products in the Unites States, a processor has to isolate all temperature deviation affected products, which must be evaluated by a competent process authority for safety as required by U.S. FDA and USDA. Often, process downtime and product quality loss could cause a significant economic loss to the food processors. Therefore, developing a real-time *lethality-based computer predictive model*, which is able to predict and trace the lethality of the critical product during process temperature deviation, is of high interest to the food canning

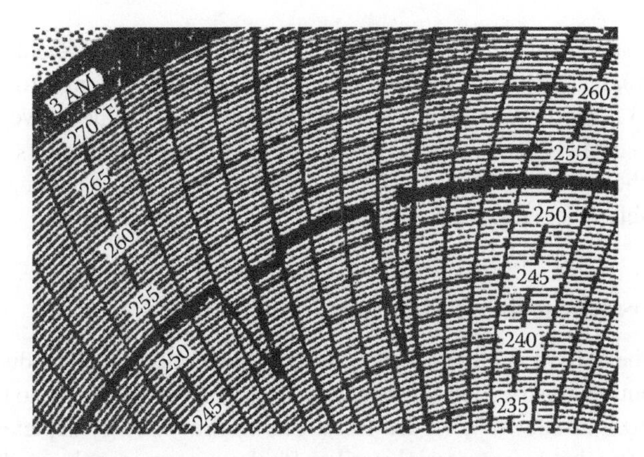

FIGURE 13.15 An industrial example of process temperature (°F) deviations in a hydrostatic sterilizer (two instance temperature deviations).

industry. The importance and benefits of using this technology have been recognized in commercial production [28].

There are a number of methods that have been discussed and proposed for the real-time lethality-based computer control of thermal processes [45,46,47]. Lethality-based computer predictive control logic for a conduction-heating product processed in a batch sterilizer has been proposed [48,49,50,51]. The method can be used for online correction of retort process temperature deviations involving conduction-heating products. Weng has further applied the lethality-based computer predictive control logic into the continuous sterilizer or pasteurizer systems [52–55]. In this section, we are going to address the principles of lethality-based *computer predictive control* of thermal process in the batch and continuous sterilizers with the capability of real-time process temperature deviation correction.

13.4.1 Illustration of Mathematical Model

The goal of thermal process control is to deliver an adequate lethality to the product being *least* processed in the sterilizer under normal and deviated process conditions. Mathematically, this can be described in Equations 13.2 through 13.4:

$$\text{min. } F_{T_{ref}}^{z} \text{ value} = f\left(F_{T_{ref}}^{z} \text{ value}(\theta)\right) \quad (\theta \in \Omega) \tag{13.2}$$

$$F_{T_{ref}}^{z} \text{ value}(\theta) = \int_{0}^{t} 10^{(T-T_{ref})/z} dt \tag{13.3}$$

$$\text{Least } F_{T_{ref}}^{z} \text{ value}(\theta_{min}) \geq \text{Target } F_{T_{ref}}^{z} \text{ value} \tag{13.4}$$

where

θ is an individual container of product
Ω is the total containers affected by process deviation
$F_{T_{ref}}^{z}$ value is the sterilization value for the microorganism of concern

The product slowest heating zone temperature T can be simulated by well-known model such as NumeriCAL model as described in the previous section.

Equation 13.2 is used to find the least $F_{T_{ref}}^{z}$ value (or Fo value for low-acid foods) container (or a group of containers), θ_{min}, which receives the minimum process among products (Ω) being processed. Equation 13.3 is used to calculate the total lethality or $F_{T_{ref}}^{z}$ value delivered to any container of product (θ) in the sterilizer. By knowing the *least* $F_{T_{ref}}^{z}$ value in the process, the model can then intelligently control the thermal process usually the process time during heating in real time to meet the required food safety level (Equation 13.4).

13.4.2 Application to the Batch Sterilizers

In batch sterilizers, the lethality delivered to each individual container of product is not uniform. A wide lethality distribution may be found in a processed batch of product, due to the temperature distribution in the sterilizer as shown in Figure 13.16. The control strategy depicted in Equations 13.2 through 13.4 is to ensure that a commercial sterility for those containers located at the slowest heating zone or the quickest cooling zone of the sterilizer (least $F_{T_{ref}}^{z}$ value, Equation 13.2) are satisfied whether or not a sterilizer temperature deviation occurred. Therefore, the least $F_{T_{ref}}^{z}$ value calculated from the containers located in *the slowest heating zone* or *the quickest cooling zone* in the sterilizer is used for the process control of the batch sterilizer system. When a process temperature deviation occurs, it is only necessary to validate or ensure the product safety for the "worst case" container in the slowest heating zone in the retort. The product safety affected by the process temperature deviation usually can be evaluated and controlled based on the retort temperature distribution profile and the recorded temperature deviation profile [52]. An extended sterilization time is often needed to ensure the food safety [48,52]. Figure 13.14b shows a typical example of using NumeriCAL model for real-time correction of process temperature deviations in a batch retort. The industry benefit of using this technology in the batch retort was discussed and recognized [28].

13.4.3 Application to the Continuous Sterilization System

In a continuous sterilizer system (e.g., hydrostatic sterilizer or reel and spiral rotary sterilizer), a uniform lethality can be delivered to the product for a tightly controlled systems, since each product sees the same process time–temperature profile. However, in case of process temperature deviations, all deviation affected carriers of containers (in hydrostatic sterilizer) or containers (in continuous rotary sterilizer) will have experienced different and unique heating time–temperature histories. As shown in Figure 13.17, for example, the entry product to the sterilization chamber first experiences the process temperature deviation and then the normal process temperature treatment. However, for the product that has already traveled to the middle of the chamber, it first experiences

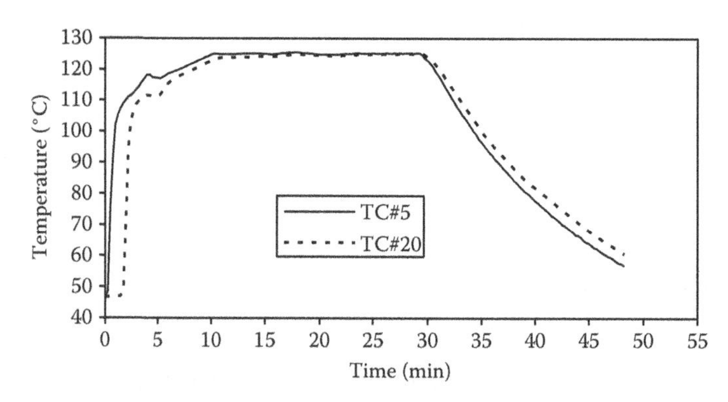

FIGURE 13.16 A typical example of the sterilizer temperature distribution in a water full immersion four basket batch sterilizer (TC #20 slowest heating zone, #5 fastest heating zone).

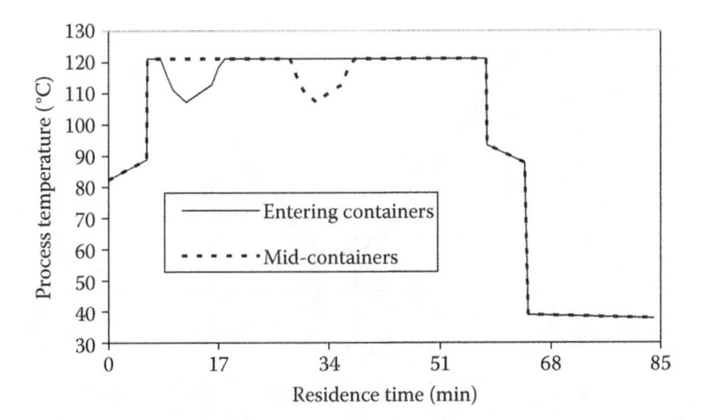

FIGURE 13.17 Two different time–temperature environmental histories for two carriers of products located at the beginning and the middle positions, respectively, in a hydrostatic sterilizer steam dome when process temperature deviation occurred.

a normal process temperature, then the deviated temperature and afterward a normal process temperature again when the deviation is cleared. As a result, a wide spread of delivered lethality values to those deviation affected products can be expected. The product which starts to significantly accumulate the lethality value upon deviation usually would suffer most. The least $F^{z}_{T_{ref}}$ value containers (Equation 13.2) need to be identified for the process control purpose [53,54,55].

Figure 13.18 shows that the total Fo-value distribution among carriers of containers affected by the process temperature deviation for a hydrostatic sterilizer. It is noted that the carrier just entering the feed leg when the temperature deviation occurred was labeled as zero (Figures 13.18 and 13.19). The total sterilization value delivered to each carrier of product was calculated by using NumeriCAL model based on its position during deviation. It was found that the total accumulated Fo value for each deviation affected carrier of containers was different, and its total Fo value strongly depends on the carrier location when deviation occurs. The minimum Fo-value carrier was #569 with total Fo value of 4.3 min. Those carriers of product from #260 to #995 were underprocessed with total

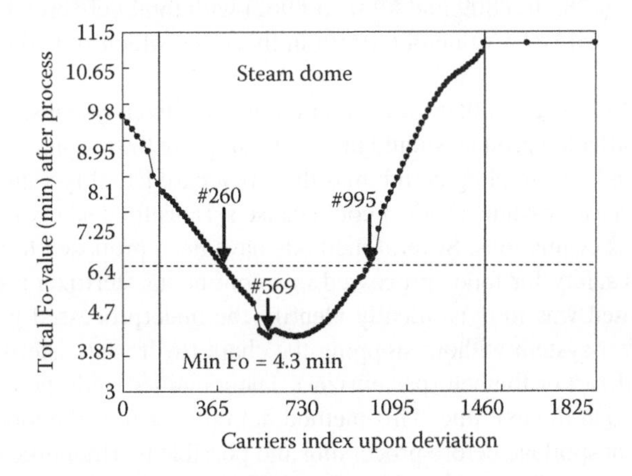

FIGURE 13.18 Total accumulated Fo-value (min) distribution among carriers of product affected by steam dome process temperature deviation (steam dome temperature dropped from 121°C to 112.8°C for 26 min, and the chain was not stopped upon deviation. Steam dome interface water level was raised by 0.508 m, and the water temperature was 90.6°C).

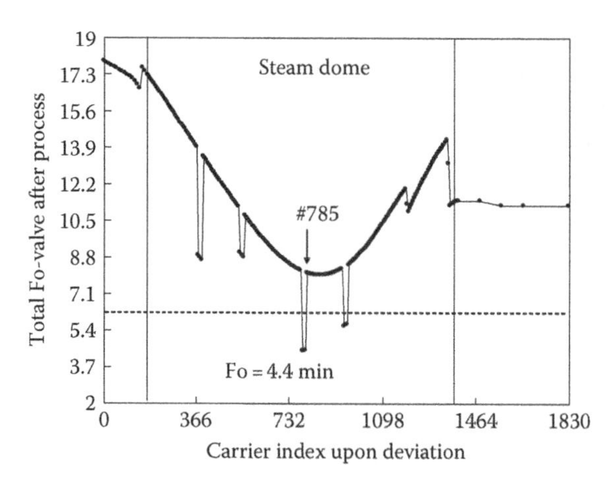

FIGURE 13.19 Total accumulated Fo-value (min) distribution among carriers of product affected by the multiple process temperature deviations (the chain was properly stopped, and the steam dome interface water level was raised by 0.6 m during deviations).

Fo-value less than the scheduled commercial sterility value of 6.0 min in this case, which should be separated from the rest of the product.

Figure 13.19 shows a case study with two consecutive process temperature deviations in a continuous hydrostatic sterilizer. The steam dome temperature first dropped from 120°C to 104.4°C for 10 min. After the process temperature was recovered for 5 min, another process temperature deviation occurred with process temperature dropped to 110°C for another 10 min. During process temperature deviations, the sterilizer chain was properly stopped and the steam dome water interface was elevated for about 0.6 m. The total lethality value delivered to each carrier of product was calculated by using NumeriCAL model based on the carrier position during deviations, and the results were plotted in Figure 13.19. As shown in Figure 13.19, the delivered total Fo values for those carriers of product affected by the process temperature deviations were all different and followed a convex type of curve with some low Fo-value pockets. The low Fo-value pockets were from the products that were covered by the raised steam dome water due to the reduced steam dome pressure during the temperature deviations. The least Fo value for all products affected by the temperature deviation was 4.4 min with the carrier index number of #785. There are two sets of carriers of products (#788 to #809 and #941 to #962) with total delivered Fo value less than the schedule commercial sterility Fo value of 6.0 min in this case, which should be separated from the rest of the product.

Figures 13.18 and 13.19 suggest that for the continuous sterilizer systems, the traditional incubation test for deviation affected product should target on sampling those containers that received least sterilization value. Random sampling could miss the worst case lethality value products. Failure to locate the true global least Fo-value product could cause serious food safety issues during both offline and online process evaluations. Several methods have been proposed to identify and *dynamically control* the food safety for foods processed in a continuous sterilizer system [53,54,55]. One of the methods proposed was to dynamically identify the underprocessed products at the exit of the continuous sterilizer system without stopping the chain (hydrostatic sterilizer) or the reel (reel and spiral rotary sterilizer) or the belt (pasteurizer). The identified underprocessed product can be mechanically separated at the exit line. This method not only ensures the food safety but also prevents potential incipient spoilage before processing and possible thermophilic spoilage in the warm water during cooling due to long system stop time. The method could be also used to ensure the product quality by avoiding overprocessed products and prevent significant economic loss due to system stop downtime.

13.5 CONCLUSIONS

The thermal processing of canned foods continues to advance to meet the customer needs particularly on the convenience and quality aspects, and thermal sterilization systems continue to advance to meet these challenges such as for improving product quality, for handling various convenience packages and geometry, and for increasing the production efficiency.

There is an increasing interest in the canning industry on using new process modeling and calculation method for ensuring food safety and for optimizing product quality and process efficiency. The importance and benefits of using proper thermal processing software such as NumeriCAL model have been more and more recognized in the food industry.

A process temperature deviation often occurs due to the failure of adequate supply of heating medium. The product affected could be underprocessed and could cause potential public health problems. Proper thermal process control with the capability of real-time process temperature deviation correction has been developed and applied for thermal processing of canned foods. The lethality-based computer predictive control method has been discussed and recently has been expanded to the continuous sterilizer systems such as hydrostatic sterilizers.

Lately we have experienced increasing interests on *aseptic processing* and packaging of low-acid foods with particles for both retail and institutional sizes. The first industrial process filing of low-acid soups containing particles processed in an aseptic processing line was accepted by U.S. FDA [56]. The major technical issues related to the aseptic processing of multiphase foods have been addressed [57]. Recognized mathematical modeling tool, such as *AseptiCAL*™ [58,59], and aseptic fillers for packing low-acid foods with particles in institutional sizes [60] and retail sizes [61] are available. In addition, various novel thermal process methods have been studied in the lab and pilot scale. The pressure-assisted and microwave-assisted thermal processes are perhaps the two most promising technologies with potential for canning industry applications. Perhaps the major challenge of the aforementioned technologies is how to increase the product throughput, so that the product quality can be recognized by the consumer and the operation cost can be comparable to the canning operations.

REFERENCES

1. CR Stumbo. *Thermobacteriology in Food Processing*, 2nd edn. Orlando, FL: Academic Press, Inc., 1973, pp. 1–6.
2. SA Elizabeth. What consumers want—and don't want—on food and beverage labels. *Food Technology* 57(11):26–36, 2003.
3. DL Downing. Sterilization systems. In: DL Downing, ed. *A Complete Course in Canning, Book I.* Baltimore, MD: CTI Publications, Inc., 1996, pp. 373–423.
4. Anon. *Canned Foods—Principles of Thermal Process Control, Acidification, and Container Closure Evaluation*, 5th edn. Washington, DC: The Food Processors Institute, 1989.
5. K De Grave. Capabilities of commercial retorting systems. In: *IFTPS European Conference*, Porto, Portugal, October, 2007.
6. A Reichert. Other agitation options for horizontally oriented containers: Gentle motion and the Shaka process. In: *IFTPS Annual Conference*, San Antonio, TX, 2011.
7. J Emanuel, R Walden. Zinetec's Shaka™ retorting process. In: *IFTPS European Conference*, Porto, Portugal, October, 2007.
8. CP Dunne. New technologies for shelf-stable foods—Successful efforts of S&T consortia. In: *Research & Development Associates Fall Forum*, November, 2009.
9. AL Brody. Advance in microwave pasteurization and sterilization. *Food Technology*, February, 2011, pp. 83–85.
10. J Tang, F Liu, SK Pathak, EE Eves II. Apparatus and method for heating objects with microwaves. US Patent Pub. No.: US 2005/0127068 A1, 2005.
11. VM (Bala) Balasubramaniam, D Farkas, EJ Turek. Preserving foods through high-pressure processing. *Food Technology*, November, 2008, pp. 32–38.

12. W Ratphitagsanti, S De Lamo-Castellvi, VM Balasubramaniam and AE Yousef. Efficacy of pressure-assisted thermal processing, in combination with organic acids, against *Bacillus amyloliquefaciens* spores suspended in deionized water and carrot puree. *Journal of Food Science* 75(1):M46–M52, 2010.
13. D Lund. Food processing: From art to engineering. *Food Technology* (September):242–247, 1989.
14. CO Ball, FCW Olson. General method of process calculation. In: CO Ball, FCW Olson, eds. *Sterilization in Food Technology*. New York: McGraw-Hill Book Company, Inc. 1957, pp. 291–312.
15. IJ Pflug. *Microbiology and Engineering of Sterilization Processes*, 7th edn. Minneapolis, MN: Environmental Sterilization Laboratory, University of Minnesota, 1990.
16. Anon. *FDA 21 CFR, Part 113—Thermally Processed Low-Acid Foods Packaged in Hermetically Sealed Containers*. Washington, DC: The Federal Register, 1991.
17. Anon. *Fish and Fisheries Products Hazards and Controls Guidance*, 4th edn. Washington, DC: FDA, 2011.
18. DR Heldman, RL Newsome. Kinetic models for microbial survival during processing. *Food Technology* 57(8):40–46, 2003.
19. V Sapru, GH Smerage, AA Teixeira, JA Lindsay. Comparison of predictive models for bacterial spore population resources to sterilization temperatures. *Journal of Food Science* 58(1):223–228, 1993.
20. M Peleg. A model of survival curves having an 'activation shoulder'. *Journal of Food Science* 67(7):2438–2443, 2002.
21. CO Ball, FCW Olson. The formula method. In: CO Ball, FCW Olson, eds. *Sterilization in Food Technology*. New York: McGraw-Hill Book Company, Inc. 1957, pp. 313–358.
22. KI Hayakawa. A critical review of mathematical procedures for determining proper heat sterilization processes. *Food Technology* (March):59–65, 1978.
23. RL Merson, RP Singh, PA Carroad. An evaluation of Ball's formula method of thermal process calculations. *Food Technology* (March):66–75, 1978.
24. JE Manson, JW Zahradnik, CR Stumbo. Evaluation of lethality and nutrient retentions of conduction-heating foods in rectangular containers. *Food Technology* 24(11):109–113, 1970.
25. AK Datta, AA Teixeira. Numerically predicted transient temperature and velocity profiles during natural convection heating of canned liquid foods. *Journal of Food Science* 53(1):191–195, 1988.
26. BM Nicolai, P Verboven, N Scheerlinck. Modeling and simulation of thermal processes. In: P Richardson, ed. *Thermal Technologies in Food Processing*. Boca Raton, FL: CRC Press, 2001, pp. 91–112.
27. JW Larkin, MR Berry. Estimating cooling process lethality for different cooling j values. *Journal of Food Science* 56(4):1063–1067, 1991.
28. Anon. Predictive control boosts process efficiency at AmeriQual foods. *Food Engineering*, September, 2001.
29. KI Hayakawa. Estimating food temperatures during various processing or handling treatments. *Journal of Food Science* 36:378–385, 1971.
30. AA Teixeira. Innovative heat transfer models: from research lab to on-line implementation. In: *Proceedings of Food Processing Automation II, ASAE*, 1992, pp. 177–184.
31. J-L Lanoiselle, Y Candau, E Debray. Predicting internal temperature of canned foods during thermal processing using a linear recursive model. *Journal of Food Science* 60(4):833–840, 1995.
32. J Noronha, M Hendrickx, A Van Loey, P Tobback. New semi-empirical approach to handle time-variable boundary conditions during sterilization of non-conductive heating foods. *Journal of Food Engineering* 24:249–268, 1995.
33. NG Stoforos, J Noronha, M Hendricks, P Tobback. Inverse superposition for calculating food product temperatures during in-container thermal processing. *Journal of Food Science* 62(2):219–224, 248, 1997.
34. CR Chen, HS Ramaswamy. Modeling and optimization of variable retort temperature (VRT) thermal processing using coupled neural networks and genetic algorithms. *Journal of Food Engineering* 53:209–220, 2002.
35. Anon. *NumeriCAL*™ users Manual—Advanced Numerical Modeling Software. Madera, CA: JBT Corporation, 2010.
36. GS Tucker. Validation of heat processes. In: P Richardson, ed. *Thermal Technologies in Food Processing*. Boca Raton, FL: CRC Press, 2001, pp.75–89.
37. JW Larkin, S Spanik. Validation and compliance of computerized control. In: *NFPA Validation of Automated Control Systems/21 CFR Part 11 Conference*, San Diego, CA, February, 2003.
38. J Sanders, E Curran. *Software Quality, a Framework for Success in Software Development and Support*. Wokingham, England: Addison-Wesley Publishing Company, 1994.

39. SV Ilyukhin, TA Haley, JW Larkin. Control system validation: Key to automated food processing. *Food Technology* 54(3):66–72, 2000.
40. MF Deniston. Validation guidelines for the automated control of food processing systems. In: *The Institute for Thermal Processing Specialists 22nd Annual Conference*, Orlando, FL, November, 2002.
41. TL Heyliger. Providing a "validatable" automated control system. In: *The Institute for Thermal Processing Specialists Annual Conference*, Washington, DC, November, 2000.
42. Z Weng. Validation methods of numerical modeling software for thermal processing. In: *The Institute for Thermal Processing Specialists Annual Conference*, Washington, DC, November, 1999.
43. T Smith, MA Tung. Comparison of formula methods for calculating thermal process lethality. *Journal of Food Science* 47:626–630, 1982.
44. MR Berry JR. Prediction of come-up time correction factors for batch-type agitating and still retorts and the influence on thermal process calculations. *Journal of Food Science* 48(4):1293–1299, 1983.
45. JEW McConnell. Effect of a drop in retort temperature upon the lethality of processed for convection heating products. *Food Technology* (February):76–78, 1952.
46. S Navankasattusas, DB Lund. Monitoring and controlling thermal processes by on-line measurement of accomplished lethality. *Food Technology* (March):79–83, 1978.
47. EB Giannoni-Succar, KI Hayakawa. Correction factor for deviant thermal processes applied to packaged heat conduction food. *Journal of Food Science* 47(2):642–646, 1982.
48. AK Datta, AA Teixeira, JE Manson. Computer-based retort control logic for on-line correction of process deviations. *Journal of Food Science* 51(2):480–507, 1986.
49. AA Teixeira, JE Manson. Computer control of batch retort operations with on-line correction of process deviations. *Food Technology* (April):85–90, 1982.
50. J Fastag, H Koide, SSH Rizvi. Variable control of a batch retort and process simulation for optimization studies. *Journal of Food Process Engineering* 19(1):1–14, 1996.
51. Z Weng. On-line process deviation correction for batch and continuous retort systems. In: *IFTPS Annual Conference*, San Antonio, TX, 2007.
52. Z Weng. Method for administering and providing on-line correction of a batch sterilization process. US Patent No. 6472008, 2002.
53. Z Weng. Controller and method for administering and providing on-line handling of deviations in a hydrostatic sterilization process. US Patent No. 6440361, 2003.
54. Z Weng. Controller and method for administering and providing on-line handling of deviations in a continuous rotary sterilization process. US Patent No. 6416711, 2003.
55. Z Weng, R Gunawardena. Controller and method for administering and providing on-line handling of deviations in a continuous oven cooking process. US Patent No. 6410066, 2003.
56. KT Higgins. The refined science of heat transfer. *Food Engineering*, August, 2005.
57. Anon. *Case Study for Condensed Cream of Potato Soup from the Aseptic Processing of Multiphase Foods*. Chicago, IL: National Center for Food Safety and Technology, 1995.
58. Z Weng. AseptiCAL software—A mathematical modeling package for multiphase foods in aseptic processing system. In: *Advanced Aseptic Processing and Packaging Workshop,* UC Davis, Davis, CA, 2000.
59. Z Weng. Value and acceptance of mathematical modeling and simulation studies in validation of aseptic processing system. In: *IFTPS Annual Conference*, San Antonio, TX, 2007.
60. DE Green, R Becker, R Hillison, K Gallagher. First aseptic filler in U.S. for low-acid products in flexible bags. *Food Processing* (August), 1988.
61. KT Higgins. New soup's on. *Food Engineering*, June 2005.

14 Thermal Processing of Ready Meals

Gary Tucker

CONTENTS

14.1 INTRODUCTION

This chapter considers *thermal processing* of *ready meals* in terms of the different types of meals on the market, their methods of *manufacture*, and the assurance of *quality* and *safety* of those meals. Exposure of foods to a time at lethal temperatures, referred to as thermal processing, is the most commonly used method to kill or control the numbers of microorganisms present within foods and on packaging material surfaces. Other methods for achieving *preservation*, such as freezing, modified atmospheres, and novel techniques will be mentioned briefly in Section 14.2 but the priority is with thermal processing techniques.

Ready meals are *multi-component foods*, typically comprising of a meat or vegetable component, and a rice, pasta or potato component. These can be mixed together in one meal, for example, in a single canned product, or the components can be separated within a *multi-compartment tray*. There are many types of ready meal, including traditional meals such as cottage pie and roast beef with Yorkshire pudding to oriental dishes with various accompaniments. The concept is that the product is a complete meal that requires minimal preparation by the consumer because the consumer receives it in a microbiologically safe condition that requires reheating only to achieve an organoleptic condition.

Thermal processing of ready meals to achieve this safe condition is complicated because of the need to ensure that all parts are processed to minimum safety criteria. However, there will be differences in the rates of heating within the meal that must be controlled so that the level of processing is not detrimental to one or more of the parts. Ready meals sold for ambient storage will be thermally sterilized, whereas those intended for chilled or frozen storage will only receive a pasteurization process. The latter process has advantages in terms of product quality because of the milder process.

The commercial market for ready meals has grown significantly over the last few years. Table 14.1 presents commercial market data for all categories of foods that receive a thermal process with market values scaled to 2003 [1,2]. Thus, the total 2003 U.K. market value of thermally processed food products was £4402 million, with ready meals contributing £1535 m. These figures include ready meals that are stored under ambient, chilled, and frozen conditions. In 2007, a report in *The Independent* newspaper suggested that Britons spent over £2 billion on ready meals last year (2006), compared with £1.35 billion in France and £1.2 billion in Germany. They predicted that the number of ready meals consumed in the United Kingdom over the next 5 years was expected to rise by over 25% to £2.6 billion.

For cultural reasons, the ready meals category is less developed in most Asia Pacific countries. A report by Emily Woon of Euromonitor International indicated that per capita consumption of ready meals in the Asia Pacific region was low in comparison with Europe and the United States, but the category was growing, on the back of changing lifestyles, retail promotion, and the introduction of new products. The retail value of the ready meals market in Asia Pacific was expected to total US$14.8 billion for 2006, an increase of about 3% over the previous year, with volume sales up by 3.6% at almost 1.9 million tones. Although much of the region is undeveloped in comparison with western countries, Japan has the world's second-largest ready meals market in value terms, after the United States. In 2005, per capita spending on ready meals in Japan was US$103, compared with US$75 in the United States and US$158 in the United Kingdom. This

TABLE 14.1
Commercial Information on the Market Size for Thermally Processed Foods (3% Inflation Assumed per Year from 1996 to 2003)

Product Category	Estimated 2003 Market (£ Million)
Cooked poultry products	359.7
Cooked other meat products	231.7
Canned, bottled fruit	208.9
Processed fruit	143.4
Spreads, dressings, pickles, sauces	755.3
Canned meat products	120.3
Pet foods	953.3
Jams, jellies	94.7
Ready meals	1535.0
Total	4402.3

is driven by increasingly hectic lifestyles, the rising number of women in the workplace and the increase in single households.

The Euromonitor report also indicated per capita spending on ready meals in Asia Pacific was only US$4 in 2005, compared to the Western Europe average of US$54. Outside Japan, the limited consumption was primarily due to the dominance of home cooking and the traditional mind-set that fresh food was better than preprepared meals. However, the ready meals sector is gaining momentum within Asia Pacific, with changing consumer lifestyles, the development of new products that cater to local tastes and aggressive promotions by retailers the main driving forces behind the growth.

The second-largest ready meals market in the Asia Pacific region is Taiwan, with estimated value sales of US$451 million in 2006, up by as much as 17% from 2005. This robust growth was driven by the strong performance of chilled ready meals, which is by far the largest and fastest growing sector, accounting for 86% of value sales. Along with the introduction of new products by manufacturers, aggressive promotion and own brand development by large retailers has been a major factor in the development of the ready meals market in Taiwan.

In China, the third-largest market in the region, thermally processed is the most popular style of ready meal, with expected retail value sales of US$370 million in 2006, an increase of 8% from the previous year. At present, other ready meal formats are virtually nonexistent in China, but canned ready meals are now approaching saturation and choice is limited. However, increasing microwave oven ownership will encourage growth in frozen and chilled ready meal sales, especially among young and single consumers, and particularly in urban centers. The development of microwave-ready meals has been high on the agenda of the national food industry and may help to trigger a new trend in Chinese cuisine.

The reasons for the lack of development in other markets, such as South Korea, are largely cultural. Consumers in these countries generally prefer to use fresh ingredients to prepare meals, and if not cooking at home will buy from foodservice outlets, such as street stalls, rather than rely on mass-produced ready meals.

Sales of ready meals globally are sure to increase over the next few years as consumer lifestyles change. Both retail and catering markets are important and have already seen tremendous growth as new processes such as *sous-vide* increase in popularity. This has opened up the chilled and frozen cabinets for ready meals, in addition to the more traditional ambient shelf space.

Sous-vide is a technology that originated in France as a mild thermal process (e.g., 70°C for 40 min) for the manufacture of high-quality foods sold to the catering sectors [3]. More recent *sous-vide* processes have targeted the psychrotrophic strains of *Clostridium botulinum* (e.g., 90°C for 10 min) that have required the process severity to be increased. Section 14.3 discusses microbiological targets and how thermal processes are validated.

Packaging formats vary significantly for ready meals and are linked with the methods of production, storage, and consumer preference. Paperboard, plastic, and aluminum trays are now commonplace for chilled and frozen ready meals, where the thermal processes are most likely to be applied to the food components before they are packaged. These packaging formats often have multiple compartments in which the components such as meat, vegetables, and rice are placed separately and each given a different thermal process. The processes given tend to be referred to as cooking steps, but there is a *microbiological kill* associated with these steps, and so it is correct to refer to this as thermal processing. The microorganism targets are discussed later in Section 14.3.

Production methods for single-component or multi-component ready meals are varied but can be categorized into those where the thermal processing step occurs prior to packaging (e.g., using heat exchangers or in high care factories) or after packaging (e.g., using retorts). The former method has the most variety in terms of the production methods, where factories making short shelf-life meals can be divided into high- and low-care (or risk) areas. Methods of ready meal manufacture are discussed in more detail in Section 14.2.

14.2 METHODS OF MANUFACTURE

In order to understand how thermal processing can be used to manufacture a complex food such as a ready meal it is important to know how microorganisms behave. In the HACCP context, food safety with respect to microorganisms is controlled by quantifying their introduction to the food, growth within that food, and survival through the production stages [4]. A factory *HACCP* plan will consider introduction, growth, and survival at all stages of manufacture.

There are categories of temperature sensitivity of microorganisms that help to define and position the thermal processes that are applied to ready meals.

- *Psychrotrophic* (cold tolerant), which can reproduce in chilled storage conditions, sometimes as low as 4°C. Having evolved to survive in extremes of cold, these are the easiest to destroy by heat.
- *Psychrophilic* (cold loving), which have an optimum growth temperature of 20°C.
- *Mesophilic* (medium range), which have an optimum growth temperature between 20°C and 44°C. These are of greatest concern with ready meals.
- *Thermophilic* (heat loving), which have an optimum growth temperature between 45°C and 60°C. In general, these organisms are only of concern with ready meals produced or stored in temperate climates.
- *Thermoduric* (heat enduring), which can survive above 70°C but cannot reproduce at these temperatures. These are of no concern to most food processes.

The middle three categories are of greatest importance in ready meal manufacture. Most of the *pathogenic microorganisms* fall within the mesophilic category, such as *Salmonella*, *Listeria*, and *Escherichia coli* O157, and so food production conditions will be designed to minimize their growth and survival during manufacture. The scheduled thermal processing conditions of *hold temperature* and *hold time* are designed specifically with the intended storage conditions that dictate which microorganisms can grow. For example, it would be unnecessary and uneconomic to fully sterilize a chilled ready meal. Thus, safe manufacture of a ready meal requires that the thermal processes are applied correctly, and the intended storage and distribution conditions are maintained for the duration of the shelf life [5].

In order to understand how microorganisms can proliferate in food products and in factory environments, it is important to understand the four stages in bacterial growth. The first two growth phases, namely, the lag phase and log phase, are important during food manufacture.

- *Lag phase*, during which the bacteria are acclimatizing to their environment, which can be several hours long.
- *Log phase*, during which reproduction occurs logarithmically for the first few hours. Conditions for growth are ideal during this period, and toxin production is most common.
- *Stationary phase*, during which the bacteria's reproduction rate is cancelled by the death rate.
- *Mortality or decline phase*, during which exhausted nutrient levels, or the level of toxic metabolites in the environment, prevent reproduction, with the results that the bacteria gradually die off.

The lag phase is critical in ready meal production because it allows the food manufacturer to complete the processing and assembly of the ready meal without surviving microorganisms germinating. *Chilled ready meal production* is a good example of this, whereby the high-risk part of the factory is likely to be held at low temperatures (10°C–12°C) in order that microorganism germination and growth are controlled. Even psychrophilic organisms require many hours to germinate at these temperatures.

The log phase is to be avoided at all costs since this can result in microorganisms doubling in numbers in short time periods (e.g., every 20–30 min if conditions permit). Toxin production is most likely to occur during the log phase.

14.2.1 Nonthermal Methods for Manufacturing Ready Meals

This chapter primarily considers thermal processing, but there are many other preservation methods that are used to manufacture ready meals [6]. For completeness, these methods are outlined in the following sections.

14.2.1.1 Freezing

Freezing of food does not render the food sterile. Although the freezing process can reduce the levels of some susceptible microorganisms, this is not significant in the context of the overall microbial quality of the food. Thus, a frozen ready meal will have been subjected to a thermal process at some stage or stages in its manufacture. At commercial freezing temperatures (–18°C to –24°C), all microbial activities are suspended, and the length of time for which the food can be kept is dependent on quality factors. It is important to note, however, that once a frozen food is defrosted, those viable microorganisms present will grow and multiply.

14.2.1.2 Chilling

Chilling is the process that lowers the food temperature to a safe storage temperature of 5°C–8°C. Chilled foods can potentially present a greater risk to public safety than frozen foods. Keeping products at a low temperature reduces the rate of microbiological and chemical deterioration of the food. In most processed chilled foods, it is microbial growth that limits the shelf life; even the slow growth rates that occur under chilled conditions will eventually result in microbial levels that can affect the food or present a potential hazard (see Table 14.2) [7]. This microbial growth can result in spoilage of the food (it may go putrid or cloudy or show the effects of fermentation), but pathogens, if present, may have the potential to grow and may show no noticeable signs of change in the food.

TABLE 14.2
Minimum Growth Temperatures for Selected Pathogens

Pathogenic Microorganism	Minimum Growth Temperature (°C)
Bacillus cereus	4.0
C. botulinum (psychrotrophic)	3.3
E. coli O157 (and other VTEC)	7.0
Listeria monocytogenes	–0.4
Salmonella species	4.0
Staphylococcus aureus	6.7
Vibrio parahaemolyticus	5.0
Yersinia enterocolitica	–1.0

Source: Betts, G. ed., *A Code of Practice for the Manufacture of Vacuum and Modified Atmosphere Packaged Chilled Foods with Particular Regard to the Risks of Botulism, Campden BRI Guideline No. 11*, Campden BRI, 1996.

To reduce microbial effects to a minimum, chilled ready meals are usually given a thermal process, sufficient to eliminate a variety of pathogens such as *Salmonella*, *Listeria*, and *E. coli* O157. A process equivalent to 70°C for 2 min is generally considered to be sufficient, but the exact process given will depend on the nature of the food. Application of hurdles to growth such as a pasteurization process and chilled storage can allow a shelf life of several days. If the pasteurization regime is more severe, for example, 90°C for 10 min [8], it is possible to extend the shelf life to between 18 and 24 days or more. The exact length depends on the suitability of the food to support microbial growth, and it is common for a company to apply the same heat process to two different foods, yet the declared shelf life of one may be 14 days, while the other may allow 20 days. The use of product ingredient(s) with low microbial count(s), ultraclean handling and filling conditions in combination with sterilized packaging (i.e., near aseptic conditions) will serve to reduce initial microbial loading of pasteurized product and thereby extend shelf life.

14.2.1.3 Drying and Water Activity Control

All microorganisms of public health significance need water to grow. Reducing the amount of water in a food that is available to the microorganism is one way of slowing or preventing growth. Thus, dried foods and ingredients, such as dried herbs and spices, will not support microbial growth and provided they are stored under dry conditions, they can have an expected shelf life of many months if not years. The shelf life of dried ready meals is usually limited by texture changes caused by moisture ingress through the packaging, with the food losing its crispness and becoming soft. Selection of suitable packaging materials is therefore critical in extending the shelf life of dried ready meals. Laminated paperboard with a plastic moisture barrier, such as polyethylene, is a common pack format for dried foods such as pot noodles.

14.2.1.4 Modifying the Atmosphere

This is a technique that is being increasingly used to extend the shelf life of fresh foods as well as bakery products, snack foods, and ready meals. Air in a package is replaced with a gas composition that will retard microbial growth and slow down the deterioration of the food. The exact composition of the gas used will depend entirely on the type of food being packaged and the biological process being controlled. Ready meals and other cook-chill products tend to use atmospheres containing 30% carbon dioxide and 70% nitrogen [9]. Modified atmosphere packaging (MAP) is generally used in combination with a mild thermal process and refrigeration to extend shelf life of ready meals toward the higher quality end of the market.

An alternative to modifying the atmosphere is vacuum packaging, where all of the gas in the package is removed. This can be a very effective way of retarding chemical changes such as oxidative rancidity development, but care needs to be taken to prevent the growth of the pathogen, *C. botulinum*, which grows under anaerobic conditions. A specific pasteurization process, referred to as the "psychrotrophic botulinum" process, is applied to the packaged food to reduce its numbers to commercially acceptable levels. By using vacuum packaging, mild heat, and chilled storage, a greatly extended shelf life can be achieved. This was the basis of "*sous-vide*" cooking, described earlier in the chapter.

14.2.2 Preservation by Thermal Processing

Breakdown by microorganisms and enzymes are the major causes of undesirable changes in foodstuffs. Both are susceptible to heat, and appropriate heating regimes can be used to reduce, inhibit, or destroy microbial and enzymatic activity. The degree of heat processing required to produce a ready meal of acceptable stability will depend on the nature of the food, its associated enzymes, the numbers and types of microorganisms, the conditions under which the processed food is stored, and other preservation techniques used [5].

Manufacture of a thermally processed ready meal can be broken down into two basic processes:

1. The food is heated to reduce the numbers to an acceptably small statistical probability of pathogenic and spoilage organisms capable of growth under the intended storage conditions.
2. The food is sealed within a hermetic package to prevent reinfection. Preservation methods, such as traditional canning, seal the food in its package before the application of heat to the packaged food product, whereas other operations such as aseptic, cook-chill, and cook-freeze heat the food prior to dispensing into its pack.

Retorting and *canning* are terms that are still widely used in the food industry to describe a wide range of thermal processes where the food is heated within the pack to achieve a *commercially sterile packaged food*. The heating usually takes place in retorts that are basically batch-type (see Figure 14.1) or continuous hot water and/or steam-heated pressure cookers (see Figure 14.2). The principal concept of food retorting is to heat a food in a hermetically sealed container, so that it is commercially sterile at ambient temperatures (see Section 14.3). In other words, so that no microbial growth can occur in the food under normal storage conditions until the package is opened [5]. Once the package is opened, the effects of retorting are lost, the food will need to be regarded as perishable, and its shelf life will depend on the nature of the food itself. Various packaging materials are encompassed within the phrase retorting, which includes not only tin-plate, but glass, plastic pots, trays, bottles and pouches, and aluminum cans.

The most heat-resistant pathogen that might survive the retorting process of low-acid ready meals is *C. botulinum*. This bacterium can form heat-resistant spores under adverse conditions, which will germinate in the absence of oxygen and produce a highly potent toxin that causes a lethal condition known as botulism. This can cause death within 7 days. In practical terms, the sterilization process must reduce the probability of a single *C. botulinum* spore surviving in a pack of low-acid product to one in one million (i.e., 1 in 10^{12}). This is called a "*botulinum cook*," and the minimum process is 3 min equivalent at 121.1°C, referred to as F_0 3 [10–12]. All retorting processes target this organism if no other effective hurdle to its growth is present. However, there is a growing trend to apply additional hurdles to microbial growth that allow the processor to use a milder heat treatment referred to as *pasteurization*.

Milk is the most widely consumed pasteurized food in the United Kingdom, and the process was first introduced commercially in the United Kingdom during the 1930s, when a treatment of

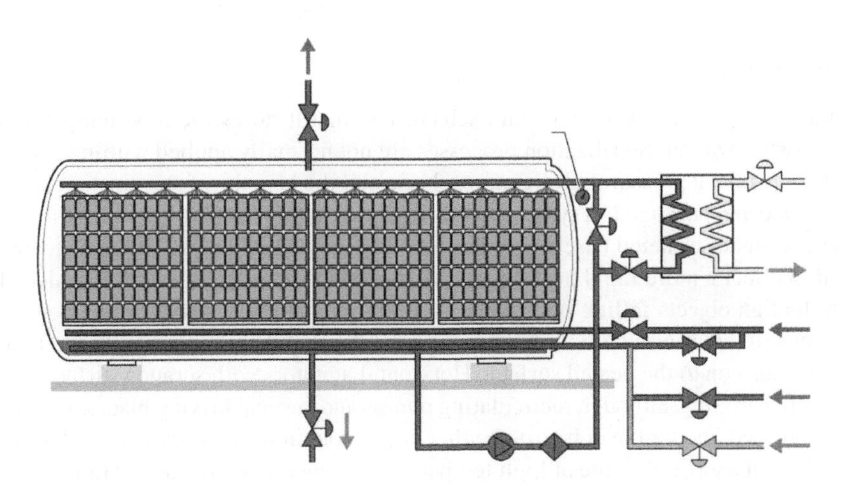

FIGURE 14.1 Example of a batch retort operating on the sprayed water principle. (Courtesy of FMC Foodtech, Chicago, IL.)

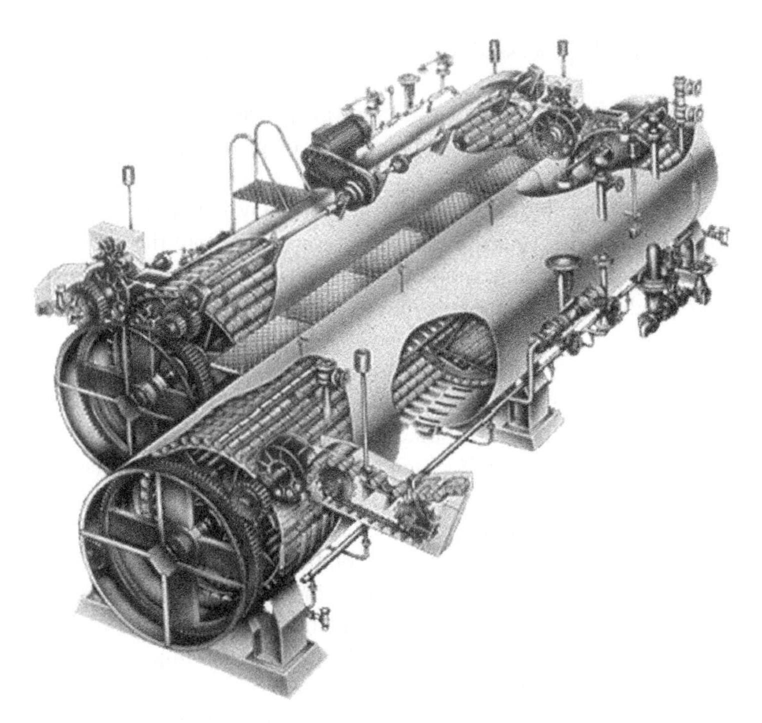

FIGURE 14.2 Reel and spiral cooker-cooler for processing cylindrical food cans. (Courtesy of FMC Foodtech, Chicago, IL.)

63°C for 30 min was used. Modern *milk pasteurization* uses an equivalent process of 72°C for 15 s. Pasteurization is nowadays used extensively in the production of many different types of food, such as fruit products, pickled vegetables, jams, and ready meals [13]. Food may be pasteurized in a sealed container (analogous to a canned food) or in a continuous process analogous to an aseptic filling operation. It is important to note that pasteurized foods are not sterile and will rely on other preservative mechanisms to ensure their extended stability for the desired length of time.

While a thermally processed ready meal for ambient storage has a shelf life of 18–24 months, a cook-freeze meal will typically have a shelf life measured in months and a cook-chill meal in days.

14.2.3 In-Vessel Systems

The vessel acts in a similar way to a heat exchanger in that it raises the food temperature to that required for pasteurization. Sterilization processes are not normally applied within vessels because of the need to operate at pressures above atmospheric and the benefits of sterilization are lost when the food is cooled and packaged. A typical vessel size is around 800–1000 kg and usually comprises a hemispherical steam-jacketed base with cylindrical sides (see Figure 14.3). Direct steam injection can be used to effect a more rapid heating rate. A hinged lid is usually present to reduce heat loss and prevent foreign objects falling into the food. With high-viscosity foods, it is essential that the food is well mixed, otherwise laminar boundary layers develop that reduce the thermal efficiencies and may assist burn on to the heated surface. Horizontal agitators with scraped surface blades offer the most effective mixing, although recirculating pumps and vertical mixing blades are alternatives.

Once pasteurized, the food can be filled either hot or cold into the containers. A *hot-fill* process will only require a short hold time at high temperature to ensure the inside container surfaces are pasteurized. This is usually achieved in a *raining water tunnel pasteurizer*, although it is possible to omit this step if

FIGURE 14.3 Jacketed vessel with horizontal scraped surface agitation for the manufacture of medium- and high-viscosity foods with particulates. (Courtesy of T. Giusti Ltd., Staffordshire, U.K.)

1. The food's acidity is high (pH < 3.8)
2. The filling temperature is above 95°C
3. The containers are prewarmed or of low heat capacity

The shelf life of a sauce of low pH will be many months if hot filled and determined by its chemistry and not by its microbiology. *Multi-component ready meals* are usually close to pH neutral and require acidification to achieve a stable pH. A ready meal packaged in a *multi-compartment tray* may contain one or more components of low pH that can be hot filled (e.g., a sweet and sour sauce). However, some or all of the other components will be pH neutral and require cold filling; which limits the chilled shelf life.

A *cold-fill* process will not guarantee commercial sterility of the container, and as such requires far greater attention to hygiene in order to minimize the introduction of microbial contamination during filling. Most cold-filled products are sold chilled and have a shelf life up to 10–12 days. This is where the concept of low- and high-care environments becomes important, in that shelf life extension beyond 10–12 days can be achieved. Many of the ready meals sold chilled or frozen are assembled in high-care areas, with their components having been manufactured in different factories and transported chilled to the assembly factory. Packaging formats for the components are varied. This method of manufacture is discussed in Section 14.4 on future trends.

14.2.4 Heat Exchangers

A heat exchanger offers the same type of thermal processing operation to that of a heated vessel in that the food is processed prior to packaging. Suitable heat exchangers for heating and cooling foods are *plate packs* for thin liquids, *tubular heat exchangers* for medium viscosity foods, and *scraped surface heat exchangers* for high viscosity foods that may contain particulates. When establishing the processing times and temperatures, it is assumed that the thermal process is delivered solely within a holding section that is usually a length of tube. Sterilization or pasteurization values are calculated from the holding tube outlet temperature and the residence time from the measured flow rate. Control of these parameters must be to high levels of accuracy because of the high temperatures and short times (*HTST*) used.

The term *UHT* (ultrahigh temperature or ultraheat treatment) is often used to describe food preservation by in-line continuous thermal processing when followed by *aseptic packing.* Aseptic packing can be carried out with metal cans, plastic pots and bottles, flexible packaging, and foil-laminated paperboard cartons. In aseptic packing, the package and food are sterilized separately and brought together and closed in a sterile environment. One potential advantage that is often quoted for UHT processing is that of enhanced food quality [14]. This is because the problem of overcooking can be reduced at the typical temperatures and holding times in a UHT process (e.g., 140°C for a few seconds). At these elevated temperatures, the lethal kill to *C. botulinum* spores is substantial, whereas the rate of the cooking reactions is less significant, in effect UHT allows extremely short process times with minimal detriment to quality [12]. This benefit is used in the production of sterilized milk and cream that would end up too brown (caramelized) with associated off-flavors if it was processed in its package.

UHT technology for ready meals has been commercialized but has not found widespread consumer acceptance, probably because the meal quality differs to that produced from conventional technology. One major drawback is that the meat and vegetable particulates heat by conduction, which is slow, and so the holding times have to be extended in order that the centers of the particulates are sterilized.

For ready meals that are processed as the complete meal, large *particulates* will almost certainly be present. This reduces the choice of conventional heat exchanger technology to two types:

1. Tubular heat exchangers with single-product tubes (monotubes) large enough for the flow diameter to be of the order of four times the size of the largest particulate (see Figure 14.4). The drawback with monotubes compared to smaller diameter multi-tubes or concentric tubes is in the reduced thermal efficiency due to the thicker product layer. However, the particulates present in the ready meal will to some extent work as "internal mixers" and will hence promote heat transfer [14].

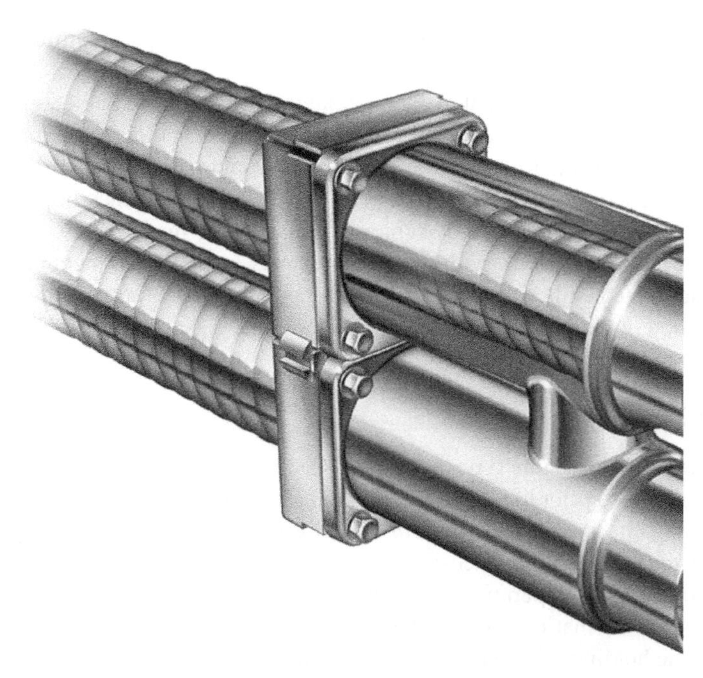

FIGURE 14.4 Monotube type tubular heat exchanger suitable for continuous processing of foods with small particulates. (Courtesy of Tetra Pak, Lausanne, Siwtzerland.)

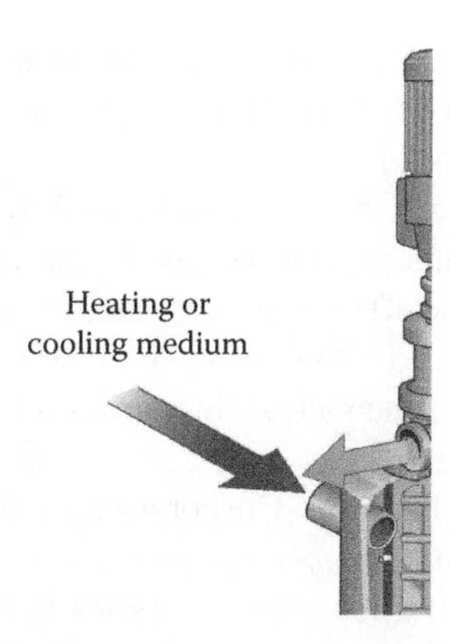

FIGURE 14.5 Scraped surface heat exchanger for processing medium- and high-viscosity foods, including those with particulates. (Courtesy of Tetra Pak, Lausanne, Switzerland.)

2. If the food viscosity is high and there is the chance of product fouling the heated or cooled surfaces of a tubular heat exchanger, the scraped surface heat exchanger must be employed (see Figure 14.5). In principle, a scraped surface heat exchanger is a monotube equipped with a rotating internal scraper. The scraper keeps the heating surface free from any deposits and also promotes turbulence. Drawbacks with scraped surface heat exchangers are the shear damage to delicate particulates and high cost of the plant.

Alternate heat exchange techniques have been explored in the food industry for ready meal production, with limited commercial success [15]. *Ohmic heating* is a technology commercialized by APV Ltd. in the 1980s in which alternating electrical current is used to heat flowing foods to pasteurization and sterilization temperatures. Once cooled, the ready meals are aseptically filled into plastic trays or bags that can be stored in ambient conditions for many months. Emmepiemme SRL entered the market more recently and has a range of equipment in commercial use.

Ohmic heating is a volumetric method, and so a food of consistent electrical resistance will heat almost instantaneously and uniformly. By adjusting the carrier liquid conductivity, it is possible to heat particulates preferentially and in doing so thermally process a ready meal with great efficiency—that is, apart from cooling, which relies on conventional tubular heat exchangers. Ohmic heating has been used to sterilize ready meals for both catering and retail markets but is no longer used in the United Kingdom for this application. Differences in the texture and flavor were the main consumer concerns with ohmically heated ready meals, so the technology did not take off for this application and was mothballed. It is now used successfully for the pasteurization of fruit products with delicate particulates in which the products are intermediates for assembly into yogurts and other dessert products.

14.2.5 In-Pack Retorting

One of the major advantages of retorting ready meals in their packages is that both food and package are thermally processed together, which allows the filled packages to be commercially sterile. The disadvantages with retorting are that the containers need to be strong enough to withstand the high temperatures and the changes in pressure differential that can occur, and that the retorts are the production bottleneck in most factories. The former disadvantage incurs extra cost with the packaging.

The traditional retorted package is the metal can processed in a steam atmosphere, although glass jars, pouches, and flexible plastic or aluminum trays can now be successfully processed in steam or hot water. These containers require an air overpressure to counteract the natural expansion of the gases present in the headspace and those released from the food as it rises in temperature. The desired effect of the overpressure is to push the lid or sides back to their original position and therefore minimize the stress on the seals.

With almost all ready meals that are thermally processed by any technology, heat transfer from the media to the product thermal center is poor because of the viscous boundary layers that develop on the inside processing surfaces. The effect is to form an insulating layer that impairs heat transfer if there is no mixing inside the packages to remove the layers. This results in lengthy processing times to achieve commercial sterility, and so few companies will process ready meals without some form of agitation.

The first equipment (introduced by FMC Corporation in the 1930s) to increase heat transfer to canned foods was the *reel and spiral cooker-cooler* (see Figure 14.2). This imposes a rotation of cans around their central axis and allows heating times to be reduced substantially (e.g., from 90 min to 15 min in the steam chamber). Cans travel on a helical reel through the steam heating and water cooling sections and in doing so rotate at high speeds. The continuous operation of the cooker-cooler allows a high throughput to be achieved and favors high-volume production. Some canned ready meals are processed in reel and spiral cooker-coolers although the high viscosity of the sauces and the high concentration of *particulates* (e.g., beans) reduce the heat transfer efficiency. Examples of ready meals processed in metal cans are meat and vegetable curries, hot pots, pasta, and pulse products.

More recently, *end-over-end (EOE) pack rotation* has become popular, where the packages are constrained in baskets or crates that rotate. Being a batch system, this offers greater production flexibility and is not restricted to cylindrical metal cans (see Figure 14.1). However, heat transfer enhancement is not as great as with the cooker-cooler, and so the throughput is lower and operating costs higher.

Ready meals in *plastic pots* and *pouches* are processed in steam/air or in water retorts where air overpressure is applied to maintain the package shape throughout the heating and cooling process. Expansion of the packages can put stresses on the heat seals of trays and pouches, which may lead to a failure in the *package integrity* and an increased chance of postprocess contamination. Low EOE rotation speeds (2–10 rpm) are often used to induce mixing but without damaging the delicate packages or seals. Some of these packs can be multi-compartment, and in these cases, the challenge is to design a thermal process that delivers a uniform level of microbial kill to all parts of the meal without overcooking any part of it.

14.3 ASSURING READY MEAL SAFETY

Although pasteurization or sterilization of the food is the desired condition, the food can be referred to as commercially sterile. By definition, *commercial sterility* or *appertization* of food is "the condition achieved by the application of heat which renders food free from viable micro-organisms, including those of known public health significance, capable of growing in the food at temperatures at which the food is likely to be held during distribution and storage" [5]. In practice, however, the food manufacturer makes a decision on the level of commercial risk with the applied thermal process because it is not possible to kill all of the microorganisms and produce a saleable product. A *pasteurization process* usually operates to 6 log reductions of the target organism [8], and this differs from fully sterilized foods where the intention is to achieve at least 12 log reductions in *C. botulinum* spores. The lower target log reductions for pasteurization are because of the reduced risks with a product having a secondary preservation hurdle to prevent microorganism growth.

The following equation is used to calculate the target process or F value from heat resistance data on the likely pathogenic or spoilage organisms present. When the F value is divided by the *decimal reduction time* (D_T) for that organism this gives the number of log reductions that will be achieved.

$$F = D_T\left(\log\frac{N_0}{N}\right) \tag{14.1}$$

where

N is the final number of organisms after a specific time–temperature history
N_0 is the initial number of organisms
D_T is the decimal reduction time at a fixed temperature (T) to reduce the number of organisms by a factor of 10 (min)

As the number and variety of ready meals increases, food companies are faced with the challenge of proving that these products are safely pasteurized or sterilized. This can sometimes be difficult if conventional temperature probe systems cannot be used to measure temperature in a food during the processing time. The categories that introduce these complexities include ready meals cooked in continuous ovens, meals with *discrete particulates* cooked in *steam-jacketed agitated vessels* or in heat exchangers. If an alternative to temperature probes is needed, the following approaches to validating microbiological process safety are the options available:

1. Microbiological methods can be used whereby cells or spores of a nonpathogenic organism, with similar temperature-induced death kinetics to the target pathogen, are embedded into an *alginate bead* [16]. The beads are made to mimic the food pieces in their thermal and physical behavior and thus pass through the process with the food. Enumeration of the surviving organisms allows the log reduction and process value to be calculated. Section 14.3.2 deals with microbiological methods.
2. Simulated trials are carried out in a laboratory where the heat transfer conditions of the process are replicated, and temperature measurement is feasible. This is more common with continuous particulate systems than with in-pack processes. Process models can then be developed that predict, for example, the temperature–time history of the critical food particulates as they travel through the heating, holding, and cooling zones of the process.
3. No validation is attempted, with the process safety being inferred from temperature probing of the bulk product or the environment. Substantial overprocessing is allowed, in order that the thermal process delivered to the product thermal center is sufficient. Chilled or frozen storage is used for the ready meal, and end-product testing for microbiological activity is usual.
4. *Time-temperature integrators* (TTIs) can be applied to gather similar process data to that from microorganisms [17,18]. Section 14.3.3 discusses TTIs because this method is one of the most exciting to have emerged in recent years.

One of the core activities involved with establishing a thermal process is the selection of "worst case conditions" likely to be experienced during normal production [19–22]. This is independent of the method chosen for process validation. The current methodology used by most food companies is to validate the microbiological process safety under worst case conditions, so that, by default, the process will be safe under normal production conditions.

To prove that the thermal process has achieved the target process value or F value during manufacture, it is necessary to conduct validation studies using an approved method. Various methods can be selected from the earlier list, and their choice depends on the costs and on the nature of the food and the process type.

It has been stated earlier that the *process validation* study should be conducted using *worst case conditions*. Therefore, by inference, it should not be possible for a normal production batch to heat more slowly than the combination of factors evaluated as worst case. To determine the worst case conditions, it is necessary to first consider the product, process, and package separately and second consider the influences of interactions. The following lists suggest the factors that should be addressed in a process validation study, although the lists are not intended to be exhaustive. This thought process, to arrive at the set of conditions that heat the slowest, not only is typical for in-pack processing but is also appropriate for continuous processing.

Product factors:

- Formulation (weight variation in ingredients e.g., high starch levels that could lead to increased viscosity)
- Fill weight (percent overfill of the key components, e.g., solids content)
- Consistency or viscosity of the liquid components (before and after processing)
- Solid components (size, shape, and weight before and after processing, potential for matting and clumping)
- Preparation methods (e.g., blanching)
- Rehydration of dried components (a complex area that should be avoided if possible)
- Heating mode (convection, conduction, and mixed or broken heating)

Container factors:

- Type (metal cans, glass jars, pouches, and semirigid containers)
- Nesting of low-profile containers (if processed in pack)
- Vacuum and headspace (residual gases with flexible containers, overpressure required to minimize stresses on pack seals)
- Orientation
- Fill method (initial temperature and the effects of delays in getting instrumented containers into the retort)
- Symmetry of pack rotation (EOE or axial)

Retort or processing system factors:

- Media type (steam, steam/air, water immersion, and raining water. Venting schedule if steam)
- Retort come-up time (should be as short as possible to minimize the quantity of heat absorbed by the product during this phase)
- Racking and dividing systems (flow of media)
- Rotation (slowest heating position usually along the retort axis)

The combination of conditions to arrive at the *worst case scenario* should be determined by a competent individual, sometimes referred to as a thermal process authority. It does not matter which validation method is used (probes, microbes, or biochemical) or whether the process is continuous, the thought process for product, package, and process conditions and interaction should be the same.

14.3.1 Temperature Probe Systems

For a *high-viscosity sauce*, typical of that in a ready meal, the position within a container that heats slowest is usually the geometric center. To validate a process for a food containing *particulates*, a large food chunk or piece is usually attached to the end of a temperature probe. Details on process validation methods can be found in numerous texts. While this will work for in-container processes, this method cannot be used for continuous processes, and so alternative methods are required.

14.3.2 Microbiological Methods for Process Validation

These are often referred to as direct methods, but they in fact rely on measuring the achieved log reductions for a process using a nonpathogenic organism and converting this to an F value using the same equation. The nonpathogenic or *marker organism* is put into the process in high numbers, so that some will survive to be counted. If there are no surviving organisms, then it is only possible to conclude that the process achieved greater than for example, 6 log reductions for a 10^6 initial loading. In this situation, there will be uncertainty as to whether the organisms died as a result of the process, during transportation to or from the factory, or if the spores germinated during the come-up time making them more susceptible to destruction at milder temperatures than for the heat-resistant spores. Hence, controlling how these tests are performed is critical and the expertise to conduct a test using encapsulated spores or organisms tends to be restricted to a limited number of microbiology laboratories. A microbiological method can be conducted using organisms distributed evenly throughout a food product (*inoculated pack*) or concentrated in small beads (*encapsulated organisms*).

The inoculated pack method is also known as the *count reduction method* and involves inoculating the entire food with organisms of known heat resistance. For ease of handling, the organisms are usually in the spore form. It is essential that some organisms survive the heat process in order that the containers can be incubated and the surviving organisms counted. The average thermal process received by a container can be calculated using the equation for F value. If the product is liquid, it is relatively easy to introduce the organisms, but for solid products it is necessary to first mix the organisms in one of the ingredients to ensure that they are dispersed evenly throughout the container. Typical levels of the inoculum are between 10^3 and 10^5 organisms per container. An alternative is to use a gas-producing organism and estimate the severity of the process by the number of blown cans.

Encapsulating spores or organisms in an *alginate bead* allows the organisms to be placed at precise locations within a container or within the food particulates. The alginate bead can be made up with a high percentage of the food material, so that the heating rate of the bead is similar to that of the food. This method has been used for *continuous processes* where the food contains *particulates* that require evaluation at their centers, and conventional temperature sensing methods cannot be used. Large numbers of alginate beads are used to determine the distribution of F values that can occur in continuous processes as a result of the distribution of particle residence times. Estimating the exact number to use in a test is not straightforward because it depends on the F-value distribution, which is not known until after the test is conducted and the results are analyzed. The number of organisms used will be greater than for an inoculated container test and can be of the order of 10^6 per bead. It is also important that not all are destroyed by the heat process otherwise it is not possible to estimate an F value.

14.3.3 Time-Temperature Integrators

These are an alternative means of process evaluation that does require live organisms to be deliberately taken into a food production environment [17,18]. A TTI can be an *enzyme* (e.g., *amylase* or peroxidase) that permanently denatures as it is heated. The reaction kinetics of the temperature-induced denaturation are designed to match those of the death kinetics for the target organism. F values for TTIs are calculated from the initial and final enzyme "activities" using a similar expression to Equation 14.1. Instead of using the initial and final number of organisms, the TTI F value uses enzyme activities. For the example where amylase is the TTI, the measurement of activity is a rate of color development when the amylase is reacted with a suitable reagent.

The breakthrough that enabled TTIs to be used for thermal process evaluation is in encapsulating the amylase solutions in silicone tubes [17]. When stored in chilled water to minimize the rate of amylase degradation, the TTI tubes can be used for up to 14 days, therefore increasing the scope for applying TTIs to factories overseas. To further extend their useable shelf life, filled TTIs can be frozen in large numbers, and TTI tubes removed as and when required. Freezing has little impact on amylase structure or on the rate at which its structure degrades by heat. It is conceivable that several hundred TTIs could be made at one time, frozen individually, and used over a period of months. This is an economical method for producing TTIs and would ensure that the kinetics for each tube are similar.

Most enzyme systems are limited to pasteurization because the enzymes are released from microorganisms that grow at moderate temperatures. One exception is an amylase from the hyperthermophilic organism *Pyrococcus furiosus*, an organism that grows in salt water pools in volcanic regions, such as Vulcano Island in Italy. The amylase it secretes is genetically designed by the microorganism to be thermostable [23]. This has been proven to work as a TTI for sterilization processes.

14.4 FUTURE TRENDS

When processing a *high-viscosity food* such as a ready meal, the limits to heat transfer are to overcome the development of boundary layers that impair heat transfer rates at the heated surfaces. Some degree of agitation is advantageous irrespective of whether it is a continuous heat exchange process or one processed in a container [24]. If particulates are introduced, then this imposes a further limitation on heat transfer rates because of the heat conduction within the particulates. Hence, most thermal process times will be considerably longer than those for foods without particulates.

Traditional thermal processes may reduce the vitamin content of food and can affect its texture, flavor, and appearance. Food manufacturers are continually looking for new ways to produce food with enhanced flavor and nutritional characteristics [13]. Much of this can be achieved by process optimization of existing processes but alternative methods are of interest that could have benefits. The development of new processes is being actively researched, to find processes that are as effective as traditional thermal systems in reducing or eliminating microorganisms but do not adversely affect the constituents of the food [6]. In addition to those mentioned in the following, ultrasound, pulsed light, and electric field and magnetic field systems are all being actively investigated.

In the United Kingdom, before any completely novel food, ingredient or process can be marketed, it has to be considered by the Advisory Committee on Novel Foods and Processes (ACNFP). The primary function of the ACNFP is to investigate the safety of novel food or process and to advise government of their findings. The European Union has also formulated "Novel Foods" legislation.

14.4.1 High-Pressure Processing

Ultrahigh-pressure processing (*UPP*) was originally considered in the 1890s, but it was not until the 1970s that Japanese food companies started to develop its commercial potential. Pressures of several thousand atmospheres (500–600 MPa) are used to kill microorganisms, but there is little evidence that high pressure is effective on spores or enzymes. Thus, chilled storage or high acidity remains essential hurdles in preventing microbial growth for these products. Jams were the first products to be produced in this way in Japan, and the process is now commercialized in Europe and the United States. Sterilization of the package is not possible using high pressure alone, and without aseptic filling, this may restrict its widespread use. There is interest in UPP for extending the shelf life of chilled ready meals [25–27].

14.4.2 Ohmic Heating

Ohmic heating achieves its preservation action via thermal effects [28], but instead of applying external heat to a food as with in-pack or heat exchangers, an electric current is applied directly to the food. The electrical resistance of the food to the current causes it to heat it up in a similar way to a light bulb filament. The advantage is that much shorter heating times can be applied than would otherwise be possible, and so the food will maintain more of its nutritional and flavor characteristics. The limitation is that ohmic cooling, or some other means of effecting rapid cooling, cannot be applied, and so cooling relies on traditional methods that are slow in comparison with ohmic heating.

Foods containing large particulates are suited to ohmic heating because the electrical properties of the particulate and carrier liquid can be designed, so that the particulate heats preferentially and instantaneously. The only commercial ohmic heater in operation in the United Kingdom (at the time of writing) is used to pasteurize fruit preparations, in which good particle definition is a key requirement. Prior to its use for fruit preparations, this ohmic heater was used for ready meal sterilization whereby the meals were manufactured to a high quality but were not a commercial success for various reasons. One such reason was that their flavor was different from that of traditional processes because of the different rates of cooking reactions.

14.4.3 Microwave Processing

Microwave processing, like ohmic heating, destroys microorganisms via thermal effects. Frequencies of 950 and 2450 Hz are used to excite polar molecules that produce thermal energy and increases temperature [29]. On the continent, a small number of microwave pasteurization units are in operation, primarily manufacturing pasta and oriental products in plastic trays. Some ready meal production uses microwaves to effect the microbial kill step. Benefits of rapid heating can result in improved quality for foods that are sensitive to thermal degradation. The technology has not received widespread adoption because of the high capital costs of the equipment and the wide distribution in temperatures across a package.

Heat generated by the microwaves pasteurizes the food and the package together, and the products are sold under chilled storage to achieve an extended shelf life. Microwave sterilization is not popular because of the need for air overpressure to maintain the shape of the flexible packages during processing. This creates complications with continuous systems in that transfer valves are required between the chambers.

14.4.4 Irradiation

Irradiation has seen much wider applications in the USA than in the United Kingdom where "public opinion" has effectively sidelined it. In the United Kingdom, there is a requirement to label food that has been irradiated or contains irradiated ingredients. In addition to killing bacterial pathogens, such as *Salmonella* on poultry, it is especially effective at destroying the microorganisms present on fresh fruit such as strawberries and thus markedly extending their shelf life. Its biggest advantage is that it has so little effect on the food itself that it is very difficult to tell if the food has been irradiated. It also has some technical limitations, in that it is not suitable for foods that are high in fat, as it can lead to the generation of off flavors. This restricts its use for sterilizing ready meals. The only commercial foods that are currently licensed for irradiation in the United Kingdom are dried herbs and spices [30], which are notoriously difficult to decontaminate by other techniques, without markedly reducing flavor. A major application for irradiation is in decontaminating packaging.

14.5 CONCLUSIONS

Changing lifestyles has ensured that consumer preference continues to increase for ready meals as convenience foods. It is likely that thermal processing of ready meals will remain as one of the key methods for their manufacture. Many thermal processing methods can be used for their production; however, two routes forward are gaining in popularity:

1. In-pack pasteurization in combination with chilled storage. This is being achieved through optimization of existing retort-based equipment and thermal processing regimes, which can be used to produce ready meals of high quality. Recent developments in overpressure retorts and packaging formats have allowed this ready meal sector to advance at a rapid rate. Retorting has the advantage of pasteurizing both ready meal and packaging together, which can extend the shelf life under chilled storage.
2. *Low-care–high-care factories* (also known as *low risk–high risk*). Ready meals of the highest quality are being manufactured using mild or *minimal heat processes* for the components. The thermal processing steps take place in the low-care parts of a factory and are followed by packaging in high-care environments. Strict control of hygiene in the high-care areas is essential in order that microorganisms are not introduced to the food. Once packaged, there is no further preservation hurdle apart from chilled storage.

Neither of the aforementioned methods can be classified as traditional thermal processes. Both involve pasteurization processes that require chilled storage as the secondary preservation hurdle. This is because the consumer trends are for higher-quality ready meals that are difficult to produce with a full sterilization process. The benefit of these mild thermal treatments is in the high quality of ready meal that can be manufactured.

REFERENCES

1. Anon. *The Food Pocket Book.* London, U.K.: NTC Publications Ltd., 1998.
2. Anon. *UK Food Market—2000 Market Review*, 12th edn. London, U.K.: Key Note Publications Ltd., 2000.
3. G Betts (ed.). *The Microbiological Safety of Sous-Vide Processing. Campden BRI Technical Manual No. 39.* Campden BRI, 1993.
4. S Leaper (ed.). *HACCP: A Practical Approach. Campden BRI Technical Manual No. 38.* Campden BRI, 1992.
5. Department of Health. *Guidelines for the Safe Production of Heat Preserved Foods.* London, U.K.: HMSO, 1994.
6. C Leadley, A Williams, and L Jones. *New Technologies in Food Preservation. Campden BRI Key Topics in Food Science and Technology—No. 8.* Campden BRI, 2003.
7. G Betts (ed.). *A Code of Practice for the Manufacture of Vacuum and Modified Atmosphere Packaged Chilled Foods with Particular Regard to the Risks of Botulism. Campden BRI Guideline No. 11.* Campden BRI, 1996.
8. J Gaze (ed.). *Pasteurisation Heat Treatments. Campden BRI Technical Manual No. 27.* Campden BRI, 1992.
9. Anon. *The Freshline Guide to Modified Atmosphere Packaging.* Air Products, 1995.
10. CO Ball and FCW Olsen. *Sterilization in Food Technology: Theory, Practice and Calculation.* New York: McGraw-Hill Book Co., 1957.
11. CR Stumbo. *Thermobacteriology in Food Processing*, 2nd edn. New York: Academic Press, 1973.
12. D Holdsworth. *Thermal Processing of Packaged Foods.* London, U.K.: Blackie Academic and Professional, 1997.
13. T Hutton (ed.). *Food Manufacturing: An Overview. Key Topics in Food Science and Technology—No. 3.* Campden BRI, 2001.
14. G Tucker and U Bolmstedt. Gently does it. *Liquid Foods International* 3(3): 15–16, 1999.
15. P Richardson (ed.). *Thermal Technologies in Food Processing.* Cambridge, U.K.: Woodhead Publishing Ltd. 2000.
16. KL Brown, CA Ayres, JE Gaze, and ME Newman. Thermal destruction of bacterial spores immobilised in food/alginate particles. *Food Microbiology* 1: 187–198, 1984.
17. GS Tucker, D Wolf. *Time-Temperature Integrators for Food Process Analysis, Modeling and Control. Campden BRI R&D Report No. 177.* Campden BRI, 2003.
18. GS Tucker, T Lambourne, JB Adams, and A Lach. Application of a biochemical time-temperature integrator to estimate pasteurisation values in continuous food processes. *Innovative Food Science and Emerging Technologies* 3: 165–174, 2002.

19. Anon. *Guidelines to the Establishment of Scheduled Heat Processes for Low-Acid Foods. Campden BRI Technical Manual No. 3*. Campden BRI, 1977.
20. N May (ed.). *Guidelines for Batch Retort Systems—Full Water Immersion—Raining/Spray Water—Steam/Air. Campden BRI Guideline No. 13*. Campden BRI, 1997.
21. N May (ed.). *Guidelines for Performing Heat Penetration Trials for Establishing Thermal Processes in Batch Retort Systems. Campden BRI Guideline No. 16*. Campden BRI, 1997.
22. N May (ed.). *Guidelines for Establishing Heat Distribution in Batch Overpressure Retort Systems. Campden BRI Guideline No. 17*. Campden BRI, 1997.
23. GS Tucker, HM Brown, PJ Fryer, PW Cox, FL Poole II, H-S Lee, and MWW Adams. A sterilisation time-temperature integrator based on amylase from the hyperthermophilic organism *Pyrococcus furiosus*. *Innovative Food Science & Emerging Technologies* 8: 63–72, 2007.
24. GS Tucker and SF Featherstone. *Essentials of Thermal Processing*. Oxford, England: Wiley-Blackwell, 2011.
25. D Sorenson, M Henchion, B Marcos, P Ward, AM Mullen, and P Allen. Consumer acceptance of high pressure processed beef-based chilled ready meals: The mediating role of food-related lifestyle factors. *Meat Science* 87(1): 81–87, 2011.
26. D Sorenson and M Henchion. Understanding consumers' cognitive structures with regard to high pressure processing: A means-end chain application to the chilled ready meals category. *Food Quality and Preference* 22(3): 271–280, 2011.
27. T Bolumar, ML Andersen, and V Orlien. Antioxidant active packaging for chicken meat processed by high pressure treatment. *Food Chemistry* 129(4): 1406–1412, doi:10.1016/j.foodchem.2011.05.082.
28. T Aymerich, PA Picouet, and JM Monfort. Decontamination technologies for meat products. *Meat Science* 78(1–2): 114–129, 2008.
29. R Vadivambal and DS Jayas. Non-uniform temperature distribution during microwave heating of food materials—A review. *Food and Bioprocess Technology* 3(2): 161–171, 2010.
30. AM Witkowska, DK Hickey, M Alonso-Gomez, and MG Wilkinson. The microbiological quality of commercial herb and spice preparations used in the formulation of a chicken supreme ready meal and microbial survival following a simulated industrial heating process. *Food Control* 22(3–4): 616–625, 2011.

15 Thermal Processing of Vegetables

Jasim Ahmed and U.S. Shivhare

CONTENTS

15.1 INTRODUCTION

Thermal processing is one of the major techniques, which aims to produce commercially sterile packaged food of optimal quality. Commercial sterility of food has been considered as the condition achieved by application of heat, which renders the food free from viable microorganisms, including those of known public health significance, capable of growing in the food at temperatures at which the food is likely to be held during distribution and storage. The production and distribution of heat-treated food with respect to intrinsic and extrinsic factors can be divided into (i) mild heat treatment (pasteurization) and (ii) severe heat treatment (sterilization) processes. The success of the either process lies in destroying all viable microorganisms together with nature of food, environmental factors, hermetic packaging, and storage temperature. The choice of processing depends on the severity of the heat treatment and on the objectives to be accomplished [1]. Due to low acidity (pH > 6.0), vegetable products are subjected to severe heat treatment during processing in order to inactivate pathogenic microorganisms. The process is commonly termed as canning. Sometimes, the canning process has been altered to treat the vegetables at lower temperature in combination with other extrinsic factors (pH, preservatives, cooling, and packaging) to reduce microbial growth and for safe handling during transportation and storage.

The ultimate objective of the production of thermal processed vegetables is to satisfy consumers. Quality of vegetable changes as it proceeds from harvest to consumer. Product quality is represented by customers' satisfaction. Quality is a human construct additive parameter comprising many properties or characteristics [2]. Quality of thermally processed vegetables encompasses sensory properties (appearance, texture, aroma, and taste), nutritive value, chemical constituents, mechanical properties, functional properties, and defects. There has been an increasing public concern toward the quality of thermally processed vegetables. The major focus on thermal processing of vegetables now is to retain maximum nutritional and sensory quality with optimum process design along with keeping the microorganisms at safe levels.

Since thermal processing is a heat intensive process resulting significant loss of color and flavor of many vegetables. Presently, extensive research is going on to incorporate nonthermal processing (high-pressure processing, ohmic heating, and pulse electric heating) into thermal processing to improve sensory quality and consumers' acceptability by reducing process temperature. It is now well established that high-pressure processing has minimum effect on quality attributes by retention of the covalent bond and instant and uniform pressure distribution throughout a food system. Therefore, a combined preservation technology would help processors to produce safe and quality products with affordable prices.

The microbial pathogens and spoilage enzymes that would render the food unfit for consumption are inactivated during heat processing. On the other hand, concentration of heat sensitive nutrients such as ascorbic acid and thiamine is greatly reduced and texture of thermally processed vegetables is excessively softened on many occasions. For example, canned vegetables may be too soft due to disintegration of cell wall materials. Fortunately, kinetics of microbial destruction and quality deterioration usually do not proceed at the same rate. Studies have shown that thermal death rates of bacteria generally proceed much faster with increased temperature (z values 10°C–15°C) than simultaneous reactions that would lead to quality loss (z values 30°C–35°C). It is therefore possible to apply the principle of high-temperature short-time (HTST), ultrahigh temperature (UHT), and aseptic packaging process for better quality retention [3].

In order to balance the benefits of thermal processing without compromising safety, a good predictive model is required. Mathematical modeling is an essential tool to predict the impact of thermal processing on quality and safety of vegetables. The degradation kinetics of quality attributes and target microorganisms during thermal processing are complex one. Mathematical models that accurately predict the progress of a chemical or biochemical reaction occurring in a homogenous liquid or semisolid phase during thermal processing, and storage are useful in many engineering applications including process optimization. Experimental studies and application of simplified models to predict and interpret kinetic parameters (reaction order, rate constant, and activation energy) have been utilized by the process industries to improve the process efficiency and quality. In thermal processing, it is possible to achieve different time–temperature combination (TTC) with the same lethal effect, but resulting in different quality losses.

The objective of this chapter is to provide an overview on quality changes during thermal processing, and quality assurance (QA) of thermally processed vegetables. Food safety principles and practices are integrated into activities identified within QA or quality control programs, or within quality management systems; therefore, these programs and systems can address both food quality and safety simultaneously [4].

15.2 CONSIDERATIONS IN THERMAL PROCESSING OF VEGETABLES

The primary objective of thermal processing of food is the destruction of microorganisms of public health concern as well as spoilage microorganisms, and enzymes. The selection of thermal processing depends on various factors like type of foods, chemical composition, type of microorganisms present in food, initial microbial load, reaction kinetics of microbial death, and nutrient loss. Vegetables differ from fruits and other foods in chemical composition and therefore require different processing conditions. The acidity of vegetables is much lower than that of fruits, and they may contain more of the heat-resistant soil organisms than fruits [5]. Furthermore, vegetables require severe heat processing to produce better flavor and texture. Some of the important factors affecting thermal processing of vegetables are types of food, pH, nature of enzymes and microorganisms, thermal resistances of enzymes and microorganisms, mode of heat transfer, and so forth.

15.2.1 Thermal Processing and Reaction Kinetics

The rate at which a substrate disappears and new product forms is commonly represented by reaction kinetics. The rate can vary with time, temperature, pressure, moisture, acidity, amount of reactants/ingredients, and environmental conditions. It represents the extent of degradation or inactivation of nutritional and sensory factors or enzymatic and microbial activity in foods during processing. Some reactions result in quality loss/degradation, whereas others produce desired color or flavor and, therefore, require optimization during processing. Estimating the reaction kinetics and kinetic parameters is one of the most challenging parts with reactions pertaining to foods [6].

The rate equation for nth-order reaction for inactivation or degradation of biological materials is given by

$$\frac{dC}{dt} = -kC^n \tag{15.1}$$

where

C is the concentration of reactant at any time t
k is the reaction rate constant, with unit of $(\text{concentration})^{1-n}/(\text{time})$
n is the order of the reaction
The negative sign indicates decrease in concentration with time

Equation 15.1 can be generalized in the following form for nth order of reaction:

$$C^{1-n} - C_0^{1-n} = (n-1)kt; \quad n > 1 \tag{15.2}$$

where C_0 is the concentration of reactant at zero time.

The reaction rate is mostly described by either zero-, first-, or second-order reaction kinetics as follows:

$$\text{Zero order:} \quad C - C_0 = -kt \tag{15.3}$$

$$\text{First order:} \quad \ln(C/C_0) = -kt \tag{15.4}$$

$$\text{Second order:} \quad 1/C - 1/C_0 = kt \tag{15.5}$$

Most of the reactions involved in processing of foods are first-order reactions [7]. In some cases, fractional reaction orders have been observed [8]. Care has to be taken to determine the order of reaction precisely, which could be used for determining the effect of thermal processing on heat labile compounds in food products.

Temperature dependence of the reaction rate constant is described by the Arrhenius equation:

$$\ln k = \frac{\ln A - E_a}{RT} \tag{15.6}$$

where

A is a constant known as preexponential factor (s^{-1})
E_a is the process activation energy (kJ mol^{-1})
T is temperature (K)
R is the universal gas constant (8.314 J mol^{-1} K^{-1})

Activation energy is the minimum energy required to initiate a reaction at molecular level and is computed from linear regression of ln(k) versus reciprocal of absolute temperature (1/T) (Figure 15.1). The magnitudes of A vary from 10^{14} to 10^{20} s^{-1} for unimolecular and from 10^4 to 10^{11} s^{-1} for bimolecular reaction, respectively. Reaction kinetic parameters of some selected vegetables are presented in Table 15.1 [9–23].

15.2.2 Microbial Inactivation Kinetics

Inactivation kinetics of microorganisms is complex and has been well described in microbiology textbooks. Thermal death rate kinetics of test microorganisms must be studied to optimize TTC

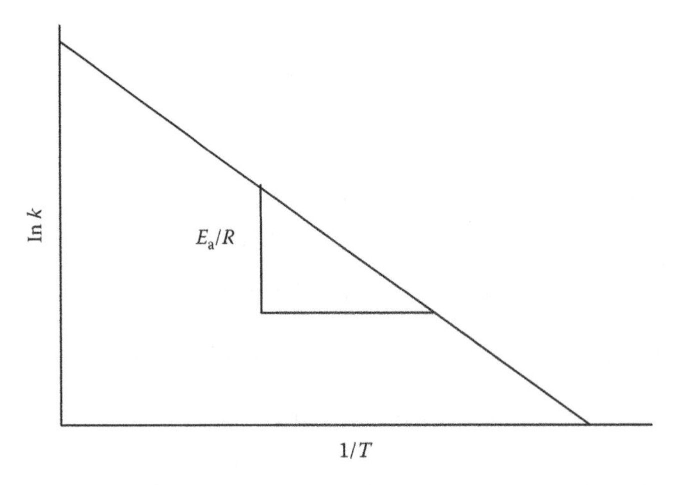

FIGURE 15.1 Temperature dependence of reaction rate constant.

TABLE 15.1
Reaction Kinetics Parameters of Selected Vegetables

Reaction Types	Vegetables	Order of Reaction	Temperature Range (°C)	Activation Energy (kJ mol⁻¹)	References
Thermal inactivation of peroxidase and lipoxygenase	Broccoli (floret)	Biphasic first	70–95	58	[10]
	Green asparagus			43–53	[10]
	Carrot			83–86	[9]
Chlorophyll degradation	Broccoli juice	First	80–120	69.0	[11]
Green color degradation	Coriander leaves	First	50–110	22.1–29.3	[12]
Chlorophyll degradation	Coriander leaves	First	80–145	11.8–95.3	[13]
Chlorophyll degradation	Mint leaves	First	80–145	16.6–41.6	[14]
Texture degradation	Potato	First	90–120	89.70	[15]
Texture softening	Potato	Second	122	116–180	[16]
Overall sensory characteristics	Vegetable puree	First	110–134	125–167	[17]
Ascorbic acid degradation)	Mushroom	First order with two degradation mechanisms	110–140	46.36–49.57	[18]
Texture softening (hardness)	Mushroom	First order with two softening mechanisms	70–100	10.32–15.22	[19]
Color (ΔE)	Pumpkin puree	First	60–100	30.39	[20]
β-Carotene				27.27	
Flow behavior				31.94	
Color (ΔE)	Tomato puree	First	100–140	162.21	[21]
	Strawberry juice	First	100–140		
	pH=2.5			92.67	
	pH=3.7			72.46	
	pH=5.0			48.05	
Texture softening (hardness)	Garlic	First	80–95	139.4	[22]
PO inactivation	Pumpkin	First	75–95	86.20	
Color (ΔE)				98.79	
Texture softening (hardness)				72.21	
PO inactivation	Carrot	First	70–90	151.4	[23]
Phenolic content		First		123.51	
Color		First			
L				187.03	
a				186.39	
b				231.16	
Texture softening (hardness)		First		151.97	

during thermal processing. Earlier evidence supports that the inactivation of microorganisms followed the first-order reaction kinetics (Equation 15.4) and represented by the following form:

$$\ln\left(\frac{N}{N_0}\right) = -kt \tag{15.7}$$

where N_0 and N represent the number of viable organisms at time zero and t, respectively.

Applying Equation 15.6 for the thermal inactivation kinetics of microbial spores at reference temperature T_{ref}, and the corresponding reference reaction rate constant k_{ref}, the following equation can be found:

$$\ln k = \ln k_{ref} - \left[\left(\frac{E_a}{R} \right) \left(\frac{1}{T} - \frac{1}{T_{ref}} \right) \right] \tag{15.8}$$

The activation energy (E_a) for bacterial spores has been reported as 500 kJ mol^{-1} [24]. The significantly higher value of E_a has been explained in various ways by researchers. The traditional view of microbial inactivation as being a process that follows first-order kinetics has recently been under renewed criticism [25]. Most of the published microbial survival curves have been described as reflecting a Weibull distribution of heat resistances rather than first-order mortality kinetics [26].

15.2.3 Effect of Time–Temperature on Microbial Death

15.2.3.1 Decimal Reduction Time (*D* Value)

Equation 15.7 indicates a linear semilogarithmic plot of N versus t. Equation 15.7 can be rewritten in common logarithmic form as:

$$\log \left(\frac{N}{N_0} \right) = \frac{-kt}{2.303} \tag{15.9a}$$

and

$$\log \left(\frac{N}{N_0} \right) = \frac{-t}{D} \tag{15.9b}$$

Equation 15.9 defines the decimal reduction time (D value), the time required to result in one decimal reduction in the survival cell population at a given temperature (Figure 15.2). In Equation 15.9,

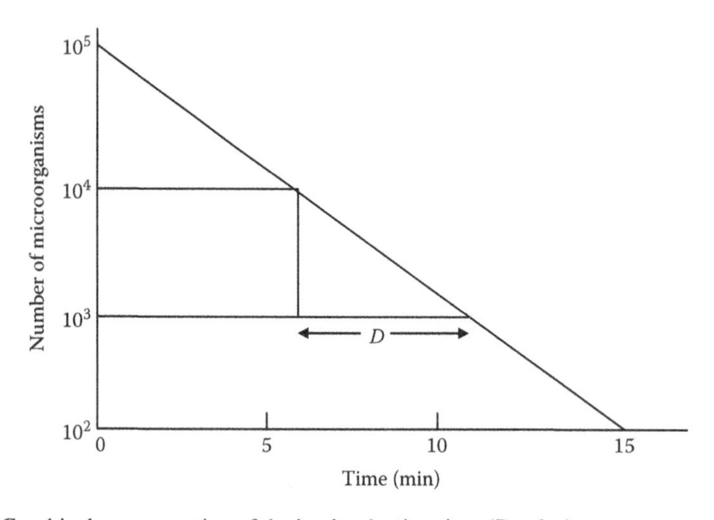

FIGURE 15.2 Graphical representation of decimal reduction time (*D* value).

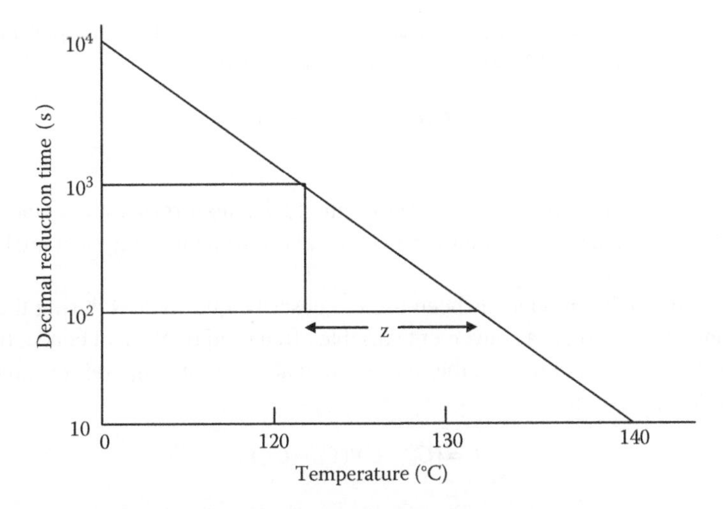

FIGURE 15.3 Graphical representation of thermal resistance constant (z value).

$D = 2.303/k$; the survival cell ($N < 1$) has been considered to be the probability of spoilage, while $N \geq 1$ implies certain spoilage.

15.2.3.2 Temperature Sensitivity and *z* Value

Temperature sensitivity of the D values is represented in terms of z value that indicates the influence of temperature on D values. The thermal resistance constant (z) is defined as the increase in temperature causing 90% reduction in D values (Figure 15.3). Different microorganisms have different z values, and even the z of the same microorganisms may vary with circumstances. Due to this reason, each food products must have its own TTC that is specific for the formulation, can size, and process parameters. Mathematically, z value is represented as

$$z = (T_2 - T_1)/[\log(D_1) - \log(D_2)] \quad (15.10)$$

where D_1 and D_2 represent decimal reduction times at temperatures T_1 and T_2, respectively.

15.2.4 Degradation Kinetics of Quality Attributes

Quality attributes of food are important from consumers' acceptability. The sensory attributes are perhaps the first criteria for acceptance or rejection. Appearance, color, flavor, and texture govern the initial acceptance of foods [27,28]. Thermal processing has significant effect on sensory attributes. However, information on the kinetic data on sensory evaluation and other properties is limited.

The kinetic parameters for degradation of a food component can be calculated by two procedures [29]. In steady-state procedure, the thermal lag (heat up or cool down) time is considered insignificant as compared to the overall processing time, and the reaction is considered to occur at constant temperature. In unsteady-state procedure, the reaction is considered to occur at a variable temperature and the data required are concentration of the degraded component and temperature profile of the sample during the heating–cooling process. Degradation of the component during the thermal lag is included in data analysis. Steady-state methods are used in most studies to estimate kinetic parameters, due to the difficulty of using temperature profile prediction models because of the natural complexity of foods.

It is evident from the work of many researchers that the kinetics of quality parameters followed the first-order reaction kinetics [8]. Equation 15.4 can be rewritten as

$$\ln(C_1/C_2) = k_c(t_2 - t_1) \quad (15.11)$$

where
C_1 and C_2 are the concentrations or physical attributes (color, texture) at times t_1 and t_2, respectively
k_c represents degradation rate constant for a chemical constituent or physical attribute

The fractional conversion model has been used to describe quality loss during thermal processing [30]. Fractional conversion is a convenient variable often used in place of concentration [31]. For irreversible first-order reaction kinetics, the rate constant at constant temperature can be determined through fractional conversion, f:

$$f = (C_0 - C)/(C_0 - C_\infty) \quad (15.12)$$

where, C_∞ is concentration at infinite time signifying that the reaction is completed. Considering fractional conversion, the first-order reaction kinetics model takes the form:

$$\ln(1 - f) = \ln[(C - C_\infty)/(C_0 - C_\infty)] = -k_c t \quad (15.13)$$

The concept of D–z has been equally valid for heat labile quality parameters. D_{ref} value generally be considered as D_c value at reference temperature T_{ref}, and denotes 90% destruction of heat labile components, the corresponding z value is designed as z_c [24]. The following equation is used to determine D values for thermolabile components.

$$\log(D/D_{ref})_c = -(T - T_{ref})/z_c \quad (15.14)$$

where the subscript c refers to a chemical compound or physical attribute.

Modeling of the real food is difficult due to complexity of the nutrients and their interaction during processing. As a result, model system under ideal condition has been considered for mathematical treatment. A complex food exhibited different rate of degradation and mechanism as compared to an isolated nutrient system. Working on thermal degradation of vegetables, most of the researchers have considered the first-order reaction kinetics for its simplicity. However, the degradation kinetics of real food is somewhat different from the first-order reaction kinetics used in the literature. The optimum temperature of a process can be affected if a first-order model instead of real kinetics occurring in the product is applied [32]. Therefore, more information is required to obtain consistent and reliable data on modeling of thermal processing.

15.3 APPLICATIONS OF THERMAL PROCESSING

Thermal processing has wide application in vegetable processing. These include blanching, cooking, evaporation, drying, extrusion, pasteurization, sterilization, microwave (MW) processing, and so forth. Due to limited scope and applicability to vegetables, the present discussion is limited to blanching, pasteurization, and sterilization only.

15.3.1 Blanching

Blanching is the treatment of vegetables in hot water or steam to inactivate oxidative enzymes. The process is extremely important if the vegetables are to be further processed by drying, freezing, or canning; otherwise, quality of the product will deteriorate during handling and storage. The thermal effect inactivates the enzymes like polyphenol oxidase (PPO), catalase, peroxidase (PO), lipoxygenase (LPO), and chlorophylase. Apart from enzymes inactivation, blanching helps to destroy microorganisms on the surface, cleans the dirt, and makes some vegetables like broccoli and spinach more compact. The process brightens the color, retards loss of vitamins, wilts vegetables, facilitates packing, and stabilizes texture, flavor, and nutritional quality (changes caused by enzymes).

Among the oxidative enzymes, it has been found that peroxidase (PO: EC 1.11.1.7) is the most heat resistant and is considered as an index of blanching adequacy. Blanching is carried out in many ways like hot water blanching, hot lye blanching, steam blanching, MW blanching, high-pressure blanching, and individual quick blanching. Hot water and steam blanching are simple; however, the leaching of nutrients (vitamins and pigments) is excessive, and the process is energy intensive. MW blanching reduces the nutrient loss [33]; however, the process may not be effective since some enzymes are difficult to inactivate. This could result in off-flavors and loss of texture and color. High-pressure blanching of vegetables has been designed on the concept of hot water or steam blanching but without the related leaching of nutrients, quality losses, and environmental effects [34]. However, the process could not achieve complete inactivation of PPO enzymes. Moist hot gas and ohmic heating have rarely been used for blanching of vegetables.

Efficacy of blanching depends on the nature of the vegetable, geometry of the sample, residence time, water quality, heat transfer conditions, and indicator enzyme activity. The proportion of resistant and labile isoenzymes and kinetic parameters are important factors to predict and optimize blanching operation. Blanching time is crucial and varies with the vegetable and sample geometry. Quality of water used to blanch vegetables may affect the texture of certain vegetables; for example, hard water could result in toughening of green beans. It is therefore advisable to check the quality of water to be used in blanching.

PO are assumed to play an important role in chlorophyll degradation, a process accompanying ripening and senescence of most fruits and vegetables. The mechanism of peroxidases in chlorophyll degradation is based on the formation of enzyme–hydrogen donor complexes [35]. In PO-mediated chlorophyll degradation, PO oxidizes the phenolic compounds with hydrogen peroxide and forms phenoxy radical. The phenoxy radical then oxidizes chlorophyll to colorless low-molecular-weight compounds through the formation of C13-hydroxychlorophyll-a and a bilirubin-like compound as an intermediate [36].

In experimental studies, thermal inactivation of PO has not been found to follow first-order kinetics [37–39]. Considering the possibility of the presence of isoenzymes at the beginning of the inactivation process, the kinetic models applied in literature are based on different mechanisms: first-order, consecutive reactions, and parallel reactions (Table 15.2) [40–45]. Details are available elsewhere [46–49].

15.3.2 Pasteurization

Pasteurization is a relatively mild thermal treatment with less impact on product quality. The process has been applied to inactivate the enzymes and pathogenic microorganisms. *Mycobacterium tuberculosis* is the widely accepted target, but recently *Listeria monocytogenes* has been considered as the test organism for pasteurization especially for the dairy products. Since the process is not severe enough to kill *Clostridium botulinum*, the pasteurized foods require refrigeration immediately after processing. Thus, pasteurized products have limited shelf life in the distribution chain. The severity and duration of heat treatment depends on nature of product, pH, initial microbial load, type of heat processing, nature of resistive microorganism, and so forth. Plate-type heat exchanger (PHE) is commonly employed to pasteurize low viscosity fluid (<5 Pa s). The cold fluid is pumped to regeneration section of PHE followed by heating at desired temperature (72°C–75°C), holding for predefined residence time (15–30 s) and finally cooled to refrigeration temperature. Viscous products can be pasteurized using scrapped surface heat exchanger whose inside surface in contact with the product is continuously scrapped by molded plastic to prevent fouling. Scope of pasteurization for vegetable products is limited because of their high pH values (Table 15.3).

TABLE 15.2
Kinetic Equations to Describe Inactivation of Peroxidase

Models	Equations	References
First order	$\ln\frac{A}{A_o} = -k \times t$	[40]
Distinct isozymes	$\frac{A}{A_o} = A_L \exp(-k_L t) + A_S \exp(-k_R t)$	[41]
Two fractions	$\frac{A}{A_o} = a \times \exp(-k_L \times t) + (1-a) \times \exp(-k_R \times t)$	[42]
Multicomponent first order	$\frac{A}{A_o} = \{\exp(-k_1 t) + r\exp(-k_2 t)\} / (1+r)$	[43]
Series	$\frac{A}{A_o} = \alpha_2 + \left[1 + \frac{\alpha_1 k_1}{k_2 - k_1} - \frac{\alpha_2 k_2}{k_2 - k_1}\right]\exp(-k_1 \times t) - \left[\frac{\alpha_1 k_1}{k_2 - k_1} - \frac{\alpha_2 k_1}{k_2 - k_1}\right]\exp(-k_2 \times t)$	[44]
*n*th order	$\frac{A}{A_o} = \{A_o^{1-n} + (n-1)kt\}^{1/(1-n)}$	[40]
Fractional conversion	$\frac{A}{A_o} = A_r + (A_o - A_r)\exp(-kt)$	[14]
Weibull distribution	$s = s_o \exp(-bt^n)$	[45]

TABLE 15.3
pH Values of Selected Vegetables

Name of Vegetables	pH	References
Asparagus	5.4–5.8	[5]
Broccoli	5.2–6.5	[5]
Cabbage	5.2–6.3	[5]
Carrots[a]	4.9–5.5	a
Cauliflower	5.7–6.5	[5]
Celery	5.5–6.0	[5]
Coriander leaves[a]	6.0–6.25	[11]
Eggplant	5.3–5.8	[5]
Lettuce	6.0–6.4	[5]
Potato[a]	5.4–5.8	a
Peas	6.0–6.2	a
Rhubarb	3.1–3.4	[5]
Spinach[a]	5.2–6.2	a
Sweet potato[a]	5.3–5.6	a
Turnip[a]	5.2–5.6	[5]
Corn[a]	6.1–6.3	a
Capsicum (green)[a]	5.1–6.0	a
Jalapeno pepper	6.0–6.60	[5]
Wax beans	6.0–6.15	a

[a] Unpublished data of J. Ahmed.

15.3.3 Sterilization

Theoretically, sterilization refers to complete inactivation of all the microorganisms present in food. However, foods are generally commercially sterilized in retorts (enclosed vessel for canning) or continuous flow system. Primary objective of the process is to destroy *C. botulinum* and its spores, and produce safe shelf-stable food. Success of the process does not depend upon inactivating microorganisms only but considering other factors associated with the inactivation of spoilage and pathogenic microorganisms. These factors are nature of foods (pH), environment (vacuum), hermetic package, and storage temperature.

15.3.3.1 Microbiological Considerations

Safety and quality retention are of major concern during thermal processing of vegetables. The process must be designed in such a way that the presence of survival microorganisms should be statistically insignificant with no health risk to the consumer. There are three types of microorganisms present in foods namely yeast, mold, and bacteria. Yeast and mold cannot withstand high temperature and therefore do not survive in thermally processed foods. However, bacteria and its spore can withstand high temperature and are of major concern. *C. botulinum*, a heat resistant, spore-forming anaerobic toxin-producing bacteria, is of major safety concern in thermally processed vegetables. The spores of *C. botulinum* do not produce toxin, while the germinating vegetative cells are capable of producing deadly neurotoxin in low-acid vegetables. Other spoilage organisms are *Clostridium sporogenes, Bacillus stearothermophillus*, and related species [50].

15.3.3.2 12D Concept

D values follow logarithmic destruction and mathematically it takes infinite time to destroy all the microorganisms present in the food. Statistical approach has therefore been used to describe the number of survival microorganisms after thermal processing to provide a margin of safety. It has been established that the minimum safe heat process given to low-acid food should decrease the population by 12 logarithmic cycles, the basis of the 12D concept or "bot cook." The process is found to be adequate even if the initial microbial load is too high. The D value for *C. botulinum* is estimated as 0.21 min at 121.1°C with a z of 10°C. In this case, application of 12D process (equivalent to 12×0.21 min) resulting in a heating period of 2.52 min at 121°C, which is known as minimum process lethality (F_o). Many low-acid foods are processed beyond the minimum value, and a F_o value of 5 min is frequently used for vegetables. The 12D process has been considered as a minimum safe standard for vegetables [51].

15.3.3.3 pH

The microbial growth depends on pH of the medium, and hence it is an important factor in determining the degree of thermal treatment. Foods are divided into two categories based on pH: acidic food (pH$<$4.5) and nonacidic food (pH$>$4.5). It has been established that *C. botulinum* cannot grow or produce toxin at pH below 4.6. Most of the vegetables are nonacidic in nature (Table 15.3) and spores of *C. botulinum* are the primary concern for vegetable processing. Therefore, it is necessary to apply the TTC sufficient to inactivate spores of *C. botulinum*. The pH can affect the D value of microorganisms in different ways. Lowering of pH reduce the heat resistance of the spores and process severity. Furthermore, the pH adjustment of the food could inhibit microbial growth especially during storage. This is considered as an extra safety measure for thermally processed foods.

15.3.3.4 Heat Transfer

Penetration of heat into the cold spot of food is confronted by, convective resistance from heating medium to outer surface of the container and from inner surface to product, as well as the resistance of the packaging material and food product. The internal resistance depends on the thermophysical properties, geometry, and product dimensions [52]. The mechanism of heat transfer through the container wall is by conduction. For metallic containers of normal thickness and thermal conductivity, there is no appreciable resistance to heat transfer. With regard to the heat transfer from the container wall into the product the mechanism largely depends on the viscosity of the food. Liquid and semiliquid foods are mainly heated by convection while solid foods are heated by conduction. However, increase in viscosity and presence of solid particles in semiliquids affect rate of heat transfer and make the process more complex.

Foods are heterogeneous in nature and varied from liquid to semisolid to solid. This heterogeneity has considerable implications in the characteristics and behavioral properties of thermally processed foods. As such, when semiliquid and solid foods are processed in still retorts, various problems are encountered. In general, still retorting is associated with slow rate of heat transfer resulting considerable loss in quality attributes. Numerous thermal processing techniques have been developed. The improved methods involve carefully engineered processes such as application of HTST principle in aseptic and agitating processing.

15.3.3.5 Sterilizers

Sterilization of vegetables is generally carried out in both batch and continuous retorts. Batch retort is suitable for small-scale operation and is available in horizontal and vertical type. The cans are stacked in crates and can be placed on the retorts by hydraulics. Product characteristics govern the choice of retorts.

Many vegetable products have high-viscosity components (vegetable stews, beans in tomato sauce, and sauce with hydrocolloids) and, therefore, require agitation to ensure better heat transfer and sterilization temperature at the center of the can. Container agitation in rotary retorts during processing provides several advantages over still-retort processing. The advantages are improved quality and reduced process time as a result of increased rate of heat penetration [53]. These involve axial or end-over-end agitation, and special retorts or modified retorts are used. For adequate sterilization of vegetables, it is important to control headspace of can, solid–liquid ratio, consistency, rotational speed of retort, process time, temperature, pressure, and operating cycles [24].

For continuous operation, hydrostatic pressure cookers are used. A typical hydrostatic cooker contains a preheat leg, a steam sterilizing chamber, and a cooling leg. These arrangements can be modified with product nature and heat treatment requirement. Saturated steam and steam–air mixture are used as heating source for the sterilization of vegetables.

15.3.3.6 Aseptic Processing

The process involves filling commercially sterilized food into a presterilized container, which is hermetically sealed under aseptic conditions. The technique is based on the principle of HTST process, which is claimed to have several advantages over conventional processing. The advantages include shorter processing time and better quality end product. In addition, it reduces energy consumption, suitable for new packaging, designing, and readily adaptable to automatic control [54]. Development of new container materials such as polymer plastics and polymer-coated paper containers have remarkable influence on the increased interest and attention devoted to aseptic processing and packaging of low-acid foods.

15.4 THERMAL PROCESSING AND QUALITY ATTRIBUTES OF VEGETABLES

15.4.1 Quality and Quality Attributes

Quality is defined as "the degree of excellence." Kramer and Twigg [55] defined quality as "the composite of those characteristics that differentiate individual units of a product, and have significance in determining the degree of acceptability by the buyer." Food quality embraces both sensory attributes that are readily perceived by the human sense and hidden attributes such as safety and nutrition that require sophisticated instrumentation to measure [56]. Quality of vegetables is based on some predefined quality parameters. These parameters have been set for product safety and reproducibility. In one way, it governs manufacture of the food according to public health significance and, on the contrary, it makes a consistent product to acquire a good market share. The processed vegetable loses some quality during processing while some quality loss occurs during storage. The major quality attributes of thermally processed vegetables are color, aroma, taste, and texture while hidden quality attributes like nutritional values and safety (chemical and microbiological) remain the most challenging concern in processed vegetables.

Color plays an important role in appearance, processing, and acceptability of vegetables. When a vegetable is exposed to light, about 4% of the incident light is reflected at the outer surface, causing specular reflectance or gloss, and the remaining 96% of incident energy is transmitted through the surface into the cellular structure of the product where it is scattered by the small interfaces within the tissue or absorbed by cellular constituents [57]. Recently imaging technology has been introduced for more precise color measurement of fruits and vegetables and multi- or hyperspectral cameras permit rapid acquisition of images at many wavelengths [2]. This type of imaging provides information about the spatial distribution of constituents (pigment, sugar, and moisture) in vegetables near the surface of the product.

Rheology, the study of the deformation and flow of matter, has been extensively applied to vegetables in an effort to understand the relationship between structure, texture, and changes taking place during processing. The mechanical properties of vegetables have been widely studied following the same technique applied for nonbiological materials. It helps to understand the mechanical behavior of vegetables to some extent, however, vegetables as a biological material differs from nonbiological materials in many respect. Various researchers have defined the term "texture" of vegetables in different ways [58]. Sensory analysis of vegetable texture in combination with mechanical measurement could represent the vegetable texture more precisely.

The application of the principle of hazard analysis and critical control points (HACCP) has done much to focus and formulize the process of demonstrating the safety of processed vegetables. It is generally agreed that verification processes are necessary to assess the effectiveness of the HACCP plan and to confirm that the food safety system once implemented adheres to the HACCP plan. An optimum TTC is an important critical control point (CCP) that has to be controlled in order to guarantee the microbiological safety of processed vegetables. The safety of sterilization process can be evaluated according to the lethality achieved and the microbiological risk alteration [59] of the target microorganisms that survive the thermal treatment.

15.4.2 Effect of Blanching on Quality of Vegetables

15.4.2.1 Enzymes

Peroxidase (PO) is considered to be the most heat-stable enzyme in vegetables since it is known that this enzyme has a relative high thermal stability and is therefore used as the indicator of blanching efficiency. Food and Drug Administration (FDA) recommends its inactivation to reduce quality loss during storage of processed foods [60]. Thermal inactivation of PO depends on the nature, thickness, geometry of the vegetables and, applied TTC. Heating times can be significantly reduced if an individual quick blanching process is followed. The temperature of the blanching step has an effect on enzymes: at a higher blanching temperature, enzyme inactivation was higher. PO seems to be more resistant to high temperatures since residual PO activity could still be detected after a high-temperature blanching (90°C for 4 min) [61]. However, it can be noticed that soaking the carrot pieces in a Ca^{2+}-solution results in a lower residual enzyme activity, compared to a low-temperature blanching (LTB; 60°C for 40 min).

The PO inactivation of savoy beet, amaranth and fenugreek was reduced to negligible amount in 1 min in hot water (95°C ± 3°C) [62] while same was achieved at 85°C in 30 s or 95°C for 15 s for spinach [63] and 99°C for 2 min for fenugreek leaves [64]. MW blanching (2 min) of artichokes inactivated PO completely without any losses of ascorbic acid while boiling water (8 min) and steam blanching (6 min) exhibited 16.7% and 28.9% loss of ascorbic acid with inactivation [65]. MW heating has also been found to be more effective than water blanching of carrot. Loss of vitamin C was lower (16.8%–31.85%) during MW—as compared to water (44.3%)—blanching [39]. There is evidence that the quality of blanched and processed food is superior if some PO activity left at the end of blanching process [66,67]. One of the problems associated with complete inactivation of PO is the presence of 1%–10% more heat-stable isoenzymes of PO in most vegetables. The complete inactivation of PO indicated overblanching of the process [68].

Though PO has been considered as an indicator for blanching process, low correlations between residual PO activity and keeping quality of frozen foods have been reported [66]. Inactivation of residual PO results in overblanching that can lead to reduced quality in frozen food and thus cause economic losses. Various researchers have advocated for consideration of other enzymes as a blanching indicator instead of PO. LPO is widely distributed in vegetables and is involved in off-flavor development and color loss [69]. Several authors have recommended that analysis of LPO activity, rather than PO, may be a more appropriate index of blanching adequacy [69,70]. Inactivation of LPO requires less heat treatment, resulting in improved color, flavor, and nutritive values [71]. Time–temperature relationship based on an 80% reduction of LPO has been recently reported [72], and the inactivation followed a first-order kinetic model.

PPO has been used as an indicator in blanching of potato. PPO is a generic term for the group of enzymes catalyzing the oxidation of phenolic compounds to produce brown pigment on cut surface of fruits and vegetables. Due to PPO, certain phenols, especially mono-, di- and polyphenols are hydroxylated in the *o*-position adjacent to an existing –OH group, further oxidized to *o*-benzoquinones and then nonenzymatically polymerized to melanins [73]. The PPO activity is related with color changes due to the formation of colored polymers. Therefore, color measurements could be considered as an indirect index of PPO activity and samples that did not show browning or other analogous colors were reasonably PPO-free [73].

Thermal resistance of enzymes is greatly affected by pH. *Bacillus licheniformis* α-amylase exhibited a higher thermal resistance between pH 6.5 and 9, while below pH 5 or above pH 11 thermal inactivation was significantly faster [74,75]. It was also noted that the stability of this enzyme at different pH values was enhanced by the addition of calcium chloride and that the optimum pH for enzyme stability was found to be 8.5 [76]. Thermal stability of LPO from asparagus was found to be independent of pH in the range 4.0–7.0 [77]. Likewise, thermal resistance of PPO from mushroom did not change significantly with pH from 5.5 to 7.5, although a slight decrease of the z value with pH was observed [78]. For horseradish, a maximum stability at pH 7 (in the range 4–10) has been reported. A study carried out on PO from asparagus revealed that the thermostability of the enzyme was affected by changes in pH (from 4.0 to 7.0), but the highest stability was observed at pH 6 [76]. However, a continuous decrease in the stability with an increase of pH was reported [79], thereby suggesting a maximum stability at mild acidic conditions.

15.4.2.2 Color and Pigment

Generally, visual color degradation of vegetables is expressed in terms of tristimulus color values (L, a, and b) individually or its combination, chroma, hue, or total color difference value. Color degradation of vegetables depends on blanching temperature, medium, and duration. While working on blanching of artichoke in three types of medium (steam for 6 min, boiling water for 8 min, and MW for 2 min), it was found that boiling water and MW-blanched samples retained maximum L and $-a$ values [64]. Though boiling water and MW blanching exhibited similar results for color values, boiling water treatment retained maximum chlorophyllaceous pigment.

The loss of color during blanching of green beans and broccoli at 40°C–96°C have been modeled by using $-a/b$ color values [80]. Initially, the color increased during processing but decreased later on. Color change was modeled by a simplified mechanism of two consecutive reactions: one that sets color and one that degrades color. Formation and degradation of visible color in vegetables is governed by processes related to the coloring compounds (like chlorophyll and chlorophyllides) irrespective of the vegetable under study [80]. The color ratio (*a*/*b*) has also been used as quality parameter for vegetables during blanching [81]. This change has been associated with the change of green color to yellow due to conversion of chlorophyll to pheophytin and further to pyropheophytin.

Blanching at high temperatures (80°C–100°C) degraded greenness (−*a* value) of vegetables faster than that of low temperature [82,83]. Both temperature and blanching time had significant effect on −*a* value but did not affect *L* and *b*. Chlorophyll *a* and *b* decreased after blanching.

The conventional perceived color is associated with hue [$\tan^{-1} b/a$] and is represented by yellow for an angle of 90°; objects with a higher hue angle are greener, with lower hue angles are more orange-red [84,85]. Blanching process reduces the hue angle for vegetables, while MW blanching of artichokes exhibited maximum hue as compared to steam or water blanching [64]. Chroma [$(a^2+b^2)^{0.5}$] decreased during blanching; however, there was no significant difference in chroma of boiling water and MW-blanched artichokes [64]. Steam blanching significantly reduced chroma.

Chlorophyll is lost during blanching process. There was 40% chlorophyll loss in savoy beet [62] while 10%–15% losses have been reported in fenugreek leaves and amaranth. During thermal processing of tomatoes, blanching resulted in increased bioavailability of lycopene from 5.1 to 9.2–9.7 mg kg^{-1}, but significant reduction in β-carotene [86]. Ohmic heating has been demonstrated to inactivate PO in less time than water blanching [87].

15.4.2.3 Texture

Blanching affects the texture of vegetables. For some vegetables, texture softening is desirable but undesirable for others like sweet potato, carrot, and jalapeno. LTB has been found to be effective for some vegetables [83,88–90] and is being practiced commercially for canned snap beans and canned cauliflower, tomato, potato, and carrot [91,92]. Maximum firming effects of these vegetables could be obtained by blanching at 55°C–80°C for times ranging from several minutes to several hours. For sweet potato, blanching at 62°C for 90 min resulted in maximum firmness [90], while high-temperature blanching disrupted cell integrity and cell adhesion, and reduced tissue rigidity. Blanching at 55°C for an hour produced maximum firmness of jalapeno [83].

Several researchers have emphasized the benefits of HTST blanching. Carrot tissues subjected to HTST (100°C, 0.58 min) retained a firmer texture than those subjected to a low temperature for long time (LTLT) (70°C, 71.10 min) [72]. Galacturonic acid and sugar content of pectin extract during blanching of carrot were determined, and immunocytochemistry experiments elucidated changes in the cell membrane. Blanching decreased galacturonic acid and total sugar contents for all treatments. Carrots subjected to HTST blanching contained higher galacturonic acid and sugars in pectins than carrots blanched by the LTLT process.

Two-stage blanching was found to be effective to maintain firmer texture of green beans. Instead of single blanching at 93°C, preliminary blanching at 70°C followed by high-temperature blanching resulted in firmer texture of green beans. During LTB, pectinase enzymes partially demethylate the pectins. This leaves −OH sites free on the pectin chain to cross-link with other pectin molecules via a calcium bridge, resulting in a firmer texture [93,94]. The beans are given high-temperature blanching in second stage that inactivates the enzymes.

Several researchers have investigated the changes in pectic substances of processed vegetables [95–98]. Results indicated that cooking-associated texture loss is often related to the dissolution of pectins. Cooking or blanching has also been found to activate enzymes [99–101]. The demethoxylation and depolymerization of pectic substances were catalyzed by pectin methylesterase (PME) or pectin polygalacturonase in blanched carrots. At 50°C–80°C, PME hydrolyzed the ester bond to yield free carboxyl groups and to release methoxyl groups. These free carboxyl groups were then cross-linked by salt bridges with calcium ions that were present in the tissue. The cross-linking bridges were often found to result in a firmer texture of fruits and vegetables [101]. The demethoxylation of pectins in blanched vegetables can also be caused by chemical saponification [101]. Chemical saponification or enzymatic deesterification produces a random deesterification of pectic polymers [101,102].

15.4.2.4 Sensory and Nutritional Quality

Ascorbic acid is one of the most heat labile nutrients and easily oxidizable by the naturally occurring enzyme system, ascorbic acid oxidase. Retention of ascorbic acid decreases with increased temperature and duration of blanching. In case of potato, significant differences in respect of ascorbic acid retention were observed during blanching at 80°C and 93°C [103]. Boiling water treatment and MW blanching of frozen beans showed same reduction of ascorbic acid content. MW-blanched broccoli retained the greatest amount of ascorbic acid compared to steam blanching.

Leaching losses are prominent during blanching. Blanching in hot water (95°C ± 3°C for 1 min) followed by potassium metabisulfite (KMS) treatment reduced leaching loss [70]. The maximum leaching loss was observed in savoy beet where 53% of β-carotene and 80% ascorbic acid (dry weight basis) lost during blanching.

Blanching of vegetable soybeans at different TTC (80°C for 30 min, 90°C for 20 min, and 100°C for 10 min) resulted in nutrients loss including sugar, vitamins B_1, B_2, and C while the loss was minimum at 100°C for 10 min [84]. Howard et al. [104] studied the effect of steam blanching on nutritional qualities of broccoli, carrot, and green beans. The maximum loss (30%) of ascorbic acid was observed in broccoli while green beans exhibited the least. The loss in carrot was about 14%. Steam blanching resulted in little or no loss in β-carotene content [105,106].

15.4.3 Effect of Pasteurization on Quality of Vegetables

Since vegetables are low-acid foods, scope of pasteurization is limited. Little information is available in the literature on pasteurization of vegetable products. Few reports are available on pureed foods and vegetable juices where the pH is lowered to ~4.0 to carry out pasteurization to retain sensory quality [107,108]. However, acidification degraded the color and pigment significantly while no significant changes were found in texture or flavor of these products.

15.4.4 Effect of Sterilization on Quality of Vegetables

Commercial sterilization is extensively used to process vegetables, which causes loss of quality during thermal processing.

15.4.4.1 Pigment and Color

The TTC used in canning process has substantial effect on pigments. Chlorophyll, carotenoids, lycopene, and xanthophyll are the predominant pigments present in vegetables. Chlorophyll is responsible for the color of green vegetables. Chlorophyll pigments have been the subject of enormous research because of their prominent function in plant physiology. Both chlorophyll *a* and *b* are the derivatives of dihydroporphyrin chelated with a centrally located magnesium atom, and they occur in an approximate ratio of 3:1 [109]. Structurally, the only difference between chlorophyll *a* and *b* is in the C-3 atom where the former contains a methyl group and the later contains a formyl group. Both these pigments differ in perceived color and thermal stability. Furthermore, the chlorophylls contain an isocyclic ring. It is believed that chlorophyll is present in the chloroplasts forming a complex with protein; however, the binding nature is not clear [110]. The central magnesium atom is easily replaced by hydrogen and thus forming pheophytins under various conditions. The chlorophyll degradation involves the loss of phytol to form chlorophyllide, loss of Mg^{2+} to form pheophytin, loss of Mg^{2+} and phytol to form pheophorbide, and loss of Mg^{2+} and carbomethoxy group to form pyropheophytin [111]. The conversion is enhanced by extended heat treatment, acidity, and storage [112]. The most common change that occurs in green vegetables during thermal processing is the conversion of chlorophyll to pheophytins, causing a color change from bright green to olive brown, which is undesirable for consumer acceptability [61,113–115].

Various methods to retain green color have been proposed. Controlling pH followed by HTST processing showed better retention of green color of vegetables [116–119]. Color retention was superior in most of these processes immediately after thermal processing but the retained chlorophyll degraded rapidly during storage. Greater stability of chlorophyll in blanched spinach puree was found when surfactants were added. However, the protective effect of surfactants lost at or above 100°C indicating its unsuitability for heat sterilized food products.

Tristumulus *L* value described the color change of mushroom [120], while *a* value represented the pigment content of many vegetables during thermal processing. The total pigment content of sweet potato, and squash correlated well with tristimulus *a* value [121,122]. The color ratio (*a/b*) is measured routinely as the quality index in tomato, orange, and red pepper processing industries [123].

15.4.4.2 Kinetics of Color Degradation of Vegetables during Thermal Processing

The color degradation kinetics of vegetables is complex, and dependable models to predict experimental color change are limited. Kinetics of pigment and color degradation of vegetables during thermal processing has been studied by several researchers [10,115,124–127]. Major finding of these studies is that both pigment and color degradation during thermal processing follows the first-order reaction kinetics.

Chlorophyll *a* degrades faster than chlorophyll *b* [116]. Degradation of chlorophyll *a* is 2.5 times faster than chlorophyll *b* at 37°C and water activity of 0.32. Thermal degradation of chlorophylls and chlorophyllides in spinach puree was studied at 100°C–145°C for a retention time of 2–25 min and 80°C–115°C for 2.5–39 min. Reaction kinetics revealed that both chlorophyll *a* and chlorophyllide *a* degraded more rapidly than the corresponding *b* form [128].

Tristimulus color ratio (*a/b*) has been used to determine reaction kinetic parameters for the discoloration of some vegetables [129]. The process activation energies for change in visual green color using rate constants determined at various temperatures. The activation energy values were found to be in the range from 33.14 to 43.38 kJ mol^{-1} for asparagus, green beans, and green peas, respectively.

Since color a value represents only major pigment and therefore does not represent the total color change of vegetables during thermal processing. In practice, any change in a value is associated with a simultaneous change in *L* and *b*. Representation of quality in term of total color may therefore be more relevant from the processing viewpoint [11,130,131]. Different combinations of tristimulus *L*, *a*, and *b* values were therefore selected to represent the color change in green vegetable [127,131]. The *La/b* was found to be the optimum combination to describe the total color degradation of pea puree [127] while combination of $L \times a \times b$ was found to adequately describe color change of many pureed foods (green chilli, red chilli, garlic, and coriander leaves) [11,107,108,130].

A fractional conversion model is currently practiced by researchers to describe color loss of vegetables during processing. The technique was exploited by many researchers [11,12,30,115] to study the kinetic parameters of visual green color loss of vegetables. They considered $-a$ value as a physical property that represented loss of visual green color and therefore, Equation 15.13 can be rearranged in the following form:

$$\ln[(-a)-(-a_{\infty})]/[(-a_{o})-(-a_{\infty})] = -kt \tag{15.15}$$

where

$-a_0$ is $-a$ value at zero time, t_0

$-a$ is $-a$ value at any time t

a_{∞} is $-a$ value at infinite time signifying complete degradation of the chlorophyll to pheophytin

Color degradation of many green vegetables (pea, green pepper, coriander leaves, and spinach) followed the first-order, while color degradation of broccoli juice followed two-step reaction kinetics (first degradation step for pheophytinization of the total chlorophyll followed by subsequent decomposition of pheophytin in second step) [81].

Degradation of both green and total color of green chilli and a mixed green leafy vegetable puree (mustard:spinach:fenugreek = 1:0.75:0.25) was studied using fractional conversion concept to determine the kinetic parameters [107,130]. Degradation of both green and total color followed the first-order reaction kinetics. The green color was represented by $-a$ value, while $L \times a \times b$ was found to adequately represent the changes of total color. Dependence of the rate constant during heat treatment obeyed the Arrhenius relationship. The activation energy values for green- and total-color degradation of green chilli were 23.04 and 25.02 kJ mol^{-1}, respectively, while the corresponding values for mixed leafy vegetables were 19.71 and 41.64 kJ mol^{-1}, respectively.

Table 15.4 reports the major findings of various researchers to relate color of vegetables with tristimulus *L*, *a*, and *b* values during thermal processing.

Color changes of many foods during processing have been transformed into a single variable, which would correlate with visual judgment and color degradation as a quality control parameter. A numerical total color difference (ΔE) technique commonly applied in paint and textile industry has been used to represent the color of food products during processing and storage [133–135]. Total color difference is the square root of the sum of the squares of the differences in each axis and is represented as

$$\Delta E = \sqrt{(L - L_{ref})^2 + (a - a_{ref})^2 + (b - b_{ref})^2} \quad (15.16)$$

Effect of thermal processing on color values has also been assessed by ΔE. Some researchers [135] advocated that ΔE values of 1 or less regarded as satisfactory for the best product, while a value greater than 3 was considered as unsatisfactory. The ΔE has been successfully applied to kinetic study of red chilli puree during thermal processing along with fractional conversion technique [30]. However, a failure of ΔE to represent color of pea and carrot puree was reported by Little [136] primarily due to large chromaticity differences.

15.4.4.3 Texture and Rheology

Textural changes occur in food during thermal processing. Softening of the tissues due to physical and chemical changes may render the food unacceptable to the consumers [137]. Thermal processing should therefore ensure a desirable texture in the product [3].

Kinetics of texture degradation optimizes the process to yield best quality product with minimal textural degradation. Recently, numerous studies have been reported in the literature on thermal softening and texture degradation kinetics of vegetables. Texture degradation kinetics is complex and different approaches have been proposed to analyze experimental data of texture degradation during thermal processing of vegetables. The literature suggests that thermal degradation kinetics of vegetables is well described by first-order reaction kinetics. Kinetic data are available on a variety of vegetables [137–139].

An alternative approach of dual mechanism first-order kinetic model has been reported while vegetables are being processed for longer period. In this concept, it is assumed that two substrates (say S_1 and S_2) participate in the textural degradation reaction consecutively resulting in two different kinetic parameters. Huang and Bourne [138] reported that the rate of thermal softening followed a biphasic behavior in which it was assumed that there was a labile component and a resistant component, and the labile component was degraded during the first heating stage. Biphasic models have been successfully applied to other vegetables like asparagus, peas, knoll-khol, and carrot [33,139]. However, the mechanism for texture changes at high temperatures may differ from the mechanism at low temperatures [137]. Thermal softening of white beans heated between 90°C and 122°C exhibited a biphasic behavior, but at temperatures 110°C, the first phase of thermal softening

TABLE 15.4
Tristimulus *L*, *a*, and *b* Values Used in Thermal Processing of Vegetables

Products	Color Value Used	Major Finding	References
Broccoli juice	*a*	Color degradation was best represented using −*a* value and fraction conversion technique was used	[10]
Coriander leaf puree	*L*, *a*, and *b*	Combination of ($L \times a \times b$) was found to represent color degradation	[11]
Jalpeno pepper	*L*, *a*, and *b*	Color parameter −*a* well correlated with process temperature to optimize blanching process of jalapeño pepper	[83]
Garlic	*L*, *a*, and *b*	Combination of ($L \times a \times b$) represented well the change of color during thermal processing	[107]
Pea puree	*A*	Green color degradation was best correlated by −*a* value. Fractional conversion technique was used	[115]
Mushroom	*L*	*L* value adequately represented thermal color degradation	[120]
Sweet potato	*a* and *b*	*a* was correlated well with pigment concentration	[121,123]
Green peas	*a* and *b*	−*a*/*b* was correlated with degree of conversion of chlorophyll	[124]
Pea puree	*L*, *a*, and *b*	Combination −(*La*/*b*) described well the color degradation of pea puree during thermal processing.	[127]
Asparagus and green beans	*a*, *b*	*a*/*b* was considered to represent color degradation	[129]
Red chilli (pepper)	*L*, *a*, and *b*	Fraction conversion model of color combination ($L \times a \times b$) adequately described color degradation of red chilli	[30]
Green peas	*a*	Color degradation was described in terms of −*a* value. Fractional conversion model was applied to represent color of green peas	[131]
Tomato	*L*, *a*, and *b*	Color combination ($L \times b/a$) exhibited high correlation with visual score of processed tomato products	[132]
Pea puree	*L*, *a*, and *b*	Hue angle ($\tan^{-1}b/a$) and combination −(*Lb*/*a*) described well the color degradation of pea puree during thermal processing	[87]
Pumpkin puree	*L*, *a*, and *b*	Combination −(*La*/*b*) described well the color degradation of pumpkin puree during thermal processing	[19]
Pumpkin	*L*, *a*, and *b*	Color parameters *L*, *a*, *b*, and ΔE well correlated with process temperature to optimize blanching process of pumpkin	[22]
Tomato puree and strawberry juice	*L*, *a*, and *b*	Combination (*La*/*b*) described well the color degradation during thermal processing	[20]
Carrot	*L*, *a*, and *b*	Color parameters *L*, *a*, and *b* well correlated with process temperature to optimize blanching process of carrot	[23]

was not seen [140]. It was pointed out that a main defect in many studies of softening has been the lack of thermal lag correction [137]. An iterative thermal lag correction procedure was proposed in which the temperature history for the center of a cylindrical container could be used to estimate the transient temperature distribution [129].

Two-step mechanisms of texture degradation have some limitation, especially in the second step where the degradation reached equilibrium softening stage and the calculated activation energy became negative. To overcome those limitations, fractional conversion technique has been employed to describe thermal softening of vegetables [14,141], and it was inferred that the first-order reaction kinetics was appropriate to describe texture degradation during thermal processing. Fractional conversion takes into account the non-zero texture property during prolonged heating, and therefore Equation 15.13 can be rearranged for texture degradation into the following forms:

$$\ln[(TP - TP_{\infty})/(TP_0 - TP_{\infty})] = -kt \tag{15.17}$$

$$TP = TP_{\infty} + (TP_0 - TP_{\infty})\exp(-kt) \tag{15.18}$$

where (TP_0) is the texture property at zero time; equilibrium texture property (TP_{∞}) is determined using the average of all measurements when the heating time is greater than 2 days to a constant TP. However, it is extremely difficult to obtain TP_{∞} for all the experiment runs as the fluctuation from sample to sample is significant. One of the biggest advantages of applying the fractional conversion technique in data reduction is that there is no longer a need to standardize the experimental protocol [141].

The geometry (size, shape, and thickness) of the sample does not significantly affect the kinetic parameters of vegetable texture [142,143]. Using potato as a model food, various researchers [144,145] studied the texture parameters using different instruments and concluded that kinetics of texture degradation were independent of testing parameters. However, contradictory reports are available in the literature while testing parameters were applied to measure food texture using various techniques. The process activation energies during thermal softening of potato [142,146,147] and texture degradation of carrots varied significantly [138,143]. The variation was explained on the basis of sample size.

The correct kinetic models of thermal degradation of vegetables texture depend on the set up of the experimental procedures; whether it is steady state (isothermal heating) or unsteady state (non-isothermal heating). The kinetic parameters obtained by these processes are significantly different. Most of the studies were carried out at steady state conditions to avoid complexity of modeling. An unsteady-state method was employed for estimating texture degradation during heating–cooling of green asparagus spears [148]. The method used a mathematical model of heat transmission for time–temperature history estimation, and a nonlinear regression of texture measurements of asparagus spears to estimate the kinetic parameters. The mathematical model and the kinetic parameters estimated could be used to design and evaluate thermal processes for green asparagus and other similar products like cucumber or corn-on-the-cob.

Texture softening is caused by the hydrolysis of pectic substances, gelatinization of starches, and partial solubilization of hemicelluloses combined with loss of cell turgor of vegetables. Heat processing will also result in changes of the cell walls, particularly the middle lamella, and gelatinization of starch. The final texture will therefore depend on the relative importance of each factor contributing to texture and the degree to which that factor has been changed by the processing method use.

Reports on LTB are available in the literature [90–92]. LTB followed by canning of vegetables resulted in higher fracturability and hardness of texture. The LTB also caused increase in gumminess and springness of canned sweet potato [90].

Many vegetables are washed, blanched, comminuted, and converted to puree or paste. The puree/paste is considered as a minimally processed food as the product is sometime acidified (~pH = 4.0) and pasteurized. These products retain color and flavor comparable to the fresh produce and are convenient to use. The rheological characteristics of these products change during thermal processing. Rheological parameters are significantly affected by process temperature. Pureed food behaves as non-Newtonian fluid with yield stress and pseudoplastic in nature [149]. Consistency index decreased with temperature while flow behavior index increased. Rheological parameters obtained from a controlled rate rheometer of some strained vegetable pureed foods during thermal processing are presented in Table 15.5. Among various rheological models, Hershel–Bulkley model (Equation 15.19) adequately described shear stress–shear rate data of vegetable purees. Presence of yield stress (τ_o) of pureed food indicates application of minimum force to initiate the flow of the product. Temperature has significant effect on τ_o and the yield stress decreased with temperature [150]:

TABLE 15.5
Rheological Characteristics of Strained Vegetables Puree

Vegetable Types	Temperature Range (°C)	Yield Stress (Pa)	Consistency Index (Pa s^n)	Flow Behavior Index (-)	Apparent Viscosity at 100 s^{-1} (Pa s)
Carrot puree[a] (7.4°Brix and pH 4.88)	20–80	1.65–3.72	5.44–11.53	0.27–0.35	0.24–0.42
Pea puree[a] (7°Brix and pH 5.92)	20–80	1.25–1.83	1.12–1.35	0.36–0.44	0.11–0.14
Corn puree[a] (14°Brix and pH 6.46)	20–80	0.43–1.18	0.58–2.06	0.26–0.30	0.02–0.05
Sweet potato[a] (11.2°Brix and pH 5.27)	20–80	1.46–2.66	0.77–1.20	0.38–0.43	0.06–0.09
Wax bean puree[a] (5.1°Brix and pH 5.05)	20–80	5.10–5.23	0.22–1.86	0.48–1.0	0.23–0.26
Pumpkin puree[b] (7.2°Brix and pH 4.35)	60–100	4.83–24.18	0.60–1.95	0.58–0.80	0.4–2.55

[a] Unpublished work of J. Ahmed.
[b] From Ref. [19].

$$\tau = \tau_o + K(\dot{\gamma})^n \tag{15.19}$$

where

τ and τ_o are shear stress and yield stress (Pa)
$\dot{\gamma}$ is shear rate (s^{-1})
K is the consistency index (Pa s^n)
n is the flow behavior index (dimensionless)

15.4.4.4 Sensory and Nutritional Quality

Nutrient losses during canning processing are significant [150]. Differences among cultivars of vegetables have been noted for most nutrients. Vitamins losses occur during commercial sterilization of vegetables. Vitamin A is comparatively stable as it is heat stable and insoluble in water, and survives during canning process. However, 30% losses during thermal processing have been reported [104]. Thiamin is heat labile and lost during sterilization process. The losses could be reduced by HTST process. Ascorbic acid is most susceptible to losses during processing due to its solubility in water and highly oxidized nature. Canning resulted in significant loss of ascorbic acid [104]. Information on degradation of nutrients during thermal processing is available in the literature [150,151]. There are about 85% losses of the heat sensitive nutrients during thermal processing. In processed foods, most significant losses occur as a result of chemical degradation. Several researchers have studied the mechanism of ascorbic acid loss and reported that it obeys the first-order reaction kinetics [152,153].

15.4.4.5 Quality Optimization

Quality optimization during thermal processing involves optimum balance between inactivation of spoilage microorganisms and retention of quality parameters. The rate of heat transfer inside a can depends upon the consistency, nature of vegetables and can size. For convection heating of vegetables, the internal mixing produces uniform cooking. However, for conduction heating, cooking starts from outside surface to inside of the can. This produces overcooking of the surfaces resulting in brown color and loss of overall acceptability. Slow heat transfer rate to the coldest region limits the applicability of the HTST process. For these situations, it is however possible to ascertain an optimum condition of TTC leading to microbiological safe food and minimized quality losses [154]. Quality optimization of thermally processed food has been well described by Holdsworth [24].

15.4.4.6 Food Safety

Food safety principles and practices have been applied to foods starting from farm produce, processed food products, ingredients, processing aids, contact surfaces, and packaging materials that are used in manufacturing the foods. Government agencies all over the world have enacted laws and regulations designed to ensure that processed foods are safe for the consumers and keep strict vigil on food safety rules. The safety of foods in international trade is governed by the World Trade Organization (WTO)/Sanitary and Phytosanitary (SPS) Agreement, which recognizes that governments have the right to reject imported foods when the health of the population is endangered. The criteria used to determine whether a food should be considered safe should be clearly conveyed to the exporting country and should be scientifically justifiable [155].

Application of HACCP to the thermal processing industries makes it convenient to identify the CCPs in the process line. Each step has to pass through CCPs and is well monitored. It provides ample opportunities to reduce chances of food hazards or contamination that could be physical, chemical, or microbiological in nature. For example, reaching of temperature at the coldest point of canned food is the most important CCP as it provides the vital information of survival of most heat resistant spores.

Each CCP reduces end-of-line sampling. This is not always possible for microbiological techniques, which take time to deliver the results. The monitoring system has always been recorded and takes corrective action. Verification of the CCPs would further reduce chances of contamination or mechanical fault during processing and assure product safety more precisely.

The worldwide acceptance of the Codex Alimentarius model has resulted in a consistent format (12 steps) for developing and implementing HACCP plans in food establishments around the world and has contributed to the recognition of HACCP systems to address food safety issues in bilateral and international trade [4].

Last few years, many thermal processing industries have implemented number of QA systems including good manufacturing practices (GMP), HACCP, International Organization of Standardization (ISO), and Total Quality Management (TQM) program in order to introduce new quality system and to produce high-quality products. However, mixed responses are available from the industries. Introduction of QA and TQM does not produce desired performance in quality. Theses implementation failures and the context of the organization require continuous adjustments of quality management activities, which should be based on assessment of quality performance [156].

15.5 CONCLUSIONS

Vegetable processing industry has undergone significant changes during the last few years in response to changing consumer demand for safe and quality products. However, canning is a traditional technology that remains the means of choice for bringing processed vegetables to the marketplace and the trend will continue because the consumers demand for high-quality safe food, which can only be manufactured by minimal thermal processes. The use of more advanced mathematical models to evaluate and predict thermal processing operation and quality retention will provide benefits to producers and customers. Thermal processing in combination with nonthermal processing technologies would produce food with better quality retention and consumers satisfactions. Introduction of TQM would ensure more product safety and smooth operation of food processing industry.

NOMENCLATURE

a	Tristimulus color value at time t measures redness and greenness
a_0	Tristimulus color value at time zero measures redness and greenness
a_∞	Tristimulus color value at time infinite measures redness and greenness
a_{ref}	Tristimulus color value measures redness and greenness considers as reference
A	Preexponential factor (s^{-1})
b	Tristimulus color value at time t measures yellowness and blueness
b_0	Tristimulus color value at time zero measures yellowness and blueness
b_∞	Tristimulus color value at infinite time measures yellowness and blueness
b_{ref}	Tristimulus color value measures yellowness and blueness considers as reference
C	Concentration of reactant/microorganisms, enzymes, or quality attributes at time t
C_0	Concentration of reactant/microorganisms, enzymes, or quality attributes at time zero
C_∞	Concentration of reactant/microorganisms, enzymes, or quality attributes at infinite time
C_1	Concentration or physical attributes (color, texture) at times t_1
C_2	Concentration or physical attributes (color, texture) at times t_2
CCP	Critical control points
D	Decimal reduction time (min)
D_1	Decimal reduction time at temperatures T_1 (min)
D_2	Decimal reduction time at temperatures T_2 (min)
D_{ref}	Decimal reduction time at reference temperature (min)
E_a	Process activation energy, kJ mol^{-1}
ΔE	Total color difference value
f	Fractional conversion
F_o	Process lethality (min)
FDA	Food and Drug Administration
GMP	Good manufacturing practices
HACCP	Hazard analysis and critical control points
HTST	High-temperature short time
ISO	International Organization of Standardization
k	Reaction rate constant, (concentration) $^{1-n}$/(time)
k_c	Degradation rate constant for a chemical constituent or physical attribute
k_{ref}	Reaction rate constant at reference temperature (concentration) $^{1-n}$/(time)
K	Consistency index (Pa s^n)
L	Tristimulus color value at time t measures lightness
L_0	Tristimulus color value at time zero measures lightness
L_∞	Tristimulus color value at infinite time measures lightness
L_{ref}	Tristimulus color value measures lightness considers as reference
LPO	Lipoxygenase
LTLT	Low temperature for long time
MW	Microwave
n	Flow behavior index
N	Number of viable microorganisms at time t
N_0	Number of viable microorganisms at time zero
PHE	Plate heat exchanger
PO	Peroxidase
PPO	Polyphenol oxidase
QA	Quality assurance
R	Universal gas constant (8.314 J mol^{-1} K^{-1})
t	Time (s)

T	Temperature of study (K)
T_1,T_2	Temperature (K)
T_{ref}	Reference temperature (K)
TP_0	Texture property at zero time
TP_∞	Texture property at infinite time
TP	Texture property at zero time
TQM	Total quality management
TTC	Time–temperature combination
UHT	Ultraheat treatment
z	Thermal resistance constant (°C)
z_c	Thermal resistance at reference temperature (°C)

Greek

τ	Shear stress (Pa)
τ_o	Yield stress (Pa)
$\dot{\gamma}$	Shear rate (s^{-1})

REFERENCES

1. DB Lund. Effects of blanching, pasteurization, and sterilization on nutrients. In: RS Harris, E Karmas, eds. *Nutritional Evaluation of Food Processing*. New York, NY: AVI Publishing, 1975, pp. 205–212.
2. JA Abott. Quality measurement of fruits and vegetables. *Postharvest Biology and Technology* 15:207–225, 1999.
3. DB Lund. Design of thermal processes for maximizing nutrient retention. *Food Technology* 31:71–78, 1977.
4. I Alli. *Food Quality Assurance: Principles and Practices*. Boca Raton, FL: CRC Press, 2003, pp. 25–38.
5. BS Luh, CE Kean. Canning of vegetables. In: BS Luh, JG Woodroof, eds. *Commercial Vegetable Processing*. Westport, CT: AVI Publishing, 1978, pp. 163–262.
6. J Ahmed, K Dolan, D Mishra. Chemical reaction kinetics pertaining to foods. In: J Ahmed, MS Rahman, eds. *Handbook of Food Process Design*. Oxford, England: Wiley-Blackwell Publication. in press, 2012.
7. RT Toledo. *Fundamentals of Food Process Engineering*. New Delhi, India: CBS Publication, 1997.
8. A Van Loey, A Francis, M Hendrickx, G Maesman, P Tobback. Kinetics of quality changes of green pea and white beans during thermal processing. *Journal of Food Engineering* 24:361–377, 1995.
9. EF Morales-Blancas, VE Chandia, L Cisneros-Zevallos. Thermal inactivation kinetics of peroxidase and lipoxygenase from broccoli, green asparagus and carrots. *Journal of Food Science* 67:146–154, 2002.
10. CA Weemeas, V Ooms, AM Loey, ME Hendrickx. Chlorophyll degradation and color loss in heated broccoli juice. *Journal of Agricultural and Food Chemistry* 47:2404–2409, 1999.
11. J Ahmed, US Shivhare, P Singh. Color kinetics and rheology of coriander leaf puree and storage characteristics of paste. *Food Chemistry* 84:605–611, 2004.
12. S Gaur Rudra, BC Sarkar, US Shivhare. Thermal degradation kinetics of chlorophyll in pureed coriander leaves. *Food and Bioprocess Technology* 1:91–99, 2008.
13. S Gaur, US Shivhare, BC Sarkar, J Ahmed. Thermal chlorophyll degradation kinetics of mint leaves puree. *International Journal of Food Properties* 10:853–865, 2007.
14. AF Rizvi, CH Tong. Fractional conversion for determining texture degradation kinetics of vegetables. *Journal of Food Science* 62:1–7, 1997.
15. MD Alvarez, W Canet. A comparison of various rheological properties for modeling the kinetics of thermal softening of potato tissue (c.v. Monalisa) by water cooking and pressure steaming. *International Journal of Food Science and Technology* 37:41–55, 2002.
16. T Ohlsson. Temperature dependence of sensory quality changes during thermal processing. *Journal of Food Science* 45:836–839, 847, 1980.
17. R Blasco, MJ Esteve, A Frigola, M Rodrigo. Ascorbic acid degradation kinetics in mushrooms in a high-temperature short-time process controlled by a thermoresistometer. *Lebensmittel-Wissenschaft und-Technologie* 37:171–175, 2004.

18. WC Ko, WC Liu, YT Tsang, CW Hsieh. Kinetics of winter mushrooms (*Flammulina velutipse*) microstructure and quality changes during thermal processing. *Journal of Food Engineering* 81:587–598, 2007.
19. D Dutta, A Dutta, U Raychaudhuri, R Chakraborty. Rheological characteristics and thermal degradation kinetics of beta-carotene in pumpkin puree. *Journal of Food Engineering* 76:538–546, 2006.
20. D Rodrigo, A van Loey, M Hendrickx. Combined thermal and high pressure colour degradation of tomato puree and strawberry juice. *Journal of Food Engineering* 79:553–560, 2007.
21. L Rejano, AH Sanchez, A Montano, FJ Casado, A de Castro. Kinetics of heat penetration and textural changes in garlic during blanching. *Journal of Food Engineering* 78:465–471, 2007.
22. EM Goncalves, J Pinheiro, M Abreu, TRS Brandao, CLM Silva. Modelling the kinetics of peroxidase inactivation, colour and textural changes of pumpkin (*Cucurbita maxima L.*) during blanching. *Journal of Food Engineering* 81:693–701, 2007.
23 EM Goncalves, J Pinheiro, M Abreu, TRS Brandao, CLM Silva. Carrot (*Daucus carota L.*) peroxidase inactivation, phenolic content and physical changes kinetics due to blanching. *Journal of Food Engineering* 97:574–581, 2010.
24. SD Holdsworth. *Thermal Processing of Packaged Foods*. London, United Kingdom: Blackie Academic & Professional, 1997, pp. 1–181.
25. MG Corradini, M Peleg. A model of non-isothermal degradation of nutrients, pigments and enzymes. *Journal of the Science of Food Agriculture* 84:217–226, 2004.
26. MAJS van Boekel. On the use of the Weibull model to describe thermal inactivation of microbial vegetative cells. *International Journal of Food Microbiology* 72:159–172, 2002.
27. OR Fennema. *Principles of Food Science Part I*. New York, NY: Marcel Dekker, 1976.
28 S Ranganna. *Handbook of Analysis and Quality Control for Fruit and Vegetable Products*. New Delhi, India: Tata McGraw-Hill Publication, 1986.
29 MK Lenz, DB Lund. Experimental procedures for determining destruction kinetics of food components. *Food Technology* 34:51, 1980.
30. J Ahmed, US Shivhare, HS Ramaswamy. A fraction conversion kinetic model for thermal degradation of color in red chilli puree and paste. *Lebensmittel-Wissenschaft und-Technologie* 35:497–503, 2002.
31 O Levenspiel. *Chemical Reaction Engineering*. New Delhi, India: Willey Eastern Publication, 1974.
32. CLM Silva. Optimization of thermal processing conditions: Effect of reaction type of kinetics. Proceedings of the Conference on Modeling of Thermal Properties and Behavior of Foods during Production, Storage and Distribution, Prague, Czech Republic, June, 1997.
33. MN Ramesh, W Wolf, D Tevini, A Bognar. Microwave blanching of vegetables. *Journal of Food Science* 67:390–398, 2002.
34. D Knorr. Process assessment of high-pressure processing of foods: An overview. In: FAR Oliveira, JC Oliveria, ME Hendrickx, D Knorr, LGM Gorris, eds. *Processing Foods: Quality Optimization and Process Assessment*. Boca Raton, FL: CRC Press, 1999, pp. 249–267.
35. HM Hemeda, BP Klein. Inactivation and regeneration of peroxidase activity in vegetable extracts treated with anti-oxidants. *Journal of Food Science* 56:68–71, 1991.
36. N Yamauchi, Y Funamoto, M Shigyo. Peroxidase-mediated chlorophyll degradation in horticultural crops. *Phytochemistry Reviews* 3:221–228, 2004.
37. J Saraiva. Effect of environment aspects on enzyme heat stability and its application in the development of time temperature integrators. PhD thesis, Escola Superior de Biotecnologia, Universidade Católica Porteguesa, Portugal, 1994.
38. JB Adams. Regeneration and kinetics of peroxidase inactivation. *Food Chemistry* 60:201–206, 1997.
39. C Soysal, Z Söylemez. Kinetics and inactivation of carrot peroxidase by heat treatment. *Journal of Food Engineering* 68:349–356, 2005.
40. L Ludikhuyze, V Ooms, C Weemaes, M Hendrickx. Kinetic study of the irreversible thermal and pressure inactivation of myrosinase from broccoli (*Brassica oleracea L. Cv. Italica*). *Journal of Agriculture and Food Chemistry* 47:1794–1800, 1999.
41. CA Weemaes, LR Ludikhuyze, I van den Broeck, M Hendrickx. Kinetics of combined pressure-temperature inactivation of avocado polyphenol oxidase. *Biotechnology and Bioengineering* 60:292–300, 1998.
42. CS Chen, MC Wu. Kinetic models for thermal inactivation of multiple pectinesterases in citrus juices. *Journal of Food Science* 63:747–750, 1998.
43. H Fujikawa, T Itoh. Characteristics of a multicomponent first-order model for thermal inactivation of microorganisms and enzymes. *International Journal of Food Microbiology* 31:263–271, 1996.
44. JP Henley, A Sadana. Categorization of enzyme deactivation using a series-type mechanism. *Enzyme and Microbial Technology* 7:50–60, 1985.

45. W Weibull. A statistical distribution of wide applicability. *Journal of Applied Mechanics* 18:293–297, 1951.
46. M Ladero, A Santos, F Garcia-Ochoa. Kinetic modeling of the thermal inactivation of and industrial β-galactosidase from *Kluyveromyces fragilis*. *Enzyme and Microbial Technology* 38:1–9, 2006.
47. L Vámos Vigyázó. Polyphenol oxidase and peroxidase in fruits and vegetables. *Critical Reviews in Food Science and Nutrition* 15:49–127, 1981.
48. WR Lencki, J Arul, RJ Neufeld. Effect of subunit dissociation, denaturation, aggregation, coagulation and decomposition on enzyme inactivation kinetics. I. First-order behavior. *Biotechnology and Bioengineering* 40:1421–1426, 1992.
49. WR Lencki, J Arul, RJ Neufeld. Effect of subunit dissociation, denaturation, aggregation, coagulation and decomposition on enzyme inactivation kinetics. II. Biphasic and grace period behavior. *Biotechnology and Bioengineering* 40: 1427–1434, 1992.
50. CR Stumbo, RS Purohit, TV Ramkrishna, DA Evans, FJ Francis. *Handbook of Lethality Guides for Low-Acid Canned Foods*. Boca Raton, FL: CRC Press, 1983, pp. 132–136.
51. GJ Banwart. *Basic Food Microbiology*, 2nd edn. New York, NY: AVI Publishing, 1989, pp. 23–31.
52. C Silva, M Hendrickx, F Oliver, P Tobback. Optimal sterilization temperatures for conduction heating foods considering finite surface heat transfer coefficients. *Journal of Food Science* 57:743–748, 1992.
53. HS Ramaswamy, C Abbatemarco, SS Sablani. Heat transfer rates in a canned food model as influenced by processing in an end-over-end rotary steam/air retort. *Journal of Food Processing and Preservation* 17:269–286, 1993.
54. RP Singh, PE Nelson. *Advances in Aseptic Processing Techniques*. London, United Kingdom: Elsevier Science Publishers Ltd., 1992, pp. 65–68.
55. A Kramer, BA Twigg. *Quality Control for the Food Industry*, 3rd edn. New York, NY: Van Nostrand Reinhold Co., 1970, pp. 15–32.
56. RL Shewfelt. What is quality? *Postharvest Biology and Technology* 15:197–200, 1999.
57. GS Birth. How light interacts with foods. In: JJ Gaffney Jr., ed. *Quality Detection in Foods*. St. Joseph, MI: American Society for Agricultural Engineering, 1976, pp. 6–11.
58. MC Bourne. *Food Texture and Viscosity: Concept and Measurement*, 2nd edn. San Diego, CA: Academic Press, 2002, pp. 1–12.
59. S Akterian, C Smout, P Tobback, M Hendrickx. Identification of CCP-limit levels for HACCP systems of thermal sterilization process using sensitivity functions. Proceeding of the Conference on Modeling of Thermal Properties and Behavior of Foods during Production, Storage and Distribution, pp. 6–12. Prague, Czech Republic, June, 1997.
60. JR Whitaker. *Principles of Enzymology for the Food Sciences*. New York, NY: Marcel Dekker Inc., 1994, pp. 615–621.
61. L Lemmens, E Tibäck, C Svelander, C Smout, L Ahrné, M Langton, M Alminger, A Van Loey, M Hendrickx. Thermal pretreatments of carrot pieces using different heating techniques: Effect on quality related aspects. *Innovative Food Science and Emerging Technologies* 10:522–529, 2009.
62. PS Negi, SK Roy. Effect of blanching and drying methods on β-carotene, ascorbic acid and chlorophyll retention of leafy vegetables. *Lebensmittel-Wissenschaft und-Technologie* 33:295–298, 2000.
63. AJ Speak, SS Speak, WHP Schreurs. Total carotenoid and β-carotene contents of Thai vegetables and the effect of processing. *Food Chemistry* 27:245–257, 1988.
64. M Bajaj, P Agarwal, KS Minhas, JS Sidhu. Effect of blanching treatment on quality characteristics of dehydrated fenugreek leaves. *Journal of Food Science and Technology* 30:196–198, 1993.
65. M Ihl, M Monslaves, V Bifani. Chlorophyllase inactivation as a measure of blanching efficacy and color retention of Artichokes (*Cynara scolymus L.*). *Lebensmittel-Wissenschaft und-Technologie* 31:50–56, 1998.
66. DC Williams, MH Lim, AO Chen, RM Pangborn, JR Whitaker. Blanching of vegetables for freezing-which indicator enzyme to choose. *Food Technology* 40:130–140, 1986.
67. B Günes, A Bayindirh. Peroxidase and lipoxygenase inactivation during blanching of green beans, green peas, and carrots. *Lebensmittel-Wissenschaft und-Technologie* 26:406–410, 1993.
68. H Böttcher. Enzyme activity and quality of frozen vegetables. I. Remaining residual activity of peroxidase. *Nahrung* 19:173–177, 1975.
69. DM Barett, C Theerakulkait. Quality indicators in blanched, frozen, stored vegetables. *Food Technology* 49:62–65, 1995.
70. MV Romero, DM Barrett. Rapid methods of lipoxygenase assay in sweet corn. *Journal of Food Science* 62:696–700, 1997.

71. RL Garrote, ER Silva, RA Bertone, RD Roa. Predicting the end point of a blanching process. *Lebensmittel-Wissenschaft und-Technologie* 37: 309–315, 2004.
72. SS Roy, TA Taylor, HL Kramer. Texture and histological structure of carrots as affected by blanching and freezing condition. *Journal of Food Science* 66:176–180, 2001.
73. C Severini, A Baiano, T De Pilli, R Romaniello, A Derossi. Prevention of enzymatic browning in sliced potatoes by blanching in boiling saline solutions. *Lebensmittel-Wissenschaft und-Technologie* 36:657–665, 2003
74. NA Saito. Thermophilic extracellular α-amylase from *Bacillus licheniformis*. *Archives of Biochemistry and Biophysics* 155:290–295, 1973.
75. SJ Tomazic, AM Klibanov. Mechanisms of irreversible thermal inactivation of Bacillus α-amylases. *The Journal of Biological Chemistry* 263:3086–3091, 1988.
76. S De Cordt, K Vanhof, J Hu, G Maesmans, M Hendrickx, P Tobback. Thermostability of soluble and immobilized a-amylase from Bacillus licheniformis. *Biotechnology and Bioengineering* 40:396–402, 1992.
77. C Ganthavorn, CW Nagel, JR Powers. Thermal inactivation of asparagus lipoxygenase and peroxidase. *Journal of Food Science* 56:47–49, 1991.
78. C Weemeas, P Rubens, S De Cordt, L Ludikhuyze, I Van Den Broeck, M Hendrickx, K Hermans, P Tobback. Temperature sensitivity and pressure resistance of mushroom polyphenoloxidase. *Journal of Food Science* 62:261–266, 1997.
79. P Lopez, J Burgos. Peroxidase stability and reactivation after heat treatment and manothermosonication. *Journal of Food Science* 60:451–455, 1995.
80. LMM Tijskens, EPHM Schijvens, ESA Bickmann. Modeling the change in color of broccoli and green beans during blanching. *Innovative Food Science and Emerging Technologies* 2:303–313, 2001.
81. ML Gunawan, SA Barringer. Green color degradation of blanched broccoli (Brassica Oleracea) due to acid and microbial growth. *Journal of Food Processing and Preservation* 24:253–263, 2000.
82. JY Song, GH An, CJ Kim. Color, texture, nutrient content, and sensory values of vegetable soybeans (Glycine max (L) Merrill) as affected by blanching. *Food Chemistry* 83:69–74, 2003.
83. A Quintero-Ramos, MC Bourne, J Barnard, A Anzaldua-Moralses. Optimization of low temperature blanching of frozen Jalapeno pepper (*Capsicum annuum*) using response surface methodology. *Journal of Food Science* 63:519–522, 1998.
84. MS Brewer, BP Klein, BK Rastogi, AK Perry. Microwave blanching effects on chemical, sensory and color characteristics of frozen green beans. *Journal of Food Quality* 17:245–259, 1994.
85. M Ihl, C Shene, E Scheuermann, V Bifani. Correlation for pigment content through colour determination using tristimulus values in a green leafy vegetable. *Journal of the Science of Food and Agriculture* 66:527–531, 1994.
86. CA Svelander, EA Tiback, LM Ahme, MIBC Langton, USO Svanberg, MAG Alminger. Processing of tomato: Impact on *in vitro* bioaccessibility of lycopene and textural properties. *Journal of the Science of Food and Agriculture* 90:1665–1672, 2010.
87. F Icier, H Yildiz, T Baysal. Peroxidase inactivation and colour changes during ohmic blanching of pea puree. *Journal of Food Engineering* 74:424–429, 2006.
88. CY Lee, MC Bourne, JP Van Buren. Effect of blanching treatments on the firmness of carrots. *Journal of Food Science* 44:615–616, 1979.
89. MC Bourne. Applications of chemical kinetic theory to the rate of thermal softening of vegetables tissue. In: JJ Jen, ed. *Quality Factors of Fruit and Vegetables*. Washington, DC: American Chemical Society, 1989, pp. 98–123.
90. VD Truong, WM Walter Jr., KL Bett. Textural properties and sensory quality of processed sweet potato as affected by low temperature blanching. *Journal of Food Science* 63:739–743, 1998.
91. A Anderson, V Gekas, I Lind, F Oliveria, R Oste. Effect of preheating on potato texture. *Critical Reviews of Food Science and Nutrition* 34:229–252, 1994.
92. DW Stanley, MC Bourne, AP Stone, WV Wismer. Low temperature blanching effects on chemistry, firmness and structure of canned green beans and carrots. *Journal of Food Science* 60:327–333, 1995.
93. JB Adams, A Robertson. Instrumental methods of quality assessment: Texture, Technical Memo No. 449. Chipping Campden, United Kingdom: Campden Food Research Association, 1987.
94. KH Moledine, M Hayder, B Oor Aikul, D Hadziyev. Pectin changes in the pre-cooking step of dehydrated mashed potato production. *Journal of the Science of Food and Agriculture* 32:1091–1102, 1981.
95. T Sajjaanantakul, JP Van Buren, DL Downing. Effect of methyl ester content on heat degradation of chelator-soluble carrot pectin. *Journal of Food Science* 54:1272–1277, 1989.

96. D Plat, N Ben-Shalom, A Levi. Changes in pectic substances in carrots during dehydration with and without blanching. *Food Chemistry* 39:1–12, 1991.
97. LC Greve, RN McArdle, JR Gohlke, JM Labavitch. Impact of heating on carrot firmness: Changes in cell wall components. *Journal of Agricultural and Food Chemistry* 42:2900–2906, 1994.
98. M Fuchigami, K Miyazaki, N Hyakumoto. Frozen carrots texture and pectic components as affected by low-temperature blanching and quick freezing. *Journal of Food Science* 60:132–136, 1995.
99. LG Bartolome, JE Hoff. Firming of potatoes: Biochemical effects of preheating. *Journal of Agricultural and Food Chemistry* 20:266–270, 1972.
100. MJH Keijbets, W Pilnik. Elimination of pectin in the presence of anions and cations. *Carbohydrate Research* 33:359–362, 1974.
101. JP Van Buren. The chemistry of texture in fruits and vegetables. *Journal of Texture Studies* 10:1–23, 1979.
102. RF McFeeters. Changes in pectin and cellulose during processing. In: T Richardson, JW Finley, eds. *Chemical Changes in Food during Processing*. Westport, CT: AVI Publishing, 1985, pp. 347–372.
103. C Arroqui, TR Rumsey, A Lopez, P Virseda. Losses by diffusion of ascorbic acid during recycled water blanching of potato tissue, *Journal of Food Engineering* 52:25–30, 2002.
104. LA Howard, AD Wong, AK Perry, BP Klein. β-carotene and ascorbic acid retention in fresh and processed vegetables. *Journal of Food Science* 64:929–936, 1999.
105. O Fennema. *Principles of Food Science: Part II*. New York, NY: Marcel Dekker, 1975, pp. 133–170.
106. M Gomez. Carotene content of some green leafy vegetables of Kenya and effects of dehydration and storage on carotene retention. *Journal of Plant Food* 3:231–244, 1981.
107. J Ahmed, US Shivhare. Thermal kinetics of color change, rheology and storage characteristics of garlic puree/paste. *Journal of Food Science* 66:754–757, 2001.
108. J Ahmed, US Shivhare, GSV Raghavan. Rheological characteristics and kinetics of color degradation of green chilli puree. *Journal of Food Engineering* 44:239–244, 2000.
109. FJ Francis. Pigments and other colorants. In: OR Fennama, ed. *Food Chemistry*, 2nd edn. New York, NY: Marcel Dekker Publication, 1985, pp. 545–585.
110. JTO Kirk, RAE Tilney-Basset. *The Plastids*, 2nd edn. Amsterdam, the Netherlands: Elsevier Publications, 1978.
111. JH von Elbe. Chemical changes in plant and animal pigments during food processing. In: OR Fennema, C Wei-Hsien, L Cheng-Yi, eds. *Role of Chemistry in Quality of Processed Foods*. Westport, CT: Westport Publication, 1986.
112. YD Lin, FM Clydesdale, FJ Francis. Organic acid profiles of thermally processed spinach puree. *Journal of Food Science* 35:641, 1970.
113. SJ Schwartz, JH von-Elbe. Kinetics of chlorophyll degradation to pyropheophytin in vegetables. *Journal of Food Science* 48:1303–1306, 1983.
114. T Rocha, A Lebart, C Marty-Audouin. Effect of pre-treatments and drying conditions on drying rate and color retention of basil. *Lebensmittel-Wissenschaft und-Technologie* 26:456–463, 1993.
115. JA Steet, CH Tong. Degradation kinetics of green color and chlorophyll in peas by colorimetry and HPLC. *Journal of Food Science* 61:924–927, 931, 1996.
116. CT Tan, FJ Francis. Effect of processing temperature on pigments and color of spinach. *Journal of Food Science* 27:232–241, 1962.
117. GJ Malecki. Retains vegetable green. *Journal of Food Engineering* 37:93, 1964.
118. FM Clydesdale, FJ Francis. Chlorophyll changes in thermally processed spinach as influenced by enzyme conversion and pH adjustment. *Food Technology* 22:135–138, 1968.
119. MS Eheart, D Odland. Use of ammonium compounds for chlorophyll retention in frozen green vegetables. *Journal of Food Science* 38:202, 1973.
120. M Rodrigo, C Calvo, C Sanchez, T Rodrigo, A Martinez. Quality of canned mushroom acidified with glucano-δ-lactone. *International Journal of Food Science and Technology* 34:161–166, 1999.
121. MW Hoover, DD Mason. Evaluation of color in sweet potato puree with the Hunter color difference meter. *Food Technology* 15:299–303, 1961.
122. FJ Francis. Relationship between flesh color and pigment content in squash. *Proceedings of American Society of Horticultural Science* 69:296, 1962.
123. EM Ahmed, LE Scott. A rapid method for the estimation of carotenoid content in sweet potato roots. *Proceedings of American Society of Horticultural Science* 80:497, 1962.
124. HJ Gold, KG Weckel. Degradation of chlorophyll to pheophytin during sterilization of canned green peas by heat. *Food Technology* 13:281–286, 1959.
125. SM Gupte, HM El-Bisi, FJ Francis. Kinetics of thermal degradation of chlorophyll in spinach puree. *Journal of Food Science* 29:379–382, 1964.

126. STAR Kajuna, WK Bilanski, GS Mittal. Color changes in bananas and plantains during storage. *Journal of Food Processing and Preservation* 22:27–40, 1998.
127. S Shin, SR Bhowmik. Thermal kinetics of colour changes in pea puree. *Journal of Food Engineering* 24:77–86, 1995.
128. FL Canjura, SJ Schwartz, RV Nunes. Degradation kinetics of chlorophylls and chlorophyllides. *Journal of Food Science* 56:1639–1643, 1991.
129. K Hayakawa, GE Timbers. Influence of heat treatment on the quality of vegetables: Changes in visual green colour. *Journal of Food Science* 42:778–791, 1977.
130. J Ahmed, A Kaur, US Shivhare. Color degradation kinetics of spinach, mustard leaves and mixed puree. *Journal of Food Science* 67:1088–1091, 2002.
131. C Smout, NE Banadda, AM Van Loey, MEG Hendrickx. Non-uniformity in lethality and quality in thermal process optimization: A case study on color degradation of green peas. *Journal of Food Science* 68:545–550, 2003.
132. JN Yeatman. Tomato products: Read tomato red? *Food Technology* 23:618–627, 1969.
133. KW Young, KJ Whittle. Colour measurement of fish minces using Hunter *L, a, b* values. *Journal of Science of Food and Agriculture* 36:383–392, 1985.
134. JB Hutchings. *Food Colour and Appearance*. London, U.K: Blackie Academic and Professional Publication, 1994.
135. KE Nanke, JG Sebranek, DG Olson. Color characteristics of irradiated aerobically packaged pork, beef and turkey. *Journal of Food Science* 64:272–278, 1999.
136. AC Little. Evaluation of single-number expressions of color difference. *Journal of Optical Society of America* 53:293–296, 1963.
137. MA Rao, DB Lund. Kinetics of thermal softening of foods: A review. *Journal of Food Processing and Preservations* 6:133–153, 1986.
138. YT Huang, M Bourne. Kinetics of thermal softening of vegetables. *Journal of Texture Studies* 14:1–9, 1983.
139. C Rodrigo, A Alavarruiz, A Martinez, A Frigola, M Rodrigo. High-temperature short-time inactivation of peroxidase by direct heating with a five channel computer controlled thermoresistor. *Journal of Food Protection* 60:967–974, 1997.
140. A Van Loey, A Fransis, M Hendrickx, G Maesmans, P Tobback. Kinetic of thermal softening of white beans evaluated by a sensory panel and FMC tenderometer. *Journal of Food Processing and Preservation* 18:407–412, 1994.
141. TR Stoneham, DB Lund, CH Tong. The use of fractional conversion technique to investigate the effects of testing parameters on texture degradation kinetics. *Journal of Food Science* 65:968–973, 2000.
142. J Loh, WM Breene. The thermal fracturability loss of edible plant tissue: Pattern and within-species variation. *Journal of Texture Studies* 12:457–460, 1981.
143. K Paulus, I Saguy. Effect of heat treatment on the quality of cooked carrots. *Journal of Food Science* 45:239–244, 1980.
144. MF Kozempel. Modeling the kinetics of cooking and precooking potatoes. *Journal of Food Science* 53:753–755, 1988.
145. K Kiysohi, O Kelko, H Yoshihiko, S Kanichi, H Hideaki. Studies on cooking rate-equation of potato and sweet potato slices. *Journal of Faculty of Fish Animal Husbandry, Hiroshima University* 17:97–102, 1978.
146. T Harada, H Tirtohusodo, K Paulus. Influence of temperature and time on cooking kinetics of potatoes. *Journal of Food Science* 50:463–468, 1985.
147. GS Mittal. Thermal softening of potatoes and carrots. *Lebensmittel-Wissenschaft und-Technologie* 27:53–258, 1994.
148. C Rodrigo, A Mateu, A Alavarruiz, F Chinesta, M Rodrigo. Kinetic parameters for thermal degradation of green asparagus texture by unsteady-state method. *Journal of Food Science* 63:126–129, 1998.
149. D Lund. Effects of heat processing on nutrients. In: E Karmas, RS Harris, eds. *Nutritional Evaluation of Food Processing*, New York, NY: AVI Publication, 1987, pp. 319–354.
150. J Ahmed. Rheological behaviour and colour changes of ginger paste during storage. *International Journal Food Science and Technology* 39:325–330, 2004.
151. DB Lund. Application of optimization in heat processing. *Food Technology* 2:97–99, 1982.
152. JR Johnson, RJ Braddock, CS Chen. Kinetics of ascorbic acid loss and nonenzymatic browning in orange juice serum: Experimental rate constants. *Journal of Food Science* 60:502–505, 1995.
153. I Saguy, IJ Kopelman, S Mizrah. Simulation of ascorbic acid stability during heat processing and concentration of grapefruit juice. *Journal of Food Process Engineering* 2:213–225, 1978.

154. AA Teixeira, JR Dixon, JW Zahradnik, GE Zinsmeister. Computer optimization of nutrient retention in the thermal processing of conduction-heated foods. *Food Technology* 23:845–850, 1969.
155. Anon. The role of food safety objectives in the management of the microbiological safety of food according to Codex documents. Document prepared for the Codex Committee on Food Hygiene, ICMSF International Commission on the Microbiological Specifications for Foods, March 2001.
156. M van der Spiegel, PA Luning, GW Ziggers, WMF Jongen. Towards a conceptual model to measure effectiveness of food quality systems. *Trends in Food Science and Technology* 14:424–431, 2003.

16 Thermal Processing of Fruits and Fruit Juices

Catherine M.G.C. Renard and Jean-François Maingonnat

CONTENTS

16.1 INTRODUCTION

The proportion of fruits that are eaten after processing has greatly increased in the last decade, with in particular a remarkable expansion and diversification of fruit juices. Canning comes a far second, while fresh-cut (outside the scope of this chapter) and purĕed products have also received renewed interest. There is a steady industrial demand for cooked chilled or frozen intermediate products, most of which are further processed, for example, in jams or frozen desserts. Concentrated juices are a commodity on the world market. As most fruits are of relatively low pH (lower than 4.5), the microbial safety of fruit products only requires mild heat treatments (pasteurization) even for long-term stability at room temperature. The limiting factors in thermal treatments of fruits are therefore more linked to physicochemical properties. The first limiting factor often is inactivation of endogenous enzymes, some of which can be relatively heat resistant. The typical example is pectin methylesterase, which is linked to cloud instability in fruit juices. The thermal treatments applied to jams are also conditioned by the physical properties of the final products and the need to reach high sugar concentrations while ensuring pectin homogeneous distribution throughout the product. In fruit juice concentrates or to a lesser extent purèed products, cumulative thermal treatments will be dominated by the evaporation steps.

At the same time, new technologies are now coming to market, either as alternative heating systems such as ohmic, radio frequency heating, or using additional physical phenomena for microbial inactivation (high-pressure processing, pulsed electric field). Juices are still the main area of interest for these new technologies, with again a difficulty to ensure sufficient enzyme inactivation for physicochemical stability. Much less experience has been accumulated on these techniques, and in spite of numerous publications in the last few years [1–4], more experimental approaches are still required at all levels. Many recent publications compare these "nonthermal" technologies with more classical pasteurization, either conventional or high temperature–short time. Major interests and innovations can lie in exploiting the potential for synergisms between heat and these physical phenomena [5], a field still scarcely explored.

Quality aspects of fruits concern their sensory properties (aroma, taste, texture, and color) and nutritional qualities. Texture and color, being amenable to physicochemical methods of (relatively) high throughput and normalized values have been the most extensively studied. Softening is an expected result of heat treatment, which can be used, for example, prior to sieving for purèed products. Only moderate heat treatments are usually applied to fruits, some of them being very sensitive to extensive texture degradation and tissue disintegration; this sensitivity is species dependent and variety dependent [6,7]. Few publications have focused on the impact of heat treatments on volatiles in fruit juices, notably orange and apple [8–10].

From a nutritional point of view, most fruits contribute primarily to sugars intake, and their interest further resides in food volume and water intake. A few fruits can contain significant amounts of starch (unripe banana or apple) or fat (avocado or olives). In terms of micronutrients, fruits and fruit products are the main source of vitamin C. They contribute to potassium and calcium ions, fibers, and provitamin A. Most data available deal with losses in vitamin C in fruit juices as a function of treatment and storage. Vitamin C is also frequently used as a processing aid, for inhibition of enzymatic browning (notably for cloudy juices and purèes); and the concentrations that are added can be much higher than those of the endogenous vitamin C.

The objective of this chapter is to provide an overview of the recent results and advances on the impact of processing on safety and quality of heat-processed fruits, and of the models that have been developed to deal with microbial safety insurance as well as nutritional or physicochemical characteristics.

16.2 MICROBIAL CONSIDERATIONS IN FRUITS THERMAL PROCESSING

The surfaces of fruits contain diverse microorganisms that are normal microflora plus the microorganisms inoculated during processing [11] and recontamination on the process lines. One of the most convenient ways to limit the microbial risk is to increase the food temperature up to a lethal value for the microorganisms. This procedure is pasteurization if the vegetative forms are destroyed and sterilization if both vegetative and spores are destroyed. Recently, minimally processed fruits and vegetables [12] or cooked and chilled [13] foods containing fruits are frequently proposed to the consumers, and the microbial safety of such products is studied.

The food products are divided into two categories based on pH: acid foods ($pH < 4.5$) and nonacid foods ($pH > 4.5$). The acid foods are less critical because *Clostridium botulinum,* a strictly anaerobic spore-forming (thus heat-resistant) and toxin-producing bacteria, cannot grow at $pH < 4.5$. Most of the fruits are acid foods, the pH varying from 1.8 for limes, 3.5 for apples, to 6.5 for melons.

The main targets of thermal treatment are reduction of microorganisms and enzyme inactivation. The effects of thermal treatment are evaluated by the microorganism destruction, the enzyme inactivation, the quality loss, etc. These food component modifications are generally modeled with the reactions kinetic concept [14].

The basis of microbial inactivation has been the assumption that the microbial mortality at constant temperature obeys the following irreversible first-order kinetics:

$$\frac{dN(t)}{dt} = -k(\vartheta)N(t) \tag{16.1}$$

where

$N(t)$ is the number of microorganism at the time t

$k(\vartheta)$ is the temperature ϑ-dependent inactivation "rate constant"

The solution of this first-order kinetic reaction is exponential and generally translated in terms of the well-known "*D* value," which is the time in minutes to reduce the microbial population by one-log cycle (base 10):

$$\log_{10}\left(\frac{N(t)}{N_0}\right) = -D(\vartheta)t \quad \text{with } D(\vartheta) = \frac{\ln(10)}{k(\vartheta)} \tag{16.2}$$

where N_0 is the initial number of microorganisms.

The temperature influence on the *D* value is traditionally assumed to be a log-linear relation, so that a plot of the *D* value versus temperature in semilogarithmic coordinates is a straight line. The popular *z* value is defined as the temperature change required to change the *D* value by a factor of 10, as given in the following:

$$\log_{10}\left(\frac{D(\vartheta)}{D_{ref}}\right) = \frac{\vartheta_{ref} - \vartheta}{z} \tag{16.3}$$

For the pathogenic bacteria (*C. botulinum*, *Bacillus cereus*, *Bacillus subtilis*, and *Clostridium perfringens*), the *z* values vary generally between 8°C and 11°C.

The reference temperatures are chosen as a function of the process design: 121.1°C for sterilization, 70°C for pasteurization; for more thermosensitive microorganisms, this reference temperature can be lower.

A very common concept is the process thermal lethality *F*, which is the product of the *D* value at the chosen temperature by the number of decimal reduction required. For example, the "12D" concept applied to *C. botulinum* at 121.1°C leads to a $12\times0.21=2.52$ min *F* value. To achieve a predetermined *F* value, the thermal history of the process should be taken into account:

$$F = \int_0^t 10^{(\vartheta-\vartheta_{ref})/zt} \tag{16.4}$$

The first-order kinetic microorganism destruction is inappropriate when the isothermal survival semilogarithmic curves are nonlinear. Such nonlinearity has been observed for different microorganisms and spores [15], and alternative models are proposed [16]. One of the simplest is derived from the cumulative Weibull distribution and is written as follows [17]:

$$\log_{10}\left(\frac{N(t)}{N_0}\right) = -\left(\frac{t}{b(\vartheta)}\right)^{\beta} \tag{16.5}$$

where $b(\vartheta)$ and β are the temperature-dependent parameters of the model. When $\beta=1$, the model is the first-order kinetic (Equation 16.2), $\beta<1$ corresponds to concave upward survival curves, and $\beta>1$ corresponds to concave downward curves [18].

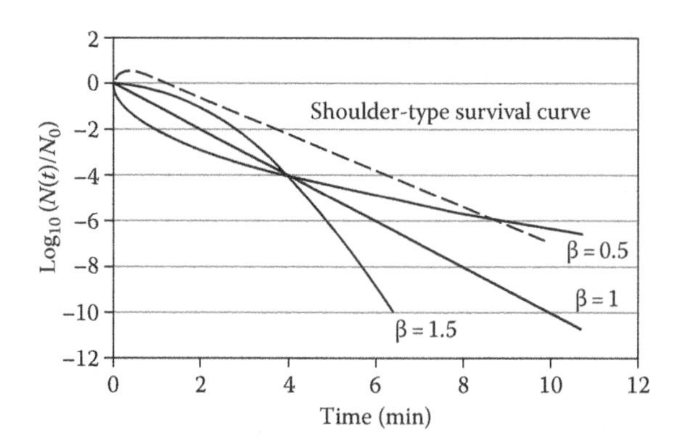

FIGURE 16.1 Simulated survival curves showing the effect of Weibull (Equation 16.5) parameter β and a shoulder-type survival curve.

The survival curves can also exhibit some "shoulders" for short treatment times [19] and tailing effects at long treatment time or a sigmoid form [20]. The shapes of these different survival curves are schematically presented in Figure 16.1, and a software tool [21] is also available to fit the different types of survival curves.

Another type of microorganism destruction model is the so-called biphasic model [22] as defined in the following:

$$\frac{N(t)}{N_0} = q\exp(-k_1\ t) + (1-q)\exp(-k_2\ t) \tag{16.6}$$

where q is a partition factor. Equation 16.6 has been applied for thermal inactivation of *C. botulinum* [23] or high-pressure effect on *Escherichia coli* in liquid whole egg [24].

The fittings of different survival curves and their influence on the heating processes have been recently reviewed [25]. Although the classical first-order kinetic model is frequently used, an optimization (quality assessments, energy saving, etc.) of the thermal processes [26] should take into account these new types of modeling.

Some thermophilic bacteria are growing even in acid foods such as fruit concentrates [27] and are encountered in process lines or orchard soils [28–30]. For example, the "*D* value" of *Acyclobacillus* in concentrated lemon product (50° brix) at 86°C may reach 69 min [31]. The fruit beverage industry applies a hot-filled-hold pasteurization process, where the product is held at 86°C–96°C for ~2 min. It is clear that this thermal treatment does not eliminate *Acyclobacillus* [29].

The majority of mold species present low heat resistance, having their vegetative structure (conidia and hyphae) easily destroyed by heat. However, some thermal resistant molds are reported in fruit products [32–34]: *Byssochlamys nivea* ($D_{85°C}$ value 34.6 min, z value 6.4°C in strawberry pulp), *Neosartorya fischeri* ($D_{85°C}$ value 19.6–29.5 min, z value 9°C in pineapple juice), and *Talaromyces flavus* ($D_{85°C}$ value 3.3 min, z value 8.2°C in strawberry pulp) [32]. Besides bacterial problems, the presence of mycotoxins represents another hazard to food safety, and patulin [35] or ochratoxin A are the most important fruit juice-associated mycotoxins. The microbial quality and safety of fruit juices was reviewed recently [36]; the authors [36] highlight the necessity of more severe heat treatments for pasteurization of exotic fruits juices, which are less acid (pH 4.5–6.8) than the classic apples or citrus (pH 1.8–4.5). In the context of an increasing demand in "freshness" and new tastes, the actual studies should consider the microbial, nutritional, and sensory aspects jointly in order to improve the quality and safety of fruit juices.

16.3 QUALITY ASSESSMENTS CONSIDERATIONS

The International Organization of Standardization defines "quality" as "the degree to which a set of inherent characteristics fulfils to requirements." This definition is clearly in close relation to consumers who can appreciate the quality of a food product as bad, poor, good, or excellent. The consumer assessment is generally subjective, but the food producers and manufacturers have to identify measurable quality attributes and have to estimate their modifications during the food storage and processing. The quality attributes generally studied for fruits are texture, color, enzyme degradation, and the nutrient contents.

The thermal treatments of foods lead to substantial modifications of the final product qualities. Although microbial and chemical safety is prevalent during food processing, the sensory attributes are certainly the first criteria for acceptance and the major cause of rejection. The attributes such as color, shape, or texture are very important for both the first acceptance and regular purchasing of the food products. Most of the studies concerning the influence of thermal treatment on quality attributes are carried out in isothermal conditions, and the kinetics of quality parameters obey an irreversible first-order reaction kinetic similar to Equation 16.1.

As the food quality attribute levels are heterogeneous from a fruit to another, the fractional conversion model is frequently used [37] associated with a first-order kinetic reaction (Equation 16.7) in which $C(t)$ is the quality attribute level at a time t and C_0 and C_∞ are, respectively, the attribute level at the beginning of the treatment and after a long time. The quality attribute level after a long time C_∞ is often temperature dependent and indicates an irreversible structure breakage of the fruit matrix. The fractional conversion model is given by

$$\frac{C(t)-C_\infty}{C_0-C_\infty}=e^{-k(\vartheta)t} \tag{16.7}$$

The kinetic parameters are greatly influenced by the temperature and the most frequently used theory in the area of food engineering is the following Arrhenius equation:

$$\frac{k}{k_{ref}}=\exp\left[\frac{E_a}{R}\left(\frac{1}{T_{ref}}-\frac{1}{T}\right)\right] \tag{16.8}$$

where T is the absolute temperature, R is the universal gas constant (8.134 J mol^{-1} K^{-1}), and E_a (kJ mol^{-1}) is the activation energy.

It can be noted that the previously mentioned z value for microorganism's destruction can be evaluated in terms of an activation energy. The activation energy of microorganism destruction varies from 8 to 15 kJ mol^{-1}, and the quality loss activation energy varies between 30 and 100 kJ mol^{-1}. Therefore, a high-temperature short-time (HTST) process can be applied to increase the product quality while ensure the microbial safety.

The quality attribute loss is also studied during the storage of canned foods, for example, vitamin C degradation in canned pineapple slices [38] obeys a first-order kinetics reaction with $k_{18.3°C}=26.975\times10^{-8}$ min^{-1} and an activation energy $E_a=11.5$ kJ mol^{-1}.

16.4 APPLICATIONS OF THERMAL PROCESSING

Thermal treatments are omnipresent in food processing: blanching, pasteurization, sterilization, cooking, drying, frying, microwave or radiofrequency or ohmic heating, etc. Moreover, a final food product is subjected to different heat treatments including the domestic cooking or cold storage. Thermal treatments are also used for disinfecting whole fruits such as mango [39] in warm water

(45°C) during 10–40 min. In this chapter, we will overview the main thermal treatments of fruits: blanching and pasteurization/sterilization.

16.4.1 Blanching

Blanching is a thermal treatment in hot water or steam aimed to inactivate oxidative enzymes naturally present in fruits and responsible for off-flavors, color change, and chemical reactions during the further processing steps and storage. This first thermal step is very important when the fruits are further processed. The influence of blanching on quality attributes [40] is generally evaluated together with the following process steps such as freezing, sterilizing, drying, and osmo-dehydration. This thermal step also helps to destroy microorganisms (bacteria, yeasts, and molds), prevent the flesh contamination when cutting [41], clean the fruits, brighten the color, and expel trapped air in the intercellular regions. The main enzymes affected by blanching are peroxidase, polyphenol oxidase, catalase, lipoxygenase, and chlorophylase; their thermal kinetic inactivation is documented [42].

Blanching is carried out by different means such as hot water, steam [43], high pressure, infrared-dry blanching [44,45], ohmic [46], fluidized bed with steam, whirling bed with a mix of hot air and steam [47], individual quick blanching system, combined with ozone [48,49]. Hot-water blanching is by far the most popular and commercially adopted process for its simplicity and economic reasons. The microbial quality of the blanching water must also be observed because the high temperature could select thermophilic bacteria. The main problem of water blanching is the leaching of important nutrients such as vitamins and pigments.

Due to the thermal diffusion in food matrices, blanching efficiency greatly depends on the size and shapes of the fruits. The thermal product conductivity governs the heat transfer in the matrix, and in the case of unsteady state, the thermal diffusivity is introduced in the heat transfer equations. The thermal diffusivities of fruits [50] vary between 1×10^{-7} and 1.8×10^{-7} m s^{-2}: the heat transfers are 1000 times faster than the mass transfer of micronutrients in fruits. It is also important to note that as the thermal transfer time depends on the fruit pieces as the size at the power 2, size reduction is interesting in terms of heat and mass transfers.

16.4.2 Pasteurization

Pasteurization is a mild treatment aiming to inactivate most of the enzymes and to inhibit the vegetative microorganism's cells, while sterilization eliminates also the spores. As mentioned previously, the thermal treatment depends on the microbial contamination. In the case of low-pH fruit products, a pasteurization process (reaching 85°C at the coldest point) allows a long shelf-life at room temperature. Different time and temperature combinations can be used.

For fruit juices, in the traditional practice, the juices are heated up to 60°C–75°C for 30 min, then filled at that temperature, closed and pasteurized at 84°C–88°C during 15–45 min depending of the size of the packaging. After this heat treatment, the products are cooled back to room temperature. HTST pasteurization is conducted at higher temperatures (>90°C) for shorter times. This can, for example, be carried out at 95°C–98°C for about 15–30 s for apple juice. Hot-fill-hold of containers ≥1 L with rapid closure gives temperatures >85°C due to thermal inertia. Ultrahigh temperature (UHT) is also applicable to juices and commonly used in larger plants.

Before this pasteurization operation, the previous operations are washing/blanching, crushing, enzyme maceration (40°C–50°C/1–2 h), pressing, centrifugation or filtration, optional second enzyme treatment, and deaeration. For citrus and especially orange juices, specific extractors are used to avoid contamination of the juice with peels, and the stabilization treatment must be applied within a few minutes of extraction to avoid pectin methylesterase action.

The fruit products are also processed in a more or less viscous liquid form or in a combination of liquid and solid phases (purées, preserves, jams, etc.). These can be treated in continuous equipment. For purées, two processes are commonly differentiated by the respective order of the sieving and cooking phases: sieving before cooking for cold-break products or after cooking for hot-break products. They result in different colors, textures, and compositions due to the possibility of enzyme activity in the cold-break products.

The process line consists of a holding tank, a mill, a heating zone, a sieve, a holding zone (in principle for a few minutes, up to 0.5 h), a pasteurization step followed by hot-fill or a cooling zone for aseptic packaging equipment. High temperatures are required during processing for fruit cooking and for adequate viscosities during pumping, so that additional heat treatments are limited. For example, for apple puree, a cooking time of 15 min at 85°C, holding tanks at 50°C–85°C, and a final pasteurization of 2–3 min at 90°C [51] then hot-fill can be sufficient to ensure stability at room temperature for months. Jams require higher temperatures for cooking and evaporation; thermal treatment under vacuum is preferred as it limits both temperature and product degradation during the concentration phase. However, temperatures must stay above pectin gelation, that is, >70°C–90°C, depending on the pectin grade.

The juices are frequently concentrated to be reused in fruit drinks. Concentration of juices is most commonly carried out by vacuum concentration, using efficient multieffect systems with the recovery of volatiles. The volatiles are later used in juice reconstitution for aroma restoration. The evaporators operate at temperatures <50°C. Redilution of the concentrated juices entails an additional pasteurization of the final product. Nonthermal alternatives (osmotic evaporation and membrane distillation) are gaining interest due to their more limited effect on juice volatiles.

16.4.3 Canned Food Products

Among the different sterilization processes, the canned food process has been one of the most widely used methods of food preservation during the twentieth century for ensuring nutritional well-being of populations. The advantages of canned foods are determinant: relative low price, storage at room temperature, acceptable nutrient contents, easy to use, and varied contents. This process consists of heating hermetically sealed food containers (cans, plastic bottles and containers, and flexible pouches) in pressurized retorts and imposing a prescribed time–temperature history [52].

In the case of fruits containing trapped air such as apricot [53], peaches [54], or plums [55], an exhausting procedure during 5–10 min at 90°C in a steam chamber is used to remove the air.

The sterilization process is achieved in batch retorts, with the cans being or not agitated, or in continuous retorts in which the cans are agitated. In these retorts, the cans are heated by steam or pressurized water, maintained at a high temperature until the whole can content is subjected to the predetermined time–temperature history. Very roughly, 15% of the process thermal efficiency is achieved during the heating phase, 15% during the cooling phase, and the rest is achieved during the holding phase. The complete time treatment depends on the size of cans and the contained product.

There are different types of retorts. The craterless retort consists in a tank in which the cans fall in hot water after which the top hatch is closed and vapor is injected up to the desired temperature. The cans are immobile and when the thermal treatment is achieved, warm water is injected in the retort, and the bottom hatch is opened to let the cans fall in the discharge cooling canal. The filling, heating, warming, and discharging are automatically carried out.

The hydrostatic sterilizers are so named because steam temperature is controlled hydrostatically by the height of the water leg, and they have self-contained structure that is often partly built outdoors. The hydrostatic sterilizers are made up of four chambers: a hydrostatic bring-up leg, a sterilizing steam section, a hydrostatic bring-down leg, and a cooling section. The cans are conveyed continuously through the different chambers by a continuous chain link, and the residence is adjusted by the speed of the conveyers.

The continuous rotary sterilizers are horizontal (indoor) rotary retorts in which the cans are conveyed by a reel while they rotate around their own axis by different means. The residence time in the sterilizer is controlled by the rotating speed of the reel. The most common systems require at least three shells in series to heat under pressure, cool under pressure, and cool at atmospheric pressure. These retorts generally accommodate a specific can size and are not flexible.

There are different types of discontinuous retorts in which the cans are contained in large baskets rotating on their own axis at different speeds or shaken at high speeds.

During the thermal treatment, the temperature in the cans is measured with special devices such as thermocouples and is modeled as a function of the operating conditions such as agitation, rotation, and transport and the contents, that is, type of fruit pieces, and syrup concentration. The heat transfer between the can wall and the content is the limiting factor of the heat transmission; the motion such as rotation and vibrations of the cans enhance this heat transfer.

The bases of the thermal treatment calculations are well documented in several books (e.g., see Ref. [52]) and are out of the scope of this chapter.

The sterilization temperature/times for canned fruits are typically 100°C/17–30 min for apricot, 93°C–95°C/25–30 min for blueberry, 100°C/20–350 min or 116°C/14–18 min for peaches, and 100°C/12 min for plums.

For high-quality orange juices, it is proposed to centrifuge the raw product and to apply two different thermal treatments for the pulp (85°C/15 s) to inactivate the pectin methylesterase and for the low pulp juice (65°C/15 s) to inactivate the microorganisms and then to blend these two products [56].

16.4.4 Continuous Sterilization and Aseptic Processing

In this technology, the product is convoyed by a volumetric pump through heat exchangers. For fruit juices, the applied process is 100°C–110°C for 0.5–1.5 min.

We can mention the problem of the particles moving in the suspending fluid. Another problem, actually not well solved, concerns the local accumulation of particles in elbows, for example, leading to plugging the process line. The treatment time is adjusted for the fastest particle in the line and the product must be poured in presterilized containers in an aseptic environment before the containers are hermetically sealed. The main problems are often encountered in the cooling zone because the product viscosity can become very high and pasty.

This technology is very elegant and energy saving but is difficult to manage, especially in aseptic filling when the product contains particles and fibers. Along with the classical tubular or scraped surface heat exchangers used for pasty or particle content fluids, new technologies such as ohmic [57] or radio frequency [58] are available for the industry. The main advantage of these techniques consists in heating in the mass both suspending fluid and particles at the same time, if their electric conductivities or their dielectric losses are almost the same, thus enhancing the quality of the final products.

16.5 QUALITY ATTRIBUTE MODIFICATIONS DURING THERMAL PROCESSING OF FRUITS

16.5.1 Texture

Texture is a very important quality attribute and is closely linked to sensory analysis [59,60]. The texture measurements of different food products by objective, subjective, and imitative tests are well described in a reference book [61]. The most used tests for fruits are puncture, double compression [7], texture profile analysis [62], Kramer cell [63], and three-point bending test with notch [64]. These tests are performed on whole fruits or pieces (slices, cubes, sticks, and cylinders).

Recently, new approaches have been developed to quantify the texture (or shape) changes during food processing. The crispiness (the number, frequency, and shape of peaks in the jagged part of a force-deformation curve) frequently studied for extruded or brittle food products [65] was applied to cucumbers [66] and apples [67]. Computer image analysis with adapted software is applied to food quality [68], including the study on damaged banana and apple tissues [69–72]. Microscopic observations give also pertinent information on the fruit tissues, for example, affected by pretreatments and freezing [73,74]. The micromechanics [75] associated with the image analysis is also a promising technique to evaluate fruit textures (apple tissue [76]) and their variations during thermal (or other) treatments. Acoustic methods are also used for the assessment of apple texture [67,77,78].

The mechanisms of texture evolution during heat processing are closely related to pectin degradation [79]. They therefore depend on the time–temperature history [80], on intrinsic parameters of the fruit tissue such as pectin content, degree of methylation, and activity of pectin degrading enzymes, and on parameters that can be adjusted during processing such as pH, the presence of cations, and notably calcium salts [81,82].

Fresh fruit texture is determined by the interaction between turgor pressure of the cells and the resistance of the cell walls that surrounds the cells. The cell walls have two functions in the living fruit: the middle lamella is responsible for cell-to-cell adhesion, and the primary cell walls for limiting cell expansion. During heat treatment, the first phenomenon that leads to texture loss is the loss of turgor pressure upon membrane destabilization. Two successive mechanisms then determine texture evolution. At low temperatures (55°C–75°C), pectin methylesterases are activated; these enzymes have high optimal temperatures and can be remarkably heat resistant [83–85]. Pectin methylesterases hydrolyze the methyl esters on the carboxylic acid at C6 of galacturonic acid, releasing methanol and free carboxyl groups. Subsequent pectin demethylation will contribute to a firmer texture by minimizing latter susceptibility to β-elimination and by enabling the formation of additional calcium cross-links. At higher temperatures, the chemical mechanisms that lead to pectin cleavage take their turn. Their efficiency will depend on pectin degree of methylation and pH. At low pHs (<4), pectin cleavage is predominantly by hydrolysis, irrespective of degree of methylation. At slightly acidic to neutral pHs, β-elimination, a mechanism specific to methylated pectins, determines pectin degradation. β-elimination demands the presence of a methylated uronic acid: highly methylated pectins are highly susceptible to β-elimination while pectic acid, totally demethylated, is impervious to this cleavage [79,81,86]. These mechanisms and their succession are shown in Figure 16.2.

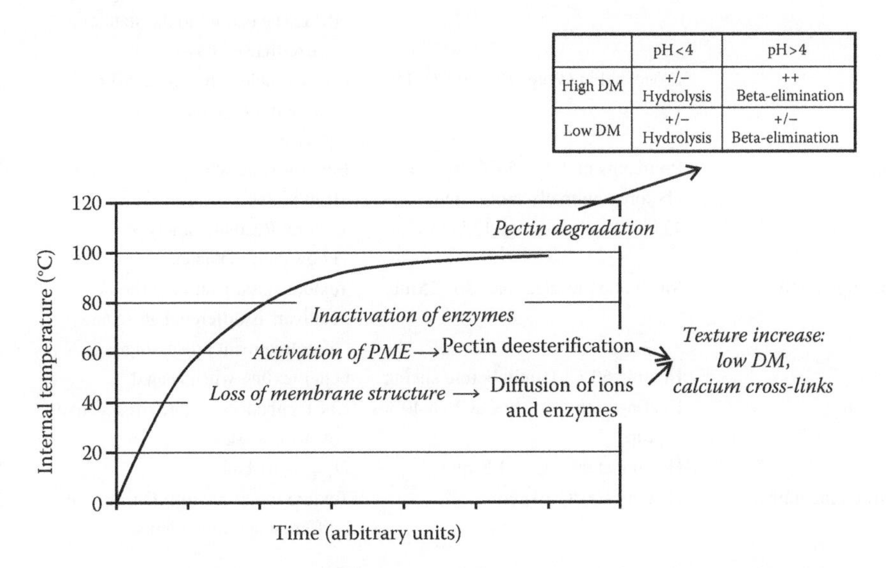

FIGURE 16.2 The two successive mechanisms involved in the vegetable texture evolution during thermal treatment.

There is therefore a potential for manipulation of texture through low-temperature blanching, enabling pectin methylesterase activation, and calcium addition [87]. This process is known as low-temperature long-time (LTLT) blanching, which results in firmer products through *in situ* activation of endogenous pectin methylesterase [88].

16.5.1.1 Thermal Softening of Fruits

In terms of reaction kinetics, the thermal softening of fruits is less documented than for vegetables [89]. Some examples of thermal softening of fruits by heat treatments are presented in Table 16.1. The D values greatly depend on the product and its variety, for canned apricots [90,93], cooked apricots [101], bananas [100], banana and plantain [92], or apples [7]. For the two later examples, the texture classification before and after heat treatments is not the same, indicating that it is not possible to predict the final product texture from texture measurements on the raw materials. Generally, the texture decrease is important (60°C–80°C) and rapid (10–20 min) when heating at the pasteurization temperatures and less important for milder treatment temperatures [90,95]. The sterilization of canned banana is also studied [97].

The same results are obtained when ohmic heating peaches slices at different frequencies [102]. When the data are available, the energy activation E_a of the first-order kinetic reaction (about 100 kJ mol^{-1}) is similar to the thermal softening of vegetables.

The texture of canned product is a crucial problem and a very often used solution for firmer products is the immersion in $CaCl_2$ solution (about 10 mM) for pears [99], peaches [54], or plums [91]. Mild

TABLE 16.1
Examples of the Thermal Softening of Fruits in Different Operating Conditions

Fruits	Treatment	Main Results	Reference
Apricots	Hot water 70°C–90°C	$D_{90°C}$ 2 min E_a 96.6 kJ mol^{-1}	[90]
Plums (different ripeness)	Canned with and without Ca^{++}	Better texture when added Ca^{++}	[91]
Plantain banana	Heating 60°C–100°C 0–30 min	Rapid texture decrease during the first 10 min	[92]
Apricots	Canned 82°C–95°C	$D_{90°C}$ 17 min E_a 116.5 kJ mol^{-1}	[93]
Apple slice	PEF and heat treatment	PEF + heat treatment disintegrates the tissue texture to the state of a freeze-thawed tissue	[94]
Guava	Processed in syrup 60°C–90°C, 1 h	Considerable softening at 90°C unlike milder temperature processing	[95]
Kiwi slices	Pretreatment 25°C–50°C 26–74 min before minimally processing	Better texture with pretreatment if fruit is ripe	[96]
Canned bananas	121°C 25 min, storage 135 days	Texture: Rasthali variety > Poowan > red banana	[97]
Apples different cultivars	Vacuum pasteurization 95°C 25 min	Texture classification of the cultivars is different before and after treatment processing	[7]
Peach slices	Heating 50°C, 10 min before slicing	better texture when heated	[98]
Canned pears	Heating with 0%–3% $CaCl_2$ in 40% syrup	$CaCl_2$ enhances texture irrespective of ripeness stage	[99]
Banana	Heating at 96.5°C 0–120 min	$D_{90°C}$ 4–10.4 min	[100]
Apricots different cultivars	Heating 100°C 10 min	Texture losses are different for the different apricot varieties	[6]

PEF, pulsed electric field.

heat treatment (50°C) was used to enhance the texture of minimally processed kiwis [96] or sliced peaches [98].

Thermal treatment, combined with other physical treatment, is used for partially disintegrating the texture, thus enhancing the juice extraction for apple (e.g., with pulsed electric fields) [94] and other materials [103]. Heat treatment (boiling during 15 min) is also used before candying plums (up to 2 months in 60–65 and 75 brix sugar syrups) [104]. In addition, a vacuum impregnation was also tested at mild temperature (30°C) for candying pineapple [105].

16.5.1.2 Rheological Behavior of Purees and Compotes

The fruit juices are generally Newtonian and poorly viscous (less than seven times the water viscosity) [106–109]. The concentrates are more viscous and exhibit a shear thinning pseudoplastic behavior [107,108,110–114]. Adding fibers drastically thickens the fruit juices [115–117], leading to highly viscous fluid exhibiting a strong yield stress.

Fruits are often prepared as semiliquid purées and compotes, which have a long storage life and are convenient. These pasty products exhibit a yield stress τ_0, a force (or stress) level required for flowing, which is a very important quality parameter for spoonability, spreadability, and sensory evaluation of such food products. The flow behavior is modeled with the following Herschel Bulkley equation [117,118] determined with rheometers. This approach is applied for apple sauces [51] or different fruit purees [119]:

$$\tau = \tau_0 + K\,\dot{\gamma}^n \tag{16.9}$$

Another interesting rheological property is the thixotropy, typically the ketchup behavior: when shaking the container, the product becomes liquid and when staying at rest (or less sheared), the yield stress is rebuilt [120]. Some purées and fruit concentrates exhibit such a rheological behavior [110,111,113,114,121–126].

The industrial rheology determination of pasty products is often made with a Bostwick consistometer [127,128], this test does not give precise rheological data but is very pertinent for formulating pasty products and control during and after processing.

16.5.1.3 Juice Cloud Stability

Pectin methylesterases are of particular interest in fruit juice processing as they are determinant for cloud stability and highly thermo (and baro) resistant, that is, they are more stable than the bacterias and yeasts present. Cloud in fruit juices are due to the presence of dissolved pectin colloids (ca. 0.1 µm) and small particles (0.5–10 µm). Cloud stability is linked to electrostatic repulsions between these particles, where negatively charged pectins surround a protein nucleus (itself positively charged at the juice pH). Pectin methylesterases in fruit juices lead to cloud destabilization [129] by the formation of sequences with consecutive free galacturonic acids on the pectin main chain, leading to a very high calcium reactivity [130] and pectin precipitation. Pectin methylesterases have been particularly studied for citrus juices and especially orange, which contains high pectin methylesterases and the juice of which should have stable cloud [129–137]. Heating of orange juices at 90°C for 1 min is sufficient to ensure their colloidal stability, but this thermal treatment must take place very rapidly after expression. Pectin methylesterases are generally deactivated following first-order kinetics; however, isoforms with differing thermal stability are present in orange, leading to biphasic kinetics [83,135]. The z values for thermolabile and thermostable orange pectin methylesterases are 10.8°C and 6.5°C for orange pulp [138], while the z values for microwave-heated orange juice are 31.1°C ($D_{60°C} = 1240$ s) and 17.6°C ($D_{60°C} = 154$ s) [139] with activations energies of 299 kJ mol^{-1} ($k_{73.5°C} = 8.5$ min^{-1}) and 532 kJ mol^{-1} ($k_{73.5°C} = 0.003$ min^{-1}) [140]. The relative proportions of the thermostable and thermolabile isoforms may vary, but the thermostable form is always the lowest, often between 5% and 10% [85,131,133,135,138,139]. It is generally considered that inactivation of the thermolabile form is sufficient for acceptable cloud stability. Efficiency of inactivation also varies with pH, with the more acidic conditions leading to faster inactivation: for thermostable orange pectin methylesterases $D_{90°C}=33$ s at pH 3.6 but 256 s at pH 4.1 [137]. Thermal stability of pectin methylesterase appears to be higher for soluble enzymes in the juice that in the pulp [133] but that was not related to isoforms distribution.

Other fruits also present pectin methylesterases with high thermal stabilities, though few studies report z values or activation energies. One of the pectin methylesterase from acerola kept its activity after 1 h at 98°C [141]. Apple pectin methylesterase presents z value of 10.1°C (with $D_{50°C}=7.4$ min) [84]. Activation energies were for banana $E_a=279$ kJ mol^{-1} ($k_{75°C}=0.55$ min^{-1}) [140], for plum (biphasic with about 60% thermolabile enzyme) $E_a=274$ kJ mol^{-1} ($k_{60°C}=0.106$ min^{-1}) and $E_a=354$ kJ mol^{-1} ($k_{60°C}=0.008$ min^{-1}) [142], and for grapefruit thermostable pectin methylesterase $E_a=329$ kJ mol^{-1} ($k_{60°C}=0.189$ min^{-1}) [143]. Influence of physical–chemical conditions was reported, with stability at pH 7>4 for plum [142] and optimal stability at pH 7 and high ionic strength for grapefruit [143]. Remarkably, high pressure and temperatures have an antagonistic effect in pectin methylesterase inactivation [83].

16.5.2 Color and Image Analysis

Color measurements by hand-held tristimulus colorimeters are reliable. These pieces of equipment give the color in different units, with the most used with food products being the L^*, a^*, b^* color space coordinates [144] or the quasi similar L, a, b system. Lightness value, L^* indicates how dark/light the sample is (varying from 0—black to 100—white), a^* is the measure of greenness/redness (varying from −60 to +60), and b^* is the measure of blueness/yellowness (varying from −60 to +60). The polar coordinate chroma or saturation C^* is an indication of how dull/vivid the product is (ranging from 0 to 60), which can be calculated from a^* and b^* Cartesian coordinates by the following equation:

$$C^* = \sqrt{a^{*2} + b^{*2}} \quad (16.10)$$

The total color difference (TCD*) is a parameter considered for the overall color difference evaluation between the reference sample (initial product or a reference ceramic plate) and the processed one. Differences in visual color can be classified based on TCD* as follows: not noticeable (0–0.5), slightly noticeable (0.5–1.5), noticeable (1.5–3.0), well visible (3.0–6.0), and great difference (6.0–12.0).

$$\text{TCD}^* = \sqrt{\left(L_0^* - L^*\right)^2 + \left(a_0^* - a^*\right)^2 + \left(b_0^* - b^*\right)^2} \quad (16.11)$$

Other parameters are calculated from L^*, a^*, and b^*. The hue angle, that is, $h^*=\tan^{-1}(b^*/a^*)$, is frequently used to characterize reddish or yellowish color. The use of a^*/b^* cannot be advocated as it reflects a faulty understanding of the color coordinates.

The browning index, defined as absorbance at 420 nm of a centrifuged and filtered juice, is also frequently used to detect the nonenzymatic browning of fruits matrices or juices.

Examples of the influence of heat treatments on fruit jams and juices are presented in Table 16.2. In Table 16.2, only TCD (or TCD*) is mentioned, which seems to be a well-recognized quality index. The most recent publications compare the thermal treatment with nonconventional ones [149,150,158–160]. The color loss kinetic is modeled with the aforementioned models [145–148,151–154], and the energy activation varies between 30 and 120 kJ mol^{-1} with an exception for the strawberry juice in which the group $L^*{\cdot}a^*/b^*$ was used [154].

TABLE 16.2
Influence of Heat Treatment on the Fruit and Puree Colors

Products	Thermal Treatment	TCD or TCD*	Comments	Reference
Peach puree	10°C–135°C up to 160 min		Fractional first-order kinetic model E_a 119 kJ mol^{-1}	[145]
Cupuacu puree	80°C–115°C up to 120 min	0–6 80°C; 0–18 115°C	Fractional first-order kinetic model E_a 36 kJ mol^{-1}	[146]
Peach puree	80°C–98°C up to 500 min	0–4 80°C; 0–18 98°C	Combined zeroth- and first-order kinetic model E_a 82 kJ mol^{-1}	[147]
Pineapple puree	70°C–110°C up to 500 min	0–20, increase with temperature	Zeroth-order kinetic E_a 83.7 kJ mol^{-1} 70°C–90°C E_a 94.4 kJ mol^{-1} 90°C–110°C	[148]
Strawberry and blackberry puree	70°C, 2 min	Strawberry 5.67 Blackberry 3.17	Comparison thermal/high pressure	[149]
Nectarine puree	85°C, 5 min	Treated/untreated: 2.5 Treated/60 storage days: 7.7	Comparison thermal/high pressure	[150]
Pineapple juice	55°C–95°C, 80 min	0–0.7 55°C; 0–1.8 65°C; 0–2.2 75°C; 0–3 85°C; 0–4 95°C	Combined kinetic model, E_a 47.3 kJ mol^{-1}	[151]
Yellow-orange cactus pear juice	75°C, 85°C, 95°C, 60 min	0–10 75°C; 0–18 85°C; 0–30 95°C	Addition of 0.1% isoascorbic acid prior to heating minimized color alteration	[152]
Purple pitaya juice	85°C 1 h, pH 4 and 6	C^a and h^a	Betacyanins in purple pitaya juice is stabilized by the addition of ascorbic, isoascorbic, citric acids	[153]
Strawberry juices	100°C–140°C, 0–120 min	(L^*a^*/b^*) parameter, kinetic fractional model	E_a 183 at pH 2.5, 168 at pH 3.7, 86 at pH 5 (kJ mol^{-1})	[154]
Apple juice	HTST 73°C, 80°C, 83°C 27 s	1.3		
Cashew apple juice	88°C, 100°C, 111°C, 121°C	0–15	Kinetic modeled with an exponential model	[155]
Cashew apple model juice	60°C, 90°C	0–6	Kinetic modeled with a biphasic model	[156]
Elderberry juice	95°C, 0–4 h	0–18	Color differences in strawberry and elderberry juices when ascorbic acid was added	[157]
Strawberry juice		0–14		
Watermelon juice	Thermosonication 25°C–45°C, amplitude 24.4–60.1 µm	4.5–5.54	Effect of process parameters on quality indices modeled with a second-order polynomial	[158]
Watermelon juice	60°C, 5–60 min	6.94–5.87	Comparison of thermal, ultraviolet-c, and high-pressure treatments	[159]
Mango nectar	60°C–85°C, 10–20 min	5.8 at optimal conditions	High-pressure homogenization + heat shock	[160]

Color degradations of fruits [153,156,157] are clearly due to thermal degradation of pigments, and the reaction kinetics of these food components are summarized in a reference book [38] for anthocyanins, betalains, and carotenoids. The reaction kinetic modeling is based on the fraction conversion model associated with a first-order kinetic reaction. The reaction kinetic constants are well documented [38,89] for different operating conditions and in different fruit matrices. The color variation kinetics is quite similar to that of pigments degradation.

Some other analyses such as near infrared (NIR) spectroscopy, middle infrared (MIR) spectroscopy [162], or hyperspectral imaging [163,164], are performed for inspecting or grading fruits and vegetables but are rarely used for determining the impact of thermal processing [165] on their quality attributes.

16.5.3 Aroma Modification

Two types of products must be distinguished when dealing with impact of thermal treatments on fruit juice aromas: the juices produced from concentrates and the more lightly heated not-from-concentrate juices. In juice concentration, normally, volatiles are recovered and concentrated for later addition back to the concentrated products. Therefore, incomplete restoration will result in the loss of volatiles, as did the former practice of mixing concentrate (without recovered volatiles) with single strength juices.

A recent review summarizes the effects of heat treatments on orange juices [9], indicating that juice volatiles are impacted by even HTST treatments due to the loss of more volatile molecules. However, this is not sufficient for detection of aroma modification by a sensory analysis. Typical cooked off-tastes are only perceived by consumers in juices from concentrates, which are subjected to concentration then pasteurization after redilution. Juices from concentrates are depleted in the more volatile molecules but contain more sulfur-containing volatiles because of thermal degradation of amino acids.

Ohmic heating of freshly squeezed orange juice resulted in slightly higher retention of limonene, myrcene, octanal, and decanal than conventional pasteurization (about 60% for all four compounds) [134]. In apple juice, conventional evaporation resulted in the loss of >95% of *trans*-2-hexenal [166]. Apple juices not from concentrate were dominated by fruity, sweet aromatics, while apple juices from concentrate were assessed as more green, fresh, and having a shampoo-like smell [10].

16.5.4 Nutritional Impact of Fruit Heat Processing

This section will focus on the micronutrients for which fruits contribute significantly to diet. These are primarily vitamin C and provitamin A carotenoids (β-, α-, and γ-carotene and β-cryptoxanthin). Fruits also contain polyphenols, which will be briefly mentioned here, as polyphenol oxidation is a factor of ascorbic acid degradation and enzymatic browning. These compounds have very different sensitivities to heat-treatment, and these sensitivities are modulated differently by physicochemical conditions, as summarized in Table 16.3.

16.5.4.1 Vitamin C

Vitamin C is present in fruits, sometimes at high concentrations as in black currant, kiwi, oranges, and lemons. It is also often added during processing to prevent enzymatic browning, in particular in apple, banana, and small red fruits. Though vitamin C probably has been the most studied vitamin in terms of loss during heat treatments, there are still gaps in the understanding (and therefore modeling) of its loss. Vitamin C is composed of two compounds, both of which carry the vitaminic property: ascorbic acid is the reduced form of vitamin C and dehydroascorbic acid is its oxidized form. Fruits may contain both forms, with ratios that vary depending on the species, the physiological state, etc.

TABLE 16.3
Main Factors of Loss and Relative Stability of Microconstituents in Fruits

	Susceptible to Oxidation	Other Chemical Degradation Mechanism	Highly Soluble (Leaching)	Concentrated in Outer Parts	Presence of Enzymes
Vitamin C	++ (ascorbate)	++ (dehydroascorbate)	+	+	+ (polyphenoloxidase)
Vitamin B9	?	+	++		
Carotenoids	+/–	+/– (isomerization)	0	+	
Dietary fibers	0	+/– (conversion to soluble fibers)	0	+	
Polyphenols	+	0	+/–	++	+ (polyphenoloxidase)

These molecules have different susceptibilities to oxidation and thermal degradation: ascorbic acid can be oxidized easily and is in particular converted to dehydroascorbic acid by coupled oxidation—reduction with the quinones of polyphenols. Oxidation of polyphenols (by polyphenoloxidase or laccase prior to heat treatment, or by autoxidation catalyzed by metals after heat treatment) thus leads to loss of ascorbic acid. Dehydroascorbic acid can be further hydrolyzed to 2,3-diketoglutaric acid, which follows the sugar degradation reaction chain [167]. Of the two moieties, ascorbic acid is thus highly susceptible to degradation by oxidation (but very stable in anaerobic conditions), while dehydroascorbic acid is susceptible to heat degradation by Strecker degradation independently of oxygen concentration [168]. The requirement for oxygen in conversion of ascorbic acid to dehydroascorbic acid explains why vitamin C degradation can deviate from the first order [169]. Oxygen solubility in water at 20°C is of 9 mg L^{-1}, that is, 0.28 mmol L^{-1} (oxygen solubility decreases with °Brix [170]). This would be sufficient to oxidize 100 mg L^{-1} of ascorbic acid. However, concentrations of ascorbic acid close to 0.5 g L^{-1} are found in citrus juices or added to cloudy apple juices. Oxygen can be a limiting reagent, especially in hermetically sealed vessels or in closed process loops. Once the pool of dissolved oxygen is exhausted, ascorbic acid degradation stops or at least becomes much slower as it is relatively resistant to Strecker degradation. However, even in anaerobic condition, all dehydroascorbic will disappear in a relatively short time due to hydrolysis and Strecker degradation. This dual mechanism explains why nonlinear models can often better predict vitamin C loss [171], and also why deaeration is a major factor for juice quality. Two reactions with three reaction rate constants are thus needed to adequately describe ascorbic acid degradation in aerobic conditions [168]:

$$AA + \tfrac{1}{2}O_2 \xleftrightarrow{K1,K2} DHA + H_2O$$

$$DHA + H_2O \xrightarrow{K3} DKG$$

where

- AA is ascorbic acid
- DHA is dehydroascorbic acid
- DKG is 2,3-diketoglutaric acid
- $K1$ is the reaction rate constant for ascorbic acid oxidation (apparent rate constant as oxygen concentration is not explicitly taken into account)
- $K2$ is the reconversion of DHA to ascorbic acid
- $K3$ is the subsequent hydrolysis of DHA

In the presence of oxygen and absence of reducing agent, oxidation of ascorbic acid is much faster than reduction of dehydroascorbic acid. Serpen and Gökmen [172] reported a half-life of 3.21 h (decreased in the presence of Fe^{3+}, increased by cysteine, a reducing agent) for a solution of ascorbic acid in pure water held at 90°C. They calculated the apparent reaction rate constants as $K1=0.218\ h^{-1}$, $K2=0.249\ 10^{-8}\ h^{-1}$, and $K3=0.218\ h^{-1}$.

However, few studies take into account the dual nature of vitamin C and the role of oxygen in ascorbic acid degradation, and conditions encountered in the laboratory (such as small volumes and large surfaces for exchange with ambient air) may not be totally representative of industrial conditions.

Losses in degassed and nondegassed tamarillo nectars kept for 10 min at 95°C were of 4% and 100%, respectively, for ascorbic acid but 75% and 80%, respectively, for dehydroascorbic acid [173]. Similar losses of vitamin C (ca. 15%) in orange juices are reported for various ohmic conditions and conventional pasteurization (90°C, 50 s) [134]. Mild pasteurization (75°C, 10 min) of a highly acidic (pH 3.2) apple puree (vitamin C coming from added acerola and lemon juice) did not decrease total vitamin C but decreased ascorbic acid by 38.5% [174]. For apricots in syrup, a loss of 18% of ascorbic acid was registered during ohmic heating (and a further loss of 28% in the first weeks of storage) [175]. For ascorbic acid, E_a of 36 kJ mol^{-1} ($z=64$°C) was found in mixed orange and clementine juice [176], $E_a=38.6$ kJ mol^{-1} (initial steps only) in orange juice under aerobic condition [177], and $E_a=47.5$ kJ mol^{-1} ($z=53$°C, with $D_{90°C}=175$ min) in rose hip pulp [178].

The other factor that may contribute to vitamin C loss is its solubility, so that it may leach during blanching and diffuse to the covering liquid during storage. Vitamin C concentrations are often higher in the outer parts of the fruit, so that peeling or sieving also contributes to its loss [179,180].

16.5.4.2 Carotenoids

Carotenoids are tetraterpenoid organic pigments characterized by their lipophilic character. They can be divided into xanthophylls (lutein) that contain oxygen, and carotenes (notably β-carotene and lycopene), which are purely hydrocarbons and thus even more lipophilic. They can also be categorized according to their nutritional status in provitamin A (β-carotene, α-carotene, γ-carotene, and β-cryptoxanthin) and nonprovitamin A carotenoids. Many studies report the impact of heat treatments on carotenoids, with contradictory results, but globally these compounds appear to be relatively stable [181,182]. There also appears to be differences in stability, with xanthophylls being less stable than the carotenes [176,183–185].

In raw natural foods, all-*trans* isomers of carotenoids are dominant but mono or poly-*cis* isomers are formed upon heating. Carotenoids are susceptible to oxidation, with first formation of epoxides followed by chain cleavage leading to apo-carotenals. Upon moderate heat treatment of tissues, there is commonly an increase in apparent carotenoid content, which is usually ascribed to easier extractability [186–189]. Prolonged heating and presence of oil seem to favor isomerization, while oxidation is favored by heat, low water activities, and of course presence of oxygen [190,191]. Carotenoid degradation follows first-order kinetics and is strongly correlated to color degradation [173,176,192,193].

Pasteurization can lead to a loss of carotenoids in orange juice, with violaxanthin and lutein being the most thermolabile; concentration of the juices further affects lutein concentrations. The provitamin A carotenoids are not significantly degraded [184]. Using microwave heating [185], z values between 10.9°C (β-carotene, with $D_{70°C}=24.6$ min) and 16.7°C (antheraxanthin, with $D_{70°C}=8.5$ min) are reported for carotenoids in orange juice. Activation energies of 110 and 156 kJ mol^{-1} (z of 22.5°C and 15.9°C) for β-carotene and β-cryptoxanthin were found in mixed orange and clementine juice [176]. Cold storage of frozen fruits, fruit purées, or concentrated fruit juices can lead to loss in carotenoids [181,194], especially in plastic containers, probably because plastic containers allow some oxygen permeability. Degradation of carotenoids is not affected by the level of dissolved oxygen in tamarillo juice [173].

The most remarkable effect of heat treatment on carotenoids in intact tissues is a marked increase of their bioavailability, noticed first *in vivo* [195–197] and studied in detail using *in vitro* models [198]. This increase in bioaccessibility/bioavailability of carotenoids appears linked to cell wall degradation, enhancing cell rupture [199–203]. High-pressure processing in contrast decreases carotenoids bioaccessibility, which can be related to pectin methylesterase activity and calcium-mediated cross-linking of the polysaccharides in the cell walls [204,205].

Carotenoids are not water soluble. In juice processing, carotenoids are linked to the pulp, and they can be present in quite high concentrations in pulpy juices such as juices made from apricot, mango or guava, which are actually comminuted rather than pressed. Furthermore, carotenoids are often concentrated in the outer tissues of plants: peeling and sieving also contribute to a decrease of their amounts in the processed products [180].

16.5.4.3 Dietary Fiber

Dietary fiber in fruits corresponds to polysaccharides (and traces of lignin) constituting their cell walls. Thermal treatments have limited impact on the total dietary fiber contents. They generally lead to a modification of the soluble/insoluble fiber balance by an increase in soluble fibers [51,206], due to pectin degradation [79]. This also leads to increased cell wall swelling.

Pressing eliminates insoluble dietary fibers from the juice, as they are retained in the pomace; the residual fibers are further eliminated upon clarification. Only minor amounts are present as cloud in cloudy juices, typically $<0.5\,g\,L^{-1}$. Peeling and sieving also retain the harder tissues, which are also the ones that are rich in dietary fibers, for example, in purées, smoothies, or pulpy juices [51]. Though the dietary fiber contents are higher than in juices, there is still a significant loss compared to whole fruits (though not necessarily when compared to fruit flesh [207]).

16.6 CONCLUSIONS

Processed fruits generally undergo mild heat treatments as their low pH is a good protection against pathogenic bacteria. A specific characteristic for fruit juices is that they may have spent long durations at only slightly elevated temperatures (40°C–50°C) for ease of processing (enzymation before or after pressing, concentration, filtration, etc.), which may result in quality modifications in particular concerning aromas. The current trends in industrial heat processing of fruits are to minimize the process intensity, both for better preservation of organoleptic and nutritional properties and for more durable processes. Fruit juices in particular are an area of active research and product development in nonthermal stabilization technologies. There has been however in the recent past increased reports of fruit juice contamination with pathogenic microorganisms such as *E. coli* O157:H7. These are linked to increased demand for fresh, unpasteurized products, and highlight the necessity for vigilance from manufacturers and public authorities.

Fruits used for processing have undergone a remarkable diversification in the last few years, including more and more exotic fruits. Some of these fruits have higher pH than the more classical citrus and pome fruits, and therefore, more attention should be paid to microbial inactivation. Among the popular fruits, mango, lychee, and sometimes pear may only require some mild acidification (typically citric acid) as their pHs are of about 4.5–5. Guava, cantaloupe, and watermelon or banana with pHs > 5 need more stringent thermal treatments.

NOMENCLATURE

a^*	Measure of greenness-redness (—)
b^*	Measure of blueness-yellowness (—)
$b(\vartheta)$	Parameter of the Equation 16.5 derived from a Weibull distribution (min)
$C(t)$	Quality attribute level at a time t
C_0	Initial quality attribute level

C_∞	Quality attribute level at equilibrium
C^*	Chroma index (—)
D_{ref}	Decimal reduction time at reference temperature (min)
$D(\vartheta)$	Decimal reduction time at temperature ϑ (min)
E_a	Activation energy (kJ mol^{-1})
$f(t)$	Fractional conversion at a time t (—)
F	Process thermal lethality (min)
F_0	Process thermal lethality at 121.1°C (min)
$k(\vartheta)$	Inactivation "rate constant" at temperature ϑ (min^{-1})
k_1, k_2	Coefficients of the biphasic model (min^{-1})
K	Parameter of the Herschel–Bulkley equation (Pa s^n)
$K1$, $K2$, $K3$	Reaction constants of ascorbic acid degradation in aerobic conditions
L^*	Lightness (—)
n	Parameter of the Herschel–Bulkley equation (—)
N_0	Initial number of spores or microorganisms
$N(t)$	Number of spores or microorganisms at time t
q	Coefficient of the biphasic model (—)
R	Universal gas constant (8.314 J mol^{-1} °C^{-1})
t	Time (min for reaction kinetics)
T	Absolute temperature (K)
TCD*	Total color difference (—)
Z	Temperature difference required for 10-fold change in D value (°C)

Greek

β	Power parameter of the Equation 16.5 derived from a Weibull distribution (—)
ϑ_{ref}	Reference temperature (°C)
$\dot{\gamma}$	Shear rate (s^{-1})
τ	Stress (Pa)
τ_0	Yield stress (Pa)

REFERENCES

1. E Rendueles et al. Microbiological food safety assessment of high hydrostatic pressure processing: A review. *LWT—Food Science and Technology* 44(5):1251–1260, 2011.
2. D Bermudez-Aguirre and GV Barbosa-Canovas. An update on high hydrostatic pressure, from the laboratory to industrial applications. *Food Engineering Reviews* 3(1):44–61, 2011.
3. G Demazeau and N Rivalain. The development of high hydrostatic pressure processes as an alternative to other pathogen reduction methods. *Journal of Applied Microbiology* 110(6):1359–1369, 2011.
4. D Knorr et al. Emerging technologies in food processing. *Annual Review of Food Science and Technology* 2:203–235, 2011.
5. MB Fox et al. Inactivation of L-plantarum in a PEF microreactor—The effect of pulse width and temperature on the inactivation. *Innovative Food Science & Emerging Technologies* 9(1):101–108, 2008.
6. CE Missang et al. Texture variation in apricot: Intra-fruit heterogeneity, impact of thinning and relation with the texture after cooking. *Food Research International* 44(1):46–53, 2011.
7. E Bourles et al. Impact of vacuum cooking process on the texture degradation of selected apple cultivars. *Journal of Food Science* 74(9):E512–E518, 2009.
8. SF Aguillar-Rosas et al. Thermal and pulsed electric fields pasteurization of apple juice: Effects on physicochemical properties and flavour compounds. *Journal of Food Engineering* 83(1):41–46, 2007.
9. PR Perez-Cacho and R Rouseff. Processing and storage effects on orange juice aroma: A review. *Journal of Agricultural and Food Chemistry* 56(21):9785–9796, 2008.
10. MP Nikfardjam and D Maier. Development of a headspace trap HRGC/MS method for the assessment of the relevance of certain aroma compounds on the sensorial characteristics of commercial apple juice. *Food Chemistry* 126(4):1926–1933, 2011.

11. A Kalia and RP Gupra. Fruit microbiology. In: YH Hui, ed. *Handbook of Fruits and Fruits Processing*. Ames, IA: Blackwell Publishing, 2006, pp. 3–28.
12. C The Nguyen and F Carlin. The microbiology of minimally processed fresh fruits and vegetables. *Critical Reviews in Food Science and Nutrition* 34:371–401, 1994.
13. F Carlin et al. Research on factors allowing a risk assessment of spore-forming pathogenic bacteria in cooked chilled foods containing vegetables: A FAIR collaborative project. *International Journal of Food Microbiology* 60(2/3):117–135, 2000.
14. MAJS van Boekel. Kinetics of inactivation of microorganisms. In: MAJS van Boekel, ed. *Kinetic Modeling in Reactions in Foods*. Boca Raton, FL:CRC Press, Taylor & Francis Group, 2009, pp. 13–1, 13–44.
15. MAJS van Boekel. On the use of the Weibull model to describe thermal inactivation of microbial vegetative cells. *International Journal of Food Microbiology* 74(1–2):139–159, 2002.
16. M Corrandini and M Peleg. The non linear kinetics of microbial inactivation and growths in foods. In: S Brul, S van Gerwen, and M Zwietering, eds. *Modelling Microorganisms in Food*. Cambridge, U.K.: Woodhead Publishing Limited, 2007, pp. 129–160.
17. P Mafart et al. On calculating sterility in thermal preservation methods: Application of the Weibull frequency distribution model. *International Journal of Food Microbiology* 72(1–2):107–113, 2002.
18. O Couvert et al. Survival curves of heated bacterial spores: Effect of environmental factors on Weibull parameters. *International Journal of Food Microbiology* 101(1):73–81, 2005.
19. M Peleg. A model of survival curves having an "activation shoulder." *Journal of Food Science* 67(7):2438–2443, 2002.
20. AH Geeraerd, CH Herremans, and JF Van Impe. Structural model requirements to describe microbial inactivation during a mild heat treatment. *International Journal of Food Microbiology* 59(3):185–209, 2000.
21. AH Geeraerd, VP Valdramidis, and JF Van Impe. GInaFiT, a freeware tool to assess non-log-linear microbial survivor curves. *International Journal of Food Microbiology* 102(1):95–105, 2005.
22. R Xiong et al. A mathematical model for bacterial inactivation. *International Journal of Food Microbiology* 46(1):45–55, 1999.
23. G Stone, B Chapman, and D Lovell. Development of a log-quadratic model to describe microbial inactivation, illustrated by thermal inactivation of Clostridium botulinum. *Applied and Environmental Microbiology* 75(22):6998–7005, 2009.
24. L Dong-Un, V Heinz, and D Knorr. Biphasic inactivation kinetics of *Escherichia coli* in liquid whole egg by high hydrostatic pressure treatments. *Biotechnology Progress* 17(6):1020–1025, 2001.
25. P Mafart et al. Quantification of spore resistance for assessment and optimization of heating processes: A never-ending story. *Food Microbiology* 27(5):568–572, 2010.
26. P Pittia et al. Safe cooking optimisation by F-value computation in a semi-automatic oven. *Food Control* 19(7):688–697, 2008.
27. CE Steyn, M Cameron, and RC Witthuhn. Occurrence of Alicyclobacillus in the fruit processing environment—A review. *International Journal of Food Microbiology* 147(1):1–11, 2011.
28. WH Groenewald, PA Gouws, and RC Witthuhn. Isolation, identification and typification of *Alicyclobacillus acidoterrestris* and *Alicyclobacillus acidocaldarius* strains from orchard soil and the fruit processing environment in South Africa. *Food Microbiology* 26(1):71–76, 2009.
29. Y Smit et al. Alicyclobacillus spoilage and isolation—A review. *Food Microbiology* 28(3):331–349, 2011.
30. FVM Silva and P Gibbs. *Alicyclobacillus acidoterrestris* spores in fruit products and design of pasteurization processes. *Trends in Food Science & Technology* 12(2):68–74, 2001.
31. MC Maldonado, C Belfiore, and AR Navarro. Temperature, soluble solids and pH effect on *Alicyclobacillus acidoterrestris* viability in lemon juice concentrate. *Journal of Industrial Microbiology & Biotechnology* 35(2):141–144, 2008.
32. FVM Silva and P Gibbs. Target selection in designing pasteurization processes for shelf-stable high-acid fruit products. *Critical Reviews in Food Science and Nutrition* 44(5):353–360, 2004.
33. BdC Martins Salomao, P Rodriguez Massaguer, and GM Falcao Aragao. Isolation and selection of heat resistant molds in the production process of apple nectar. *Ciencia e Tecnologia de Alimentos* 28(1):116–121, 2008.
34. M Zimmermann et al. Modeling the influence of water activity and ascospore age on the growth of Neosartorya fischeri in pineapple juice. *LWT—Food Science and Technology* 44(1):239–243, 2011.
35. AdS Sant'Ana, A Rosenthal, and PRd Massaguer. The fate of patulin in apple juice processing: A review. *Food Research International* 41(5):441–453, 2008.

36. AA Lima Tribst, Ad Souza Sant'Ana, and PRd Massaguer. Review: Microbiological quality and safety of fruit juices—Past, present and future perspectives. *Critical Reviews in Microbiology* 35(4):310–339, 2009.
37. TR Stoneham, DB Lund, and CH Tong. The use of fractional conversion technique to investigate the effects of testing parameters on texture degradation kinetics. *Journal of Food Science* 65(6):968–973, 2000.
38. R Villota and JG Hawkes. Reaction kinetics in food systems. In: L Heldman, ed. *Handbook of Food Engineering*. Boca Raton, FL: CRC Press, 2007, pp. 125–286.
39. Y Kim, JK Brecht, and ST Talcott. Antioxidant phytochemical and fruit quality changes in mango (Mangifera indica L.) following hot water immersion and controlled atmosphere storage. *Food Chemistry* 105(4):1327–1334, 2007.
40. C Noe-Aguilar et al. Blanching at low temperatures: A thermal bioprocess applied to fruits and vegetables to improve textural quality. *Food Science and Biotechnology* 13(1):104–108, 2004.
41. EB Solomon et al. Thermal inactivation of salmonella on cantaloupes using hot water. *Journal of Food Science* 71(2):M25–M30, 2006.
42. JB Adams. Review: Enzyme inactivation during heat processing of food-stuffs. *International Journal of Food Science & Technology* 26(1):1–20, 1991.
43. LN Gerschenson, AM Rojas, and AG Marangoni. Effects of processing on kiwi fruit dynamic rheological behaviour and tissue structure. *Food Research International* 34(1):1–6, 2001.
44. Z Yi and P Zhongli. Processing and quality characteristics of apple slices under simultaneous infrared dry-blanching and dehydration with continuous heating. *Journal of Food Engineering* 90(4):441–452, 2009.
45. Z Yi et al. Processing and quality characteristics of apple slices processed under simultaneous infrared dry-blanching and dehydration with intermittent heating. *Journal of Food Engineering* 97(1):8–16, 2010.
46. H Yildiz, H Bozkurt, and F Icier. Ohmic and conventional heating of pomegranate juice: Effects on rheology, color, and total phenolics. *Food Science and Technology International* 15(5):503–512, 2009.
47. S Mukherjee and PK Chattopadhyay. Whirling bed blanching of potato cubes and its effects on product quality. *Journal of Food Engineering* 78(1):52–60, 2007.
48. EMC Alexandre et al. Influence of aqueous ozone, blanching and combined treatments on microbial load of red bell peppers, strawberries and watercress. *Journal of Food Engineering* 105(2):277–282, 2011.
49. MV Selma et al. Effect of gaseous ozone and hot water on microbial and sensory quality of cantaloupe and potential transference of *Escherichia coli* O157:H7 during cutting. *Food Microbiology* 25(1):162–168, 2008.
50. RP Singh. Heating and cooling processes for foods. In: DR Heldman, DB Lund, eds. *Handbook of Food Engineering*. Boca Raton, FL: CRC Press, Taylor & Francis Group, 2007, pp. 397–426.
51. M Colin-Henrion et al. Instrumental and sensory characterisation of industrially processed applesauces. *Journal of the Science of Food and Agriculture* 89(9):1508–1518, 2009.
52. A Teixeira. Thermal processing of canned foods. In: DR Heldman and DB Lund, eds. *Handbook of Food Engineering*. Boca Raton, FL: CRC Press, 2007, pp. 745–797.
53. M Siddiq. Apricots. In: HY Hui, ed. *Handbook of Fruits and Fruits Processing*. Ames, IA: Blackwell Publishing, 2006, pp. 279–291.
54. M Siddiq. Peach and nectarine. In: HY Hui, ed. *Handbook of Fruits and Fruits Processing*. Ames, IA: Blackwell Publishing, 2006, pp. 519–531.
55. M Siddiq. Plums and prunes. In: HY Hui, ed. *Handbook of Fruits and Fruits Processing*. Ames, IA: Blackwell Publishing, 2006, pp. 553–564.
56. JV Carbonell et al. Chilled orange juices stabilized by centrifugation and differential heat treatments applied to low pulp and pulpy fractions. *Innovative Food Science and Emerging Technologies* 12(3):315–319, 2011.
57. C Chen, K Abdelrahim, and I Beckerich. Sensitivity analysis of continuous ohmic heating process for multiphase foods. *Journal of Food Engineering* 98(2):257–265, 2010.
58. F Marra, Z Lu, and JG Lyng. Radio frequency treatment of foods: Review of recent advances. *Journal of Food Engineering* 91(4):497–508, 2009.
59. Nd Belie et al. Use of physico-chemical methods for assessment of sensory changes in carrot texture and sweetness during cooking. *Journal of Texture Studies* 33(5):367–388, 2002.
60. K Roininen et al. Perceived eating difficulties and preferences for various textures of raw and cooked carrots in young and elderly subjects. *Journal of Sensory Studies* 18(6):437–451, 2003.
61. M Bourne, ed. *Food Texture and Viscosity: Concepts and Measurements*, 2nd edn. London, U.K.: Academic Press, 2002.

62. E Madieta, R Symoneaux, and E Mehinagic. Textural properties of fruit affected by experimental conditions in TPA tests: An RSM approach. *International Journal of Food Science & Technology* 46(5):1044–1052, 2011.
63. MB Sousa et al. Effect of processing on the texture and sensory attributes of raspberry (cv. Heritage) and blackberry (cv. Thornfree). *Journal of Food Engineering* 78(1):9–21, 2007.
64. FR Harker et al. Instrumental measurement of apple texture: A comparison of the single-edge notched bend test and the penetrometer. *Postharvest Biology and Technology* 39(2):185–192, 2006.
65. G Roudaut et al. Crispness: A critical review on sensory and material science approaches. *Trends in Food Science & Technology* 13(6–7):217–227, 2002.
66. Y Yoshioka et al. Quantifying cucumber fruit crispness by mechanical measurement. *Breeding Science* 59(2):139–147, 2009.
67. F Costa et al. Assessment of apple (Malus* domestica Borkh.) fruit texture by a combined acoustic-mechanical profiling strategy. *Postharvest Biology and Technology* 61(1):21–28, 2011.
68. C Zheng, D-W Sun, and L Zheng. Recent developments and applications of image features for food quality evaluation and inspection—A review. *Trends in Food Science & Technology* 17(12):642–655, 2006.
69. R Yoruk et al. Machine vision analysis of antibrowning potency for oxalic acid: A comparative investigation on banana and apple. *Journal of Food Science* 69(6):E281–E289, 2004.
70. R Quevedo et al. Determination of senescent spotting in banana (Musa cavendish) using fractal texture Fourier image. *Journal of Food Engineering* 84(4):509–515, 2008.
71. R Quevedo et al. Description of the kinetic enzymatic browning in banana (Musa cavendish) slices using non-uniform color information from digital images. *Food Research International* 42(9):1309–1314, 2009.
72. L Fernández, C Castillero, and JM Aguilera. An application of image analysis to dehydration of apple discs. *Journal of Food Engineering* 67(1–2):185–193, 2005.
73. J Suutarinen et al. The effects of calcium chloride and sucrose prefreezing treatments on the structure of strawberry tissues. *Lebensmittel-Wissenschaft und-Technologie* 33(2):89–102, 2000.
74. Sv Buggenhout et al. Minimizing texture loss of frozen strawberries: Effect of infusion with pectinmethylesterase and calcium combined with different freezing conditions and effect of subsequent storage/thawing conditions. *European Food Research and Technology* 223(3):395–404, 2006.
75. HK Mebatsion et al. Modelling fruit (micro)structures, why and how? *Trends in Food Science & Technology* 19(2):59–66, 2008.
76. ML Oey et al. Effect of turgor on micromechanical and structural properties of apple tissue: A quantitative analysis. *Postharvest Biology and Technology* 44(3):240–247, 2007.
77. A Zdunek et al. New contact acoustic emission detector for texture evaluation of apples. *Journal of Food Engineering* 99(1):83–91, 2010.
78. A Zdunek et al. Evaluation of apple texture with contact acoustic emission detector: A study on performance of calibration models. *Journal of Food Engineering* 106(1):80–87, 2011.
79. AGJ Voragen et al. Pectins. In: AM Stephen and I Dea, eds. *Food Polysaccharides*. New York, NY: Marcel Dekker Inc., 1995, pp. 287–339.
80. GE Anthon and DM Barrett. Characterization of the temperature activation of pectin methylesterase in green beans and tomatoes. *Journal of Agricultural and Food Chemistry* 54(1):204–211, 2006.
81. SM Krall and RF McFeeters. Pectin hydrolysis: Effect of temperature, degree of methylation, pH, and calcium on hydrolysis rates. *Journal of Agricultural and Food Chemistry* 46(4):1311–1315, 1998.
82. T Sajjaanantakul, JP Van Buren, and DL Downing. Effect of cations on heat degradation of chelator-soluble carrot pectin. *Carbohydrate Polymers* 20(3):207–214, 1993.
83. T Duvetter et al. Pectins in processed fruit and vegetables: Part I—Stability and catalytic activity of pectinases. *Comprehensive Reviews in Food Science and Food Safety* 8(2):75–85, 2009.
84. JM Denes, A Baron, and JF Drilleau. Purification, properties and heat inactivation of pectin methylesterase from apple (cv Golden Delicious). *Journal of the Science of Food and Agriculture* 80(10):1503–1509, 2000.
85. AR Hirsch et al. Impact of minimal heat-processing on pectin methylesterase and peroxidase activity in freshly squeezed Citrus juices. *European Food Research and Technology* 232(1):71–81, 2011.
86. T Sajjaanantakul, JP Van Buren, and DL Downing. Effect of methyl ester content on heat degradation of chelator-soluble carrot pectin. *Journal of Food Science* 54(5):1272–1277, 1989.
87. ACK Sato, EJ Sanjinez-Argandona, and RL Cunha. The effect of addition of calcium and processing temperature on the quality of guava in syrup. *International Journal of Food Science and Technology* 41(4):417–424, 2006.

88. CN Aguilar et al. Blanching at low temperatures: A thermal bioprocess applied to fruits and vegetables to improve textural quality. *Food Science and Biotechnology* 13(1):104–108, 2004.
89. J Ahmed and US Shivhare. Thermal processing of vegetables. In: DW Sun, ed. *Thermal Food Processing*. Boca Raton, FL: CPC Press, Taylor & Francis Group, 2006, pp. 387–422.
90. P Varoquaux, M Souty, and F Varoquaux. Water blanching of whole apricots. *Sciences Des Aliments* 6(4):591–600, 1986.
91. IAG Weinert, J Solms, and F Escher. Quality of canned plums with varying degrees of ripeness. II. Texture measurements and sensory evaluation of texture and colour. *Lebensmittel-Wissenschaft und -Technologie* 23(2):117–121, 1990.
92. B Qi, KG Moore, and J Orchard. Effect of cooking on banana and plantain texture. *Journal of Agricultural and Food Chemistry* 48(9):4221–4226, 2000.
93. CG Mallidis and C Katsaboxakis. Effect of thermal processing on the texture of canned apricots. *International Journal of Food Science & Technology* 37(5):569–572, 2002.
94. NI Lebovka, I Praporscic, and E Vorobiev. Effect of moderate thermal and pulsed electric field treatments on textural properties of carrots, potatoes and apples. *Innovative Food Science and Emerging Technologies* 5(1):9–16, 2004.
95. AC Kawazoe Sato, R Lopes da Cunha, and EJ Sanjinez-Argandona. Evaluation of colour, texture and mass transfer during guava in syrup processing. *Brazilian Journal of Food Technology* 8(2):149–156, 2005.
96. S Beirao-da-Costa et al. Effects of maturity stage and mild heat treatments on quality of minimally processed kiwifruit. *Journal of Food Engineering* 76(4):616–625, 2006.
97. A Karthiayani and CT Devadas. Textural properties of canned banana. *Journal of Food Science and Technology* 44(4):440–442, 2007.
98. A Koukounaras, G Diamantidis, and E Sfakiotakis. The effect of heat treatment on quality retention of fresh-cut peach. *Postharvest Biology and Technology* 48(1):30–36, 2008.
99. AH Rather et al. Effect of fruit ripeness and calcium chloride on the quality of canned Bartlett pears. *Advances in Food Sciences* 31(2):109–113, 2009.
100. O Gibert et al. A kinetic approach to textural changes of different banana genotypes (Musa sp.) cooked in boiling water in relation to starch gelatinization. *Journal of Food Engineering* 98(4):471–479, 2010.
101. C Ella Missang et al. Texture variation in apricot: Intra-fruit heterogeneity, impact of thinning and relation with the texture after cooking. *Food Research International* 44(1):46–53, 2011.
102. MV Shynkaryk et al. Ohmic heating of peaches in the wide range of frequencies (50 hz to 1 mhz). *Journal of Food Science* 75(7):E493–E500, 2010.
103. E Vorobiev and N Lebovka. Enhanced extraction from solid foods and biosuspensions by pulsed electrical energy. *Food Engineering Reviews* 2(2):95–108, 2010.
104. C Nunes et al. Effect of candying on microstructure and texture of plums (Prunus domestica L.). *LWT—Food Science and Technology* 41(10):1776–1783, 2008.
105. JM Barat et al. Pineapple candying at mild temperature by applying vacuum impregnation. *Journal of Food Science* 67(8):3046–3052, 2002.
106. J Giner et al. Rheology of clarified cherry juices. *Journal of Food Engineering* 30(1/2):147–154, 1996.
107. CA Zuritz et al. Density, viscosity and coefficient of thermal expansion of clear grape juice at different soluble solid concentrations and temperatures. *Journal of Food Engineering* 71(2):143–149, 2005.
108. A Altan and M Maskan. Rheological behavior of pomegranate (Punica granatum L.) juice and concentrate. *Journal of Texture Studies* 36(1):68–77, 2005.
109. R Ibarz et al. Flow behavior of clarified orange juice at low temperatures. *Journal of Texture Studies* 40(4):445–456, 2009.
110. NL Chin et al. Modelling of rheological behaviour of pummelo juice concentrates using master-curve. *Journal of Food Engineering* 93(2):134–140, 2009.
111. V Falguera and A Ibarz. A new model to describe flow behaviour of concentrated orange juice. *Food Biophysics* 5(2):114–119, 2010.
112. D Gabriele, M Migliori, and Bd Cindio. Innovation in fig syrup production process: A rheological approach. *International Journal of Food Science & Technology* 45(9):1947–1955, 2010.
113. D Bergamaschi Guedes, A Mota Ramos, and MD Martins Silva Diniz. Effect of temperature and concentration on the physical properties of watermelon pulp. *Brazilian Journal of Food Technology* 13(4):279–285, 2010.
114. AM Sharoba and MF Ramadan. Rheological behavior and physicochemical characteristics of goldenberry (Physalis peruviana) juice as affected by enzymatic treatment. *Journal of Food Processing and Preservation* 35(2):201–219, 2011.

115. N Grigelmo-Miguel, A Ibarz-Ribas, and O Martin-Belloso. Flow properties of orange dietary fiber suspensions. *Journal of Texture Studies* 30(3):245–257, 1999.
116. N Grigelmo-Miguel, A Ibarz-Ribas, and O Martin-Belloso. Rheology of peach dietary fibre suspensions. *Journal of Food Engineering* 39(1):91–99, 1999.
117. MK Krokida, ZB Maroulis, and GD Saravacos. Rheological properties of fluid fruit and vegetable puree products: Compilation of literature data. *International Journal of Food Properties* 4(2):179–200, 2001.
118. R Fügel, R Carle, and A Schieber. Quality and authenticity control of fruit purées, fruit preparations and jams—A review. *Trends in Food Science & Technology* 16(10):433–441, 2005.
119. R Maceiras, E Álvarez, and MA Cancela. Rheological properties of fruit purees: Effect of cooking. *Journal of Food Engineering* 80(3):763–769, 2007.
120. JF Maingonnat, L Muller, and JC Leuliet. Modelling the build-up of a thixotropic fluid under viscosimetric and mixing conditions. *Journal of Food Engineering* 71(3):265–272, 2005.
121. JE Lozano and A Ibarz. Thixotrophic behaviour of concentrated fruit pulps. *Lebensmittel-Wissenschaft und -Technologie* 27(1):16–18, 1994.
122. V Hegedusic, T Lovric, and A Parmac. Influence of phase transition (freezing and thawing) on thermophysical and rheological properties of apple puree-like products. *Acta Alimentaria* 22(4):337–349, 1993.
123. AM Ramos and A Ibarz. Thixotropy of orange concentrate and quince puree. *Journal of Texture Studies* 29(3):313–324, 1998.
124. S Bhattacharya. Yield stress and time-dependent rheological properties of mango pulp. *Journal of Food Science* 64(6):1029–1033, 1999.
125. V Speiciene et al. Rheological properties of currant purees and jams: Effect of composition and method of proceeding. *Journal of Food, Agriculture & Environment* 6(3–4):162–166, 2008.
126. B Santanu, US Shivhare, and GSV Raghavan. Time dependent rheological characteristics of pineapple jam. *International Journal of Food Engineering* 3(3), 2007.
127. RR Milczarek and KL McCarthy. Relationship between the Bostwick measurement and fluid properties. *Journal of Texture Studies* 37(6):640–654, 2006.
128. KL McCarthy, RF Sacher, and TC Garvey. Relationship between rheological behavior and Bostwick measurement during manufacture of ketchup. *Journal of Texture Studies* 39(5):480–495, 2008.
129. S Croak and M Corredig. The role of pectin in orange juice stabilization: Effect of pectin methylesterase and pectinase activity on the size of cloud particles. *Food Hydrocolloids* 20(7):961–965, 2006.
130. H Lee et al. De-esterification pattern of Valencia orange pectinmethylesterases and characterization of modified pectins. *Journal of the Science of Food and Agriculture* 88(12):2102–2110, 2008.
131. J Carbonell et al. Pectin methylesterase activity in juices from mandarins, oranges and hybrids. *European Food Research and Technology* 222(1):83–87, 2006.
132. LSFCA Collet et al. A kinetic study on pectinesterase inactivation during continuous pasteurization of orange juice. *Journal of Food Engineering* 69(1):125–129, 2005.
133. B Ingallinera et al. Effects of thermal treatments on pectinesterase activity determined in blood oranges juices. *Enzyme and Microbial Technology* 36(2–3):258–263, 2005.
134. S Leizerson and E Shimoni. Effect of ultrahigh-temperature continuous ohmic heating treatment on fresh orange juice. *Journal of Agricultural and Food Chemistry* 53(9):3519–3524, 2005.
135. F Sampedro, D Rodrigo, and M Hendrickx. Inactivation kinetics of pectin methyl esterase under combined thermal-high pressure treatment in an orange juice-milk beverage. *Journal of Food Engineering* 86(1):133–139, 2008.
136. E Sentandreu et al. Effects of heat treatment conditions on fresh taste and on pectinmethylesterase activity of chilled mandarin and orange juices. *Food Science and Technology International* 11(3):217–222, 2005.
137. TB Tribess and CC Tadini. Inactivation kinetics of pectin methyl-esterase in orange juice as a function of pH and temperature/time process conditions. *Journal of the Science of Food and Agriculture* 86(9):1328–1335, 2006.
138. L Wicker and F Temelli. Heat inactivation of pectinesterase in orange juice pulp. *Journal of Food Science* 53(1):162–164, 1988.
139. S Tajchakavit and HS Ramaswamy. Thermal vs. microwave inactivation kinetics of pectin methylesterase in orange juice under batch mode heating conditions. *Lebensmittel-Wissenschaft und-Technologie* 30(1):85–93, 1997.
140. A Espachs-Barroso et al. Inactivation of plant pectin methylesterase by thermal or high intensity pulsed electric field treatments. *Innovative Food Science and Emerging Technologies* 7(1–2):40–48, 2006.

141. de SA Assis, ABG Martins, and OMM de Faria Oliveira. Purification and characterization of pectin methylesterase from acerola (Malpighia glabra L.). *Journal of the Science of Food and Agriculture* 87(10):1845–1849, 2007.
142. CS Nunes et al. Thermal and high-pressure stability of purified pectin methylesterase from plums (Prunus Domestica). *Journal of Food Biochemistry* 30(2):138–154, 2006.
143. Y Guiavarc'h et al. Purification, characterization, thermal and high-pressure inactivation of a pectin methylesterase from white grapefruit (Citrus paradisi). *Innovative Food Science & Emerging Technologies* 6(4):363–371, 2005.
144. FJ Francis. Colorimetric properties of foods. In: MA Rao, SSH Rizvi, and AK Datta, eds. *Engineering Properties of Foods*. Boca Raton, FL: CRC Press, Taylor & Francis Group, 2005, pp. 703–732.
145. IMLB Avila and CLM Silva. Modelling kinetics of thermal degradation of colour in peach puree. *Journal of Food Engineering* 39(2):161–166, 1999.
146. FM Silva and CLM Silva. Colour changes in thermally processed cupuacu (Theobroma grandiflorum) puree: Critical times and kinetics modelling. *International Journal of Food Science & Technology* 34(1):87–94, 1999.
147. S Garza et al. Non-enzymatic browning in peach puree during heating. *Food Research International* 32(5):335–343, 1999.
148. C Benjar and N Athapol. Color degradation kinetics of pineapple puree during thermal processing. *LWT—Food Science and Technology* 40(2):300–306, 2007.
149. A Patras et al. Impact of high pressure processing on total antioxidant activity, phenolic, ascorbic acid, anthocyanin content and colour of strawberry and blackberry purees. *Innovative Food Science and Emerging Technologies* 10(3):308–313, 2009.
150. J Garcia-Parra et al. Effect of thermal and high-pressure processing on the nutritional value and quality attributes of a nectarine puree with industrial origin during the refrigerated storage. *Journal of Food Science* 76(4):C618–C625, 2011.
151. M Rattanathanalerk, N Chiewchan, and W Srichumpoung. Effect of thermal processing on the quality loss of pineapple juice. *Journal of Food Engineering* 66(2):259–265, 2005.
152. MR Mosshammer et al. Impact of thermal treatment and storage on color of yellow-orange cactus pear (Opuntia ficus-indica L. Mill. cv. "Gialla") juices. *Journal of Food Science* 71(7):C400–C406, 2006.
153. KM Herbach et al. Effects of processing and storage on juice colour and betacyanin stability of purple pitaya (Hylocereus polyrhizus) juice. *European Food Research and Technology* 224(5):649–658, 2007.
154. D Rodrigo, Av Loey, and M Hendrickx. Combined thermal and high pressure colour degradation of tomato puree and strawberry juice. *Journal of Food Engineering* 79(2):553–560, 2007.
155. LF Damasceno et al. Non-enzymatic browning in clarified cashew apple juice during thermal treatment: Kinetics and process control. *Food Chemistry* 106(1):172–179, 2008.
156. L Queiroz Zepka et al. Thermal degradation kinetics of carotenoids in a cashew apple juice model and its impact on the system color. *Journal of Agricultural and Food Chemistry* 57(17):7841–7845, 2009.
157. E Sadilova et al. Matrix dependent impact of sugar and ascorbic acid addition on color and anthocyanin stability of black carrot, elderberry and strawberry single strength and from concentrate juices upon thermal treatment. *Food Research International* 42(8):1023–1033, 2009.
158. A Rawson et al. Effect of thermosonication on bioactive compounds in watermelon juice. *Food Research International* 44(5):1168–1173, 2011.
159. Z Chao et al. Comparison of thermal, ultraviolet-c, and high pressure treatments on quality parameters of watermelon juice. *Food Chemistry* 126(1):254–260, 2011.
160. AAL Tribst et al. Quality of mango nectar processed by high-pressure homogenization with optimized heat treatment. *Journal of Food Science* 76(2):M106–M110, 2011.
161. J Ahmed, US Shivhare, and K Mandeep. Thermal colour degradation kinetics of mango puree. *International Journal of Food Properties* 5(2):359–366, 2002.
162. I Scibisz et al. Mid-infrared spectroscopy as a tool for rapid determination of internal quality parameters in tomato. *Food Chemistry* 125(4):1390–1397, 2011.
163. AA Gowen et al. Hyperspectral imaging—An emerging process analytical tool for food quality and safety control. *Trends in Food Science & Technology* 18(12):590–598, 2007.
164. Q Jianwei and L Renfu. Measurement of the optical properties of fruits and vegetables using spatially resolved hyperspectral diffuse reflectance imaging technique. *Postharvest Biology and Technology* 49(3):355–365, 2008.
165. N Nguyen Do Trong et al. Prediction of optimal cooking time for boiled potatoes by hyperspectral imaging. *Journal of Food Engineering* 105(4):617–624, 2011.

166. P Onsekizoglu, KS Bahceci, and MJ Acar. Clarification and the concentration of apple juice using membrane processes: A comparative quality assessment. *Journal of Membrane Science* 352(1–2):160–165, 2010.
167. JP Yuan and F Chen. Degradation of ascorbic acid in aqueous solution. *Journal of Agricultural and Food Chemistry* 46(12):5078–5082, 1998.
168. R Garcia-Torres et al. Effects of dissolved oxygen in fruit juices and methods of removal. *Comprehensive Reviews in Food Science and Food Safety* 8(4):409–423, 2009.
169. I Oley et al. Temperature and pressure stability of L-ascorbic acid and/or [6s] 5-methyltetrahydrofolic acid: A kinetic study. *European Food Research and Technology* 223(1):71–77, 2006.
170. GD Sadler, J Roberts, and J Cornell. Determination of oxygen solubility in liquid foods using dissolved-oxygen electrode. *Journal of Food Science* 53(5):1493–1496, 1988.
171. MG Corradini and M Peleg. Prediction of vitamin loss during non-isothermal heat processes and storage with non-linear kinetic models. *Trends in Food Science & Technology* 17:24–34, 2006.
172. A Serpen and V Gökmen. Reversible degradation kinetics of ascorbic acid under reducing and oxidizing conditions. *Food Chemistry* 104(2):721–725, 2007.
173. C Mertz et al. Characterization and thermal lability of carotenoids and vitamin C of tamarillo fruit (Solanum betaceum Cav.). *Food Chemistry* 119(2):653–659, 2010.
174. A Landl et al. Effect of high pressure processing on the quality of acidified Granny Smith apple purée product. *Innovative Food Science & Emerging Technologies* 11(4):557–564, 2010.
175. G Pataro, G Donsì, and G Ferrari. Aseptic processing of apricots in syrup by means of a continuous pilot scale ohmic unit. *LWT—Food Science and Technology* 44(6):1546–1554, 2011.
176. C Dhuique-Mayer et al. Thermal degradation of antioxidant micronutrients in Citrus juice: Kinetics and newly formed compounds. *Journal of Agricultural and Food Chemistry* 55(10):4209–4216, 2007.
177. MC Manso et al. Modelling ascorbic acid thermal degradation and browning in orange juice under aerobic conditions. *International Journal of Food Science and Technology* 36(3):303–312, 2001.
178. M Karhan et al. Kinetic modeling of anaerobic thermal degradation of ascorbic acid in rose hip (Rosa canina L) pulp. *Journal of Food Quality* 27(5):311–319, 2004.
179. MW Davey, K Kenis, and J Keulemans. Genetic control of fruit vitamin C contents. *Plant Physiology* 142(1):343–351, 2006.
180. RK Toor and GP Savage. Antioxidant activity in different fractions of tomatoes. *Food Research International* 38(5):487–494, 2005.
181. AL Vasquez-Caicedo et al. Effects of thermal processing and fruit matrix on beta-carotene stability and enzyme inactivation during transformation of mangoes into puree and nectar. *Food Chemistry* 102(4):1172–1186, 2007.
182. HS Lee and GA Coates. Thermal pasteurization effects on color of red grapefruit juices. *Journal of Food Science* 64(4):663–666, 1999.
183. L Cinquanta et al. Effect on orange juice of batch pasteurization in an improved pilot-scale microwave oven. *Journal of Food Science* 75(1):E46–E50, 2010.
184. JJT Gama and CM de Sylos. Effect of thermal pasteurization and concentration on carotenoid composition of Brazilian Valencia orange juice. *Food Chemistry* 100(4):1686–1690, 2007.
185. A Fratianni, L Cinquanta, and G Panfili. Degradation of carotenoids in orange juice during microwave heating. *LWT—Food Science and Technology* 43(6):867–871, 2010.
186. CH Lin and BH Chen. Stability of carotenoids in tomato juice during processing. *European Food Research and Technology* 221(3–4):274–280, 2005.
187. C Seybold et al. Changes in contents of carotenoids and vitamin E during tomato processing. *Journal of Agricultural and Food Chemistry* 52(23):7005–7010, 2004.
188. AA Abushita, HG Daood, and PA Biacs. Change in carotenoids and antioxidant vitamins in tomato as a function of varietal and technological factors. *Journal of Agricultural and Food Chemistry* 48(6):2075–2081, 2000.
189. V Dewanto et al. Thermal processing enhances the nutritional value of tomatoes by increasing total antioxidant activity. *Journal of Agricultural and Food Chemistry* 50(10):3010–3014, 2002.
190. E Mayer-Miebach et al. Thermal processing of carrots: Lycopene stability and isomerisation with regard to antioxidant potential. *Food Research International* 38(8–9):1103–1108, 2005.
191. RK Toor and GP Savage. Effect of semi-drying on the antioxidant components of tomatoes. *Food Chemistry* 94(1):90–97, 2006.
192. D Dutta, UR Chaudhuri, and R Chakraborty. Effect of thermal treatment on the beta-carotene content, colour and textural properties of pumpkin. *Journal of Food Science and Technology-Mysore* 43(6):607–611, 2006.

193. SK Sharma and M le Maguer. Kinetics of lycopene degradation in tomato pulp solids under different processing and storage conditions. *Food Research International* 29(3/4):309–315, 1996.
194. WW Fish and AR Davis. The effects of frozen storage conditions on lycopene stability in watermelon tissue. *Journal of Agricultural and Food Chemistry* 51(12):3582–3585, 2003.
195. W Stahl and H Sies. Uptake of lycopene and its geometrical-isomers is greater from heat-processed than from unprocessed tomato juice in humans. *Journal of Nutrition* 122(11):2161–2166, 1992.
196. K Frohlich et al. Effects of ingestion of tomatoes, tomato juice and tomato puree on contents of lycopene isomers, tocopherols and ascorbic acid in human plasma as well as on lycopene isomer pattern. *British Journal of Nutrition* 95(4):734–741, 2006.
197. KH van het Hof et al. Carotenoid bioavailability in humans from tomatoes processed in different ways determined from the carotenoid response in the triglyceride-rich lipoprotein fraction of plasma after a single consumption and in plasma after four days of consumption. *Journal of Nutrition* 130(5):1189–1196, 2000.
198. DB Rodriguez-Amaya. Quantitative analysis, *in vitro* assessment of bioavailability and antioxidant activity of food carotenoids—A review. *Journal of Food Composition and Analysis* 23(7):726–740, 2010.
199. EA Tydeman et al. Effect of carrot (*Daucus carota*) microstructure on carotene bioaccessibilty in the upper gastrointestinal tract. 1. *In vitro* simulations of carrot digestion. *Journal of Agricultural and Food Chemistry* 58(17):9847–9854, 2010.
200. EA Tydeman et al. Effect of carrot (*Daucus carota*) microstructure on carotene bioaccessibility in the upper gastrointestinal tract. 2. *In vivo* digestions. *Journal of Agricultural and Food Chemistry* 58(17):9855–9860, 2010.
201. RM Faulks and S Southon. Challenges to understanding and measuring carotenoid bioavailability. *Biochimica Et Biophysica Acta—Molecular Basis of Disease* 1740(2):95–100, 2005.
202. CA Svelander et al. Processing of tomato: Impact on *in vitro* bioaccessibility of lycopene and textural properties. *Journal of the Science of Food and Agriculture* 90(10):1665–1672, 2010.
203. O Livny et al. beta-carotene bioavailability from differently processed carrot meals in human ileostomy volunteers. *European Journal of Nutrition* 42(6):338–345, 2003.
204. S Van Buggenhout et al. *In vitro* approaches to estimate the effect of food processing on carotenoid bioavailability need thorough understanding of process induced microstructural changes. *Trends in Food Science & Technology* 21(12):607–618, 2010.
205. S Van Buggenhout et al. Pectins in processed fruits and vegetables: Part III—Texture engineering. *Comprehensive Reviews in Food Science and Food Safety* 8(2):105–117, 2009.
206. CMGC Renard. Effects of conventional boiling on the polyphenols and cell walls of pears. *Journal of the Science of Food and Agriculture* 85(2):310–318, 2005.
207. C Le Bourvellec et al. Phenolic and fiber composition of applesauce is close to that of apple flesh. *Journal of Food Composition and Analysis* 24:537–547, 2011.

Part III

Innovations in Thermal Food Processes

17 Aseptic Processing and Packaging

Min Liu and John D. Floros

CONTENTS

17.1 INTRODUCTION

Commercial sterilization has been used widely for food preservation during the twentieth century. It produces safe and shelf-stable foods with good nutritional and sensory attributes and has contributed significantly to the well-being of the world's population [1]. Aseptic processing and packaging (APP) is one of the two major commercial sterilization methods, also called "in-flow" sterilization. The other commonly used method is "in-container" sterilization, which generally refers to conventional canning processes [2]. The term "commercial sterilization" is defined as the absence of any microorganisms capable of reproducing in foods under nonrefrigerated conditions of storage and distribution, instead of the absolute absence of every existing form of microbial life [1].

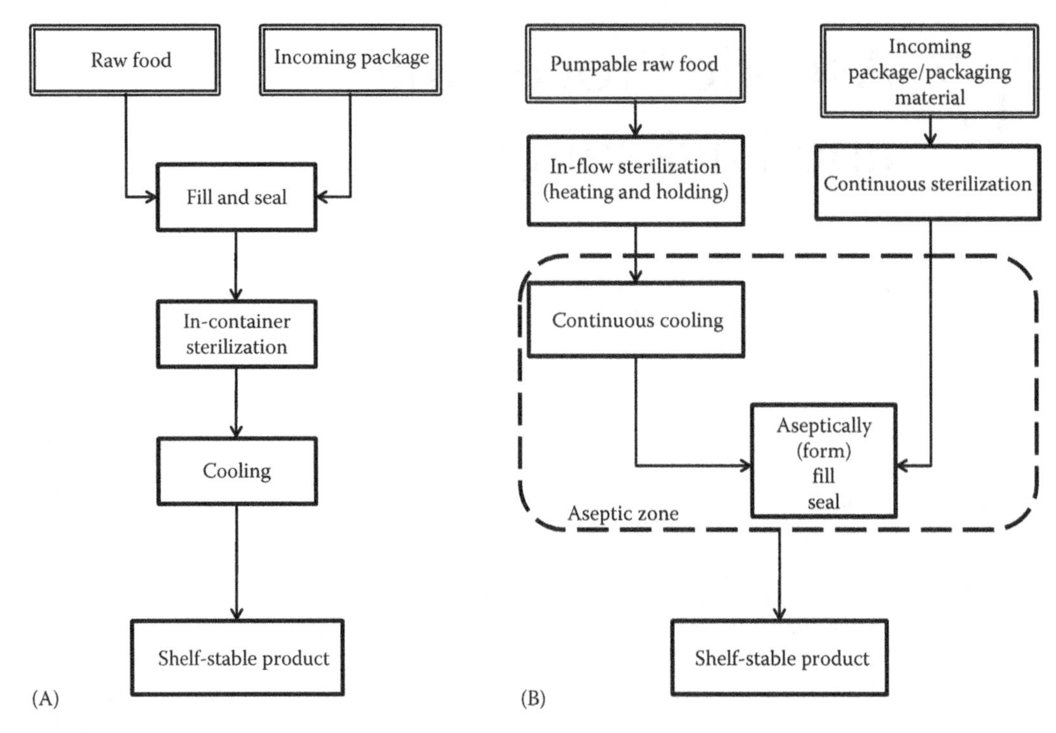

FIGURE 17.1 A comparison between (A) conventional thermal processing and packaging ("in-container" sterilization) and (B) APP ("in-flow" sterilization). (Based on the information in Floros, J.D. et al., Aseptic packaging technology, in P.E. Nelson, ed., *Principles of Aseptic Processing and Packaging*, 3rd edn., pp. 101–133, Purdue University Press, West Lafayette, IN, 2010.)

In conventional canning process, the product is placed into a container, such as glass bottles or metal cans, before sealing and sterilization [2]. However, in APP, the product and the package (or packaging material) are continuously sterilized separately (Figure 17.1). The commercially sterile and cooled product is then filled into the presterilized package in a sterile environment and hermetically sealed to produce a shelf-stable final product with extended shelf life but without the need of refrigerated storage [3]. APP has become a commercial success worldwide [4], because it minimizes quality losses in foods as compared to conventional retorting systems with long and slow heating (Figure 17.2). Meanwhile, APP allows the marketing of shelf-stable foods in more economical, lightweight, and easy to transport packaging systems, such as plastic bottles, cups and trays, or paper-based cartons, which cannot withstand normal retort processing conditions [4]. APP has been applied to almost any pumpable low- or high-acid food, and examples include juices/juice drinks, fruit and vegetable purees and concentrates, milk/dairy products, flavored drinks, liquid eggs, puddings, mineral water, tomato products, edible oils, wine, soy sauce, and many others [3,7].

The first aseptically processed milk in metal cans was developed in Denmark by Jonas Nielsen in 1913, almost 100 years after 1810, when Nicolas Appert discovered the canning process [1]. In 1917, M.E. Dunkley filed the first U.S. patent related to aseptic processing that described an invention of sterilizing cans and lids with saturated steam; the cans were then filled with presterilized material [1]. In 1933, the American Can Company developed a filling machine called the heat-cool-fill (HCF) system. However, the first commercially accepted aseptic canning system was the Dole system in 1951 [8]. This system was based on William M. Martin's patent [8] filed in the 1940s,which described an apparatus and method for preserving products in sealed containers by filling and

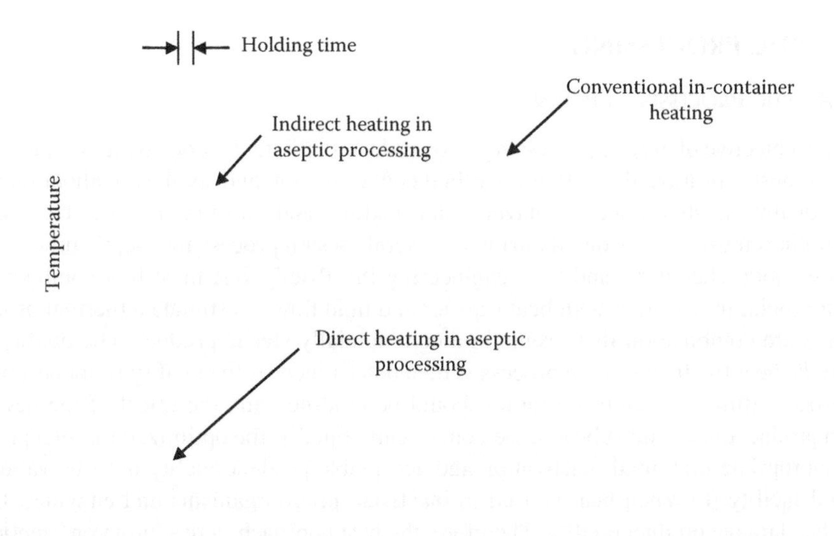

FIGURE 17.2 Comparison of product heating curves for aseptic processing and conventional "in-container" processing. (Based on the information in Lopez, A., Aseptic processing and packaging, in A. Lopez, ed., *A Complete Course in Canning,* pp. 160–211, The Canning Trade, Inc., Baltimore, MD, 1987; Singh, R.P. and Heldman, D.R., *Introduction to Food Engineering,* 4th edn., pp. 429–440, Academic Press/Elsevier, Burlington, MA, 2009.)

sealing under aseptic conditions. Consumer-size aseptic packages as we know them today were first developed in the 1960s in Europe for the milk industry by Tetra Pak [3]. The United States did not use APP until much later, after the U.S. Food and Drug Administration (FDA) approved the use of hydrogen peroxide as a sterilant for aseptic packages in 1981 [8]. Since then, the process has advanced rapidly, and the earlier Dole system is no longer the predominant choice. The newly developed systems are capable of filling high-and low-acid fluids into less expensive packaging alternatives such as plastic or laminated containers. The first successful patent filing of an aseptic process with particulates was by the Campbell Soup Company in the mid-1980s [8]. However, aseptic processing of low-acid particulate foods continues to be a major field of research in academia and development in industry. In 2010, more than 660 aseptic packaging systems were installed in the United States and many more worldwide [9].

The advantages of APP include the following [3,5,7,8,10,11]: (a) pumpable foods can be heated rapidly to a very high temperature (e.g., 300°F/150°C), resulting in very short process times (also called holding times) to achieve commercial sterility, and then quickly cooled, without the long come-up and cool-down times that are unavoidable during conventional canning processes (Figure 17.2). The result is a high-temperature short-time (HTST) process or ultrahigh-temperature (UHT) process that preserves higher nutritional quality and better sensory characteristics in foods. (b) A variety of packages can be used in aseptic packaging systems, ranging from traditional glass bottles and aluminum cans to the semirigid and flexible polymeric packages or composites of plastic/paperboard/metal; these lightweight packages are easier to make, transport, and handle, resulting in lower energy use and transportation costs. (c) Room temperature can be used throughout storage and distribution, thus reducing energy consumption as compared to a refrigerated supply chain. (d) APP allows various package shapes and sizes to be used with functionalities such as microwavability, convenience, easy opening and pouring, and great consumer appeal. (e) The process is faster and the overall cost is lower, because less energy is needed for sterilization, processing, packaging, storage, and transportation.

17.2 ASEPTIC PROCESSING

17.2.1 Aseptic Processing Design

The ultimate objective of aseptic processing is to produce shelf-stable foods with extended shelf life. The process must minimize the risk of microbial contamination and spoilage without refrigeration and retain quality attributes such as flavor, color, texture, and nutritional value. Therefore, three major disciplines must be considered during the overall design process for aseptic processing: food microbiology, food chemistry, and food engineering [8]. Briefly, one must first combine the principles of microbial inactivation with heat transfer and fluid flow to estimate a thermal process time and temperature combination that results in a commercially sterile product. The quality destruction due to the heat treatment of the process is then determined. If the quality is not acceptable, the thermal process (time/temperature profile) should be modified and the effect of the new thermal process on product quality must be assessed once again. Finally, the optimized thermal process that achieves appropriate microbial inactivation and acceptable product quality must be validated in a commercial facility [8].When heat is used to inactivate microorganisms and enzymes, this same heat can also damage product quality. Therefore, the best approach of designing on aseptic process is to optimize the temperature and time combination in order to maximize retention of food quality, while achieving commercial sterility.

17.2.2 Microbiology of Aseptic Processing

The extent of thermal treatment needed for commercial sterilization depends on the microorganisms that survive and grow in a product. The most heat-resistant microorganism, known as target microorganism, is chosen to evaluate commercial sterilization conditions. The design of an aseptic process follows many of the same principles as those used in the design of a canning process [8]. These principles include knowledge of thermal inactivation parameters for microbial cells and spores, and thermal destruction kinetics for quality attributes and enzymes [8]. The well-known $F^{z}_{T_{ref}}$ value for canning processes, or sterilization value, can also be used for the evaluation and calculation of aseptic processes. The $F^{z}_{T_{ref}}$ value is the time required to kill a known amount of microorganisms with a known z-value in a given food at some reference temperature (T_{ref}). The kinetics of thermal destruction of microorganisms can be described as a first-order reaction under lethal temperatures [12] and can be defined by

$$F^{z}_{T_{ref}} = \int_{0}^{t} 10^{(T-T_{ref})/z}\, dt, \tag{17.1}$$

where

T is the temperature of the product at the slowest-heating point at time t (known as the cold spot)

T_{ref} is the reference temperature for the microorganism of concern, and z value is given by

$$z = \frac{T_1 - T_2}{\log_{10} D_{T_1} - \log_{10} D_{T_2}}, \tag{17.2}$$

where z value is the change in temperature from T_1 to T_2 needed to change the D value by a factor of 10 from D_{T1} to D_{T2}. The D value is the time needed for a microbial population to be reduced by 90%, or 1-log cycle at a constant heating temperature T, and is given by

$$D_T = \frac{\log_{10} N_0 - \log_{10} N}{t}, \quad (17.3)$$

where

N_0 is the initial number of microorganisms

N is the number of microorganisms surviving after time *t* at temperature T

The F value is referred to as F_0 value, when the reference temperature is 121°C (250°F) and the z value is 10°C (18°F). Different foods have different F_0 values, because the thermal resistance of microorganisms can be influenced by a variety of factors. F_0 values have been established for many foods. For example, low-acid foods (pH>4.6 and water activity>0.85) must receive a treatment resulting in a 12-log reduction of *Clostridium botulinum*, and the D value for *C. botulinum* at 121°C in a low-acid food is 0.25 min; thus, the F_0 value is 12×0.25 min=3 min, which indicates that such a thermal process is equivalent to a heat treatment of 3 min at 121°C. Many combinations of time and temperature can yield an F_0 value of 3 min, such as low temperature long time (LTLT), HTST, or UHT. However, according to the FDA, the whole thermal process for in-container sterilization including the heating and cooling time is used to calculate the F value, but for aseptic processing, only the holding time and temperature are allowed toward the calculation of the F value, even if the heating and cooling sections may contribute some thermal destruction of microorganisms. This is because the heating and cooling processes are very rapid, and the temperature distribution during heating and cooling is uncertain (Figure 17.2) [13].

17.2.3 Chemistry of Aseptic Processing

The destruction of color, flavor, and nutrients, and the inactivation of enzymes follow similar kinetics to that of the inactivation of microorganisms. Two parameters are important in thermal destruction kinetics: (1) the destruction rate at a given temperature and (2) the sensitivity of the destruction rate to temperature [14]. These two parameters correspond to the D value and z value of microbial inactivation, respectively, and the destruction of nutrients in food products is quantified by the cook value (C), which has been defined analogously to the F value as

$$C = \int_0^t 10^{(T-T_{ref})/z} dt \quad (17.4)$$

A cooking index C_0 is also defined similar to F_0, but the reference temperature is 100°C, rather than 121°C, and the z value here is 33°C instead of 10°C. D values or destruction rates for nutrients are usually much greater than those for microorganism (Figure 17.3). This makes thermal processing feasible by reducing bacteria populations very fast, while retaining food quality. On the other hand, the z value for destruction of nutrients and quality factors (e.g., vitamins, colors, and flavors) is also much higher than that of microorganisms. This implies that the difference between destruction rates of nutrients and microorganisms at higher temperatures is much larger than that at lower temperatures (Figure 17.3). In general, the z values for microorganisms are 5°C–15°C, while for nutrients and other quality factors are 20°C–40°C or even higher [18,19]. Thus, HTST and UHT processes are more effective in retaining nutrients and quality than LTLT processing. On the other hand, the C_0 values associated with LTLT processes with an F_0 value = 10 are in the order of 30–300 but only around 1–10 for HTST at higher temperature (130°C–145°C) [1]. This forms the basis for aseptic processing [17], which can take full advantage of HTST and UHT processing as the food can be heated to 130°C–145°C over a very short time using indirect or direct heating (Figure 17.2) while the in-container processing needs much longer heating time to reach only 120°C.

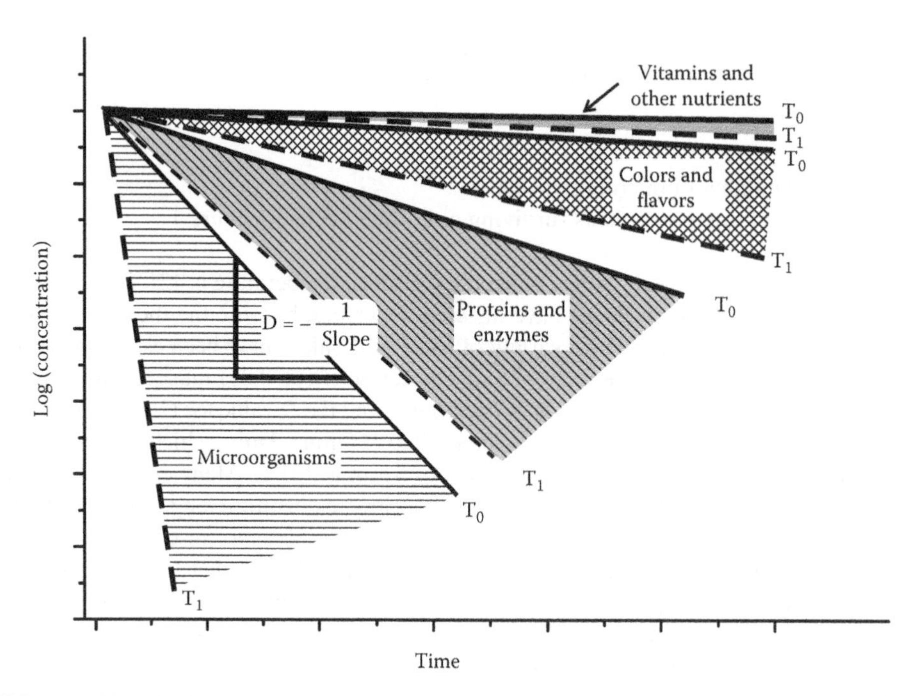

FIGURE 17.3 Hypothetical examples of thermal destruction of microorganisms, enzymes, colors, and vitamins at T_0 and T_1 (higher than T_0). The shaded regions represent D values for each quality characteristics or inactivation of microorganisms during thermal processing. (Based on the information in Dock, L.L. and Floros, J.D., Thermal and non-thermal preservation methods, in M. Schmidl, and T. Labuza, eds., *Science and Technology of Functional Foods,* Aspen Publishers, 1999; Jelen, P., *Canadian Inst. Food Sci. Technol. J.,* 16(3), 159, 1983; Jay, J.M. et al., *Modern Food Microbiology,* 7th edn., pp. 420–431, Springer, New York, 2005; Ashirifie-Gogofio, J. and Floros, J.D., Aseptic processing and packaging, in Da-Wen Sun, ed., *Handbook of Food Safety Engineering*, Oxford, England, Wiley-Blackwell, 2011.)

17.2.4 Elements of an Aseptic Processing System

A basic aseptic processing system consists of the following essential elements [8,13]: raw product supply tank, pumps with flow control, deaerator, heating section, holding tube, cooling section, back pressure valve, aseptic surge tank, and aseptic packaging system (Figure 17.4).

17.2.4.1 Flow Control

APP relies on the continuous flow of products and packages [8]. The continuous flow of a product is achieved by special pumps, a critical consideration in the design of any aseptic processing system. A metering pump is required to maintain the required rate of product flow according to FDA regulations. Pumps may be classified as centrifugal or positive displacement. A positive displacement pump is used as a metering pump more often than a centrifugal pump, because the former maintains a more consistent flow as pressure fluctuates due to the opening/closing of divert valves or fouling in the heat exchanger. However, centrifugal pumps can act as metering pumps when used with a closed-loop control system with a flow meter and a variable frequency drive [8]. There are two types of positive displacement pumps: rotary and reciprocating. Selection of the type of positive displacement pumps depends upon product characteristics such as viscosity, presence and size of particulates, pressure drop in the system, and cost [8]. For example, rotary positive displacement pumps are usually the most economical choice for processing homogeneous fluids or fluids with small particles (<3 mm) and the pressure drop in the system is less than 150 psi (around 10 atm). However, a reciprocating piston pump or a progressive-cavity pump must be used, if the product contained particles larger than 3 mm and up to about 8 cm in size [8]. The pump must prevent

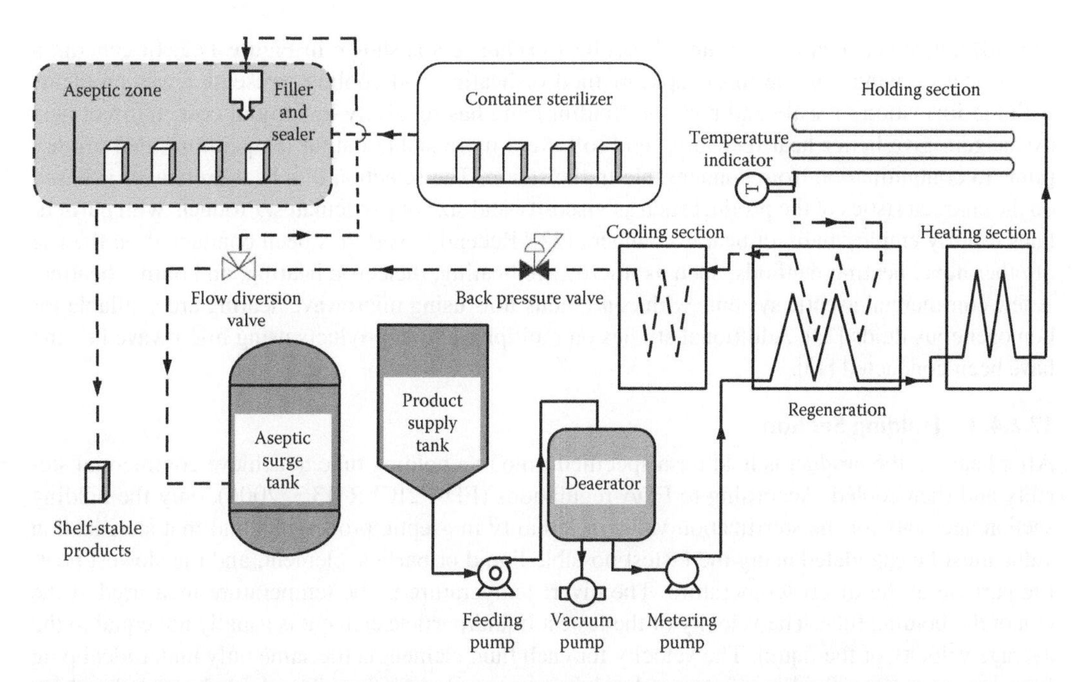

FIGURE 17.4 Schematic of an aseptic processing system. (Based on the information in Floros, J.D. et al., Aseptic packaging technology, in P.E. Nelson, ed., *Principles of Aseptic Processing and Packaging*, 3rd edn., pp. 101–133, Purdue University Press, West Lafayette, IN, 2010; Kumar, P., *Aseptic Processing of a Low-acid Multiphase Food Product using a Continuous Flow Microwave System,* North Carolina State University, Raleigh, NC, 2006.)

product recontamination by bacteria during operation, if it is used in the sterile zone of an aseptic system. Typically, these aseptic pumps have steam tracings around the seals and gaskets to maintain sterility in all critical areas [8].

17.2.4.2 Heating and Cooling Sections

Heating the product to the holding temperature and subsequently cooling it before packaging are carried out continuously using heat exchangers. These heat exchangers can be classified into two groups [13,20,21]:

a. Direct heat exchangers, where steam is condensed in the product for heating, and later evaporation removes vapor to cool. Direct heating can take place by the following:
   - *Steam injection* (steam into product): For homogeneous and high-viscosity products sensitive to shear stress such as creams, desserts, and viscous sauces.
   - *Steam infusion* (product into steam): Also for homogeneous and high-viscosity products sensitive to shear stress such as creams and desserts.
b. Indirect heat exchangers, where the product and the heat exchanging fluid are separated by the heating surface. Indirect heating can be done by the following:
   - *Plate heat exchangers*: For homogeneous and low-viscosity products containing small particles up to 5 mm, such as dairy and juice products. Plate heat exchangers are frequently used for energy regeneration to reduce the cost of heating and cooling (Figure 17.4).
   - *Tubular heat exchangers*: For homogeneous and high-viscosity products containing larger particles up to 10 mm, such as soups and fruit purees.
   - *Scraped surface heat exchangers*: For nonhomogeneous and/or highly viscous products containing very large particles up to 15 mm such as diced fruit preserves and soups.

The difference between indirect and direct heat exchangers is shown in Figure 17.2. In general, a direct heat exchanger is the most rapid method of heating and cooling, presents fewer problems with the formation of scale and burn-on (fouling) and has relatively low initial cost. Indirect heat exchangers give better heat recovery, tend to have a more stable output temperature, and are not prone to contamination from condensable in the steam. The selection of a heat exchanger is based on the characteristics of the product such as viscosity and size of particulates. Products with particulates usually employ indirect heat exchangers [22]. Recently, work has been conducted on the use of other novel heating methods, such as microwave heating, dielectric heating, and ohmic heating. A few commercial aseptic systems with continuous flow using microwave heating are available for homogeneous fluids [23]. Additional studies on multiphase food products using microwave heating have been conducted [13].

17.2.4.3 Holding Section

After heating, the product is held for a specified time in a holding tube to achieve commercial sterility and then cooled. According to FDA regulations (FDA 21CFR113.3, 2006), only the holding section accounts for the sterilization value or lethality in aseptic processing, and that sterilization value must be calculated using the fastest possible liquid or particle element, and the slowest heating particle at the divert temperature. The divert temperature is the temperature measured at the exit of the holding tube. The velocity of the fastest liquid/particle element is usually not equal to the average velocity of the liquid. The velocity for each fluid element is the same only under ideal plug flow. However, most fluids are under either laminar or turbulent flow. The fluid elements for these flows have a velocity profile shown in Figure 17.5. The velocity of the fastest element for developed laminar flow is twice that of the average flow velocity. This factor may be less than 2, when rheology and flow conditions are more accurately known. The velocity profile results in different lengths of time that each particular element spends in the heat exchanger or the holding tube. Therefore, one must be able to determine the length of time that various elements spend in the system, in other words, their residence time distribution (RTD) must be calculated, in order to estimate the lethality received by each element.

Since the holding tube is the heart of the process and ensures commercial sterility, several precautions must be taken [8]. First, the holding tube must be inclined upward with a gradient of at least 2 cm/m (0.25 inch/foot) in order to eliminate air pockets and ensure product draining. Second, the holding tube must follow good sanitary design criteria, such as a smooth interior without

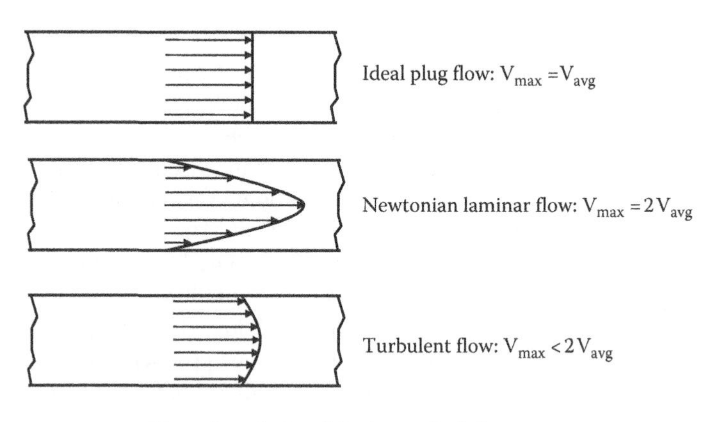

FIGURE 17.5 Velocity profiles of flows. (Based on the information in Singh, R.P. and Heldman, D.R., *Introduction to Food Engineering,* 4th edn., pp. 429–440, Academic Press/Elsevier, Burlington, MA, 2009.)

protrusions, easily disassembled for inspection, cleaning, and a fail-safe system for reassembly, so that the tube cannot be shortened or changed in diameter. Third, the temperature of the tube should be maintained as constant as possible. It should not be exposed to condensate drip or drafts of cold air. An insulation layer can be placed on the holding tube to limit heat losses, but heat cannot be applied at any point along the tube. Finally, the holding tube must be maintained under isobaric conditions well above the vapor pressure of the product at the process temperature to prevent flashing or boiling. This is usually accomplished by a back-pressure valve.

17.2.4.4 Back Pressure

A back-pressure valve is necessary in aseptic systems to prevent flashing of the product in the holding tube or heat exchanger, because the product is usually heated above the normal boiling temperature of water at atmospheric pressure [8]. The back pressure can be located anywhere after the holding tube. Alternative methods are needed if the product contains particles because the back-pressure valve will most likely destroy the particles. Such methods include the use of a second positive displacement pump running slightly slower than the metering pump or a pressurized aseptic surge tank.

17.2.4.5 Deaerators

Most aseptically processed products must be deaerated prior to thermal processing and packaging. The purpose of deaeration is to remove the entrapped air and minimize oxidative reactions that may reduce the quality of the product during processing and storage [13]. Generally, the deaerator is a vessel maintained at a certain vacuum using a vacuum pump. The location of deaerator depends on the sensitivity of the product to temperature. It is more economical to operate the deaerator at higher temperature; thus, the deaerator is often located after preheating or regeneration, if the product does not lose volatile flavors. Otherwise, the product must be deaerated before any thermal treatment.

17.2.4.6 Aseptic Surge Tank

Aseptic surge tanks are used to accumulate the sterile product prior to packaging. They provide flexibility to the process, when the flow rate of sterile product is not compatible with packaging capacity or when the packaging system is not operational due to malfunctions [8,13]. The capacity of an aseptic surge tank may vary from 100 gallons to several thousand gallons.

17.3 ASEPTIC PACKAGING

17.3.1 Packaging Functions

Containment, protection and preservation of the food, and properly informing the consumer are the basic functions of food packaging [24,25]. Development of an effective package begins by considering the general packaging functions and the protective role that packaging usually assumes in any product/package/environment system [3]. It is evident that every food constituent is susceptible to some environmental factors, and these environmental factors can cause food deterioration. For example, food lipids and vitamins could be oxidized and destroyed by atmospheric oxygen, if the oxygen barrier properties of the package were not adequate for the desired product shelf life. Problems with oxidation are minimal when foods are packaged in metal or glass container. However, aseptic processing takes advantage of less expensive and lightweight polymeric containers, which also have measurable oxygen permeation. Therefore, oxygen barrier properties need serious attention, when one chooses the type of a package for aseptic products [26]. Other examples include lipid rancidity and vitamin destruction caused by light, nutritional quality loss, and organoleptic changes due to undesired water vapor permeation, etc. [3]. Overall, consideration of barrier and sealing properties, chemical composition and compatibility with the contained food, and physical structure are essential when selecting packages for aseptic food products.

Besides considering the basic functions of a package mentioned earlier, several other special considerations are needed for aseptic packaging [3]. For example, the package must be commercially sterilized before filling. This requires the establishment of a closed, aseptic zone/system.

17.3.2 Packaging Materials

Aseptic packaging can utilize less expensive and lighter-weight packaging materials such as plastic polymers, in addition to metal or glass containers commonly used in conventional canning. However, no single polymer has all the essential properties or meets all the packaging requirements for aseptic packaging [3]. Therefore, a composite multilayered structure is often used to meet package design needs. For example, polyvinylidene chloride (PVDC) and ethylene vinyl alcohol (EVOH) have good barrier properties; high-density polyethylene (HDPE), low-density polyethylene (LDPE), and polypropylene (PP) have good sealing properties; and polyethylene teraphthalate (PET) or nylon has good mechanical and heat resistant properties. Therefore, composites using combinations of these three groups can create a heat-resistant rigid container with good sealing strength and gas barrier properties [3].

The selection of a packaging material depends on the nature of the product, the cost of both the product and the package, and the preference of the consumer. Generally, there are four types of aseptic packaging [3]: (1) rigid containers, such as metal cans, metal drums, glass bottles, and composites cans; (2) paperboard containers, such as web-fed or roll-fed paper/foil/plastic cartons and preformed/machine-erected cartons; (3) semirigid plastic containers, such as web-fed thermoformed or preformed cups, tubs, and trays; and (4) flexible plastic containers, such as blow molded bottle, pouches, sachets, bag-in-box, and bag-in-drum. Each type of the container has its own properties and characteristics. The most widespread consumer package for aseptic products is the paperboard laminated carton [1]. The structure of a typical paperboard carton is shown in Figure 17.6. The paper provides mechanical strength, good heat resistance, and light protection, but it is coated with polymers due to its sensitivity to moisture. The aluminum foil layer acts as an excellent gas, water, and light barrier, especially when laminated between two plastic layers, and is very thermally stable [27]. All of these properties have made the paper/foil/plastic laminations very popular in aseptic packaging. Several kinds of semirigid plastic containers are now aseptically produced. HDPE and PP are the two most common thermoplastics used, and pigments are often added to better protect the contents from light [27]. A current trend is the use of PET bottles for aseptically filled beverages, which allow consumers to see the product unlike in traditional aseptic juice box designs [3].

17.3.3 Sterilization in Aseptic Packaging

Sterilization of packaging and food contact surfaces is essential in aseptic systems. The required microbial count reduction (D values) for the sterilization of the food contact packaging material

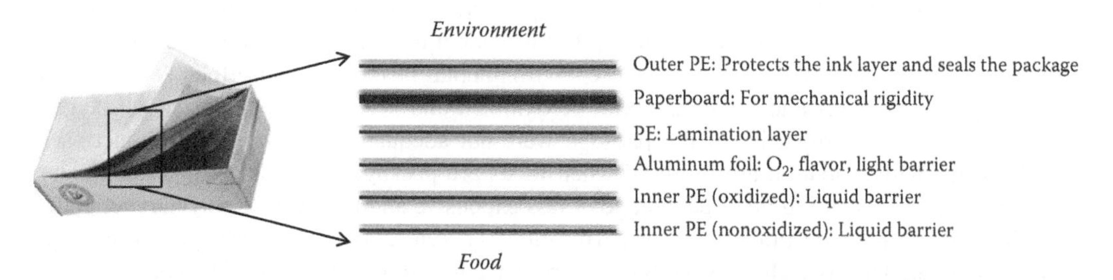

FIGURE 17.6 Typical structure of a paperboard-laminated carton for aseptic packaging. (Based on the information in Robertson, G.L., *Food Packaging: Principles and Practice,* pp. 255–270, Taylor & Francis/CRC Press, New York, 2006; Floros, J.D. et al., Aseptic packaging technology, in P.E. Nelson, ed., *Principles of Aseptic Processing and Packaging*, 3rd edn., pp. 101–133, Purdue University Press, West Lafayette, IN, 2010.)

surface depends on the type of the product, its desired shelf life, and the likely storage temperature [1]. A minimum of four decimal reductions in bacterial spores is required for acidic products (pH<4.5), while a decimal reduction of six is required for sterile, neutral, low-acid products (pH>4.5). A full 12-decimal reduction process must be given, if *C. botulinum* were able to grow in the product [1]. Care should be taken to minimize the initial microbial load on the packaging material. Packages and food contact surfaces for aseptic systems may also get contaminated by contact with air, machines, or individuals [28]. Therefore, the packaging materials should be produced, transported, and stored under conditions that are as free from microorganisms as possible. Four main sterilization processes for packaging material are in industrial use, either individually or in combination [3]:

- *Mechanical processes*: water rinsing/flushing, air blasting, brushing, and ultrasound. Mechanical processes are used primarily for precleaning and reduction of the initial microbial load.
- *Heat*: saturated steam, superheated steam, hot air, mixture of hot air and steam, and heat from extrusion during packaging formation. High temperature is normally used; thus, it is not compatible with heat-sensitive plastics.
- *Irradiation processes*: ionizing rays, infrared rays, and ultraviolet rays. They can be used mostly for heat-sensitive packaging materials or combine with other methods.
- *Chemical processes*: hydrogen peroxide (H_2O_2) is the most popular sterilization method for aseptic packaging, because it is relatively fast and efficient.
- *Combinations*: hydrogen peroxide/heat and hydrogen peroxide/UV. Hydrogen peroxide/heat is very popular for the same reasons of hydrogen peroxide alone, except that heat can improve the efficiency further and/or dry the package. A synergistic effect between H_2O_2 and UV is achieved.

17.3.4 Aseptic Packaging Systems

The equipment for aseptic packaging can be broadly classified into six basic categories [3,5,10,17,24,29,30].

a. *Fill and seal.* Aseptic canning systems and systems using glass or preformed plastic containers (thermoformed, injection, or blow molded) belong to this group. The preformed containers are sterilized, filled, and then sealed in a sterile environment. Examples include Dole, Serac, Remy Metal Box (Freshfill), Gasti, Crosscheck, Hamba, Ampack, and Remy [3,17,24,30].
b. *Erect, fill, and seal.* A knocked-down blank of incoming material is erected, sterilized, filled, and sealed. Examples include KHS, SIGPKL-Combibloc, and LiquiPak [3,17,24,30].
c. *Form, fill, and seal.* In this system, a roll of incoming material is sterilized, formed, filled, and sealed in an aseptic environment. Typical systems include carton and pouch-forming equipment with examples such as Tetra-Pak, International Paper, Asepak, Impaco, Astepo, DuPont, Prodo-Pak, and Thimmonier [3,5,17,24,29,30].
d. *Thermoform, fill, and seal.* Roll stock is sterilized or a sterile surface is provided. The material is then heated, thermoformed, filled, and sealed in a sterile environment. Examples include Benco, Bosch, Conoffast, Thermoforming USA, and IWRA-Hassia [3,5,17,24,30].
e. *Blow mold, fill, and seal.* The extrudable material is first blow molded in this process. The product is then filled into the package and sealed before the molds are opened, or the sterile bottle is closed in the mold and delivered to a separate operation, where the sterile container is opened, filled, and sealed in an aseptic environment. The process can produce bottles and other semirigid containers. Examples include Rommellag, Holopack, Remy, Serac, Bottlepack, and Sidel [3,17,30].

f. *Bulk packaging and storage systems.* Presterilized (or in-line sterilized) bags are aseptically filled and sealed. Drums and totes are first sterilized and then filled and sealed under sterile conditions. The larger size storage tanks are usually sterilized by chemical disinfectants (e.g., iodophores) and/or heat and specially designed valves allow aseptic filling and closing. Examples include bag-in-box systems such as Scholle, FMC FoodTech-FranRica, DuPont Star Asept, Liqui-Box, ELPO, and Rapak; drum systems such as Scholle, FMC FoodTech-FranRica, and Cherry-Burrell; tank and railcar systems such as FMC FoodTech-FranRica and Enerfab; and the Fischer Group for large ships [3,5,17,24,30].

17.3.5 Selection of Aseptic Packaging Systems

The demand for an increased variety of aseptically processed foods is growing, and the desire to include larger food particles, such as vegetable pieces in soups or fruit dices in yogurts, is increasing [3]. All products have specific chemical and physical properties and shelf-life requirements. In turn, this requires specific protection and preservation methods provided by packaging, such as oxygen barrier to minimize nutrient loss from oxidation reactions. The selection of a package for aseptic processing must be done carefully; otherwise, the package will either fail to provide the required functions for the food or cause negative impact on the quality of the food. For example, the package may absorb odors, flavors, and colors from the food. Customers' demands, such as volume and size of the package, and transparency of the package, cannot be ignored when selecting a packaging system. Flexibility of an aseptic system is also a very important factor, particularly the number of products that a system can process/produce. As the initial capital cost for an aseptic system is usually higher than other methods, food manufacturers must continually look for ways to add value to their products. Therefore, the desire to process various products can arise [3]. This may lead to situations where food manufacturers want to process and fill several products with their aseptic systems, when the systems were not originally intended for such use. In other words, many factors must be considered in selecting a packaging system for a given application. It is worth noting that all considerations for the selection of an aseptic packaging system are interrelated and should be viewed as components of a single system.

17.3.6 Aseptic Bulk Storage

Aseptic processing systems can deliver their commercially sterile products to a variety of packaging systems. Aseptic packaging systems can fill packages as small as 50 mL or as large as 300-gallon bag-in-box. Aseptic bulk storage refers to processing products during a harvest season for later re-packaging or other use. The size of the storage tanks can be up to 1 million gallons. Aseptic bulk storage was initiated and has become successful for several reasons [31]. Seasonality was the first motivating factor, but better quality products and global competition have also been the driving forces through the years.

Dr. Phil Nelson of Purdue University and his group were the first to work on aseptic bulk storage for tomatoes in the late 1960s [31]. The first aseptic bulk storage was then developed and patented by Purdue University in 1972 and was licensed exclusively to Bishopric company for commercialization [32]. The first commercial system, consisting of eight 40,000-gallon tanks, was installed in 1973 at NAAS Foods, Portland, IN, for the storage of single-strength chopped tomato. Later that year, eight 80,000-gallon tanks were installed at Kikko Foods in Haronamachi, Japan, for storage of tomato puree. Since then, 40,000- and 80,000-gallon tanks have been installed at numerous sites/companies for aseptic storage of grape juice, tomato concentrates, apple products, and other foods [33].

All large tanks must be sterilized before any sterile product is placed inside [31]. The first step in such a sterilization cycle is a presterilization operation to reduce the initial microbial load using an acid or base wash, brushing with mechanical brushes or other methods (i.e., ultrasound), and a final water rinse. The second step is the actual sterilization operation. This is achieved by steam or liquid chemicals (i.e., iodoforms). The sterilant is introduced into the tank from the bottom as the top manifold is still open. When the tank is full of sterilant, the top is closed and the sterilization begins. After the sterilization cycle is completed, the condensates or the liquid sterilants are pressurized out of the tank by sterile air or sterile nitrogen introduced through sterile high-efficiency particulate air (HEPA) filters at the top of the tank. Finally, sterile product comes into the tank from the bottom, through specially designed valves, while the pressure in the tank is always maintained above atmospheric.

Another important property of aseptic bulk storage is mobility [31]. This allows very large aseptic containers to move across countries and around the world. As a result, large tanker ships that can carry enormous amounts of sterile foods across oceans have been developed. On the ground, several aseptic containers have been used for transportation, including truck trailers, piggyback trailers, and rail cars.

Pouch and bag-in-box systems are used for the aseptic bulk storage in addition to tankers. Several such systems are available today, ranging from very small size pouch applications (i.e., ELPO, FranRica, Liqui-Box, Scholle) to very large size pouch applications (i.e., Asepak, Bosch, DuPont, Inpaco) [31]. The aseptic bag-in-box systems generally start with a bag that has been presterilized by radiation, whereas the pouch systems start with packaging material that is first sterilized, then aseptically made into a pouch, and finally filled. Laminated materials are now utilized to construct strong boxes replacing wooden and paperboard boxes for bag-in-box applications.

Aseptic bulk storage techniques have replaced the traditional methods of concentration or freezing for preserving commodity products like tomato paste or orange juice. They are now used throughout the world to transport huge quantities of foods.

17.4 ESTABLISHMENT, VALIDATION, AND REGULATION OF APP

Aseptic systems are unique in many ways. One characteristic that separates aseptic systems from most other thermal processing systems is the need to document the process adequately. Documentation is necessary to satisfy requirements for the product, the aseptic surge tank (if any), the adequacy of the sterilization process for packages, and the adequacy of the sterilization process for filling and packaging equipment. Regulatory agencies require registration of processing plants, and the filing of thermal processes and sterilization procedures before products can be distributed from companies that manufacture shelf-stable, low-acid foods [33]. The time–temperature combinations required to commercially sterilize the product, along with the list of factors critical to achieve and maintain sterility in the processing and packaging system, referred to as the scheduled process, is provided by the processor and must be filed with the appropriate regulatory agency. The scheduled process must be established by a qualified person referred to as a "process authority" and kept on file by the processor.

In order to design a time–temperature process that provides commercially sterile products and establish a scheduled process, the thermal resistance of specific spoilage organisms and/or organisms of potential public health significance found in the food should be considered by the process authority. Other factors such as product flow rate, flow characteristics of the product as it passes through the holding tube, and the system's design/operation must also be considered. The following section will discuss the critical factors for processing and packaging equipment.

17.4.1 Critical Factors for the Scheduled Process

For every scheduled process, a list of critical factors must be identified by the process authority. It will include any item that must be controlled and monitored to assure product sterility. The selection and quantification of critical factors, to be part of the scheduled process, begins during the initial equipment design review and continues through the testing and start-up period [33]. First, potential critical factors are identified, instrumentation and control functions are reviewed, and the regulatory compliance is evaluated during the design review. Once equipments are installed, testing will then be conducted to verify proper operation. This work begins with testing the preproduction sterilization cycle to assure that all product contact surfaces are sterilized properly. The list of potential critical factors identified during the preinstallation review may be modified during testing. As part of this testing, temperature monitoring and control points are established, and the sequence of the sterilization operation is finalized. Examples of critical factors include time and temperature during the presterilization cycle of equipment, maintenance of a minimum pressure differential within regenerative heat exchangers, product flow rates, and time and temperature of product sterilization.

17.4.2 Critical Factors for Packages and Packaging Equipment

Critical factors need to be identified for packages as well as packaging equipment. These critical factors must also be included in the scheduled process, and they are identified to achieve and maintain sterility within packages and packaging equipment. For example, packaging equipment currently used in the United States for low-acid foods employ either hydrogen peroxide with heat, peroxyacetic acid-based sterilants with heat, ionizing radiation, or heat alone to sterilize the aseptic zone and/or the packaging materials. Therefore, the critical factors usually include chemical consumption rate, sterilant, concentration, and contact time and temperature, and define the specific machine conditions for the biological sterility testing and challenge studies that follow. As the testing proceeds, the critical factors or their specific values may be modified on the basis of test results.

Validation of an aseptic system should be done once the scheduled process is established. Challenge studies with inoculated packs of product or simulated product should be conducted for confirmation of proper system operation. These challenge studies are conducted in batches by inoculating product packs with an appropriate test organism and then processing them at the maximum flow rate specified in the scheduled thermal process. The automatic controls and safety devices built into the processing and packaging systems should also be challenged to verify proper function. Validation of presterilization of packaging machines and sterilization of packages is also important [34]. This normally involves microbial challenge testing and package integrity tests. For filling and packaging machines, the microbial challenge testing to be conducted will depend on the sterilization agent or agents to be used and the type of product to be packaged. Package integrity tests are classified as either destructive or nondestructive methods [3], many of which are described in the *Annual Book of ASTM Standards* [35] and the *Bacteriological Analytical Manual* (*BAM*) [36]. Destructive methods involve tests that partially or completely destroy the packages and include the following tests [37,38]: bubble test (ASTM E 515), electrolytic test (BAM), dye test (ASTM F 1585 and F 1929), burst test (ASTM F 1140 and F 2045), microbial challenge test (BAM), and teardown procedures (BAM). Nondestructive tests are designed not to damage the package or the contents and include the following methods [37–39]: visual inspection (BAM), pressure difference (BAM and ASTM D 3078), ultrasonic (BAM), and infrared thermography methods [3].

There are three primary sets of regulatory requirements for food safety, which are applicable to aseptic food processing and packaging operations in the United States. All food products manufactured and/or sold in the United States fall under the regulatory jurisdiction of either the FDA or the Food Safety and Inspection Service (FSIS) of the U.S. Department of Agriculture (USDA) depending on the proportion of meat or poultry in the food product. Regulation of the commercial production of all foods other than meat, poultry, and processed egg products falls under the purview of the FDA. Production of formulated meat or poultry products containing at least 3% raw meat or 2% cooked poultry falls under the regulatory jurisdiction of the USDA's FSIS [40], as described in 21 CFR Parts 108, 110, and 113 (FDA) and 9 CFR Parts 308, 318, 320, 327, and 381 (USDA) [41]. Aseptic operations involving milk or milk products must also comply with the applicable provisions of the Grade A Pasteurized Milk Ordinance (PMO). There are currently some major differences and many other less significant variations among these sets of regulations. Therefore, it is imperative that food processors become thoroughly familiar with each applicable set of regulations. From a global perspective, aseptic regulations are dynamic in nature with significant differences among countries. For example, European regulations use spoilage data in evaluating aseptic systems, whereas FDA requires microbiological challenge tests and chemical tests to approve aseptic systems [30]. International processors must understand the important regulatory differences between the United States and other countries.

17.5 CONCLUSIONS

APP is an established and major commercial sterilization method for preserving shelf-stable foods with many advantages. APP requires less energy for processing, packaging, transportation, and storage and can deliver high quality, safe, and convenient food products with a smaller environmental footprint. Successful application of APP requires the highest quality of incoming materials, strict control of processing conditions, extended knowledge of food processing and packaging equipment components, and effective management of all postprocess operations. APP is a relatively new technology, being used commercially for only the last 50 years. Therefore, it still faces many challenges. For example, APP of multiphase foods containing large particulates presents many problems that are now being investigated by research. Aseptic packaging is an essential part of any aseptic system. New and emerging packaging materials with better gas barrier properties can improve food quality and extend product shelf life. Packaging materials used in aseptic packaging are not subjected to intense heat; thus, new packaging materials may be easily adopted in APP. In the future, even bio-based polymers may be considered for APP. In addition to enhancing the properties of polymeric films, active packaging may also be the solution for problems caused by permeated oxygen. Oxygen scavenging films have shown excellent results with regard to ascorbic acid oxidation as compared to regular plastic films and glass containers. Overall, the global market for aseptic products will continue to grow and APP will evolve to meet the ever-changing customer needs, particularly safety, cost, convenience, quality, and environmental stewardship.

REFERENCES

1. GL Robertson. *Food Packaging: Principles and Practice*. New York: Taylor & Francis/CRC Press, 2006, pp. 255–270.
2. ZJ Weng. Thermal processing of canned foods. In: DW Sun, ed. *Thermal Food Processing: New Technologies and Quality Issues*. Boca Raton, FL: CRC/Taylor & Francis, 2006, pp. 353–362.
3. JD Floros, I Weiss, and LJ Mauer. Aseptic packaging technology. In: PE Nelson, ed. *Principles of Aseptic Processing and Packaging*, 3rd edn. West Lafayette, IN: Purdue University Press, 2010, pp. 101–133.
4. A Teixeira. Thermal processing of canned foods. In: DR Heldman and DB Lund, eds. *Handbook of Food Engineering*. Boca Raton, FL: CRC Press/Taylor & Francis, 2007, pp. 784–790.
5. A Lopez. Aseptic processing and packaging. In: A Lopez, ed. *A Complete Course in Canning*. Baltimore, MD: The Canning Trade, Inc., 1987, pp. 160–211.
6. RP Singh and DR Heldman. *Introduction to Food Engineering,* 4th edn. Burlington, MA: Academic Press/Elsevier, 2009, pp. 429–440.
7. JM Krochta. Food packaging. In: DR Heldman and DB Lund, eds. *Handbook of Food Engineering*. Boca Raton, FL: CRC Press/Taylor & Francis, 2007, p. 900.
8. MT Morgan, DB Lund, and RK Singh. Design of the aseptic processing system. In: PE Nelson, ed. *Principles of Aseptic Processing and Packaging*, 3rd edn. West Lafayette, IN: Purdue University Press, 2010, pp. 3–29.
9. T Szemplenski. Aseptic packaging in the United States: Systems, installations & trends for food & beverage markets. *Packaging Strategies*, 2010. Available at: http://www.packstrat.com/PS/Home/Files/PDFs/2010AsepticPackInUSbrochure.pdf. Accessed on August 28, 2011.
10. AC Hersom. Aseptic processing and packaging of food. *Food Review International* 1(2): 215–270, 1985.
11. H Reuter. *Aseptic Packaging of Food*. Lancaster, PA: Technomic Publishing Co., Inc., 1989.

12. D Lund. Heat transfer in foods. In: M Karel, OR Fennema, and DB Lund, eds. *Physical Principles of Food Preservation*. New York: Marcel Dekker, 1975, pp. 50–91.
13. P Kumar. *Aseptic Processing of a Low-acid Multiphase Food Product using a Continuous Flow Microwave System*. Raleigh, NC: North Carolina State University, 2006.
14. LL Dock and JD Floros. Thermal and non-thermal preservation methods. In: M Schmidl and T Labuza, eds. *Science and Technology of Functional Foods*. Gaithersburg, MD: Aspen Publishers, 1999, pp. 49–88.
15. P Jelen. Review of basic technical principles and current research in UHT processing of foods. *Canadian Institute of Food Science and Technology Journal* 16(3): 159–166, 1983.
16. JM Jay, MJ Loessner, and DA Golden. *Modern Food Microbiology*, 7th edn. New York: Springer, 2005, pp. 420–431.
17. J Ashirifie-Gogofio, and JD Floros. Aseptic processing and packaging. In: Da-Wen Sun, ed. *Handbook of Food Safety Engineering*, Oxford, England, Wiley Blackwell, 2011, pp. 524–542.
18. S Nielsen, B Ismail, and GD Sadler. Chemistry of aseptically processed foods. In: PE Nelson, ed. *Principles of Aseptic Processing and Packaging*, 3rd edn. West Lafayette, IN: Purdue University Press, 2010, pp. 75–100.
19. D Lund. Heat processing. In: M Karel, OR Fennema, and DB Lund, eds. *Physical Principles of Food Preservation*. New York: Marcel Dekker, 1975.
20. MJ Lewis and NJ Heppell. *Continuous Thermal Processing of Foods: Pasteurization and UHT Sterilization*. Greenwood Village, CT: Aspen Publishers, 2000, pp. 59–127.
21. PJ Skudder. Ohmic heating. In: EMA Willhoft, ed. *Aseptic Processing and Packaging of Particulate Foods*. London, U.K.: Blackie Academic & Professional, 1993, pp. 74–89.
22. JD Kelder, PM Coronel, and PM Bongers. Aseptic processing of liquid foods containing solid particulates. In: R Simpson, ed. *Engineering Aspects of Thermal Food Processing*. Boca Raton, FL: CRC Press, 2008, pp. 49–72.
23. P Coronel, J Simunovic, KP Sandeep, and GD Cartwright. Aseptic processing of sweet potato purees using a continuous flow microwave system. *Journal of Food Science* 70(9): E531–E536, 2005.
24. FA Paine and HY Paine. *A Handbook of Food Packaging*. London, U.K.: Leonard Hill, 1983.
25. JAG Rees and J Bettison. *Processing and Packaging of Heat Preserved Foods*. London, U.K.: Blackie Academic and professional, 1991.
26. RT Toledo. Post-processing changes in aseptically packed beverages. *Journal of Agricultural and Food Chemistry* 34(3): 405–408, 1986.
27. U Strole. Carton laminates for aseptic packaging. In: H Reuter, ed. *Aseptic Packaging of Food*. Lancaster, PA: Technomic Publ. Co., Inc., 1989, pp. 221–226.
28. DS Lee, KL Yam, and L Piergiovanni. Thermally preserved food packaging: Retortable and aseptic. In: DS Lee, KL Yam, and L Piergiovanni, eds. *Food Packaging Science and Technology*. Boca Raton, FL: CRC Press, 2008, pp. 379–391.
29. BR Harte, JR Giacin, T Imai, JB Konczal, and H Hoojjat. Effect of sorption of organic volatiles on the mechanical properties of sealant films. In: HD Henyon, ed. *Food Packaging Technology, ASTM STP 1113*. Philadelphia, PA: American Society for Testing and Materials, 1991, pp. 18–30.
30. JRD David, RH Graves, and VR Carlson. *Aseptic Processing and Packaging of Food: A Food Industry Perspective*. Boca Raton, FL: CRC Press, 1996, pp. 7–200.
31. JD Floros, M Ozdemir, and PE Nelson. Trends in aseptic bulk storage and packaging. *Food, Cosmetics and Drug Packaging* 21: 236–239, 1998.
32. NH Mermelsteinm. Aseptic bulk storage and transportation. *Food Technology* 54(4): 107–111, 2000.
33. DI Chandarana, JA Unverferth, RP Knap, MF Deniston, KL Wiese, and B Shafer. Establishment the aseptic processing and packaging operation. In: PE Nelson, ed. *Principles of Aseptic Processing and Packaging*, 3rd edn. West Lafayette, IN: Purdue University Press, 2010, pp. 135–150.
34. G Moruzzi, WE Garthright, and JD Floros. Aseptic packaging machine pre-sterilisation and package sterilisation: Statistical aspects of microbiological validation. *Food Control* 11(1): 57–66, 2000.
35. Annual book of American Society for Testing and Materials (ASTM) Standards. West Conshohocken, PA: ASTM International, 1998–2002. Available at: http://global.ihs.com
36. Bacteriological Analytical Manual, Chapter 22. Food and Drug Administration (United States), Center for Food Safety and Applied Nutrition, 2001. Available at: http://www.fda.gov/Food/ScienceResearch/LaboratoryMethods/BacteriologicalAnalyticalManualBAM/default.htm. Accessed on August 28, 2011.
37. V Gnanasekharan and JD Floros. Package integrity evaluation: Criteria for selecting a method—Part I. *Packaging Technology and Engineering* 3(6): 44–48, 1994.
38. V Gnanasekharan and JD Floros. Package integrity evaluation: Criteria for selecting a method—Part II. *Packaging Technology and Engineering* 3(7): 67–72, 1994.

39. V Gnanasekharan and JD Floros. A theoretical perspective on the minimum leak size for package integrity evaluation. In: BA Blakistone and CL Harper, eds. *Plastic Package Integrity Testing—Assuring Seal Quality*. Herndon, VA: Institute of Packaging Professionals, 1995, pp. 55–65.
40. LM Weddig. Federal regulation of aseptic processing and packaging. In: PE Nelson, ed. *Principles of Aseptic Processing and Packaging*, 3rd edn. West Lafayette, IN: Purdue University Press, 2010, pp. 151–155.
41. DI Chandarana. Acceptance procedures for aseptically processed particulate foods. In: RK Singh and PE Nelson, eds. *Advances in Aseptic Processing Technologies*. London, U.K.: Elsevier Applied Science, 1992, pp. 261–278.

18 Ohmic Heating for Food Processing

António Augusto Vicente, Inês de Castro, José António Teixeira, and Ricardo Nuno Pereira

CONTENTS

18.1 INTRODUCTION

Ohmic heating (OH) (also called *Joule heating*, electrical resistance heating, direct electrical resistance heating, electroheating, or *electroconductive heating*) is defined as a process where electric currents are passed through foods to heat them (Figure 18.1). Heat is internally generated due to electrical resistance [1]. OH is distinguished from other electrical heating methods by (a) the presence of *electrodes* contacting the foods (in microwave and inductive heating *electrodes* are absent), (b) the frequency applied (unrestricted, except for the specially assigned radio or microwave frequency range), and (c) waveform (also unrestricted, although typically sinusoidal).

OH concept is well known, and various attempts have been made to use it in *food processing*. A successful application of electricity in food processing was developed in the nineteenth century to pasteurize milk [2]. This *pasteurization* method was called "electropure process" and by 1938, it was used in ~50 milk pasteurizers in five U.S. states and served about 50,000 consumers [3]. This application was abandoned apparently due to high processing costs [1]. Also, other applications were abandoned because of the short supply of inert materials needed for the *electrodes*, although electroconductive *thawing* was an exception [4].

However, research on ohmic applications in *fruits*, *vegetables*, *meat products*, and *surimi* has been undertaken by several authors, more recently [5–8].

Aseptic processing is considerably developed in the food industry especially for liquid foods which are processed predominantly by heat exchangers. Food technologists and the food industry are interested in extending this technology to more complex foods such as highly viscous, low-acid, and particulate-containing foods. In fact, most of the technologies actually used rely on conductive, convective, and radiative *heat transfer*. Their application to *particulate foods* is limited by the time required to ensure the *sterilization* of the center of larger particles, often causing overcooking of the surrounding volume. Consequently, product safety is achieved at the expense of quality.

In order to maximize food quality high-temperature short-time (HTST) processes have been used. This process is based on reactions that reduce the level of bacterial spores with lower activation energy than ones responsible for the loss of quality (e.g., reduction in vitamins and color degradation).

OH technology has gained interest recently because the products are of a superior quality to those processed by conventional technologies [8–10]. Moreover, the ohmic heater assembly can be incorporated into a complete product *sterilization* or cooking process. Among the advantages claimed for this technology are uniformity of heating and improvements in quality with minimal structural, nutritional, or organoleptic changes [11]. The potential applications are very wide and include, for example, blanching, evaporation, dehydration, fermentation [12], *pasteurization*, and *sterilization*.

OH is currently used in Europe, Asia, and North America to produce a variety of high-quality, low- and high-acid products containing particulates. A large number of additional applications are being developed for this technology as shall be seen further in this chapter.

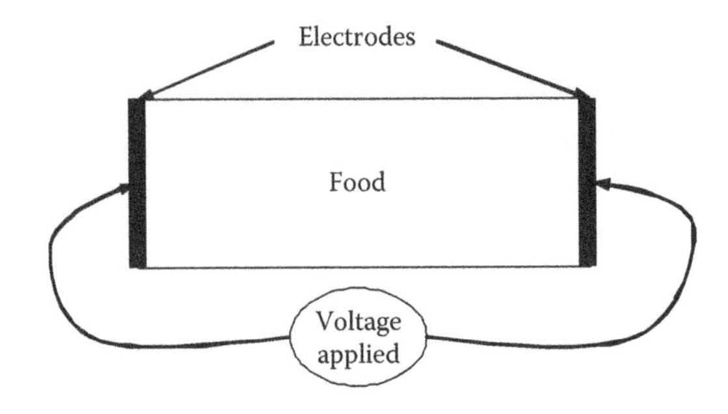

FIGURE 18.1 The principle of ohmic heating (OH).

A consortium of 25 partners from industry (food processors, equipment manufacturers, and ingredient suppliers), academia (food science, engineering, microbiology, and economics), and government was formed in 1992 in the United States in order to overcome the challenges still remaining about regulatory evaluation and formula optimization.

A 5 kW of pilot-scale *continuous* flow ohmic system (APV Baker, Ltd., Crawley, United Kingdom) was evaluated and a wide variety of shelf-stable (low and high acid) products were developed. These included broccoli and cheese (broccoli florets in Cheddar cheese sauce), shrimp gumbo (shrimp, okra, and tomatoes in a savory sauce), strawberries in glaze, oriental chicken (chicken, waterchestnuts, corn, peapods, bamboo shoots, mushrooms, carrots, and red peppers in Savoy sauce), and pasta primavera (shrimp, *surimi*, carrots, green beans, and mushrooms in Alfredo sauce). Texture, color, and nutrient retention of these products were found to be equal to, or higher than, those processed by traditional methods such as freezing, retorting, and aseptic processing. The consortium concluded that the technology was technically and economically viable [9].

The *OH* system allows for the production of new, high-added-value, shelf-stable products with a quality previously unattainable with alternative *sterilization* techniques, especially for *particulate foods*. Its major advantages are as follows [9]:

- *Continuous* production without heat-transfer surfaces
- Rapid and uniform treatment of liquid and solid phases with minimal heat damages and nutrient losses (e.g., unlike microwave heating, which has a finite penetration depth into solid materials)
- Ideal process for shear-sensitive products because of low flow velocity
- Optimization of capital investment and product safety as a result of high solids loading
- Reduced fouling when compared to conventional heating
- Better and simpler process control with reduced maintenance costs
- Environmentally friendly system

Some of the disadvantages accounting for *OH* are the higher initial operational costs and the lack of information or validation procedures for this technology.

The following sections will introduce several aspects of the *OH* technology, with emphasis on modeling and on the application to industrial processes.

18.2 MODELING OHMIC HEATING

The main objective of *OH* is to increase the temperature of food materials to a point at which the food is considered adequately processed. Also, the so-called *cold spot* must be identified. The most effective manner requires the exact measurement of the temperature profile in the food material when heating is applied. *Models* describing the effect of the main operating parameters on temperature of food materials (and especially of the system's cold spot) are required.

The *models* are mathematical representations of physical and chemical phenomena such as the laws of conservation of mass, energy, and momentum. They can be applied to generate the results of virtual experiments. This saves time and money and can be the only way to obtain the relevant data. The more complex a *model*, in terms of structure and the equations used, the more accurate the results. Other equations may be required to express particular observations or conditions or simply to establish boundary conditions. It is not possible to consider all the phenomena involved in the process to be modeled. Instead, assumptions are usually made to simplify the set of equations needed to solve the problems without oversimplifications. Another important issue is the parameters of the *model*. These must be evaluated as accurately as possible to avoid jeopardizing all the modeling effort. In fact, even if a *model* is correctly built, the solutions obtained from it can be very far from reality if the parameters involved are not properly determined.

In order to *model* the *OH* process (section D), all the aforementioned equations, parameters, and assumptions have to be taken into account, which will be discussed in the following sections.

18.2.1 Basic Equations

18.2.1.1 Electric Field

In order to generate heat in an *OH* system, an *electric field* must be applied to the food. The *electric field* (voltage distribution) is a function of the electrode and system geometry, *electrical conductivity*, and also of the applied voltage [1]. The *electric field* is determined by the solution of Laplace's equation:

$$\nabla(\sigma \cdot \nabla V) = 0 \tag{18.1}$$

where
σ is the electrical conductivity
∇V is the *voltage gradient*

This equation has been obtained combining Ohm's law with the continuity equation for electric current [13], and differs from the usual form of Laplace's equation:

$$\nabla^2 V = 0 \tag{18.2}$$

because σ is a function of both position and temperature, as mentioned before.

In order to solve Equation 18.1, boundary conditions specific for each case must be established. The solution has been obtained by de Alwis and Fryer [1] for a static ohmic heater containing a single particle, using as boundary conditions: (a) a uniform voltage on the *electrodes* or (b) no current flux across the boundary elsewhere. For a more general case of many different particles flowing in a fluid composed of several liquid phases (e.g., vegetable soup, where different vegetable solid pieces are dip in a fluid broth with at least an aqueous and a lipid phase), the mathematical solution for Equation 18.1 is, to our knowledge, still unknown. In these cases, the prediction of the *electric field* has been based on semiempirical *models* (see, e.g., Sastry and Palaniappan [14] or Sastry [15]).

The determination of the *electric field* is one of the most challenging subjects of the modeling effort in the *OH* technology.

18.2.1.2 Heat Generation

In order to ohmically heat a food, it is necessary to pass electrical current through it. The heat generated in the food by that current ($\dot{Q}$) is proportional to the square of its intensity (I), the proportionality constant being the electrical resistance (R), thus yielding

$$\dot{Q} = R \cdot I^2 \tag{18.3}$$

Alternatively, if both electrical conductivity (σ) and voltage gradient (∇V) are known, it is possible to write as follows:

$$\dot{Q} = |\nabla V|^2 \cdot \sigma \tag{18.4}$$

where σ is a function of position and temperature. The dependence of position is because foods are not necessarily homogeneous materials, the limiting scenarios being foods containing particles (e.g., *vegetables* soup) and that of a reasonably homogeneous liquid (e.g., orange juice). The relation of σ with temperature is usually well described by a straight line of the type [5]:

$$\sigma_T = \sigma_{ref} \cdot \left[1 + m \cdot (T - T_{ref})\right] \quad (18.5)$$

where

σ_T is the *electrical conductivity* at temperature T
σ_{ref} is the *electrical conductivity* at a reference temperature, T_{ref}
m is the temperature coefficient

18.2.1.3 Energy Balance

The next step for model is to determine the temperature distribution of the fluid and the solid phases (if present). This is done by establishing an *energy balance* for the fluid phase and solid phase [16]. Both the equations for the fluid and solid phases are established based on the knowledge of the predominant energy (mainly heat) transfer phenomena occurring during *OH*. These are schematically outlined in Figure 18.2 together with the relevant mass and momentum transfer phenomena, z and r, which are the axial and radial coordinates, respectively.

According to Figure 18.2, for a *continuous* process (the more general case) and for the fluid phase (the system considered here is a tubular heater—which is the most common geometry—where the

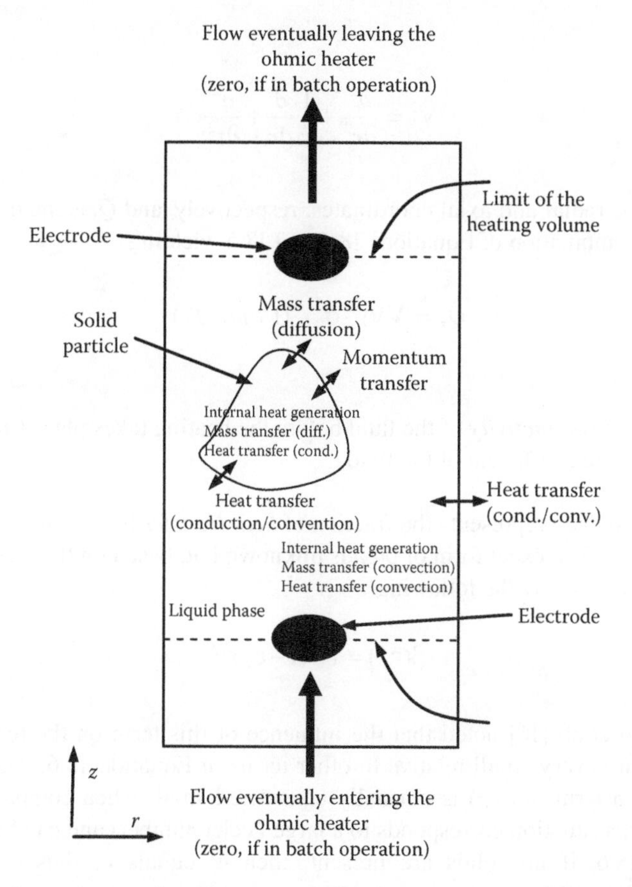

FIGURE 18.2 Main heat, mass, and momentum transfer phenomena occurring during OH. The presence (continuous) or absence (batch) of a mass flow through the heater will alter the hydrodynamic conditions and thus the relative importance of the phenomena.

electrodes are placed along the heater, thus creating a longitudinal voltage field; in any case, the application of the following equation to the case of a radial voltage field—i.e., where the *electrodes* are placed on the walls of the tube—should be straightforward), the *energy balance* is

$$\rho_f \cdot C_{pf} \cdot \bar{v}_z \cdot \varepsilon_f \cdot \frac{\partial T_f}{\partial z} = \beta(\varepsilon_f) \cdot \nabla(k_f \cdot \nabla T_f) - \frac{n_p \cdot A_p \cdot h_{fp}}{V_{sys}} \cdot (T_f - T_{ps}) + \frac{\dot{Q}_f}{V_{sys}} \tag{18.6}$$

where

ρ_f is the fluid density
C_{pf} is the fluid specific heat
$\bar{v}_z$ is the mean fluid velocity profile in the axial direction
ε_f is the fluid volume fraction in the heater
T_f is the fluid temperature
k_f is the fluid thermal conductivity
n_p is the number of solid particles in the considered volume
A_p is the area of each solid particle
$_{fp}$ is the fluid-particle *heat transfer* coefficient
V_{sys} is the volume of the system between the *electrodes*
T_{ps} is the temperature at the surface of the solid particle

being also

$$\nabla^2 = \frac{\partial^2}{\partial r^2} + \frac{1}{r}\frac{\partial}{\partial r} + \frac{\partial^2}{\partial z^2} \tag{18.7}$$

where r and z are the radial and axial coordinates, respectively, and $\dot{Q}_f$ is the heat generated in the fluid, given by the combination of Equations 18.4 and 18.5, yielding

$$\dot{Q}_f = |\nabla V|^2 \cdot \sigma_{0f} \cdot (1 + m_f \cdot T_f) \tag{18.8}$$

where

σ_{0f} is the *electrical conductivity* of the fluid before the heating takes place (initial value of σ_f)
m_f is the temperature coefficient of the fluid

Still in Equation 18.6, $\beta(\varepsilon_f)$ represents the fraction of *heat transfer* by *conduction* through the mixture in the fluid phase. The exact form of $\beta(\varepsilon_f)$ is unknown but, based on the Kopelman *model* [17], its expression can be given by the following:

$$\beta(\varepsilon_f) = 1 - (1 - \varepsilon_f)^{2/3} \tag{18.9}$$

In any case, Orangi et al. [16] noted that the influence of this term on the result is marginal, as the conduction term is very small relative to other terms in Equation 18.6. Also in this equation, the axial conduction term ($\partial^2/\partial z^2$) is normally considered small when compared with the radial *conduction* term (this situation corresponds to a large Peclet number) and can be neglected.

For Equation 18.6, if no solids are present, then n_p equals 0, thus canceling the terms $n_p \cdot A_p \cdot h_{fp} \cdot (T_f - T_{ps})$.

Also for Equation 18.6, the value of $\bar{v}_z$ (the mean fluid velocity profile in the axial direction) can be obtained assuming plug flow (an assumption generally valid for high solids concentration) or considering the more general case of a flow with a velocity gradient (assuming a power law fluid behavior), which can be given by the following [16]:

$$v_z = \left(\frac{3n+1}{n+1}\right)\cdot \overline{v}_z \cdot \left[1-\left(\frac{r}{R}\right)^{n+1/n}\right] \tag{18.10}$$

where n is the flow behavior index. In this case, v_z will replace $\overline{v}_z$ in Equation 18.6.

In order to obtain the solution of Equation 18.6, initial and boundary conditions must now be established. As initial condition, the following applies:

$$T_f = T_0, \text{ at } z = 0 \tag{18.11}$$

For the boundary condition at $r = R$ (R being the radius of the heater tube), it is necessary to consider the balance between conductive and convective *heat transfer* on the outer surface of the tube:

$$-k_f \cdot \nabla T_f \cdot \vec{n}\Big|_w = U \cdot (T_f - T_{air})\Big|_w \tag{18.12}$$

where

$\vec{n}$ is the unit vector normal to the surface of the tube's wall (w)
U is the overall *heat transfer* coefficient
T_{air} is the temperature of the air surrounding the walls of the heater

This balance must then be applied to Equation 18.6 to obtain the boundary condition at $r = R$.

Another boundary condition must be established on the axis ($r = 0$) which can be obtained by applying Equation 18.6 when $r \to 0$.

The *energy balance* for the solid phase is as follows (assuming spherical particles, Figure 18.2):

$$\rho_p \cdot C_{pp} \cdot \frac{\partial T_p}{\partial t} = \nabla(k_p \cdot \nabla T_p) + \frac{\dot{Q}_p}{V_p} \tag{18.13}$$

where

ρ_p is the particle density
C_{pp} is the particle specific heat
T_p is the particle temperature
t is the time
k_p is the particle thermal conductivity
V_p is the volume of the particle
T_{ps} is the temperature at the surface of the solid particle
$\dot{Q}_p$ is the heat generated in the particle, given by the combination of Equations 18.4 and 18.5, yielding

$$\dot{Q}_p = |\nabla V|^2 \cdot \sigma_{0p} \cdot (1 + m_p \cdot T_p) \tag{18.14}$$

where

σ_{0p} is the *electrical conductivity* of the particle before the heating takes place (initial value of σ_p)
m_p is the temperature coefficient of the particle

Similar to what has been made with Equation 18.6, initial and boundary conditions must now be established for Equation 18.13. As initial condition, the following holds:

$$T_p = T_{0p}, \text{ at } t = 0 \tag{18.15}$$

where T_{0p} is the initial temperature of the particle.

The boundary condition is established at the surface (*s*) of the particle, where it is again necessary to consider the balance between conductive and convective *heat transfer*, given by the following:

$$-k_p \cdot \nabla T_p \cdot \vec{n}\Big|_s = h_{fp} \cdot (T_{ps} - T_f)\Big|_s \tag{18.16}$$

If there are more than one solid phases present, this *energy balance* must be applied to each of them.

18.2.1.4 Mass and Momentum Balance

In order to completely describe the system, it is necessary to solve *mass and momentum balance* equations in three dimensions. This includes the solutions of those balances for both the solid phase (or phases, if several are present) and the liquid phase (or phases, if several are present) and constitutes substantial modeling and computational efforts [1,16,18]. The hydrodynamics of the overall system (*batch* or *continuous*) must be well characterized in order to define fully the velocity field of all phases present in the ohmic heater. This is related to the determination of the quickest particle (or portion of fluid) and is crucial in determining the safety of the heat treatment. There are commercial software programs, based on computational fluid dynamics, that are able to address these issues. These are being used to obtain the flow field in a *continuous* ohmic heater to use this information to obtain the temperature distribution of the solid and of the liquid phases in a two-phase flow (Castro et al., unpublished results). However, these are trials and alternatives have been used.

The equations described earlier will form the basis for the development of *models* to predict the *heating rate* of fluids and fluids containing solid particles. According to de Alwis et al. [19], the *heating rate* is a function of (a) *electrical conductivity* of the food constituents (of both fluid and solid phases), (b) particle geometry and size, (c) *particle orientation*, and (d) thermal variation of physical properties. Sastry and Palaniappan [14] added particle's volume fraction to this list. While the particles' size, geometry, orientation, and volume fraction are parameters relatively easy to measure and control and there are extensive data available on the temperature dependence of the main physical property, *electrical conductivity* (especially particle's electrical conductivity) is a critical point [15].

18.2.2 Main Parameter: *Electrical Conductivity*

Probably, the most important parameter in *OH* modeling is σ [20]. Some of the characteristics of this property are summarized as follows:

- σ can be anisotropic (varies in different directions).
- Changes in the value of σ reflect changes in the matrix structure, for example, during starch gelatinization or cell lysis.
- The value of σ that is not suitable for ohmic processing can be suitably modified, for example, by blanching.

However, perhaps the most striking feature of σ is its dependence on temperature, as it has been shown to increase with increasing temperature (T). This observation is due to a variable opposition

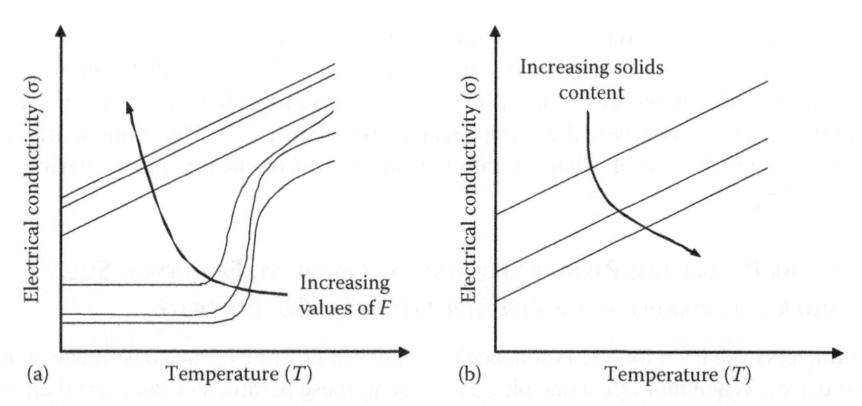

FIGURE 18.3 (a) Variation of σ with T for solids and for several electric field strength values (F), showing that the relationship between σ and T becomes linear for increasing values of F (the lower limit of F—i.e., $F = 0\ \mathrm{V \cdot m^{-1}}$—represents conventional heating); (b) variation of σ with T for liquids, showing that σ has always a linear relationship with T, with the value of σ decreasing for increasing concentration of nonpolar (thus nonconductive) constituents (e.g., soluble solids) (Adapted from Sastry, S.K. and Palaniappan, S., *Food Technol.*, 45(12), 64, 1992.)

(drag force) to the movement of the ions responsible for conducting the electricity in food materials: for higher temperatures, that opposition is less important than for lower temperatures [21].

For most solid foods, σ increases sharply with temperature at around 60°C and this has been attributed to the breakdown of cell wall materials [22], releasing ionic compounds to the bulk medium. Figure 18.3a represents this dependence for several *electric field* (F) strength values, demonstrating that the relationship between σ and T becomes linear for increasing values of F (the lower limit of F—i.e., $F = 0\ \mathrm{V \cdot m^{-1}}$—represents conventional heating). This effect has been demonstrated by, among others, Palaniappan and Sastry [5], who suggested that it may be due to electro-osmotic effects which could increase the effective conductivity at low temperatures. This means that, for normal *OH* conditions, the relationship between σ and T can be represented by Equation 18.5.

For liquid foods, σ always has a linear relationship with T; however, the value of σ decreases if nonpolar (thus nonconductive) constituents are present (Figure 18.3b). If those constituents are generally regarded as the "solid content" of the liquid (e.g., the pulp content of juice), it is possible to write, for example, for tomato and orange juices [6],

$$\sigma_T = \sigma_{ref} \cdot \left[1 + K_1 \cdot (T - T_{ref})\right] - K_2 \cdot S \quad (18.17)$$

where

K_1 and K_2 are constants
S is the dissolved solids content

A similar influence of solids concentration on σ has also been reported by other authors [8,23], but in this cases, the solids were particles of significant sizes suspended in a fluid. This situation differs from the one given earlier because the value of σ is not that of the fluid but a combined value for the mixture fluid + particles, the so-called *effective electrical conductivity* (σ_{eff}). In fact, not only solid content but also solid size will influence the mixture σ_{eff}.

Another major factor affecting σ is the ionic content of the food: the higher the ionic content, the higher the value of σ [24]. This has been demonstrated by several authors [5,25–27].

σ is also a function of the frequency at which it is measured [28] and evidence also shows that the value of σ of a food heated electrically is different from that cooked by conventional heating [20]. In fact, Wang and Sastry [7] demonstrated that samples preheated by either conventional or *OH* had a higher *heating rate* due to increased σ, and that the values found for σ were different in each case. It has been mentioned that σ *versus* T is usually a linear relationship. However, this is not always the case. Several examples exist where the curve became nonlinear [27,29]. This nonlinearity has been attributed to the equilibration of solutes within the samples during heating, thus altering the value of the observed σ_{eff}.

18.2.3 Other Parameters: Particle Orientation, Geometry, Frequency, Size, and Concentration—the *Effective Electrical Conductivity*

Other parameters have been found to influence the values of *OH* rate of the constituents of a food as mentioned before. When more than one phase is present, these parameters can exert their influence *via* the effective conductivity of the mixture (σ_{eff}) which is the case of *particle size* and concentration, or they can directly influence the *heating rate* of the different constituents, which is the case of *particle orientation* and geometry.

Particle geometry was demonstrated to be of importance only when the aspect ratio of the solid particle is far from unity [30]. If this is the case, and for a static *OH* system containing a fluid and a single elongated particle with a lower conductivity than the fluid, de Alwis et al. [19] showed that when the particle was placed with its longer axis parallel to the *electric field* (Figure 18.4a), then it would heat slower than the fluid. However, if such a particle would be placed with its longer axis perpendicular to the *electric field* (see Figure 18.4b), then the particle would heat faster than the fluid. Exactly the reverse would occur if the particle had a higher conductivity than the fluid (Figure 18.4c and d). This has been confirmed by mathematical [1,31]. The cases documented by Figure 18.4a and d are referred by Sastry and Salengke [32] as being worst-case scenarios. The particle will be underheated compared to the fluid, and it may be underprocessed. In the case of Figure 18.4a, underheating may result from the major part of the current bypassing the particle, thus heating more the surrounding fluid. Alternatively (Figure 18.4d), underheating may result from the particle transmitting most of the current thus creating a low field gradient within the particle and low current densities in its vicinity. A similar situation will be that of a cluster of particles (more conductive than the fluid) blocking the cross section of the heater. However, the authors conclude that both the latter cases are avoidable in an industrial application (e.g., controlling *particle size* or on-line sensing the σ of the product to divert possible particle clumps from the product stream prior to heating), the former being the one that may cause greater concern. In practice, if a single low conductivity particle enters the system, there is potential for underprocessing that particle, once its *cold spot* would have a significantly lower temperature than that of the fluid [15]. Such a particle might be a fat globule which, if carrying a microbial load, could present a serious risk for the safety of the product being processed. Usually, in order to minimize undesirable electrochemical reactions between *electrodes* and food product, an increase in the alternating *electric field* frequency is required [33,34]. There have been several efforts in order to evaluate the impact of *electric field* frequency during *OH* processing of vegetable food. *Electrical conductivity* and temperature rise of turnip under *OH* at four frequencies (4, 10, 25, and 60 Hz) were evaluated, and it has been shown that the heating rate increased with decreasing frequency [35]. Much in the same way, the lowering of electric field frequency for the moderate electric field (MEF) isothermal processing of red beet resulted in higher tissue damage and improved the mass transfer processes [28,36]. More recently, it has been shown that the combination of high-frequency *OH* (200 kHz) with *pasteurization* at moderate temperature followed by rapid cooling minimizes texture degradation of peach tissue [37]. These latter results provided new information on the impact of *electric field* frequency on *OH*, which is useful for *OH* process design and modeling.

Most of the conclusions presented earlier have been drawn for a static ohmic heater. What happens then in a *continuous*-flow ohmic heater? In this case, the fluid will be agitated, and

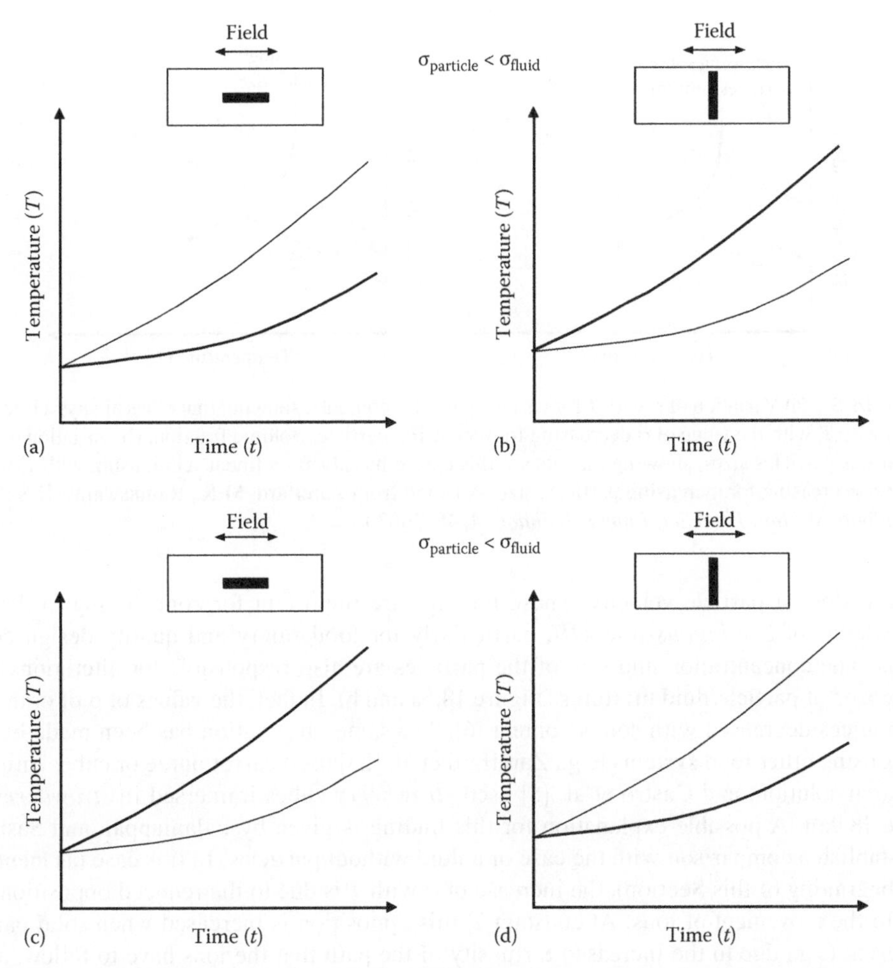

FIGURE 18.4 Heating curves of a single solid particle suspended in a fluid contained in a static ohmic heater (▬ particle; — fluid): (a) and (b) when the particle has a lower electrical conductivity than the fluid and its longer axis is parallel to the electric field (a) or perpendicular to the electric field (b); (c) and (d) when the particle has a higher electrical conductivity than the fluid and its longer axis is parallel to the electric field (c) or perpendicular to the electric field (d).

therefore, the fluid-particle *heat transfer* will be higher than in the static situation. Sastry and Salengke [32] have shown that the static situation is not always related to the worst-case scenario. In fact, when the solid is less conductive than the fluid, the worst situation will be that of a mixed fluid (continuous situation). Nevertheless, if the solid is more conductive than the fluid, then the worst case will be that of a static fluid (static situation). Evidently, none of the rheological properties of the fluid have great influence, and viscous fluids tend to make a *continuous* system behave as a static system [38]. Sarang et al. [39] have studied the residence time distribution (RTD) of particulate foods in a *continuous* flow *OH*, using radio-frequency identification technology. In particular, the effect of solids' concentration (30%, 40%, 50%, 60%, 70%, and 80%, v/v) and rotational speed of the agitators (55, 40, and 25 rpm) inside the ohmic heaters was evaluated and RTD of potato in starch solution, in a pilot-scale ohmic heater was determined for each experimental condition. Results show that maximum particle residence time increased with the rotational speed of the agitators ($P < 0.05$), and no particular trend was observed between the mean residence time and the solid concentration. The velocity of the fastest particle was 1.62 times the mean product velocity while the product mean velocity was two

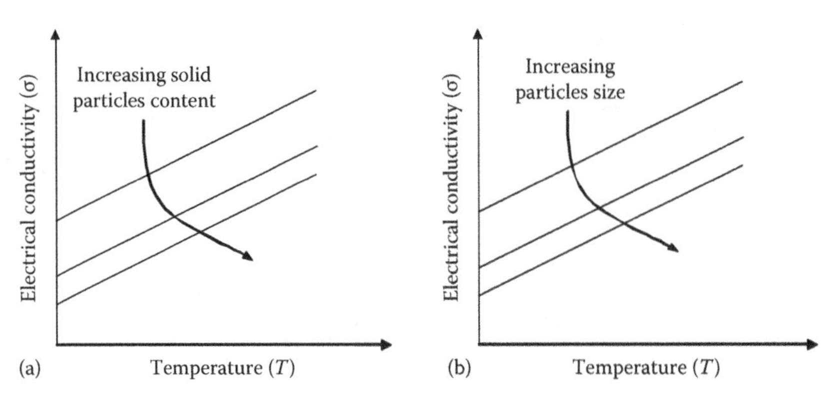

FIGURE 18.5 (a) Variation of σ with T for various particles contents, showing that σ has always a linear relationship with T, with the value of σ decreasing for increasing particles volume fraction; (b) variation of σ with T for various particles sizes, showing that also in this case σ has always a linear relationship with T, with the value of σ decreasing for increasing particles size (Adapted from Zareifard, M.R., Ramaswamy, H.S., Trigui, M., Marcotte, M., *Inn. Food Sci. Emerg. Technol.*, 4, 45, 2003.)

times the slowest particle velocity. These findings are important for contributing to the body of knowledge of *continuous*-flow *OH*, particularly for food safety and quality design considerations. The concentration and size of the particles are also responsible for alterations of the *heating rate* of particle/fluid mixtures (Figure 18.5a and b). In fact, the values of σ of orange and tomato juices decreased with solids content [6]. The same observation has been made by other authors using other food systems [e.g., Zareifard et al. [23] used carrot puree or cubes immersed in a starch solution, and Castro et al. [8] used *strawberry* cubes immersed in *strawberry* pulp (Figure 18.9a)]. A possible explanation for this finding is given by Palaniappan and Sastry [6] who establish a comparison with the case of a fluid without particles. In this case (as mentioned in the beginning of this Section), the increase of σ with T is due to the reduced opposition (drag force) to the movement of ions. At constant T, this opposition is increased when solid particles are present (e.g., due to the increased tortuosity of the path that the ions have to follow, among other effects), and this may be the reason for the observed decreasing trend of σ with increasing particles content (Figure 18.5a), which is in line with what happened with dissolved solids concentration (see Figure 18.3b). This same explanation may be applied to justify the decreasing of σ when *particle size* increases (Figure 18.5b). This effect has been observed by several authors in various systems [6,8,23].

If a solid particle with a lower σ than the fluid, in which the particle is immersed and is subjected to *OH*, it will heat slower than the fluid (assuming a particle with an aspect ratio close to unity or aligned as in Figure 18.4a). However, if a number of such particles are present, simulations have shown and experiments have confirmed [14] that although the *heating rate* of the particles is initially lower than that of the fluid, *heating rate* of the particles overtakes that of the fluid (Figure 18.6a). This can be explained in light of what has been represented in Figure 18.4b, where a particle with a lower σ blocking the current *conduction* would heat faster than the fluid; similarly, as the particles concentration increases, they will increasingly block the *conduction* paths forcing a greater proportion of the total current to flow through them, thus showing higher *heat generation* rates than those of the fluid. This means that the particle's σ, although lower than the fluid's σ at low temperatures, will increase faster (Figure 18.6b), overtaking the latter for higher temperatures therefore justifying the shape of the heating curves in Figure 18.6a.

Attempts have been made to determine the (overall) σ_{eff} of particle/fluid mixtures, which is a function of the conductivity and solids fraction of the component phases. The first relation was determined by Maxwell [20] and is valid for a dilute dispersion of spheres with a volume fraction ε_p and *electrical conductivity* σ_p immersed in a fluid with electrical conductivity σ_f:

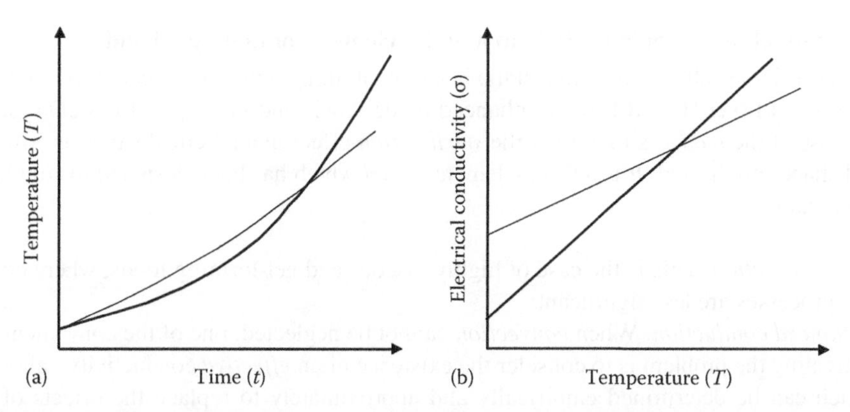

FIGURE 18.6 Behavior of a mixture of a fluid with a high concentration of particles (these having a value of σ lower than that of the fluid) (—— particles; —— fluid); (a) heating of particles and fluid (Adapted from Sastry, S.K. and Palaniappan, S., *J. Food Eng.*, 15, 241, 1992) and (b) corresponding variation of σ with *T* (Adapted from Zareifard, M.R., Ramaswamy, H.S., Trigui, M., Marcotte, M., *Inn. Food Sci. Emerg. Technol.*, 4, 45, 2003.)

$$\sigma_{eff} = \sigma_f \cdot \frac{1 - 2 \cdot \Phi \cdot \varepsilon_p}{1 + \Phi \cdot \varepsilon_p} \tag{18.18}$$

where

$$\Phi = \frac{\sigma_f - \sigma_p}{2 \cdot \sigma_f - \sigma_p}.$$

Modifications of this equation have been proposed, and there are a number of different *models* for σ_{eff} [26]. However, the main problem persists because σ_{eff} will vary with the position and orientation of the particles in the mixture. Kopelman [17] proposed the following *model*:

$$\sigma_{eff} = \frac{\sigma_f \cdot (1 - C)}{(1 - C) \cdot \left(1 - \varepsilon_p^{1/3}\right)} \tag{18.19}$$

where

$$C = \varepsilon_p^{2/3} \cdot \left(1 - \frac{\sigma_p}{\sigma_f}\right).$$

These models (of which Equation 18.19 is an example) can be of use if they are representative in the conditions under which they are to be applied. If this is the case, then it is possible to use σ_{eff} to estimate the temperature increase of a mixture when it is subjected to *OH* by a very simple calculation.

18.2.4 Models

The *models* described in the following sections are the ones that are either most quoted in the literature or those that, in our opinion, constituted important landmarks in the modeling effort in *OH*. Generally, all assume that the physical properties of the phases in presence are constant with respect to the temperature, with the obvious exception being *electrical conductivity*. Considering the coexistence of several solid particles immersed in a fluid, it is also generally assumed that the particles have short interaction times with the surrounding particles and with the walls of the heater.

18.2.4.1 Model for a Single Particle in a Static Heater Containing a Fluid

A *model* for a single solid particle in a static heater containing a fluid has been developed initially by de Alwis and Fryer [1] and further enhanced by de Alwis and Fryer [40] for a 2D system. The main purpose of the *model* is to predict the *sterilization* effect in a practical situation, where complex food shapes are heated. It is a finite-element *model* which has been designed to simulate three types of situations:

1. *Zero convection*. This is the case of highly viscous and gel-forming foods, where convective processes are less significant.
2. *Enhanced conduction*. When *convection* cannot be neglected, one of the convenient ways of treating the problem is to consider the existence of an *effective* conductivity value (σ_{eff}), which can be determined empirically and approximately to replace the effects of both convection and conduction.
3. *Well-stirred liquid*. In a previous work [19], where an unstirred fluid was used, no hot spots were noticed when very low viscosity mixtures were heated, indicating that rapid convective mixing was taking place. In this case, a well-stirred liquid condition can be applied.

To obtain the *electric field* distribution, Equation 18.1 was written for a 2D situation, yielding

$$\frac{\partial}{\partial x}\left(\sigma_x \cdot \frac{\partial V}{dx}\right) + \frac{\partial}{\partial y}\left(\sigma_y \cdot \frac{\partial V}{dy}\right) = 0 \tag{18.20}$$

where x and y are the space coordinates of the system, and σ_x and σ_y the values of the σ in the x- and y-direction, respectively. The solution of this equation was obtained using boundary conditions of (a) a uniform voltage on the *electrodes* or (b) no current flux elsewhere across the boundary.

For *heat generation*, a network theory approach was used in which each triangular element was considered an isolated network, with nodal voltages known by solution of Equation 18.20. *Heat generation* in the particle was thus found by the following:

$$\dot{Q} = \sum_{i=1}^{i=3} V_i \cdot I_i \tag{18.21}$$

where i is the node number.

Finally, the temperature distribution has been found by means of an *energy balance*, similar to that presented in Equation 18.13:

$$\rho \cdot C_p \cdot \frac{\partial T}{\partial t} = \frac{\partial}{\partial x}\left(k_x \cdot \frac{\partial T}{\partial x}\right) + \frac{\partial}{\partial y}\left(k_y \cdot \frac{\partial T}{\partial y}\right) + \frac{\dot{Q}}{V_{sys}} \tag{18.22}$$

where

k_x and k_y are the values for the thermal conductivity in the x- and y-directions
V_{sys} is the volume of the relevant part of the system (either solid or liquid)

The *model* was validated by comparing the results of the simulations to those from pieces of lamb. The *model* has also been used to simulate the heating of particles with different shapes and under different orientations, the results of which have been presented previously.

Later on, Zhang and Fryer [41] extended the work subsequently by using commercial software to simulate the behavior of a two-phase system containing different numbers of particles. A review article is available [20]. The same commercial software was also used by Fu and Hsieh [31] to simulate the temperature distribution in a 2D *OH* system. The simulation results were compared with experiments made in an ohmic heater that was especially designed to meet the main assumptions of the model (a 2D heating system containing a fluid and a single solid particle). They concluded that the boundary conditions and physical property values were crucial to obtain a reliable simulation.

18.2.4.2 Model for Multiparticle Mixtures in a Static Heater Containing a Fluid

The case of a mixture of particles in a static heater containing a fluid was studied [14]. This is a 3D finite element *model* aiming at predicting the temperature of the mixture of a liquid in which various particles are suspended. The most accurate solution has to be obtained considering the existence of a *particle size* distribution, having the *heat-transfer* problem to be solved for each particle. Such solution is unlikely to be applied in practice due to the extreme computational effort needed. Instead, an "average size" particle can be assumed, with cubic geometry.

Again, the electric field problem has to be solved. A solution for Equation 18.1, which would be difficult for a mixture of several particles since it would need knowledge of the location and properties of every particle at all points in time instead, the authors obtained a estimation of the electric field distribution using a circuit theory-based approach. Two cases were analyzed:

1. *Mixtures of several particles involving large particle populations*. Such a mixture is considered to be composed of a *continuous* (fluid) and a discrete (particles) phase. The equivalent electrical circuit is that of parallel fluid (R_{fP}) and particles (R_{pP}) resistances, and in series with a liquid resistance (R_{fS}); therefore, the total resistance (R) is

$$R = R_{fS} + \frac{R_{fP} \cdot R_{pP}}{R_{fP} + R_{pP}} \quad (18.23)$$

and

$$R_{fS} = \frac{l_{fS}}{A_{fS} \cdot \sigma_f} \quad (18.24)$$

$$R_{pP} = \frac{l_{pP}}{A_{pP} \cdot \sigma_p} \quad (18.25)$$

$$R_{fP} = \frac{l_{fP}}{A_{fP} \cdot \sigma_f} \quad (18.26)$$

where l_{fS}, l_{fP}, and l_{pP} are the lengths of each phase, which are related to the length of the heater (l) and to each other by

$$l = l_{fS} + l_{fP} \quad (18.27)$$

and by

$$l_{pP} = l_{fP} \quad (18.28)$$

and for A_{fS}, A_{fP}, and A_{pP} (the cross-sectional areas of each phase), they are related to the cross-sectional area of the heater (A) and to each other by

$$A = A_{fS} = A_{pP} + A_{fP} \tag{18.29}$$

In a similar way to Kopelman [17], it was assumed that the cross-sectional area and length of the discrete phase (particles) could be estimated from the volume fraction of that phase (ε_p) by

$$A_{pP} = A \cdot \varepsilon_p^{2/3} \tag{18.30}$$

and

$$l_{pP} = l \cdot \varepsilon_p^{1/3} \tag{18.31}$$

Finally, the *electric field* was calculated using

$$V = I \cdot R(x) \tag{18.32}$$

where $R(x)$ is the resistance as calculated by Equation 18.23 up to position x, where the voltage (V) is to be calculated.

2. *Mixtures with relatively small numbers of particles.* In this situation, the orientation and location of the particles are easily known, thus the equivalent resistance can be calculated separately for zones with or without particles. For the former, the equivalent resistance (R) can be calculated as in Equation 18.23, while for the latter, the equivalent resistance simply equals R_{fS}, as in Equation 18.24. The electric field can then be calculated using Equation 18.32.

The next problem is *heat generation*. This has been solved for this *model* as explained previously by using Equations 18.3 through 18.5.

Finally, the *energy balance* must be established for the fluid phase and the particles, to allow the determination of temperatures of each phase during the heating process. For the fluid phase, Equation 18.6 was slightly modified (Equation 18.6 is valid for the more general case of a continuous ohmic heater) for usage in a static (*batch*) ohmic heater, which yields

$$\rho_f \cdot V_f \cdot C_{pf} \cdot \frac{T_f^{n+1} - T_f^n}{\Delta t} = -U \cdot A_w \cdot \left(\overline{T}_f - \overline{T}_{air}\right) - n_p \cdot A_p \cdot h_{fp} \cdot \left(\overline{T}_f - \overline{T}_{ps}\right) + \dot{Q}_f \tag{18.33}$$

where

V_f is the volume of the fluid
n is the time step index
$\overline{T}_{air}$ is the average temperature of the air surrounding the heater and

$$\overline{T}_f = \frac{T_f^{n+1} - T_f^n}{2} \tag{18.34}$$

$$\overline{T}_{ps} = \frac{T_{ps}^{n+1} - T_{ps}^n}{2} \tag{18.35}$$

For the particles, Equation 18.13 holds with Equation 18.16 as the boundary condition.

The simulations were carried out and compared to the experimental results obtained with potato cubes of various sizes heated in a static ohmic heater containing phosphate solutions of various concentrations to simulate fluid phases with different electrical conductivities. This mathematical *model* was considered to be in agreement with the experiments. Critical parameters were found to be the (a) conductivities of liquid and solid phases and (b) volume fraction of each phase. This *model* has been compared with the one presented in the previous section in terms of their capability of predicting the behavior of the static ohmic heater in a worst-case scenario [32]. The assessment of which *model* is more conservative depends on whether the particles are more or less conductive than the fluid.

18.2.4.3 Model for Multiparticle Mixtures in a Continuous Flow Heater Containing a Fluid

The *model* described in previous section for a static heater has been extended to the situation of a *continuous* flow heater by Sastry [15]. Equations and assumptions were essentially the same (Equations 18.23 through 18.35), in the introduction of fluid and particle flow in and out of the system. Both fluid and particles will increase in temperature (and thus conductivity) during their path through the heater, and therefore, the voltage drop must be calculated not for the whole heater but separately for each of the incremental sections into which the heater must be divided for the calculations. As a consequence of this, Equations 18.23 through 18.32 have to be applied separately for each of those sections, in order to determine the *electric field*. Subsequently, Equations 18.3 through 18.5 need to be applied for the *heat generation*, and finally, Equations 18.33 through 18.35 are applied to resolve the thermal problem.

The simulations performed with this *model* allowed several important aspects of *continuous* processing with *OH* technology to be emphasized:

- Although when a multiparticle mixture of low conductivity is present, these particles tend to heat faster than the fluid, if an isolated low conductivity particle crosses the system, it can be underprocessed.
- The RTD and the fluid-particle *heat transfer* coefficient are crucial aspects to consider when designing a continuous heater.

These results were confirmed and extended by Orangi et al. [16], who studied a similar problem when investigating the *continuous* flow *sterilization* of solid–liquid food mixtures by *OH*.

18.2.4.4 Other Models

Other *models* of the behavior of particle/fluid mixtures subjected to *OH* have been developed. Benabderrahmane and Pain [42] presented a *model* based on the principle of a mean slip velocity between the fluid and the particles in plug flow (the Slip Phase *Model*). The existence of a difference between the velocity of the fluid and that of the particles has been demonstrated by Lareo et al. [43], Lareo [44], Liu et al. [45], Fairhurst [46], and Fairhurst and Pain [47] and is therefore considered in this *model*. It also takes into consideration internal *heat generation*, convective *heat transfer*, and heat *conduction* within the solid particles, thus following closely the scheme outlined in Figure 18.2, although the heater walls are assumed to be adiabatic. One of the main contributions of this *model* is that it demonstrates the importance of the *temperature gradient* inside the particles. It also confirmed results obtained earlier by other authors (see, e.g., Sastry [15]) showing that for a mixture of particles in a fluid, the process critical point is situated in the fluid and not in the particles. For conventional heating processes, the critical point is in the particles. Other in-depth problems have been addressed that arose during the modeling effort during the previous two decades. Lacey [48,49] and Lacey et al. [50], who studied the problem of thermal runaway while modeling *OH*, first by considering only *heat transfer* by *conduction* and second including heat *convection*, which was found to dominate heat conduction [51]. In the same line, a diffusion-*convection* problem was addressed by Kavallaris and Tzanetis [52], having the Heaviside function representing the food resistivity (which is the inverse of the conductivity). Effects of local *electric field* on the heat transfer *model* of a continuous flow *OH* process were also assessed [53]. The *model* fluid used was a mixture of reconstituted skimmed milk and whey protein concentrate (WC) solution. Two-dimensional simulations of laminar flow *OH* were carried out using two different *models* to determine the volumetric heating rate; the first *model* considered the local *electric field* by solving the Laplace equation while the other assumed an average voltage gradient applied between the two *electrodes*. Despite the difference in the wall temperature shown by the two *models*, the centerline fluid temperature and the average outlet temperature were considered to be essentially similar, demonstrating that volumetric or bulk heating dominates in the *OH* process. The predicted outer electrode wall temperature and average outlet temperature have agreed reasonably well with the measured values.

18.3 THERMAL AND NONTHERMAL EFFECTS OF OH

This section reveals the research efforts to determine the effects of *OH*, both thermal and nonthermal, on microbes and on food constituents such as enzymes and vitamins. The information on proteins is limited to the studies made with *surimi* and its gel-forming ability due to the presence of the myosin heavy chain (see, e.g., Refs. [54,55]).

18.3.1 Microbial Kinetics

Microbial inactivation in foodstuffs is predominantly carried out by thermal processes, and the thermal inactivation kinetics of most of the target microorganisms is well studied. The need to reduce processing time and the increasing interest in using *OH* as an alternative heating technology to conventional *heat transfer* during commercial processes of *sterilization* or *pasteurization* was the impetus for the study of nonthermal mechanisms of *microbial inactivation*.

The destruction of microorganisms by nonthermal effects such as electricity is still not well understood and generates some controversy. Little work has been done in this field. Moreover, most of the published results do not refer to the sample temperature or cannot eliminate temperature as a variable parameter.

The application of *OH* in the fermentation by *Lactobacillus acidophilus*, a lactic acid bacterium used in the dairy industry and with human health implications [12] was studied. The lid of the fermentation vessel was equipped with ports for thermocouple, pH probe, inoculation, water circulation coil, medium circulation, and two stainless steel plate *electrodes*. Metal surfaces were coated with epoxylite for electrical insulation and inertness. Temperature control was carried out either under conventional (by *continuous* water circulation) or *OH* (a constant low voltage of 15 V or a high voltage of 40 V was applied), at different temperatures (30°C, 35°C, and 40°C). The results indicated that the lag phase for fermentations at 30°C was significantly lower (18-fold) under low-voltage ohmic conditions, which was also the lowest lag period of all the conditions tested. Although additional investigation is needed to explain the shortening of lag phase, this may be due to the improvement of absorption of nutrients and minimization of the inhibitory action of fresh medium. The application of an electrical field may induce pore formation in membranes (similar to the electroporation mechanism used to transform cells in molecular biology studies) allowing a faster and efficient transport of the nutrients into the cells thus decreasing lag phase. The minimum generation time for *L. acidophilus* was not affected by the heating method. Small differences in the final pH of the fermentation between the two methods were also reported. The consumption of glucose and the release of lactic acid were not significantly affected by the heating method. On the other hand, the production of lacidin A, an antimicrobial protein (bacteriocin) by the bacterium, decreased when fermentation occurred under ohmic conditions. The electrical current measurements follow approximately the changes in growth of *L. acidophilus* during the fermentation, and these changes in σ could, perhaps, be used to monitor fermentations. This study provides evidence that *OH* in the food industry may be useful to shorten the time for processing yogurt and cheese production [12], among other applications.

The kinetics of inactivation of *Bacillus subtilis* spores by *continuous* or intermittent ohmic and conventional heating were studied by Cho et al. [56] to determine if electricity had an additional effect on the killing of this microorganism at single- and double-stage heating treatments. Experiments were conducted in an ohmic fermentor for temperatures ranging from 88°C to 97°C (see Table 18.1). Spores heated at 92.3°C had significantly lower decimal reduction time (D) values when using ohmic rather than conventional heating. These results indicate that electricity has an additional killing effect against bacterial spores. The dependence on temperature of the D (z value) and the activation energy (E_a) were not significantly affected indicating that electricity affects the death rate but not the temperature dependency of the spore inactivation process.

Palaniappan et al. [57] indicated that electricity did not influence inactivation kinetics, but the application of a nonlethal *electric field* reduces the intensity of the subsequent thermal treatment. This implies that the *electric field* lowers the heat resistance of microorganisms. Microbial death during *OH* was mainly attributed to thermal effects, while the nonthermal effects were insignificant (Table 18.2).

The effect of an *electric field* on the thermal inactivation kinetics of a highly heat-resistant microorganism *Byssochlamys fulva* has been studied. This is a thermotolerant, ascospore producing, filamentous fungus and was investigated by Castro et al. [73]. It can also produce the important mycotoxin, patulin. *B. fulva* death kinetics were determined in an industrial *strawberry* pulp (14.5°Brix, $pH = 4.0$). The experimental D values for *B. fulva* obtained under *OH* (D_{oh}) conditions were half the ones obtained for conventional heating (D_{conv}) ($T = 85$°C, $D_{conv} = 7.23$; $D_{oh} = 3.27$). Unsurprisingly, these results are not consistent with those obtained for *B. subtilis* and *Zygosaccharomyces bailii* (a bacterium and yeast, respectively) and the nonthermal inactivation mechanism needs to be studied in more detail, in terms of the effect of the *electric field* on the membrane integrity of ascospores and enzymes participating in ascospores activation. Effects of *OH* on patulin production in food by *B. fulva* and degradation of patulin require investigation. Nonthermal effects of *OH* on the heat resistance of the *Escherichia coli*, which frequently contaminates dairy products when their manufacture conditions are unsanitary, were studied in goat

TABLE 18.1
Kinetic Constants and Thermal Inactivation Parameters for *B. subtilis* Spores under Conventional and Ohmic Heating (OH)

Temperature (°C)	D_{conv} (min^{-1})	D_{oh} (min^{-1})	k_{0conv} (s^{-1})	k_{0oh} (s^{-1})
88.0	32.8	30.2	0.00117	0.001271
92.3	9.87	8.55	0.003889	0.004489
95.0	5.06	–	0.007586	–
95.5	–	4.38	–	0.008763
97.0	3.05	–	0.012585	–
99.1	–	1.76	–	0.021809
z (°C)	8.74	9.16	–	–
E_a (kJ·mol^{-1})	–	–	292.88	282.42

Source: Cho, H.Y. et al., *Biotechnol. Bioeng.*, 62, 368, 1999.

TABLE 18.2
Kinetic Constants and Thermal Inactivation Parameters for *Zygosaccharomyces bailii* under Conventional and Ohmic Heating (OH)

Temperature (°C)	D_{conv} (min^{-1})	D_{oh} (min^{-1})	k_{0conv} (s^{-1})	k_{0oh} (s^{-1})
49.8	294.6	274.0	0.008	0.009
52.3	149.7	113.0	0.016	0.021
55.8	47.21	43.11	0.049	0.054
58.8	16.88	17.84	0.137	0.130
z (°C)	7.19	7.68	–	–
E_a (kJ·mol^{-1})	–	–	123.97	116.19

Source: Palaniappan, S. et al., *Biotechnol. Bioeng.*, 39, 225, 1992.

milk, and death kinetics using conventional and *OH* were compared [58]. The microorganism's inactivation was faster when the *OH* was applied, indicating that, in addition to the thermal effect, the presence of an electric field provided a nonthermal killing effect over vegetative cells of *E. coli*. Sun et al. [59] have studied the effects of *OH* (internal heating by electric current) and conventional heating (external heating by hot water) on viable aerobes and *Streptococcus thermophilus* 2646 in milk under identical temperature history conditions. It was found that both the microbial counts and the calculated decimal reduction time (*D* value) resulting from *OH* were significantly lower than those resulting from conventional heating. Death kinetics (*D and z* values) corresponding to thermal inactivation of spores of *Bacillus licheniformis* was studied in cloudberry jam [58]. Results have shown that the time required for thermal treatment was reduced with *OH* treatment, indicating that, in addition to the thermal effect, the presence of an electric field provided a nonthermal killing effect over spores of *B. licheniformis*. Authors concluded that electrical current affected the death rate but did not affect the temperature dependency of the spore inactivation process. A nonthermal *model* was proposed by Machado et al. [60] that successfully describes the *E. coli* death kinetics under MEF treatment. Authors have demonstrated that *electric fields* are involved in the inactivation of *E. coli*, without possible synergistic temperature effects, concluding that MEF ranging from 50 to 220 V·cm^{-1} treatment holds potential for *sterilization* of thermolabile products by itself or as a complement of the traditional heat-dependent techniques. Therefore, it is clear that by reducing the time required or also the temperature for inactivation of microorganisms, due to the possible existence of nonthermal effects, the use of *OH* could diminish negative thermal effects of heating treatments, opening a new perspective for shorter, less aggressive aseptic processing. However, food safety plans must be fully validated before an industrial implementation and new critical limits must be established regarding critical control points monitoring. Furthermore, online monitoring of the process would become a function of time temperature and *electric field* instead of just time and temperature binomial.

Data on nonthermal effects are scarce and more studies are needed to determine, for example, the effect of electricity on the physiological characteristics of microbes, changes in glycosylation degree of proteins and lipids, and other elements that can affect the heat resistance of microorganisms. The differences between microorganisms such as bacteria, filamentous fungi and yeast need to be fully recognized.

The effects of *OH* on fermentations using immobilized cells or high cell density systems should also be further investigated, namely, in terms of substrate/metabolite diffusional limitations, lag phase duration and efficiency of metabolite production. The effect on toxic metabolite production by microorganisms and degradation in food requires investigation in general.

18.3.2 Enzyme Degradation Kinetics

The industrial application of the *OH* technology is fully dependent on its validation with experimental data in order to evaluate the effects of the *electric field* on microorganisms, enzymes toxins, and biological tissues.

Enzymes are used in the food industry for (a) improving food quality (e.g., texture and flavor), (b) recovery of by-products, and (c) achieving higher yields of extraction [61,62]. Enzymes may also have negative effects on food quality such as production of off-odors, tastes, and altering texture. Control is required in many *food processing* steps to promote/inhibit enzymatic activity during processing. Studies on the degradation kinetics of enzymes have been conducted to determine the effects of novel processing technologies, such as pulse *electric fields* [63–66] and high hydrostatic pressure [67,68], on *enzyme inactivation* kinetics.

The effects of *OH* on enzyme activity have not been investigated extensively. However, the sensitivity of several related enzymes toward *OH* has been studied by the authors (Castro et al., [72]) lipoxigenase (*LOX*), *polyphenoloxidase* (*PPO*), *pectinase* (*PEC*), *alkaline phosphatase* (*ALP*), and *β-galactosidase* (*β-GAL*), which is summarized in Table 18.3, included the origins of the selected enzymes, the media used for the enzyme activity assays and the methods for those assays. For both ohmic and conventional processes, and for similar samples' thermal histories, the corresponding D values were determined, leading to the calculation of the z value, the activation energy (E_a) and the pre-exponential factor (k_0) of Arrhenius equation for each enzyme (Equations 18.36 through 18.38):

$$\frac{\log C_A - \log C_{Ao}}{t} = \frac{1}{D} \tag{18.36}$$

$$\frac{\log D_2 - \log D_1}{T_2 - T_1} = \frac{1}{Z} \tag{18.37}$$

$$k(t) = k_0 \exp^{\left(-\frac{E_a}{RT}\right)} \tag{18.38}$$

Inactivation of ALP, PEC, and peroxidase (POD) during *OH* of milk, fruit, and vegetable juices, respectively, was carried out at several incubation temperatures by Jakób et al. [69]. For both ohmic and conventional processes, and for similar samples' thermal histories, experimental data of enzyme

TABLE 18.3
Overview of Enzymatic Systems Tested, Media Used, and Activity Determination Methods

Enzyme	Origin	Medium	Activity Measurement
Lipoxygenase (LOX)	Soybean (Sigma)	Tris-HCl buffer (pH 9)	Spectrophotometric at 234 nm and 25°C; substrate: linoleic acid
Polyphenoloxidase (PPO)	Apples (cv. Golden delicious)	Phosphate buffer (pH 6)	Spectrophotometric at 395 nm and 30°C; substrate: catechol
Pectinase (PEC)	Fermentation (Novo enzymes)	Citrate buffer (pH 4.5)	Spectrophotometric at 276 nm and 40°C; substrate: polygalacturonic acid
Alkaline phosphatase (ALP)	Raw milk	Milk	Spectrophotometric at 410 nm and 37°C; substrate: *p*-nitrophenylphosphate
β-galactosidase (β-GAL)	Genetically modified *S. cerevisiae* with β-GAL gene of *A. niger*	Fermentation broth	Spectrophotometric at 405 nm and 65°C; substrate: *p*-nitrophenylgalactopyranosyde (pNPG)

inactivation were modeled simultaneously using the so-called multitemperature evaluation. The parameters of the individual *models*, such as rate constants and activation energies, were estimated.

LOX are nonheme but iron-containing dioxygenases which catalyze the dioxygenation of polyunsaturated fatty acids containing a *cis*,*cis*-1,4-pentadiene-conjugated double bond system such as linoleate and linolenate, to yield hydroperoxides. Garotte et al. [70] evaluated the sensory characteristics of blanched vegetable purées to which isolated enzymes had been added and found that *LOX* was the most active enzyme in aroma deterioration in English green peas and green beans. *LOX* is distributed widely in *vegetables* and evidence is accumulating to support its involvement in off-flavor development and color loss [70]. To optimize the blanching process of *vegetables*, it is essential to establish the kinetic *model* for the inactivation of this indicator enzyme. *LOX* is also important in the baking industry because it interacts with a gluten side chain, in dough making the gluten more hydrophobic and subsequently stronger. With stronger gluten, the dough has better gas retention properties and increased tolerance to mixing.

Table 18.4 presents the kinetic parameters obtained for the thermal degradation of *LOX*. This demonstrates that the *electric field* has an additional effect on the *LOX* inactivation with much lower D values. This means that for the same inactivation degree the time required for thermal treatment is much lower, thus reducing negative thermal effects in the other food components.

PEC hydrolyzes the linkages that bind the small building blocks of galacturonic acid together in pectic substances, producing smaller molecules. Its main functions in foods are (a) to clarify juices and wines, (b) to reduce viscosity, (c) to accelerate the rate of filtration, (d) to prevent pectin gel formation of fruit juice from the fruit, and (e) to improve color extraction, for example, from grape skin [61].

The kinetic parameters for *PEC* inactivation are presented in Table 18.5, and it is clear that the *electric field* does not have any influence in the inactivation kinetics. Both conventional and ohmic processing D and z values are identical. The inactivation of exogenous PEC in apple juice and cloudberry jams has been found to follow first-order kinetics, and the differences between the activation energy values for ohmic and conventional heating were within the error of estimation [69]. Results have also revealed that pre-exponential factors in the Arrhenius equation were always higher for *OH*; the factor was about 30%–40%, which corresponded to a temperature increment of about 0.8°C. However, authors concluded that almost the full extent of activity loss with *OH* was caused by the Joule heat effect. Only a very small contribution of effects such as modification of the enzyme surface charge or enzyme environment by ionization of solution components due to the presence of an electric field could be inferred from these results.

TABLE 18.4
Kinetic Constants and Thermal Inactivation Parameters for the Thermal Degradation of LOX

Temperature (°C)	60	62	65	68	70	75	78
D_{conv} (min)	117.8	20.44	10.03	5.43	3.76	0.99	0.77
D_{oh} (min)	6.92	5.38	–	1.57	–	0.58	0.47
z_{conv} (°C)			10.83				
z_{oh} (°C)			14.75				
k_{0conv} (s^{-1})			0.9971				
k_{0_oh} (s^{-1})			0.9971				
E_{a_conv} ($J \cdot mol^{-1}$)			0.7574×10^{-3}				
E_{a_oh} ($J \cdot mol^{-1}$)			0.9686×10^{-3}				

Source: Castro, I. et al., *J. Food Sci.*, 69(9), C696, 2004.

TABLE 18.5
Kinetic Constants and Thermal Inactivation Parameters for the Thermal Degradation of PEC

Temperature (°C)	60	62	65	68	70	75
D_{conv} (min)	14.00	9.79	3.41	1.56	0.75	–
D_{oh} (min)	15.00	8.70	3.57	1.67	0.83	0.52
z_{conv} (°C)			8.11			
z_{oh} (°C)			7.75			
k_{0_conv} (s^{-1})			0.997			
k_{0_oh} (s^{-1})			0.997			
E_{a_conv} ($J{\cdot}mol^{-1}$)			0.782×10^{-3}			
E_{a_oh} ($J{\cdot}mol^{-1}$)			0.560×10^{-3}			

Source: Castro, I. et al., *J. Food Sci.*, 69(9), C696, 2004.

The enzyme responsible for enzymatic browning, *orthodiphenol oxygen oxidoreductase*, is also known as catecholase, tyrosinase, phenolase, and PPO [61,71]. *PPO* is present in most foods and participates only in the beginning phase of oxidation as it catalyzes the change of monophenols to diphenols which are then changed to highly reactive, colored *o*-quinones. These then react with other o-quinones, amino acids, reducing sugars, etc., to form polymers that precipitate, resulting in the appearance of a dark color [71]. There are various phenols present in *fruits* and *vegetables* that are simultaneously used as substrates for *PPO*. Oxidative browning occurs only when the tissues are disrupted or destroyed and the compounds come into contact with air and with each other. The browning of *fruits* and *vegetables* is undesirable and so careful handling is required to avoid tissue injury. However, browning is important to flavor and color development (e.g., cocoa and tea), and so the tissues of these plants are deliberately damaged.

The effects of the *electric field* on the thermal inactivation kinetics of *PPO* (Table 18.6) are similar to the ones described for *LOX*. An enhanced *enzyme inactivation* is obtained when an *electric field* is present, thus reducing inactivation time. It was hypothesized that the electric field may have removed the metallic prosthetic groups, present in both enzymes LOX and PPO thus causing the enhancement of activity loss [72].

TABLE 18.6
Kinetic Constants and Thermal Inactivation Parameters for the Thermal Degradation of PPO

Temperature (°C)	70	75	80	85	90	95
D_{conv} (min)	193.0	61.61	23.74	10.30	3.15	1.51
D_{oh} (min)	–	19.37	6.79	3.52	0.92	0.58
z_{conv} (°C)			11.85			
z_{oh} (°C)			12.77			
k_{0_conv} (s^{-1})			0.997			
k_{0_oh} (s^{-1})			0.997			
E_{a_conv} ($J{\cdot}mol^{-1}$)			0.777×10^{-3}			
E_{a_oh} ($J{\cdot}mol^{-1}$)			0.815×10^{-3}			

Source: Castro, I. et al., *J. Food Sci.*, 69(9), C696, 2004.

In another study, the effects of voltage gradient, temperature, and holding time on PPO activity during *OH* were also investigated at pasteurization temperatures range in fresh squeezed grape juice [73]. It was observed that the critical temperature at which the enzyme deactivation starts was lower at $40\,V\cdot cm^{-1}$ than that at 20 or $30\,V\cdot cm^{-1}$. Moreover, enzyme activity was found to be significantly lower at $40\,V\cdot cm^{-1}$ than at 20 or $30\,V\cdot cm^{-1}$ at 70°C and 80°C. These effects were attributed to the faster increase in electrical conductivity at higher voltage gradients.

ALP is an enzyme used as an indicator of the effectiveness of the milk thermal processing. For this enzyme, conventional and ohmic inactivation mechanisms are similar, and the results obtained do not demonstrate a clear additional effect of the *electric field* on the *enzyme inactivation* (Table 18.7). These results are somehow in accordance with ones found by Jakób et al. [69] who have studied the ALP activity loss on bovine and caprine milk under *OH* in the temperature range of 52°C–64°C. In this study, the principal conclusion was that activity loss under *OH* was mainly caused by the Joule heat effect in accordance with the conclusions drawn for PEC inactivation.

The use of the enzyme β-*GAL* in lactose containing foodstuffs may be an area of interest. It may be useful for people who are lactose intolerant and cannot eat milk and dairy products. When the intestine produces little or no lactase, lactose (the milk sugar) is not digested and moves into the colon, where bacteria ferment it, producing hydrogen, carbon dioxide, and organic acids. The results of this fermentation are diarrhea, flatulence (gas), and abdominal discomfort. However, conventional and ohmic inactivation mechanisms are similar, and the results obtained (Table 18.8) do not show a clear additional effect of the *electric field* on the *enzyme inactivation* [72].

POD catalyzes a large number of reactions in vegetable and plant tissues, causing browning of foods and leading to discoloration, off-flavors, and nutritional damage. It is also known that the thermal stability of vegetable PODs varies considerably with the source and origin [74]. POD inactivation experiments were carried out for several vegetable mixtures [69]. The temperature range applied was 58°C–74°C for broccoli and potato mixtures and 62°C–78°C for a carrot mixture. Results revealed that during the early stages of heating, minimal differences were observed between *OH* and conventional heating. However, after for prolonged heating times, POD inactivation proceeded faster under *OH* than under conventional heating. Furthermore, it has been concluded that despite activation energies of the irreversible reaction were essentially the same for ohmic and conventional heating, the rate constants at the reference temperature were different; *OH* may have caused a significant entropic effect on the rate of inactivation of POD. Authors have stated that this implies that a cause of the decreased stability does not lie in the modification of enzyme tertiary

TABLE 18.7
Kinetic Constants and Thermal Inactivation Parameters for the Thermal Degradation of ALP

Temperature (°C)	55	60	62	65	67	70
D_{conv} (min)	35.46	19.00	4.41	3.54	1.96	0.91
D_{oh} (min)	31.75	11.00	4.06	2.89	1.28	0.89
z_{conv} (°C)			9.00			
z_{oh} (°C)			9.34			
k_{0_conv} (s^{-1})			0.997			
k_{0_oh} (s^{-1})			0.997			
E_{a_conv} ($J\cdot mol^{-1}$)			0.649×10^{-3}			
E_{a_oh} ($J\cdot mol^{-1}$)			0.675×10^{-3}			

Source: Castro, I. et al., *J. Food Sci.*, 69(9), C696, 2004.

TABLE 18.8
Kinetic Constants and Thermal Inactivation Parameters for the Thermal Degradation of β-GAL

Temperature (°C)	65	70	72	75	78	80
D_{conv} (min)	182.0	33.90	12.89	2.99	0.52	0.50
D_{oh} (min)	–	12.24	9.70	2.77	0.64	0.28
z_{conv} (°C)			5.12			
z_{oh} (°C)			5.13			
k_{0_conv} (s^{-1})			0.997			
k_{0_oh} (s^{-1})			0.997			
E_{a_conv} ($J \cdot mol^{-1}$)			0.370×10^{-3}			
E_{a_oh} ($J \cdot mol^{-1}$)			0.350×10^{-3}			

Source: Castro, I. et al., *J. Food Sci.*, 69(9), C696, 2004.

structure by the electric field but can be explained by the affect of modified environment due to the increased number of ions and their different distributions around enzyme molecules.

In summary, *OH* can enhance the rate of inactivation of enzymes present in food materials. However, no general behavior or trends are observed, and the effect of *OH* and nonthermal effect due to the presence of an *electric field* seem to be both dependent on the enzyme and probably also on the food system. The *electric field* reduces the *D* values for *LOX* and *PPO* while *z* and E_a values are not greatly affected. POD also seemed to be prone to destabilization electric fields during OH, which promoted changes on thermodynamic properties of *enzyme inactivation*. In the case of PEC and ALP, a slight decrease of enzyme stability occurred, and joule heat effect was predominant.

18.3.3 Ascorbic Acid Degradation Kinetics under OH

Ascorbic acid (vitamin C) is frequently used as a food preservative or a vitamin supplement. Industrially its main function is to prevent browning and discoloration, thus enhancing the shelf life of several products. Ascorbic acid is, for example, an effective inhibitor of POD in *fruits* such as kiwi.

It is known to be thermolabile and its degradation mechanism is specific to the particular system in which it is integrated. Degradation depends on whether aerobic or anaerobic pathways are employed [75]. The aerobic degradation pathway is related to the presence of oxygen (either in the headspace or dissolved) [76]. The anaerobic pathway is mainly driven by the storage temperature, with lower temperature storage being the only way to minimize the degradation rate [77]. When oxygen is present, the contribution of anaerobic degradation to the total vitamin C loss is small compared to aerobic degradation [78]. Several studies have been made to determine the kinetic parameters of thermal degradation of ascorbic acid in food systems under conventional and *OH* conditions (Table 18.9). It must be stressed that thermal degradation kinetics are variable and depend on the products being processed. In summary, from the limited amount of data available comparing the degradation of ascorbic acid by conventional heating and *OH* demonstrate no statistically significant differences.

Castro et al. [29] studied *ascorbic acid* degradation kinetics in industrial *strawberry* pulps, concluding that it followed first-order kinetics for conventional and *OH* treatments. The obtained kinetic parameters were identical, for the temperature range of 60°C–97°C and for the two types of heating processes. This leads to the conclusion that the presence of lower values of the *electric field*

TABLE 18.9
Kinetic Constants and Thermal Inactivation Parameters for the Degradation of Ascorbic Acid in Several Fruits

Products	Temperature Range (°C)	pH	°Brix	E_a (kJ mol^{-1})	K_0 (s^{-1})	z (°C)	$D_{75°C}$ (min)	Sources
Strawberry pulp	60–97 (conventional heating)	4.0	14.5	21.36	0.15	46.7	175	[29]
Strawberry pulp	60–97 (OH)	4.0	14.5	21.05	0.14	46.7	169	[29]
Grapefruit juice	61.0–96.0 (conventional heating)	3.05	11.2	21.0	3.90×10^{-2}	49.3	1354	[79]
Grapefruit juice	60.0–91.0 (conventional heating)	3.05	31.2	22.0	6.10×10^{-2}	45.0	1228	[79]
Lime	20.0–92.0 (conventional heating)	5.92	6.3	58.1	1.55×10^{4}	35.8	1186	[80]
Lemon	20.0–92.0 (conventional heating)	2.94	6.0	46.5	3.59×10^{2}	44.6	949	[80]
Tangerine	20.0–92.0 (conventional heating)	4.10	13.4	44.6	2.25×10^{2}	46.5	771	[80]
Grapefruit	20.0–92.0 (conventional heating)	3.54	11.2	56.9	9.29×10^{3}	36.5	1276	[80]
Orange juice	70.3–97.6 (conventional heating)	3.60	12.5	128.3	3.23×10^{13}	19.0	24110	[81]
Orange juice	70.3–97.6 (conventional heating)	3.60	36.7	97.4	1.62×10^{9}	24.9	10447	[81]
Orange juice	65–90 (conventional heating)	Not reported	Not reported	12.6	3.30×10^{4}	Not reported	Not reported	[82]
Orange juice	65–90 (OH)	Not reported	Not reported	12.5	3.26×10^{4}	Not reported	Not reported	[82]

strength (<20 V·cm^{-1}) do not affect the *ascorbic acid* degradation. Lima [83] drew similar conclusions for orange juice systems.

The *OH* technology will be extremely useful in minimizing thermal gradients (and thus overprocessed volumes) and consequently minimizing ascorbic acid degradation (and other thermolabile compounds). The final products obtained are expected to be of higher nutritional and organoleptic quality. This may also be extremely important in producing functional foods.

18.3.4 Whey Proteins

Whey proteins are increasingly used as nutritional and functional ingredient in several foods formulations, particularly in the form of dry whey, whey concentrate (WC), and whey protein isolate

(WPI). The major proteins in whey are β-lactoglobulin (β-Lg), which constitutes more than 50% of the total protein in whey, and α-lactalbumin (α-Lac) and bovine serum albumin (BSA), which represent 20% and 5%, respectively. The functionality of whey proteins is determined by changes in their physical and chemical properties during the production process that commonly includes thermal processing such as, preheating, *pasteurization*, and *sterilization*. The effects of heat treatment on denaturation of whey proteins in milk and whey systems has been extensively studied in the literature, once it is the most commonly applied in food processing [83]. It is known that heating treatments at elevated temperatures (>60°C) result in thermal denaturation of globular whey proteins, which is assumed generally to be a process consisting of at least two steps: a partial unfolding of the native protein and a subsequent aggregation of unfolded molecules [84]. Pereira et al. have studied the effects of *OH* on whey protein denaturation. The results showed that *OH* determined lower concentration of free sulfydryls (unfolding) and less aggregation (irreversible denaturation) in heated whey protein solutions, when compared with conventional heating performed under similar thermal histories. Less whey protein denaturation occurs during direct heating than during indirect heating [85,86]. During *OH*, heat is generated directly within the food sample (volumetric or direct heating) and hence the problems associated with *heat transfer* surfaces are eliminated [87]. In fact, uniform heating and the absence of hot surfaces minimize overprocessed volumes. *OH* may have determined lower whey protein denaturation rates interfering with the interactions between whey proteins. There is no information concerning nonthermal effects of *OH* on whey proteins unfolding and aggregation mechanisms. Nonthermal effects of electric fields can be related with conformational disturbances on tertiary protein structure due to rearrangement of hydrogen bonds, hydrophobic interactions, and ionic bonds [73]. The presence of electric fields may affect ionic movement in the medium and influence biochemical reactions by changing molecular spacing and affecting the rates of interchain reactions. Therefore, the presence of MEF of low frequency can also dictate distinctive interactions and associations of denatured proteins that need to be further studied and verified.

18.4 BASIC OHMIC HEATER CONFIGURATIONS

This section presents the basic configurations for ohmic heaters but it does not imply that these are the only ones available. In fact other possibilities exist, including some under development, which are not considered here. The fundamental requirements for *OH* equipment for *food processing* are a pair of *electrodes*, a container for the food to be processed and an alternating power supply.

The ohmic heater can be integrated into a *batch* or *continuous process*. The most typical configuration for the ohmic heater is that of a horizontal cylinder with one electrode placed in each extremity for a *batch* process (Figure 18.7a). For a *continuous* process, the design of the ohmic heater can be more variable, depending on the manufacturer. It can range from a simple tube with pairs of opposing *electrodes* mounted on the tube walls opposite to each other (Figure 18.7b) to coaxial tubes acting as *electrodes* with the food flowing between (Figure 18.7c), or a vertical tube with the *electrodes* embodied at regular intervals (Figure 18.7d). As the *electric field* is perpendicular to the food flow for the equipment represented in Figure 18.7b and c, these configurations are often called *cross-field*. If the electric field is parallel to the food flow in the (Figure 18.7d) the configuration obtained is termed *in-field*.

Ideally, it is possible to consider that in the cross-field configuration, the *electric field* strength is constant. For the in-field configuration, σ will increase, and therefore, the field strength experienced by the product will increase as it approaches the outlet. The product will heat during its path through the heater. To minimize this effect, when multiple *electrodes* are used in series (Figure 18.7d), they are spaced to account for the increase of σ of food with temperature. Therefore, as the product approaches the outlet of the heater, a lower value of the electric field strength is needed, and this is accomplished by increasing the spacing between each pair of *electrodes*.

The choice of the best configuration will obviously depend on the food being processed and the objectives of the process (e.g., cooking, *pasteurization*, *sterilization*).

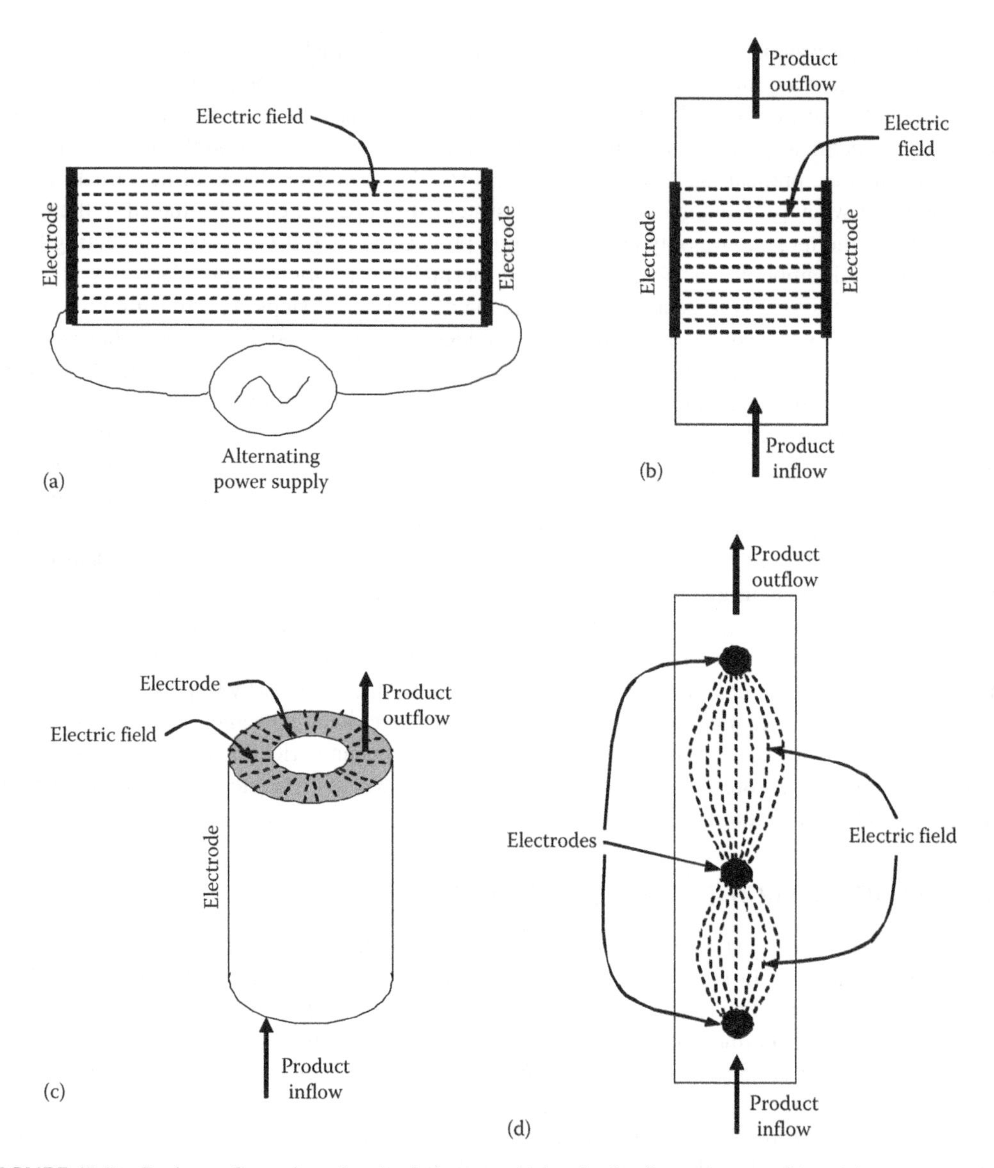

FIGURE 18.7 Basic configurations for ohmic heaters: (a) batch; (b–d) continuous. (b) and (c) correspond to the cross-field configuration while (d) corresponds to the *in-field* configuration, depending on if the electric field lines are perpendicular or parallel to the direction of the flow of the food, respectively.

Batch processes are typically used to cook, for example, *meat products* [88], while *continuous* processes are more appropriate for viscous fluids or fluids with particulates.

18.5 APPLICATIONS

The *OH* technology provides new, high-added-value, shelf-stable products with a quality unachievable by the traditional processing technologies. The process can be used in (a) the thermal processing of high-acid food products such as tomato-based sauces, (b) *pasteurization* of whole liquid egg [9], (c) fish pastes, (d) *meat products* processing as an alternative cooking to the traditional smokehouse, and (e) the extraction of specific cell metabolites (macromolecules, sugars, and pigments) via MEF treatment, where heat and *electric field* work together to increase the permeability.

18.5.1 Meat Products

The first experiments using *OH* in *meat products* were made in Finland in the 1970s, but operational difficulties led to the abortion of the project. Later on, in the 1990s, a project for cooking liver pâté and hams was carried out in France by the Meat Institute Development Association (ADIV) and Électricité de France (EDF).

One *OH* application is *thawing* process, when the traditional *thawing* methods cannot provide *high-quality products*. Wang et al. [89] applied the technology to frozen meat samples, in a liquid-contact *thawing* method. The results demonstrated a uniform and quicker *thawing* process. Also, meat properties such as color and pH were not changed significantly, and the final products achieved a good *thawing* quality. These results demonstrate the potential uses of *OH* in contact *thawing*, especially for *meat products*. Although the initial batch results were very promising, the attempts to develop a prototype version for a *continuous* cooking process were unsuccessful.

Canadian scientists [88] have been working on the ohmic cooking of *meat products* (*sausages* and *ham*) with promising results in batch operations. Brine-cured *meat products* were found to be admirably suited to *OH* having extremely reduced cooking times (e.g., a *ham* weighing 1 kg was cooked in less than 2 min). However, these flash cooking times did not reduce the bacterial load to levels that guaranteed product safety. *Pasteurization* had to be taken into consideration and a time–temperature profile was established. However, under these conditions, no noticeable changes were reported in the products' taste, texture, or shelf life.

The knowledge of *electrical conductivity* of meats and nonmeat ingredients is essential in the formulation of meat products destined for *OH*, once intrinsic levels of electrically conductive materials can be dramatically altered by the nature of the ingredients added in solid–liquid food mixtures. Sarang et al. [90] developed a blanching method for increasing the ionic content in chicken chow mein consisting of a sauce and different solid components (celery, water chestnuts, mushrooms, bean sprouts, and chicken) and adjusting the electrical conductivities of the solid components to that of sauce. In this study, chicken chow mein samples containing blanched particulate were compared with untreated samples with respect to *OH* uniformity at 60 Hz up to 140°C. It has been observed that following pretreatment, it may be possible to uniformly heat the entire product during *OH* and that such pretreatment does not compromise product quality. Zell et al. [91] compared σ values of nonmeat ingredients in solution/suspension and when incorporated in a meat product at typical usage levels. In this study, the influence of fat incorporation, fiber orientation, and salt injection on σ of whole beef muscle was also evaluated during *OH* process. The protocols developed would be useful for preparation of other multicomponent *meat products* for *OH sterilization*.

18.5.2 Fruit and Vegetable Products

Strawberry fruit jams are extremely important for the Portuguese fruit jam industry because they account for most of its sales (~90%). The search for alternative processing technologies leading to higher quality products is one of the main goals, and the economic viability of this technology depends on the possibility of applying it to most of the products processed by this industry [8]. In the study by Castro et al. [8], several *strawberry*-based products were tested by *OH*. The obtained results showed that, for most of the products, high *heating rates* could be achieved, despite the significant differences of σ between the products stet test. Also, the increase of the applied *electric field* could increase the *heating rate* (Figure 18.8a and b).

The suitability of *OH* for *strawberry* products having different solids concentrations, or Brix values, was also tested [8]. σ was shown to decrease with the increase of solids content in a mixture of particles with a bimodal *particle size* distribution, but the decrease was more significant for the bigger particles tested (Figure 18.9a). The results also suggest that for higher solids content (>20% w/w) and sugar contents more than 40.0°Brix, σ is too low to use in the conventional ohmic heaters and a new design is required (Figure 18.9b).

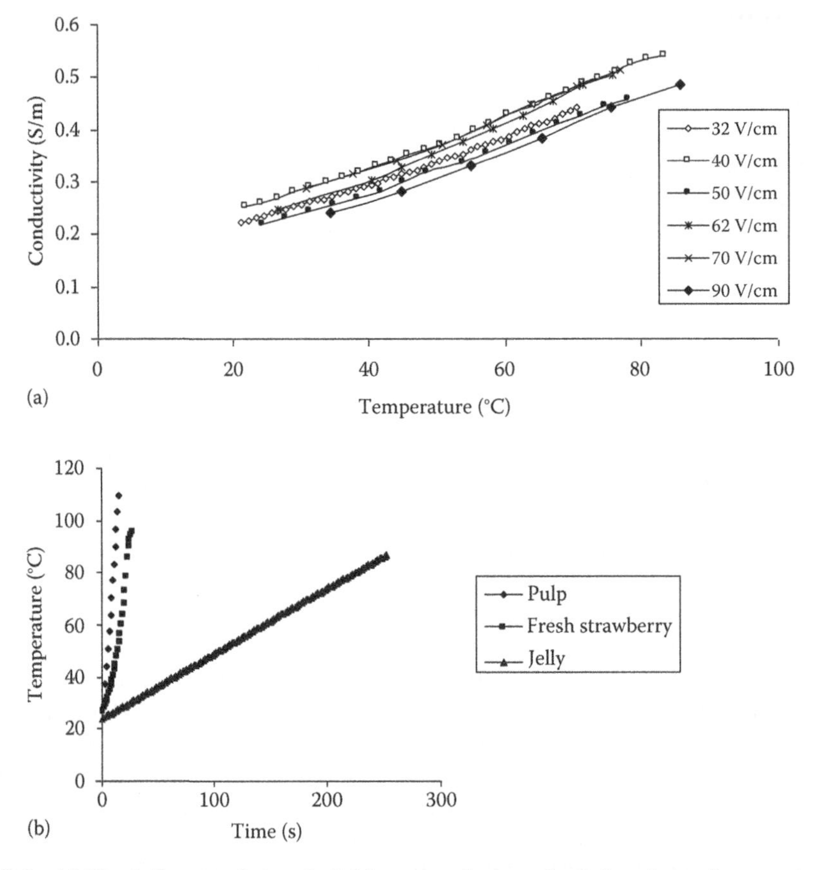

FIGURE 18.8 (a) The influence of electric field on electrical conductivity of strawberry pulp; (b) differences in heating rates of three strawberry based products. (From Castro, I. et al., *Innovat. Food Sci. Emerg. Technol.*, 5, 27, 2004.)

Most of the commercially available *vegetables* are submitted to more than one thermal treatment (e.g., blanching, *sterilization*). Wang and Sastry [7] studied the effects of an ohmic pretreatment and found no significant changes in the moisture content of the final products. This technology might be an alternative to conventional blanching treatments. It has already been demonstrated that the texture of *fruits* and *vegetables* is greatly influenced by temperature and thermal processing [92,93]. The firmness of processed *vegetables* (e.g., canned cauliflower, asparagus) can be improved by low-temperature, long-time pretreatments. Eliot et al. [94] studied the influence of precooking by *OH* on the firmness of *cauliflower*. The experimental data showed that *OH* combined with low-temperature precooking in saline solutions offers a viable solution to HTST sterilization of cauliflower florets. A similar study was also performed with potato cubes [95,96] and concluded that an ohmic pretreatment prevented loss of firmness when compared to a conventional pretreatment (50% in some cases).

OH appears to be an effective method for enhancement of processes controlled by mass transfer since it affects the integrity of biological tissue by solubilizing the pectic substances constituting the cellular wall and by providing electroporation of cell membranes [97,98]. Diffusion is enhanced, electrical conductivity changes are more linear during heating, and moisture migrates more easily out of the tissue [99,100]. Ohmic pretreatment has been found useful in drying and juice extraction processes and in sugar extraction from beets [101,102]. Allali et al. [103] studied the effect of blanching by *OH* on the kinetics of osmotic dehydration of strawberries and found that ohmic blanching permits the effective damage of cells by combination of electrical and thermal effects,

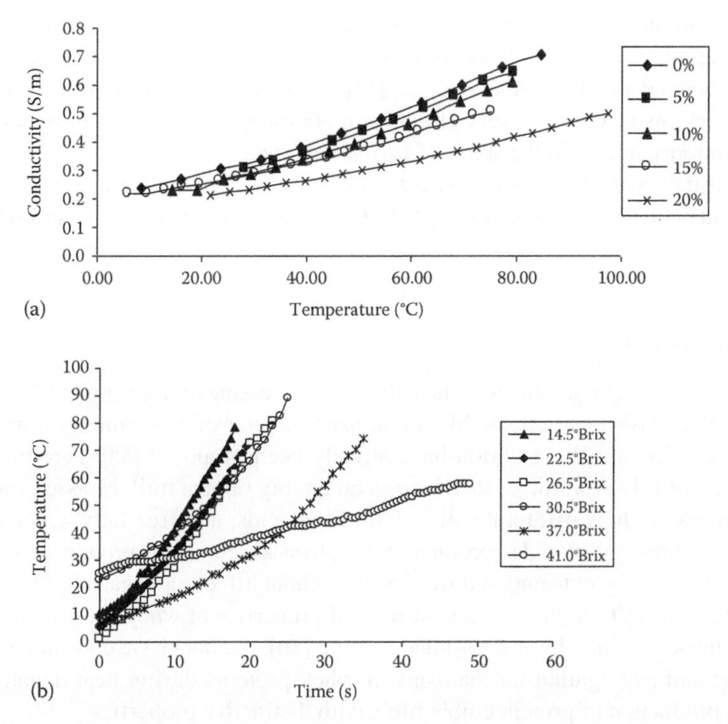

FIGURE 18.9 (a) The influence of solid particles concentration on electrical conductivity of strawberry pulp; (b) differences in heating rates of strawberry pulps with different Brix values. (From Castro, I. et al., *Innovat. Food Sci. Emerg. Technol.*, 5, 27, 2004.)

thus resulting in an important enhancement of water and sugar transfers during osmotic dehydration of strawberries.

The presence of MEF during *OH* may induce electropermeabilization and thus increase the permeability of the cell membrane of plant tissues. MEF processing is defined as a process of controlled, possibly reversible permeabilization, characterized by the use of electric fields (typically from 1 to 100 $V \cdot cm^{-1}$), with arbitrary waveform electric fields, with or without the attendant *OH* effects [104]. For example, *OH* of raw potato tissue causes an increase in conductivity at a lower temperature than conventional heating; for the same final temperature of heating, *OH* yields greater permeabilization of the potato tissue than conventional heating; the gradual change in dielectric properties that was seen over the same temperature range for ohmically heated samples suggests that the presence of an *electric field* was responsible for the membrane changes produced [105].

In another study, Sensoy and Sastry [105] have compared extraction via MEF with that via hot water of black (fermented) tea leaves and dried and fresh mint leaves and investigated effects of electric field strength and frequency on the extraction of fresh mint leaves. Authors showed that electrical breakdown is achieved even at low electric field strengths and that MEF treatment shows potential to be used as an alternative to conventional heating for extraction from cellular materials.

18.5.3 Seafood Products

Most of the scientific literature concerning seafood products deals with *surimi*. *Surimi* is stabilized myofibrillar proteins from fish muscle, which is used in several Japanese food products. The textural properties of the products treated by *OH* were found to be superior to those heated in a 90°C water bath. Also, an increase in shear stress and shear strain of *surimi* gels was found when ohmic technology was applied [106] and, in addition, a superior gel quality was achieved. Higher *heating rates* are not beneficial to *surimi* manufacture [54]. Instead, slow *heating rates* produce stronger gels. Ohmic processing of *surimi* is very effective in obtaining a wide range of linear *heating rates* which plays an important role in the study of *surimi* gelation.

It is expected that with the *continuous* research and development of new electrode materials and of equipment with innovative design, several other food applications using OH will soon be possible.

18.5.4 Dairy Products

OH technology offers great potential to reduce the overprocessing of high-protein liquid foods products, such as milk or protein solutions. Modeling *heat transfer* of a continuous flow *OH* of reconstituted skimmed milk and WC solution have already been proposed [53]. Pereira et al. [58,107] studied the effects of HTST ohmic *pasteurization* on quality of goat milk by assessing physical and chemical properties, such as pH, total solids, total fatty acids, and free fatty acids, and have concluded that this technology provides products with physical–chemical properties similar to those of the products obtained by conventional treatment, without affecting negatively the quality of goat milk flavor. Effects of *OH* on physical and structural properties of whey protein edible films were compared with those obtained by conventional heating [86]. Authors have concluded that *OH* influenced unfolding and aggregation mechanisms of whey proteins during heat denaturation, which determined the production of protein edible films with distinctive properties.

18.6 ECONOMICS

Although there are some *OH* facilities in operation around the world, it is not widespread and the data on commercial *OH* operations are scarce. Several applications of *OH* are still under study and various products are being processed in pilot-scale equipment, but not at industrial scale. As a consequence, processing costs were not yet fully assessed.

With the objective of determining the economic viability of this technology, a comparison of *OH* with other *food-processing* methods was made by Allen et al. [108]. In their study, several assumptions were made about the products, facilities, and components of costs. Estimated values for labor, energy, quality control, packing, maintenance, and equipment depreciation were provided by well-trained food technologists and researchers. The product and production characteristics of the food products under analysis are summarized in Tables 18.10 and 18.11.

The main difference between the two types of ohmic heaters used in this study, OH 75 and OH 300, as their power is 75 and 300 kW, respectively.

The conclusions withdrawn from this study show the importance of the type of product and of the power of the ohmic heater on the economics of the process. The OH 300 appears to be a viable alternative to conventional processing technologies when processing low-acid foods. It may also be a useful technology for high-acid foods, but only if *high-quality products* are required.

Studies conducted at the Agri-Food Canada's Food Research and Development Centre (FRDC) with *meat products*, where the traditional smokehouse cooking was replaced by an ohmic process, indicated that energy savings of at least 70% could be achieved. In fact, the specific energy consumption required to cook a brine-cured meat product was on the order of 210 and 258 $kJ \cdot kg^{-1}$, compared to 859 $kJ \cdot kg^{-1}$ for conventional cooking [88]. In an industrial production context, energy savings would undoubtedly be even greater, as considerably less energy would be lost because of optimally designed equipment. Also, time savings would probably be considerable (a few minutes with ohmic cooking as oppose to a week with the traditional smoking house), indicating that ohmic cooking is an economic method for making excellent *meat products*.

TABLE 18.10
Low-Acid Formulated Products: Product and Production Characteristics for Various Processing Technologies

	Processing Technology		
	Retorting	Freezing	Continuous-Flow OH
Studied products	Fresh or frozen meat and vegetables	Fresh or frozen meat and vegetables	Fresh or frozen meat and vegetables
Carrier	Gravy or sauce	Gravy or sauce	Gravy or sauce
Package	Microwavable tray	Microwavable tray	Microwavable tray
Product storage	Shelf stable	Frozen	Shelf stable
Product shelf life	About 1 year	About 1 year	About 1 year
Annual production	5.44 million Kg	5.44 million Kg	OH 75: 3.18 million kg[a] OH 300: 12.70 million kg[a]

Source: Allen, K. et al., *Food Technol.*, 50, 269, 1996.
[a] Per main ohmic heater.

TABLE 18.11
High-Acid Formulated Products: Product and Production Characteristics for Various Processing Technologies

	Processing Technology		
	Freezing	Conventional Thermal Heating (Tubular Heat Exchanger)	Continuous-Flow OH
Studied products	Fresh fruit	Fresh fruit	Fresh or frozen fruit
Carrier	Sugar syrup	Sugar syrup	Sugar and starch syrup
Package	Plastic tubs	Bag-in-box	Bag-in-box
Product storage	Frozen	Shelf stable	Shelf stable
Product shelf life	About 1 year	1 year	1 year
Annual production	15.88 million kg	15.88 million kg	OH 75: 1.58 million kg[a] OH 300: 6.3 million kg[a]

Source: Allen, K. et al., *Food Technol.*, 50, 269, 1996.
[a] Per main ohmic heater.

Studies on the economic viability of *OH* for *continuous* aseptic processing of *strawberry* pulps and *strawberry* particulate jams (high-acid products) are currently being made in Portugal, but no further details are available at this time.

Economic viability and detailed costing analysis need to be made on a case-by-case basis. To obtain more accurate results on the profitability of this technology when applied to specific facilities, it is necessary to consider full production schedules, startup costs, as well as considerations on product type [108]. Such actions must be taken in order to establish the commercial feasibility in this technology.

18.7 OVERVIEW, PROBLEMS, AND FUTURE CHALLENGES

Recently, several thermal and nonthermal technologies have been proposed for *food processing*, having as their main objective to contribute to the production of safer and higher quality food products. *OH* is one of these technologies presenting as its main advantage that *heat generation* occurs inside the material being processed. It is emphasized here that the internal *heat generation* induced by *OH* eliminates the problems associated with heat *conduction* in food materials, preventing the overcooking typical of conventional thermal *food processing*. Moreover, as the heating process is generally faster than that with heat exchangers, the killing of microbial cultures is obtained with a less pronounced degradation of most food components. Therefore, *OH* is indicated for the processing of foods containing particulate materials and foods with non-Newtonian rheological properties. The characterization, modeling, and control of all the phenomena associated with the application of *OH* in *food processing* will lead to the achievement of the main goal when heating foods—to have an effective control on the temperature profile of food materials.

Several applications have been proposed and developed for *OH* in relation to microbial control to replace the heating methods for food *pasteurization*, *sterilization*, and *food processing*—blanching, evaporation, dehydration, fermentation and extraction—and several units are already working at the industrial scale.

Various reactions must be considered to occur in a food matrix that includes in its composition several components with different electrical, heat, mass, and momentum transfer properties. Obviously, each individual component and their interactions will play a decisive role in the efficiency of the *OH* process. Such a role may be played on the thermal killing of the microbial population or on the modification of the functional, nutritional, and organoleptic properties of foods. This reasoning clearly points out the complexity associated with thermal processing of foods and in particular with *OH* thermal processing.

Several additional points associated with the fundamentals of ohmic processing must be addressed. In the particular case of a *continuous* flow ohmic heater, the flow properties of the food components must be fully characterized and related to the temperature profiles of those components. These phenomena may be considered from a macroscopic point of view. However, microscopic phenomena also take place. At this level, disruption of cell membranes, structural changes in food components (including enzyme inactivation) and mass transfer phenomena must be considered.

Also, novel applications and a deeper knowledge of current applications must be obtained. Obviously, further research and development work on *OH* in these areas are essential.

Research is required in the following areas:

1. To elucidate on the relative importance of electric current properties and the corresponding temperature values on the killing of microbes and in particular resistant structures (e.g., spores)
2. To evaluate the effect of *OH* on microbial toxins (e.g., mycotoxins)
3. To characterize the effect of *OH* on the nutritive, organoleptic, and functional properties of foods (this work should be initiated by studying the effects on individual food components, including water, and should be extended to whole food products; at this point, it must be emphasized that each food product requires specific treatments)
4. To characterize the flow of the food components when being processed in an ohmic heater by developing adequate hydrodynamic *models*
5. To develop methods that will allow for a more precise mapping of temperatures on foods submitted to *OH*
6. To develop *models* that can adequately describe ohmic processing of foods
7. To implement these *models* so that an adequate control of the rate of heating can be achieved, thus minimizing the thermal degradation effects on desirable product attributes but maintaining a safe product

In terms of process development, the more relevant points to be addressed are as follows:

- Food preservation (with respect to spores and filamentous fungi)
- *Food processing*—application of *OH* to *particulate foods* (food purees containing particles, food suspensions, etc.) and foods with a now-Newtonian rheological behavior

- *Food processing*—integration of *OH* in existing food industries as an alternative to conventional preheating, in those cases where higher quality products and more efficient processing are to be obtained
- Food *thawing* and water removal processes (evaporation and dehydration)
- Application to fermentation processes; this includes how electrical current affects microbial metabolism and membrane and cell-wall transport properties
- Extraction and purification processes—the possibility of applying an electrical field to existing processes to increase the yields and purification of biological macromolecules
- To develop alternative designs for ohmic heaters, adapted to specific food processes
- To extend the use of *OH* to other processes other than food processes

18.8 CONCLUSIONS

OH is a *food processing* technology that presents several advantages when compared to existing technologies. It can improve food quality and is a clean technology. Also the economic advantages of its use have been demonstrated.

The application of *OH* for *food processing* is not well characterized and not all its potentialities have been fully exploited due to the complexity of the phenomena occurring during *OH* processing and the complexity of food materials.

In conclusion, if attention is paid to the phenomena occurring during *OH*, there are (a) electrical current generation and transport, (b) *heat generation* and transport, and (c) eventually mass and momentum transport. These phenomena are intimately associated and interact with each other. Also, there are several consequences occurring as a result of the electrical current passage and heat being generated because of the electrical current. These reactions include thermal and electrical *microbial killing*, nutrients modification, and *enzyme inactivation*.

This technology will gain an increasing importance in food processing as a deeper understanding of the fundamental science is achieved.

NOMENCLATURE

A_{fP}	Cross-sectional area of parallel fluid (m^2)
A_{fS}	Cross-sectional area of serial fluid (m^2)
A_p	Area of each solid particle (m^2)
A_{pP}	Cross-sectional area of parallel particles (m^2)
C_A	Activity (U)
C_{A0}	Initial activity (U)
C_{pf}	Fluid-specific heat ($J \cdot kg^{-1} \cdot K^{-1}$)
C_{pp}	Particle-specific heat ($J \cdot kg^{-1} \cdot K^{-1}$)
D	Decimal reduction time (s)
D_{conv}	Decimal reduction time, under conventional heating conditions (s)
D_{oh}	Decimal reduction time, under ohmic heating conditions (s)
E_a	Activation energy ($J \cdot mol^{-1}$)
F	Electric field ($V \cdot m^{-1}$)
I	Current intensity (A)
$\dot{Q}$	Heat generated ($J \cdot s^{-1}$)
$\dot{Q}_f$	Heat generated in the fluid ($J \cdot s^{-1}$)
$\dot{Q}_p$	Heat generated in the particle ($J \cdot s^{-1}$)
R	Electrical resistance (Ω)
R	Radius of the heater tube (m)
R_{fP}	Parallel resistance of fluid (Ω)
R_{fS}	Serial resistance of fluid (Ω)

R_{pP}	Parallel resistance of particles (Ω)
S	Dissolved solids content ($kg \cdot m^{-3}$)
T_{0p}	Initial temperature of the particle (K)
T_{air}	Temperature of the air surrounding the walls of the heater (K)
T_f	Fluid temperature (K)
T_p	Particle temperature (K)
T_{ps}	Temperature at the surface of the solid particle (K)
T_{ps}	Temperature at the surface of the solid particle (K)
T_{ref}	Reference temperature (K)
U	Overall heat transfer coefficient ($W \cdot m^{-2} \cdot K^{-1}$)
V	Voltage (V)
V_f	Volume of the fluid (m^3)
V_p	Volume of the particle (m^3)
V_{sys}	Volume of the relevant part of the system (m^3)
V_{sys}	Volume of the system between the electrodes (m^3)
Z	Dependence on temperature of D value (K)
h_{fp}	Fluid-particle heat transfer coefficient ($W \cdot m^{-2} \cdot K^{-1}$)
i	Node number (-)
k	Frequency constant (s^{-1})
k_0	Preexponential factor (s^{-1})
k_f	Fluid thermal conductivity ($W \cdot m^{-1} \cdot K^{-1}$)
k_p	Particle thermal conductivity ($W \cdot m^{-1} \cdot K^{-1}$)
k_x	Thermal conductivity in the x direction ($W \cdot m^{-1} \cdot K^{-1}$)
k_y	Thermal conductivity in the y direction ($W \cdot m^{-1} \cdot K^{-1}$)
l_{fP}	Length of parallel fluid (m)
l_{fS}	Length of serial fluid (m)
l_{pP}	Length of parallel particles (m)
m	Temperature coefficient ($S \cdot m^{-1} \cdot K^{-1}$)
m_f	Temperature coefficient of the fluid ($S \cdot m^{-1} \cdot K^{-1}$)
m_p	Temperature coefficient of the particle ($S \cdot m^{-1} \cdot K^{-1}$)
n	Flow behavior index (-)
n	Time step index (-)
$\vec{n}$	Unit vector normal to the surface of the tube's wall (-)
n_p	Number of solid particles in the considered volume (-)
r	Radial coordinates (-)
t	Time (s)
$\bar{v}_z$	Mean fluid velocity profile in the axial direction ($m \cdot s^{-1}$)
$\bar{v}_z$	Mean fluid velocity profile in the axial direction ($m \cdot s^{-1}$)
w	Tube's wall (-)
x, y	Space coordinates (-)
z	Axial coordinates (-)
∇V	Voltage gradient (V)
$\beta(\varepsilon_f)$	Fraction of heat transfer by conduction through the mixture in the fluid phase (-)
ε_f	Fluid volume fraction in the heater (-)
ε_p	Volume fraction (-)
ρ_f	Fluid density ($kg \cdot m^{-3}$)
ρ_p	Particle density ($kg \cdot m^{-3}$)
σ_{0f}	Electrical conductivity of the fluid before the heating takes place ($S \cdot m^{-1}$)
σ_{0p}	Electrical conductivity of the particle before the heating takes place ($S \cdot m^{-1}$)
σ_{eff}	Effective electrical conductivity ($S \cdot m^{-1}$)
σ_{ref}	Electrical conductivity at a reference temperature ($S \cdot m^{-1}$)

σ_T	Electrical conductivity at temperature T ($S \cdot m^{-1}$)
σ_x	Electrical conductivity in the x-direction ($S \cdot m^{-1}$)
σ_y	Electrical conductivity in the y-direction ($S \cdot m^{-1}$)
σ	Electrical conductivity ($S \cdot m^{-1}$)
ALP	Alkaline phosphatase
β-GAL	β-galactosidase
HTST	High-temperature short-time
LOX	Lipoxigenase
MEF	Moderate electric field
POD	Peroxidase
OH	Ohmic heating
PEC	Pectinase
PPO	polyphenoloxidase

REFERENCES

1. AAP de Alwis, PJ Fryer. A finite-element analysis of heat generation and transfer during OH of food. *Chemical Engineering Science* 45(6):1547–1559, 1990.
2. BE Getchel. Electric pasteurization of milk. *Agriculture Engineering* 16(10):408–410, 1935.
3. BD Moses. Electric pasteurization of milk. *Agriculture Engineering* 19:525–526, 1938.
4. S Mizrahi, I Kopelman, J Perlaman. Blanching by electroconductive heating. *Journal of Food Technology* 10:281–288, 1975.
5. S Palaniappan, SK Sastry. Electrical conductivities of selected solid foods during OH. *Journal of Food Process Engineering* 14:221–236, 1991.
6. S Palaniappan, SK Sastry. Electrical conductivity of selected juices: Influences of temperature, solids content, applied voltage, and particle size. *Journal of Food Process Engineering* 14:247–260, 1991.
7. WC Wang, SK Sastry. Changes in electrical conductivity of selected vegetables during multiple thermal treatments. *Journal of Food Process Engineering* 20:499–516, 1997.
8. I Castro, JA Teixeira, AA Vicente. The influence of field strength, sugar and solid content on electrical conductivity of strawberry products. *Journal of Food Process Engineering* 26:17–29, 2003.
9. DL Parrott. Use of OH for aseptic processing of food particulates. *Food Technology* 45(12):68–72, 1992.
10. HJ Kim, YM Choi, TCS Yang, IA Taub, P Tempest, P Skudder, G Tucker, DL Parrott. Validation of OH for quality enhancement of food products. *Food Technology* 50:253–261, 1996.
11. PJ Skuder. OH in food processing. *Asian Food Journal* 4(4):10–11, 1989.
12. HY Cho, AE Yousef, SK Sastry. Growth kinetics of *Lactobacillus acidophilus* under ohmic heating. *Biotechnology and Bioengineering* 49:334–340, 1996.
13. WH Hayt. *Engineering Electromagnetics*, 4th edn. New York, NY: McGraw-Hill, 1981.
14. SK Sastry, S Palaniappan. Mathematical modelling and experimental studies on ohmic heating of liquid-particle mixtures in a static heater. *Journal of Food Engineering* 15:241–261, 1992.
15. SK Sastry. A model for heating of liquid-particle mixtures in a continuous flow ohmic heater. *Journal of Food Process Engineering* 15:263–278, 1992.
16. S Orangi, SK Sastry, Q Li. A numerical investigation of electroconductive heating in solid-liquid mixtures. *International Journal of Heat and Mass Transfer* 41:2211–2220, 1998.
17. IJ Kopelman. Transient heat transfer and thermal properties in food systems. PhD dissertation, Michigan State University, East Lansing, MI, 1966.
18. SK Sastry, Q Li. Modelling the ohmic heating of foods. *Food Technology* 50(5):246–249, 1996.
19. AAP de Alwis, K Halden, PJ Fryer. Shape and conductivity effects in the ohmic heating of foods. *Chemical Engineering Research and Design* 67:159–168, 1989.
20. PJ Fryer, Z Li. Electrical resistance heating of foods. *Journal of Food Science and Technology* 4:364–369, 1993.
21. Anon. Theory and application of electrolytic conductivity measurement. The Foxboro Company, Foxboro, MA, 1982 (unpublished).
22. SK Sastry, S Palaniappan. Ohmic heating of liquid-particle mixtures. *Food Technology* 45(12):64–67, 1992.
23. MR Zareifard, HS Ramaswamy, M Trigui, M Marcotte. Ohmic heating behaviour and electrical conductivity of two-phase food systems. *Innovative Food Science and Emerging Technologies* 4:45–55, 2003.

24. SK Sastry, JT Barach. Ohmic and inductive heating. *Journal of Food Science* 65:42–46, 2000.
25. I Castro, JA Teixeira, AA Vicente. The influence of food additives on the electrical conductivity of a strawberry pulp. Proceedings of the 32nd Annual Food Science and Technology Research Conference, University College Cork, Cork, Ireland, 2002.
26. S Palaniappan, SK Sastry. Modelling of electrical conductivity of liquid-particle mixtures. *Food and Bioproducts Processing, Part C: Transactions of the Institution of Chemical Engineers* 69:167–174, 1991.
27. WC Wang, SK Sastry. Salt diffusion into vegetable tissue as a pretreatment for ohmic heating: Electrical conductivity profiles and vacuum infusion studies. *Journal of Food Engineering* 20:299–309, 1993.
28. M Lima, SK Sastry. The effects of ohmic frequency on hot-air drying and juice yield. *Journal of Food Engineering* 41:115–119, 1999.
29. I Castro, JA Teixeira, S Salengke, SK Sastry, AA Vicente. Ohmic heating of strawberry products: Electrical conductivity measurements and degradation kinetics. *Innovative Food Science and Emerging Technologies* 5:27–36, 2004.
30. JW Larkin, SH Spinak. Safety considerations for ohmically heated, aseptically processed, multiphase low-acid food products. *Food Technology* 50(5):242–245, 1996.
31. WR Fu, CC Hsieh. Simulation and verification of two-dimensional ohmic heating in static system. *Journal of Food Science* 64:946–949, 1999.
32. SK Sastry, S Salengke. Ohmic heating of solid-liquid mixtures: A comparison of mathematical models under worst-case heating conditions. *Journal of Food Process Engineering* 29:441–458, 1998.
33. C Amatore, M Berthou, S Hebert. Fundamental principles of electrochemical ohmic heating of solutions. *Journal of Electroanalytical Chemistry* 457:191–203, 1998.
34. C Samaranayake, SK Sastry. Electrode and pH effects on electrochemical reactions during ohmic heating. *Journal of Electroanalytical Chemistry* 577(1):125–35, 2005.
35. M Lima, B Heskitt, SK Sastry. The effect of frequency and wave form on the electrical conductivity-temperature profiles of turnip tissue. *Journal of Food Process Engineering* 22:41–54, 1999.
36. S Kulshrestha, SK Sastry. Frequency and voltage effects on enhanced diffusion during moderate electric field (MEF) treatment. *Innovative Food Science & Emerging Technologies* 4(2):189–94, 2003.
37. MV Shynkaryk, J Taehyun, VB Alvarez, SK Sastry. Ohmic heating of peaches in the wide range of frequencies (50 Hz to 1 MHz). *Journal of Food Science* 75:E493–E500, 2010.
38. WG Khalaf, SK Sastry. Effect of fluid viscosity on the ohmic heating rate of solid-liquid mixtures. *Journal of Food Engineering* 27:145–158, 1996.
39. S Sarang, B Heskitt, P Tulsiyan, SK Sastry. Residence time distribution (RTD) of particulate foods in a continuous flow pilot-scale ohmic heater. *Journal of Food Science* 74:E322–E327, 2009.
40. AAP de Alwis, PJ Fryer. The use of network theory in the finite-element analysis of coupled thermo-electric fields. *Communications of Applied Numerical Methods* 6:61–66, 1990.
41. L Zhang, PJ Fryer. Models for the electrical heating of solid-liquid food mixtures. *Chemical Engineering Science* 48:633–642, 1993.
42. Y Benabderrahmane, JP Pain. Thermal behaviour of a solid/liquid mixture in an ohmic heating sterilizer—slip phase model. *Chemical Engineering Science* 55:1371–1384, 2000.
43. C Lareo, CA Branch, PJ Fryer. The flow behaviour of solid-liquid food mixtures. *Proceedings of Institute of Chemical Engineers Symposium: Food Process Engineering* 1994, pp. 203–210.
44. C Lareo. The vertical flow of solid-liquid food mixtures. PhD dissertation, Department of Chemical Engineering, University of Cambridge, Cambridge, U.K., 1995.
45. S Liu, JP Pain, J Procter, AAP de Alwis, PJ Fryer. An experimental study of particle flow velocities in solid-liquid food mixtures. *Chemical Engineering Communications* 124:97–114, 1993.
46. PG Fairhurst. Contribution to the study of the flow behaviour of large nearly neutrally buoyant spheres in non-Newtonian media: Application to HTST processing. PhD dissertation, Département de Génie Chimique, Université de Technologie de Compiègne, Compiègne, France, 1998.
47. PG Fairhurst, JP Pain. Passage time distributions for high solid fraction solid-liquid food mixtures in horizontal flow: Unimodal size particle distributions. *Journal of Food Engineering* 39:345–357, 1999.
48. AA Lacey. Thermal runaway in a non-local problem modelling ohmic heating. Part I: Model derivation and some special cases. *European Journal of Applied Mathematics* 6:127–144, 1995.
49. AA Lacey. Thermal runaway in a non-local problem modelling ohmic heating. Part II: General proof of blow-up and asymptotes of runaway. *European Journal of Applied Mathematics* 6:201–224, 1995.
50. AA Lacey, DE Tzanetis, PM Vlamos. Behaviour of a non-local reactive convection problem modelling ohmic heating of food. *The Quarterly Journal of Mechanics and Applied Mathematics* 52:623–644, 1999.

51. CP Please, DW Schwendeman, PS Hagan. Ohmic heating of foods during aseptic processing. *IMA Journal of Mathematics for Business and Industry* 5:283–301, 1994.
52. NI Kavallaris, DE Tzanetis. An ohmic heating non-local diffusion-convection problem for the Heaviside function. *ANZIAM Journal* 44(E):E114–E142, 2002.
53. HJ Tham, XD Chen, BR Young, GG Duffy. Ohmic heating of dairy fluids—Effects of local electric field on temperature distribution. *Asia-Pacific Journal of Chemical Engineering* 4:751–758, 2009.
54. J Yongsawatdigul, JW Park. Linear heating rate affects gelation of Alaska pollock and pacific whiting surimi. *Journal of Food Science* 61(1):149–153, 1996.
55. J Yongsawatdigul, JW Park, E Kolbe. Degradation kinetics of myosin heavy chain of pacific whiting surimi. *Journal of Food Science* 62(4):724–728, 1997.
56. HY Cho, AE Yousef, SK Sastry. Kinetics of inactivation of *Bacillus subtilis* by continuous or intermittent ohmic and conventional heating. *Biotechnology and Bioengineering* 62:368–372, 1999.
57. S Palaniappan, SK Sastry, ER Richter. Effects of electroconductive heat treatment and electrical pretreatment on thermal death kinetics of selected microorganisms. *Biotechnology and Bioengineering* 39:225–232, 1992.
58. RN Pereira, J Martins, C Mateus, JA Teixeira, AA Vicente. Death kinetics of in goat milk and in cloudberry jam treated by ohmic heating. *Chemical Papers* 61(2):121–126, 2007.
59. HX Sun, S Kawamura, JI Himoto, K Itoh, T Wada, T Kimura. Effects of ohmic heating on microbial counts and denaturation of proteins in milk. *Food Science and Technology Research* 14 (2):117–123, 2008.
60. LF Machado, RN Pereira, RC Martins, JA Teixeira, AA Vicente. Moderate electric fields can inactivate *Escherichia coli* at room temperature. *Journal of Food Engineering* 96(4):520–527, 2010.
61. LP Somogyi, HS Ramaswamy, YH Hui. Biology, principles and applications. In: LP Somogyi, HS Ramaswamy, YH Hui, eds. *Processing Fruits: Science and Technology, Vol. 1*. Lancaster, PA: Technomic Publishing Co., 1996.
62. A van Loey, B Verachtert, M Hendrickx. Effects of high electric field pulses on enzymes. *Trends in Food Science & Technology* 12:94–102, 2002.
63. GV Barbosa-Cánovas, UR Pothakamury, E Palou, BG Swanson. Biological effects and applications of pulsed electric fields for the preservation of foods. In: GV Barbosa-Cánovas, UR Pothakamury, E Palou, BG Swanson, eds. *Nonthermal Preservation of Foods*. New York, NY: Marcel Dekker, 1998.
64. AJ Castro, BG Swanson, GV Barbosa-Cánovas, R Meyer. Pulsed electric field modification of milk alkaline phosphatase activity. In: GV Barbosa-Cánovas, QH Zhang, eds. *Pulsed Electric Fields in Food Processing: Fundamental Aspects and Applications, Vol. 3*. Lancaster, PA: Technomic Publishing Co., 2001, pp. 65–83.
65. SY Ho, GS Mittal, JD Cross. Effects of high field electric pulses on the activity of selected enzymes. *Journal of Food Engineering* 31:69–84, 1997.
66. T Grahl, H Markl. Killing of microorganisms by pulsed electric fields. *Applied Microbiology and Biotechnology* 45:148–157, 1996.
67. L Ludikhuyze. High pressure technology in food processing and preservation: A kinetic case study on the combined effect of pressure and temperature on enzymes. PhD dissertation. Katholieke Universiteit Leuven, Leuven, Belgium, 1998.
68. S Denys, AM van Loey, ME Hendrickx. A modelling approach for evaluating process uniformity during batch high hydrostatic pressure processing: Combination of a numerical heat transfer model and enzyme inactivation kinetics. *Innovative Food Science and Emerging Technologies* 1:5–19, 2000.
69. A Jakób, J Bryjak, H Wójtowicz, V Illeová, J Annus, M Polakovic. Inactivation kinetics of food enzymes during ohmic heating. *Food Chemistry* 123:369–376, 2010.
70. RL Garotte, ER Silva, RA Bertone. Kinetic parameters for thermal inactivation of cut green beans lipoxygenase using unsteady-state methods. *Latin American Applied Research* 33:87–90, 2003.
71. L Kim, MS Hwang, ES Kim. Extraction, partial characterization, and inhibition patterns of polyphenol oxidase from burdock (*Arctium lappa*). In: CY Lee, JR Whitaker, eds. *Enzymatic Browning and It's Prevention*. Washington, DC: American Chemical Society Symposium Series 600, 1995, pp. 267–276.
72. I Castro, B Macedo, J Teixeira, AA Vicente. The effect of electric field on important food-processing enzymes: Comparison of inactivation kinetics under conventional and ohmic heating. *Journal of Food Science* 69(9):C696–C701, 2004.
73. F Icier, H Yildiz, T Baysal. Polyphenoloxidase deactivation kinetics during ohmic heating of grape juice. *Journal of Food Engineering* 85:410–417.
74. DS Robinson. Peroxidases and their significance in fruits and vegetables. In: PF Fox, ed. *Food Enzymology, Vol. 1*. Amsterdam, The Netherlands: Elsevier Applied Science, 1991, pp. 399–425.

75. S Tannenbaum. Ascorbic acid. In: O Fennema, ed. *Principles of Food Science. Part I. Food Chemistry*, 2nd edn. New York, NY: Marcel Dekker, 1976.
76. Tetra Pak. *The Orange Book*. Lund, Sweden: Tetra Pak Processing Systems AB, 1998.
77. DJ Trammell, DE Dalsis, RT Malone. Effect of oxygen on taste, ascorbic acid loss and browning for HTST-pasteurized, single strength orange juice. *Journal of Food Science* 61(3):477–489, 1986.
78. R Villota, JG Hawkes. Reaction kinetics in food systems. In: D Heldman, D Lund, eds. *Handbook of Engineering*. New York, NY: Marcel Dekker, 1992.
79. I Saguy, IJ Kopelman, S Mizrah. Simulation of ascorbic acid stability during heat processing and concentration of grapefruit juice. *Journal of Food Process Engineering* 2:213–225, 1978.
80. D Alvarado, NP Viteri. Efecto de la temperatura sobre la degradaci6n aerobica de vitamina C en jugos de frutas cítricas. *Archivo Latinoamericano de Nutricion* 39(4):601–612, 1989.
81. JR Johnson, RJ Braddock, CS Chen. Kinetics of ascorbic acid loss and nonenzymatic browning in orange juice serum: Experimental rate constants. *Journal of Food Science* 60(3):502–505, 1995.
82. M Lima. Ascorbic acid degradation kinetics and mass transfer effects in biological tissue during ohmic heating. PhD dissertation, Ohio State University, Columbus, OH, 1996.
83. I Nicorescu, C Loisel, C Vial, A Riaublanc, G Djelveh, G Cuvelier, J Legrand. Combined effect of dynamic heat treatment and ionic strength on denaturation and aggregation of whey proteins—Part I. *Food Research International* 41(7):707–713, 2008.
84. RN Pereira, BWS Souza, MA Cerqueira, JA Teixeira, AA Vicente. Effects of electric fields on protein unfolding and aggregation: Influence on edible films formation. *Biomacromolecules* 11(11):2912–2918, 2010.
85. PS Patrick, HE Swaisgood. Sulfhydryl and disulfide groups in skim milk as affected by direct ultra-high-temperature heating and subsequent storage. *Journal of Dairy Science* 59:594–600, 1976.
86. RLJ Lyster, TC Wyeth, AG Perkin, H Burton. Comparison of milks processed by the direct and indirect methods of ultra-high-temperature sterilization: V. Denaturation of the whey proteins. *Journal of Dairy Research* 38:403–408, 1971.
87. B Bansal, XD Chen, SXQ Lin. Skim milk fouling during ohmic heating. In: H Müller-Steinhagen, R Malayeri, AP Watkinson, eds. *6th International Conference on Heat Exchanger Fouling and Cleaning—Challenges and Opportunities*; Engineering Conferences International Symposium Series; Kloster Irsee: Ostallgäu, Germany, 2005, Vol. RP2, pp. 133–140.
88. G Piette, C Brodeur. Ohmic cooking for meat products: The heat is on! *Le Monde Alimentaire* 5(6):22–24, 2001.
89. WC Wang, JI Chen, HH Hua. Study of liquid-contact thawing by ohmic heating. *Proceedings of IFT Annual Meeting and Food Expo*, Anaheim, California, 2002.
90. S Sarang, SK Sastry, J Gaines, TCS Yang, P Dunne. Product formulation for ohmic heating: Blanching as a pretreatment method to improve uniformity in heating of solid–liquid food mixtures. *Journal of Food Science* 72:E227–E234, 2007.
91. M Zell, JG Lyng, DA Cronin, DJ Morgan. Ohmic heating of meats: Electrical conductivities of whole meats and processed meat ingredients. *Meat Science* 83:563–570, 2009.
92. C Hoogzand, JJ Doesburg. Effect of blanching on texture and pectin of canned cauliflower. *Food Technology* 15:160–163, 1961.
93. LG Bartolome, JH Hoff. Firming of potatoes: Biochemical effects of pre-heating. *Journal of Agriculture and Food Chemistry* 20:266–270, 1972.
94. SC Eliot, A Goullieux, JP Pain. Processing of cauliflower by ohmic heating: Influence of precooking on firmness. *Journal of the Science of Food and Agriculture* 79:1406–1412, 1999.
95. SC Eliot, A Goullieux, JP Pain. Combined effects of blanching pretreatments and ohmic heating on the texture of potato cubes. *Science des Aliments* 19:111–117, 1999.
96. SC Eliot, A Goullieux. Application of the firming effect of low-temperature long-time precooking to ohmic heating of potatoes. *Science des Aliments* 20:265–280, 2000.
97. NI Lebovka, I Praporscic, E Vorobiev. Combined treatment of apples by pulsed electrical fields and by heating at moderate temperature. *Journal of Food Engineering* 65:211–217, 2004.
98. I Praporscic, NI Lebovka, S Ghnimi, E Vorobiev. Ohmically heated, enhanced expression of juice from soft vegetable tissues. *Biosystems Engineering* 93:199–204, 2006.
99. K Halden, AAP de Alwis, PJ Fryer. Changes in electrical conductivity of foods during ohmic heating. *International Journal of Food Science and Technology* 25:9–25, 1990.
100. M Lima, BF Heskitt, SK Sastry. Diffusion of beet dye during electrical and conventional heating at steady-state temperature. *Journal of Food Process Engineering* 24(5):331–340, 2001.

101. WC Wang, SK Sastry. Study of electro-thermal effects on juice yield from cellular tissues. Paper presented at the 1998 IFT Annual Meeting, June 20–24, Atlanta, GA.
102. LA Fedorenchenko, IS Guly, IG Bazhal, LD Bobrovnik, NS Karpovich, YA Zabrodskaya, TA Martynenko. The effects of a low-voltage electric field on the electrical resistivity of beet tissue during extraction process. *Sakharnaeiia Promyshlennost* 2:23–24, 1983.
103. H Allali, L Marchal, E Vorobiev. Blanching of strawberries by ohmic heating: Effects on the kinetics of mass transfer during osmotic dehydration. *Food Bioprocess Technology* 3:406–414, 2010.
104. SK Sastry, I Sensoy. Extraction using moderate electric fields. *Journal of Food Science* 69:FEP7–FEP13, 2004.
105. SA Kulshrestha, SK Sastry. Low-frequency dielectric changes in cellular food material from ohmic heating: Effect of end point temperature. *Innovative Food Science and Emerging Technologies* 7:257–262, 2006.
106. J Yongsawatdigul, JW Park, E Kolbe. Electric conductivity of Pacific whiting surimi paste during ohmic heating. *Journal of Food Science* 60(5):922–935, 1995.
107. RN Pereira, RC Martins, AA Vicente. Goat milk free fatty acid characterization during conventional and ohmic heating pasteurization. *Journal of Dairy Science* 91(8):2925–2937, 2008.
108. K Allen, V Eidman, J Kinsey. An economic-engineering study of ohmic food processing. *Food Technology* 50:269–273, 1996.

19 Radio Frequency Dielectric Heating

Yanyun Zhao and Qingyue Ling

CONTENTS

19.1 INTRODUCTION

Radio frequency (RF) dielectric heating is a heating technology that allows for rapid, uniform heating throughout a medium. This technology generates heat energy within the product and throughout its mass simultaneously due to the frictional interactions of polar dielectric molecules rotating in response to an externally applied AC electric field [1,2]. *RF dielectric heating* offers several advantages over conventional heating methods in food application, including saving energy by increasing heat efficiency; achieving rapid and even heating; reducing checking, the uneven stresses in the product as a result of evening the product moisture profile; avoiding pollution as there are no by-products of combustion; increasing production without an increase in overall plant length; saving floor space as efficient heat transfer results in faster product transfer and reduced oven length; and automatically compensating for variations in product moisture. In addition, this technology can be easily adapted during implementation to be compatible with automated production batch and/or continuous flow processing [1–4].

RF heating of foods dates back to the 1940s. The early efforts attempted to apply RF energy to cook various processed meat products, to heat bread, and to dehydrate and blanch vegetables [1,5,6]. However, the work resulted in very few commercial installations, primarily because of the high overall operating costs associated with using RF energy in the 1940s. By the 1960s, studies on the application of RF energy in foods focused mainly on thawing frozen products, which resulted in several commercial production lines [7–9]. The next generation of commercial applications of RF energy in the food industry was postbake drying of cookies and snack foods, starting in the late 1980s [3,4,10,11]. The use of RF heating resulted in reduced energy consumption and improved product quality, both considered great advantages over the traditional baking oven. By the 1990s, due to increased food safety requirements, great attention has been given to the use of RF energy for pasteurization and sterilization of foodstuffs for providing high-quality and safety food products. However, commercial RF dielectric heating systems for the purpose of food pasteurization or sterilization are not known to be in use in the United States due to reasons such as the inability to ensure sterilization of the entire package, lack of suitable packaging materials, and unfavorable economics when compared with conventional thermal processing technologies. By the 2000s, due to the availability of more powerful commercial computer simulation tools or software of electromagnetic field and thermal transfer, more research work has been done to understand effectiveness of RF dielectric heating of various foodstuffs in terms of heating uniformity, shape and size, and RF heating system design. At the same period, RF heating of raw food items has been studied for the purpose of insect control or disinfestation to improve quality of foods. In addition, more research on pasteurization and sterilization of various foods has been conducted, demonstrating more promising applications of RF dielectric heating technologies in the near future.

This chapter will consider the major technology aspects of RF dielectric heating, discuss the potential use of *mathematical modeling and computer simulation* for the design of the RF heating system, *RF applications* in food *pasteurization and sterilization*, and *disinfestation*, and provide insights on the major technical challenges in *RF pasteurization and sterilization* of foods, especially ready-to-eat, packaged items.

19.2 TECHNOLOGY OF RF DIELECTRIC HEATING

RF heating is a subset of several electromagnetic-based methods of heating materials. Table 19.1 lists the most common electromagnetic-based heating methods, the critical parameters that most affect heating rate, and the common *frequency range* of applications.

The technology of RF heating involves applying a high-voltage AC signal to a set of parallel electrodes set up as a capacitor (Figure 19.1). The medium to be heated is sandwiched between the electrodes, and an AC displacement current flows through the medium. As a result, polar molecules in the medium align and rotate in opposite fashion to match the applied AC electric field. Heating occurs as polar molecules interact with neighboring molecules, resulting in lattice and frictional losses as they rotate. The higher the frequency of the alternating field, the greater the energy imparted to the medium, until the frequency is so high that rotating molecules can no longer keep up with the external field due to lattice limitations. The frequency at which lattice limitations occur is called the "*Debye resonance*." It is the frequency at which maximum energy can be imparted to a medium for a given electric field strength. Other higher frequency limitations include the reduction of field penetration into a medium as frequency is increased in the presence of high electrical conductivity and/or high dielectric loss components. A typical existing design for a dielectric heating system is shown in Figure 19.2.

RF dielectric heating differs from higher frequency electromagnetic radiative dielectric heating (e.g., microwave ovens). With RF heating, the wavelength of the chosen frequency is large compared to the dimensions of the sample being heated. With electromagnetic radiative heating, the wavelength is comparable, or even smaller, than the dimensions of the sample being heated. An

TABLE 19.1
Electromagnetic-Based Heating Methods

Heating Method	Key Parameters	Frequency Range	Common Frequencies
Ohmic (lower f) (electric field)	σ, E	<1 MHz	50 Hz, 60 Hz
Capacitive dielectric (medium f) (electric field)	f, ε'', E	100 kHz < f < 100 MHz or Wavelength >> sample size	10 MHz, 27 MHz, 39 MHz, and others
Radiative dielectric (higher f) (electric field)	f, ε'', E	100 MHz < f < 100 GHz or Wavelength ≤ sample size	915 MHz, 2.45 GHz, 5.8 GHz, 24.124 GHz, and others
Inductive/ohmic combination (lower f) (magnetic field)	Hysteresis losses f (B–H curves, f) and eddy current losses $f(H, \sigma)$	50 Hz < f < 1 MHz	50 Hz, 60 Hz, 1–50 kHz, 450 kHz, and others
Inductive (medium f) (magnetic field)	f, μ'', H	1 MHz < f < 100 MHz or Wavelength >> sample size	50 MHz and others
Radiative magnetic (higher f) (magnetic field)	f, μ'', H	100 MHz < f < 100 GHz or Wavelength ≤ sample size	915 MHz, 2.45 GHz, 5.8 GHz, 24.124 GHz, and others

Source: From Ishii, T.K., *Handbook of Microwave Technology: Vol. 2. Applications*, Academic Press, San Diego, CA, 1995.

Note: σ = electrical conductivity (S/m)
ε'' = electric permittivity (F/m)
μ'' = magnetic permeability (H/m)
f = electric field frequency (Hz)
E = RMS electric field intensity (V/m)
H = RMS magnetic field intensity (A/m)
B = magnetic flux density (W/m^2)

example of RF dielectric heating would be two large parallel electrodes placed on opposite sides of a sample with an AC displacement current flowing through it. An example of electromagnetic radiative heating would be a metal chamber with resonant electromagnetic standing wave modes, as in a microwave oven. Thus, RF dielectric heating method offers the advantages of providing more uniform heating over the sample geometry due to deeper wave penetration into the sample, as well as simple uniform field patterns (as opposed to the complex nonuniform standing wave patterns in a microwave oven). RF dielectric heating operates at frequencies low enough to use standard power grid tubes, which are both lower cost (for a given power level) and allow for generally much higher power generation levels than microwave tubes.

RF heating also differs from lower frequency ohmic heating. RF heating depends on dielectric losses; ohmic heating relies on direct ohmic conduction losses in a medium and requires the electrodes to contact the medium directly (i.e., the current cannot penetrate a plastic film or air gap). RF dielectric heating methods also offer advantages over low-frequency ohmic heating, including the ability to heat a medium that is enclosed inside an insulating plastic package or surrounded by an air or deionized water barrier (i.e., the electrodes do not have to contact the medium directly). The performance of RF dielectric heating is therefore also less dependent on the product making a smooth contact to the electrodes. RF dielectric heating methods are not dependent on the presence of DC electrical conductivity and can heat insulators as long as they display polar dielectric losses. Examples of electromagnetic heating (microwave ovens), ohmic heating, and RF dielectric heating systems are shown in Figure 19.3.

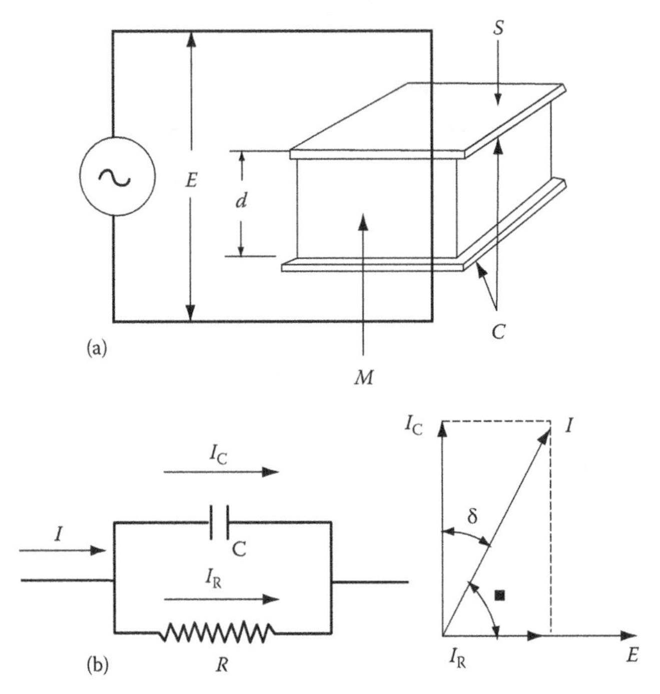

FIGURE 19.1 Diagram of RF dielectric heating (C=capacitor, d=distance between two electrodes, E=electric supply, M=magnetic field, and R=resistor). (a) Dielectric heating diagram and (b) dielectric heating: equivalent circuit diagram. (From Zhao, Y. et al., *J. Food Process. Eng.*, 23, 25, 2000.)

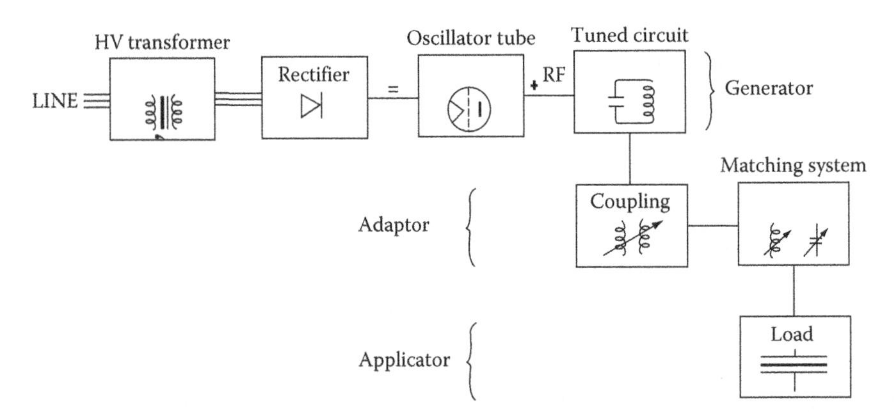

FIGURE 19.2 Block diagram of a typical capacitive (RF) dielectric heating system. (From Zhao, Y. et al., *J. Food Process. Eng.*, 23, 25, 2000.)

19.3 DIELECTRIC PROPERTIES OF FOOD MATERIALS IN RADIO FREQUENCY RANGE

Knowledge of *dielectric properties* of food materials in RF range is essential in guiding further development, improvement, and scale-up of RF dielectric heating in the food industry. The two dielectric properties of greatest interest are the *energy storage* and *loss terms* of the *electric permittivity* (ε' and ε'', respectively). Permittivity describes dielectric properties that influence reflection of electromagnetic waves at interfaces and the attenuation of the wave energy within material. The complex relative permittivity ε of a material can be expressed in the following complex form:

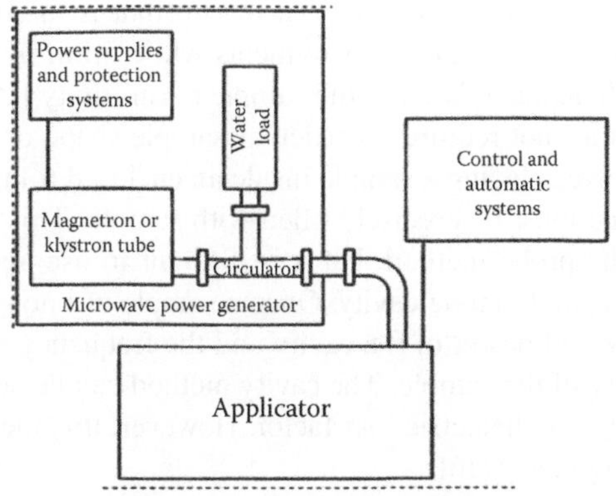

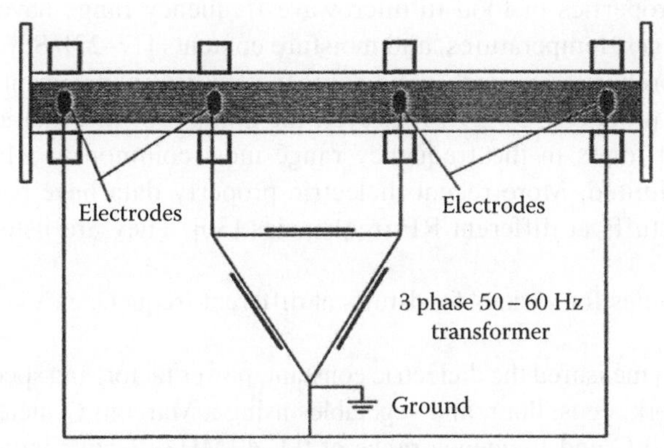

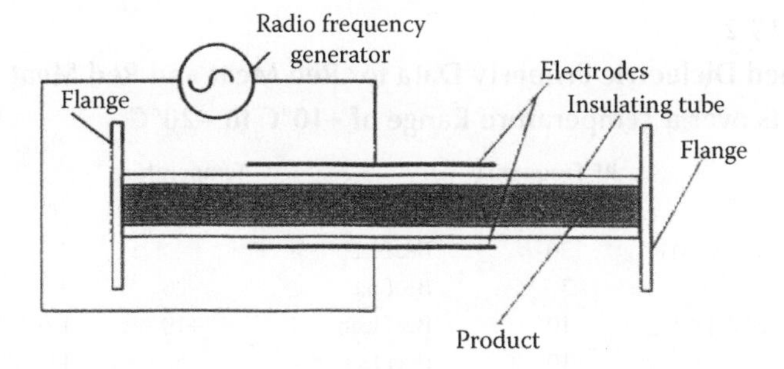

FIGURE 19.3 RF dielectric heating as compared with microwave and ohmic heating. (From Zhao, Y. et al., *J. Food Process. Eng.*, 23, 25, 2000.)

$$\varepsilon = \varepsilon' - j\varepsilon'' \tag{19.1}$$

The real part ε' is referred to as the dielectric constant and represents stored energy when the material is exposed to an electric field, while the dielectric loss factor ε'', which is the imaginary part, influences energy absorption and attenuation, and $j = \sqrt{-1}$.

The three most popular methods for *measuring dielectric properties* of foods and commodities are open-ended coaxial probe, transmission line, and resonant cavity method [14]. The probe method is based on a coaxial line ending abruptly at the tip that is in contact with the material being tested. This method offers broadband measurements while minimizing sample disturbance. The measured reflection coefficient is related to the sample permittivity [15]. The probe method is the easiest to use because it does not require a particular sample shape or special containers. The transmission line method involves placing a sample inside an enclosed transmission line. The cross section of the transmission line must be precisely filled with sample. This method is usually more accurate and sensitive than the probe method, but it is difficult to use and time consuming. The resonant cavity method uses a single-mode cavity. Once a sample of known geometry is placed in the cavity, the changes in reflected power of the cavity and the frequency of resonance are used to compute the dielectric property of the sample. The cavity method can be accurate and is especially suited for samples with a very low dielectric loss factor. However, this method provides dielectric properties at only one fixed frequency [16].

Frequency, temperature, salt content, moisture content, and the state of moisture (frozen, free, or bound) are the major factors that influence dielectric properties of food materials. Many studies on dielectric properties of food in microwave frequency range have been reported for different frequency ranges, temperatures, and moisture contents [17–23]. Several comprehensive reviews on dielectric properties provide good sources of useful experimental data from many foods and agricultural products [23–29]. However, the amount of information available on the dielectric properties of foods in the frequency range most common to RF dielectric heating (1–200 MHz) is very limited. More recent dielectric property data have been published since 2000 for various foodstuffs at different RF frequencies [30]. They are listed in Table 19.2 for further reference.

Some of reported studies for various foodstuffs at different frequencies and temperature ranges are summarized later.

Ede and Haddow [34] measured the dielectric constant, power factor, and specific conductivity of foods including beef, pork, eggs, flour, and vegetables using a Marconi Q-meter over the temperature range of –39°C–100°C and frequency range of 0.1–40 MHz. Results indicated that for foods that are poor conductors, the electrical conductivity varied very little with frequency. For foods

TABLE 19.2
Published Dielectric Property Data for Red Meat and Red Meat Products over a Temperature Range of +10°C to –20°C

Source	RF Frequency (MHz)	Products	Temperature (°C)	ε'	ε''
Ede and Haddow [31]	20	Beef fat	0	7	3.5
	3	Beef fat	–20	3	0.21
Bengtsson et al. [32]	10	Beef lean	+10	101	808
	10	Beef lean	–5	18	25.2
	10	Beef lean	–10	13	6.5
	35	Beef lean	+10	79	237
	35	Beef lean	–5	13	7.8
	35	Beef lean	–10	9	3.6
Tran and Stuchly [17]	100	Beef	+1.2	63.6	115.3
Zhang et al. [33]	27.12	Pork luncheon	+5	46.6	526.7
Farag et al. [24]	27.12	Beef lean	–1	68.7	241.9
		50:50 Mixture	–1	37.3	92.4
		Beef fat	*–1*	*18.3*	*26.1*

Source: Modified from Nelson, S.O., *Trans. ASAE,* 16(2), 384, 1973.

that are poor dielectrics, the variation in electrical conductivity with frequency was high. They also found that for most foods, the electrical conductivity decreased with reduced temperature and showed a sharp drop at the freezing point.

Bengtsson et al. [32] measured the dielectric constant and loss tangent for lean beef and codfish over the temperature range of −25°C–10°C and frequency range of 10–200 MHz using a Boonton RX-meter. They found that the values of dielectric constant and loss tangent decreased with increased frequency, and also showed a sharp increase at defrosting temperatures (~1°C–2°C). Variation in dielectric properties caused by variability in raw material and by frozen storage was relatively small for meat and codfish. Dielectric properties were quite similar for lean meat, lean fish, and herring, but of much lower value for fats.

Kent and Jason [33] elucidated the nature of the mechanisms that influenced the dielectric properties of frozen fish muscle (cod and haddock) and measured the dielectric properties over the frequency range of 0.1 Hz–10 GHz. The results revealed that the time dependence of the dielectric properties were attributable to changes in the distribution of water between the solid and liquid states.

Tran and Stuchly [17] measured dielectric properties of raw beef (meat and liver), chicken, and salmon at three temperatures between 1.2°C and 64°C in the frequency range of 100–2500 MHz using an open-ended coaxial line sensor and an automatic network analyzer. This method has proved to be very suitable for measuring high water content substances.

Kent [28] published the dielectric properties of cooked and raw fish including codfish, herring, haddock, and sprat at selected frequencies from 10 Hz to 10 GHz and selected temperatures from −78.5°C to 90°C.

Cole et al. [35] used time domain spectroscopy (TDS) to study dielectric properties of water/oil and oil/water microemulsions from 10 MHz to 3 GHz. Merabet and Bose [36] used time domain reflectometry (TDR) to measure the dielectric constant and loss factor of water in the radio and microwave frequencies (10 MHz–8 GHz). Puranik et al. [37] used TDR to determine the dielectric properties of a honey–water mixture in the frequency range of 10 MHz–10 GHz at 25°C. Looan and Smulders [38] also used TDR to measure the dielectric permittivity of foodstuffs (including solutions of mono-, di-, and polysaccharides) in the frequency range of 20 kHz–5 GHz, considering the influences of species, composition, and temperature.

Ikediala et al. [39] studied dielectric properties of four apple cultivars between 30 and 3000 MHz at 5°C–55°C, using the open-ended coaxial-line probe technique. Dielectric constant of apples decreased with frequency and decreased slightly with increasing temperature. The dielectric loss factor increased linearly with temperature in the RF range but was nearly constant at the microwave frequencies. Minimum dielectric loss factor of apples was observed at about 915 MHz. Dielectric constant and loss factor were not influenced by cultivar, pulp section, or degree of ripeness of apples.

Wang et al. [40] determined the dielectric properties of a whey protein gel, a liquid whey protein mixture, and a macaroni and cheese product and their constituents by using a custom-built temperature-controlled test cell and an Agilent 4291B impedance analyzer over a temperature range from 20°C to 121.1°C, at frequencies of 27, 40, 915, and 1800 MHz. As temperature increased, dielectric constants of whey protein products increased at 27 and 40 MHz, but decreased at 915 and 1800 MHz. Dielectric loss factors of whey protein products increased sharply with increasing temperatures at 27 and 40 MHz, but increased mildly at 915 and 1800 MHz. Similar results were observed with macaroni and cheese. The penetration depths of electromagnetic energy at 27 and 40 MHz were about four times as great as those at the microwave frequencies 915 and 1800 MHz in all tested samples.

Wang et al. [41] also measured dielectric properties of six fruit and nut commodities between 1 and 1800 MHz using an open-ended coaxial-line probe technique and at temperatures between 20°C and 60°C. The dielectric loss factor of fresh fruits decreased with increasing frequency at constant temperatures. The loss factor of fresh fruits increased almost linearly with increasing temperature at 27 MHz. Both dielectric constant and loss factor of nuts were very low compared to those of fresh fruits. The temperature effect on dielectric properties of nuts was not significant at 27 MHz.

Farag et al. [24] evaluated the dielectric properties of three different beef meat blends (lean, fat, and 50:50 mixture) over a range of temperature from −18°C to +10°C at frequency of 27.12 MHz. Overall, while dielectric constant (ε′) and dielectric loss factor (ε″) values at 27.12 MHz increased significantly as from −18°C to +10°C for lean beef and 50:50 mixture, the dielectric values for fat did not increase significantly over the temperature range. The most significant increase in dielectric properties was found from −5°C to −1°C or thawing temperature range. For instance, lean beef ε′ changed from 36.0 to 68.8 and ε″ values from 81.4 to 241.9, respectively, as temperature changed from −5°C to −1°C.

Sacilik et al. [42] evaluated the dielectric properties of flaxseeds at different moisture content (6%–22%) and bulk densities (587–723 kg m^{-3}) over a frequency range of 50 kHz–10 MHz using a parallel-plate capacitor sample holder method. The study showed that the relative permittivity ε′ and loss factor ε″ were greatly influenced by the moisture content and bulk density of flaxseeds as well as operating frequency. They increased as moisture content and bulk density increased at a given frequency. The moisture content, however, was the most significant factor affecting the dielectric property. Based on the measured data, the researchers developed empirical models to predict the relative permittivity and loss factor with moisture content, bulk density, and frequency.

Luigni et al. [43] measured dielectric constant (ε′) and loss factor (ε″) of fresh hen eggs in the frequency range of 20–1800 MHz using an open-ended coaxial probe after the eggs were stored at room temperature for 1, 2, 4, 8, and 15 days. The dielectric properties of hen eggs were measured on albumen and yolk separately. The results indicated that both dielectric constant and loss factor of albumen were decreased as frequency increased from 20 to 1800 MHz although more significant decrease for loss factor. For yolk, similar relationship existed between the dielectric properties and frequency although both the dielectric constant and loss factor are consistently smaller than that of albumen. Finally, the researchers found that the dielectric constant of albumen at 20 MHz was significantly influenced by storage time. The dielectric constant of yolk increased continuously with storage time and the loss factor of yolk was also characterized by a general and significant increase in the storage period of 15 days.

19.4 MATHEMATICAL MODELING AND COMPUTER SIMULATION OF RF DIELECTRIC HEATING OF FOODSTUFFS

Due to the complexity of the RF system where the heating pattern depends on a large number of factors, simulation-based design can save significant time and resources in developing microbiologically safe processes. Such simulation-based design can drastically reduce the number of experiments needed to predict the location of cold points and the time–temperature history at these locations for the actual food and equipment combinations. State-of-the-art commercial software simulating the electromagnetic and heat transfer properties has been used for microwave food process design. Such software can provide a comprehensive insight into the heating process by showing interior power absorption in a 3D object. Simulation-based design can allow the process and equipment designers to judiciously choose proper combinations of food and process parameters in an efficient manner, reducing some of the time and expenses in prototype building.

19.4.1 Mathematical Modeling

Based on mass and heat transfer in processed foods, mathematical electromagnetic models for dielectrics can be adapted to model the composite food, packaging, and water barrier-layered arrangement. These models can be combined with a 2D or 3D electric field finite-difference method (*FDM*) or *finite element method* (*FEM*) model program described by Ishii [12] and Roussy and Pearce [44] to account for material heterogeneity, field penetration variations, and temperature distribution in whole materials. In addition, the time domain in the modeling needs to be considered. Two currently employed techniques of time domain approach to RF heating circuit modeling are the transmission-line matrix (TLM) method developed by Johns [45] and the finite-different time-domain (FDTD) method formulated by Yee [46]. Both techniques are extensively used to analyze electromagnetic structures of arbitrary geometry. FDTD is heavily favored by the scattering problem and radiation, while TLM is used mostly by researchers interested in guided propagation problems. Although they have been developed independently, both methods have been extensively applied to solve similar electromagnetic field diffusion and network problems in the time domain.

19.4.1.1 Governing Equations

The heat transfer is considered due to conduction within the food material, convection at the material surface and heat generation due to RF heating. The guiding heat transfer equation in an electromagnetic field is

$$\rho C_p \frac{\partial T}{\partial t} = \nabla(k\nabla T) + Q \tag{19.2}$$

where the RF power absorption density "Q" delivered to the food sample for a given electric field intensity "E" can be described as [44,47]

$$Q = 2\pi f \varepsilon_0 \varepsilon_r'' E^2 \tag{19.3}$$

where
$\varepsilon_0 \varepsilon_r'' = \varepsilon''$
ε_0 is the permittivity of free space, 8.854×10^{-12} F/m
The subscript r designates the relative energy loss term of permittivity

The electric field intensity, E, is governed by the electromagnetic field and affected by the dielectric properties of samples
Zhao et al. [13] described the modeling of electromagnetic field, and their determination is discussed in the modeling approach of this chapter. Equation 19.2 can be written as Equation 19.4 in 3D rectangular coordinate:

$$\rho C_p \frac{\partial T}{\partial t} = \frac{\partial}{\partial x}\left(k\frac{\partial T}{\partial x}\right) + \frac{\partial}{\partial y}\left(k\frac{\partial T}{\partial y}\right) + \frac{\partial}{\partial z}\left(k\frac{\partial T}{\partial z}\right) + Q \tag{19.4}$$

Heat generation is a function of temperature and moisture at a particular location (x,y,z). Time–temperature profiles of the food materials in an RF heating system can be predicted using the aforementioned models with known thermal and dielectric properties of food samples as described in Equation 19.2.

In general, mathematical modeling can play a crucial role in the design and optimization of the product and processing parameters during RF heating. The model has to take into account conductive heat transfer, internal heat generation, and convective and evaporative heat losses at the boundaries. The composite model can be developed by combining heat transfer models with dielectric property models in 2D or 3D node. If the material is heterogeneous, assumption about that heterogeneity has to be made to be able to create 2D or 3D finite element model, based on known mesh or grid of electromagnetic field values. In addition, a food sample is usually packaged in plastic or other materials, as well as surrounded by an air layer. Those factors add the complexity of model, which becomes a composite model of multiple dielectrics.

The heating time t_h for a given temperature rise (ΔT) is then given by Equation 19.5 [48]:

$$t_h = \frac{C_P \rho \Delta T}{E^2 \omega \varepsilon''} = \frac{C_P \rho \Delta T}{P_V} \quad (19.5)$$

where C_p = specific heat of the medium (J/kg°C); P_v = maximum power per unit volume (W/m^3); ω = angular frequency = $2\pi f$ (rad/s); ρ = density of medium (kg/m^3).

The penetration capacity or depth of an electromagnetic energy field to a dielectric material is determined by its dielectric properties and wavelength (λ_0) in free space. The penetration depth (d_p) can be estimated according to Hippel [49]:

$$d_p = C/(2\pi f \sqrt{}(2\varepsilon'(\sqrt{}(1 + (\varepsilon''/\varepsilon')^2) - 1))) \quad (19.6)$$

where

- c is the speed of light in free space (3×10^8 m/s)
- ε' is referred to as the dielectric constant and represents stored energy when the material is exposed to an electric field
- ε'' is the dielectric loss factor, which influences energy absorption and attenuation
- f is the frequency in Hz

When loss tangent $\tan \delta = \varepsilon''/\varepsilon'$ is low (i.e., far less than 1), penetration depth can be simplified by the following equation [50]:

$$d_p = \frac{c}{2\pi f \sqrt{\varepsilon'} \tan \delta} \quad \text{or} \quad \frac{c\sqrt{\varepsilon'}}{2\pi f \varepsilon''} \quad (19.7)$$

19.4.1.2 Modeling Approach

Mathematical electromagnetic models for layered composite dielectrics can be adapted to model the composite food, packaging, and air/water barrier-layered arrangement. These models can then be combined with a 2D or 3D electric field FEM model or FDM model program such as described by Roussy and Pearce [44] and Ishii [12] to account for any material heterogeneities and field penetration variations. Examples of electromagnetic field FEM programs for complex composite geometrical structures include the High Frequency Structure Simulator (HFSS) and the Maxwell Extractor electromagnetic field solver programs developed by Ansoft, Inc., Pittsburgh, PA; FEMLAB by Comsol, Los Angeles, CA; ANSYS from Ansys Corp., Canonsburg, PA; and CST from Microwave Studio, Darmstadt, Germany. These programs can solve for electromagnetic field quantities in a FEM mesh or grid (2D), or in tetrahedral or other shaped finite element blocks (3D) based on imported arrays of electrical permittivity (both storage and loss terms), electrical conductivity, and magnetic permeability tensors for all the constituent components. They can also be based on mechanical drawings (e.g., AutoCAD.DXF or SolidWorks.SLDDRW format) of the complex geometrical structures such as packaged foods.

A separate simulation run will need to be conducted for every value of frequency as well as for every value of temperature since the dielectric terms are functions of both. The electric field solutions are exported to a program that calculates the power delivered and heat generated for a given time increment in each finite element block. These arrays of heat generation values can then be used as heat source terms in a FEM or FDM heat transfer conservation of energy program. The energy program also accounts for thermal fluxes between adjacent finite element blocks based on modeled values for thermal conductivity between blocks, for the thermal heat capacity of the blocks, and for various boundary conditions between the blocks. Then the true temperature rise vs. time can be solved for each FEM block with a 2D model.

The overall *computer simulation* will iterate cyclically back and forth between the electromagnetic field program and the heat transfer program. Each cycle will correspond to some time increment, and the solution for temperature rise in each FEM block will be used to influence the imported dielectric and conductivity properties for each FEM block in the next run of the electromagnetic field. This overall simulation can then be used to predict the temperature rise versus time of all the various constituents of the composite-packaged samples. By containing the temperature dependence of the dielectric properties of the food and packaging constituents, this model will aid in predicting thermal runaway or hot spots when different heterogeneities are assumed and small thermal mass pockets of lossy dielectric materials exist (e.g., water or food trapped in package seams). The level of complexity of the model will depend on assumptions relating to material homogeneity, composite constituent geometry, and field penetration depth nonuniformity.

19.4.2 Computer Simulation of RF Heating of Foodstuffs

Computer simulation of RF dielectric heating of foodstuffs has been studied since the 2000s. The type of food stuffs has been simulated varied from vegetable seeds, fruits, and meats to study the effects of dielectric properties and electrodes shapes and sizes on heating uniformity and power distribution that affect quality of foodstuffs.

In 2003, Yang et al. [50] investigated the application of a 3D finite-element computer program package, TLM-FOOD-HEATING (FAUSTUS Scientific Corp., Victoria, Canada), on the simulation of RF dielectric heating of radish and alfalfa seeds. The flow diagram of the TLM-Food-Heating program is illustrated in Figure 19.4. Seeds placed inside polystyrene boxes were described as a rectangle, where nonuniform boundary conditions presented complications in the simulation. Temperature- and moisture-dependent thermal properties and temperature- and frequency-dependent dielectric properties of seeds were considered in the modeling. The time–temperature profiles of seeds at different locations within the rectangle seeds boxes were simulated. The model was then validated using experimental data. Simulated temperature distribution in seeds was consistent with the measured results. The absolute temperature differences between simulated and measured values in radish seeds packed in a 100×25×50mm rectangle seeds box were 1.8°C, 1.1°C, 8.9°C, and 13.6°C in the center, top, edge, and bottom locations, respectively, when seeds reached about 80°C and were 0.9°C, 2.4°C, 7.75°C, and 14.3°C, respectively, in alfalfa seeds box in the same conditions.

In 2005, Marra et al. [51] conducted computer simulation of temperature distribution inside cylindrical meat batters at different levels of RF output powers of 100, 200, 300, and 400W. RF heating of meat batters was modeled as a multiphysics problem and was built based on a quasi-static electromagnetic equation coupled with heat transfer by conduction, plus a generation term that is function of local E-field. Electromagnetic and heat equations were simultaneously solved using FEMLAB 3.1 software from Comsol. The goodness of fit of the model was evaluated and validated by comparing numerical results with measured temperature profiles for the meat batters that were heated using a custom built 50Ω, 600W RF Oven. The results showed good agreement between simulation results and experiment data. Due to different heating rates occurred within the meat samples, uneven temperature distribution in samples was observed. The higher heating rate occurred at the bottom of the samples. Nonuniform temperature distribution was worsened as RF output power increased from 100 to 400W. The study indicated the

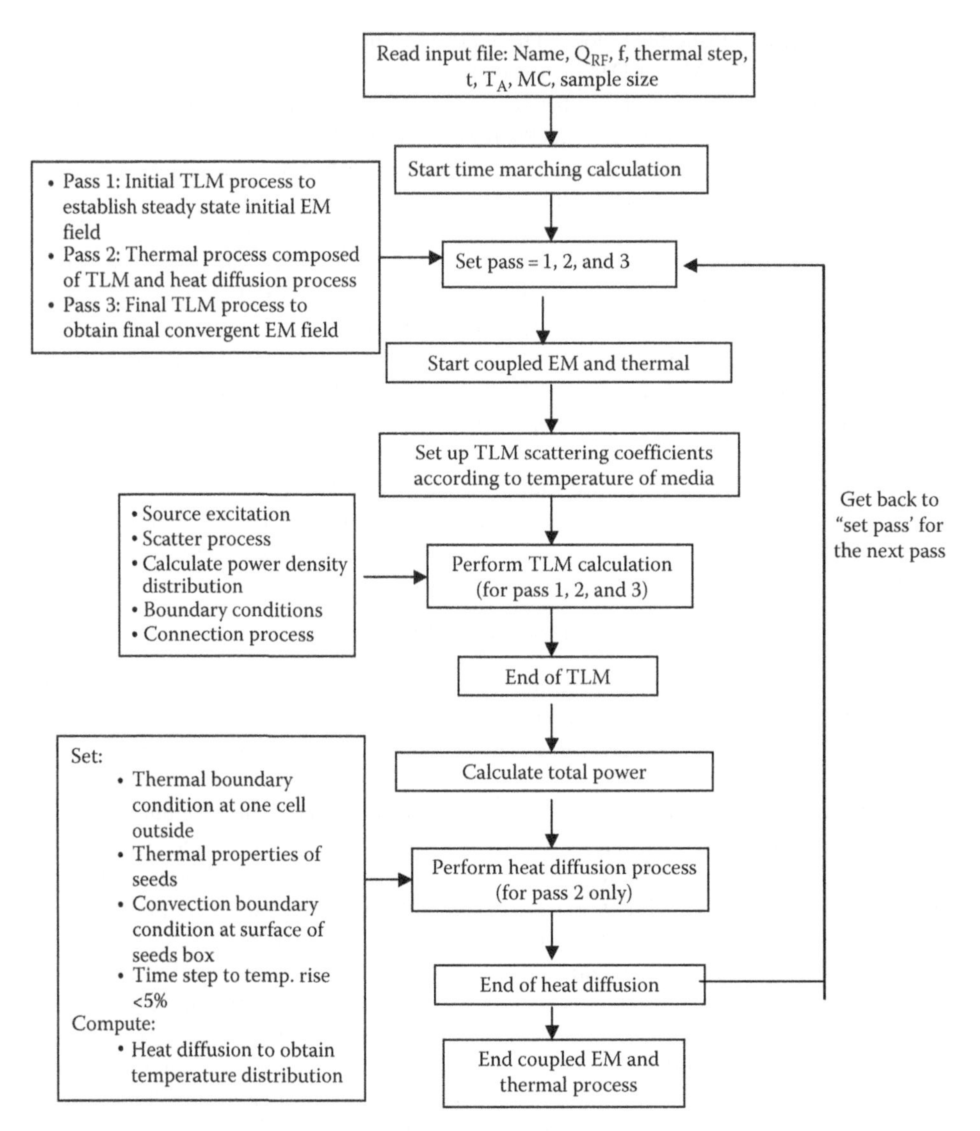

FIGURE 19.4 Flow diagram of the TLM-Food-Heating program. (From Yang, J. et al., *J. Food Process. Eng.*, 26(3), 239, 2003.)

future research to study the effect of different domain geometries and shapes and dimensions on temperature distribution.

Romano and Marra [52] conducted the follow-up research in 2006 to examine the effects of sample shape and orientation of meat batters on heating rate and temperature distribution. Three sample shapes of cube, cylinders, and sphere were simulated in this study. The cubic length of the cube is 7.35 cm; the cylinder has a diameter of 7.97 cm and height of 7.97 cm; and the sphere diameter is 9.12 cm. A 3D multiphysics mathematical model has been developed, for which the specific absorbed power, appearing as source term in the heat conduction equation, was obtained by solving Gauss law, stated for electro-quasi-static conduction in heterogeneous materials with nonuniform permittivity. A numerical solution has been carried out again using FEMLAB 3.1, a FEM-based software. The RF heating simulation was conducted considering an RF oven consisting of a cubic chamber with electrically insulated walls and of two parallel rectangular upper and lower electrodes. The results showed significant impact of the sample shape on heating rate and on temperature distribution within samples. Cubic-shaped samples exhibited a fast and more even heating with a good absorption of power. In case of cylindrically shaped samples, a vertical orientation showed better heating rate and more even temperature distribution than horizontally oriented cylinders. Spherical shapes are the least favored with nonuniform temperature distribution and lowest power absorption.

In 2007, Birla et al. [53] conducted computer simulation of heating uniformity and temperature distribution within a model apple (1% gel, spherical shape) in air and immersed in water. Computer simulation was done using FEMLAB 3.2 from COMSOL Multiphysics, Burlington, MA. It was used to solve the coupled electromagnetic, momentum, and transport equations. The primary mode of RF energy interaction was modeled as a quasi-static electrical field between two electrodes. A 12 kW parallel plate RF heating system with operating frequency of 27.12 MHz, Strayfield Fastran E-200 from Strayfield International Limited, Woringham, United Kingdom, was used to validate computer simulation. The results showed that uniform heating of spherical model fruit in air would not be possible. For spherical model fruit immersed in water, heating rate was significantly improved since the presence of water greatly enhanced power coupling and it only took half time of heating in air to reach 50°C temperature. However, temperature distribution within the spherical apple is not uniform across the whole volume, with lower temperature at the surface and higher temperature at the center of the sphere. The temperature could vary from 40°C at the surface to 65°C at the center. To overcome this problem, the researchers created rotation and movement of fruit during heating, and it evened out the effect of proximity of the sphere to electrodes and container walls. However, it created uneven heating at different horizontal positions. It was found that horizontal and vertical model fruit positions with respect to electrodes significantly influence the heating patterns inside the model fruit. The study suggested that movement and rotation of the sphere object during heating is the only plausible solution for improving heating uniformity.

The most recent computer simulation has been done on heating of dry food materials by Tiwari et al. [54] in 2010. A computer simulation model was developed to heat dry foods such as grains, wheat flours, nuts, herbs, spices, and bakery products using RF heating treatment. Hard red spring wheat flour was selected as a representative dry food material for computer simulation to study RF heating characteristics to improve heating uniformity of dry food materials based on the validated computer simulation model. FEM-based commercial software, FEMLAB v3.4 was used to solve coupled electromagnetic and heat transfer equations. A 12 kW, 27.12 MHz parallel plate RF heating system (Strayfield Fastran with E-200) was used to validate the simulation model. The wheat flour was filled in three polypropylene trays (300 × 220 × 60 mm) stacked one on top of other. It was designed to facilitate temperature mapping at multiple layers. The simulation results showed that temperature values at the mid-layers were highest, followed by those of top and bottom layers. Corners were heated more than centers at each layer. Simulation results confirmed that nonuniform distribution of RF power density resulted in nonuniformity in the sample. Therefore, analysis of RF power distribution inside food materials can be a starting step to understand the complex RF heating process and to predict temperature distribution in dry food materials.

19.5 RF DIELECTRIC HEATING OF FOODSTUFFS

RF dielectric heating has been used in the food industry with proven processes available for a wide range of applications (Table 19.3), including heating packaged bread [62–64], blanching vegetables [5], thawing frozen foods [7,8,88,89], baking and postdrying snack foods [3,4,10,11], and pasteurizing and sterilizing processed meat products [56,57,59–61,65,82–87].

19.5.1 Pasteurization and Sterilization

The interests in using RF dielectric heating to pasteurize and sterilize foodstuffs have increased as a result of its capability for achieving rapid and uniform heating. It is anticipated that RF dielectric

TABLE 19.3
Summary of Applications of Capacitive (RF) Dielectric Heating in Food Materials

Applications	Frequency (MHz)	Processing Temperature (°C)	Processing Time	Advantages	Food Items	References
Blanching of vegetables	150	100	3 min	Limiting loss of nutritive value	Fruits and vegetables	[5]
Baking and postbaking	27.12	~100	—	Rapid drying, improved color and flavor	Crackers, cookies, and bread	[3,4,10,11]
Cooking	27	~70–80	2–13 min	Reduced cooking time, increased cooking yield	Meat products	[55–61]
Heating	14–17	~52–66	20–59 s	Killing mold, refreshing stale bread	Bread	[62–64]
Disinfection	27	50–60	5–20 min	Killing insects, energy efficiency, less impacts on quality and germination rate of seeds	Grains, in-shell walnuts, apples, and legumes	[65–81]
Pasteurizing and sterilizing of foodstuffs	9 27 60	~80	~10 min	Improved sensory quality, reduction of juice losses	Cured ham, sausage emulsion, meat loaf, military ration, poly tray food, eggs, in-shell almonds, and seeds	[50,56,57, 59,60, 61,63,65, 82–87]
Thawing of frozen foods	14–17 36–40 35	>1.67	2–15 min 34 min	Maintaining color and flavor of thawed foods, faster thawing, less drip loss	Egg, fruits, vegetables, fish, beef blocks, sausage, pies, and bacon	[7,8,88,89]

Source: Modified from Ishii, T.K., *Handbook of Microwave Technology: Vol. 2. Applications*, Academic Press, San Diego, CA, 1995.

heating will probably serve as an improved means of producing higher-quality, shelf-stable foods for civilian and military use [61].

The study of RF pasteurization of meats dates back to the 1950s. Pircon et al. [83] described a process for diathermal (energy generated by combination of the ionic and molecular mechanism) sterilization of boned ham at 9 MHz. Based on fundamental studies of capacitance and conductivity of cured luncheon meats, they designed test cells (into which hams were placed in Pyrex tubes between steel electrodes) using an expansion follower plate for the "hot" electrode to take up pressure caused by expansion of the meat. Due to the high conductivity of the meat, the tubes had to be made long and narrow. Experiments were also conducted with plastic-walled cans with concentrically corrugated metal caps as the electrodes. An industrial model 15 kW oscillator with 9 kV across the plates and operating at 9 MHz was used. They achieved energy conversion efficiency of 56.6%, and reported that ~2.7 kg pieces of ham could be heated to sterilization temperature (~80°C) in about 10 min. However, the process had not been used commercially, probably because of high costs and problems with designing processing cells or cans suitable for large-scale operation in the 1950s. Also, after processing, the ham had to be transferred under aseptic conditions to separate containers for supplementary surface heat treatment.

In the late 1960s, Bengtsson and Green [82] developed continuous RF pasteurization of cured hams packaged in Cryovac casings at 60 MHz. They used a generator of 1 kW output and a conveyor for feeding the material between the electrodes of the load condenser. The frequency could be shifted between 35 and 60 MHz by a simple modification of the circuit. They obtained power efficiency of about 25% at 60 MHz, and with knowledge of the dielectric properties of the material, the power efficiency could have been improved by proper design of the generator. Results showed that for lean 0.91 kg hams, treatment time to reach the desired central temperature of 80°C could be reduced to 1/3 by heating in a condenser tunnel operating at 60 MHz. Results also showed substantial reduction of juice losses and a tendency to improve sensory quality compared with hot water processing.

In the 1990s, Houben et al. [55–57] described RF pasteurization of moving sausage emulsions. They performed stationary heating tests at 27 MHz with sausage emulsions of varying formulations stuffed in tubes (inner diameter 50 mm) made of different materials. The integrated 27 MHz heating station consisted of two Colpitt generators with power levels of 25 kW and 10 kW, both equipped with a heating unit. They found that polytetrafluoroethylene (PTFE), polycarbonate, and borosilicate glass performed well as tube materials, but both polyvinyl chloride (PVC) and polyvinylidene chloride (PVDC) heated selectively and therefore could not be used in RF heating. They also found that energy absorption in the working load could be drastically increased by surrounding the tube with demineralized water instead of air. The time required to heat a sausage emulsion from 15°C to 80°C at a mass flow of 120 kg/h was about 2 min. By using RF energy, a heating rate of up to 40°C/min was achieved, compared to about 1°C/min in the center of a sausage during a conventional process (product diameter of 50 mm). Overall energy efficiency (power effectively absorbed in the mass to be heated divided by the power supplied from the mains to the generator) was around 30%, a value the authors thought could probably be improved especially designing the generator-heating unit combination. Sausage products heated using RF energy had a good appearance, smooth surface, and did not show moisture or fat release.

Proctor Strayfield at Horsham, PA, developed a "Magnatube" pasteurization system [2] that demonstrated success in the cooking/sterilization of scrambled eggs, as well as in the creation of a "skinless" meat loaf from a pumped slurry using a vertical 100 mm diameter tube system. The slurry entered from a meat pump at the base of the tube, passed up through the electrode system, and emerged as a skinless "meat loaf" at the top of the heating tube. An extremely uniform temperature distribution across the heated product was achieved, with very good yield.

Foodborne illness is a significant burden. About 48 million (1 in 6 Americans) get sick each year, 128,000 are hospitalized, and 3,000 die [84]. To meet this increased challenge for food industry, more studies have been carried out to examine effectiveness of RF dielectric heating for pasteurization and sterilization of various foodstuffs including beefs, eggs, ham, meat emulsion, bread, and in-shell almonds [63,85–87,90,91]. The potential for controlling human pathogens on alfalfa seeds, used in production of sprouts by RF heating at 39 MHz was studied by experimental exposure of alfalfa seeds which was artificially contaminated with *Salmonella*, *E. coli* O157:H7, and *Listeria monocytogenes* [50]. Significant reductions in the populations of all three pathogens were achieved without reductions in seed germination. However, the desired levels of reduction in bacterial populations (5-log CFU/g) were not achieved without sever reductions in germination of the alfalfa seed. Bacterial reductions appeared better when seed moisture content at time of treatment was in the 5%–7% (wet base) range than at higher moisture levels. The study also confirmed that seeds can tolerate higher temperature without damage to viability when moisture contents at the time of dielectric heating are lower. The study concluded that RF dielectric heating treatments can increase the germination of alfalfa seed lots with substantial hard-seed content, thus increasing sprouting yield, without abrasion of the seed coat, and the process might be considered for this purpose as well as for reduction of human pathogen population in sprouting seeds.

The effectiveness of RF heating on inactivation of *E. coli* in ground beef and its effect on the shelf-stability were investigated with a comparison to hot water bath by Guo et al. [90] in 2004. An RF heater of 1.5 kW at a frequency of 27.12 MHz was used to heat the ground beef samples inoculated with *E. coli* K12 until the center temperature of each sample reached 72°C. The beef was shaped into a plastic container with a dimension of 9.5 cm in diameter and 15 cm high. The same size samples were heated in a water bath with starting temperature of 65°C and ending temperature of 90°C until the center of temperature reached 72°C. The average heating time for the RF heating and water bath are 4.25 and 150.33 min, respectively. The heat-treated samples were then stored at 4°C for up to 30 days for a shelf-life study. Although both RF and water-bath heating had similar effect of reducing *E. coli* counts and maintaining of the shelf-life of 30 days, RF heating significantly reduced the heating time and generated more uniform heating than water bath.

Sterilization of scrambled eggs using RF heating was conducted to produce safe and shelf-stable egg products and compared with conventional retort sterilization [85]. *Clostridium sporogenes* spores were used as a surrogate to validate the RF sterilization process to control *Clostridium botulinum*. Scrambled eggs for RF and retort sterilization were prepared on a stainless griddle at 135°C for 2 min before they were placed in the 6 lb capacity polymeric tray (24.5 × 23.5 × 4.5 cm) and vacuum sealed with a laminated foil lid. For RF sterilization, a tray of scrambled eggs was placed in a 6 kW, 27.12 MHz pilot-scale RF heating system. The temperature of the heated eggs was measured at four locations of upper and lower centers as well as two corners. Heating was conducted based on the temperature profile at the least heated spot to reach sterilization value (F_0) of about 4.6 min. For retort sterilization, the tray of scrambled eggs was heated using circulating hot water at 125°C under the overpressure of 206.8 kPa. The processing time was about 80 min to provide an F_0 of about 4.6 min at the core center of the tray. After heating by either RF or retort method, the egg trays were cooled in a vessel with circulating water at 21°C for about 20 min to allow the temperature at the center of eggs to be lowered to 80°C. Evaluation of quality was done between RF and retort treated scrambled eggs after both types of samples were processed and stored at 4°C for 7 days. The results of heating and quality evaluation showed that RF heating reduced the processing time to one-third that of conventional retort process to reach the same lethality level for *C. sporogenes* spores with better quality of scrambled eggs due to more uniform heating and reduced thermal processing time.

Liu et al. [87] explored the application of RF energy in conjunction with conventional hot air treatment for control of mold in prepackaged bread loaf. A 6 kW, 27.12 MHz RF system was used to develop treatment parameters based on minimum time–temperature conditions that were required for 4-log reduction of *Penicillium citrinum* spores while yielding acceptable bread quality. Sliced enriched white bread samples were inoculated with 3.2×10^6 CFU/g of pure mold strain before heat treatment. The inoculated bread samples packaged in Ziploc bags were then placed in the RF system and heated by combined RF energy and hot air system till the highest temperature in bread reached 53°C, 58°C, 63°C, or 68°C. The treated samples were then stored for a storage test, and the quality of bread was analyzed for moisture content, water activity, and firmness. Compared with untreated bread samples, combined RF and hot air-treated samples (4.5 min to 58°C) had a longer storage life of 66 days versus 38 days for untreated samples while there is no significant difference in sample qualities in terms of moisture content, water activity, and firmness during the storage.

In 2010, RF dielectric heating was applied to pasteurize in-shell almonds to control *Salmonella* without causing a substantial loss of product quality by Gao et al. [63]. Washed in-shell almonds (1.7 kg) were heated using a 6 kW, 27.12 MHz pilot scale RF system with hot air heating at 55°C, followed by forced room air cooling. The results showed that almond temperature above 75°C at 23% moisture contents for 2–4 min RF heating could achieve 5-log reduction of *Salmonella*. The RF treatment process for 20 min reduced the moisture content to 5.7% w.b. Almond qualities (fatty acid, peroxide value, kernel skin, and core color) were not affected by the RF treatment after the accelerated shelf-life test for 20 days at 35°C simulating 2 years of real storage at 4°C.

Another interesting application of RF heating was conducted to control brown rot on peaches and nectarines by Casals et al. [92]. A serous postharvest disease affecting peaches and nectarines, brown rot by *Monilinia* spp., is treated using RF dielectric heating, an alternative for nonchemical fungicide treatment in Spain. Fruit used in this study included "Red Jim" and "Big Orange" nectarines and "Merryl O'Henry," "Placido, and Summer Rich" peaches. Before RF heating treatment, all fruits were inoculated with isolate of *M. fructicola*. A 15 kW, 27.12 RF system was used to heat the fruit samples on a conveyor belt in-between the two RF heating electrodes with a adjustable gap of 65–205 mm for varying the amount of RF power. The results showed that the RF treatment significantly reduced the incidence of brow rot in "summer Rich" peaches inoculated at 10^3, 10^4, and 10^5 conidia mL^{-1}. The same RF treatment did not reduce incidence of brown rot for nectarines at given treatment power with electrode gap of 17 mm and exposure time from 11.25 to 22.5 min when internal temperature reached to 62°C at the end of RF heating.

Despite the research, commercial RF dielectric heating systems for the purpose of food pasteurization or sterilization are not known to be in use. The reasons for the lack of success in commercial operation are complexity, expense, nonuniformity of heating, inability to ensure sterilization of the entire package, lack of suitable packaging materials, and unfavorable economics when compared to prepared frozen foods in the United States [93].

19.5.2 Cooking

There has been limited research on RF cooking of foodstuffs to improve quality and reduce cooking time compared to conventional cooking using hot water bath and steam [55–57]. Until the 2000s, a few studies have investigated the effect of RF heating on the quality of meat products, compared to conventional water bath.

In 2002, Laycock et al. [58] conducted RF cooking of three different meat products. The effect of RF heating on the quality, heating rate, and temperature history of three types of meat products (ground, comminuted, and muscle) was investigated after heating to center temperature of 72°C in a 1.5 kW, 27.12 MHz RF heating system. The results were compared with that obtained from heating in a water bath. RF processed meat products resulted in lower juice losses and acceptable color and texture. The comminuted and muscle meat samples exhibited improved texture while ground beef was very chew and elastic. The results indicated that ground beef achieved highest power efficiency (60.17%), followed by comminuted meat (44.41%), and muscle (43.38%). Cooking time for three meat products was 5.83, 13.5, and 13.25 min for ground beef, comminuted meat, and muscle, respectively, significantly reduced compared to 151, 130, and 109 min in water bath. It is concluded that the well-mixed comminuted and ground meat products appeared to be most promising for RF cooking.

The same research group [59] conducted a follow-up research to further study the effect of RF heating on the texture, color, and sensory properties of a pork-based white pudding. An output power of 450 W from a customized RF batch oven was used to heat all meat samples. To avoid arcing during RF cooking, white pudding samples was cased in a polyethylene cell and placed in a purposed built circulating water cell. There were four cooking protocols used in this study to compare the quality of the meat samples: (1) steam cooked to 73°C for 2 min; (2) water bath cooked 73°C for 2 min; (3) RF cooked to 73°C for 2 min; and (4) RF cooked to 73°C for 2 min followed by an extended holding time. All heat-treated white pudding samples were evaluated for texture, color, and sensory. The results showed that RF-heated puddings were not significantly different from water bath- and steam oven-heated samples with regard to color, texture, and expressible fluid. The results of a sensory similarity test with 60 panelists indicated no detected difference between puddings cooked by RF and conventional methods. However, RF heating reduced cooking time to 7 min 40 s as compared to 33 min for conventional cooking. This research concluded that RF cooking could be very suitable for industrial heating of encased white puddings.

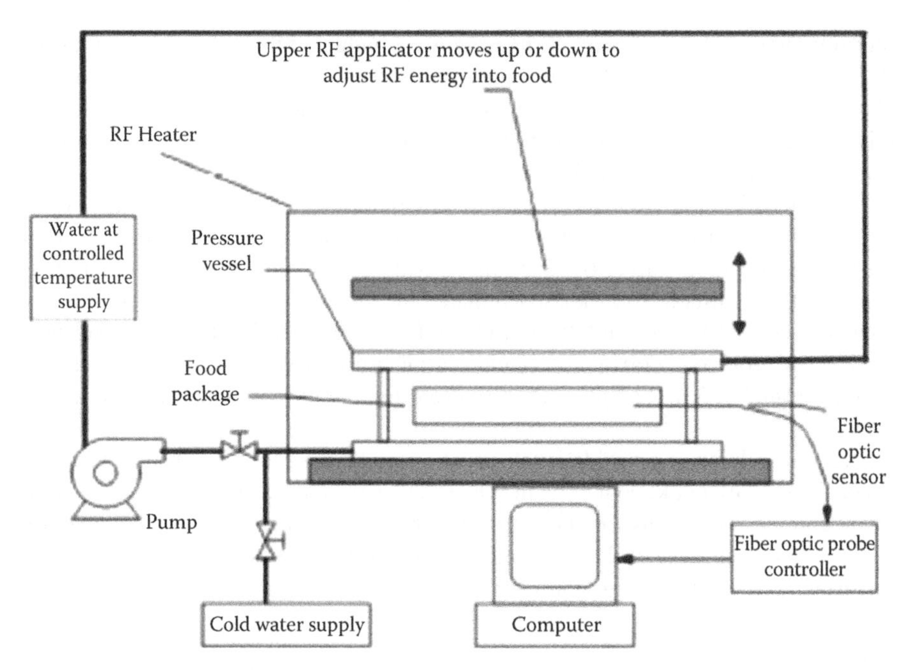

FIGURE 19.5 Simplified schematic diagram of the RF sterilization system developed at Washington State University (Pullman, WA). (From Wang, Y. et al., *J. Food Sci.*, 68(2), 539, 2003.)

In 2003, a study was conducted with a pilot-scale 6kW, 27MHz RF system (COMBI 6-S; Strayfield Fastran, United Kingdom) to investigate the effectiveness in shortening process time and in improving quality for foods sealed in 6lb military-ration polymeric trays [61]. The applicators and the tuning system were modified to improve the electromagnetic field uniformity between the two plate applicators. A pressurized vessel was developed to provide an overpressure of up to 0.276MPa gauge that allows foods in large polymeric trays to be heated up to 135°C without bursting (Figure 19.5). The system consisted of two exchangers: one used steam for heating, and the other used tap water for cooling. A surge tank was used to help maintain an overpressure with compressed air. The circulating water temperature could be varied from 20°C to 130°C, an overpressure from 0.136 to 0.204MPa gauge (20–30 psig). To reduce fringe effect at the interface between the side of the food package and the air in the RF applicators, low-conductivity water was used to immerse the food tray to approximately match the dielectric properties of the food. Together, the water and package present a very flat surface to the imposed field, and push the boundary with air away from the food package. This is expected to reduce the effect of variations in package thickness that could lead to nonuniform heating in the food. Temperature-controlled water from water-conditioning system is used to match the temperature of the heated food to prevent cooling of the package surface. Chemical marker *M-1* was used to evaluate heating uniformity in 20% whey protein gels as a model food, and macaroni and cheese was processed to assess the influence of RF process on product quality.

Another study of RF cooking of beef rolls was conducted by Tang et al. [60] in 2005. Chemical, physical, and sensory aspects of quality were compared on encased beef rolls cooked in a steam oven (80°C) and an RF heating oven (500W, 27.12MHz) under circulating water at 80°C. The RF heating reduced cooking time to 23% and 31% of steam cooking times, respectively, in noninjected meat (PG1) and in rolls prepared with curing brines (PG2, PG3, and PG4). Compared to steam, cooking yields were significantly higher and toughness of texture were significantly lower for RF-cooked PG1 rolls and for meat injected with brines containing water binding dielectrically inactive additives (PG4) but not for brined rolls lacking the latter ingredients (PG2 and PG3). Sensory analysis from 50 untrained panelists indicated no texture difference between RF-cooked samples and steam-cooked samples of PG1 and PG4.

19.5.3 Disinfestation

RF heating energy has long been used as an alternative method for nonchemical pest control as early as 1929 due to its large penetration depths, heating uniformity, and differential or targeted heating of insects in foodstuffs [65–69]. Most early research on RF insect control has been focused on stored grains [67,68]. Although most of these studies showed that RF energy could be effective for insect control, the method was not cost effective when compared to inexpensive chemical fumigation in use that time [68]. Until the mid-1990s, due to the more restrict control and requirement on chemical fumigation, RF heating has been further investigated on various horticulture and food products including in-shell walnuts, rice, apple, and legumes [70–81]. The following are brief summary of those recent studies.

The effectiveness of RF heating on insect control for in-shell walnut has been studied since the early 2000s. Several pilot-scale studies have been conducted to rapidly deliver thermal energy to walnuts to control three important insects infesting walnuts: codling moth [71], Indian meal moth [73], and navel orangeworm [74]. RF energy directly interacts with the kernel inside the shell resulting in a fast temperature ramp to a lethal level of 53°C–57°C for insect pests [75]. In order to transfer pilot-scale research results to large-scale industrial implementation, studies were conducted to investigate the effects of walnut orientation on temperature distribution inside the kernel, differential heating between open and closed shell walnuts, quality difference between actual stored walnuts and walnuts subjected to accelerated storage life tests, and energy cost for large-scale treatment [76]. The results indicated that temperatures of vertically oriented walnuts were 7.4°C higher than those of horizontally oriented walnuts. The open shell walnuts heated much faster in RF systems than closed shell walnuts after 1.5 min of blanching. The accelerated storage life tests at 35°C for 10–20 days provided accurate simulations of actual storage at 4°C for 1–2 year periods, respectively, based on similar peroxide values and free fat acid values. The energy cost of RF treating 4540 kg/h was 0.23 cents/kg. Based on the earlier results, an industrial-scale 27 MHz, 25 kW RF system was used to conduct a large-scale treatment on in-shell walnuts [77,78] to investigate the technical feasibility of using RF energy in commercial postharvest insect control in in-shell walnuts. Nonuniform vertical temperature distribution was measured in the RF unit, indicating that mixing and circulated hot air were needed to obtain the required treatment uniformity. The average energy efficiency of the RF system was estimated to be 79.5% when heating walnuts at 1562 kg/h. The RF treatment of achieving an average walnut surface temperature of 60°C, and minimum temperature of 52°C for 5 min caused 100% mortality of fifth-instar navel orangeworm larvae, the most heat tolerant target pest, in both unwashed and air-dried walnuts. Walnut quality in terms of kernel color, peroxide values and free fat acid values were similar to untreated controls after 20 days at 35°C storage condition, simulating 2 years of storage under commercial storage condition of 4°C.

In 2005, Wang et al. studied postharvest treatment to control codling moth in fresh apples using water-assisted RF heating [79]. The combined treatment included preheating in warm water at 45°C for 30 min and RF heating to 48°C in a hydraulic fruit mover placed between two parallel plate electrodes in a 12 kW, 27.12 MHz RF system. After RF treatment, apples were transferred to a 48°C hot water bath and held for 5, 10, 15, and 20 min before being cooled in ice water for 30 min. The mortality of codling moth in the infested apples was evaluated, and the quality of apples was assessed by measuring weight loss, firmness, color, soluble solids content, and titratable acidity after 0, 7, and 30 days of storage at 4°C. The results showed that the treatment at 48°C for 15 min was the most practical and effective for both insect control and apple quality.

Lagunas et al. [80] studied nonchemical disinfestations of rice using RF energy. Natural and laboratory infested dried rice containing all life cycles of lesser grain borers (*Rhyzopertha dominica*) and Angoumois grain moths (*Sitotroga cerealella*) were used to evaluate efficacy of RF power treatment. RF treatment at 55°C–60°C for 5 min resulted in >99% mortality of all biological stages of *Sitotroga cerealella* (grain moth). RF treatment at 50°C for *Sitotroga cerealella* with a 2 h tempering or 60°C for *R. dominica* with 1 h tempering provided 100% control. No moisture loss or changes in milling quality of the treated rice were observed.

In 2010, the effectiveness of RF disinfestation treatment for legumes was investigated by Wang et al. [81]. A pilot-scale 27.12 MHz, 6 kW RF system was used to thermally treat chickpea, green pea, and lentil samples together with a customized hot air system. A 3 kg legume samples were put in a plastic container (25.5 cm × 15 cm × 10 cm) and placed on the convey belt moving at a speed of 0.56 m/min in between the two electrode plates. Hot air at 60°C was supplied into the RF cavity through an air distribution box. Chickpea, green pea, and lentil samples in plastic containers were RF heated for 5–7 min to reach 60°C center temperature and were held in 60°C hot air for 10 min. After heat treatment, legume samples were spread in a 1 cm thick layer on a tray and cooled by forced room air for 15, 12, and 18 min for chickpea, green pea, and lentil, respectively. After treatment, the quality of legumes was evaluated for weight loss, moisture content, color, and germination. The results showed that RF heat treatment sharply reduced the heating time as compared to hot air heating. RF heating uniformity was greatly improved by 60°C forced air and movement of the container between the electrodes. An RF treatment protocol was developed as an effective disinfestations treatment for legumes with RF heating 60°C held for 10 min in hot air followed by forced room air cooling in a 1 cm layer. RF treatments had little effect on any of the measured quality parameters. But hot air treatments, on the other hand, reduced the sample weight and moisture content significantly for chickpea and lentil.

19.6 TECHNICAL CHALLENGES AND ENERGY EFFICACY OF RF APPLICATIONS IN FOODSTUFFS

A number of potential problems need to be addressed before the RF technology can be successfully commercialized to food pasteurization and sterilization, especially packaged foods. The major technical challenges include avoiding dielectric breakdown (arcing) and thermal runaway heating, which can lead to both packaging failure and product destruction. In addition, economic constraints exist that inhibit the use of the technology in its present form.

The effectiveness of RF dielectric heating is highly dependent on the dielectric properties of product, which are in turn strong functions of both frequency and temperature. The energy efficiency and/or heating rate will be maximized at or near the location in frequency of a "Debye resonance"—the frequency at which the dielectric loss factor is at the maximum for a particular material. Multiple Debye resonance might occur in a product made up of many different materials, causing the packaging materials to overheat or even burn, resulting in packaging failure.

19.6.1 Radio Frequency Dielectric Breakdown (Arcing) and Thermal Runaway Heating

One potential problem of RF dielectric heating is the potential for *thermal runaway or hot spots* in a heterogeneous medium. Sanders [88] mentioned that in most applications of dielectric heating to food, runaway heating is unavoidable if contacts with both electrodes are maintained. Dielectric loss factors are often strong functions of temperature. Thus, small pockets of lossy dielectric food materials (e.g., a small thermal mass trapped in the seams of a package) might heat very rapidly, burning and then melting the package. Another problem is the potential for dielectric breakdown (arcing) if the electric field strength across the sample is very high. This can be complicated by surface irregularities and edge effects of the surface in contact with the electrodes [88].

Cathcart et al. [62] found that while heating packaged bread using RF power, arcing occurred between the upper electrode and the edges of the cylindrical loaf, resulting in some burning at those points. This difficulty was overcome by rounding the edges of the loaf so that there were no corners adjacent to the electrodes. Moyer and Stotz [5] indicated that when blanching vegetables at 7, 10, and 30 MHz, arcing and burning occurred between the upper electrode and the vegetables, and between the individual particles, due to the high voltages at the electrodes. By decreasing voltage (lowering the electric field across a given sample thickness) while increasing frequency to 150 MHz to maintain the same output power density in the medium, the arcing problem was solved. Sanders [88] further demonstrated the significance of using an air gap between the top electrode and the upper material surface to lower the voltage and therefore the electric field across the sample thickness. By introducing an air gap, some of the voltage between the electrodes appears in this gap, the exact amount depending on the relative heights of the gap and of the material, and on the dielectric properties of the material. The effective voltage is reduced by a factor f_p [82]:

$$f_p = \frac{1}{\sqrt{\left[\frac{\varepsilon_r \tan\delta\, h_a}{h_s}\right]^2 + \left[\frac{\varepsilon_r h_a}{h_s} + 1\right]^2}} \tag{19.8}$$

where

ε_r is the relative permittivity (or dielectric constant, $\varepsilon_r = \varepsilon'/\varepsilon_0$)

h_a, h_s are heights of the air gap and the material (m), respectively

Sanders [88] also noted that arcing could be controlled by immersing the product in deionized water or surrounding it with ice, thus minimizing field distortions due to any irregularities, and again lowering the voltage or effective electric field across the sample thickness. Dielectric breakdown (arcing) and thermal runaway heating were also observed during preliminary tests to heat-packaged *surimi* seafood in a commercial RF oven [94].

19.6.2 Packaging Failure

When using RF power to heat packaged food products, it was found that certain packaging materials failed in an RF field. When heating packaged bread, Cathcart et al. [62] found that some wrapping, including wax and glassing paper, were unsatisfactory. Wax paper softened when bread showed an internal temperature above 38°C and glassing papers became tacky between 52°C and 60°C. However, they found that cellophane is a satisfactory wrapping material because it did not break or become unsealed in any of the heat treatments employed. As stated before, Houben et al. [56,57,61] found that PTFE, polycarbonate, and borosilicate glass performed well as tube materials when using RF power to pasteurize sausage emulsion; however, both PVC and PVDC heated selectively (due to their polar molecular structures) and therefore could not be used in RF heating. Burning of packaging material was also observed under some conditions during preliminary tests with RF heating of vacuum packaged surimi seafood [94].

Thus, a major challenge in the use of RF power to pasteurize and sterilize packaged food products is avoiding packaging failure or modification by selecting satisfactory packaging materials. This requires a full understanding of the dielectric properties of packaging material over a range of frequencies and temperatures. Moreover, it is important to avoid any factors that may cause high local power densities, such as the concentration of a small amount of a food/water within the seam of the packaging material. This could result in a small thermal mass combined with a large dielectric loss factor being exposed to a potentially large localized voltage gradient or electric field.

Two patent technologies may help to solve some of the problems that RF heating faces in food pasteurization and sterilization [94,95]. Frequency could be controlled to match Debye resonances of the dominant constituents of the medium, adjusting appropriately with temperature, and continually controlling electric field strengths [94]. Product geometry and packaging materials could be optimized to prevent dielectric breakdown (arcing), although packaging methodologies would have to ensure that food could be kept out of the seams. The electrode system could also be developed to lower the field strengths on the packaging seams. To prevent or reduce the risk of thermal runaway, a gridded electrode system could be developed with an infrared scanner to monitor the overall medium. This would add the capability of heating subcomponents of the medium independently with different field strengths, or even switching off some of the electrode sections in different duty cycles to prevent "hot spots" [94,95].

19.6.3 Energy Efficiency and Economic Analysis

Many concerns have been raised about the potential high overall operational cost of using RF dielectric heating in food processing, which has delayed the commercialization of this technology. Although no complete investigation on economic analysis has been conducted, several studies show estimated data of energy efficiency and cost [1,7,57,82,83,88].

Kinn [1] illustrated the cost in dollars per hour for operating various size standard RF generators and the cost of RF power in dollars per kilowatt for the various generators. Pircon et al. [83] reported that an average energy conversion efficiency (comparing the thermal energy output to the electrical energy input) of 56.6% can be achieved for a 15 kW and 29 MHz generator. An approximate cost analysis indicated an electricity cost of $0.00722/kg of meat compared to a steam cost of $0.0026/kg of meat, based on a 910 kg batch. Pircon et al. [83] believed that because of the sliding cost scale for electricity, a large volume of meat would bring the costs closer together. In addition, a continuous dielectric process would not involve any losses between batches as would in a steam process. Jason and Sanders [7] did some economic analysis of thawing using RF energy and concluded that the cost of dielectric thawing is about one-eighth that of freezing, and only a very small fraction of total cost. Sanders [88] reported that a value of 0.2 kW/kg has been shown to be a suitable (though not necessarily the maximum) power density for thawing blocks of meat. Bengtsson and Green [82] reported that using RF power to pasteurize ham, the highest power efficiency that could be obtained (with salted ham in the generator-electrode combination used) was 25%, but 50%–60% would be required in an industrial application. They believed that with knowledge of the dielectric properties of the materials to be heated, this higher level of power efficiency could be achieved through proper design of the generator. In the RF pasteurization of sausage emulsions, Houben et al. [57] found that optimum overall energy efficiency, defined as power effectively absorbed in the mass to be heated divided by the power supplied from the mains to the generator, was somewhat higher than 30%, and could probably be improved by careful design of the generator-heating unit combination.

New technology and quantity production in manufacturing facilities have reduced the cost of equipment and operations. Many of these RF applications may now be considerably economically sound. In addition, RF dielectric heating has the potential to improve the quality of food products, and this improvement must in some way be evaluated in the economic study. Newer technologies available for RF dielectric heater application systems, such as synthesizers, broadband RF amplifiers, and tunable matching networks, offer the promise of greater energy efficiency due to the ability of the system to tune to the optimum "Debye resonance" frequencies of a heated medium. In addition, greater control of the applied voltage and resulting electric field will lead to the reduction of arcing and thermal runaway problems inherent in older RF heating systems. Most of the published economic studies conducted on RF heating focused on older technologies under greatly different economic circumstances. For that reason, there is a need for an updated economic analysis on RF heating that takes into account the latest technological improvements, as well as the economic conditions in place in today's marketplace.

19.7 CONCLUSIONS

RF dielectric heating technology has vast potential applications in the food industry. The rapid and uniform heating patterns of RF energy are very attractive for providing safe, high-quality food products. To maximize the performance of an RF dielectric heating system and to move the technology to commercial pasteurization and sterilization of foodstuffs, following areas of research are necessary:

- To understand more about the effects of food formulation on heating patterns
- To study the effects of equipment design factors, such as frequency to achieve better uniformity of heating
- To develop variable frequency RF system for improved uniformity of heating
- To understand the factors affecting heating patterns, including qualitative changes occurring with frequency changes
- To monitor and real-time adjust for process deviations in RF processing

NOMENCLATURE

Symbol definition

C_p	Specific heat (kJ/kg°C)
d_p	Penetration depth
E	Strength of electric field of the wave at the location permittivity or dielectric
f	Frequency of the radio frequency waves (Hz)
h	Heat transfer coefficient (W/m²K)
h_a, h_s	Heights of the air gap and the material (m) in Equation 19.7
j	Constant and $j=\sqrt{-1}$
k	Thermal conductivity (W/m°C)
P_v	Maximum power per unit volume (W/m³) in Equation 19.5
Q	RF power absorption density (W) in Equation 19.2
T	Temperature in the food samples (°C)
∇T	Temperature gradient
$\partial T/\partial n$	Temperature gradient in n direction (K/m)
t	Time (s)
(x,y,z)	Three dimension in x, y, and z directions (m)
α	Thermal diffusivity (m²/s)
ε_0	Permittivity of free space in Equation 19.3, 8.854×10^{-12} F/m
ε'	Energy storage term of permittivity
ε''	Energy loss term of permittivity: $\varepsilon_0\varepsilon_r''=\varepsilon''$, the subscript r designates the relative
ε_r	Relative permittivity (or dielectric constant, $\varepsilon_r=\varepsilon'/\varepsilon_0$) in Equation 19.7
ρ	Density (kg/m³)
ω	Angular frequency $=2\pi f$ (rad/s) in Equation 19.5
δ	Penetration depth (skin depth) δ at which the field has dropped to 37% of its surface level in Equation 19.6

REFERENCES

1. TP Kinn. Basic theory and limitations of high frequency heating equipment. *Food Technology* 1:161–173, 1947.
2. AL Koral. *Proctor-Strayfield Magnatube Radio Frequency Tube Heating System*. Horsham, PA: Proctor Strayfield, A Division of Proctor & Schwartz, Inc., 1990.
3. Anonymous. An array of new applications are evolving for radio frequency drying. *Food Engineering* 59(5):180, 1987.

4. Anonymous. RF improves industrial drying and baking processes. *Processing Engineering* 70:33, 1989.
5. JC Moyer, E Stotz. The blanching of vegetables by electronics. *Food Technology* 1:252–257, 1947.
6. BE Proctor, SA Goldblith. Radar energy for rapid food cooking and blanching, and its effect on vitamin content. *Food Technology* 2:95–104, 1948.
7. AC Jason, HR Sanders. Dielectric thawing of fish. I. Experiments with frozen herrings. *Food Technology* 16(6):101–106, 1962.
8. AC Jason, HR Sanders. Dielectric thawing of fish. II. Experiments with frozen white fish. *Food Technology* 16(6):107–112, 1962.
9. Anonymous. Radio waves cut meat defrosting time. *Processing Engineering* 73:21, 1992.
10. J Rice. RF technology sharpens bakery's competitive edge. *Food Processing* 6:18–24, 1993.
11. NH Mermelstein. Microwave and radiofrequency drying. *Food Technology* 52(11):84–86, 1998.
12. TK Ishii. *Handbook of Microwave Technology: Vol. 2. Applications*. San Diego, CA: Academic Press, 1995.
13. Y Zhao, B Flugstad, E Kolbe, JW Park, JH Wells. Using capacitive (radio frequency) dielectric heating in food processing and preservation—A review. *Journal of Food Processing Engineering* 23:25–55, 2000.
14. T Ohlsson. Temperature dependence of sensory quality changes during thermal processing. *Journal of Food Science* 45:836–839, 847, 1980.
15. NL Sheen, IM Woodhead. An open-ended coaxial probe for broad-band permittivity measurement of agricultural products. *Journal of Agricultural Engineering Research* 74:193–202, 1999.
16. DS Engelder, CR Buffler. Measuring dielectric properties of food products at microwave frequencies. *Microwave World* 12:6–15, 1991.
17. VN Tran, SS Stuchly. Dielectric properties of beef, beef liver, chicken and salmon at frequencies from 100 to 250 MHz. *Journal of Microwave Power* 22:29–33, 1987.
18. R Seaman, J Seals. Fruit pulp and skin dielectric properties for 150 MHZ to 6400 MHz. *Journal of Microwave Power and Electromagnetic Energy* 26(2):72–81, 1991.
19. RK Calay, M Newborough, D Probert, PS Calay. Predictive equations for the dielectric properties of foods. *International Journal of Food Science and Technology* 29(6):699–713, 1994.
20. AG Herve, J Tang, L Leudecke, H Feng. Dielectric properties of cottage cheese and surface treatment using microwave. *Journal of Food Engineering* 37(4):389–410. 1998.
21. PA Berbert, DM Queiroz, EF Sousa, MB Molina, EC Melo, LRD Faroni. Dielectric properties of parchment coffee. *Journal of Agricultural Engineering Research* 80:65–81. 2001.
22. A Garcia, JL Torres, E Prieto, M De Blas. Dielectric properties of grape juice at 0.2 and 3 GHz. *Journal of Food Engineering* 48:203–211, 2001.
23. H Feng, J Tang, RP Cavalieri. Dielectric properties of dehydrated apples as affected by moisture and temperature. *Transactions of the ASAE* 45(1):129–135, 2002.
24. KW Farag, JG Lyng, DJ Morgan, DA Cronin. Dielectric and thermophysical properties of different beef meat blends over a temperature range of −18 to +10°C. *Meat Science* 79:740–747, 2008.
25. SO Nelson. Electric properties of agricultural products. A critical review. *Transactions of the ASAE* 16(2):384–400, 1973.
26. NN Mohsenin. *Electromagnetic Radiation Properties of Foods and Agricultural Products*. New York, NY: Gordon & Breach Science Publishers, 1984.
27. RE Mudgett. Electrical properties of foods. In: RV Decareau, ed. *Microwaves in the Food Processing Industry*. New York, NY: Academic Press, 1985, pp. 15–37.
28. M Kent. *Electrical and Dielectric Properties of Food Materials*. Essex, U.K.: Science and Technology Publishers, 1987.
29. KR Foster, HP Schwan. Dielectric properties of tissues and biological materials: A critical review. *CRC Critical Review—Biomedical Engineering* 17(1):25–104, 1989.
30. F Marra, L Zhang, JG Lyng. Radio frequency treatment of foods: review of recent advances. *Journal of Food Engineering* 91:497–508, 2009.
31. S Ryynanen. The electromagnetic properties of food materials: A review of the basic principles. *Journal of Food Engineering* 29(4):409–429, 1995.
32. NE Bengtsson, J Melin, K Remi, S Soderlind. Measurements of the dielectric properties of frozen and defrosted meat and fish in the frequency range 10–200 MHz. *Journal of Science in Food and Agriculture* 14:592–604, 1963.
33. L Zhang, JG Lyng, N Brunton, D Morgan, B McKenna. Dielectric and thermophysical properties of meat batters over a temperature range of 5–85°C. *Meat Science* 68:173–184, 2004.
34. AJ Ede, RR Haddow. The electrical properties of food at high frequencies. *Food Manufacture* 26:156–160, 1951.

35. RH Cole, G Delbos, P Winsor IV. Study of dielectric properties of water/oil and oil/water microemulsions by time domain and resonance cavity methods. *Journal of Physical chemistry* 89:3338–3343, 1985.
36. M Merabet, TK Bose. Dielectric measurements of water in the radio and microwave frequencies by time domain reflectometry. *Journal of Physical Chemistry* 92:6149–6150, 1988.
37. S Puranik, A Kumbharkhane, S Mehrotra. Dielectric properties of honey-water mixture between 10 MHz to 10 GHz using time domain technique. *Journal of Microwave Power and Electromagnetic Energy* 26(4):196–201, 1991.
38. WKP Looan, PE Smulders. Time-domain reflectometry in carbohydrate solutions. *Journal of Food Engineering* 26:319–328, 1995.
39. JN Ikediala, J Tang, SR Drake, LG Neven. Dielectric properties of apple cultivars and codling moth larvae. *Transactions of the ASAE* 43(5):1175–1184, 2000.
40. Y Wang, TD Wig, J Tang, LM Hallberg. Dielectric properties of foods relevant to RF and microwave pasteurization and sterilization. *Journal of Food Engineering* 57(3):257–268, 2003.
41. S Wang, J Tang, RP Cavalieri, DC Davis. Differential heating of insects in dried nuts and fruits associated with radio frequency and microwave treatments. *Transactions of the ASAE* 46(4):1175–1184, 2003.
42. K Sacilik, C Tarimci, A Colak. Dielectric properties of flaxseeds as affected by moisture content and bilk density in the radio frequency range. *Biosystems Engineering* 93(2):153–160, 2005.
43. R Luigni, A AL-Shami, G Mikhaylenko, J Tang. Dielectric characterization of hen eggs during storage. *Journal of Food Engineering* 82:450–459, 2007.
44. G Roussy, JA Pearce. *Foundations and Industrial Applications of Microwaves and Radio Frequency Fields. Physical and Chemical Processes*. New York, NY: John Wiley & Sons, 1995.
45. PB Johns. On the relationship between TLM and finite difference methods for Maxwell's equations. *IEEE Transactions of Microwave Theory Technology MTT* 35(1), 60–61, 1987.
46. KS Yee. Numerical solution of initial boundary value problems involving Maxwell's equations. *IEEE Transactions of Antennas and Propagation AP* 14(3):302–307, 1966.
47. AK Datta. Fundamentals of heat and moisture transport for microwaveable food product and process development. In: AK Datta, RC Anatheswaran, eds. *Handbook of Microwave Technology for Food Applications*. New York, NY: Marcel Dekker, 2000, pp. 115–172.
48. M Orfeuil. *Electric Process Heating: Technologies/Equipment/Applications*. Columbus, OH: Battelle Press, 1987.
49. AR Von Hippel. *Dielectric Properties and Waves*. New York, NY: John Wiley & Sons, 1954.
50. J Yang, Y Zhao, JH Wells. Computer simulation of vegetable sprout seeds in capacitive radio frequency (RF) dielectric heating. *Journal of Food Processing Engineering* 26(3):239–263, 2003.
51. F Marra, J Lyng, V Bomano, B McKenna. Radio-frequency heating of foodstuff: Solution and validation of a mathematical model. *Journal of Food Engineering* 79:998–1006, 2007.
52. V Romano, F Marra. A numerical analysis of radio frequency heating of regular shaped foodstuffs. *Journal of Food Engineering* 84:449–457, 2008.
53. SL Birla, S Wang, J Tang. Computer simulation of radio frequency heating of model fruit immersed in water. *Journal of Food Engineering* 84:270–280, 2008.
54. G Tiwari, S Wang, J Tang, SL Birla. Computer simulation model development and validation for radio frequency (RF) heating of dry food materials. *Journal of Food Engineering* 105:48–55, 2011.
55. JH Houben, PS Van Roon, B Krol. Continuous radio frequency pasteurization of sausage emulsions. Trends in modern meat technology, 2. *Proceedings of the International Symposium,* Den Dolder, the Netherlands, 1987.
56. JH Houben, PS Van Roon, B Krol. Radio frequency pasteurization of moving sausage emulsions. In: P Zeuthen, JC Cheftel, C Eriksson, TR Gormley, P Linko, eds. *Thermal Processing and Quality of Foods*. Gaithersburg, MD: Aspen Publishers, Inc., 1990, pp. 1.171–1.177.
57. J Houben, L Schoenmakers, E van Putten, P van Roon, B Krol. Radio-frequency pasteurization of sausage emulsions as a continuous process. *Journal Microwave Power and Electromagnetic Energy* 26(4):202–205, 1991.
58. L Laycock, P Piyasena, GS Mittal. Radio frequency cooking of ground, comminuted and muscle meat products. *Meat Science* 65:959–965, 2003.
59. NP Brunton, JG Lyng, W Li, DA Cronin, D Morgan, B McKenna. Effect of radio frequency heating on the texture, color and sensory properties of a comminuted pork meat product. *Food Research International* 38:337–344, 2005.
60. X Tang, JG Lyng, DA Cronin, C Durand. Radio frequency heating of beef rolls from biceps femoris muscle. *Meat Science* 72:467–474, 2006.

61. Y Wang, TD Wig, J Tang, LM Hallberg. Sterilization of foodstuffs using radio frequency heating. *Journal of Food Science* 68(2):539–544, 2003.
62. WH Cathcart, JJ Parker, HG Beattie. The treatment of packaged bread with high frequency heat. *Food Technology* 1:174–177, 1947.
63. M Gao, J Tang, R Villa-Rojas, Y Wang, S Wang. Pasteurization process development for controlling Salmonella in in-shell almonds using radio frequency energy. *Journal of Food Science* 104:299–306, 2011.
64. NE Bengtsson. Electronic defrosting of meat and fish at 35 MHz and 2450 MHz. A laboratory comparison. *Proceedings of the 5th International Congress of Electro-Heat*, Wiesbaden, Germany, 1963.
65. TJ Headlee, RC Burdette. Some facts relative to the effect of high frequency radio waves on insect activity. *Journal of the New York Entomological Society* 37:59–64, 1929.
66. H Frings. Factors determining the effects of radio frequency electromagnetic fields and materials they infest. *Journal of Economic Entomology* 45:396–408, 1952.
67. SO Nelson, WK Whitney. Radio frequency electric fields for stored grain insect control. *Transactions of the ASAE* 3:133–137, 144, 1960.
68. SO Nelson, 1996. Review and assessment of radio frequency and microwave energy for stored-grain insect control. *Transactions of the ASAE* 39:1475–1484, 1996.
69. SO Nelson, JA Payne. RF dielectric heating for pecan weevil control. *Transactions of the ASAE* 31:456–458, 1982.
70. GJ Hallman, JL Sharp. *Radio Frequency Heat Treatment: Quarantine Treatment for Pests of Food Plants*. San Francisco, CA: Westview Press, pp. 165–170, 1994.
71. S Wang, JN Ikediala, J Tang, JD Hansen, E Mitcham, R Mao, B Swanson. Radio frequency treatments to control codling moth in in-shell walnuts. *Postharvest Biology and Technology* 22:29–38, 2001.
72. S Wang, J Tang, JA Johnson, E Matcham, JD Hansen, RP Cavalieri, J Bower, B Biasi. Process protocols based on radio frequency energy to control field and storage pests in in-shell walnuts. *Postharvest Biology and Technology* 26:265–273, 2002.
73. JA Johnson, S Wang, J Tang. Thermal death kinetics of fifth-instar *Plodia interpunctella. Journal of Economic Entomology* 96:519–524, 2003.
74. S Wang, J Tang, JA Johnson, JD Hansen. Thermal death kinetics of fifth-instar *Amyelois transitella* (Walker) larvae. *Journal of Stored Products Research* 38:427–440, 2002.
75. S Wang, J Tang, RP Cavalieri, D Davis. Differential heating of insects in dried nuts and fruits associated with radio frequency and microwave treatments. *Transactions of the ASAE* 46:1175–1182, 2003.
76. S Wang, J Tang, T Sun, EJ Mitcham, T Koral, SL Birla. Considerations in design of commercial radio frequency treatments for postharvest pest control in in-shell walnuts. *Journal of Food Engineering* 77:304–312, 2006.
77. S Wang, M Monzon, JA Johnson, EJ Mitcham, J Tang. Industrial-scale radio frequency treatments for insect control in walnuts I: Heating uniformity and energy efficiency. *Postharvest Biology and Technology* 45:240–246, 2006.
78. A Wang, M Monzon, JA Johnson, EJ Mitcham, J Tang. Industrial-scale radio frequency treatments for insect control in walnuts II: Insect mortality and product quality. *Postharvest Biology and Technology* 45:247–253, 2006.
79. S Wang, SL Birla, J Tang, JD Hansen. Postharvest treatment to control codling moth in fresh apples using water assisted radio frequency heating. *Postharvest Biology and Technology* 40:89–96, 2006.
80. MC Lagunas-Solar, Z Pan, NX Zeng, TD Truong, E Khir, KSP Amaratunga. Application of radio frequency power for non-chemical disinfestations of rough rice with full retention of quality attributes. *Applied Engineering in Agriculture* 23(5):647–654, 2007.
81. S Wang, G Tiwari, S Jiao, JA Johson, J Tang. Developing postharvest disinfestations treatment for legumes using radio frequency energy. *Biosystems Engineering* 105:341–349, 2010.
82. NE Bengtsson, W Green. Radio-frequency pasteurization of cured hams. *Journal of Food Science* 35:681–687, 1970.
83. LJ Pircon, P Loquercio, DM Doty. High-frequency heating as a unit operation in meat processing. *Agriculture and Food Chemistry* 1(13):844–847, 1953.
84. CDC report. CDC estimates of foodborne illness in the United States. CEC 2011 Estimates: Findings. http://www.cdc.gov/foodborneburden/2011-foodborne-estimates.html (last accessed on January 5, 2012.)
85. K Luechapattanaporn, Y Wang, J Wang, J Tang, LM Hallberg, CP Dune. Sterilization of scrambled eggs in military polymeric trays by radio frequency energy. *Journal of Food Science* 71(4): 288–294, 2005.
86. L Zhang, JG Lyng, NP Brunton. Quality of radio frequency heated pork leg and shoulder ham. *Journal of Food Engineering* 75:275–287, 2006.

87. Y Liu, J Tang, Z Mao, JH Mah, S Jiao, S Wang. Quality and mold control of enriched white bread by combined radio frequency and hot air treatment. *Journal of Food Engineering* 104:492–498, 2011.
88. HR Sanders. Dielectric thawing of meat and meat products. *Journal of Food Technology* 1:183–192, 1966.
89. SO Nelson, CY Lu, LR Beuchat, MA Harrison. Radio-frequency heating of alfalfa seed for reducing human pathogens. *Transactions of the ASAE* 45(6):1937–1942, 2002.
90. Q Guo, P Piyasena, GS Mittal, W Si, J Gong. Efficacy of radio frequency cooking in the reduction of *Escherichia coli* and shelf stability of ground beef. *Food Microbiology* 23:112–118, 2006.
91. JG Lyng, DA Cronin, NP Brunton, W Li, X Gu. An examination of factors affecting radio frequency heating of an encased meat emulsion. *Meat Science* 75:470–479, 2007.
92. C Casals, I Vinas, A Landl, P Picouet, R Torres, J Usall. Application of radio frequency heating to control brown rot on peaches and nectarines. *Postharvest Biology and Technology* 58:218–224, 2010.
93. US FDA Center for Food Safety and Applied Nutrition. Kinetics of microbial inactivation for alternative food processing technologies. A report of the Institute of Food Technologists for the Food and Drug Administration of the U.S. Department of Health and Human Services. Washington, DC: USDA, 2000.
94. B Flugstad, E Kolbe, J Park, JH Wells, Y Zhao. Capacitive (RF) dielectric heating system for pasteurization, sterilization and thawing of various foods. US Patent # 6303166. Issued October 16, 2001.
95. Q Ling, E Kolbe, Y Zhao, B. Flugstad, JA Park. Variable frequency automated capacitive radio frequency (RF) dielectric heating system. US Patent # 6657173. Issued December 2, 2003.

20 Infrared Heating

Noboru Sakai and Weijie Mao

CONTENTS

20.1 INTRODUCTION

In thermal processing for the food industry, conduction and convective heat transfer have mainly been used; for radiant heat transfer, infrared radiation (IR) of short wavelength such as solar light and an IR lamp has also been used. Recently, interest in heating by far-infrared radiation (FIR) has increased due to the development of commercial FIR heaters, which have high emissivity in the IR region of long wavelength.

It is considered that meat and fish cooked by charcoal or bread baked in a stone oven are more delicious than those cooked or baked in a standard gas oven. However, these heating methods are not suitable in industrial production due to cost. Heating using FIR heaters is a heating method that is similar to that of a charcoal or stone oven, and thus, it is capable of producing more flavorful foods. In addition, recent developments in FIR equipment offer rapid and economical methods for manufacturing food products with high organoleptic and nutritional value.

In FIR heating, heat is supplied to the food by electromagnetic radiation from the FIR heaters. The rate of energy transfer between the heater and the food depends on the temperature difference between the heater and food. The FIR energy emitted from the heater passes through the air and absorbed by the food; the energy is then converted into heat by interactions with molecules in the food. Heat passes through the food from the surface layer by conduction. The related process of near-infrared radiation (NIR) heating is based on the same principle, using the appropriate wavelength. In contrast, in conventional heating the heat is mainly supplied to the surface of the food by convection from circulating hot air and products of combustion.

The difference in the heating mechanism between IR and conventional heating makes a difference in cost and in the quality of the products. The features of IR heating for food processing are as follows:

- Efficient heat transfer to the food reduces the processing time and energy costs.
- The air in the equipment is not heated, and consequently, the ambient temperature can be kept at normal levels.
- It is possible to design compact and automatic constructions with high controllability and safety.
- Exact heating control is required because there is a danger of overheating due to the rapid heating rates.

20.2 BASIC PRINCIPLES OF IR HEATING

IR energy is a form of electromagnetic energy. It is transmitted as a wave, which penetrates the food and is then converted to heat. IR radiation is classified as the region of wavelength between visible light (0.38–0.78 μm) and microwaves (MW) (1–1000 mm). Furthermore, IR radiation is divided into three classes [1] according to wavelength, as shown in Figure 20.1. However, the classification presented in Figure 20.1 has not been universally established. The wavelengths of the electromagnetic energy emitted from industrial FIR heaters are mainly in the range of 2.5–30 μm [2]. Hence, FIR heating usually means irradiative heating at a wavelength of 2.5–30 μm. The peak wavelength, at which the maximum energy is irradiated from the heater, is determined by the heater temperature. This relationship is described by the following basic laws.

20.2.1 PLANK'S LAW

When a black body is heated to a temperature *T*, electromagnetic energy is emitted from the surface. This energy has a definite distribution that is represented by Plank's law. Plank's law gives the energy flux irradiated from the black body in the wavelength range of λ to $\lambda + d\lambda$:

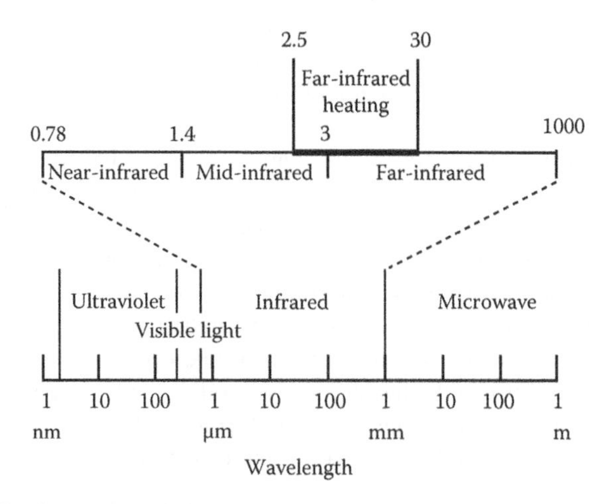

FIGURE 20.1 Classification of IR radiation.

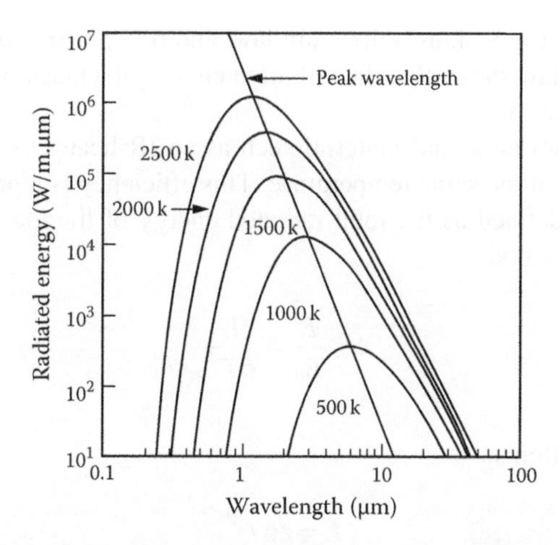

FIGURE 20.2 Spectral characteristics of black body radiation at different temperature.

$$E_\lambda = \frac{2\pi c^2 h}{\lambda^5}\frac{d\lambda}{\exp(ch/\kappa\lambda T)-1} = \frac{C_1 d\lambda}{\lambda^5\left[\exp(C_2/\lambda T)-1\right]} \tag{20.1}$$

$$C_1 = 3.7413\times10^{-16}\,\text{Wm}^2,\ C_2 = 1.4388\times10^{-2}\,\text{mK}$$

where

h is the Plank's constant
c is the velocity of light
κ is the Boltzmann's constant
T is the absolute temperature in Kelvin (K)

A plot of Equation 20.1 is given in Figure 20.2 and shows that the radiation energy increases as the temperature of a black body rises.

20.2.2 Wien's Displacement Law

It is proven from Figure 20.2 that the peak wavelength, λ_{max}, in which the radiation energy becomes a maximum, shortens with the rise in temperature. The relationship between λ_{max} and T can be determined by differentiating Equation 20.1 with respect to λ at constant T and setting the result equal to zero:

$$\lambda_{max} = \frac{2898}{T} \tag{20.2}$$

20.2.3 Stefan–Boltzmann Law

The total emissive energy radiated from a black body with temperature T is given by the integral of Equation 20.1 over all wavelengths:

$$E = \int_0^\infty E_\lambda\, d\lambda = \frac{2\pi\kappa^4 T^4}{c^2 h^3}\left(\frac{\pi^4}{15}\right) \equiv \sigma T^4 \tag{20.3}$$

This equation is known as the Stefan–Boltzmann law, and σ (=5.670×10^{-8} W/mK^{4}) is the Stefan–Boltzmann constant. This law shows that the radiation energy of a black body is proportional to the fourth power of its temperature.

The energy emitted from an actual material such as an IR heater or food is smaller than that emitted from a black body at the same temperature. This efficiency is represented as the emissivity of the material, which is defined as the total radiated energy of the material divided by the total radiated energy of the black body:

$$\varepsilon = \frac{E}{E_b} = \frac{E}{\sigma T^4} \tag{20.4}$$

The equation can be rewritten as

$$E = \varepsilon \sigma T^4 \tag{20.5}$$

where ε is a number between 0 and 1, and materials or bodies obeying this equation are called gray bodies.

A black body also absorbs the energy according to the same law, Equation 20.1, and gray bodies absorb a fraction of the quantity that a black body would absorb, corresponding to their absorptivity, β. Under thermodynamic equilibrium, the monochromatic emissivity and absorptivity of a body are equal from Kirchhoff's law. The fraction of the incident energy that is not absorbed is reflected, corresponding to the reflectivity, r_f. The relationship between these fractions is shown by the following equation:

$$\varepsilon = \beta = 1 - r_f \tag{20.6}$$

As the temperature of the heater is raised, the amount of radiant energy delivered to the food increases and the peak wavelength decreases. Since the maximum running temperature of FIR heaters, which are usually made from fine ceramics (Figure 20.3a), is 600–950 K, the peak wavelength is 3–5 μm [3]. In NIR heating, the maximum temperature for a short-wave radiator such as a tungsten filament lamp (Figure 20.3b) is 2400–2500 K, which corresponds to a peak wavelength of 1.1–1.3 μm [4].

When radiant electromagnetic energy impinges upon a food surface, it may induce changes in the electronic, vibrational, and rotational states of atoms and molecules. The kinds of mechanisms for energy absorption are determined by the wavelength range of the incident irradiative energy [2]: changes in the electronic state correspond to wavelengths in the range of 0.2–0.7 μm (ultraviolet [UV] and visible light); changes in the vibrational state correspond to wavelengths in the range of 2.5–100 μm (FIR); and changes in the rotational state correspond to wavelengths above 100 μm (MWs). Food is generally composed of water and organic compounds such as carbohydrates,

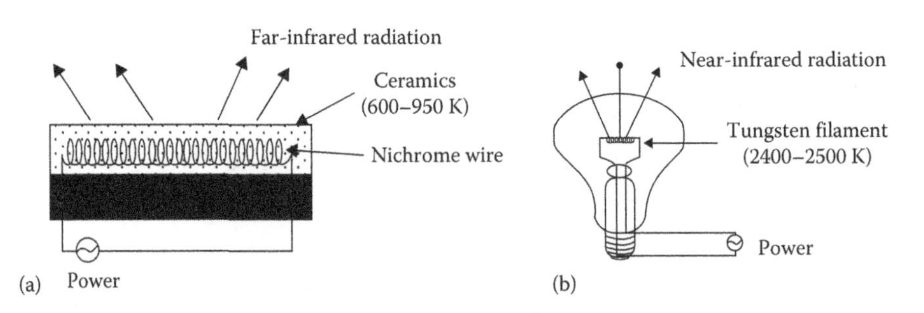

FIGURE 20.3 Schematic diagrams of FIR heater and IR ramp. (a) Ceramic FIR heater. (b) IR ramp.

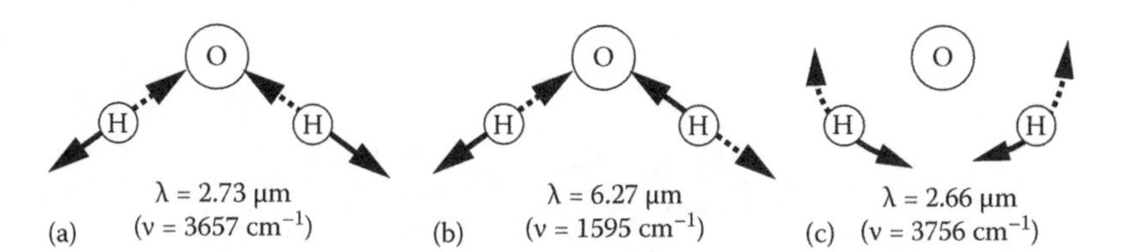

FIGURE 20.4 Normal vibration of water: (a) symmetrical stretching vibration, (b) asymmetrical stretching vibration, and (c) variable angle vibration.

proteins, and fat, and the IR absorption characteristics of these materials are important. The normal vibration of water is represented as follows [2]: (1) the symmetrical stretching vibration, (2) the antisymmetric stretching vibration, and (3) the symmetrical deformation vibration, as shown in Figure 20.4. These vibrational frequencies, namely, the wave number, are shown in Figure 20.4. The IR energy in proportion to these frequencies is efficiently absorbed in a body. Food therefore absorbs the IR energy at wavelengths greater than 2.5 μm most efficiently through the mechanism of changes in the molecular vibrational state, which can lead to heating.

20.3 ATTENUATION FACTOR AND PERMEABILITY

The penetration of IR energy into the food is an important factor to discuss with regard to IR heating.

The incident energy is partly reflected and transmitted at the air/food boundary. The energy transmitted into the food is attenuated exponentially with the penetration distance. The attenuation factor determines the energy absorption within the food as a function of depth from the surface of the food, as described by Lambert's law:

$$I_\lambda = I_{\lambda 0} \exp\left(-\alpha_\lambda x\right) \tag{20.7}$$

where

I_λ is the energy flux at the wavelength of λ

α_λ is the spectral attenuation factor

The attenuation factor of water, which is shown in Figure 20.5, changes remarkably with wavelength. For example, $\alpha_\lambda = 0.355$, 10,800, 2138 cm^{-1} when $\lambda = 1$, 3, 6 μm [5], respectively.

The spectral transmittance, determined as a ratio of the incident energy and its transmission, is as follows:

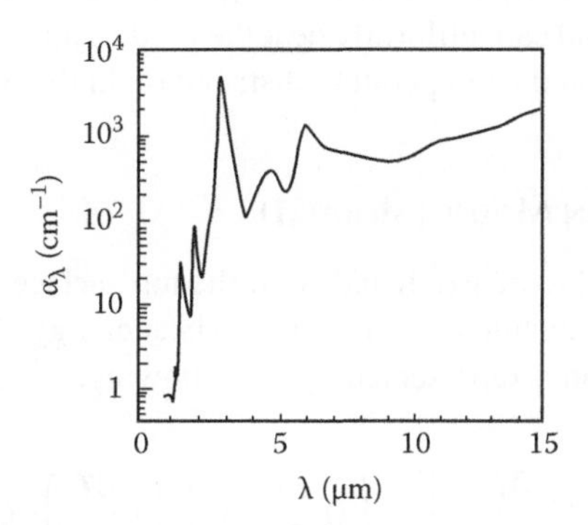

FIGURE 20.5 Absorption coefficients of water.

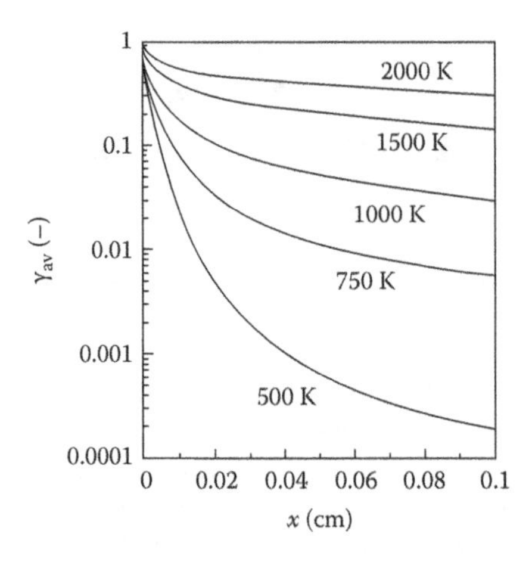

FIGURE 20.6 Average permeability of energy emitted from different temperature radiator.

$$\gamma_\lambda = \frac{I_\lambda}{I_{\lambda 0}} = \exp(-\alpha_\lambda x) \tag{20.8}$$

From the previous equation, the penetration distances, in which the spectral transmittance becomes 0.01, are 12.97 cm, 4.26×10^{-4} cm, and 2.15×10^{-3} cm, for $\lambda=1$, 3, and 6 μm, respectively. These results indicate that the permeability of the IR energy is greatly dependent on the wavelength.

Since the energy irradiated from an IR heater has distribution, the average permeability must be calculated by integrating the equation over the full wavelength:

$$\gamma_{av}(x) = \frac{\int_0^\infty I_\lambda \, d\lambda}{\int_0^\infty I_{\lambda 0} \, d\lambda} = \frac{\int_0^\infty \exp(-\alpha_\lambda x) \, d\lambda}{\sigma T^4} \tag{20.9}$$

The average permeability changes corresponding to the heater temperature, since the emitted energy $I_{\lambda 0}$ changes with temperature. Figure 20.6 shows the calculated average permeability in assuming the attenuation factor to be the value of water. It is proven from Figure 20.6 that the permeability of IR decreases as the radiator temperature is reduced.

As the features of FIR, there are reports of two different standpoints: (1) the energy is almost entirely absorbed in the surface vicinity and the penetration can be neglected and (2) the energy permeates to the food inside and can uniformly heat the food. Using two different models, the effect of the permeability of the IR on the temperature distribution in the food can be shown [6].

20.3.1 Energy Penetration Model (Model 1)

The IR energy was considered to be irradiated from the top surface of a cylindrical sample and to penetrate in the axial direction, while it is converted to heat energy. A heat transfer equation with a term of internal heat generation is represented by the following:

$$\rho C_p \frac{\partial T}{\partial t} = \frac{1}{r}\frac{\partial}{\partial r}\left(kr\frac{\partial T}{\partial r}\right) + \frac{\partial}{\partial z}\left(k\frac{\partial T}{\partial z}\right) + Q \tag{20.10}$$

where Q is the internal heat generation term, which can be described by the following based on Lambert's law:

$$Q = -\frac{dI}{dx} = \int_0^\infty \alpha_\lambda I_{\lambda 0} \exp(-\alpha_\lambda) d\lambda \tag{20.11}$$

The surface energy flux with a wavelength of λ, $I_{\lambda 0}$, is represented by the following based on Plank's law:

$$I_{\lambda 0} = \phi \left[\frac{2\pi c^2 h}{\lambda^5} \frac{d\lambda}{\exp(ch/\kappa\lambda T_h) - 1} - \frac{2\pi c^2 h}{\lambda^5} \frac{d\lambda}{\exp(ch/\kappa\lambda T_f) - 1} \right] \tag{20.12}$$

where ϕ is the overall absorption coefficient, which can be estimated from a radiation shape factor, F_{21}, the emissivity of the IR heater, ε_1, and the emissivity of food, ε_2, as follows [6]:

$$\phi = \varepsilon_1 \varepsilon_2 F_{21} \tag{20.13}$$

20.3.2 Surface Absorption Model (Model 2)

It is assumed that all of the irradiated energy is converted to thermal energy at the surface without permeating to the inside of the food. The fundamental equation is represented by the simple heat conduction equation:

$$\rho C_p \frac{\partial T}{\partial t} = \frac{1}{r} \frac{\partial}{\partial r} \left(k_t r \frac{\partial T}{\partial r} \right) + \frac{\partial}{\partial z} \left(k_t \frac{\partial T}{\partial z} \right) \tag{20.14}$$

The heat generation term by IR radiation is included in the boundary condition using the Stefan–Boltzmann law:

$$k_t \frac{\partial T}{\partial z} = -h_t (T - T_a) + \sigma\phi (T_h^4 - T^4) \tag{20.15}$$

To compare the two models, numerical calculation was performed by using a finite element method to solve the fundamental equation. Figure 20.7 shows the temperature history of the cylindrical sample at the points of axial direction. The calculation conditions were as follows: the IR heater temperature, 500 K; the ambient temperature, 293 K; the initial sample temperature of sample, 293 K; and the overall absorption coefficient, 0.4. The lines and points in Figure 20.7 represent the calculated temperature based on the energy penetration model (Model 1) and that based on the surface absorption model (Model 2), respectively. In the energy penetration model, the attenuation factor was adopted from the value of water. The calculation results according to the two models almost agree, though there are some differences in the surface temperatures. These results indicate that at the temperature range for which FIR is mainly irradiated, the heat source term can be included in the boundary condition. This boundary condition could be used in the modeling of heat transfer in the food heated by FIR [7].

In assuming the IR heater temperature to be 1500 K, the temperature histories of a cylindrical sample are shown in Figure 20.8. The overall absorption coefficient was assumed to be 0.02 so that the temperature rise may be similar to that of 500 K. In the case of a high-temperature radiator such

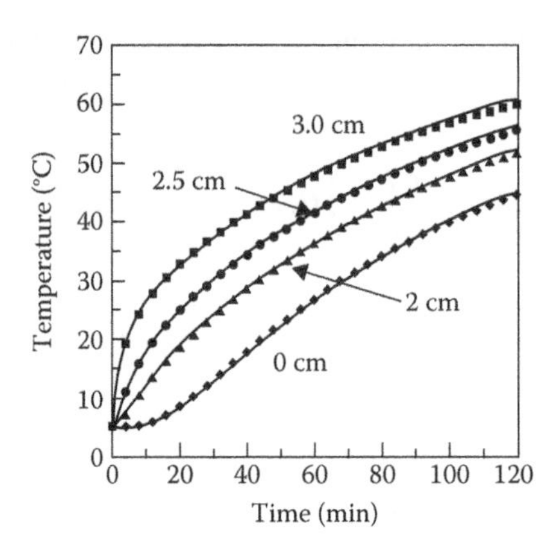

FIGURE 20.7 Comparison of calculated temperature histories between the two models at 500 K (——: Model 1, ■●▲◆: Model 2).

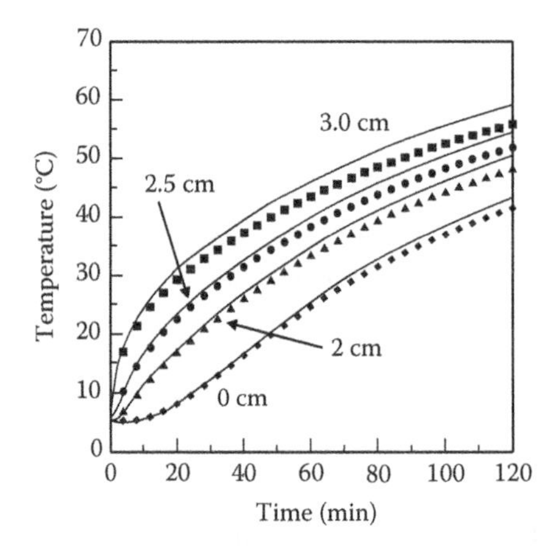

FIGURE 20.8 Comparison of calculated temperature histories between the two models at 1500 K (——: Model 1, ■●▲◆: Model 2).

as an IR lamp, the overall absorption coefficient is small because the shape of the heat source is mainly a thin wire (filament), and the radiating area is small. In the penetration model (Model 1), the surface temperature is lower than that of Model 2, and the difference between the central and surface temperatures decreases since the energy permeates the food.

This difference in the permeability of IR greatly affects the quality of the food during the heating process. In the cooking or baking process, it is important to properly control physical changes such as crust formation, protein denaturation, and fat melting as well as the burning colors of the surface and the moisture profile of the food. The rapid surface heating of foods by FIR can be used to seal in moisture (or water), fat, and aroma compounds without burning the surface, and results in products with highly acceptable sensory characteristics.

20.4 APPLICATION

20.4.1 Baking, Roasting, and Grilling

Recently, novel technologies such as microwave–infrared combination baking (MIR) and air jet impingement–infrared radiation combination (AJIR) have been used for baking bread [8,9], cake [10], and rice cakes [11]. By using MIR baking, the time-saving advantage of MW baking was combined with the surface browning advantage of IR heating.

20.4.1.1 Cake Baking

Sumnu et al. [12] compared the effects of different baking methods: MW, IR, and the MIR. An MIR oven (Advantium oven TM, General Electric Company, Louisville, KY) was used for this study, as shown in Figure 20.9. For MW baking, the cake samples were baked with only the MW operating mode at 50% power for at 50% power for 3.25, 3.50, 3.75, and 4.00 min, using a 760 W MW oven. For IR baking, the cake samples were baked with only the halogen lamp operating mode at 50% power for 9.0, 9.5, 10.0, and 10.5 min and at 70% power for 7.0, 7.5, 8.0, and 8.5 min. Halogen lamp heating provides near IR radiation (wavelength 0.7–5 μm). For the MIR baking, cake samples were baked at 50% MW power and 50% and 70% IR power levels for 3.0, 4.0, 5.0, and 6.0 min. Surface color was an important factor for evaluating the effect of baking and was measured using a Minolta Color Reader (CR-10, Japan) using the IE, L^*, a^*, and b^* color scale. Total color change (ΔE) was calculated from Equation 20.16:

$$\Delta E = \left[\left(L^* - L_0^*\right)^2 + \left(a^* - a_0^*\right)^2 + \left(b^* - b_0^*\right)^2\right]^{1/2} \tag{20.16}$$

where subscript 0 represents the initial value. The IR power was found to be a significant factor affecting the ΔE value of the cakes (Figure 20.10). Similarly, when IR heating was combined with MW heating, the IR power was found to be a significant factor in affecting the ΔE value of cakes. Higher IR power browned the surface of cakes within a shorter time. The specific volume of baked cake samples was determined by the rape seed displacement method. The specific volume of cakes increased as the baking time increased (Figure 20.11). cakes baked by MWs had less specific volume than the IR-baked cakes. Since IR heating provided focusing of radiation at the surface, the crust forming at the surface of the samples retarded the expansion of the crumb. Crumb firmness of cake samples was measured using a universal testing machine (Lloyd Instruments LR 30K, United Kingdom) after the cake samples had been cooled for 1 h and compressed by 25% of their original thicknesses at a speed of 55 mm/min. Very low firmness values in cakes baked in an IR oven showed that the cake structure was not developed yet, and the cake was still an unbaked form (Figure 20.12). Although the crumb texture was very soft, a very thick crust was observed in these cakes. By combining MWs with IR heating, the firmness of MW-baked cakes

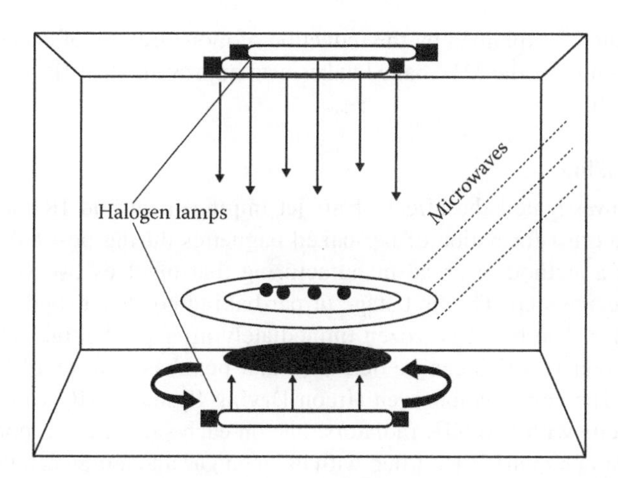

FIGURE 20.9 Schematic diagram of an MIR oven. (From Sumnu, G. et al., *J. Food Eng.*, 71, 150, 2005.)

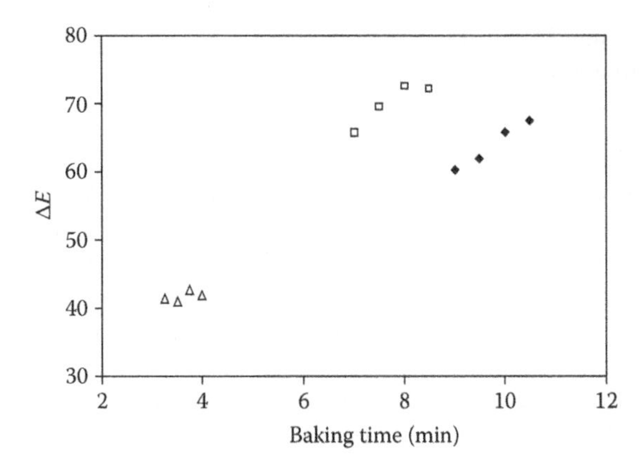

FIGURE 20.10 Changes in color (ΔE value) of cakes during baking by different methods. (△): MW baking at 50% power, (□): IR baking at 50% power, (♦): IR baking at 70% power. (From Sumnu, G. et al., *J. Food Eng.*, 71, 150, 2005.)

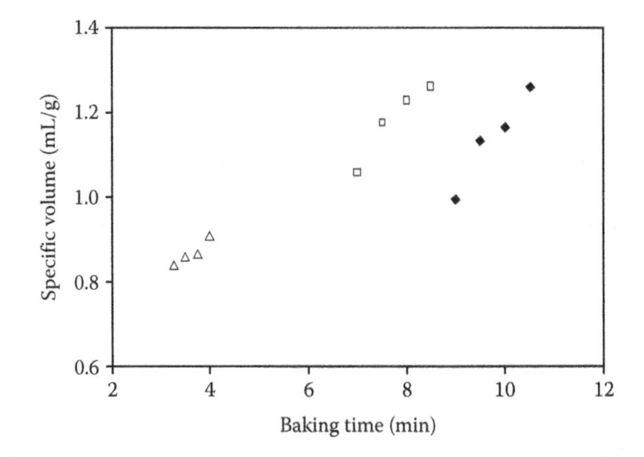

FIGURE 20.11 Changes in specific volume of cakes during baking by different methods. (△): MW baking at 50% power, (□): IR baking at 50% power, (♦): IR baking at 70% power. (From Sumnu, G. et al., *J. Food Eng.*, 71, 150, 2005.)

was reduced. This can be explained by the reduction of moisture loss of cakes in the presence of IR radiation. The quality of the MW-baked cakes was improved when IR heating was combined with MW heating [12].

20.4.1.2 Bread Baking

Olsson et al. [13] investigated the effect of air jet impingement and IR radiation (alone or in combination) on the crust formation of par-baked baguettes during post-baking. Par-baking (or partially baking) is a method of bread manufacturing that involves two stages of baking with an intermediate freezing step. The first stage of par-baking creates a rigid bread structure with a minimal crust color. The bread is frozen immediately after par-baking. The par-baked bread is a semifinished product, which can be delivered and post-baked at retail locations to provide fresh-baked bread. The combination oven (Ircon Drying Systems AB, Vänersborg, Sweden) is composed of a cassette with two NIR radiators, one on each side of a slot nozzle (Figure 20.13). The radiators consist of a quartz tube filled with halogen gas and a tungsten thread. To create the impinging air jet, a fan and a heater are connected to the oven. The maximum average velocity

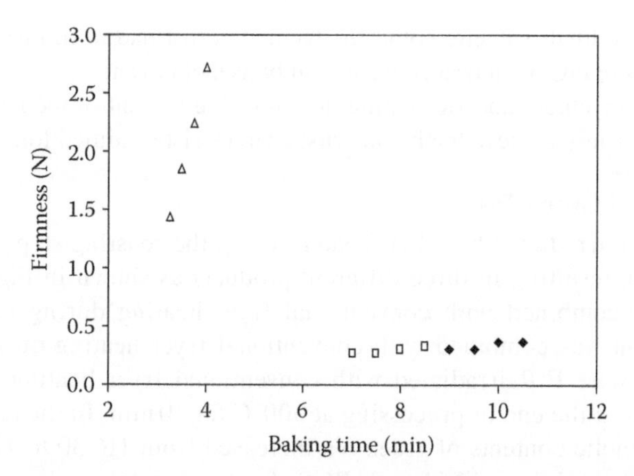

FIGURE 20.12 Changes in specific volume of cakes during baking by different methods. (△): MW baking at 50% power, (□): IR baking at 50% power, (◆): IR baking at 70% power. (From Sumnu, G. et al., *J. Food Eng.*, 71, 150, 2005.)

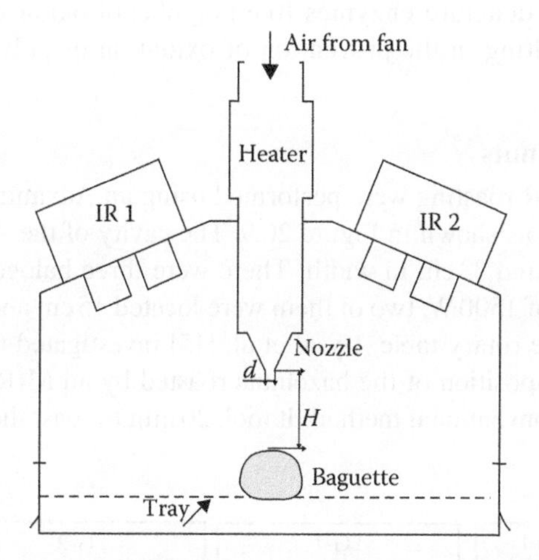

FIGURE 20.13 The IR-impingement combination oven. (From Olsson, E.E.M. et al., *J. Food Sci.*, 70, 484, 2005.)

of the jet exit is about 7.5 m/s. The maximum temperature is over 2100°C, which corresponds to a wavelength peak of 1.2 μm. The crust color, crust thickness, total water loss, and heating time were examined.

IR radiation and air jet impingement are rapid heating methods that increase the rate of color development and shorten the total heating time. The crumb temperature rose to a high temperature earlier in the case of IR heating, most likely due to the penetration properties of the IR radiation [13]. The high heat transfer rates in impingement resulted in high mass transfer rates. Lower total water loss was also generally observed for IR-heated bread than for the impingement-heated bread [13].

IR heating caused slow initial color development, but after a lag period, the crust temperature reached a sufficient temperature and color, the color development was rapid. In general, higher color development was observed for IR heating than impingement heating, and both were higher than for conventional heating. IR in combination with impingement heating was found to increase the color development rate and shorten the heating time when the air jet had sufficiently high temperature and velocity.

To achieve the same final baguette color, the baguette crust had to thinner when heated by IR than when heated by impingement or a conventional household oven.

By combining impingement and IR heating, it is possible to control the heat and mass transfer processes and consequently achieve the bread crust characteristics aimed for.

20.4.1.3 Roasting of Green Tea

Green tea leaves were irradiated by a FIR heater during the roasting step and irradiated again after the drying step, resulting in three different products as shown in Figure 20.14. In FG-1, FIR irradiation was combined with conventional fryer heating during the roasting step. In FG-2, FIR irradiation was combined with conventional fryer heating during the drying step. In FG-3, tea leaves were FIR irradiated with conventional fryer heating during the roasting step, and additionally at the end of processing at 300°C for 10 min. In the technological process (FG-3), the total phenolic contents of green tea increased from 116.30 to 171.77 mg/g; the total flavanol contents increased from 17.54 to 24.76 mg/g; the ascorbic acid contents also increased from 3.07 to 4.20 mg/g; and the amounts of epicatechin gallate and epigallocatechin gallate were significantly increased from 2.41 to 4.59 mg/mL and 20.61 to 28.54 mg/mL, respectively, compared with the nonirradiated green tea (control). These results indicate that the FIR irradiation in the roasting step might denature enzymes like polyphenol oxidases (PPO) more effectively than general heating, resulting in the prevention of oxidation or polymerization of polyphenol compounds [14].

20.4.1.4 Roasting Hazelnuts

Both MW roasting and MIR roasting were performed using an Advantium oven (General Electric Company, Louisville, KY), as shown in Figure 20.9. The cavity of the Advantium oven was 21 cm in height, 48 cm in length, and 33 cm in width. There were three halogen lamps in the oven, each having a maximum power of 1500 W, two of them were located 15 cm above the food surface, while the other was just under the rotary table. Uysal et al. [15] investigated the color, texture, moisture content, and fatty acid composition of the hazelnuts roasted by an MIR oven and compared them with those roasted by the conventional method. It took 20 min to roast the hazelnuts conventionally,

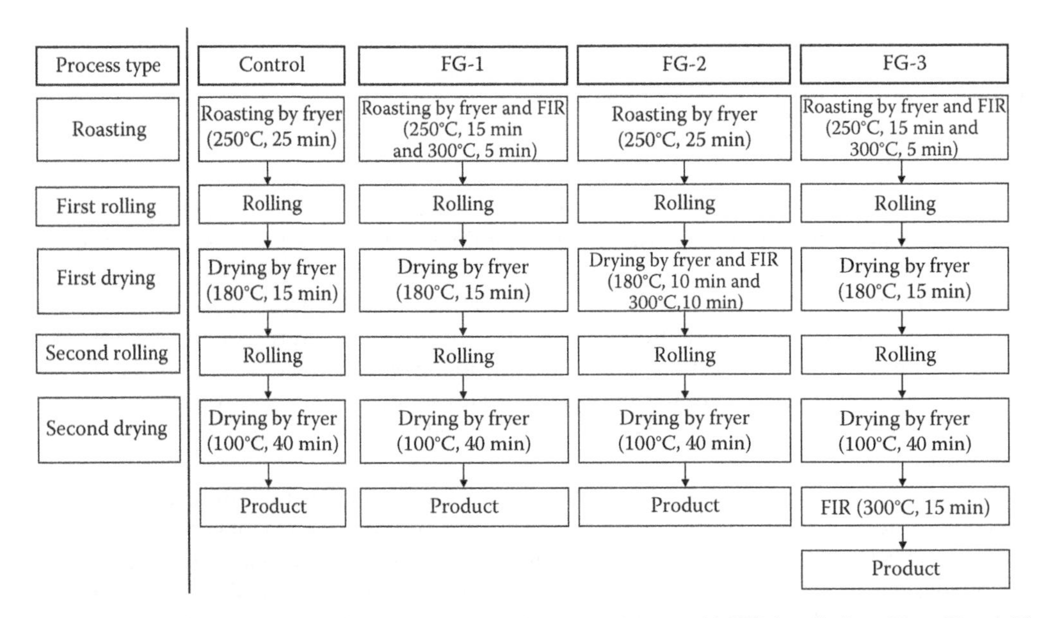

FIGURE 20.14 Diagram of manufacturing process of green tea leaves with FIR irradiation. (From Uysal, N. et al., *J. Food Eng.*, 90, 255, 2009.)

but just 2.5 min when using the MIR oven. It was possible to obtain a similar quality and fatty acid composition by using conventional and MIR ovens.

20.4.1.5 Grilling Meat Patties

Brackman et al. studied the influence of combined IR-grilling/hot air cooking processing conditions on the selected quality and sensory attributes of hamburgers [16]. The purpose of grilling is to develop a crust-like texture and color, while the remainder of the cooking process is performed in a convection oven. For grilling, a lab-scale IR-grill (Afoheat, Veurne, Belgium), consisting of a propane fired metal-fiber burner with a maximum power of 16 kW, was used. The relation between the duration of IR-grilling and the weight and fat loss of the meat patties was studied. During the short grilling stage water and fat are partly lost as shown in Figure 20.15. The longer the grilling stage, the higher the temperature of the hamburger at the point where it exits the grill and enters the convection oven, thus effectively reducing the cooking time, which was defined as the time required to reach a minimum of 72°C in the center of the meat patty's. Finally, the longer the grilling stage, the more developed the crust prior to convection cooking, thus potentially preventing moisture and fat loss during the subsequent convection cooking stage. Compared to only the grilling of the hamburgers, the effect of single-sided grilling time on the moisture, total weight and fat loss is reduced when grilling is followed by convection cooking (Figure 20.16) [16].

20.4.2 Drying

IR drying has been investigated as a potential method for obtaining high-quality dried foodstuffs including fruits [17,18], vegetables [19], and grains [20]. Chua and Chou compared both intermittent MW and IR drying for potatoes and carrots and found that step variation of the IR intensity during intermittent drying reduced the color change of the product compared to constant radiative, intermittent IR drying [20]. Figure 20.17 shows the declining step variation of the IR intensity for $\alpha = 1/4$, where $\alpha = T_{on}/(T_{on} + T_{off})$, and T_{on} and T_{off} represent the on-period and off-period of IR lamps, respectively. Conversely, a step-up IR intensity profile can be employed by reversing the varying intensity direction. It can be observed from Figure 20.18 that using a variable step-down IR intensity drying schedule resulted in benefit by giving a shorter drying time for the carrot samples and was thus beneficial. Figure 20.19 shows the color degradation (ΔE) for carrot samples under step-up, step-down, and constant IR radiation with $\alpha = 1/4$. The results for the carrot samples indicated that

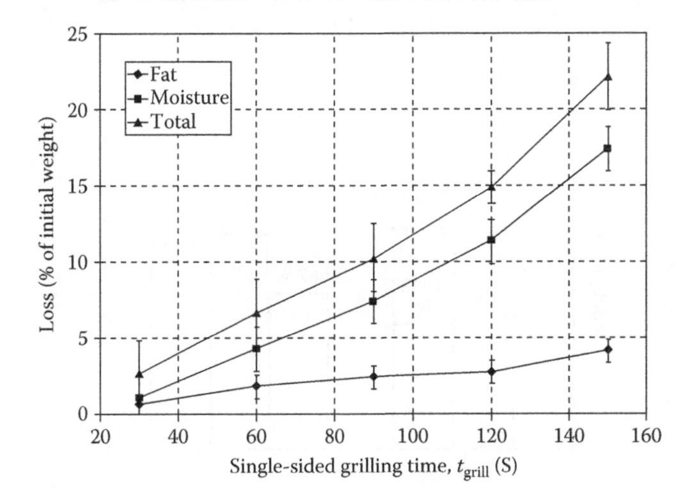

FIGURE 20.15 Influence of total grilling time (s) on fat, moisture and total weight loss (% of initial weight) of the meat patties. (From Braeckman, L. et al., *J. Food Eng.*, 93, 437, 2009.)

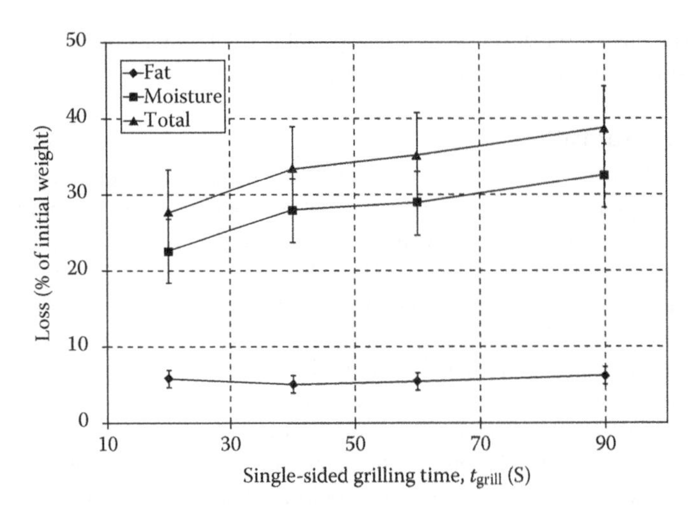

FIGURE 20.16 Influence of the single-sided grilling time (s) on fat, moisture, and total weight loss (% of initial weight) of the meat patties during combined grilling and subsequent cooking in a convection oven. (From Braeckman, L. et al., *J. Food Eng.*, 93, 437, 2009.)

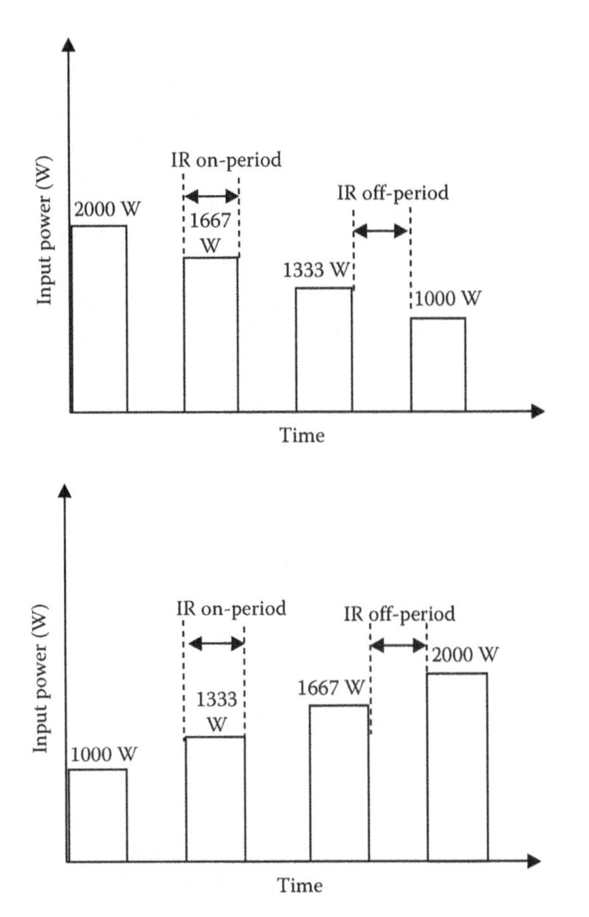

FIGURE 20.17 Step-down and step-up regulation of IR intensity. (From Chua, K.J. and Chou, S.K., *Int. J. Food Sci. Technol.*, 40, 23, 2005.)

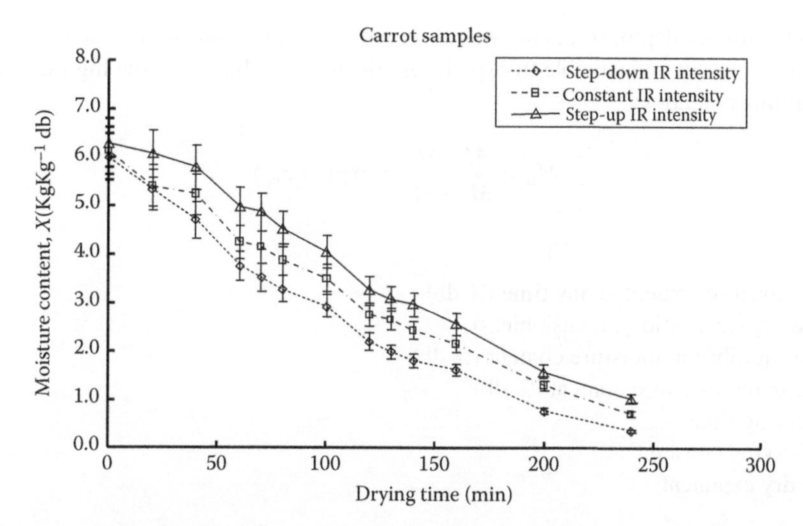

FIGURE 20.18 Drying kinetics of the agro-samples under stepwise variation of IR intensity. (From Chua, K.J. and Chou, S.K., *Int. J. Food Sci. Technol.*, 40, 23, 2005.)

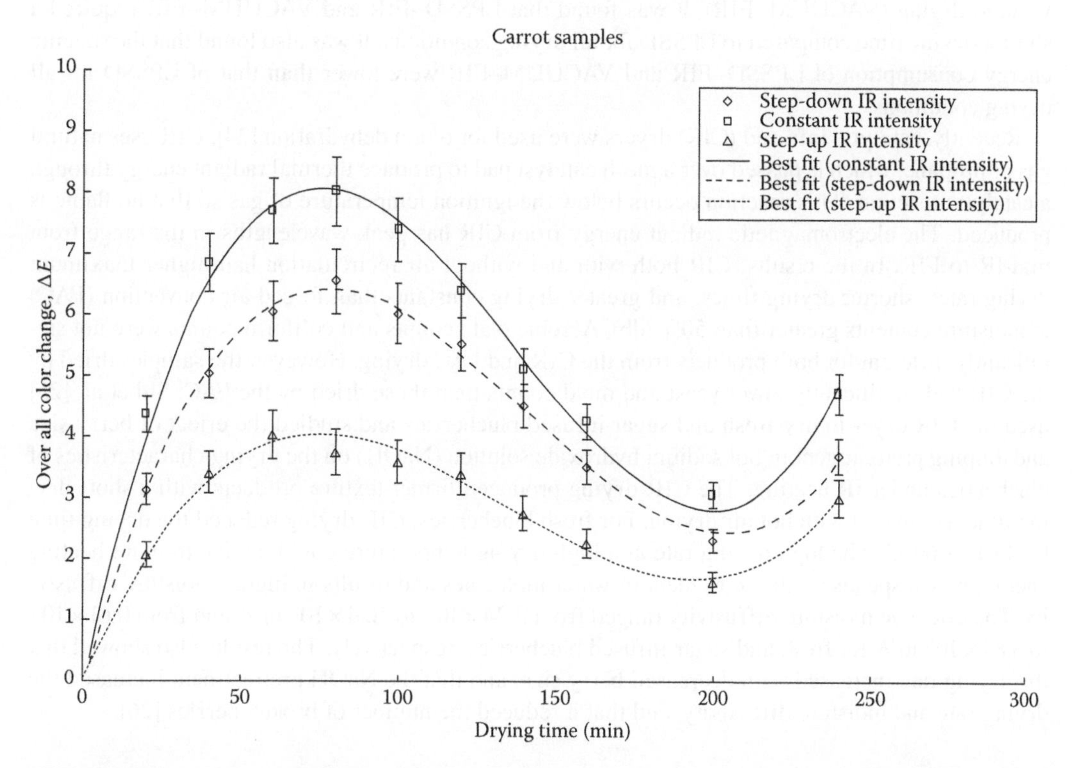

FIGURE 20.19 Net color change (ΔE) of the agro-samples under stepwise variation of IR intensity (with 95% confidence intervals indicated). (From Chua, K.J. and Chou, S.K., *Int. J. Food Sci. Technol.*, 40, 23, 2005.)

the color reduction arising from employing step-down IR intensity drying and step-up IR intensity drying was approximately 15.2% and 18.1%, respectively [21].

Although IR heating provides a rapid means of heating and drying, it is attractive only for surface heating application. Das et al. [22] developed a prototype vibratory dryer coupled with an IR heating source to drying high-moisture paddy. The drying rate was found to be dependent on both radiation intensity and grain bed depth. IR drying was observed to be falling rate process and the Page Model (Equation 20.17) was found suitable to explain the IR drying behavior of the high-moisture paddy under a vibrating condition:

$$M_R = \frac{M - M_e}{M_0 - M_e} = \exp\left(-k_3 t_{n_1}\right) \tag{20.17}$$

where

M is the moisture content at any time (% db)
M_R is the moisture ratio (dimensionless)
M_e is the equilibrium moisture content (% db)
M_0 is the initial moisture content (% db)
t is the drying time
k_3 is the drying constant
n_1 is the dry exponent

Nimmol et al. [23] investigated the use of a drying system combining low-pressure superheated steam and FIR (LPSSD–FIR) for banana and compared the drying system with combined FIR and vacuum drying (VACUUM–FIR). It was found that LPSSD–FIR and VACUUM–FIR required a shorter drying time compared to LPSSD for all drying conditions. It was also found that the specific energy consumption of LPSSD–FIR and VACUUM–FIR were lower than that of LPSSD for all drying conditions.

Recently, catalytic infrared (CIR) dryers were used for onion dehydration [24], CIR uses natural gas or propane, which is passed over a mesh catalyst pad to produce thermal radiant energy through a catalytic reaction. This reaction occurs below the ignition temperature of gas so that no flame is produced. The electromagnetic radiant energy from CIR has peak wavelengths in the range from mid-IR to FIR. In the results, CIR both with and without air recirculation had higher maximum drying rates, shorter drying times, and greater drying constants than forced air convection (FAC) at moisture contents greater than 50% (db). Aerobic plate counts and coliform counts were not significantly different for both products from the CIR and FAC drying. However, the samples dried by the CIR had significantly lower yeast and mold counts than those dried by the FAC. Shi et al. [25] used the CIR dryer to dry fresh and sugar-infused blueberries and studied the effect of berry size and dipping pretreatment in hot sodium hydroxide solution (NaOH) on the drying characteristics of blueberries under IR heating. The CIR drying produced firmer-texture products with a short drying time, compared with hot air drying. For fresh blueberries, CIR drying reduced the drying time by 44% at 60°C. The high drying rate at a high drying temperature could be due to more heating energy, which speeds up the movement of water molecules and results in higher moisture diffusivity. The effective moisture diffusivity ranged from 2.24×10^{10} to $16.4 \times 10^{10}\,m^2/s$ and from 0.61×10^{10} to $3.84 \times 10^{10}\,m^2/s$ for fresh and sugar-infused blueberries, respectively. The results also showed that the drying rate increased with decreased berry size, and that the NaOH pretreatment increased the drying rate and moisture diffusivity, and that it reduced the number of broken berries [26].

20.4.3 Dry-Blanching

A new simultaneous infrared dry-blanching and dehydration (SIRDBD) method utilizes IR dry heat to blanch fruits and vegetables and simultaneously remove a certain amount of moisture [27]. It is a new approach for blanching and partially dehydrating fruits and vegetables, such as onions [28], cashew kernels [29], and apples [30] resulting in high product quality. The processing parameters, radiation intensity, and slice thickness showed significant effects on the process characteristics and product quality of apple slices [31]. Thinner slices lead to faster inactivation of enzymes and quicker moisture reduction [32]. High radiation intensity and/or thin slices had a faster increase of product surface and center temperatures, and quicker moisture reduction and enzyme inactivation than with low radiation intensity and/or thick slices. Peroxidase (POD) was more heat resistant than PPO. The surface color change of the apple slices during the process was primarily due to the decrease of the L^* value and increase of the a^* value, corresponding to the enzymatic browning that occurred during the process, where L^* and a^* stand for whiteness or brightness/darkness and redness/greenness, respectively. For prolonged heating, continuous heating under high radiation intensity could result in severe discoloration of the apple slices.

20.4.4 Pasteurization

IR heating has also been widely used for pasteurizing foods, such as strawberry [33] and turkey frankfurters [34] to reduce or eliminate pathogens or other spoilage microorganisms. Huang and Sites [35] developed an IR pasteurization process with automatic temperature control for inactivation of surface-contaminated *Listeria monocytogenes* on ready-to-eat meats such as hotdogs. The temperature of the hotdog surfaces was effectively controlled by the feedback control mechanism during the IR pasteurization process; overcooking and burning of the hotdogs was prevented. IR was explored as a technique for decontaminating paprika powder [36,37]. For powder with an a_w value of 0.88 heated by near-IR of 20 kW/m^2 to 95°C–100°C, the load of *Bacillus cereus* spores was reduced by 4.5 $\log_{10}$ CFU/g within 6 min; the final spore concentration remained approximately 2 $\log_{10}$ CFU/g due to higher D values on the surface. Therefore, IR could be used to reduce microbial numbers in the paprika powder [37].

Compared with thermal conductive heating, FIR heating was more effective for inactivating the vegetative bacteria cells, such as *Escherichia coli* and *Staphylococcus aureus*. Hashimoto, Sawai, Igarashi, and Shimizu [38] observed that the bacteria were impaired and killed by FIR even when the bulk of the suspension was kept below the lethal temperature. They have investigated the effects of FIR radiation on *E. coli* and *S. aureus* in phosphate-buffered saline under the condition that the bulk of the suspension was kept below 40°C by cooling. It should be noted that FIR heating is more effective for pasteurization than NIR [39], whereas both FIR and NIR irradiation caused the inactivation of *Bacillus subtilis* spores [40]. For both types of IR irradiation, the inactivation effect increased with decreasing spore suspension volume; that is, inactivation increased with a decrease in the thickness of the liquid layer. However, the pattern of the decrease differed greatly for both types of irradiation. Irradiation to spores using the NIR heater resulted in a gradual decrease in the number of colonies as irradiation time increased (Figure 20.20b). In contrast, a rapid increase in colony counts was observed at the beginning of the FIR irradiation (Figure 20.20a). The increase in the colony counts of the spores shown in Figure 20.20a is called activation (heat activation). FIR irradiation caused spore activation at a bulk temperature of only 40°C–60°C, whereas this temperature range is extremely low for heat activation. By contrast, thermal conductive heating at 75°C for 1–2 h did not cause heat activation of *B. subtilis* as shown in Figure 20.20a. Hashimoto et al. also examined the germination, and they showed that the germination rate of spores markedly increased after FIR irradiation as shown in Figure 20.21. In contrast, the germination rate for spores irradiated by the NIR heater was almost the same as that of the nonirradiated spores (irradiation time of zero). These results suggest that FIR irradiation activates *B. subtilis* spores, resulting in an increase in colony counts. Figure 20.22 shows the transmission IR spectrum of *B. subtilis* spores by microinfrared spectroscopy. Large absorption bands appeared at around 1650 cm^{-1} (6.0 mm) and 1550 cm^{-1} (6.4 mm), which are similar to the main wavelengths of the FIR heater. Many bioactive substances have absorptions around 6 mm. The amide I band around 1650 cm^{-1} (6.0 mm) and the amide II band around 1550 cm^{-1} (6.4 mm) [41] provide information about the average secondary structure of the protein and its a-helical character, respectively. Dipicolinic acid (DPA), which is synthesized during the late phase of spore formation and accumulates in the spore, has an absorption band at 1570 cm^{-1} [42].The main wavelength of the FIR heater was 3–7 µm, corresponding to the absorption IR spectra of the spore. The radiation energy produced by the FIR heater was easily absorbed by the spore cell components and activated the bioactive substances involved in germination; once the spores

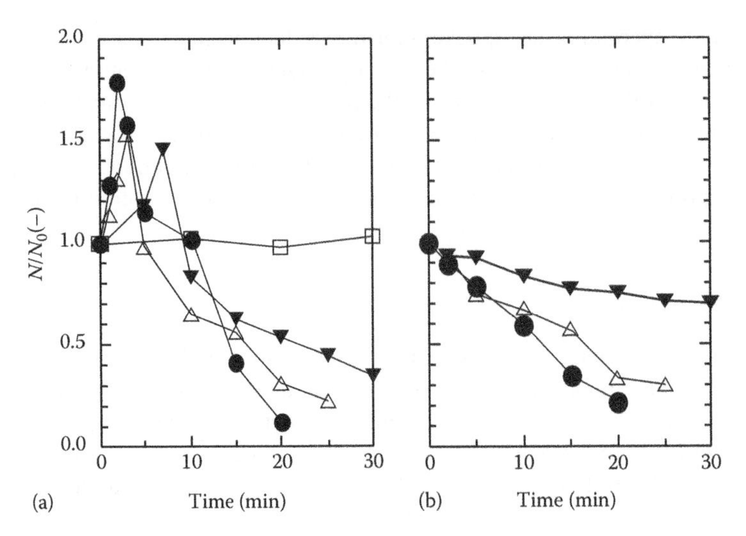

FIGURE 20.20 Effect of irradiation on *B. subtilis* colony counts. (a) FIR and (b) NIR heater irradiation with spore suspension volumes of (•) 20 mL, (Δ) 40 mL, and (▾) 60 mL. Thermal conductive heating at 75°C (□). The average initial concentration was 2.2×10^7 CFU mL^{-1}. (From Sawai, J. et al., *Int. Biodeterior. Biodegrad.*, 63, 196, 2009.)

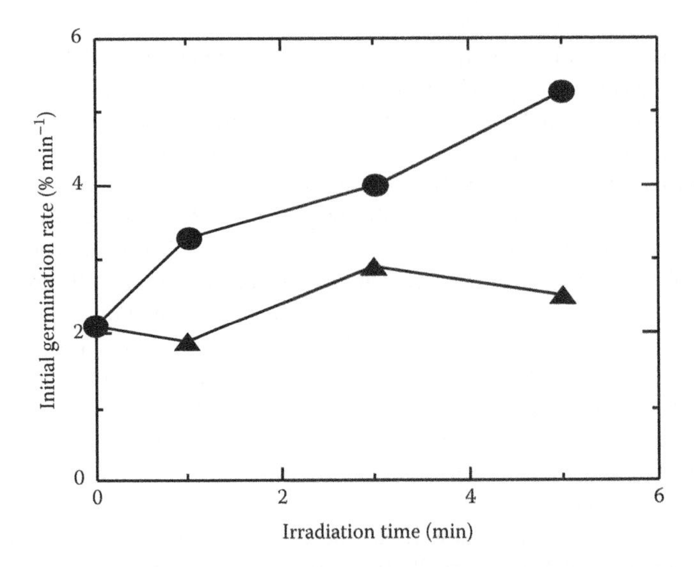

FIGURE 20.21 Germination rate of *B. subtilis* spores irradiated by (C) FIR heater and (:) NIR heater (spore suspension volume: 20 mL). (From Sawai, J. et al., *Int. Biodeterior. Biodegrad.*, 63, 196, 2009.)

germinate, their resistance to heat and chemical treatments is lost. Compared with direct inactivation of spores, promoting spore germination before sterilization allows for lower sterilization temperatures and chemical concentrations thereby eliminating product damage [43].

Recently, a new combination of IR heating and UV irradiation was used for the surface decontamination of various agricultural products [42]. Hamanaka reported that the combination of IR and UV irradiation was effective in the inactivation of mold spores [44]. It is well known that UV radiation has a pasteurization effect, and it is used for the pasteurization of water, air, food containers, etc. UV radiation has sufficient quantum energy for exciting and ionizing organic molecules; the DNA of the microorganism may be directly damaged by the germicidal action, which does not

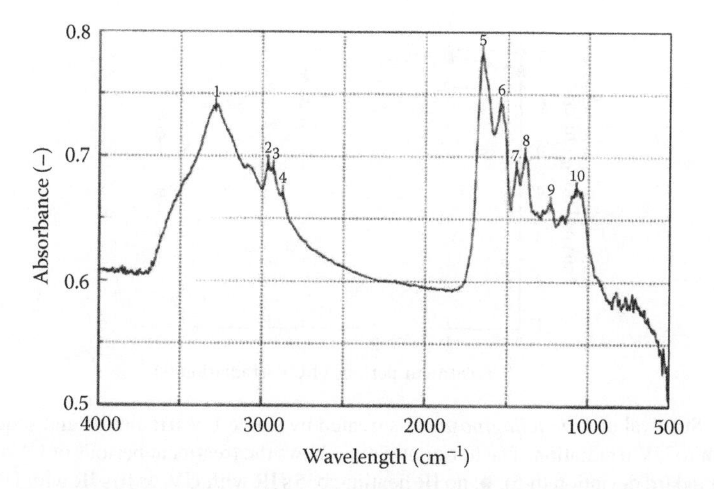

FIGURE 20.22 Microinfrared spectra of the *B. subtilis* spores used in this study. 1: O–H stretching vibration (near 3300 cm^{-1}); 2–4: C–H stretching (near 2900 cm^{-1}); 5: amide I band (near 1650 cm^{-1}); and 6: amide II band (near 1550 cm^{-1}). (From Sawai, J. et al., *Int. Biodeterior. Biodegrad.*, 63, 196, 2009.)

originate from heat. However, a relatively long time is needed to reduce the microbial counts to the required levels. Quality deterioration and acceleration of physiological responses might be caused by these long periods of treatment with UV irradiation even if the reduction of microbial counts to the required levels was accomplished. IR heating was applied to a variety of agricultural commodities (wheat and soybean [45], almonds [46]), effectively reducing the microbial counts in foods. The percentage of damaged fig fruits by mold growth was suppressed by single and sequential treatment, compared with that in the control sample as shown in Figure 20.23. The combination of IR heating and UV irradiation was also found to be effective in the inactivation of fig-related *Rhodotorula mucilaginosa* cells. Figure 20.24 shows the survival data for *R. mucilaginosa* cells after IR heating and sequential treatment with IR heating prior to UV irradiation (IR–UV). Figure 20.25 shows the

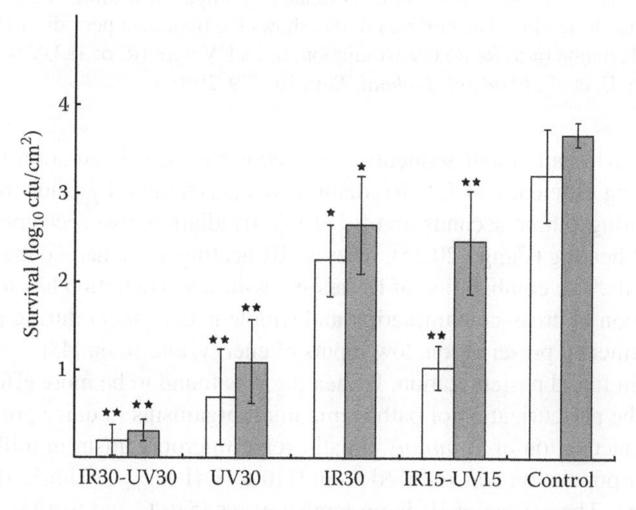

FIGURE 20.23 Fungi survival on fig fruit surface following single and sequential treatments of IR heating and UV irradiation. Error bars denote the standard deviation (n-3). Asterisks (*,**) denote the significant differences between control (C) and treated sample by Student's *t*-test ($p<0.05, <0.01$). The numbers shown with IR or UV on the horizontal axis are the treatment periods in seconds, □: day zero, ■: 2 days in storage. (From Sawai, J. et al., *Int. Biodeterior. Biodegrad.*, 63, 196, 2009.)

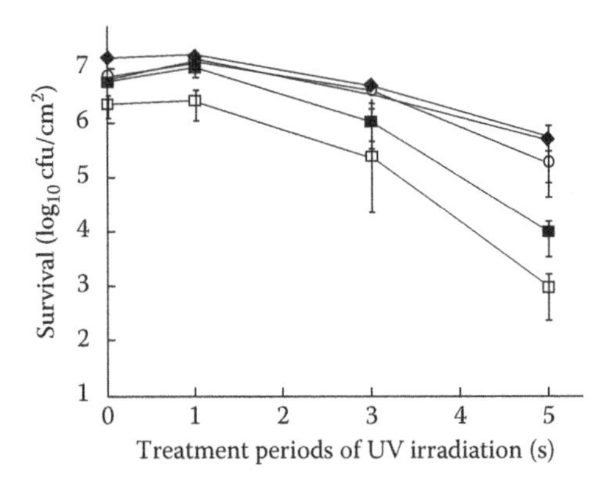

FIGURE 20.24 Survival of *R. mucilaginosa* cells treated by single UV irradiation and sequential treatment of IR heating prior to UV irradiation. The horizontal axis shows the treatment periods of UV irradiation. Error bars denote the standard deviation (n-5). ◆: no IR heating, ○: 5 s IR with UV, ○: 10 s IR with UV, ■: 20 s IR with UV, □: 30 s IR with UV. (From Hamanaka, D. et al., *Food Sci. Technol. Res.*, 16, 279, 2010.)

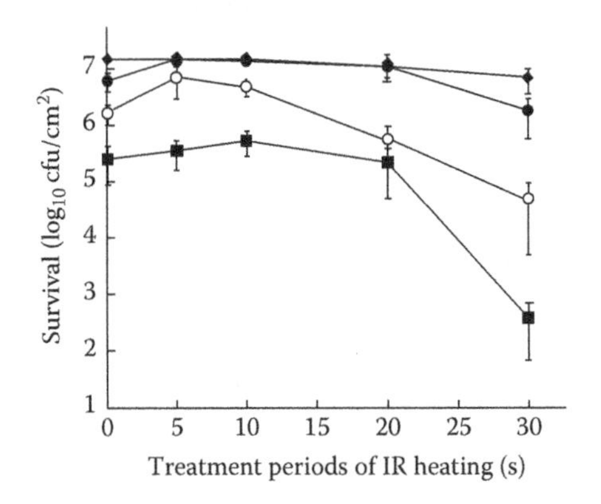

FIGURE 20.25 Survival of *R. mucilaginosa* cells treated by single IR heating and sequential treatment of UV irradiation prior to IR heating. The horizontal axis shows the treatment periods of IR heating. Error bars denote the standard deviation (n-5). ◆: no UV irradiation, ○: 1 s UV with IR, ○: 3 s UV with IR, ■: 5 s UV with IR. (From Hamanaka, D. et al., *Food Sci. Technol. Res.*, 16, 279, 2010.)

survivals after UV irradiation and sequential treatment with UV irradiation prior to IR heating (UV–IR). The killing efficiency of UV irradiation was accelerated by prior treatment with more than 20 s of IR heating. Three seconds and 5 s of UV irradiation also accelerated the killing efficiency of 30 s of IR heating (Figure 20.25), whereas IR heating alone had no inactivating effects on *R. mucilaginosa* cells. The combination of IR heating with UV irradiation has some advantages not only in the prevention of cross-contamination and simple management during processing but also in terms of environmental preservation, low inputs of energy, and so on [43].

IR is also used in liquid pasteurization. IR heating was found to be more effective than conventional heating for the pasteurization of pathogenic microorganisms in dairy products. The efficacy of IR heating for inactivation of *S. aureus*, a pathogenic microorganism in milk, was investigated [42]. The *S. aureus* population was reduced from 0.10 to 8.41 $\log_{10}$ CFU/mL, depending upon the treatment conditions. The effects of IR lamp temperatures (536°C and 619°C), volumes of treated milk samples (3, 5, and 7 mL), and treatment times (1, 2, and 4 min) on the lethality of targeted microbes were found to be statistically significant. Complete inactivation of *S. aureus* was obtained in the case within 4 min at a 619°C lamp temperature, resulting in an 8.41 $\log_{10}$ CFU/mL reduction. Some of the injured cells were able to repair in the enrichment culture. Further investigation of IR

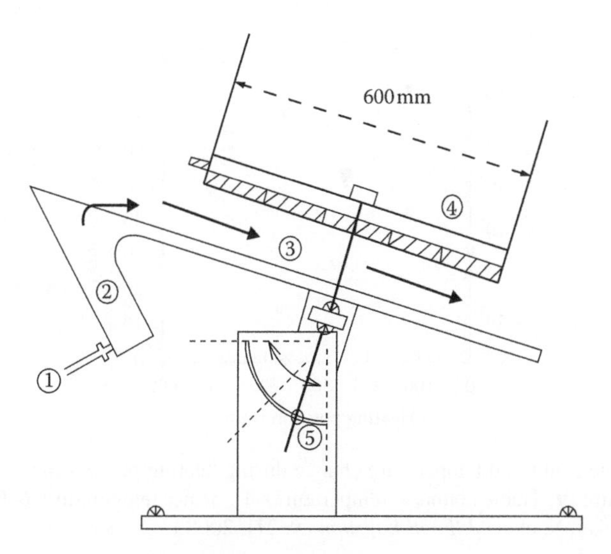

FIGURE 20.26 Experimental apparatus. (1) Inlet of sample, (2) tank, (3) heating line, (4) FIR heater, and (5) angle regulator of heating line. (From Sakai, N. et al., *J. Food Eng. Jpn.*, 9, 271, 2008.)

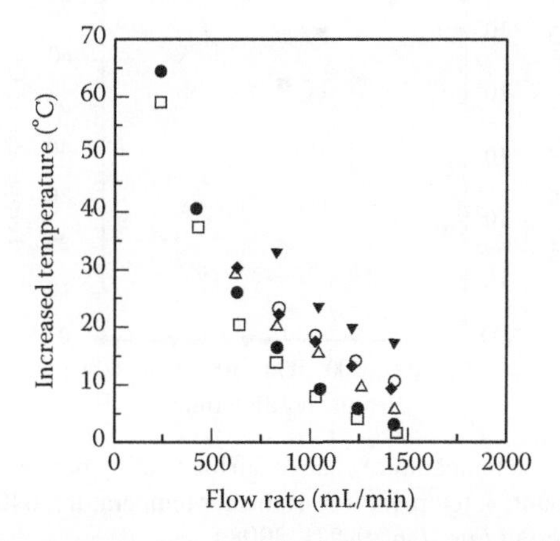

FIGURE 20.27 Relationship between flow rate of liquid and its increased temperature. Angle of the heating unit (□: 0°, •:1°, Δ: 2°, ♦: 3°, o: 4°, ▾: 5°). Initial liquid temperature: 18°C, Supplied electricity: 3.2 kW. (From Sakai, N. et al., *J. Food Eng. Jpn.*, 9, 271, 2008.)

heating for longer treatment times (>4 min) indicated that there was no growth observed following the enrichment culture in most cases for treatment at the lamp temperature of 619°C. IR heating has a potential for effective inactivation of *S. aureus* in milk [47].

Sakai et al. [48] developed pasteurization equipment using FIR for liquid food. Figure 20.26 shows a schematic diagram of the equipment. The equipment is composed of five parts: a inlet of sample, a tank, a heating line, an FIR heater, and an angle regulator of the heating line. The capacity of the tank was 2 L. The width of the trough-like heating line was 50 mm, the length was 855 mm, and the depth was 30 mm. The length of the heater was 600 mm and the width 120 mm. The emissivity of the heater was 0.85. The temperature of the liquid food changed with the radiation intensity (electricity supplied), the angle of the incline, and the flow rate. At any angle, with decreasing flow rate, the temperature increased, while at the same flow rate, the larger the angle, the higher the temperature became (Figure 20.27). Since heating by IR can directly heat liquid food, the temperature of the surface of the food was high. However, because the liquid food is irradiated by FIR from the upper of the liquid food, the temperature on a stainless side is lower than that of sample, stainless steel plate which has contacted

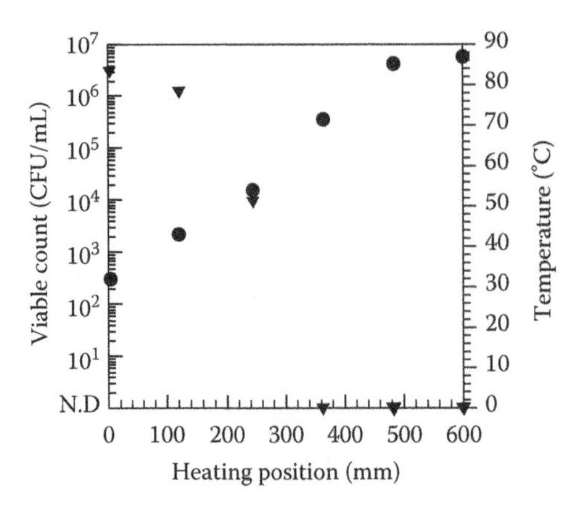

FIGURE 20.28 Viable count and temperature change during heating process in the case of *T. halophilus*. Angle of the heating unit (▾: viable count, •: temperature): 1°, heater temperature: 640°C, sample flow rate: 240 mL/min. (From Sakai, N. et al., *J. Food Eng. Jpn.*, 9, 271, 2008.)

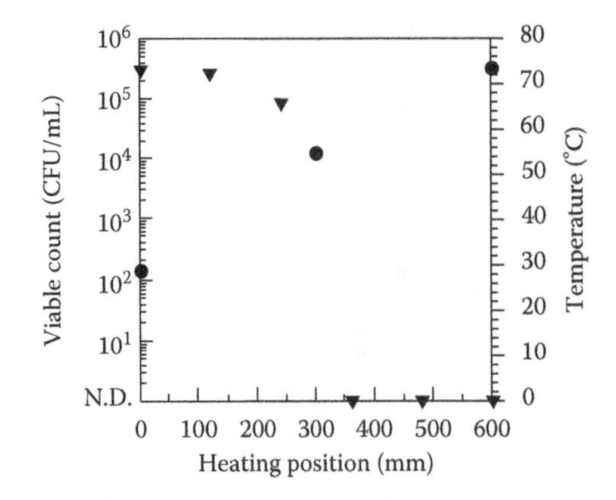

FIGURE 20.29 Viable count and temperature change during heating process in the case of *Z. rouxii*. Angle of the heating unit (▾: viable count, •: temperature): 1°, heater temperature: 640°C, sample flow rate: 240 mL/min. (From Sakai, N. et al., *J. Food Eng. Jpn.*, 9, 271, 2008.)

the food would be maintained a low temperature. As a result, it seems to be able to reduce the adhesion of the fouling in the stainless steel plate, which was an important problem in the conventional heat exchanger. *Tetragenococcus halophilus* and *Zygosaccharomyces rouxii* were chosen as the target microorganisms. The viable count of *T. halophilus* and *Z. rouxii* is shown in Figures 20.28 and 20.29, respectively. The horizontal axis is the distance for the liquid flow; the symbol of "•" is the experimental temperature of the sample, which increased almost linearly as the liquid moved downstream. The initial viable count of *T. halophilus* was about 1×10^6 CFU/mL and that of *Z. rouxii* was about 2×10^5 CFU/mL. The viable bacterial count had not begun to decrease at the position of 120 mm, after the temperature of the sample was over 40°C, and the viable bacterial count decreased quickly. In addition, at a later position, no bacteria were detected. In conclusion, the FIR system is effective in the pasteurization of liquid food.

20.5 CONCLUSIONS

IR heating applications are expected to glow as a result of the increasing demand for safe food products of high nutritional value and organoleptic quality. There are many examples of IR heating as being superior to conventional heating. However, there are few theoretical explanations of the IR effects, and there is therefore a need for basic research on the mechanism of energy transfer and its effect on changes in taste and the nutritional components of the food. Furthermore, in order to improve the IR heating process design, the practical development of quantitative models is needed to predict temperature and product quality as a function of temperature. Finally, innovation of IR heating equipment combined with MW or ohmic heating will lead to increased energy efficiency and higher quality products in the food industry.

NOMENCLATURE

c Velocity of light (m/s)
E Emissive power (W/m^2)
F_{21} Radiation shape factor (—)
h Plank's constant (mK)
h_t Surface heat transfer coefficient (W/m^{2}K)
I_λ Energy flux (W/m^2)
k_t Thermal conductivity (W/mK)
M Moisture content (%)
Q Internal heat gene (W/m^3)
r_f Reflectivity (—)
t Time (s)
T Temperature (K)
α_γ Spectral attenuation factor (μm^{-1})
β Absorptivity (—)
γ Permeability (—)
ε Emissivity (—)
κ Boltzmann's constant (J/K)

REFERENCES

1. T Omori. The basis and feature of infrared ray. In: T Omori, ed. *Bio-Electromagnetics and Its Application*. Tokyo, Japan: Fuji Techno-System, 1992, pp. 36–43.
2. M Shimizu, A Hashimoto, H Igarashi. Far-infrared radiation technology. *New Food Industry* 33(8): 23–30, 1991.
3. M Sugiyama. Utilization of far infrared radiation to food industry. *Food Industry ("Shokuhin Kogyo")* 29(10): 26–34, 1986.
4. N Sasamori. Principles of utilization of far infrared radiation and measurement of equipment performance. *Japan Food Science* 27(8): 23–29, 1988.
5. MW Irvine, JB Pollack. Infrared optical properties of water and ice. *Icarus* 8: 324–360, 1968.
6. N Sakai, A Fujii, T Hanzawa. Heat transfer analysis in a food heated by far-infrared radiation. *Nippon Shokuhin Kogyo Gakkaishi* 40(7): 469–477, 1993.
7. N Shilton, P Malikarjunan, P Sheridan. Modeling of heat transfer and evaporative mass loss during the cooking of beef patties using far-infrared radiation. *Journal of Food Engineering* 55: 217–222, 2002.
8. AK Datta, S Sahin, G Sumnu, SO Keskin. Porous media characterization of breads baked using novel heating modes. *Journal of Food Engineering* 79: 106–116, 2007.
9. O Sakiyan, G Sumnu, S Sahin, V Meda. Investigation of dielectric properties of different cake formulations during microwave and infrared—Microwave combination baking. *Journal of Food Science* 72: 205–213, 2007.
10. SO Keskin, G Sumnu, S Sahin. A study on the effects of different gums on dielectric properties and quality of breads baked in infrared-microwave combination oven. *European Food Research and Technology* 224: 329–334, 2007.
11. G Sumnu, AK Datta, S Sahin, SO Keskin, V Rakesh. Transport and related properties of breads baked using various heating modes. *Journal of Food Engineering* 78: 1382–1387, 2007.

12. G Sumnu, S Sahin, M Sevimli. Microwave, infrared and infrared-microwave combination baking of cakes. *Journal of Food Engineering* 71: 150–155, 2005.
13. EEM Olsson, AC Tragardh, LMA Hrne. Effect of near-infrared radiation and jet impingement heat transfer on crust formation of bread. *Journal of Food Science* 70: 484–491, 2005.
14. J Park, JM Lee, YJ Cho, C Kim, C Kim, KC Nam, SC Lee. Effect of far-infrared heater on the physicochemical characteristics of green tea during processing. *Journal of Food Biochemistry* 33: 149–162, 2009.
15. N Uysal, G Sumnu, S Sahin. Optimization of microwave—Infrared roasting of hazelnut. *Journal of Food Engineering* 90: 255–261, 2009.
16. L Braeckman, F Ronsse, PC Hidalgo, J Pieters. Influence of combined IR-grilling and hot air cooking conditions on moisture and fat content, texture and colour attributes of meat patties. *Journal of Food Engineering* 93: 437–443, 2009.
17. ZL Pan, C Shih, TH Mchugh, E Hirschberg. Study of banana dehydration using sequential infrared radiation heating and freeze-drying. *LWT-Food Science and Technology* 41: 1944–1951, 2008.
18. Y Zhu, ZL Pan. Effect of dipping treatments on color stabilization and texture of apple cubes for infrared dry-blanching process. *Journal of Food Processing and Preservation* 31: 632–648, 2007.
19. T Baysal, F Icier, S Ersus, H Yildiz. Effects of microwave and infrared drying on the quality of carrot and garlic. *European Food Research and Technology* 218: 68–73, 2003.
20. ZL Pan, R Khir, KLB Garber, ET Champagne, JF Thompson, A Salim, BR Hartsough, S Mohamed. Drying characteristics and quality of rough rice under infrared radiation heating. *Transactions of the ASABE* 54: 203–210, 2010.
21. KJ Chua, SK Chou. A comparative study between intermittent microwave and infrared drying of bioproducts. *International Journal of Food Science and Technology* 40: 23–39, 2005.
22. I Das, SK Das, S Bal. Drying kinetics of high moisture paddy undergoing vibration-assisted infrared (IR) drying. *Journal of Food Engineering* 95: 166–171, 2009.
23. C Nimmol, S Devahastin, T Swasdisevi, S Soponronnarit. Drying and heat transfer behavior of banana undergoing combined low-pressure superheated steam and far-infrared radiation drying. *Applied Thermal Engineering* 27: 2483–2494, 2007.
24. MM Gabel, Z Pan, KSP Amaratunga, LJ Harris, JF Thompson. Catalytic infrared dehydration of onions. *Journal of Food Science* 71: 351–357, 2006.
25. J Shi, ZL Pan, TH Mchugh, D Wood, E Hirschberg, D Olson. Drying and quality characteristics of fresh and sugar-infused blueberries dried with infrared radiation heating. *LWT-Food Science and Technology* 41: 1962–1972, 2008.
26. J Shi, Z Pan, TH Mchugh, D Wood, Y Zhu, RJ Avena, B Ehirschberg. Effect of berry size and sodium hydroxide pretreatment on the drying characteristics of blueberries under infrared radiation heating. *Journal of Food Science* 73, 259–265, 2008.
27. Y Zhu, ZL Pan, TH Mchugh, DM Barrett. Processing and quality characteristics of apple slices processed under simultaneous infrared dry-blanching and dehydration with intermittent heating. *Journal of Food Engineering* 97: 8–16, 2010.
28. GP Sharma, RC Verma, P Pathare. Mathematical modeling of infrared radiation thin layer drying of onion slices. *Journal of Food Engineering* 71: 282–286, 2005.
29. HU Hebbar, NK Rastogi. Mass transfer during infrared drying of cashew kernel. *Journal of Food Engineering* 47: 1–5, 2001.
30. Y Zhu, ZL Pan, TH McHugh, DM Barrett. Processing and quality characteristics of apple slices processed under simultaneous infrared dry-blanching and dehydration with intermittent heating. *Journal of Food Engineering* 97: 8–16, 2010.
31. Y Zhu, ZL Pan. Processing and quality characteristics of apple slices under simultaneous infrared dry-blanching and dehydration with continuous heating. *Journal of Food Engineering* 90: 441–452, 2009.
32. Y Zhu, ZL Pan. Effect of dipping treatments on color stabilization and texture of apple cubes for infrared dry-blanching process. *Journal of Food Processing and Preservation* 31: 632–648, 2007.
33. F Tanaka, P Verboven, N Scheerlinck, K Morita, K Iwasaki, B Nicolaı. Investigation of far infrared radiation heating as an alternative technique for surface decontamination of strawberry. *Journal of Food Engineering* 79: 445–452, 2007.
34. L Huang. Infrared surface pasteurization of Turkey frankfurters. *Innovative Food Science and Emerging Technologies* 5: 345–351, 2004.
35. L Huang, J Sites. Elimination of *Listeria monocytogenes* on hotdogs by infrared surface treatment. *Journal of Food Science* 73: 27–31, 2008.

36. N Staack, L Ahrne, E Borch, D Knorr. Effect of infrared heating on quality and microbial decontamination in paprika powder. *Journal of Food Engineering* 86: 17–24, 2008.
37. N Staack, L Ahrne, E Borch, D Knorr. Effects of temperature, pH, and controlled water activity on inactivation of spores of *Bacillus cereus* in paprika powder by near-IR radiation. *Journal of Food Engineering* 89: 319–324, 2008.
38. A Hashimoto, J Sawai, H Igarashi, M Shimizu. Effect of far-infrared irradiation of pasteurization of bacteria suspended in liquid medium below lethal temperature. *Journal of Chemical Engineering of Japan* 25: 275–281, 1992.
39. A Hashimoto, J Sawai, H Igarashi, M Shimizu. Irradiation power effect on pasteurization of bacteria on or within wet-solid medium. *Journal of Chemical Engineering of Japan* 26: 331–333, 1993.
40. J Sawai, K Matsumoto, T Saito, Y Isomura, R Wada. Heat activation and germination—Promotion of *Bacillus subtilis* spores by infrared irradiation. *International Biodeterioration & Biodegradation* 63: 196–200, 2009.
41. SM Ede, LM Hafner, PM Fredericks. Structural changes in the cells of some bacteria during population growth: A Fourier transform infrared-attenuated total reflectance study. *Applied Spectroscopy* 58: 317–322, 2004.
42. DL Perkins, CR Lovell, BV Bronk, B Setlow, P Setlow, ML Myrick. Effects of autoclaving on bacterial endospores studied by Fourier transform infrared microspectroscopy. *Applied Spectroscopy* 58: 749–753, 2004.
43. D Hamanaka, N Norimura, N Baba, K Mano, M Kakiuchi, F Tanaka, T Uchino. Surface decontamination of fig fruit by combination of infrared radiation heating with ultraviolet irradiation. *Food Control* 22: 375–380, 2011.
44. D Hamanaka, F Tanaka, T Uchino, GG Atungulu. Effect of combining infrared heating with ultraviolet irradiation on inactivation of mold spores. *Food Science and Technology Research* 16: 279–284, 2010.
45. D Hamanaka, T Uchino, W Hu, E Yasunaga, HM Sorour. Effects of infrared radiation on the disinfection for wheat and soybean. *Journal of Japanese Society of Agricultural Machinery* 65: 64–70, 2003.
46. J Yang, G Bingol, Z Pan, MT Brandl, TH Mchugh, H Wang. Infrared heating for dry-roasting and pasteurization of almonds. *Journal of Food Engineering* 101: 273–280, 2010.
47. K Kathiravan, I Joseph, D Ali. Efficacy of infrared heat treatment for inactivation of *Staphylococcus aureus* in milk. *Journal of Food Process Engineering* 31: 798–816, 2008.
48. N Sakai, Y Oshimai, Y Yamanaka, M Fukuoka. Development of liquid food pasteurization equipment using far-infrared radiation heating. *Journal of Food Engineering of Japan* 9: 271–276, 2008.

21 Microwave Heating

Servet Gulum Sumnu and Serpil Sahin

CONTENTS

21.1 INTRODUCTION

Microwaves are electromagnetic waves that have a frequency range from 300 MHz to 30 GHz. Although microwave spectrum is wide, only certain frequencies are allowed in food processing due to possibility of interaction of microwaves with telecommunication devices. Frequencies of 915 and 2450 MHz are used in industrial applications and home-type microwave ovens, respectively [1].

There are two major mechanisms of interaction of foods with microwaves. One is ionic conduction and the other is dipolar rotation. In ionic conduction, the positively charged ions will accelerate in the direction of the electric field in microwave oven, while negatively charged ions will move in the opposite direction [1]. Since electric field is changing its direction depending on frequency, the direction of the motion of the positive and negative ions changes. Moving particles collide with the adjacent particles and set them into more agitated motion. Thus, temperature of the particle increases. In the presence of polar molecules, heat generation in microwave oven takes place by the mechanism of dipolar rotation. When polar molecules interact with electric field, they try to orient themselves in the direction of the field and they collide with the other molecules. The change in direction of the field results in further collision since they try to line up with the reversed directions. Thus, thermal agitation and heating take place.

This chapter will give information on the properties of foods affecting microwave heating and interaction of microwave heating with food components. Moreover, the recent advances in microwave food processes such as baking, drying, extraction, pasteurization, sterilization, and thawing will be discussed.

21.2 FOOD PROPERTIES AFFECTING MICROWAVE HEATING

Dielectric properties are the main food properties affecting microwave heating. Thermal food properties such as thermal conductivity and specific heat capacity have secondary effect.

21.2.1 Dielectric Properties of Foods

The dielectric properties can be categorized into two types, which are dielectric constant and dielectric loss factor. Dielectric constant (ε') is the ability of a material to store microwave energy while dielectric loss factor (ε'') is the ability of a food to dissipate microwave energy into heat.

Depending on the magnitude of dielectric properties, it can be decided whether a food will be heated faster or not in microwave oven. In addition, it is possible to understand whether heat will be distributed.

Dielectric constant and loss factor data are important for microwave heating and microwave food processing and especially for process modeling studies.

The rate of heat generation per unit volume (Q) at a location inside the food during microwave heating can be characterized by the following equation. This equation shows that the higher the loss factor, the more microwave absorption will take place:

$$Q = 2\pi\varepsilon_0\varepsilon'' f E^2 \tag{21.1}$$

where ε_0 is dielectric constant of free space, ε'' is dielectric loss factor of the food, f is frequency of oven (Hz), and E is the root-mean-squared (rms) value of the electric field intensity (V/m) [2].

Dielectric properties of foods depend on composition, temperature, and bulk density of food and also frequency of microwave oven. Dielectric properties of foods have been recently reviewed by various researchers [3–5]. Table 21.1 shows the differences in dielectric properties of different food materials at microwave frequencies.

21.2.1.1 Effects of Composition on Dielectric Properties

Moisture and salt content are the major components in foods, which affect microwave heating. The higher the moisture content, the higher the dielectric properties will be observed. It has been recently shown that increasing the moisture content increases dielectric properties of cake batter [10].

At microwave frequencies, dielectric loss factor is composed of two components, which are dipolar loss and ionic loss as given in the following [1]:

$$\varepsilon'' = \varepsilon''_d + \varepsilon''_\sigma \tag{21.2}$$

where

ε''_d is dipolar loss factor

ε''_σ is ionic loss factor

Dipole loss results from the relaxation of water molecules [17]. Ionic loss results mainly from ionic conduction. In the presence of salt, ionic loss factor is the most dominant component. The

TABLE 21.1
Recent Dielectric Properties Data of Various Foods Measured at Microwave Frequencies

Food	Temperature (°C)	Frequency (MHZ)	ε′	ε″	References
Apple	24	915	52.2 ± 2.5	6.2 ± 5.0	[6]
Apple	24	2450	50.4 ± 2.4	8.9 ± 0.7	[6]
Beef (lean)	23	2450	43.7	13.7	[7]
Bread dough (hamburger)	23	2450	21.02	9.2	[8]
Bread (hamburger)	23	2450	3.14	0.32	[8]
Butter (unsalted)	30	915	25.8 ± 1.1	4.9 ± 0.3	[9]
Butter (unsalted)	30	2450	24.5 ± 0.9	4.3 ± 0.3	[9]
Butter (salted)	30	915	12.5 ± 0.8	36.4 ± 1.3	[9]
Butter (salted)	30	2450	9.0 ± 0.3	15.5 ± 0.7	[9]
Cake batter (Madeira)	20	2450	32.6 ± 0.5	12.6 ± 0.3	[10]
Chicken breast	23 ± 1	2450	49.0	16.1	[7]
Egg (whole)	20	915	56.4 ± 0.9	16.3 ± 0.9	[10]
Egg (whole)	20	2450	54.0 ± 1.0	13.5 ± 1.1	[10]
Grape juice (red)	20	3000	65.36 ± 0.21	9.976 ± 0.139	[11]
Honey	20	915	11.9 ± 0.2	5.2 ± 0.1	[12]
Honey	20	2450	9.2 ± 0.2	3.7 ± 0.1	[12]
Milk	22	915	68.1 ± 0.0	14.3 ± 0.2	[13]
Milk	22	2450	65.9 ± 0.0	14.3 ± 0.1	[13]
Rice cake batter	25	2450	18.27	8.91	[14]
Pink salmon (middle part)	20	915	57.0 ± 0.6	22.8 ± 1.2	[15]
Tomato (cherry)	20	2450	71.7 ± 0.5	16.2 ± 0.5	[16]
Turkey breast	23 ± 1	2450	56.3	18.0	[7]
Wine (red)	20	3000	66.63 ± 0.16	19.867 ± 0.133	[11]

addition of salt to the system decreases dielectric constant since the dissolved ions in salt bind water molecules, which results in a decrease in dielectric constant [1]. On the other hand, the increase in salt concentration of food increases loss factor which can be due to the electrophoretic migration of dissolved salts [9]. As can be seen in Figure 21.1, the increase in salt content increases the loss factor [9]. The same trend was observed in meat products [18]. Loss factor of salted hydrocolloid model systems was also positively affected by salt concentration [19].

Dielectric properties of carbohydrates and proteins by themselves are low. The interaction of these components with water in the system affects the extent of microwave heat ability. Turabi et al. (2010) [14] showed that addition of different types of gums to rice cake formulations affected dielectric constant and loss factor differently. Higher concentrations of starches were shown to decrease dielectric loss factor since greater water retention of hydrocolloid decrease the amount of free water content in the system [19]. Dielectric constant of soybean protein isolate was found to be a function of frequency, temperature, concentration, and pH. Dielectric properties of soy protein isolate dispersion changed significantly at 90°C as a result of protein denaturation [20]. The variation in pH affected electrostatic charges of protein molecules significantly, which also affected dielectric properties.

Dielectric properties of fats and oils are very low since they are nonpolar. The addition of fat to food system reduces the dielectric properties due to the dilution effects of fat. Increase in fat content from 12.4% to 29.7% decreased dielectric constant in comminuted meat blend but the reduction in loss factor was not significant [18]. On the contrary, it was shown that an addition of fat to cake formulation increased dielectric constant and loss factor of microwave baked cakes which may be explained by the decrease in porosity of cakes in the presence of fat [21].

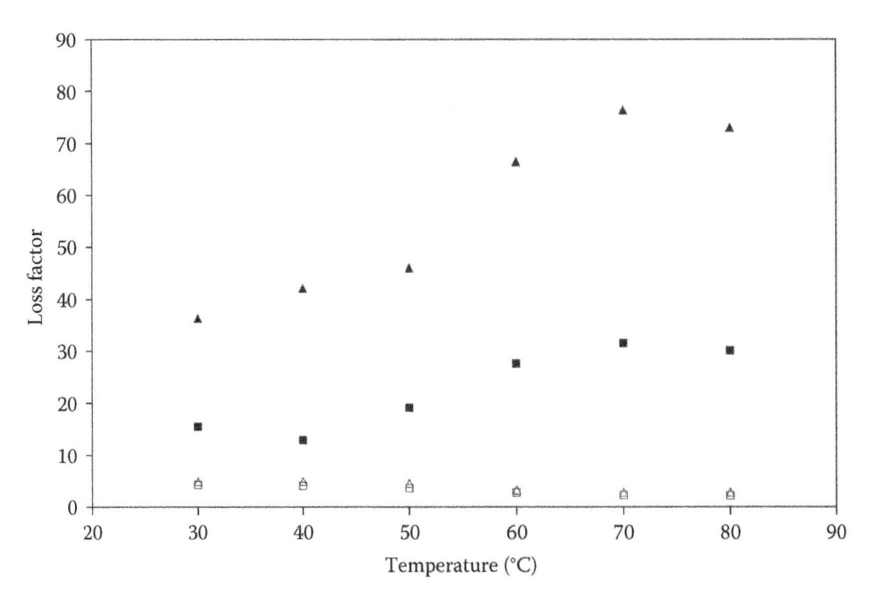

FIGURE 21.1 Variation of dielectric loss factor of salted and unsalted butter at different temperatures and frequencies. (▲) Salted butter at 915 MHz, (Δ) unsalted butter at 915 MHz, (■) salted butter at 2450 MHz, (□) unsalted butter at 2450 MHz. (Data obtained from *J. Food Eng.*, 82, Ahmed, J., Ramaswamy, H.S., and Raghavan, V.G.S., Dielectric properties of butter in the microwave frequency range as affected by salt and temperature, 351–358, Copyright 2007, with permission from Elsevier.)

21.2.1.2 Effects of Temperature on Dielectric Properties

Variation of dielectric properties with temperature depends on the presence of bound or free water in the system [1]. For powdered materials, the increase in temperature increases both dielectric constant and loss factor. Higher temperatures increase the mobility of water, which results in higher dielectric properties. Ice that has dielectric constant of 3.2 and loss factor of 0.0029 is almost transparent to microwaves [22]. Thus, dielectric properties of frozen foods are significantly low. Dielectric properties increase during thawing, which increases the risk of thermal runaway. Faster thawing of a part of food increases loss factor of that part due to the high dielectric loss factor of free water. This results in more microwave absorption, more uneven heating, and thermal runway [22]. It has been recently shown that dielectric constant and loss factor of frozen potato puree increased significantly during thawing [23]. Both dielectric parameters were found to be very low (3–10 for dielectric constant and 0–3 for loss factor) up to −10°C and level off at above −2°C.

If sample contains free water, dielectric properties are known to decrease as temperature increases. This can be explained by the presence of less hydrogen bonding and more intense movements, which requires less energy to overcome the intermolecular bonds at higher temperatures. Thus, dielectric properties decrease [1].

Dielectric constant of pure honey (containing no more than 18% water) increased with temperature [12]. For honey diluted with water, the variation of dielectric constant with temperature was a function of frequency. Dielectric constant of diluted honey decreased with increasing temperature at lower frequencies and increased with temperature at higher frequencies. The differences can be explained by the presence of bound and free water content. Water is present mainly in bound form in pure honey. In diluted honey, free water plays a dominant role as compared to bound water.

In the presence of higher salt concentration where ionic loss is dominating, dielectric loss factor increases as temperature increases due to the decreased viscosity of the liquid and increased mobility of the ions. As can be seen in Figure 21.1, the effect of temperature on dielectric loss factor of salted and unsalted butter is different [9]. As temperature increased, loss factor of unsalted butter decreased but that of salted butter increased (Figure 21.1).

21.2.1.3 Effects of Frequency on Dielectric Properties

As discussed before, dielectric loss factor is composed of two components: dipole loss and ionic loss. Ionic conductivity plays a major role for frequencies less than 200MHz [5]. If microwave frequencies are considered, both ionic conduction and dipolar rotation of free water are significant [24]. The contribution of ionic loss to the total dielectric loss factor is less at 2450MHz than at 915MHz [22].

The frequency dependence of dielectric properties of fruits and vegetables has been studied by Nelson and coworkers [25]. They found that dielectric constant and loss factor shows different trends with frequency change. The dielectric constant decreases with frequency in the range of 0.2–20GHz. On the other hand, loss factor may decrease or increase with frequency. The loss factor decreases as frequency increases from 0.2GHz, reaches a minimum in the region between 1 and 3GHz, and then increases. At lower frequencies, dielectric behavior is dominated by ionic conductivity. Bound water relaxation and free water relaxation are the mechanisms responsible at higher frequencies.

Ahmed et al. [9] studied the effects of frequency on dielectric properties of salted and unsalted butter. The dielectric constant of unsalted butter was found to be independent of frequency but dielectric constant of salted butter decreased as frequency increased. On the other hand, loss factor of both salted and unsalted butter decreased as frequency increased. Dielectric properties of different flours (green pea, lentil, and soybean flour) were also shown to decrease as frequency increased [26]. Liu and coworkers studied variation of loss factor of bread within the frequency ranges of 10–1800MHz [17]. The ionic conduction was found to be the main factor within all frequency ranges, while the effect of dipole loss was not significant at low frequencies but has medium effect at high frequencies.

21.2.1.4 Effects of Porosity on Dielectric Properties

Dielectric properties are expected to be less for highly porous foods since pores are filled with air and dielectric properties of air are very low. The dielectric constant of air is 1 and dielectric loss factor of air is almost 0. This affects dielectric properties of porous products. Cakes containing protein-based fat replacer had the highest dielectric constant and loss factor [21]. Addition of fat to cake formulations increased dielectric properties of cakes, which was explained by the decrease in porosity.

Figure 21.2 shows that both dielectric constant and loss factor of bread decreased during baking in different ovens [8]. The decrease in dielectric properties of breads during baking was related to the increase in porosity. Moisture content of breads decreases but their porosity increases during baking. The effect of porosity on dielectric properties was found to be dominant as compared to the effect of moisture content. The decrease in dielectric properties of cakes during baking have been observed by other researchers too [21]. Porosity of apples was shown to affect dielectric properties [27]. Apples which were vacuum impregnated by isotonic solution showed higher loss factor values since substitution of air inside apple by isotonic solution decreased the porosity of apples.

21.2.2 Thermal Properties of Foods

Thermal conductivity and specific heat capacity are the thermal properties that affect microwave heating of foods. Thermal conductivity is important when the size of foods is large or when the processing time is long [28]. After microwaves are absorbed by foods, heat is transferred by conduction. In order to achieve uniform heating, low powers are used so that heat conduction can reduce nonuniform temperature distribution.

Dielectric properties of oil are known to be very low. However, it has been stated that oil is heated faster in microwave oven as compared to water because of the low specific heat of oil. It is not correct to consider it as a general statement since it was shown that the effect of specific heat on heat generation in microwave oven was a function of sample size [29].

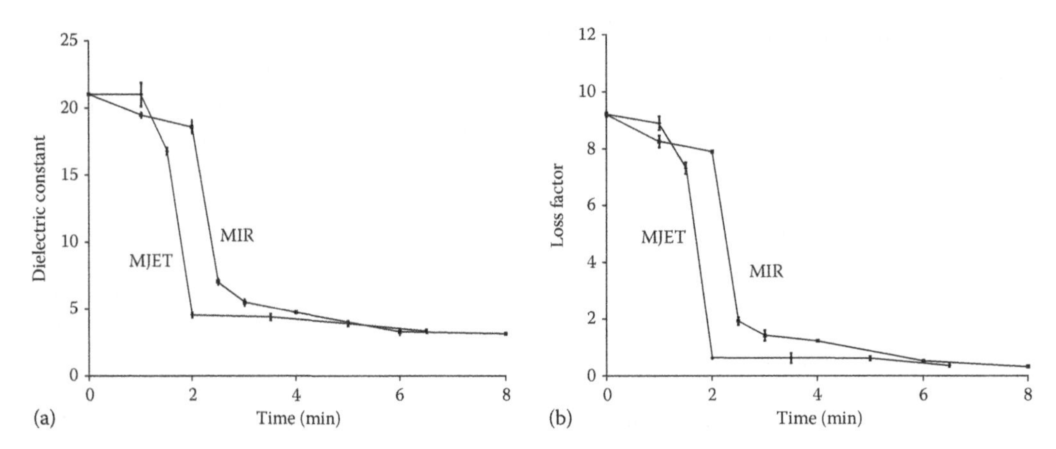

FIGURE 21.2 Variation of dielectric properties of bread during baking in different ovens. (a) Dielectric constant and (b) dielectric loss factor. MJET: microwave-jet impingement combination; MIR: microwave-infrared combination. (From *J. Food Eng.*, 78, Sumnu, G., Datta, A.K., Sahin, S., Keskin, S.O., and Rakesh, V., Transport and related properties of breads baked using various heating modes, 1382–1387, Copyright 2007, with permission from Elsevier.)

21.3 INTERACTION OF FOOD COMPONENTS WITH MICROWAVES

Studying the interaction of starch, protein, and fat content of foods with microwaves are critical to improve the quality of microwave processed foods and to develop new microwavable foods.

21.3.1 STARCH

Starch is the main ingredient in baked products. Products baked in microwave oven are known to have less starch gelatinization due to short processing time [30]. It has been recently shown that gelatinization mechanisms were the same for both microwave and conventional heating methods [31]. Starch granules began to swell at the beginning of gelatinization. When 80°C was reached, most of the granules were swollen and distorted. As heating was continued to 95°C starch granules were disintegrated and melted together. Zylema and coworkers [32] had previously reported that gelatinization mechanism for microwave heating was the same as conventional heating. The effects of microwave heating on corn starch gelatinization were studied by Sakonidou et al. [33], and it was observed that starch gelatinization cannot be completed during microwave heating. This can be explained by limited starch–water interaction during the short microwave heating period. Sakiyan et al. [21] also showed that there was insufficient starch gelatinization in the presence of microwave heating. The gelatinization degrees of cakes baked in microwave oven and conventional oven ranged from 55% to 78% and from 85% to 93%, respectively.

When noodle dough was heated in microwave oven to study starch gelatinization, it was found that starch did not gelatinize, although the required gelatinization temperatures were reached [34]. The fast heating mechanism was responsible for the insufficient starch gelatinization. However, when intermittent heating was applied, it was possible to achieve sufficient gelatinization at the surface of dough.

Microwave heating reduced crystallinity, solubility, and swelling characteristics of wheat and corn starches [35]. The physicochemical properties of waxy corn starch subjected to microwave heating remained almost constant. Microwave heating increased gelatinization temperatures compared to unheated starch.

Anderson and Guraya [36] investigated the effects of microwave heating on digestibility, pasting, and morphological properties of waxy and nonwaxy starch. Digestibility and physical properties of starches were not affected significantly in the presence of microwave heating. Setback viscosity of microwave-heated samples increased. This suggested a reassociation of amylopectin branch chains during microwave heating.

The effects of cooking methods on starch hydrolysis kinetics and resistant starch content of rice and soybeans have been studied [37]. As cooking methods, autoclave, electric cooker, microwave oven, and stone pot were used. The highest starch hydrolysis constant was determined for cooking with autoclave, followed by electric cooker and then by microwave oven. This result showed that less starch was gelatinized during microwave heating. Resistant starch was found to be inversely correlated with starch hydrolysis constants.

21.3.2 Proteins

Although the exact mechanism is not known, microwave-induced gluten changes have been previously reported [38]. Functional properties of microwave-heated gluten have recently been studied [39]. Microwave heating did not change electrophoretic patterns of gluten significantly as compared to unheated gluten. Microwave heating decreased solubility and emulsifying capacity but increased foam stability values of gluten samples.

The effects of microwave heating on viscoelastic property and microstructure of soybean isolate have been recently studied [40]. Soy protein suspensions heated with microwave at different times had larger pores than samples heated in a boiling bath. Similar results were observed for whey protein isolate gels [41]. It was found that whey protein isolate gel heated by microwaves had larger pores than samples heated by conventional means.

When protein denaturation of microwave pasteurized egg white was compared with that of water bath pasteurized ones, it was found that using microwave pasteurization resulted in less protein denaturation [42].

21.3.3 Oils

The effects of microwaves on oil quality have been studied for a long time by various researchers. Albi et al. [43] compared different methods (microwave heating, conventional heating, and exposure of microwave energy without heating) on physical and chemical properties of fats. The temperature of samples heated by microwave and conventional means were 170°C ± 10°C and 180°C ± 2°C, respectively. To observe whether nonthermal effect of microwave on oil quality was present, the temperature of microwave-treated samples was kept below 40°C. Density, viscosity, and squalene content of oil heated by microwaves were more adversely affected as compared to conventional heating. On the other hand, treating samples with and without microwaves did not result in any quality change in oils as compared to original untreated samples. Microwave treatment of oil without heating produced no oil alterations in terms of acid value, peroxide value, dielectric constant, and glyceridic polar compound [44]. This showed that no nonthermal effects of microwaves were observed. Caponio et al. [45] also showed that in microwave-heated samples, polar compounds of triglyceride oligopolymers and the oxidized triglycerides were determined to be significantly higher than those obtained in conventional heating. More oxidative degradation was found in microwave-heated samples. On the other hand, the diglyceride levels did not vary significantly between the heating methods. In contrast to other studies, Chu and Hsu showed that microwave-heated soybean oil had better oil quality in terms of lower acid and peroxide values than that heated by conventional heating [46]. This can be explained by short microwave processing time.

Differential scanning calorimetry (DSC) has been recently used for assessing the deterioration effect of microwave heating on olive oils [47,48]. It was possible to discriminate different oil samples by using their thermal properties upon cooling [48].

Tocopherol, chlorophyll, and carotene content of oil were found to decrease during microwave heating since these compounds were thermally labile [49]. Similar results could be observed in conventional heating, but in this study, the results obtained by microwave heating were not compared with conventional heating.

Valli et al. [50] compared the effects of conventional and microwave heating on phenolic profiles and antioxidant capacity of olive oil. The compounds of 3,4-dihydroxyphenylethanol linked to elenolic acid, elenolic acid, and *p*-hydroxyphenylethanol linked to elenolic acid decreased after heating by both methods. Dialdehydic form of elenolic acid and *p*-hydroxyphenyl ethanol linked to dialdehydic form of elenolic acid increased during heating, but the increase was more significant for conventional heating. On the other hand, no significant differences were found between these methods in terms of antioxidant capacity.

21.4 MICROWAVE PROCESSING OF FOODS

In recent years, there has been much research on microwave. Microwave radiation is used in the food industry for reheating, drying, baking, disinfestation, extraction, tempering/thawing, as well as for the inactivation of microorganisms in foods. The progress made in the literature in different microwave food processes will be summarized in the following sections:

21.4.1 Microwave Reheating

The most common application of microwave especially at home is reheating of foods. Reheating meals in the microwave can be both convenient and time effective. Almost everything can be reheated in a microwave oven. The effects of different reheating methods on the quality characteristics of precooked ground pork patties with different combinations of salt (1.2%) and phosphate (0.0.3%) have been studied by Choi et al. [51]. The reheating rate by microwave oven was faster than by electric grill and decreased with increase in salt and addition of phosphate. Cooking loss and reduction in patty diameter after reheating by microwave oven were higher than by electric grill, and these values decreased with increasing salt/phosphate levels.

It was found that microwave reheating of ready-to-eat surimi-based shrimp-imitation product may produce undesired textural changes [52]. During microwave heating, considerable product shrinkage and thus increase in density occurs. This results in significant increase in product toughness.

Undesired texture development is also observed in microwave reheating of breads. Reheating of breads using microwave ovens considerably toughened the crumb texture when internal boiling was induced as compared to conventional reheating [53]. It was concluded that moisture loss had a relatively minor toughening effect, and the main reason of toughening was leaching of amylose out of the granules in the case of sustained boiling during microwave heating, as compared to conventional heating. The free amylose solution is pushed by the generated steam pressure toward the air-cell wall interface. Upon cooling, the amylose undergoes rapid phase changes; thus, toughening occurs. In order to minimize this problem, internal boiling should be minimized, starch concentration should be below the threshold level and complex forming agents can be used to interfere with the amylose phase.

Sogginess may be observed during microwave reheating because of higher rate of moisture loss and lower temperature of air surrounding the food. There are some studies to prevent this problem. In the study of Chen et al. [54], mackerel nuggets were coated with hydroxy propyl methyl cellulose (HPMC) prior to battering or coated with batter containing HPMC in order to improve crust crispness through the formation of a thermal gel barrier preventing the diffusion of water from meat into the crust during microwave reheating. Diffusion of water from fish meat into the crust was inhibited during microwave reheating in HPMC-coated products. As a result, crispness of HPMC-coated mackerel nuggets that had been reheated in a microwave was improved as compared to that of non-HPMC-coated nuggets.

21.4.2 Microwave Drying

Drying is one of the most important preservation methods in food processing. In drying, moisture is removed up to the desirable level, at which microbial spoilage and deteriorative chemical reactions are avoided or greatly minimized.

Conventional hot air drying may result in serious damage to flavor, color, and nutrients of the treated material and gives unacceptably poor quality product since it involves exposure of food to high temperature for long times. Using microwave reduces the drying time of the biological products without quality degradation.

Pinchukova et al. [55] found that microwave irradiation is less energy consuming and was particularly useful for effective drying of thermally unstable materials in short periods of time. Drying time of purslane leaves decreased and effective moisture diffusivity values increased by increasing the microwave output power [56]. Increasing the sample load resulted in increase in drying time and decrease in effective moisture diffusivity. Similarly, in microwave drying of celery leaves, drying time decreased from 34 to 8 min by increasing the microwave output power from 180 to 900 W, and the drying time increased from 25 to 49 min by increasing the sample load from 25 to 100 g [57]. Effective mass diffusivity values for drying of different food materials can be seen in Table 21.2.

Microwave power affects not only the drying rate but also the quality of dried products. It was shown that rehydration capacity of the coriander (*Coriandrum satium L.*) samples decreased as microwave power increased [60]. In addition, retention of vitamin C increased by 87.9%, density increased by 14.6%, extent of browning increased by 8.7%, while soluble solids content decreased by 12.9% when microwave power level of 2.5 W/g was used for drying of Chinese jujube fruit as compared to the lower microwave power level of 1.2 W/g [63].

TABLE 21.2
Effective Mass Diffusivity Values for Drying of Different Food Materials

Samples	Drying Condition	Effective Diffusivity (m^2/s)	References
Purslane leaves	Microwave drying (180–900 W)	$5.913\times10^{-11}-1.872\times10^{-10}$	[56]
	Microwave drying with sample load from 25 to 100 g	$9.889\times10^{-11}-3.292\times10^{-11}$	
Bamboo shoot slices	Microwave drying (140–350 W)	$4.153\times10^{-10}-22.835\times10^{-10}$	[58]
Onion slices	Sun drying	8.339×10^{-10}	[59]
	Oven drying at 50°C	7.468×10^{-10}	
	Oven drying at 70°C	1.554×10^{-9}	
	Microwave drying at 210 W	4.009×10^{-8}	
	Microwave drying at 700 W	4.869×10^{-8}	
Coriander (*Coriandrum satium L.*)	Microwave drying (180–360 W)	6.3×10^{-11}–2.19×10^{-10}	[60]
Finger-root (*Boesenbergia pandurata*)	Hot air drying, 60°C	0.2073×10^{-10}	[61]
	Hot air drying, 70°C	0.4106×10^{-10}	
	Microwave vacuum drying (13.3 kPa) at 2880 W	5.7910×10^{-10}	
	Microwave vacuum drying (13.3 kPa) at 3360 W	6.8767×10^{-10}	
Parboiled wheat	Spouted bed drying (50°C–90°C)	1.44×10^{-10}–3.32×10^{-10}	[62]
	Microwave-assisted spouted bed drying (288–624 W)	5.06×10^{-10}–11.3×10^{-10}	

Recently, microwave drying of basil leaves (*Ocimum Basillium L.*) [64] and tomatoes [65] has been studied. A real-time aroma-monitoring system to control microwave drying process was designed by using carrot and apple as test samples [66].

Heat and mass transfer in foods during microwave drying was studied both experimentally and theoretically [67]. In the mathematical model, the variation of physical properties with composition, structure, and temperature was considered. Heat generation was calculated by using the approximation of Lambert's law. Predicted model for surface and center temperatures and moisture loss during drying were in good agreement with the experimental data.

Less uniform heating is the main problem of microwave heating. Therefore, new trends in the food drying industry lead to the combination of different drying technologies. Microwave can be combined with convective drying to improve uniformity. Kouchakzadeh and Shafeei [68] studied microwave-convective drying of pistachios. In microwave, air and combined microwave-air drying methods, drying times of nettle leaves were in the ranges of 6–14, 40–120, and 2–12 min, while the energy consumption was 0.07–0.10, 0.25–0.34, and 0.08–0.16 kWh, respectively [69].

Vacuum freeze drying is a method of dehydration of frozen materials by sublimation under vacuum with a sufficient heat supply. It yields the best quality products. However, the major problem with conventional freeze drying is the long drying time required, which in turn leads to higher energy consumption and higher equipment costs.

Microwave and freeze drying can be combined to achieve faster drying along with high-quality product at lower costs. The technique that combines freeze drying and microwave is called microwave freeze drying (MFD). It is one of the most promising techniques to accelerate drying and to enhance overall quality. In MFD, microwaves heat the material volumetrically in a vacuum environment, and the rate of freeze drying increases significantly.

The temperature of the material external surface, ice core, and sublimation interface is kept as high as possible by means of microwave energy in MFD process, and in consequence, maximum process efficiency is provided. In MFD, microwave is used as a heat source to supply sublimation heat so that the drying time is shortened greatly. MFD is practically a rapid, simple, efficient, economic, and novel dehydration technique, which can be used for drying of foodstuffs. The quality of foods dried using this method is comparable with the ones dried using freeze drier with over 96% saving on time and enormous amount on energy [70]. In another study, MFD time was about half of the conventional freeze drying time in drying of sea cucumber (*Stichopus japonicus*) and similar quality products could be obtained in both methods [71]. MFD was also used for drying of potato slices [72] and cabbage [73].

However, there are problems in the application of MFD technology. When water is frozen, the dielectric constant and dielectric loss factor decrease sharply. Therefore, the drying rate cannot be remarkably enhanced when ordinary MFD is applied for dehydration of high moisture containing food materials. MFD process can be enhanced significantly by the addition of dielectric materials with high loss factor to the porous or liquid materials to be dried. In the research performed by Wang et al. [74] and Wang and Chen [75], silicon carbide (SiC) was selected as the dielectric material, and the skim milk was used as the representative solid material in the aqueous solution to be freeze dried. The drying time was greatly reduced compared to cases without the aid of the dielectric material.

Another potential way to increase the rate of MFD is to change dielectric properties of materials. MFD is used to treat instant soups [76], and it was found that food ingredients can improve the dielectric constant of instant soup, which improves the efficiency of microwave absorption. The dielectric properties of food materials are affected by their chemical composition such as moisture, sugar, salt, protein, and ash contents. Several researchers have shown that osmotic dehydration (OD) prior to microwave-assisted drying leads to shorter drying time, in consequence lower energy consumption and better quality dried product [77–79] since dielectric properties change. OD involves partial removal of water from food, such as fruits and vegetables, by immersing them in hypertonic solutions. It is often used as a pretreatment of drying processes to reduce energy consumption and improve product quality. During OD simultaneous counter-current mass transfer processes, namely, diffusion of water from the food to the solution and diffusion of the solute from the solution to the food occurs. Leaching of natural solutes from the food into the solution also occurs, but it is quantitatively negligible.

Combination of microwave with vacuum is another alternative method for drying. Microwave vacuum drying (MVD) combines the advantages of rapid volumetric heating by microwaves and low-temperature evaporation of moisture with faster moisture removal by vacuum. It provides better drying option because vacuum lowers the process temperatures and increases the rate of evaporation, resulting in shorter drying times and improved product quality.

MVD has been widely used in the food industry. Several food products, such as beetroots [80], finger-root (*Boesenbergia pandurata*) [61], and apple slices [81] have been successfully dried by MVD and MVD reduced drying time significantly in comparison with convective method.

In general, microwave vacuum dried samples were better than hot air dried ones in terms of rehydration capacity [61] because of porous structure. Combined drying of hot air and microwave-vacuum resulted in a dried mushrooms of superior quality when compared with the hot air-dried ones, exhibiting lower overall color variation, higher porosity, greater rehydration capacity, and softer texture [82]. In addition, Markowski et al. [83] found that MVD resulted in puffed potato particles characterized by a porous microstructure with a network of open cavities; however, hot air drying yielded potato particles with compacted cells characterized by very few open micro-cavities. Lower vacuum level and microwave power level are recommended to reduce the loss of volatiles in MVD of rosemary [84]. Setiady et al. [85] reported that cells of potatoes dried using MVD remained intact after rehydration and retained suitable textural and color properties. In addition, sensory panelists ranked MVD potatoes ahead of those dried using freeze drying and hot air drying. Vacuum microwave-treated beetroots as well as freeze dried ones exhibited lower compressive strength, better rehydration capacity, and higher antioxidant capacity than those dehydrated by convective method [80].

Microwave drying has several advantages such as increasing the drying rate, improving the energy efficiency, and product quality as mentioned before. However, it has a limitation of uneven heating. In microwave drying, this may result from nonuniform moisture distribution in the drying material and/or nonuniform distribution of the electromagnetic field in the microwave cavity. In the case of food matrices-containing solid and liquid phases, the differences in dielectric and thermo-physical properties between these phases influence the heating characteristics.

Combination of microwave and fluidized bed drying compensate some of the drawbacks of each other. Agitation of particles in the drying bed facilitates heat and mass transfer due to constantly renewed boundary layer at the particle surface. Heat and mass transfer rates between the particles and drying medium are higher because of the large contact area between the particles and drying medium and good mixing of solids. The uniformity of the temperature among the particles and nearly uniform moisture distribution throughout the bed can be provided by well mixing due to fluidization in fluidized bed drying, and the drying times can be reduced by using microwave energy. Microwave-assisted fluidized bed dryer has been recently studied for drying of shelled corn [86].

The spouted bed technique is a modified form of conventional fluidization techniques which facilitates agitation of relatively coarse particles such as wheat grains. Microwave-assisted spouted bed drying could achieve a good expansion ratio, breaking force, and rehydration ratio in drying of potato cubes [87]. Microwave-assisted spouted bed drying was also used for drying of parboiled wheat for bulgur production [62]. It was observed that as microwave intensity increased, drying rate increased significantly (Figure 21.3). This can be due to higher steam pressure developed inside the particles at higher microwave intensity. Relatively, large amounts of internal heating when microwave was used results in higher moisture vapor generation inside the grain. This increase in vapor pressure promotes the transfer of moisture to the outside and creates higher drying rates than conventional drying.

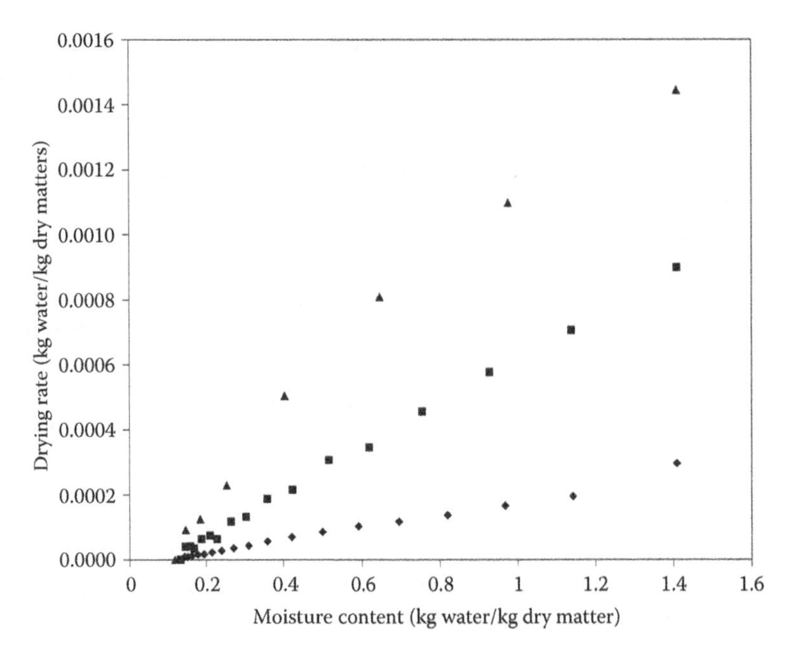

FIGURE 21.3 Drying rate curves for spouted bed drying and microwave-assisted spouted bed drying at 50°C. (♦) Spouted bed only, (■) 3.5 W/g (db), and (▲) 7.5 W/g (db). (From *Food Bioprod. Process.*, Kahyaoglu, L.N., Sahin, S., and Sumnu, G., Spouted bed and microwave-assisted spouted bed drying of parboiled wheat, in print, Copyright 2012, with permission from Elsevier.)

More porous structure was observed in wheat samples dried in microwave-assisted spouted bed compared to air dried ones [88]. Sphericity and bulk density were higher when high temperature was combined with high microwave power.

In addition to microwave-assisted fluidized or spouted bed dryers, there are some designs to improve uniformity of heating in microwave ovens. A rotary device was built in a home-type microwave oven to improve uniformity [89], and this was tried for drying of soy bean.

Different combinations of drying methods were compared for drying of carrot pieces [90]. Carrot pieces dried using MFD method were better than MVD and microwave-enhanced spouted bed dried products with respect to rehydration ratio and retention of carotene and vitamin C. However, the energy consumption in MFD method was the highest. In addition, MVD and microwave-enhanced spouted bed drying were much faster than MFD. Similarly, microwave freeze dried potato/banana restructured chip samples had better rehydration ratio than microwave vacuum dried samples [91]. In addition, the hardness of microwave vacuum dried samples was about three times higher than microwave freeze dried ones.

Freeze dried blueberries had higher antioxidant activity followed by the combination of hot air-microwave vacuum, microwave-vacuum, and hot air drying methods [92]. Carotene retention of carrot slices and the vitamin C retention of apple slices dried by combined microwave vacuum and freeze drying method were close to those of freeze dried ones and much better than those of conventional hot air dried ones [93]. Rehydration capacity, color retention, and texture of carrot and apple chips dried in combination method were similar to those of freeze dried ones.

Microwave can also be combined with infrared (IR) to decrease drying time. Microwave and IR-assisted microwave (IR-microwave) drying was used for the production of bread crumbs [94].

Apart from combining different drying techniques, drying efficiency can be improved using intermittent drying. Intermittent convective drying is drying with different drying rates in several periods. It allows more time for the internal moisture to diffuse to the surface part of the material, when no heat is supplied or a reduced level of heating is provided, or when the humidity of drying air is increased. Comparison of the drying kinetics showed that intermittent microwave drying was an effective method for shortening the drying time as compared to the intermittent IR drying and convective-microwave drying [95].

21.4.3 Microwave Baking

The studies in recent years about baking are mostly related to microwave baking of cakes and using combination baking for different products. In the study of Garcia-Zaragoza et al. [96], the glycemic index of pound cake baked in a two-cycle microwave toaster and a conventional oven was compared. No significant differences ($P > 0.05$) were found in the products baked in microwave oven and conventional oven with respect to hydrolysis index and glycemic index.

Cells of the crumbs of pound cakes baked in microwave were 20% larger as compared to the ones baked in conventional oven [97]. However, shape factor was 0.81 for both microwave and conventionally baked crumbs, and crumbs from both oven types were similar in appearance. Conventionally baked product had a greater amount of protein matrix; however, the matrix structure of crumb of microwave-baked pound cakes was comparable with that of conventionally baked ones. It was concluded that pound cake dough with its high content of fat, sugar, and moisture can be baked using microwave.

In another study, cakes baked in the microwave oven at 250 W showed improved textural properties (springiness, moisture content, and firmness) as compared to cakes baked in the convective oven [98]. In addition, microwave baking allowed for up to 93% reduction in baking time, in comparison with convective baking of Madeira cake batter. The variation in sample center temperature exhibited two distinct stages; a short "warming-up" period, and a "constant temperature (approximately 103°C)" stage. A temperature profile through the cake depth revealed that the surface temperature was lower than the center.

Faster staling in microwave-baked products as compared to the conventionally baked ones is an important problem that should be solved for the implication of microwave for baking. There are some studies to retard staling in microwave-baked products. In the study of Seyhun et al. [99], different types of starches (corn, potato, waxy corn, amylomaize, and pregelatinized starches) were added to cake formulations to retard staling of microwave-baked cakes. Starches, except amylomaize, were effective in reducing firmness during storage. The most effective starch type for retardation of staling of microwave-baked cakes, was pregelatinized starch. More amylose was leached in control cakes, which do not have additional starch, during microwave baking than conventionally baked cakes.

Although baking time is significantly reduced in microwave oven, there are some quality defects in microwave-baked products such as firm and tough texture and lack of color and crust [100]. Therefore, hybrid ovens which combine microwave with other heating methods can be recommended. Quality of cakes baked in microwave, IR, IR-microwave combination and conventional ovens were compared in terms of weight loss, specific volume, firmness, and color of cakes [101]. Cakes baked in a microwave oven had the lowest quality. However, IR-microwave combination oven produced cakes having similar firmness and color values with conventionally baked ones.

IR-microwave combination baking can also be used for baking of cookies [102]. The spread ratio of conventionally baked cookies was significantly lower than that of the cookies baked in IR-microwave oven. IR power, microwave power, and baking time were found to have a significant effect on the spread ratio of the cookies.

The effects of gums on macro- and microstructure of breads baked in IR-microwave combination and conventional oven were investigated [103]. Heating mechanism was found to be effective on pore structure development. Breads baked in IR-microwave combination oven were more porous than conventionally baked ones (Figures 21.4 and 21.5). Pores of conventionally baked breads had spherical, oval-like shapes while pores of breads baked in IR-microwave combination oven form

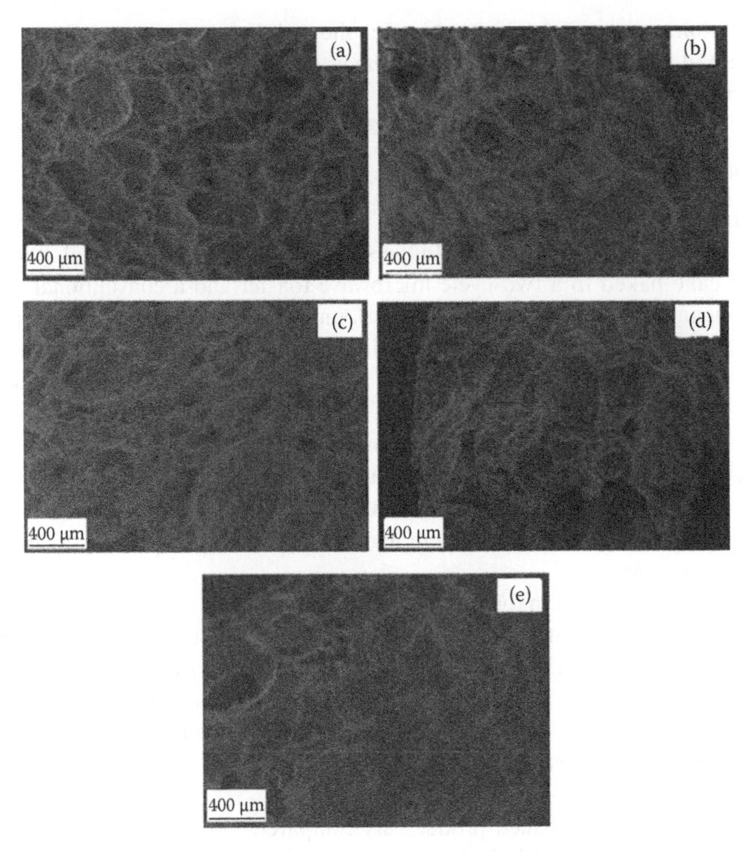

FIGURE 21.4 Microstructure at 70× magnification of (a) no gum, (b) xanthan, (c) guar, (d) xanthan-guar blend, and (e) κ-carrageenan added breads baked in conventional oven. (From *Food Hydrocolloids*, 23, Ozkoc, S.O., Sumnu, G., and Sahin, S., 2182–2189, Copyright 2009, with permission from Elsevier.)

channels. Control breads (containing no gum) baked in IR-microwave combination oven had larger pores than those baked in conventional oven. The addition of xanthan gum improved uniformity of breads baked in IR-microwave combination oven. There are differences in microstructure of breads baked in different ovens. Most of the starch granules were distorted in conventionally baked breads. On the other hand, in IR-microwave-baked breads, both granular and deformed starch structures were observed.

IR-microwave oven can also be used to produce gluten-free baked products. Baking conditions in IR-microwave combination oven and formulations of gluten-free cakes made from rice flour were optimized using response surface methodology [104]. The optimum conditions and formulations can be seen in Table 21.3. Cakes baked in IR-microwave oven had similar color and firmness values to those of conventional oven (Table 21.4).

Staling of breads baked in different ovens (microwave, IR-microwave combination, and conventional) was investigated [105]. Retrogradation enthalpies of microwave-baked breads were the highest (Figure 21.6). Retrogradation enthalpies of samples baked in IR-microwave combination oven were less than the values of microwave-baked breads but more than the values of conventionally baked ones. This shows that combination baking partially solved rapid staling problem. The hardness, setback viscosity, and crystallinity values of microwave-baked samples were also found to be the highest among other heating modes. Using IR-microwave combination heating made it possible to produce breads with similar staling degrees as conventionally baked ones in terms of retrogradation enthalpy and FTIR outputs related to starch retrogradation.

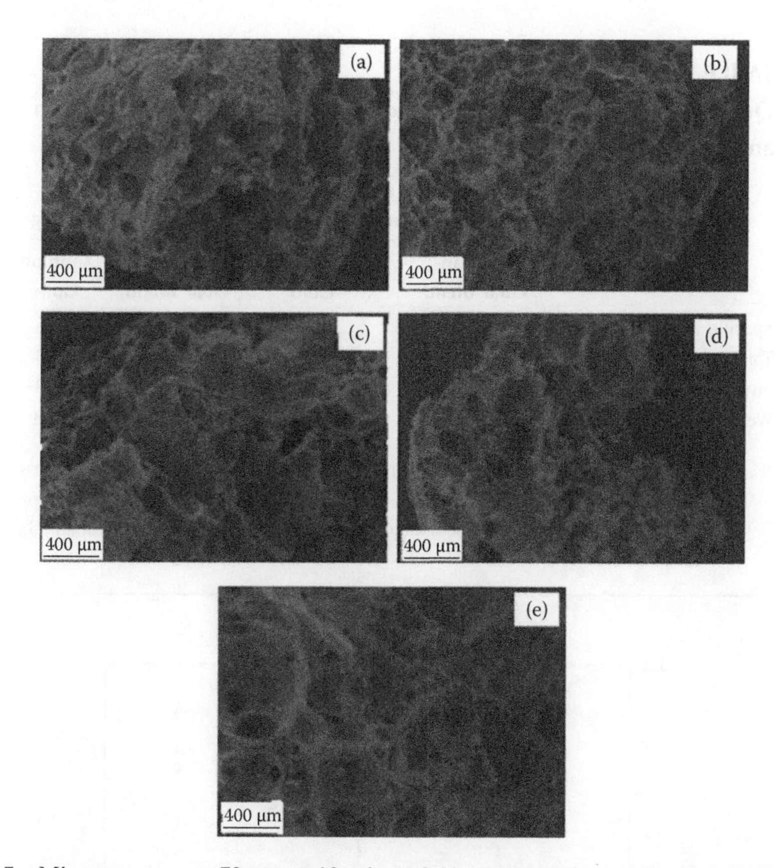

FIGURE 21.5 Microstructure at 70× magnification of (a) no gum, (b) xanthan, (c) guar, (d) xanthan-guar blend, and (e) κ-carrageenan added breads baked in IR-microwave combination oven. (From *Food Hydrocolloids*, 23, Ozkoc, S.O., Sumnu, G., and Sahin, S., 2182–2189, Copyright 2009, with permission from Elsevier.)

TABLE 21.3
Optimum Formulations and Baking Conditions in Infrared-Microwave Combination Oven for Rice Cakes Containing Xanthan Gum and Xanthan-Guar Gum Blend

	Xanthan-Guar Gum Blend		Xanthan Gum	
	Coded	Uncoded	Coded	Uncoded
Emulsifier content (%)	0.56	4.68	0.88	5.28
Halogen lamp power (%)	1.00	70	0.08	60
Baking time (min)	−1.00	7	−0.99	7

Source: With kind permission from Springer Science+Business Media: *Food Bioprocess Techn.*, Optimization of baking of rice cakes in infrared-microwave combination oven by response surface methodology, 1, 2008, 64–73, Turabi, E., Sumnu, G., and Sahin, S.

TABLE 21.4
Quality Characteristics of the Cakes Baked in Conventional Oven and at Optimum Conditions of IR-Microwave Combination Oven

	IR-Microwave Combination Baking		Conventional Baking	
	Xanthan-Guar Blend	Xanthan Gum	Xanthan-Guar Blend	Xanthan Gum
Specific volume (mL/g)	1.57	1.75	1.30	1.38
Total color change (ΔE)	32.18	30.56	26.43	25.34
Firmness (N)	2.58	2.32	2.35	2.13
Weight loss (%)	13.69	11.94	10.65	7.70

Source: With kind permission from Springer Science+Business Media: *Food Bioprocess Techn.*, Optimization of baking of rice cakes in infrared-microwave combination oven by response surface methodology, 1, 2008, 64–73, Turabi, E., Sumnu, G., and Sahin, S.

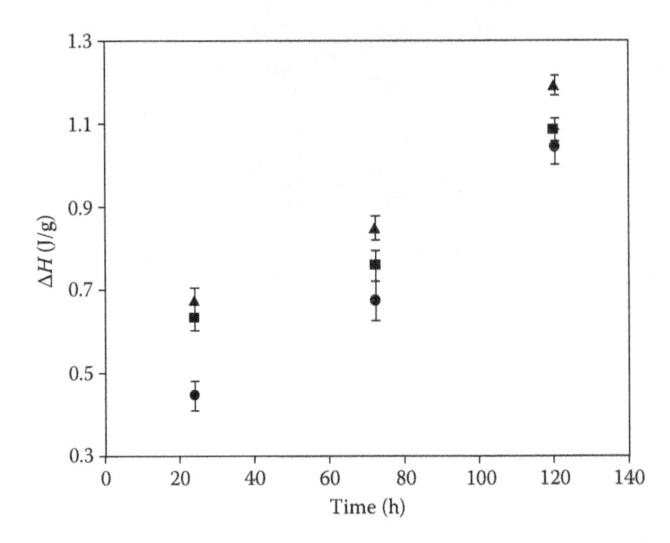

FIGURE 21.6 Variation in retrogradation enthalpy of baked in different ovens during storage (▲) microwave oven, (■) infrared-microwave combination oven, and (•) conventional oven. (With kind permission from Springer Science+Business Media: *Eur. Food Res. Technol.*, Investigation of physicochemical properties of breads baked in microwave and infrared-microwave combination ovens during storage, 228, 2009, 883–893, Ozkoc, S.O., Sumnu, G., Sahin, S., and Turabi, E.)

21.4.4 Microwave Disinfestation

Since losses of grain due to insect infestation are high, disinfestation of grain is very important for the safe storage of grain in the legume industry. Fumigation with methyl bromide (MeBr) is commonly used for postharvest insect control. MeBr fumigation is only practical at treatment temperatures ≥5°C, with lower treatment temperatures requiring higher doses or extended exposure times [106]. This might not be easy. It is required to develop a practical alternative method to control insect pests in legumes to have a minimum impact on product quality and environment. Heat disinfestation treatments are relatively easy to apply, leave no chemical residues, and may offer some fungicidal activity. However, it is difficult to accomplish disinfestation using conventional hot air heating methods without causing deleterious effects to product quality. Low heating rates also may increase the thermotolerance of the targeted insects. Thus, there has been an increasing interest in developing advanced thermal treatments for postharvest insect control in legumes.

Microwaves can be used to control insects in stored cereals and cereal products. The use of microwaves to kill insects is based on the dielectric heating effect produced in grain, which is a relatively poor conductor of electricity. Since this heating depends upon the electrical properties of the material, there is a possibility of advantageous selective heating in mixtures of different substances. The major advantage of using microwave energy is that no chemical residues are left in the food. Control of insect infestation by microwave power has been studied by many researchers [107–109].

Microwave can be used in disinfestation of grains. In a study by Phillips et al. [107], flowing grain containing different insect species at different life stages were exposed to microwave to determine the most tolerant ones to microwaves. Among the species of lesser grain borer, there is rice weevil or red flour beetle. The most susceptible insects were pupae of the rice weevil, and the most tolerant were larvae of the red flour beetle and eggs of the lesser grain borer. No insects survived treatment, and grain showed no market quality loss as determined in milling and baking studies.

The mortality of three common species of stored-grain insects, namely, *Tribolium castaneum*, *Cryptolestes ferrugineus*, and *Sitophilus granarius*, has been studied [108]. Infested wheat samples having different moisture contents were exposed to microwave energy at different power levels, for different times. Complete kill of adults of all three species and of postembryonic stages of *T. castaneum* was achieved at 500 W with an exposure time of 28 s. Mortality of insects increased with either power or exposure time or both. Germination of wheat kernels was lower after treatment with microwave energy. There was no significant difference in the quality of grain protein, flour protein, flour yield, flour ash, and loaf volume of wheat treated with microwave energy.

Similar study has been conducted for rye infested with *T. castaneum* (Herbst) [109]. Eggs were the most vulnerable while adults were the least vulnerable. Subsequent tests showed that microwaves reduced rye germination and flour yield. No effect of treatment on grain protein content, falling number, sodium dodecyl sulfate sedimentation volume, mixograph and farinograph dough development times, or baking properties was detected.

21.4.5 Microwave-Assisted Extraction

In recent years, microwave-assisted extraction (MAE) has been considered as an alternative to conventional extraction methods due to its several advantages such as shorter extraction time, less amount of solvent usage, and higher extraction rate. In MAE, both solvent and the plant materials are heated by microwaves depending on their dielectric properties. Water molecules inside the plant sample are interacted with microwaves by dipolar rotation mechanism, and the plant is heated. Internal pressure inside the plant increases and ruptures the cell wall. This enhances the leaching out of the active material from the plant which increases the extraction rate. The efficiency of microwave heating depends on the dissipation factor of solvents. The dissipation factor is defined as the ratio of dielectric loss factor to dielectric constant which represents the ability of the solvent to absorb microwave energy and transfer it as heat to the surrounding molecules [110].

In Table 21.5, the dielectric constants and dissipation factors of the solvents used in MAE is given. As can be seen in Table 21.5, methanol and 2-proponal are the two solvents with the highest dissipation factor. Therefore, it is expected that extraction time will be shortened when these solvents are used. On the other hand, hexane is almost transparent to microwaves so almost no heat is produced during the extraction.

MAE was used to extract phenolic compounds from flax seed [113], citrus peels [114], sour cherry pomace [115], green tea [116], and beans [117]. MAE of flax seed resulted in higher phenolic yield as compared to conventional extraction, and the extraction time was reduced significantly [113]. Hayat et al. [114] used three different methods (microwave, ultrasound, and rotary)

TABLE 21.5
Dielectric Constants and Dissipation Factors of Some Solvents at 20°C Which Are Commonly Used in MAE

Solvents	Dielectric Constant	Dissipation Factor ×10⁴	References
Acetone	20.7	5555	[111]
Ethanol	24.3	2500	[112]
Hexane	1.89	0.1	[111]
Methanol	32.6	6400	[111,112]
2-Propanol	19.9	6700	[112]
Water	78.3	1570	[111,112]

on extraction of phenolic compounds from citrus peels. Microwave extraction was found to be the best method to achieve higher yields and higher antioxidant activity in a shorter time. In the MAE of phenolic compound from sour cherry pomace [115], it was found that solvent type, microwave power, solvent to solid ratio, and extraction time affected total phenolic content significantly. As can be seen in Figure 21.7, the highest phenolic content was reached with the mixture of ethanol–water (1:1 v/v) followed by ethanol and water for both microwave power levels. As explained before, the difference in the effects of solvents on total phenolic content is related to the dielectric properties. The higher effectiveness of ethanol–water mixture was associated with higher dissipation factor. In another study, it was also shown that solvent type was the most important parameter for the recovery of anthocyanins from grape skins by MAE [118]. When the effects of MAE conditions on tea polyphenols were studied, the most important factor affecting extraction rate was found to be the extraction time which was followed by microwave intensity [116]. Total phenolic contents of beans determined by MAE with water at 100°C were two to three times as much as those determined by conventional extraction with water at the same temperature [117].

MAE was also used to extract fat from cacao powder and nibs [119]. The solvent hexane used in extraction is almost transparent to microwaves. However, heating of cacao can take place since

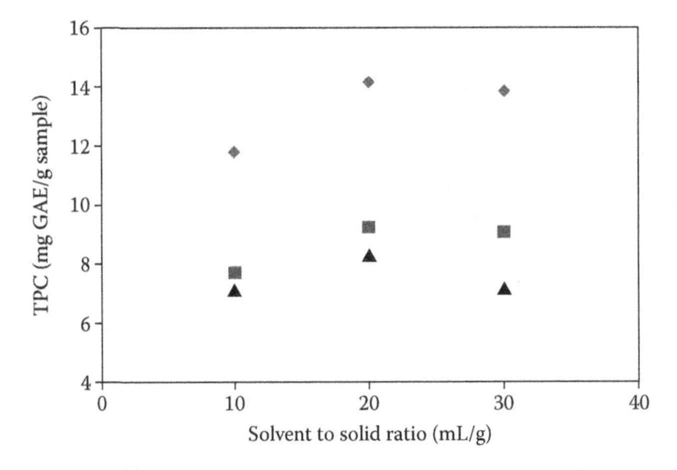

FIGURE 21.7 The effects of solvent type and different solvent to solid ratios on totol phenolic content of sour cherry pomace extracts obtained using 400 W microwave power level and 14 min. (♦) Ethanol-water, (■) ethanol, and (▲) water. (From Simsek, M., Microwave extraction of phenolic compounds from fruit pomaces, MSc Thesis, Middle East Technical University, Ankara, Turkey, 2010.)

moisture content of cacao is interacted by microwaves. Thus, fat is diffused to the surrounding solvent which solubilizes fat rapidly.

There are two types of commercial MAE systems which are closed-vessel systems (under controlled pressure and temperature) or open-vessel systems (under atmospheric pressure) [120]. The maximum temperature of sample in open-vessel systems is determined by the boiling point of the solvent. High temperatures under high pressures can be attained in closed systems.

MAE can also be used without the addition of any solvent. In solvent-free microwave extraction (SFME), microwave hydrodiffusion and gravity (MHG), and pressurized solvent-free microwave extraction (PSFME) methods, no solvents are added and extraction takes place at atmospheric pressure. SFME was developed by Lucchesi et al. [121] for the isolation and concentration of essential oils in plant materials. In this method, microwave heating is combined with dry distillation. Fresh or rehydrated plant is placed in the microwave oven which is connected to a Clevenger apparatus. As explained in MAE, the internal heating of moisture inside the plant results in cell burst. Essential oil is freed and drawn by the water inside the plant by distillation. The vapor passes through the condenser and condenses while the distillate is collected continuously. This method has been used to extract essential oils from different spices or fruits such as oregano [122], laurel [123], *Rosmarinus officinalis L.* [124], *Thymus vulgaris L.* [125], and mango [126].

The principle of MHG is based on diffusion of biomaterials from plants as a result of microwave heating and collection of extract by gravity [127]. MHG was used to extract essential oil from citrus peels [128], dried caraway [129], and flavanols from onion [130]. The essential oil obtained from rosemary by MHG and hydrodistillation was similar in their composition, but MHG was shown to be quicker (15 min) than hydrodistillation (180 min) [128]. In addition, MHG was more energy efficient (0.2 kWh) than hydrodistillation (3 kWh).

PSFME is a new method where plant material is introduced to the closed system without any solvent addition. The moisture inside the plant absorbs microwaves, and internal heating takes place, resulting in a pressure increase inside the cells. The high pressure breaks the cell walls, and compounds are released from the plant [131]. PSME has been recently tried on extraction of phenolic compounds from berries. In PSFME, berry extract with the highest phenolic content and antioxidant activity were obtained. In addition, in PSFME, the compounds of quercetin and isorhamnetin were obtained which could not be extracted by other methods such as pressurized liquid extraction, pressing, and maceration.

Microwave steam diffusion has been recently compared with steam diffusion for extraction of essential oil from orange peel [132]. Yield and composition of essential oil extracted by microwave steam diffusion were comparable with steam diffusion. However, combining microwaves with saturated steam enhances the release of essential oil present inside cells of plant tissues. Microwave steam diffusion was shown to have advantages of savings in time and energy, cleanness, and reduction in waste water as compared steam diffusion.

21.4.6 Microwave Tempering/Thawing

Microwave thawing has the advantages of fast thawing, reduction in microbial growth, chemical deterioration and drip loss, less space requirements, and low cost [133].

Microwaves can easily thaw food with a small size but there are problems due to the uneven heating if microwaves are used in the industry to thaw a very large amount of frozen foods. Thus, the major restriction in microwave thawing is runaway heating. Runaway heating takes place due to uneven temperature distribution within the food. Since there are significant differences in dielectric properties of ice and water, unfrozen part of food containing water will be heated faster than unfrozen parts. In order to avoid runaway heating, microwave tempering instead of thawing is usually preferred.

Tempering is defined as increasing the temperature of a frozen food to temperature below freezing point (–5°C to –2°C) at which the food will still be firm but can be further processed [134].

In microwave tempering, all parts of frozen food will be at a temperature below the freezing point which reduces the nonuniform heat distribution within the food [135].

There are some studies in literature where microwave thawing of foods have been modeled [133,136,137]. Modeling temperature distribution inside the frozen potato puree during tempering in combined IR-microwave oven has been recently studied [23]. It has been found that the increase in microwave power level and IR power decreased tempering times. Combining microwaves with IR changed the heating profile of microwave tempering. In another study, in order to prevent runaway heating, continuous microwave power was combined with air convection and microwave power cycles were applied together with refrigerated air to keep surface temperature low [137]. A 3D mathematical model was developed to solve the unsteady state heat and mass transfer equations for microwave thawing.

Besides modeling studies, effects of microwave thawing on quality of foods have been studied. Oz [138] compared the effects of different thawing methods (microwave oven and refrigerator) on heterocyclic aromatic amine contents of beef cooked by different methods. Thawing methods did not have a significant effect on the heterocyclic aromatic amine contents of samples.

Quality changes of anchovy and bluefish thawed by different methods (refrigerator, microwave, and running water) were compared [139]. Microwave thawing prolonged the spoilage of fish. The spoilage of samples thawed in water and a refrigerator took place after the third thawing process. However, microwave-thawed samples spoilt after the fourth thawing process. In another study where the effect of different thawing methods (microwave and refrigerator) on quality of shrimp were studied, it was found that microwave thawing resulted in higher thawing loss than thawing in refrigerator [140]. This may be due to rapid internal heating, which enhanced the vaporization of water from the samples, in microwave thawing. Moreover, the frozen foods contain frozen and unfrozen part and a nonuniform distribution of lipids which have different dielectric properties. This may result in localized heating while the other parts remain frozen. Thus, drip loss increased. However, it was found that thawing method did not affect the texture of shrimps significantly.

The effect of microwave thawing on quality of breads prepared using different flour types and gums was investigated [141]. After thawing, a significant shrinkage was observed, a porosity reduction in all samples, a major softening of the crust in control samples, a viscous crumb in most samples, a more rubbery crust especially in samples containing guar gum were observed. A more rubbery crust especially in samples containing guar gum was also noticed, which can be ascribed to water-binding capacity of gums or to gums/water interactions and/or to inhibition effect of gluten-starch interactions of gums influencing the plasticity of the bread.

21.4.7 Microwave Pasteurization

The use of microwave processing technology is rapidly becoming a viable method for pasteurization and sterilization of high- and low-acid foods. Previous studies have utilized continuous flow microwave processing to aseptically processed and packaged fruit and vegetable purees in large flexible containers [142], as well as in-pack sterilization of prepared meals [143]. Retention of product color and antioxidant activities has been reported using 915 MHz microwaves [142,144,145].

Nonuniform temperature distribution and control of processing parameters are the major drawbacks of microwave pasteurization and sterilization. Kumar et al. [146] studied methods to overcome this problem and found that static mixers, installed at the exit of each of the microwave applicators, improved temperature uniformity for both green pea and carrot purees.

Pasteurization of mechanically tenderized beef inoculated with *E. coli* O157:H7 has been studied using temperature controlled two-stage microwave heating [147]. In the primary heating stage, the sample surface temperature was increased to an initial temperature set point. In the secondary heating stage, the heating was continued with a small fraction of microwave power. Secondary heating stage was more effective for inactivation of *E. coli* and background microflora. Complete inactivation of *E. coli* and background microflora was observed with heating at temperatures above 70°C for more than 1 min.

It was possible to reduce *Listeria monocytogenes* by 7-log in inoculated beef frankfurters using 600 W microwave oven for 12–15 min [148]. Huang and Sites [149] demonstrated the feasibility of a proportional integral differential controlled microwave heating process for in-package pasteurization of frankfurters. Overall rate of inactivation of *Listeria monocytogenes* was 30%–75% higher with microwave in-package pasteurization as compared to water immersion heating at the same surface temperatures.

Microbial contamination of eggs, particularly with pathogenic bacteria is of increasing concern globally. As a result, the research on the shell egg pasteurization has drawn the attention of researchers. Functional properties of egg white after in-shell pasteurization by microwave and by immersion in hot water were compared [42]. Microwave pasteurized in-shell eggs had nearly 20% higher enthalpy of denaturation, four times more viscous, 50% more stable foam with 30% less foam density, and 78% clearer as compared to the hot water pasteurized in-shell egg white.

Continuous flow microwave pasteurization system was designed and evaluated for apple ciders from cold press and hot press [150]. Process lethality was verified based on inoculation of *Escherichia coli* 25922 in apple cider, in which the pasteurization process resulted in 5 log reduction. Continuous flow microwave system was also used in the study of Kumar et al. [151] in which the feasibility of aseptic processing of salsa con queso was shown.

Chen et al. [152] developed a numerical model to couple electromagnetic- and thermal-field calculations in packaged foods moving in pressurized 915 MHz microwave cavities with circulating water at above 120°C. The simulation model was validated with a pilot-scale microwave system using direct temperature measurement data and indirect color patterns in whey protein gels *via* formation of the thermally induced chemical marker M-2 (4-hydroxy-5-methy-3(2H)-furanone) which is a chemical compound formed as a result of Millard reaction.

Two-magnetron-in-series heating system was developed to increase heating capacity [153]. It was concluded that two-magnetron-in-series heating system could be applied to the small-scale pasteurization of liquid food. Its heating performance was increased by twofold as compared to that of the two single-magnetron heating systems, resulting in energy saving of 7%.

Cinquanta et al. [154] studied the effects of orange juice batch pasteurization using improved pilot-scale microwave oven on activity of pectin methyl esterase enzyme, color, carotenoid, and vitamin C contents have been studied. As an indicator enzyme for orange juice pasteurization, the inactivation of heat sensitive fraction of pectin methyl-esterase enzyme showed a z value of 22.1°C. Microwave pasteurization at 70°C for 1 min reduced carotenoid content by about 13%. Among the seven carotenoid compounds quantified, zeazanthin and β-carotene content decreased by about 26%, while no differences were observed for β-cryptoxanthin. The decrease in vitamin C content after microwave heating was slight. Degradation of carotenoids in orange juice during microwave heating at different time/temperature combinations has also been studied by Fratianni et al [155]. The z value for total carotenoids was 14.2°C. Among the carotenoid compounds, antheraxanthin was the less temperature dependent with a z value of 16.7°C, while β-carotene was the most temperature dependent with a z value of 10.9°C.

D values associated alkaline phosphatase inactivation under microwave heating were an order of magnitude lower than under conventional thermal heating [156]. In other words, alkaline phosphatase inactivation occurred much faster under microwave heating condition than under conventional heating thereby confirming the existence of enhanced thermal effects from microwave.

The inactivation kinetics of enzymes polyphenol oxidase and peroxidase was also studied for batch microwave treatment of green coconut water [157]. Thermal inactivation of these enzymes during microwave processing was significantly faster in comparison with conventional processes reported in literature.

21.5 CONCLUSIONS

Microwave heating has the advantages of faster heating, high energy efficiency, space savings, selective heating, and obtaining highly nutritious products. Microwave can be used for reheating, drying, baking, disinfestation, extraction, thawing/tempering, and pasteurization. One of the drawbacks of microwave heating is nonuniform heating of foods. There is also the problem of lack of color and flavor formation, tough texture, and rapid staling in microwave-baked products. Combination of microwave with other heating methods can be used to overcome these drawbacks. In drying microwave can be combined with convective air drying, freeze drying, vacuum drying, spouted, or fluidized bed drying. By the addition of dielectric materials with high loss factor or by changing the dielectric properties of food material to be dried, microwave drying rate can be enhanced. In baking, IR-microwave ovens are commonly used to improve product quality. Microwave disinfestation prevents the risk of chemical residues left in the food. SFME has the advantages of savings in time and energy, cleanness, and reduction in waste water. In recent years, different designs of SFME have been developed to improve microwave extraction efficiency. There is still need for research on microwave processing to widen the application areas.

REFERENCES

1. S Sahin, SG Sumnu. *Physical Properties of Foods*. New York, NY: Springer, 2006, pp. 173–174, 178.
2. AC Metaxas, RJ Meredith. *Industrial Microwave Heating*. London: Peter Peregrinus, 1983, p. 72.
3. MS Venkatesh, GSV Raghavan. An overview of microwave processing and dielectric properties of agri-food materials. *Biosystems Engineering* 88:1–18, 2004.
4. AK Datta, G Sumnu, GSV Raghavan. Dielectric properties of foods. In: MA Rao, SSH Rizvi, AK Datta, eds. *Engineering Properties of Foods*, 3rd edn. Boca Raton, FL: CRC Press Taylor & Francis Group, 2005, pp. 501–565.
5. ME Sosa-Marales, L Valerio-Junco, A Lopez-Malo, HS Garcia. Dielectric properties of foods: Reported data in the 21st century and their potential applications. *LWT-Food Science and Technology* 43:1169–1179, 2010.
6. W Guo, X Zhu, SO Nerlson, R Yue, H Liu, Y Liu. Maturity effects on dielectric properties of apples from 10 to 4500 MHz. *LWT-Food Science and Technology* 44:224–230, 2011.
7. JG Lyng, L Zhang, NP Brunton. A survey of the dielectric properties of meat and ingredients used in meat product manufacture. *Meat Science* 89:589–602, 2005.
8. G Sumnu, AK Datta, S Sahin, SO Keskin, V Rakesh. Transport and related properties of breads baked using various heating modes. *Journal of Food Engineering* 78:1382–1387, 2007.
9. J Ahmed, HS Ramaswamy, VGS Raghavan. Dielectric properties of butter in the microwave frequency range as affected by salt and temperature. *Journal of Food Engineering* 82:351–358, 2007.
10. AH Al-Muhtaseb, MA Hararah, EK Megahey, WAM McMinn, TRA Magee. Dielectric properties of microwave baked cake and its constituents over a frequency range of 0.915–2.450 GHz. *Journal of Food Engineering* 98:84–92, 2010.
11. A Garcia, JL Torres, M De Blas, A De Francisco, R Illanes. Dielectric characteristics of grape juice and wine. *Biosystems Engineering* 88:343–349, 2004.
12. W Guo, Y Liu, X Zhu, S Wang. Temperature dependent dielectric properties of honey associated with dielectric heating. *Journal of Food Engineering* 102:209–216, 2011.
13. W Guo, X Zhu, H Liu, R Yue, S Wang. Effects of milk concentration and freshness on microwave dielectric properties. *Journal of Food Engineering* 99:344–350, 2010.
14. E Turabi, M Regier, G Sumnu, S Sahin, M Rother. Dielectric and thermal properties of rice cake formulations containing different gum types. *International Journal of Food Properties* 13:1199–1206, 2010.
15. Y Wang, JM Tang, B Rasco, F Kong, S Wang. Dielectric properties of salmon fillets as a function of temperature and composition. *Journal of Food Engineering* 87:236–246, 2008.
16. R De los Reyes, A Heredia, P Fito, E De los Reyes, A Andrés. Dielectric spectroscopy of osmotic solutions and osmotically dehydrated tomato products. *Journal of Food Engineering* 80:1218–1225, 2007.
17. Y Liu, J Tang, Z Mao. Analysis of bread loss factor using Debye equations. *Journal of Food Engineering* 93:453–459, 2009.

18. L Zhang, JG Lyng, NP Brunton. The effect of fat, water and salt on the thermal and dielectric properties of meat batter and its temperature following microwave or radio frequency heating. *Journal of Food Engineering* 80:142–151, 2007.
19. LM Guardeno, JM Catalá-Civera, P Plaza-González, T Sanz, A Salvador, SM Fiszman, I Hernando. Dielectrical, microstructural and flow properties of sauce model systems based on starch, gums and salt. *Journal of Food Engineering* 98:34–43, 2010.
20. J Ahmed, HS Ramaswamy, GSV Raghavan. Dielectric properties of soyprotein isolate dispersions as a function of concentration, temperature and PH. *LWT-Food Science and Technology* 41:71–81, 2008.
21. O Sakiyan, G Sumnu, S Sahin, V Meda. Investigation of dielectric properties of different cake formulations during microwave and infrared-microwave combination baking. *Journal of Food Science* 72:E205–E213, 2007.
22. J Tang, F Hao, M Lau. Microwave heating in food processing. In: XH Yang, J Tang, eds. *Advances in Bioprocessing Engineering*. Hackensack, NJ: World Scientific Publishing Co., 2002, pp. 1–44.
23. N Seyhun, H Ramaswamy, G Sumnu, S Sahin, J Ahmed. Comparison and modeling of microwave tempering and infrared assisted microwave tempering of frozen puree. *Journal of Food Engineering* 92:339–344, 2009.
24. S Ryynanen. The electromagnetic properties of food materials—A review of basic principles. *Journal of Food Engineering* 26:409–429, 1995.
25. SO Nelson, WR Forbus, KC Lawrence. Microwave permittivities of fresh fruits and vegetables from 0.2 to 20GHz. *Transactions of the ASAE* 37:183–189, 1994.
26. W Gua, S Wang, G Tiwari, JA Johnson, J Tang. Temperature and moisture dependent dielectric properties of legume flour associated with dielectric heating. *LWT-Food Science and Technology* 43:193–201, 2010.
27. ME Martin-Esparza, N Martinez-Navarrete, A Chiralt, P Fito. Dielectric behaviour of apple (var. Granny Smith) at different moisture contents effect of vacuum impregnation. *Journal of Food Engineering* 77:51–56, 2006.
28. RF Schiffmann. Food product development for microwave processing. *Food Technology* 40:94–98, 1986.
29. SA Barringer, EA Davis, J Gordon, KG Ayappa, HT Davis. Effect of sample size on the microwave heating rate: Oil vs. water. *American Institute of Chemical Engineers Journal* 40:1433–1439, 1994.
30. G Sumnu, S Sahin. Baking using microwave processing. In: M Regier, H Schubert, eds. *Microwave Processing of Foods*. Cambridge, U.K.: Woodhead Publishing, 2005, pp. 119–142.
31. C Bilbao-Sáinz, M Butler, T Weaver, J Bent. Wheat starch gelatinization under microwave irradiation and conduction heating. *Carbohydrate Polymers* 69:224–232, 2007.
32. BJ Zylema, JA Grider, J Gordon, EA Davis. Model wheat starch systems heated by microwave irradiation and conduction with equalized heating times. *Cereal Chemistry* 62:447–453, 1985.
33. EP Sakonidou, TD Karapantsios, SN Raphaelides. Mass transfer limitations during starch gelatinization. *Carbohydrate Polymers* 53:53–61, 2003.
34. C Xue, N Sakai, M Fukuoka. Use of microwave heating to control the degree of starch gelatinization in noodles. *Journal of Food Engineering* 87:357–362, 2008.
35. G Lewandowicz, T Janowski, J Fornal. Effect of microwave radiation on physico-chemical properties and structure of cereal starches. *Carbohydrate Polymers* 42:193–199, 2000.
36. AK Anderson, HS Guraya. Effects of microwave heat-moisture treatment on properties of waxy and non-waxy rice starches. *Food Chemistry* 97:318–323, 2006.
37. SH Han, SW Lee, C Rhee. Effects of cooking methods on starch hydrolysis kinetics and digestion resistant fractions of rice and soybean. *European Food Research and Technology* 227:1315–1321, 2008.
38. Y Yin, CE Walker. A quality comparison of breads baked by conventional versus non-conventional ovens: A review. *Journal of the Science of Food and Agriculture* 67:283–291, 1995.
39. E Yalcin, O Sakiyan, G Sumnu, S Celik, H Koksel. Functional properties of microwave treated wheat gluten. *European Food Research and Technology* 227:1411–1417, 2008.
40. HH Liu, MI Kuo. Effect of microwave heating on viscoelastic property and microstructure of soy protein isolate gel. *Journal of Texture Studies* 42:1–9, 2011.
41. W Gustaw, S Mleko. Gelation of whey proteins by microwave heating. *Milchwissenschaft* 62:439–442, 2007.
42. SRS Dev, V Orsat, Y Gariepy, GSV Raghavan, C Ruiz-Feria. Selected post heating properties of microwave or hot water heated egg white for in shell pasteurization. *International Journal of Food Properties* 13:778–788, 2010.
43. T Albi, A Lanzon, A Guinda, MC Perez-Camino, M Leon. Microwave and conventional heating effects on physical and chemical parameters of edible fats. *Journal of Agricultural and Food Chemistry* 45:3000–3003, 1997.

44. T Albi, A Lanzon, A Guinda, M Leon, MC Perez-Camino. Microwave and conventional heating effects on thermoxidative degradation of edible fats. *Journal of Agricultural and Food Chemistry* 45:3795–3798, 1997.
45. F Caponio, A Pasqualone, T Gomes. Effects of conventional and microwave heating on the degradation of olive oil. *European Food Research and Technology* 215:114–117, 2002.
46. YH Chu, HF Hsu. Comparative studies of different heat treatments on quality of fried shallots and their frying oils. *Food Chemistry* 75:37–42, 2001.
47. E Chiavaro, C Barnaba, E Vittadini, MT Rodriguez-Estrada, L Cerretani, A Bendini. Microwave heating of different commercial categories of olive oil: Part II. effect on thermal properties. *Food Chemistry* 115:1393–1400, 2009.
48. E Chiavaro, MT Rodriguez-Estrada, A Bendini, M Rinaldia, L Cerretanic. Differential scanning calorimetry thermal properties and oxidative stability indices of microwave heated extra virgin olive oils. *Journal of the Science and Agriculture* 91:198–206, 2011.
49. R Malherio, I Oliveira, M Vilas-Boas, S Falcao, A Bento, JA Pereira. Effect of microwave heating with different exposure times on physical and chemical parameters of olive oil. *Food and Chemical Toxicology* 47:92–97, 2009.
50. E Valli, A Bendini, L Cerretani, SP Fu, A Segura-Carreter, MA Cremonini. Effects of heating on virgin olive oils and their blends: Focus on modifications of phenolic fraction. *Journal of Agricultural and Food Chemistry* 58:8158–8166, 2010.
51. JH Choi, JY Jeong, HY Kim, KI An, CJ Kim. The effects of electric grill and microwave oven reheating methods on the quality characteristics of precooked ground pork patties with different NaCl and phosphate levels. *Korean Journal for Food Science of Animal Resources* 28:535–542, 2008.
52. M Uzzan, E Kesselman, O Ramon, IJ Kopelman, S Mizrahi. Mechanism of textural changes induced by microwave reheating of a surimi shrimp imitation. *Journal of Food Engineering* 74:279–284, 2006.
53. M Uzzan, O Ramon, IJ Kopelman, E Kesselman, S Mizrahi. Mechanism of crumb toughening in bread-like products by microwave reheating. *Journal of Agricultural and Food Chemistry* 55:6553–6560, 2007.
54. CL Chen, PY Li, WH Hu, MH Lan, MJ Chen, HH Chen. Using HPMC to improve crust crispness in microwave-reheated battered mackerel nuggets: Water barrier effect of HPMC. *Food Hydrocolloids* 22:1337–1344, 2008.
55. NA Pinchukova, AY Voloshko, OV Shyshkin, VA Chebanov, BHP Van de Kruijs, JCL Arts, MHCL Dressen, J Meuldijk, JAJM Vekemans, LA Hulshof. A facile microwave-mediated drying process of thermally unstable/labile products. *Organic Process Research & Development* 14:1130–1139, 2010.
56. E Demirhan, B Ozbek. Drying kinetics and effective moisture diffusivity of purslane undergoing microwave heat treatment. *Korean Journal of Chemical Engineering* 27:1377–1383, 2010.
57. E Demirhan, B Ozbek. Thin-layer drying characteristics and modeling of celery leaves undergoing microwave treatment. *Chemical Engineering Communications* 198:957–975, 2011.
58. LM Bal, A Kar, S Satya, SN Naik. Drying kinetics and effective moisture diffusivity of bamboo shoot slices undergoing microwave drying. *International Journal of Food Science and Technology* 45:2321–2328, 2010.
59. D Arslan, MM Ozcan. Study the effect of sun, oven and microwave drying on quality of onion slices. *LWT-Food Science and Technology* 43:1121–1127, 2010.
60. A Sarimeseli. Microwave drying characteristics of coriander (*Coriandrum sativum L.*) leaves. *Energy Conversion and Management* 52:1449–1453, 2011.
61. N Therdthai, H Northongkom. Characterization of hot air drying and microwave vacuum drying of fingerroot (*Boesenbergia pandurata*). *International Journal of Food Science and Technology* 46:601–607, 2011.
62. LN Kahyaoglu, S Sahin, G Sumnu. Spouted bed and microwave-assisted spouted bed drying of parboiled wheat. *Food and Bioproducts Processing*, In print: 2012. doi:10.1016/j.fbp.2011.06.003.
63. SZ Fang, ZF Wang, XS Hu, H Li, WR Long, R Wang. Shrinkage and quality characteristics of whole fruit of Chinese jujube (Zizyphus jujuba Miller) in microwave drying source. *International Journal of Food Science and Technology* 45:2463–2469, 2010.
64. E Demirhan, B Ozbek. Microwave-drying characteristics of basil. *Journal of Food Process Engineering* 34:476–494, 2010.
65. Y Ando, T Orikasa, T Shiina, I Sotome, S Isobe, Y Muramatsu, A Tagawa. Application of microwave to drying and blanching of tomatoes. *Journal of the Japanese Society for Food Science and Technology-Nippon Shokuhin Kagaku Kogaku Kaishi* 57:191–197, 2010.
66. GSV Raghavan, ZF Li, N Wang, Y Gariepy. Control of microwave drying process through aroma monitoring. *Drying Technology* 28:591–599, 2010.

67. JR Arballo, LA Campanone, RH Mascheroni. Modeling of microwave drying of fruits. *Drying Technology* 28:1178–1184, 2010.
68. A Kouchakzadeh, S Shafeei. Modeling of microwave-convective drying of pistachios. *Energy Conversion and Management* 51:2012–2015, 2010.
69. I Alibas. Determination of drying parameters, ascorbic acid contents and color characteristics of nettle leaves during microwave-, air- and combined microwave-air-drying. *Journal of Food Process Engineering* 33:213–233, 2010.
70. S Abbasi, S Azari. Novel microwave-freeze drying of onion slices. *International Journal of Food Science and Technology* 44:974–979, 2009.
71. X Duan, M Zhang, AS Mujumdar, SJ Wang. Microwave freeze drying of sea cucumber (*Stichopus japonicus*). *Journal of Food Engineering* 96:491–497, 2010.
72. R Wang, M Zhang, AS Mujumdar. Effects of vacuum and microwave freeze drying on microstructure and quality of potato slices. *Journal of Food Engineering* 101:131–139, 2010.
73. X Duan, M Zhang, AS Mujumdar. Studies on the microwave freeze drying technique and sterilization characteristics of cabbage. *Drying Technology* 25:1725–1731, 2007.
74. W Wang, GH Chen, FR Gao. Effect of dielectric material on microwave freeze drying of skim milk. *Drying Technology* 23:317–340, 2005.
75. W Wang, GH Chen. Heat and mass transfer model of dielectric-material-assisted microwave freeze-drying of skim milk with hygroscopic effect. *Chemical Engineering Science* 60:6542–6550, 2005.
76. R Wang, M Zhang, AS Mujumdar. Effect of food ingredient on microwave freeze drying of instant vegetable soup. *LWT-Food Science and Technology* 43:1144–1150, 2010.
77. X Duan, M Zhang, XL Li, AS Mujumdar. Ultrasonically enhanced osmotic pretreatment of sea cucumber prior to microwave freeze drying. *Drying Technology* 26:420–426, 2008.
78. R Wang, M Zhang, AS Mujumdar. Effect of osmotic dehydration on microwave freeze-drying characteristics and quality of potato chips. *Drying Technology* 28:798–806, 2010.
79. RM Borquez, ER Canales, JP Redon. Osmotic dehydration of raspberries with vacuum pretreatment followed by microwave-vacuum drying. *Journal of Food Engineering* 99:121–127, 2010.
80. A Figiel. Drying kinetics and quality of beetroots dehydrated by combination of convective and vacuum-microwave methods. *Journal of Food Engineering* 98:461–470, 2010.
81. QH Han, LJ Yin, SJ Li, BN Yang, JW Ma. Optimization of process parameters for microwave vacuum drying of apple slices using response surface method. *Drying Technology* 28:523–532, 2010.
82. D Argyropoulos, A Heindl, J Muller. Assessment of convection, hot-air combined with microwave-vacuum and freeze-drying methods for mushrooms with regard to product quality. *International Journal of Food Science and Technology* 46:333–342, 2011.
83. M Markowski, J Bondaruk, W Błaszczak. Rehydration behavior of vacuum-microwave-dried potato cubes. *Drying Technology* 27:296–305, 2009.
84. A Calin-Sanchez, A Szumny, A Figiel, K Jaloszynski, M Adamski, AA carbonell-Barachina. Effects of vacuum level and microwave power on rosemary volatile composition during vacuum-microwave drying. *Journal of Food Engineering* 103:219–227, 2011.
85. D Setiady, J Tang, F Younce, BA Swanson, BA Rasco, CD Clary. Porosity, color, texture, and microscopic structure of Russet potatoes dried using microwave vacuum, heated air and freeze drying. *Applied Engineering in Agriculture* 25:719–724, 2009.
86. L Momenzadeh, A Zomorodian, D Mowla. Experimental and theoretical investigation of shelled corn drying in a microwave-assisted fluidized bed dryer using Artificial Neural Network. *Food and Bioproducts Processing* 89:15–21, 2011.
87. WQ Yan, M Zhang, LL Huang, JM Tang, AS Mujumdar, JC Sund. Study of the optimisation of puffing characteristics of potato cubes by spouted bed drying enhanced with microwave. *Journal of the Science of Food and Agriculture* 90:1300–1307, 2010.
88. LN Kahyaoglu, S Sahin, G Sumnu. Physical properties of parboiled wheat and bulgur produced using spouted bed and microwave assisted spouted bed drying. *Journal of Food Engineering* 98:159–169, 2010.
89. RF Wang, ZY Li, WG Su, JS Ye. Comparison of microwave drying of soybean in static and rotary conditions. *International Journal of Food Engineering* 6:1–10, 2010.
90. WQ Yan, M Zhang, LL Huang, JM Tang, AS Mujumdar, JC Sun. Studies on different combined microwave drying of carrot pieces. *International Journal of Food Science and Technology* 45:2141–2148, 2010.
91. H Jiang, M Zhang, AS Mujumdar, RX Lim. Comparison of the effect of microwave freeze drying and microwave vacuum drying upon the process and quality characteristics of potato/banana re-structured chips. *International Journal of Food Science and Technology* 46:570–576, 2011.

92. EI Mejia-Meza, JA Yanez, NM Davies, B Rasco, F Younce, CM Remsberg, C Clary. Improving nutritional value of dried blueberries (*Vaccinium corymbosum L.*) combining microwave-vacuum, hot-air drying and freeze drying technologies. *International Journal of Food Engineering* 4:1–6, 2008.
93. ZW Cui, CY Li, CF Song, Y Song. Combined microwave-vacuum and freeze drying of carrot and apple chips. *Drying Technology* 26:1517–1523, 2008.
94. S Tireki, G Sumnu, A Esin. Production of bread crumbs by infrared-assisted microwave drying. *European Food Research and Technology* 222:8–14, 2006.
95. KJ Chua, SK Chou. A comparative study between intermittent microwave and infrared drying of bioproducts more options. *International Journal of Food Science and Technology* 40:23–39, 2005.
96. FJ Garcia-Zaragoza, ME Sanchez-Pardo, A Ortiz-Moreno, LA Bello-Perez. *In vitro* starch digestibility and expected glycemic index of pound cakes baked in two-cycle microwave-toaster and conventional oven. *International Journal of Food Sciences and Nutrition* 61:680–689, 2010.
97. ME Sanchez-Pardo, A Ortiz-Moreno, R Mora-Escobedo, JJ Chanona-Perez, H Necoechea-Mondragon. Comparison of crumb microstructure from pound cakes baked in a microwave or conventional oven. *LWT-Food Science and Technology* 41:620–627, 2008.
98. EK Megahey, WAM McMinn, TRA Magee. Experimental study of microwave baking of Madeira cake batter. *Food and Bioproducts Processing* 831:277–287, 2005.
99. N Seyhun, G Sumnu, S Sahin. Effects of different starch types on retardation of staling of microwave-baked cakes. *Food and Bioproducts Processing* 83:1–5, 2005.
100. G Sumnu. A review on microwave baking of foods. *International Journal of Food Science and Technology* 36:117–127, 2001.
101. G Sumnu, S Sahin, M Sevimli. Microwave, infrared and infrared-microwave combination baking of cakes. *Journal of Food Engineering* 71:150–155, 2005.
102. SO Keskin, S Ozturk, H Koksel, G Sumnu. Halogen lamp-microwave combination baking of cookies. *European Food Research and Technology* 220:546–551, 2005.
103. SO Ozkoc, G Sumnu, S Sahin, E Turabi. Investigation of physicochemical properties of breads baked in microwave and infrared-microwave combination ovens during storage. *European Food Research and Technology* 228:883–893, 2009.
104. E Turabi, G Sumnu, S Sahin. Optimization of baking of rice cakes in infrared-microwave combination oven by response surface methodology. *Food and Bioprocess Technology* 1:64–73, 2008.
105. SO Ozkoc, G Sumnu, S Sahin, E Turabi. Investigation of physicochemical properties of breads baked in microwave and infrared-microwave combination ovens during storage. *European Food Research and Technology* 228:883–893, 2009.
106. S Wang, G Tiwari, S Jiao, JA Johnson, J Tang. Developing postharvest disinfestation treatments for legumes using radio frequency energy more options. *Biosystems Engineering* 105:341–349, 2010.
107. TW Phillips, S Halverson, T Bigelow, G Mbata, W Halverson, M Payton, S Forester, P Ryas-Duarte. Microwave treatment of flowing grain for disinfestation of stored-product insects. In: *8th International Working Conference on Stored Product Protection (IWCSPP)*, New York, 2003, pp. 626–628.
108. R Vadivambal, DS Jayas, NDG White. Wheat disinfestation using microwave energy. *Journal of Stored Products Research* 43:508–514, 2007.
109. R Vadivambal, DS Jayas, NDG White. Controlling life stages of *Tribolium castaneum* (Coleoptera: Tenebrionidae) in stored rye using microwave energy. *Canadian Entomologist* 142:369–377, 2010.
110. O Zuloaga, N Etxebarria, LA Fernandez, JM Madariaga. Optimization and comparison of microwave assisted extraction and Soxhlet extraction for the determination of polychlorinatedbiphenyls in soil samples using an experimental design approach. *Talanta* 50:345–357, 1999.
111. SC Verma, S Nigam, CL Jain, P Pant, MM Padhi. Microwave-assisted extraction of gallic acid in leaves of Eucalyptus x hybrida Maiden and its quantitative determination by HPTLC. *Der Chemica Sinica* 2:268–277, 2011.
112. V Mandal, Y Mohan, S Hemalatha. Microwave assisted extraction-an innovative and promising extraction toll for medicinal plant research. *Pharmacognosy Reviews* 1:7–18, 2007.
113. V Beejmohun, O Fliniaux, E Grand, F Lamblin, L Bensaddek, P Christen, J Kovensky, MA Fliniaux, F Mesnard. Microwave-assisted extraction of the main phenolic compounds in flaxseed. *Phytochemical Analysis* 18:275–282, 2007.
114. K Hayat, S Hussain, S Abbas, U Farooq, BM Ding, SQ Xia, CS Jia, XM Zhang, WS Xia. Optimized microwave-assisted extraction of phenolic acids from citrus mandarin peels and evaluation of antioxidant activity *in vitro*. *Separation and Purification Technology* 70:63–70, 2009.
115. M Simsek. Microwave extraction of phenolic compounds from fruit pomaces. MSc Thesis, Middle East Technical University, Ankara, Turkey, 2010.

116. DC Li, JG Jiang. Optimization of the microwave-assisted extraction conditions of tea polyphenols from green tea. *International Journal of Food Sciences and Nutrition* 61:837–845, 2010.
117. N Sutivisedsak, HN Cheng, JL Willett, WC Lesch, RR Tangsrud, A Biswas. Microwave-assisted extraction of phenolics from bean *(Phaseolus vulgaris L.)*. *Food Research International* 43:516–519, 2010.
118. A Liazid, RF Guerrero, E Cantos, M Palma, CG Barroso. Microwave assisted extraction of anthocyanins from grape skins. *Food Chemistry* 124:1238–1243, 2011.
119. S Elkhori, JRJ Pare, JMR Belanger, E Perez. The microwave-assisted process (MAP): Extraction and determination of fat from cocoa powder and cocoa nibs. *Journal of Food Engineering* 79:1110–1114, 2007.
120. B Kaufmann, P Christen. Recent extraction techniques for natural products: Microwave-assisted extraction and pressurized solvent extraction. *Phytochemical Analysis* 13:105–113, 2002.
121. ME Lucchesi, F Chamat, J Smadja. An original solvent free microwave extraction of essential oils from spices. *Flavour and Fragrance Journal* 19:134–138, 2004.
122. B Bayramoglu, S Sahin, G Sumnu. Solvent-free microwave extraction of essential oil from oregano. *Journal of Food Engineering* 88:535–540, 2008.
123. B Bayramoglu, S Sahin, G Sumnu. Extraction of essential oil from Laurel Leaves by using microwaves. *Separation Science and Technology* 44:722–733, 2009.
124. OO Okoh, AP Sadimenko, AJ Afolayan. Comparative evaluation of the antibacterial activities of the essential oils of *Rosmarinus officinalis L.* obtained by hydrodistillation and solvent free microwave extraction methods. *Food Chemistry* 120:308–312, 2010.
125. MT Golmakani, K Rezaei. Comparison of microwave-assisted hydro distillation with the traditional hydrodistillation method in the extraction of essential oils from *Thymus vulgaris L. Food Chemistry* 109:925–930, 2008.
126. HW Wang, YQ Liu, SL Wei, ZJ Yan, KA Lu. Comparison of microwave-assisted and conventional hydrodistillation in the extraction of essential oils from mango (*Mangifera indica L.*) flowers. *Molecules* 15:7715–7723, 2010.
127. MA Vian, X Fernandez, F Visinoni, F Chemat. Microwave hydrodiffusion and gravity, a new technique for extraction of essential oils. *Journal of Chromatography A* 1190:14–17, 2008.
128. N Bousbia, MA Vian, MA Ferhat, BY Meklati, F Chemat. A new process for extraction of essential oil from Citrus peels: Microwave hydrodiffusion and gravity. *Journal of Food Engineering* 90:409–413, 2009.
129. A Farhat, AS Fabiano-Tixier, F Visinoni, M Romdhane, F Chemat. A surprising method for green extraction of essential oil from dry spices: Microwave dry-diffusion and gravity. *Journal of Chromatography A* 1217:7345–7350, 2010.
130. H Zill-e, MA Vian, JF Maingonnat, F Chemat. Clean recovery of antioxidant flavonoids from onions: Optimising solvent free microwave extraction method. *Journal of Chromatography A* 1216:7700–7707, 2009.
131. T Michel, E Destandau, C Elfakir. Evaluation of a simple and promising method for extraction of antioxidants from sea buckthorn (*Hippophae rhamnoides L.*) berries: Pressurised solvent-free microwave assisted extraction. *Food Chemistry* 126:1380–1386, 2011.
132. A Farhat, AS Fabiano-Tixier, M El Maataoui, JF Maingonnat, M Romdhane, F Chemat. Microwave steam diffusion for extraction of essential oil from orange peel: Kinetic data, extract's global yield and mechanism. *Food Chemistry* 125:255–261, 2011.
133. BJ Taher, MM Farid. Cyclic microwave thawing of frozen meat: Experimental and theoretical investigations. *Chemical Engineering and Processing* 40:379–389, 2001.
134. SJ James. Food refrigeration and thermal processing at Laangford, UK: 32 years of research. *Food and Bioproducts Processing* 77:261–280, 1999.
135. MD Schaefer. Microwave tempering of shrimp with susceptors. MS Thesis, Virginia Polytechnic Institute and State University, Blacksburg, VA, 1999.
136. CM Liu, QZ Wang, N Sakai. Power and temperature distribution during microwave thawing, simulated by using Maxwell's equations and Lambert's law. *International Journal of Food Sciences and Technology* 40:9–21, 2005.
137. LA Campanone, NE Zaritzky. Mathematical modeling and simulation of microwave thawing of large solid foods under different operating conditions. *Food and Bioprocess Technology* 3:813–825, 2010.
138. F Oz. The effect of different thawing methods on heterocyclic aromatic amine contents of beef chops cooked by different methods. *Journal of Animal and Veterinary Advances* 9:2327–2332, 2010.
139. T Baygar, O Ozden, D Ucok. The effect of the freezing-thawing process on fish quality. *Turkish Journal of Veterinary and Animal Sciences* 28:173–178, 2004.

140. S Boonsumrej, S Chaiwanichsiri, S Tantratian, T Suzuki, R Takai. Effects of freezing and thawing on the quality changes of tiger shrimp (*Penaeus monodon*) frozen by air-blast and cryogenic freezing. *Journal of Food Engineering* 80:292–299, 2007.
141. IG Mandala, K Sotirakoglou. Effect of frozen storage and microwave reheating on some physical attributes of fresh bread containing hydrocolloids. *Food Hydrocolloids* 19:709–719, 2005.
142. P Coronel, VD Truong, J Simunovic, KP Sandeep, GD Cartwright. Aseptic processing of sweetpotato purees using a continuous flow microwave system. *Journal of Food Science* 70:531–536, 2005.
143. Z Tang, G Mikhaylenko, F Liu, JH Mah, R Pandit, F Younce, J Tang. Microwave sterilization of sliced beef in gravy in 7-oz trays. *Journal of Food Engineering* 89:375–383, 2008.
144. LE Steed, VD Truong, J Simunovic, KP Sandeep, P Kumar, GD Cartwright, KR Swartzel. Continuous flow microwave-assisted processing and aseptic packaging of purple-fleshed sweetpotato purees. *Journal of Food Science* 73:E455–E462, 2008.
145. T Sun, J Tang, JR Powers. Antioxidant activity and quality of asparagus as affected by microwave-circulated water combination and conventional sterilization. *Journal of Food Chemistry* 100:813–819, 2007.
146. P Kumar, P Coronel, VD Truong, J Simunovic, KR Swartzel, KP Sandeep, G Cartwright. Overcoming issues associated with the scale-up of a continuous flow microwave system for aseptic processing of vegetable purees. *Food Research International* 41:454–461, 2008.
147. LH Huang, J Sites. New automated microwave heating process for cooking and pasteurization of microwaveable foods containing raw meats. *Journal of Food Science* 75:E110–E115, 2010.
148. LH Huang. Computer-controlled microwave heating to in-package pasteurize beef frankfurters for elimination of Listeria monocytogenes. *Journal of Food Process Engineering* 28:453–477, 2005.
149. LH Huang, J Sites. Automatic control of a microwave heating process for in-package pasteurization of beef frankfurters. *Journal of Food Engineering* 80:226–233, 2007.
150. TS Gentry, JS Roberts. Design and evaluation of a continuous flow microwave pasteurization system for apple cider. *LWT-Food Science and Technology* 38:227–238, 2005.
151. P Kumar, P Coronel, J Simunovic, KP Sandeep. Feasibility of aseptic processing of a low-acid multiphase food product (salsa con queso) using a continuous flow microwave system. *Journal of Food Science* 72:E121–E124, 2007.
152. H Chen, J Tang, F Liu. Simulation model for moving food packages in microwave heating processes using conformal FDTD method. *Journal of Food Engineering* 88:294–305, 2008.
153. YJ Kim, SW Lim, JK Chun. Serial flow microwave thermal process system for liquid foods. *Food Science and Biotechnology* 14:446–449, 2005.
154. L Cinquanta, D Albanese, G Cuccurullo, M Di Matteo. Effect on orange juice of batch pasteurization in an improved pilot scale microwave oven. *Journal of Food Science* 75:E46–E50, 2010.
155. A Fratianni, L Cinquanta, G Panfili. Degradation of carotenoids in orange juice during microwave heating. *LWT-Food Science and Technology* 43:867–871, 2010.
156. MG Lin, HS Ramaswamy. Evaluation of phosphatase inactivation kinetics in milk under continuous flow microwave and conventional heating conditions. *International Journal of Food Properties* 14:110–123, 2011.
157. KN Matsui, JAW Gut, PV de Oliveira, CC Tadini. Inactivation kinetics of polyphenol oxidase and peroxidase in green coconut water by microwave processing. *Journal of Food Engineering* 88:169–176, 2008.

22 Combination Treatment of Pressure and Mild Heating

Takashi Okazaki and Yujin Shigeta

CONTENTS

22.1 INTRODUCTION

Hydrostatic pressure (HP) was first applied to sterilization of milk as a food processing method over a century ago.[1] Since that time, many researchers have experimented on the effects of HP on microorganisms, denaturation of protein, enzymatic reaction, and so on. In particular, a large amount of information about microorganisms such as sterilization, spore germination, and growth inhibition has been accumulated for more than a century. Various techniques were developed from these HP fundamental studies, and some of them have already been put into practice as innovative food processing methods. In addition, the HP treatment combined with heating can not only enforce the effectiveness of the HP treatment and the heating treatment but also lead to the development of innovative techniques. For example, Okazaki and Shigeta have developed two techniques, that is, a pressure-shucked oyster (PSO) machine[2] and a microbe-controlling instrument by HP.[3] These techniques of HP are considerably affected by temperature. The magnitude of necessary pressures to shuck oysters is at 400 MPa when the treatment temperature is at 10°C, but the magnitude can be reduced to 80 MPa when the temperature rises to 40°C. Such a temperature difference between 10°C and 40°C has a big effect on shucking oysters.

Fish sauce consumed mainly in Southeast Asia is a special type of seasoning produced by autolysis of fish, which is significantly accelerated in mild heating such as at 50°C. Since mild heating at 50°C accelerates the autolysis as well as occurrence of its decomposition, the autolysis is produced in a low or room temperature range. Moreover, a lot of salt is added to the autolysate to avoid its decomposition. Okazaki and Shigeta have also developed an innovative production method of fish sauce by the combination of HP and mild heating, using the growth inhibition of microorganisms due to HP.[4,5] This method can produce fish sauce in quite a short time (48 h) without its decomposition compared to the traditional method (taking more than a year), which requires adding salt to fish sauce.

In another example, inactivation of bacterial spores is also affected considerably by temperature. When bacterial spores are killed by HP at 400 MPa, the temperature difference between 25°C and 65°C strongly influences the inactivation rate.[5] In addition to these instances, almost all the phenomena on HP will be necessary to consider temperature as an important factor. Therefore, in this chapter, some of the newest information on pressure-assisted heating will be reviewed.

22.2 PRESSURE-SHUCKED OYSTER

Okazaki and Shigeta[7] had developed a method of shucking oysters using HP in a cooperative study with Hiroshima University and Yanmor Ltd. (an agricultural engineering company) since 1997. Before starting the development of the process, some patents and studies about shucking oysters were considered, and utilization of HP was selected for the direction of the development. In the submitted patent,[7] it is reported that the HP treatment alone in a pressure range of 100–400 MPa detaches the adductor muscle from the shells. To make sure that the adductor muscle is separated from shells, an experiment shown in Table 22.1 was carried out. Oysters are packed in a container with sea water and pressurized in a pressure range of 100–400 MPa at 10°C. The pressurization at 400 MPa separates the adductor muscle from shells completely in the treatments from 5 to 15 min, but that of 300 MPa for 5–10 min does not detach some part of the adductor muscle from the shells and those of 100 and 200 MPa for 5–15 min did not detach it at all. The sphere designated in the patent is in a pressure range of 100–400 MPa, and furthermore, it is also reported in the USA that oysters are shucked by HP of 311 MPa.[9] Accordingly, it is shown that a higher range of HP is needed to shuck oysters by HP treatment alone.

Table 22.2 shows the effect of temperatures on PSOs. The pressure magnitude of shucked oysters is considerably affected by the temperatures. The magnitude of pressure needed to shuck oysters is

TABLE 22.1
Situation of Shucking Oysters at 10°C

	Treatment Time (min)		
Pressure (MPa)	**5**	**10**	**15**
100	×	×	×
200	×	×	Δ
300	Δ	○	◎
400	◎	◎	◎

×, no shucked; Δ, a little shucked; ○, shucked but not separated; ◎, completely shucked and separated.

TABLE 22.2
The Effect of Temperature on Shucked Oysters by HP

	Temperature (°C)			
Pressure (MPa)	**10**	**20**	**30**	**40**
100	×[a]	×	Δ[a]	○[a]
200	×	Δ	○	◎[a]
300	○	◎	◎	—
400	◎	◎	—	—

[a] Table 22.1 is referred.

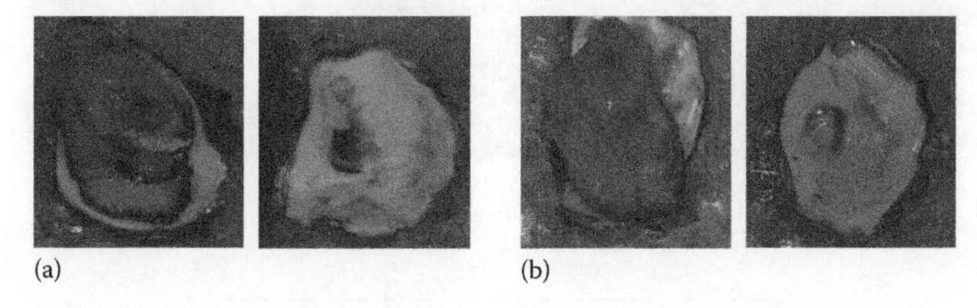

FIGURE 22.1 Photographs of oysters shucked by hands (a) and by pressure (b).

400 MPa at 10°C and 100 MPa at 40°C, respectively. Furthermore, some tests were carried out, and finally, the best condition capable of completely shucking oysters was decided as the following: oysters are first held for 15 min at 41°C to raise the oyster temperature in advance and are then treated at 85 MPa at 42°C for 5 min. These oysters shucked by HP are almost the same appearance as those shucked by hand (Figure 22.1).

22.2.1 Shucking Machine by HP

The treatment process of shucked oyster by HP is shown in Figure 22.2. First some oysters are adjusted to approximately 41°C by holding them in the seawater pool at 41°C for 15 min, and

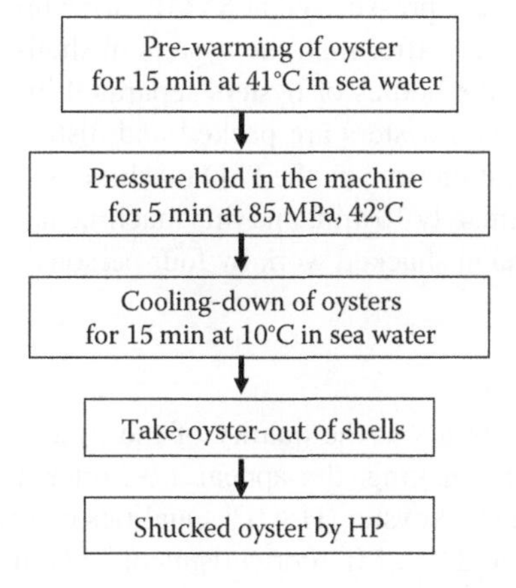

FIGURE 22.2 Procedures of PSOs.

FIGURE 22.3 Instrument of PSOs installed in Japan.

TABLE 22.3
Operation Procedure of PSO Machine

Procedure	Operation Time (min)
Setting oysters	2
Pour of sea water	0.2
Elevation of pressure	0.2
Hold of pressure	5
Exhaust of sea water	2.2
Take-out of oysters	1
Total time	10.7

successively the adjusted oysters are pressurized at 85 MPa for 5 min at 42°C in an oyster-shucking machine (Figure 22.3). Immediately after that, the oysters in shells are held in a seawater pool for 5 min to cool the oysters. Next, the bodies of oysters separated in shells are dropped into another seawater pool. Finally, the shucked oysters are packed and distributed in markets as safe oysters without shell particles. The operation time on the PSO machine is shown in Table 22.3. Total operation time in one cycle is approximately 10 min, and this machine has the capability to shuck 1500 kg of oysters in 8 h, equivalent to hand shucked work by four persons.

22.2.2 Quality of PSOs

There is, however, no information about the quality of the oysters shucked by HP at a moderate temperature. Immediately after shucking, the appearance, odor, taste, and texture is almost the same in both PSO and hand shucked oyster (HSO).[10] Qualities of PSO and HSO after shucking are compared with each other (Table 22.4). Microorganisms of PSO are less than those of HSO in an order immediately after being shucked, but after 7 days both microbes increase in 10^4 CFU/g to the

TABLE 22.4
Quality Comparison of PSOs and HSOs

Measured Qualities	PSO	HSO
pH	6.25	6.30
Microorganisms (CFU/g)	1.3×10^{2}	1.1×10^{3}
D-activity[a] (μg/g)	425	920
Taurine (mg/100 g)	435	505
Glycogen (g/100 g)	1.35	1.62
Total amino acid (g/100)	1.21	1.22

[a] Triphenylformazan (μg) formed in gill (g) is indicated as dehydrogenase activity (D-activity).

same extent. PSO's pH declines somewhat faster than HSO's, and both become below pH 6.0 in 7 days. Taurine and glycogen, both of which are components peculiar to oysters, do not change for 10 days in PSO and HSO. PSO's dehydrogenase activity (D-activity), whose value is used for freshness of oyster, is half in HSO's during the storage. Therefore, the pH and D-activity of PSO show that the bodies of oysters are somewhat damaged by HP at 41°C.

22.2.3 Advantages of PSOs

There are some advantages of PSO and its method. Many of the HSO contain some particles of shell, which are hidden between gills and/or adhere to an adductor muscle. These particles are considered as undesirable foreign substance of foods in Japan, owing to the risk of cutting the inside of people's mouths. The PSOs contain no particles of shells in the bodies of oysters (Figure 22.1). Furthermore, some of the oysters' bodies are damaged by mistake when oysters are shucked by hands using knives, and they cannot be sold in market; PSOs, however, have no damage. In addition, although it is difficult to shuck the oysters of small or curved shells, such difficulty can be overcome by using the PSO method. Therefore, PSO should be introduced as an innovative way of shucking oysters in the future.

22.3 AUTOLYSIS OF UNSALTED FISH PROTEIN UNDER PRESSURIZATION

Fish sauces have traditionally been produced in Southeast Asian countries and in some local regions in Japan. Many kinds of processed foods and dishes are seasoned with fish sauce as a flavor enhancer, and the demand for sauce has increased remarkably in recent years, owing to its potential for extensive use. A large quantity of salt is, however, added to the fish sauce to prevent its rancidity. In some cases, the salt concentration of fish sauce is over 25%, so its use is limited because of strong salty taste and consumer's preference for reducing salt intake. Moreover, the addition of a great deal of salt inhibits the activity of proteinase and peptidase in fish, which leads to a long autolysis period. In many cases, it takes from half a year to 2 years to complete the production of fish sauce. Therefore, a reduction in the concentration of salt is desirable to extend its use and for health as well as for shortening the production period. A large number of researchers have tried to reduce the salt concentration of fish sauce.

Some reports show that the addition of fish guts or proteinases accelerates the autolysis of fish proteins and reduces the autolysis time at a salt concentration <20%. In addition, maintaining lower or higher temperatures than room temperature is also performed during the autolysis period in order to control the growth of microbes. This treatment can decrease the amount of added salt. Knowing the previous situation, and for the purpose of producing a fish sauce without adding salt, Okazaki and Shigeta[11] attempted to use HP. A pressure above approximately 60 MPa inhibits the growth of any heterotrophic microbes.[4,5] On the contrary, pressure above approximately 2 kbar (200 MPa) inactivates proteinases like chymotrypsin.[12] Carboxypeptidase is also inactivated at 405 MPa.[13] From these studies, it was inferred that the growth of microorganisms causing rancidity could be inhibited while enzymes involved in autolysis would not be inactivated under pressures ranging from 60 to 200–400 MPa. If fish is maintained at an optimal temperature for the action of proteinases contained in fish, autolysis will progress rapidly without rancidity. Moreover, proteinases in fish will work more effectively owing to the lack of possible inhibition by salt. To substantiate such an idea, some studies were conducted,[11,14,15] and the results are shown in this section.

22.3.1 Autolysis Condition under HP

The autolysis of fish progresses due to the proteinases and peptidases contained in its pyloric ceca, stomach, muscle, and so on. Thus, it is necessary to find the optimal conditions of those enzymes to work as rapidly as possible. Table 22.5 shows the effect of pressure on the autolysis in a pressure range of 50–250 MPa. The total nitrogen (T-N) and formol nitrogen (F-N) in pressure autolytic extract (PAE) are 2.4–2.7 g/100 g and 1.1–1.3 g/100 g in the pressure range, respectively. The quality of fish sauce is generally evaluated by T-N and F-N concentration, and if their values are higher than 2.0 and 1.0, the quality is evaluated as good, respectively. The difference among T-Ns in each pressure is not great, so the optimal pressure for autolysis cannot be confirmed in the pressure range. In the case of small shrimps, the pressure revealing the highest T-N seems to exist in around 200 MPa (not shown). It is, however, important to autolyze fish at pressures as low as possible for practical reasons and to adopt the pressure at which no growth of any microbes will occur. A pressure range capable of inhibiting growth of microorganism is reported in some previous studies. Aoyama[4] shows that the growth of some microorganisms were inhibited at 40–60 MPa in an optimal growth temperature, and ZoBell and Johnson[5] indicate the pressure range of 50–60 MPa. Considering these pressure conditions, the pressure for autolysis is set to 60 MPa though it is somewhat outside of the optimal pressure to autolyze.

The effect of temperature on autolysis of anchovies under HP was tested in a range of 20°C–70°C, and the optimal temperature of their autolysis lies in around 50°C (Figure 22.4). In the case of small shrimps, the temperature revealing the highest T-N was around 55°C (not shown). Noda et al. investigated the properties of some proteinases contained in sardine, showing that the optimal temperature for the activity was in the range of 40°C–55°C.[16] Two aminopeptidases in the internal organs of sardine were stable up to 50°C.[17] Therefore, the optimal temperature for autolysis of fish is thought to exist in around 50°C–55°C even under HP.

Figure 22.5 shows changes in T-N, F-N, and ratio of F-N/T-N during autolysis of anchovy at 50°C at 60 MPa. In addition, the photographs of anchovies can be seen in Figure 22.6. The bodies

TABLE 22.5
Effects of Pressure on T-N and F-N for PAE at 50°C for 24 h

Pressure (MPa)	T-N (g/100 g)	F-N (g/100 g)
50	2.58	1.44
100	2.64	1.24
150	2.74	1.30
200	2.64	1.29
250	2.42	1.28

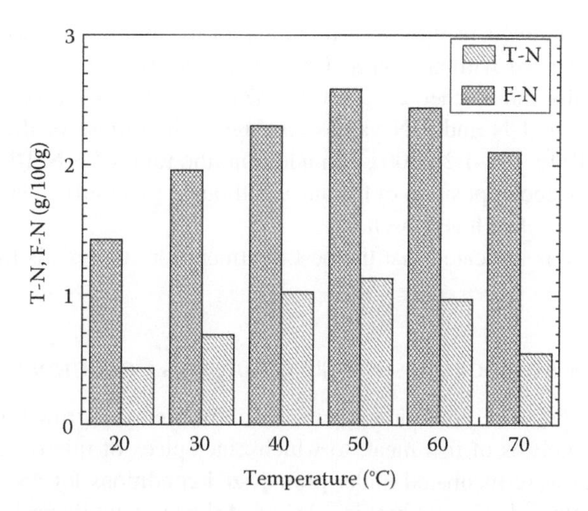

FIGURE 22.4 Effects of temperature on T-N (total nitrogen) and F-N (formol nitrogen) at 60 MPa for 24 h.

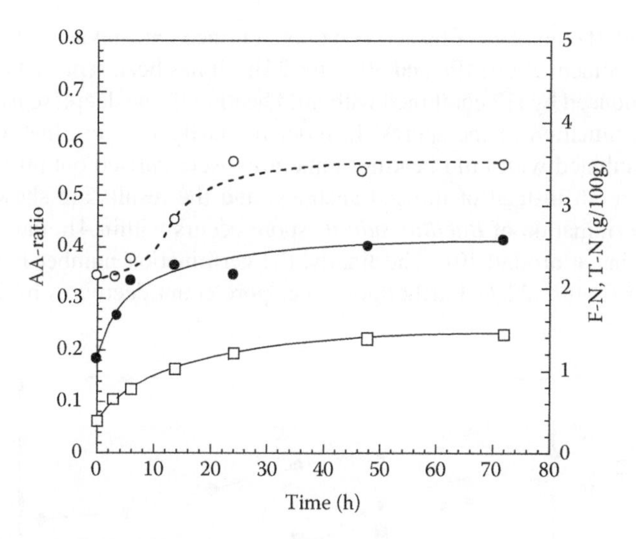

FIGURE 22.5 Changes in F-N, T-N, and amino acid ratio (AA-ratio, F-N/T-N), during anchovy autolysis under pressurization under 60 MPa at 50°C. □, F-N; •, T-N; o, AA-ratio.

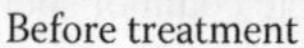
Before treatment

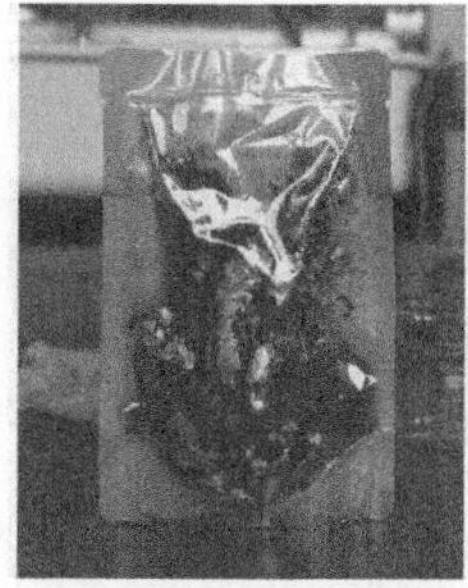
After treatment for 12 h

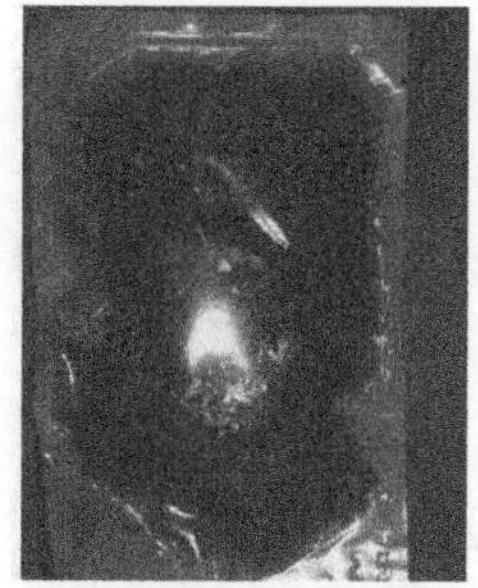
After treatment for 24 h

FIGURE 22.6 Progress of anchovy autolysis under HP at 60 MPa and 50°C.

of anchovies greatly crumbled in the first 6h compared to initial ones, and after 24h, the bodies were completely resolved without bones. T-N was greater than 2.0g/100g, and thereafter, the value increased gradually and reached 2.6g/100g in 48h. F-N also increased gradually and reached 1.4g/100g in 48h. These T-N and F-N values are higher than those of the commercially available fish sauce (1.2g/100g, 0.4–1.2g/100g). In addition, the ratio of F-N/T-N is over 0.55 in 24h. Generally, the degree of decomposition of fish meat is thought to be sufficient when the ratio is over 0.5, so the treatment time of 48h is chosen.

From these results, it is indicated that the best treatment for autolysis is for 48h at 60MPa and 50°C.

22.3.2 Behavior of Microorganisms during Autolysis Condition under HP

It is reported that the pressure condition capable of inhibiting the growth of some microorganisms was applied to autolysis of fish meat, to which nine spices of microorganisms were added.[4] The minced anchovies were incubated under pressurized conditions for 48h at 30°C–55°C. They were decomposed at atmospheric pressure in all runs of the tests, while anchovies autolyzed under pressurization retained all the sound situations, and then no microorganisms can be detected. These results reveal that the safety of the innovative way of autolysis without addition of salt was confirmed.

On the other hand, the number of bacterial spores in minced anchovies considerably decreased after the pressure treatment at 60MPa and 50°C for 24h.[4] It has been reported that germination of bacterial spores is induced by HP combined with mild heating.[18] The decrease is, therefore, thought to be caused by germination of the spores. In order to clarify the germination and inactivation of spores by HP combined with mild heating, some tests were carried out utilizing rich nutrients such as in glucose broth instead of minced anchovy, and the results are shown in Figures 22.7 and 22.8.[19,20] The germination of *Bacillus subtilis* spore occurs within 4h, inactivation of the germinated spores begins at around 10h, and finally, the germination number is in agreement with inactivation number (Figure 22.7). Furthermore, the spore count decreases by 2–3 log cycles in a

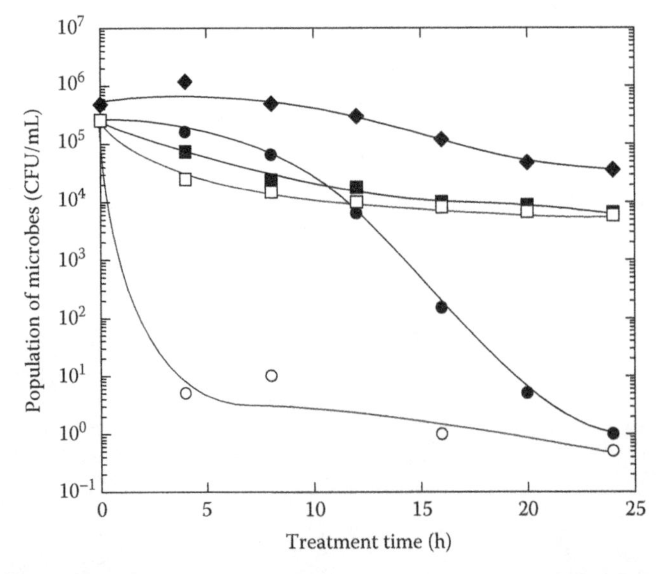

FIGURE 22.7 Changes in viable counts and spore counts of *B. subtilis* spores and vegetative cells in the presence or absence of nutrients at 40°C under pressurization at 60MPa. ●, viable count of spore in nutrients; ○, ungerminated spore count in nutrients; □, viable count of vegetative cells in nutrients; ◆, viable count of spore in no-nutrients; ■, ungerminated spore count in no-nutrients.

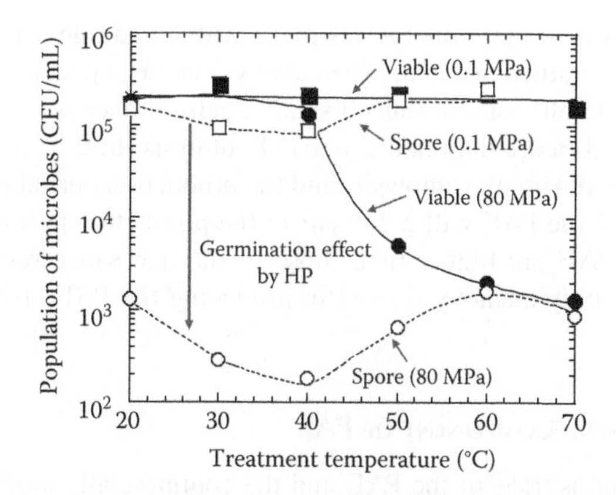

FIGURE 22.8 Effect of temperature on germination and inactivation of *B. subtilis* spores in nutrients. ■, viable count (0.1 MPa); □, spore count (0.1 MPa); ●, viable count (80 MPa); ○, spore count (80 MPa); ×, initial spore count.

temperature range of 20°C–70°C at 80 MPa in an hour and is in agreement with viable count above 60°C (Figure 22.8). Probably, the germinated spores will be killed at temperatures above 50°C due to losing their heat resistance, although they are alive at temperatures below 40°C. This reveals that germination and inactivation of bacterial spores will occur successively during the autolysis of anchovies.

Therefore, the growth of microorganisms is not only inhibited, but also bacterial spores are germinated and inactivated during autolysis condition under HP. The pressure-induced germination is also introduced in detail in Sections 22.4.3 and 22.4.4.

22.3.3 Quality of PAE

The quality of PAE is shown in Table 22.6. The commercially available fish sauce and soy sauce are also listed in the same table. The NaCl content of the PAE is much lower than those of the fish

TABLE 22.6
Chemical Components of PAE and Commercially Available Fish Sauce and Soy Sauce

		Fish Sauce			
	PAE	Nampla	Japanese 1	Japanese 2	Soy Sauce
pH	6.6	5.4	5.2	5.5	5.4
NaCl (%)	0.6	20.9	14.9	22.0	13.6
T-N (%)	2.62	1.74	1.92	1.24	1.41
F-N (%)	1.42	1.06	1.13	0.86	0.74
F-N/T-N	0.54	0.61	0.59	0.69	0.45
Glu	1.47	1.22	1.13	0.92	1.23
Ala	1.00	0.78	0.82	0.72	0.79
Asp	0.88	0.84	0.56	0.69	0.19
Tau	0.39	0.18	0.26	0.32	0
Total amino acid	11.56	8.45	9.60	6.76	6.95

Japanese 1, commercial product made from sardine in Japan; Japanese 2, commercial product made from shrimp in Japan.

sauce and soy sauce. This is because the PAE is made without adding salt, and the traces of NaCl in PAE are due to those contained originally in anchovy. The pH value of PAE is higher than those for the fish sauce. The pH value of fish sauce usually declines because of the growth of microorganisms over a long period (more than half a year) of autolysis. In the production of this PAE, no microorganisms can grow during the autolysis, and the production period is extremely short (48 h), so that the pH decrease of the PAE will not occur in the production period. T-N (2.6 g/100 g) and F-N (1.4 g/100 g) of the PAE are higher than those of the fish sauce reported (T-N: 1.6 g/100 g, F-N: 0.4–1.2 g/100 g). As only anchovy is used in producing the PAE, T-N and F-N will become concentrated.

22.3.4 Free Amino Acid Component of PAE

The free amino acid compositions of the PAE and the commercially available fish sauce and soy sauce are shown in Table 22.6. The sum of the amino acids in the PAE (11.6 g/100 g) is much higher than those of fish sauce and soy sauce. The most-contained amino acids are Glu (1.47 g/100 g), Ala (0.98 g/100 g), and Asp (0.88 g/100 g), which are known to be favorable taste-active amino acids, so that this PAE is expected to have a good taste. Except for these, Leu (1.26 g/100 g) and Lys (1.24 g/100 g) are also contained much in the PAE. In contrast, nucleotides (AMP, IMP, and GMP) are not detected in the PAE like fish sauce. Thus, the taste of the PAE is not to be attributed to nucleotides.

Since the NaCl content of PAE is low, it is difficult to compare it with commercially available fish sauce by sensory evaluation. Therefore, three PAE samples containing 0%, 6%, and 15% NaCl were prepared and compared with fish sauce containing 15% NaCl (Figure 22.9). The result shows

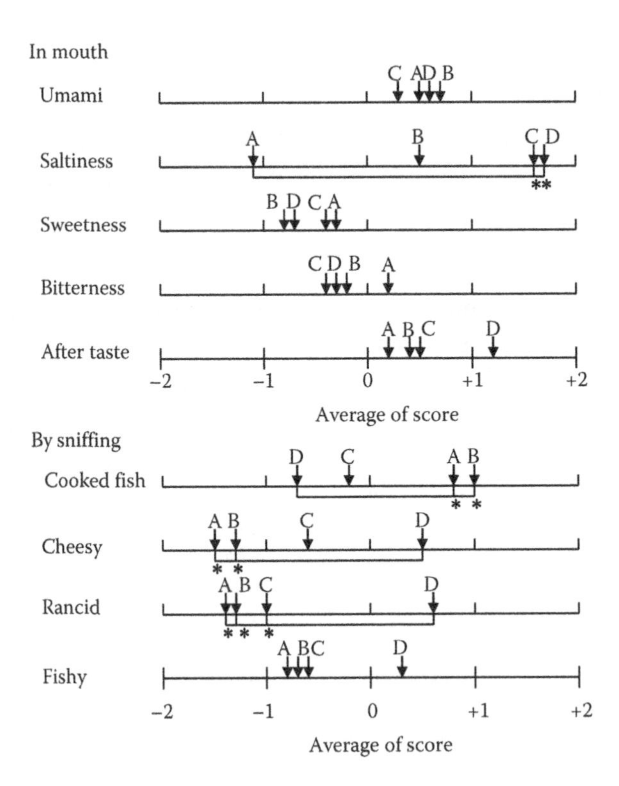

FIGURE 22.9 Sensory evaluation of PAE, and commercially available fish sauce by the scoring method. A, 0% PAE; B, 6% NaCl PAE; C, 15% NaCl PAE; D, 15% NaCl fish sauce. *Significant between two samples ($p<0.05$).

that there is no significant difference among the four samples (i.e., "umami," "sweetness," "bitterness," and "after-taste"). Though it is shown that the taste of peptides and amino acids changes in the absence or presence of salt, such differences are not found in the sensory test. However, as for saltiness, the taste of 0% and 6% NaCl PAE was significantly less salty than the 15% NaCl PAE and fish sauce. This is because the NaCl content of the PAE is much less than fish sauce. By sniffing, the differences of the odor are also evident. The 0% and 6% NaCl PAE have an odor like cooked fish, which is stronger than the fish sauce ($p<0.05$). "Cheesy" odor and "rancid" odor of PAE are significantly lower than those of fish sauce ($p<0.05$). Because this PAE can be produced in a short period (48 h) with no contact with oxygen, no oxidation of oily components in fish occurs during the pressurized treatment. So the odor of PAE was considerably different from fish sauce.

22.3.5 Application of Other Marine Materials of Autolysis under Pressurization

There are some traditional salted foods produced with autolysis in Japan, except for anchovy fish sauce. Okazaki and Shigeta have also applied the autolysis under HP to two typical Japanese foods that are made from livers of cuttlefishes[14] and guts of sea cucumbers,[15] respectively. The fish sauce made from cuttlefish livers is autolyzed under the condition of 60 MPa and 50°C for 48 h, and the composition of its amino acids are shown in Table 22.7. Compared to raw materials, it is revealed that T-N, F-N, and their T-N/F-N ratio have high values. The typical amino acids such as Glu, Asp, Ala, and Tau are also concentrated. In addition, a fish paste called Shiokara in Japan, which is made from guts of sea cucumbers, is also autolyzed at 60 MPa at 30°C for 24 h. Compared to raw materials and the conventional method, T-N, F-N, and their T-N/F-N ratio have high values

TABLE 22.7
Amino Acid Components of Squid Liver Extract Produced by Autolytic Hydrolysis Treatment under Pressure Treatment

	Raw Squid Lever	Squid Liver Extract
Tau	460	960
Asp	640	1,860
Thr	300	930
Ser	290	910
Glu	660	2,020
Gly	290	680
Ala	410	1,080
Val	390	1,120
Cys	140	130
Met	210	570
Ile	390	1,080
Leu	610	1,500
Tyr	450	140
Phe	380	830
Lys	570	2,120
His	130	370
Arg	240	n.d.
Total FAA	6,570	16,270

FAA, free amino acids
n.d., <10 mg/100 g.

TABLE 22.8
Comparison of Analytical Data in Sea Cucumber's Guts Processed by Conventional Method with Those by HP Method

	Raw	Conventional	Pressurized
pH	6.6	6.3	6.3
NaCl (g/100 g)	1.6	6.1	1.5
F-N (mg/100 g)	70	200	360
T-N(mg/100 g)	210	430	760
Mesophilic bacteria count (CFU/g)	5.9×10^2	3.2×10^2	8.8×10^3
Psychrotrophic bacteria count (CFU/g)	6.4×10^3	1.5×10^3	8.6×10^2

Raw, raw sea cucumber's guts; conventional, raw sea cucumber's guts ripened for 7 days at 5°C with 5% salt; pressurized, raw sea cucumber's guts ripened for 24 h at 30°C without salt; T-N, total nitrogen; F-N, formol nitrogen; mesophilic, number of colonies after 48 h at 35°C; psychrotrophic, number of colonies after 7 days at 7°C.

(Table 22.8). The typical amino acids such as Glu, Asp, Leu, Lys, Ala, and Tau are higher than untreated guts.[15]

As fish sauces and fish pastes have traditionally been produced in Southeast Asian countries and these products are consumed all over the world as one of the seasoning ingredients, if the quantity of salt added in these products needs to be controlled, autolysis under HP will be a useful method to produce these products. This new method is called "autolysis under pressurization."[11] This technique can be applied to the autolysis of other fish species as shown earlier as well as the hydrolysis of proteins with proteinase and other type of enzymes. Furthermore, to carry out the autolysis under pressurization, a special pressure machine, whose capacity is from 50 mL to 300 L, had already been developed in Japan (Figure 22.10).

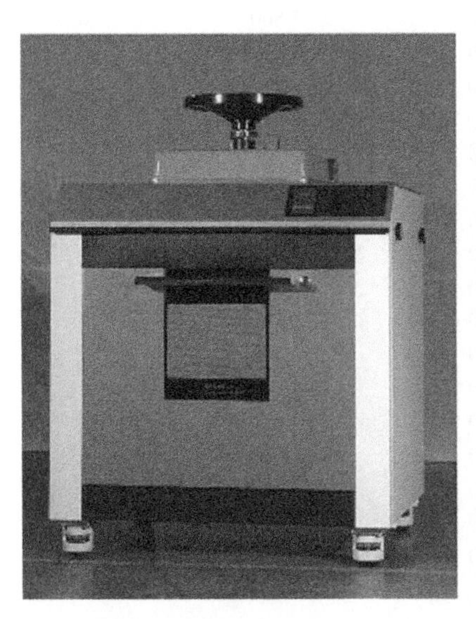

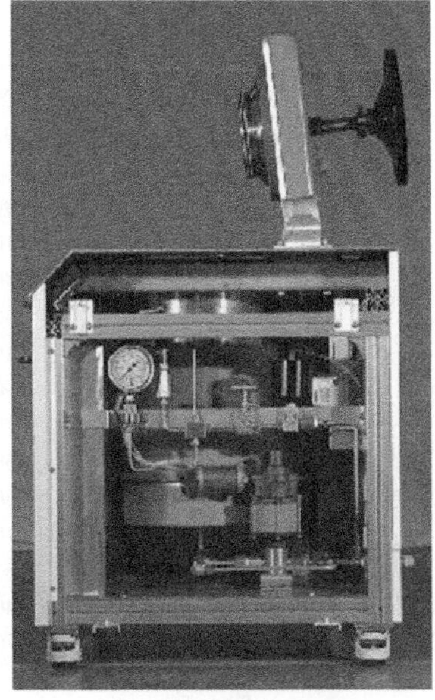

FIGURE 22.10 Pressure instrument newly developed to carry out autolysis under pressure, called "Marugoto Ekisu" in Japanese. (Courtesy of Toyo Koatsu, Hiroshima, Japan.)

22.4 INACTIVATION OF BACTERIAL SPORES BY PRESSURE-ASSISTED HEATING

Almost all foods and their ingredients are naturally contaminated by many kinds of microorganisms, some of which, especially bacterial spores frequently caused spoilage of foods because of their heat resistance. The general method to completely kill the bacterial spores is moist heat at temperatures higher than 100°C in food industries. Some of the bacterial spores in special species are known to survive even in the treatments at 120°C for 40 min. Such heat treatments lead to deteriorate their favorable functions like flavor, color, textures, and nutritive components. HP treatment is expected to be one of the innovative ways of killing bacterial spores without those deteriorations.[21,22] Several high value-added products with improved qualities and shelf life have been commercialized such as vegetable, fruit, meat, and fish-based products in Europe, the United States, and Japan.

In order to kill bacterial spores, numerous researchers have extensively investigated their inactivation treatments by HP.[18,23,24] However, it has come to be known that it is difficult to inactivate bacterial spores completely by HP treatment alone. Nakayama et al.[25] reported that the spores of six *Bacillus* strains are not inactivated even at 981 MPa for 40 min at 5°C–10°C. Combination treatments of heating and HP are shown to be an effective way of enhancing the inactivation of bacterial spores by many researchers.[26–36]

Okazaki and Shigeta[19,20,36,37] have investigated combination treatment of heating and HP, and germination of spore induced by HP in order to achieve the effective inactivation of bacterial spores without deterioration of food qualities. In addition, the pressure-induced germination of spores is applied to the extension of shelf life of foods. Some of their efforts will be introduced in this chapter.

22.4.1 Compact Pressure Vessel for Pressure Treatment

The thermal treatments combined with pressure are carried out using the compact pressure vessel shown in Figure 22.11. HP is introduced to the pressure vessel from the right side through a pressure pump, and a sheath thermocouple (1.5 mm in diameter) is inserted from the left side. In the case of liquid sample like spore solution, 2 mL solution is poured into a sample container (4 × 100 mm silicon rubber sealed at one end) without air bubbles, and the open end of the container is sealed with a silicon

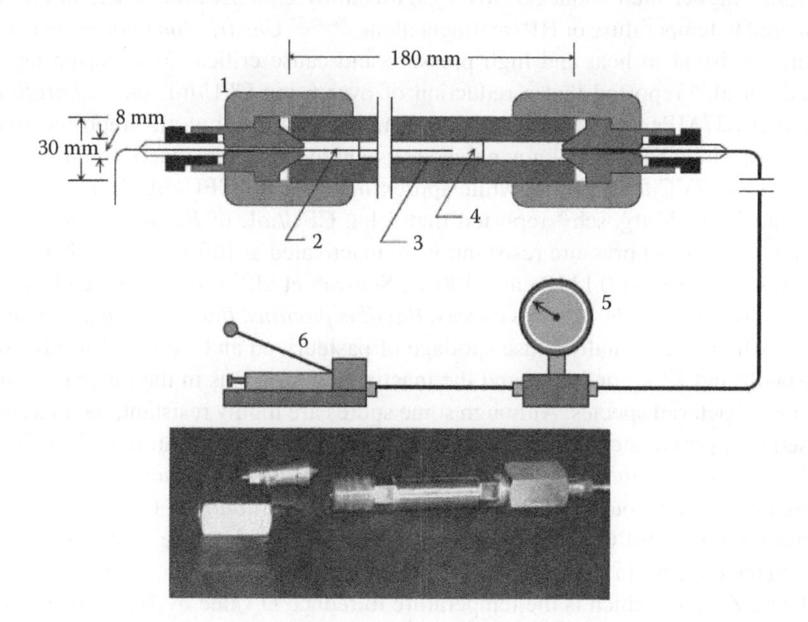

FIGURE 22.11 A schematic diagram of experimental apparatus. 1, pressure vessel; 2, thermocouple; 3, silicon tube; 4, sample; 5, pressure gauge; 6, pressure pump.

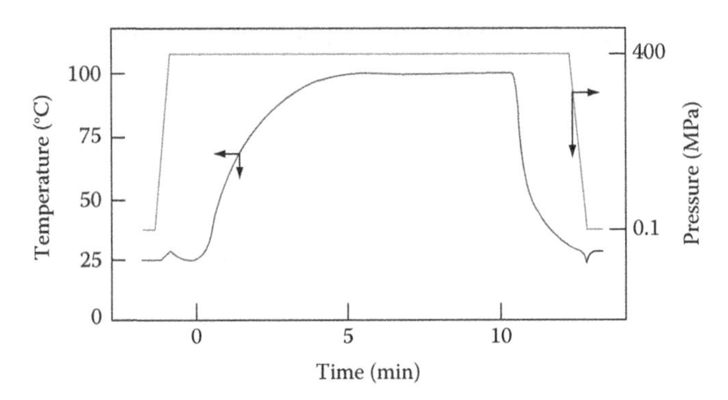

FIGURE 22.12 An example of temperature and pressure history curve. The treatment condition is at 100°C at 400 MPa for 10 min.

rubber plug. The container filled with the solution is inserted into the vessel. After the sample is in place, HP is applied to the vessel, and the pressure vessel is maintained in a water bath at 25°C for 5 min to ensure the initial temperature. Then, the pressure treatments combined with heating start, at the same time the pressure vessel is placed in a heating bath heated in a temperature range of 90°C–120°C.

As a typical example, the history of curves of pressure and temperature on a 10 min treatment at 100°C and 400 MPa is shown in Figure 22.12.

It can be seen that each treatment time contains rising time for 5 min without cooling time. When a packed sample is pressurized up to 400 MPa, the sample temperature increased by 8°C due to adiabatic compressive heating and after a few minutes the temperature returned to the initial one, at which the heating of the sample is started.

22.4.2 Combination of Elevated Temperature and High Hydrostatic Pressure

Combination of elevated temperature and high hydrostatic pressure (HHP), which is defined as a pressure range higher than about 300 MPa, significantly enhances the inactivation of bacterial spores compared to temperature or HP treatment alone.[26–32,36] *Clostridium botulinum* is well known to have spores resistant to heat and high pressure and cause critical food poisoning for human beings. Reddy et al.[26] reported that a reduction of over 6 log CFU/mL on *C. botulinum* spores was observed at 827 MPa and 75°C for 20–30 min. Cléry-Barraud et al.[27] reported that *Bacillus anthracis* spores, which have been known to be resistant to heat, were inactivated up to 9 log CFU/mL at 500 MPa and 75°C for 360 min, while approximately 2 log CFU/mL decrease was observed at 500 MPa and 20°C. Margosch[28] reported that 3 log CFU/mL of *B. amyloliquefaciens* spores, which is known to be most pressure resistant, were inactivated at 100°C and 800 MPa, whereas no inactivation was observed at 0.1 MPa and 100°C. Scurrah et al.[32] showed that eight *Bacillus* spp. (40 isolates) such as *B. subtilis*, *Bacillus cereus*, *Bacillus pumilus*, *Bacillus coagulans*, and *Bacillus licheniformis*, which occasionally cause spoilage of pasteurized and sterilized foods, were inactivated at 600 MPa and 72°C for 1 min, and the inactivation ratio was in the range of 0–6 log CFU/mL according to bacterial species. Although some spores are highly resistant, the inactivation ratio was increased to approximately 3–6 log CFU/mL by increasing temperature to 95°C.[32]

Survival curves of *B. subtilis* in combination treatment of elevated temperature and HHP are shown in Figure 22.13a through c. The survival curves of *B. subtilis* at 0.1, 100, and 200 MPa in the temperature range of 90°C–110°C were linear (r = 0.999–0.985). The death rates of the spores increased with the increase in temperature. The D value, which is the time to reduce the spore number by 1/10, and Z value, which is the temperature to reduce D value by 1/10, calculated by linear regression are shown in Table 22.9. It can be seen that higher pressure reduced the D values at 100°C and 105°C. On the other hand, the Z values at 100 and 200 MPa were approximately twice those at

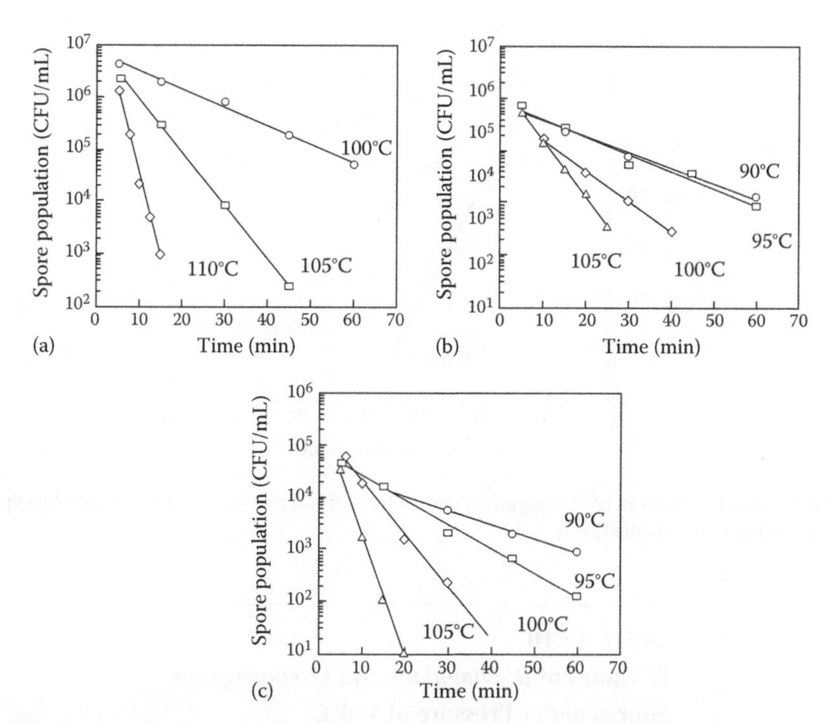

FIGURE 22.13 Survival curves of *B. subtilis* spores in a temperature range of 90°C–110°C at (a) atmospheric pressure, (b) 100 MPa, and (c) 400 MPa.

TABLE 22.9
Parametric Values Obtained from Survival Curves of *B. subtilis* Spores Used in This Study

	D Values (mm)					
Pressure (MPa)	**90**	**95**	**100**	**105**	**110**	**Z Values (°C)**
0.1			28.3	10.0	3.2	10.6
100	27.2	24.7	14.3	7.8		20.0
200	35.9	21.4	9.9	5.5		17.1

D value is the time required to reduce the number of spores by 1/10 at the temperature shown.
Z value is the temperature to reduce D value by 1/10.
D value and Z value are shown in Table 22.10.

ambient pressure and those reported before (Z = 7°C–12°C)[38] This might show that pressure reduces the temperature effect on thermal death rate.

Figure 22.14 shows the death behaviors of *B. coagulans* spores by thermal treatments (110°C) combined with HP in the pressure range of 0.1–400 MPa. Pressure increase enhanced the inactivation of the spores. The D values calculated from the survival curves are listed in Table 22.10. Though the D values at 100 MPa were almost the same as those at 0.1 MPa, at pressures higher than 200 MPa, the D values became considerably lower. The D values at 200 and 400 MPa were respectively 1/3 and 1/8 times lower than that at 0.1 MPa. The survival curves of *B. coagulans* spores by thermal treatments combined with 400 MPa become straight lines in a temperature range

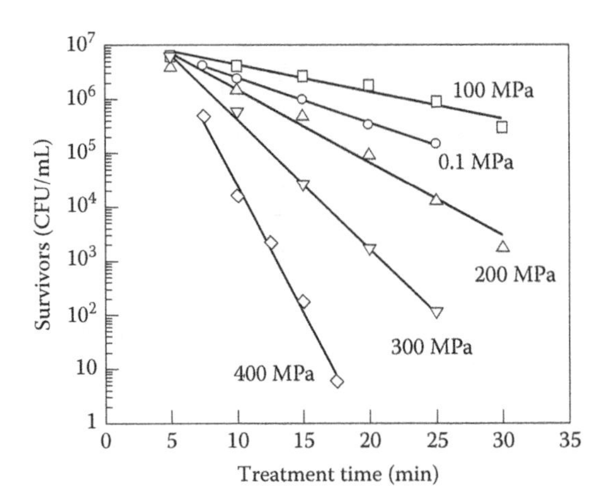

FIGURE 22.14 Death behaviors of *B. coagulans* spores by thermal treatments at 110°C combined with pressure in a pressure range of 0.1–400 MPa.

TABLE 22.10
D Values of *B. coagulans* and *C. sporogenes* Spores under Pressure at 110°C

	$D_{110°C}$ Values (min)	
Pressure (MPa)	***B. coagulans***	***C. sporogenes***
0.1	17.8	17.7
100	19.1	20.0
200	7.4	3.7
300	4.2	2.3
400	2.1	1.3

D value is shown in Table 22.9.

of 50°C–110°C (r = 0.999–0.987), as shown in Figure 22.15. The death rates of the spores increased gradually from 50°C to 90°C and thereafter exponentially rose from 90°C. The Z values calculated from these curves are shown in Figure 22.16. The D values gently decreased when the temperature rose from 50°C to 90°C, but exponentially decreased from 90°C. The Z values at 400 MPa calculated from the D values were 39.1°C in the temperature range of 50°C to 90°C, and 12.1°C in the temperature range of 90°C to 110°C, whereas the Z values at 0.1 MPa was 6.7°C in the temperature range of 104°C to 117°C. Thus, temperature is expected to be effective when higher than 90°C at 400 MPa. However, it must be considered that the Z values at 400 MPa are approximately twice as large as those at 0.1 MPa.

The survival curves of *Bacilllus stearothermophilus* spores at 113°C in the pressure range of 0.1–400 MPa are shown in Figure 22.17. Inactivation ratio of *B. stearothermophilus* spores was also enhanced by combination with elevated temperature and HHP as previously reported.[29] The shape of each survival curve proved to be a downward convex. It is reported that survival curves of bacterial spores under HHP do not obey a first-order rate equation, as was often previously reported.[30,36,39] This means that there is a possibility of being unable to inactivate the spores completely even by the simultaneous treatments of pressure and heating. Furthermore, as reported by some researchers, a small number of superdormant spores survive after such combination treatment.[27,28] In the case of

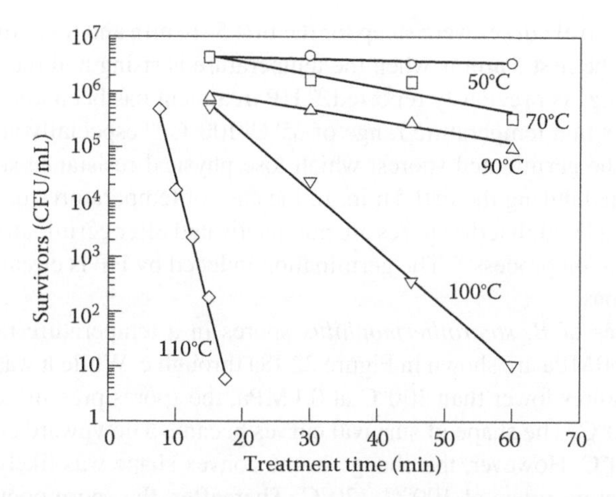

FIGURE 22.15 Survival curves of *B. coagulans* spores by thermal treatments in a temperature range of 50°C–110°C combined with pressure at 400 MPa.

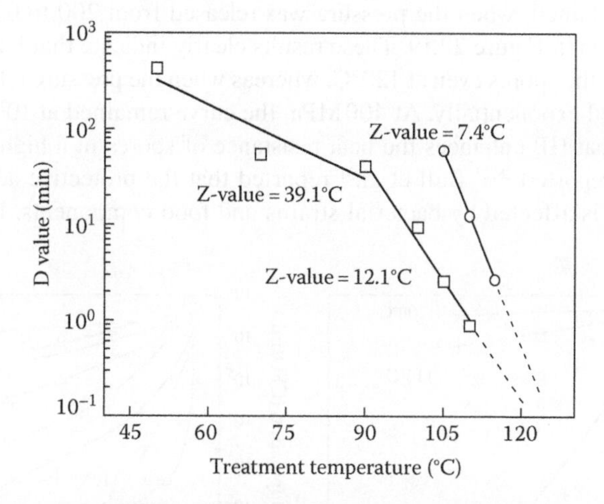

FIGURE 22.16 Comparison of D values of *B. coagulans* spores at 400 MPa with those at atmospheric pressure in a temperature range of 50°C–110°C. □, 400 MPa; ○, atmospheric pressure.

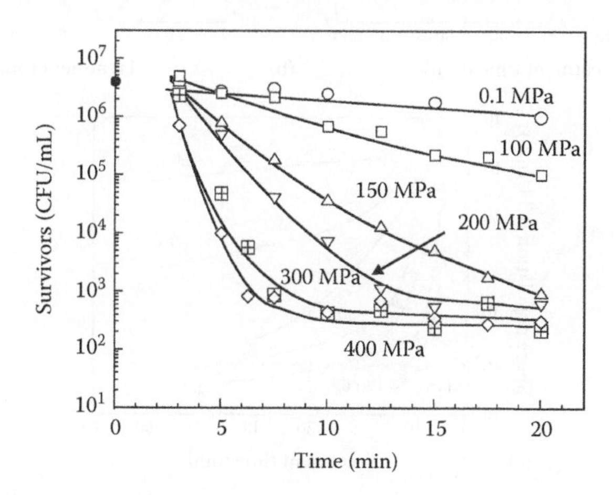

FIGURE 22.17 Survival curves of *B. stearothermophilus* spores by thermal treatments at 113°C combined with pressure in the range of 0.1–400 MPa. ●, initial spore number.

200–400 MPa, the survival curves were steep for the first 5–10 min and thereafter flattened out. One of the reasons is that the first 5 min is when the temperature is rising and the spores are efficiently killed after germinating, as previously reported.[30] HP treatment has been known to induce the germination of the spores in a temperature range of 35°C–100°C,[40] especially at 70°C–95°C.[30] Thus, it is considered that the germinated spores, which lose physical resistance such as to heat and to pressure, are inactivated during the first 5 min. In the case of temperatures of >95°C and pressures of >100 MPa, it is considered that the spores are not inactivated after germinating, but inactivated by bypassing the germination process.[30] The germination induced by HP is circumstantially discussed in the following sections.

The survival curves of *B. stearothermophilus* spores in a temperature range of 50°C–120°C under 0.1, 200, and 400 MPa are shown in Figure 22.18a through c. While it was most difficult to kill the spores at temperatures lower than 100°C at 0.1 MPa, the spores pressurized at 400 MPa began to be inactivated at 60°C. The shape of survival curves became a downward convex, particularly in the range of 70°C–90°C. However, the change in the convex shape was likely to terminate within 10 min in the temperature range of 100°C–120°C. Thereafter, the spore populations remained on the order of about 10^2 CFU/mL for 50 min (Figure 22.18c). The influence of HHP (200 MPa and 400 MPa) on survival curves of *B. stearothermophilus* spores at 120°C is shown in Figure 22.19. The survival curve obtained, when the pressure was released from 200 to 0.2 MPa after a 10 min treatment, is also shown in Figure 22.19. These results clearly indicate that HP at 200 MPa was not able to fully inactivate the spores even at 120°C, whereas when the pressure released to 0.2 MPa, the spores were inactivated exponentially. At 400 MPa, the curve remained at 10^2 CFU/mL for 50 min. This might indicate that HP enhances the heat resistance of spores in a higher temperature range as other researchers reported.[28,29] Bull et al.[31] reported that the protective action of HHP as well as inactivation action is affected by bacterial strains and food components. It is reported that HP

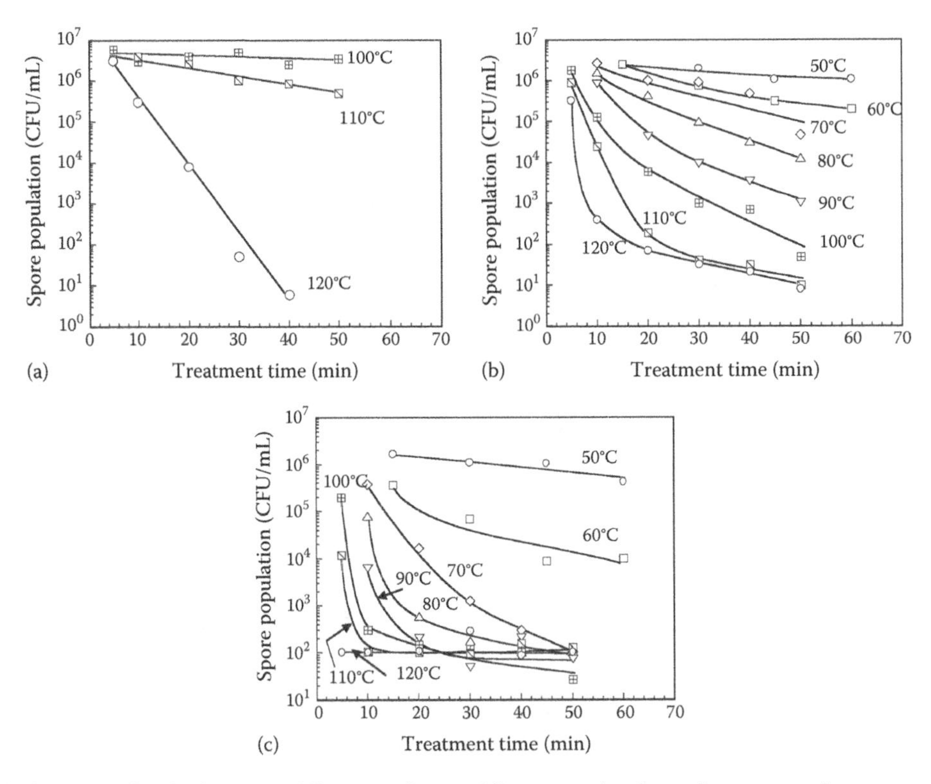

FIGURE 22.18 Survival curves of *B. stearothermophilus* spores by thermal treatments in a temperature range of 50°C–120°C combined with pressure at (a) 0.1 MPa, (b) 200 MPa, and (c) 400 MPa.

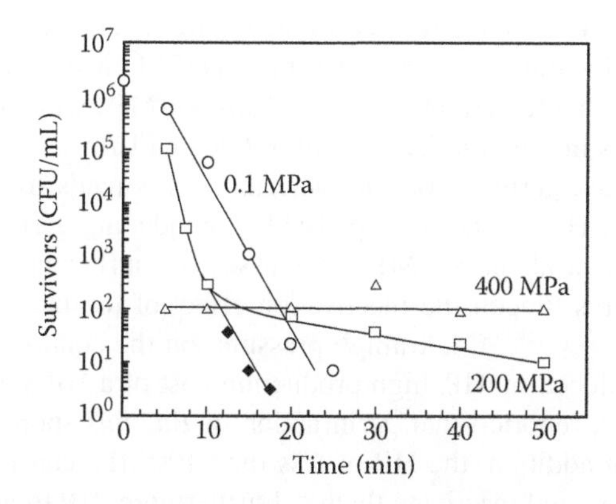

FIGURE 22.19 Comparison of survival curves of *B. stearothermophilus* spores by thermal treatment at 120°C combined with pressure. ◆, Pressure was released to be 0.1 MPa after the 10 min thermal treatment combined with 200 MPa.

suppresses thermal denaturation of *B. subtilis* DNA, and the effect increases from 122 to 162 MPa.[41] Black et al.[18] mentioned the involvement of superdormant spores. Enhancing heat resistance of the spores by HHP might be related to the same phenomenon as reported earlier.[18,41]

In addition, adiabatic compressive heating, especially when the sample temperature is high, affects critically the inactivation of bacterial spores in combined process of elevated temperature and HHP. Therefore, it is important to know that the increasing rate of water, which is the main component of foods, is approximately 3°C/100 MPa due to adiabatic compressive heating. Krebbers[42] reported that temperature of tomato puree, whose initial temperature is 90°C, reached 121°C during combined process of heat and HHP at 700 MPa, 90°C for 30 s. Adiabatic compressive heating instantaneously and uniformly increases the temperature of subjected food, and adiabatic expansive cooling also instantaneously decreases the subjected food temperature regardless of size, shape, and phase, though the loss of heat toward a pressure vessel is considered to control the temperature exactly.[43] This rapid and uniform heating and cooling action compared to traditional heating is one of the advantages of pressure-assisted heating, which might be useful for avoiding thermal deterioration of foods.

Therefore, the combination treatment of heating and HHP can effectively inactivate heat-resistant spores at lower temperatures than those treated on traditional heating and so will be able to reduce deterioration of foods due to heating. However, for processed foods like long-life products preserved at normal temperature, it must be considered that there is a risk that a small amount of superdormant spores will survive and some protective effect due to HHP will work on heat-resistant spores.

22.4.3 Combination of Mild Heating and Moderate HP-Induced Germination of Bacterial Spores

Inactivation of bacterial spores by HP treatment has been discussed in terms of spore germination by numerous researchers.[18,27,33,35,44,45] HP treatment combined with mild heating greatly induces the germination of bacterial spores. It has been reported that HP treatment also induces the germination of dormant spores that are hard to germinate with nutrients.[46] Because germinated spores lose their resistance to heat, HP, and toxic chemicals, the HP-induced germination has a potential ability for sterilization method by combination or subsequent heating. First, the HP-induced germination of bacterial spores has been reported by Clouston and Wills.[44] They showed that the germination of *B. pumilus* spores occurred at pressures exceeding 500 MPa and was the prerequisite for inactivation by compression. Moerman[35] reported that several *Bacillus* sp. and *Clostridium* sp. spores, which are inoculated in pork marengo (a kind of stew), are inactivated up to 1.5–4 log CFU/mL by HP treatment at 400 MPa and 50°C, and microbial shelf life is extended by the treatment. van Opstal[33] reveals that the combined HP treatment of 100–600 MPa with heating at 40°C could germinate the *B. cereus* spores in milk on the order of 3–6 log CFU/mL.

The effect of HP-induced germination on bacterial spores tends to increase according to an increase with pressure, but HHP is not indispensable for inducing germination.[33,34,47,48] Moderate HP, whose range is less than about 300 MPa, possesses the sufficient effect of HP-induced germination on bacterial spores though the inactivation effect of the moderate HP on the bacterial spores is hardly expected.[19,20,37,45,47,48] Ultrahigh pressure, on the contrary, has disadvantages such as protein denaturation induced by HP, high production cost of a HP apparatus.[22] In 1970, Gould and Sale, for the first time, reported that germination of *Bacillus* spore is induced in a pressure range of 25–100 MPa.[45] In addition, the HP of less than 100 MPa can induce the germination of *B. stearothermophilus* spore and inactivate them.[40] Furthermore, HP treatment in a pressure range of 90–350 MPa induces the germination of *B. subtilis* spore.[48]

Figure 22.20 shows the germination and inactivation of *B. subtilis* spores induced by moderate HP combined with heating in phosphate buffer (pH 7.0). As shown in Figure 22.20, the germination of 1–4 log CFU/mL is induced by relatively low pressure even at 20–100 MPa, and its optimum temperature is found to be in the temperature range of 40°C–50°C. At less than 40°C, inactivation of the spore is hardly observed, but higher than 50°C, the inactivation ratio begins to coincide with germination ratio. The germination ratio, however, declines at higher temperatures. Thus, once the HP treatment is done at moderate temperatures that are appropriate for spore germination, the temperature suitable for thermal inactivation should be elevated to inactivate the germinated spores. This two-step treatment is thought to be effective for spore inactivation.

However, it should be noted that all the spores are not germinated in the HP treatment combined with mild heating as many researchers have already reported.[27,36] Even at ultrahigh pressure, it is difficult to kill completely all the bacterial spores by HP-induced germination. Since the superdormant spores survive slightly after the treatment, the HP treatment combined with heating is inappropriate for long-life foods, which are distributed in market at the ambient temperature. However, on the chilled foods, in particular, a reduction of bacterial spores in it by the HP treatment combined with mild heating decrease in initial bacterial number, leads to the significant shelf-life extension

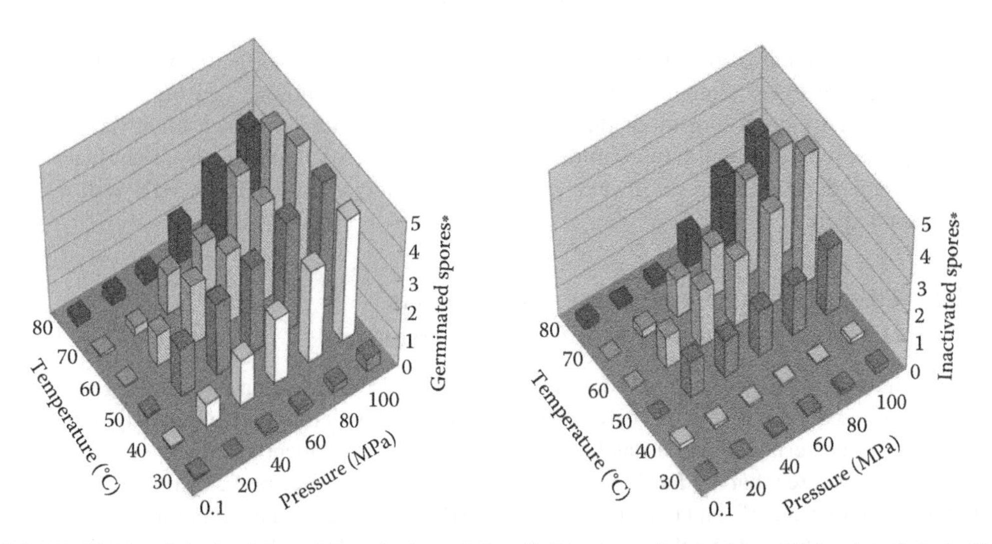

FIGURE 22.20 Germination and inactivation of *B. subtilis* spores induced by HP in phosphate buffer. *Germinated spores and inactivated spores are the mean Log (N_0/N_1) and Log (N_0/N_2), respectively, where N_0 is the initial number of spores, N_1 is the number of viable colonies after HP and heat treatment, and N_2 is the number of viable colonies after HP treatments.

of the chilled foods. Especially, *B. cereus*, which is thermostable pathogenic bacterium, is well germinated by moderate HP,[33,34,37] so that the moderate HP treatment combined with mild heating has a potential for reducing the risk of food-borne illness. Furthermore, this results in a reduction in the use of food preservatives. With the development of a chilled food chain, various kinds of chilled foods have been developed in recent years. Thus, the moderate HP treatment combined with mild heating should be appropriate for reducing bacterial spores that are hazardous microbes for food preservation and should also inhibit deterioration of foods. Furthermore, the cost of the device is expected to be less expensive than a HHP machine.

22.4.4 Influence of Environmental Factors on Moderate HP-Induced Germination

It has been reported that some environmental factors surrounding bacterial spore affect germination of bacterial spores considerably, not only under atmospheric pressure but also even under HP. Three main factors, nutrients, pH, and water activity (a_w), which affect germination of bacterial spores, are introduced in this section.

22.4.4.1 Influence of Germinants

It is well known that germination of bacterial spores is induced by germinants such as amino acids and sugars.[49] In addition, these germinants work considerably for the germination under moderate HP treatment.[19,20,33,37,45,51] Gould and Sale[45] revealed that addition of L-Ala accelerates germination of *B. subtilis*, *B. coagulans*, and *B. cereus* spores in the moderate HP treatment. van Opstal[33] also reported that germination ratio of *B. cereus* spores is 1.5–3.0 log cycle higher in milk than in phosphate buffer. This phenomenon means that moderate HP-induced germination has a potential for pasteurization of foods, which contain nutrients including these germinants. The accelerating effect of nutrients on germination of *B. subtilis* spores by moderate HP treatment, using glucose broth as a model of liquid food is shown in Figure 22.21. Glucose broth, which contains rich nutrients, induced the germination of spores, especially in the lower pressure range of 20–60 MPa more remarkably than that in phosphate buffer that contains no germinants. On the contrary, the accelerating effect of glucose broth on the germination is diminishing with the increase in pressure, and the difference of germination ratio in glucose broth and phosphate buffer is slight at 200 MPa (data not shown). Gould and Sale[45] showed that the germination initiated by lower pressure (<101 MPa) is markedly affected by the L-Ala, and the influence of L-Ala decreases with the increase of pressure

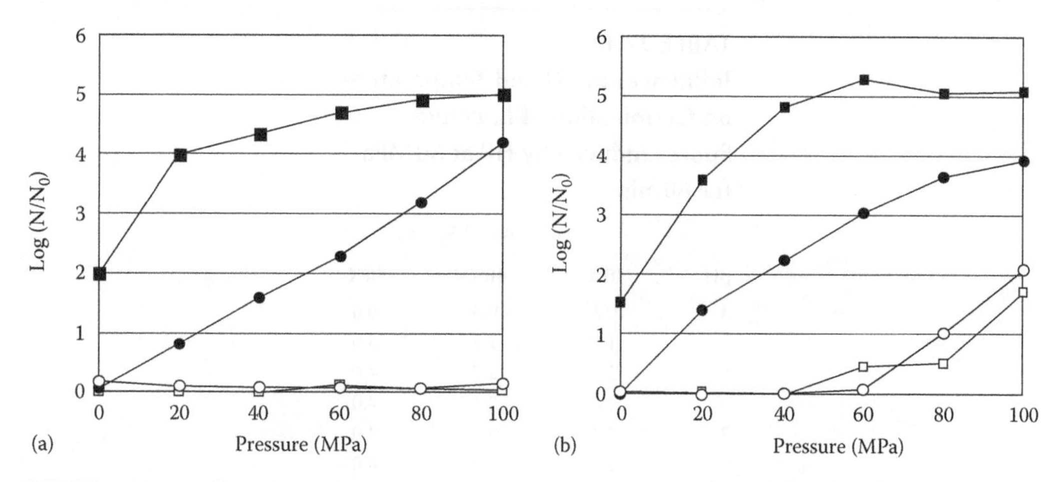

FIGURE 22.21 Effect of pressure on germination and inactivation of *Bacillus* spores induced by pressure at 40°C for 60 min. (a) *B. subtilis* and (b) *B. cereus*. ●, germinated spores in phosphate buffer; ○, inactivated spores in phosphate buffer; ■, germinated spores in glucose broth; □, inactivated spores in glucose broth.

and becomes negligible between 101 and 203 MPa. Raso[51] indicated that addition of L-Ala accelerated the germination at 250 MPa, 25°C, but did not at 690 MPa, 40°C. On the contrary, it has been reported that the germination ratio of the spores was higher in milk at 300 and 600 MPa than that in phosphate buffer.[33]

The mechanism of germination of bacterial spores induced by HP treatment was not clarified completely, but several researchers have studied about the mechanisms.[18,50,52–54] Wuytack et al.[52] suggested that the germination pathway of *B. subtilis* at 100 MPa shares steps with the pathway induced by nutrient germinant such as L-Ala and AGFK (asparagine, glucose, fructose, potassium ions), in addition, the pathway of the germination at HHP such as 600 MPa avoids the pathway of nutrient germinant working at lower pressure such as 100 MPa, and the germination mechanisms are different from those at lower pressures.

As stated earlier, since some germinants can induce germination of bacterial spores at relatively low hydrostatic pressure (LHP), the foods originally containing germinants can be treated by LHP such as 20–60 MPa, which leads to germinate effectively the spores in the foods and kills them. If the foods contain no germinants, then some germinants can be added to them.

22.4.4.2 Influence of pH

The germination of bacterial spores is significantly affected by pH, so that it is important to investigate the influence of pH on the HP-induced germination to apply for sterilization of various types of foods. Table 22.11 shows the HP-induced germination of *B. cereus* spores in a pH range of 3–10. At 60 MPa and 30°C–50°C, the germination of *B. cereus* spores occurred frequently in the pH range of 5–9. Gould and Sale[45] reported that the optimal pH range of germination for *B. cereus* initiated by HP at 100 MPa was broader than that of germination initiated by nutrients at ambient pressure. At acidic region (pH 3), germination is less than 1 log CFU/mL in all the temperatures, which shows that HP-induced germination of *B. cereus* spores occurs little in this acidic region. Raso et al.[51] also reported that the germination of *B. cereus* spores induced by HP treatment at 250 MPa, 25°C, pH 6–7.8 is greater than the germination at pH 3.5. It is reported that germination and inactivation of *B. cereus* spores due to HP treatment in the pressure range of 150–600 MPa is more effective in a neutral region than in an acidic region.[34]

The germination behavior of *B. subtilis* spores induced by moderate HP is shown in Table 22.12. The germination hardly occurs at pH 5 in all the temperatures tested. The germination works well at 50°C when the pH value becomes higher than pH 5. In acidic region at lower than pH 5, the

TABLE 22.11
Influences of pH and Temperature on Germination of *B. cereus* Spores Induced by HP at 60 MPa for 60 min

	Germinated Spores		
pH	30°C	40°C	50°C
3	0.3	0.3	0.6
4	1.0	1.7	3.0
5	2.1	2.7	4.0
6	2.4	3.1	4.0
7	2.3	2.9	4.0
8	2.3	2.9	4.0
9	1.9	2.8	4.0
10	2.0	2.5	3.0

TABLE 22.12
Influences of pH and Temperature on Germination of *B. subtilis* Spores Induced by HP at 60 MPa for 60 min

	Germinated Spores		
pH	30°C	40°C	50°C
3	0.2	0.5	1.6
4	0.1	0.2	0.5
5	0.2	0.2	0.6
6	0.1	1.2	1.9
7	0.1	1.2	3.0
8	0.1	0.8	3.8
9	0.1	0.3	3.5
10	0.1	0.5	3.1

significant reduction of spores is observed. Stewart et al.[55] reported that greater inactivation of *B. subtilis* and *Clostridium sporogenes* spores by HP treatment at 404 MPa and 45°C was observed when the pH value was lowered from pH 7 to 4. This reduction is not inhibited by phenol, which is a germination inhibitor; therefore, it is considered that the reduction is not related to germination (data not shown). It is well known that bacterial spores become more sensitive to heat at low pH than neutral pH.[56] In addition, Wuytack and Michiels[55] reported that the heat resistance of *B. subtilis* spores under HP decreases at low pH (<5) and the low pH inhibits the germination induced by HP at 100 MPa and suggested that the decrease is ascribed to permeability of protons into the spore. The reduction seemed to be due to inactivation by combination effect such as heat, low pH, and pressure.

The optimum pH for germination induced by HP lies in around neutral pH, but the germination behavior is different among bacterial species. For example, the spore of *Alicyclobacillus acidoterrestris* that is attributed to an odor in acidic fruit products is germinated by HP even in a low acidic side.[57] Thus, pH dependence of spore germination is also thought to affect remarkably the HP-induced germination. Furthermore, it is reported that pH dependence of spore inactivation is influenced by protein and sugar.[58,59] When HP-induced germination is applied to various foods, it should be noticed that pH and bacterial species bring in different results. Therefore, more information should be accumulated in the future in order to put the HP technology in practice.

22.4.4.3 Influence of a_w

In addition to pH, a_w is another important environmental factor for microbiologic control. In general, a_w is thought to affect various aspects of microorganisms, for instance, germination of spores, their outgrowth, and growth. When the surrounding a_w of bacterial spores becomes lower, not only their heat resistance becomes higher but also their germination and inactivation are inhibited at ambient pressure.[60,61] Under HP, the a_w also affects the germination and inactivation of bacterial spores. Sale et al.[62] reported about inactivation of *B. coagulans* spores by HP treatment at 101 MPa and 65°C, in the presence of sucrose and NaCl, and showed that the inactivation is inhibited at a_w of 0.96 with NaCl but was hardly inhibited at the same a_w with sucrose. Raso et al.[51] reported that the initiation of germination and the inactivation of *B. cereus* spores induced by HP of 250 MPa, 25°C, and 690 MPa, 40°C, are inhibited in a_w of 0.92 with sucrose. The germination behavior of *B. subtilis* and *B. cereus* spores at 60 MPa, 40°C–60°C, at a_w of 0.89 and 0.94, with sucrose and NaCl are shown in Figure 22.22. In both sucrose and NaCl, the germination ratio tends to decrease

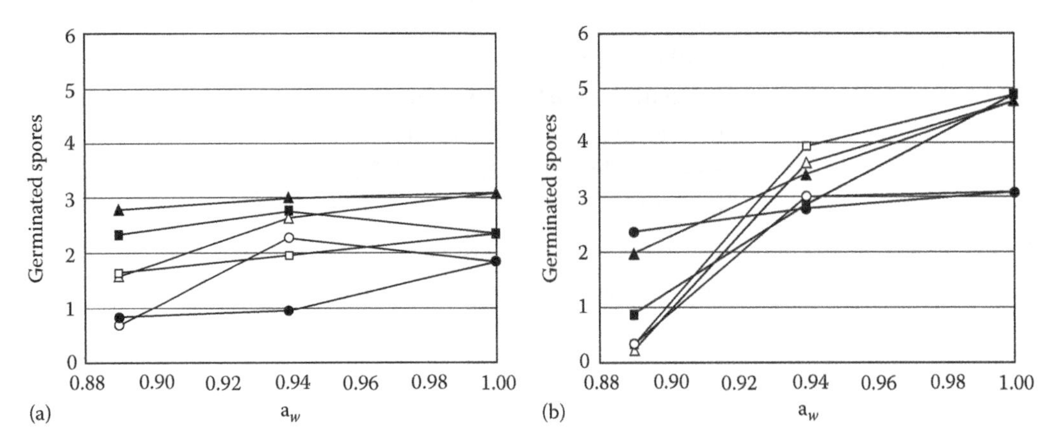

FIGURE 22.22 Relationship of water activity (a_w) and germination of *Bacillus* spores induced by HP at 60MPa for 60min. (a) *B. subtilis* and (b) *B. cereus*. ●, NaCl at 40°C; ○, sucrose at 40°C; ▲, NaCl at 50°C; △, sucrose at 50°C; ■, NaCl at 60°C; □, sucrose at 60°C.

with lowering of a_w. In the case of *B. subtilis*, the inhibitory effect of NaCl on germination became smaller than those of sucrose at 50°C and 60°C. In *B. cereus*, the inhibitory effect of NaCl was slightly larger than that of sucrose at a_w of 0.94 but was conversely smaller than sucrose at a_w of 0.89. Sale et al.[62] reported that ionic solutions (NaCl and calcium chloride) inhibit the inactivation of *B. coagulans* spores at 100MPa and 65°C, but nonionic solutions (sucrose and glycerol) rarely have the inactivation effect. Raso et al.[51] showed that *B. cereus* spores are more resistant to the germination at 250MPa, 25°C, and 690MPa, 40°C, in the presence of 0.5–4M sucrose (a_w of 0.99–0.92), when compared to no sucrose added. From these results, it is indicated that the HP-induced germination is inhibited at lower a_w adjusted by both NaCl and sucrose, but the inhibitory behavior of sucrose and NaCl is different, depending on the bacterial species and treatment conditions. On the contrary, at 40°C, sucrose enhances the HP-induced germination of *B. subtilis* spores at a_w of 0.94. As shown in Figure 22.21, nutrients such as sugar, amino acids, and inorganic ions promote the germination of spores under HP. Glucose is one of the major germinants or cogerminants, and sucrose is a germinant of *B. subtilis* spores. Sucrose might enhance the HP-induced germination in certain concentrations.

It has been reported that the germination of *Bacillus* sp. spores was inhibited by 10%–15% NaCl at ambient pressure.[63] It is, however, notable that HP treatment combined with mild heating induces the germination even in the high concentrations of NaCl and sucrose. These results indicated that HP-induced germination has a potential for reducing method of bacterial spores in the broad a_w range. However, it should be emphasized that the germination and the inactivation effect of bacterial spores are affected by bacterial species and composition of foods and foodstuffs. Treatment conditions such as pressure, temperature, and pH also affect the germination and the inactivation. In addition, the condition of sporulation such as temperature, pH, a_w, and minerals affected the germination.[33,51,64] In order to reinforce the confidence of food pasteurization to which HP-induced germination is applied, it is also effective to combine the pasteurization with some hurdle technologies.[66] For example, when preservatives like nisin are used with the pressure-assisted heating together, the inactivation of bacterial spores as well as other microbes can be enhanced.[47,66] Further investigation will be needed to apply this technology to practical sterilization.

22.5 CONCLUSIONS

As shown in this chapter, the combination of HP and heating is an effective way of enhancing the effectiveness of pressure and heating, respectively. On the shucking oysters, the combination can lower the magnitude of pressure by which oysters can be shucked, compared to pressure alone. The reduction of the pressure not only can reduce the cost manufacturing of the PSO machine but also give durability to the machine. As a result, the machine could be put into practice in market. The PSOs are almost the same as HSOs in appearance, though the reduction of freshness can be somewhat observed. Since the effectiveness and safety of the machine are superior to the demerits of freshness, the take-up of the machine in the industry will be expected in the future.

The combination treatment of growth inhibition of microorganisms by HP and the effect of activating some enzymes by mild heating can produce fish sauce in a quite short time without adding salt to them. The combination treatment utilizes autolysis of fish containing originally proteinases and is also applied to other food materials that have less proteinases or no proteinase if some commercial proteinase is added to them. The combination treatment machine has already been developed in the Japanese market.

Using pressure-assisted heating enables effective inactivation of bacterial spores in foods, thus preserving their qualities. However, it should be noted that a small amount of the superdormant spores can survive. However, the reduction of the spore will be useful for extension of shelf life in chilled foods, delicatessens, and so on; and the reduction of the initial microbe number in them will lead to the reduced use of food preservatives.

It should be noticed that environmental factors such as temperature, pH, and a_w affect the HP-induced germination and inactivation. Therefore, in order to promote the use of pressure-assisted heating, food components, their microflora, and their distribution conditions should be considered together.

REFERENCES

1. B H Hite. The effect of pressure in the preservation of milk. *West Virginia University Agricultural Experiment Station, Bulletin* 58: 15–35, 1899.
2. A Murokoshi, K Namba. The way of shucking bivalves. Japanese Patent Toku-Gan-Hei 10-340234, 1998.
3. T Okazaki. The way of seasoning under pressure. Japanese Patent 3475328.
4. Y Aoyama, Y Shigeta, T Okazaki, Y Hagura, K Suzuki. Growth inhibition of microorganisms by hydrostatic pressure. *Food Science and Technology Research* 10: 268–272, 2004.
5. C E ZoBell, F H Johnson. The influence of hydrostatic pressure on the growth and viability of terrestrial and marine bacteria. *Journal of Bacteriology* 57: 179–189, 1949.
6. T Okazaki, T Yoneda, K Suzuki. Combined effects of temperature and pressure on sterilization of *Bacillus subtilis* spores. *Nippon Shokuhin Kogyo Gakkaishi* 41: 536–541, 1994.
7. A Murokoshi. Development of a device for shucking oysters. *Nippon Suisan Gakkaishi* 70: 671–673, 2004.
8. Y Miura, I Hatsukade. The way of producing shucking bivalves. Japanese Patent Toku-Gan-Hei 3-127839, 1991.
9. H He, R M Adams, D E Farkas, M T Morrissey. Use of high-pressure processing for oyster shucking and shelf-life extension. *Journal of Food Science* 67: 640–645, 2002.
10. A Murokoshi. Tehnical materials in Marino-forum 21, no. 45, 2001.
11. T Okazaki, Y Shigeta, Y Aoyama, K Namba. Autolysis of unsalted fish protein under pressurization. *Fisheries Science* 69: 1255–1260, 2003.
12. Y Taniguchi, K Suzuki. Pressure inactivation of α-chymotrypsin. *Journal of Physical Chemistry* 87: 5185–5193, 1983.
13. T Ohmori, T Shigehisa, S Taji, R Hayashi. Biochemical effects of high hydrostatic pressure on the lysosome and proteases involved in it. *Bioscience, Biotechnology, and Biochemistry* 56, 1285–1288, 1992.
14. Y Shigeta, Y Aoyama, T Okazaki, T Matsui, K Namba. Growth inhibition of microorganisms by hydrostatic pressure in autolytic hydrolysis of unsalted squid liver. *Nippon Shokuhin Kagaku Kogaku Kaishi* 55: 117–120, 2007.
15. T Okazaki, Y Shigeta, Y Aoyama. Autolysis of unsalted guts of sea cucumber under hydrostatic pressure. *Nippon Suisan Gakkaishi* 73: 734–738, 2007.

16. M Noda, K Murakami. Purification and characterization of two acid proteinases from the stomach. *Biochimica et Biophysica Acta*, 658, 27–34, 1981.
17. V V Thanh, M Noda, K Murakami. Purification and some properties of two aminopeptidases from sardines. *Agricultural and Biological Chemistry* 47: 2453–2459, 1983.
18. E P Black, P Setlow, A D Hocking, C M Stewart, A L Kelly, D G Hoover. Response of spores to high-pressure processing. *Comprehensive Reviews in Food Science and Food Safety* 6(4): 103–119, 2007.
19. Y Aoyama, Y Shigeta, T Okazaki, Y Hagura, K Suzuki. Non-thermal inactivation of *Bacillus* spores by pressure-holding. *Food Science and Technology Research* 11(3): 324–327, 2005.
20. Y Aoyama, Y Shigeta, T Okazaki, Y Hagura, K Suzuki. Germination and inactivation of *Bacillus subtilis* spores under combined conditions of hydrostatic pressure and medium temperature. *Food Science and Technology Research* 11(1): 101–105, 2005.
21. V M Balasubramaniam, D Frakas. High-pressure food processing. *Food Science and Technology International* 14(5): 413–418, 2008.
22. T Norton, D W Sun. Recent advances in the use of high pressure as an effective processing technique in the food industry. *Food and Bioprocess Technology* 1(1): 2–34, 2008.
23. J A Torres, G Velazquez. Commercial opportunities and research challenges in the high pressure processing of foods. *Journal of Food Engineering* 67(1–2): 95–112, 2005.
24. H Zhang, G S Mittal. Effects of high-pressure processing (HPP) on bacterial spores: An overview. *Food Reviews International* 24: 330–351, 2008.
25. A Nakayama, Y Yano, S Kobayashi, M Ishikawa, K Sakai. Comparison of pressure resistances of spores of six *Bacillus* strains with their heat resistances. *Applied and Environmental Microbiology* 62(10): 3897–3900, 1996.
26. N R Reddy, R C Tetzloff, H M Solomon, J W Larkin. Inactivation of *Clostridium botulinum* nonproteolytic type B spores by high pressure processing at moderate to elevated high temperatures. *Innovative Food Science and Emerging Technologies* 7(3): 169–175, 2006.
27. C Cléry-Barraud, A Gaubert, P Masson, D Vidal. Combined effects of high hydrostatic pressure and temperature for inactivation of *Bacillus anthracis* spores. *Applied and Environmental Microbiology* 70(1): 635–637, 2004.
28. D Margosch, M A Ehrmann, R Buckow, V Heinz, R F Vogel, M G Gänzle. High-pressure-mediated survival of *Clostridium botulinum* and *Bacillus amyloliquefaciens* endospores at high temperature. *Applied and Environmental Microbiology* 72(5): 3476–3481, 2006.
29. E Patazca, T Koutchma, H S Ramaswamy. Inactivation kinetics of *Geobacillus stearothermophilus* spores in water using high-pressure processing at elevated temperatures. *Journal Food Science* 71(3): M110–M116, 2006.
30. E Ananta, V Heinz, O Schlüter, D Knorr. Kinetic studies on high-pressure inactivation of *Bacillus stearothermophilus* spores suspended in food matrices. *Innovative Food Science and Emerging Technologies* 2(4): 261–272, 2001.
31. M K Bull, S A Olivier, R J van Diepenbeek, F Kormelink, B Chapman. Synergistic inactivation of spores of proteolytic *Clostridium botulinum* strains by high pressure and heat is strain and product dependent. *Applied and Environmental Microbiology* 75(2): 434–445, 2009.
32. K J Scurrah, R E Robertson, H M Craven, L E Pearce, E A Szabo. Inactivation of *Bacillus* spores in reconstituted skim milk by combined high pressure and heat treatment. *Journal of Applied Microbiology* 101: 172–180, 2006.
33. I van Opstal, C F Bagamboula, S C Vanmuysen, E Y Wuytack, C W Michiels. Inactivation of *Bacillus cereus* spores in milk by mild pressure and heat treatments. *International Journal of Food Microbiology* 92(2): 227–234, 2004.
34. S Oh, M J Moon. Inactivation of *Bacillus cereus* spores by high hydrostatic pressure at different temperatures. *Journal of Food Protection* 66(4): 599–603, 2003.
35. F Moerman. High hydrostatic pressure inactivation of vegetative microorganisms, aerobic and anaerobic spores in pork Marengo, a low acidic particulate food product. *Meat Science* 69(2): 225–232, 2005.
36. T Okazaki, K Kakugawa, T Yoneda, K Suzuki. Inactivation behavior of heat-resistant bacterial spores by thermal treatments combined with high hydrostatic pressure. *Food Science and Technology Research* 6(3): 204–207, 2000.
37. Y Shigeta, Y Aoyama, T Okazaki, T Hagura, K Suzuki. Hydrostatic pressure-induced germination and inactivation of *Bacillus* spores in the presence or absence of nutrients. *Food Science and Technology Research* 13(3): 193–199, 2007.

38. H A C Thijssen, P J A M Kerkhof. Effect of temperature-moisture content history during processing on food quality. In: T Hoyem, O Kvale, Eds. *Physical, Chemical and Biological Changes in Food Caused by Thermal Processing*. London, U.K.: Applied Science Publishers, 1977, pp. 10–30.
39. K Kimura, M Ida, Y Yoshida, K Ohki, M Onomoto. Application of high pressure for sterilization of low acid food. In: R Hayashi, C Balny, Eds. *High Pressure Bioscience and Biotechnology*. Kyoto, Japan: Elsevier Science BV, 1996, pp. 429–432.
40. S Furukawa, I Hayakawa. Investigation of desirable hydrostatic pressure required to sterilize *Bacillus stearothermophilus* IFO 12550 spores and its sterilization properties in glucose, sodium chloride and ethanol solutions. *Food Research International* 33(10): 901–905, 2000.
41. C G Heden, T Lindahl, I Toplin. The stability of deoxyribonucleic acid solutions under high pressure. *Acta Chemica Scandinavica* 18: 1150–1156, 1964.
42. B Krebbers, A M Matser, S W Hoogerwerf, R Moezelaar, M M M Tomassen, R W van den Berg. Combined high-pressure and thermal treatments for processing of tomato puree: Evaluation of microbial inactivation and quality parameters. *Innovative Food Science and Emerging Technologies* 4: 377–385, 2003.
43. Y Shao, S Zhu, H Ramaswamy, M Marcotte. Compression heating and temperature control for high-pressure destruction of bacterial spores: An experimental method for kinetics evaluation. *Food Bioprocess Technology* 3: 71–78, 2010.
44. J G Clouston, P A Wills. Initiation of Germination and Inactivation of *Bacillus pumilus* spores by hydrostatic pressure. *Journal of Bacteriology* 97(2): 684–690, 1969.
45. G W Gould, A J H Sale. Initiation and germination of bacterial spores by hydrostatic pressure. *Journal of General Microbiology* 60: 335–346, 1970.
46. J Wei, I M Shah, S Ghosh, J Dworkin, D G Hoover, P Setlow. Superdormant spores of *Bacillus* species germinate normally with high pressure, peptidoglycan fragments, and bryostatin. *Journal of Bacteriology* 192(5): 1455–1458, 2010.
47. E P Black, M Linton, R D McCall, W Curran, G F Fitzgerald, A L Kelly, M F Patterson. The combined effects of high pressure and nisin on germination and inactivation of *Bacillus* spores in milk. *Journal of Applied Microbiology* 105: 78–87, 2008.
48. H N T Minh, P Dantigny, J M P Cornet, P Gervais. Germination and inactivation of *Bacillus subtilis* spores induced by moderate hydrostatic pressure. *Biotechnology and Bioengineering* 107(5): 876–883, 2010.
49. Y Yasuda, K Tochikubo. Relation between D-glucose and L-and D-alanine in the initiation of germination of *Bacillus subtilis* spore. *Microbiology and Immunology* 28(2): 197–207, 1984.
50. E Y Wuytack, S Boven, C W Michiels. Comparative study of pressure-induced germination of *Bacillus subtilis* spores at low and high pressures. *Applied and Environmental Microbiology* 64(9): 3220–3224, 1998.
51. J Raso, M M Góngora-Nietoa, G V Barbosa-Cánovas, B G Swanson. Influence of several environmental factors on the initiation of germination and inactivation of *Bacillus cereus* by high hydrostatic pressure. *International Journal of Food Microbiology* 44: 125–132, 1998.
52. E Y Wuytack, J Soons, F Poschet, C W Michiels. Comparative study of pressure-and nutrient-induced germination of *Bacillus subtilis* spores. *Applied and Environmental Microbiology* 66(1): 257–261, 2000.
53. E P Black, K Koziol-Dube, D Guan, J Wei, B Setlow, D E Cortezzo, D G Hoover, P Setlow. Factors influencing germination of *Bacillus subtilis* spores via activation of nutrient receptors by high pressure. *Applied and Environmental Microbiology* 71(10): 5879–5887, 2005.
54. J Wei, P Setlow, D G Hoover. Effects of moderately high pressure plus heat on the germination and inactivation of *Bacillus cereus* spores lacking proteins involved in germination. *Letters in Applied Microbiology* 49: 646–651, 2009.
55. C M Stewart, C P Dunne, A Sikes, D G Hoover. Sensitivity of spores of *Bacillus subtilis* and *Clostridium sporogenes* PA 3679 to combinations of high hydrostatic pressure and other processing parameters. *Innovative Food Science and Emerging Technologies* 1: 49–56, 2000.
56. A Palop, J Raso, R Pagán, S Condón, F J Sala. Influence of pH on heat resistance of spores of *Bacillus coagulans* in buffer and homogenized foods. *International Journal of Food Microbiology* 46(3): 243–249, 1999.
57. S Y Lee, R H Dougherty, D H Kang. Inhibitory effects of high pressure and heat on *Alicyclobacillus acidoterrestris* spores in apple juice. *Applied and Environmental Microbiology* 68(8): 4158–4161, 2002.
58. Y L Gao, X R Ju. Statistical prediction of effects of food composition on reduction of *Bacillus subtilis* As 1.1731 spores suspended in food matrices treated with high pressure. *Journal of Food Engineering* 82: 68–76, 2007.

59. Y L Gao, X R Ju, W F Qiu, H H Jiang. Investigation of the effects of food constituents on *Bacillus subtilis* reduction during high pressure and moderate temperature. *Food Control* 18(10): 1250–1257, 2007.
60. W G Murrell, W J Scott. The heat resistance of bacterial spores at various water activities. *Journal of General Microbiology* 43: 411–425, 1966.
61. A C Baird-Parker, B Freame. Combined effect of water activity, pH and temperature on the growth of *Clostridium botulinum* from spore and vegetative cell inocula. *Journal of Applied Microbiology* 30(3): 420–429, 1967.
62. A J H Sale, W Gould, W A Hamilton. Inactivation of bacterial spores by hydrostatic pressure. *Journal of General Microbiology* 60: 323–334, 1970.
63. F K Cook, M D Pierson. Inhibition of bacterial spores by antimicrobials. *Food Technology* 37(11): 115–126, 1983.
64. N Igura, Y Kamimura, M S Islam, M Shimoda, I Hayakawa. Effects of minerals on resistance of *Bacillus subtilis* spores to heat and hydrostatic pressure. *Applied and Environmental Microbiology* 69(10): 6307–6310, 2003.
65. L Leistner. Further developments in the utilization of hurdle technology for food preservation. *Journal of Food Engineering* 22: 421–432, 1994.
66. E Y Wuytack, C W Michiels. A study on the effects of high pressure and heat on *Bacillus subtilis* spores at low pH. *International Journal of Food Microbiology* 64: 333–341, 2001.

23 pH-Assisted Thermal Processing

Alfredo Palop and Antonio Martínez Lopez

CONTENTS

23.1 INTRODUCTION

Heat sterilization ensures food safety and preservability through bacterial destruction, but it also involves the application of intense treatments that cause additional food quality losses. In many cases, bacterial destruction is not necessary for food preservation, and bacterial growth can be inhibited by controlling environmental factors that affect viability. In this case, microbes will not be destroyed, but they will not be able to grow, and the preservation techniques used, much less intense, will affect food quality much less.

Acid pH has the advantage, over other environmental factors that affect microbial viability, that it also decreases microbial *heat resistance*. Hence, acidified foods can be heat treated at lower temperatures, keeping the same levels of microbial destruction. Moreover, acid pHs hamper the *recovery* and growth of the survivors to the heat treatment. *Microorganisms* show their maximum heat resistance at neutral pH, decreasing with the increase in acidification. The extent of the effect of acid pH on heat resistance depends on the microorganism being tested and on the heating temperature. It is also influenced by the *type of acid* and its concentration.

The inhibitory effect of *acid pH* on microbial growth has long been used to preserve foods from spoilage. Indeed, *acidification* has been used as a food preservation technique through thousands of years in human history, naturally by fermentation or artificially by adding weak acids. Fermented milks were produced for the first time long time ago. Probably when humans started to tame dairy species and to use their milks, lactic bacteria from soil or plants contaminated accidentally the milk, expanding to the recipients used to collect and preserve it and eventually installing on them. The mechanism of action of this preservation technique, though unknown by our ancestors, was simply by a microbial lactic fermentation that reduced the pH value of the milk and was able to extend itself the preservation of milk. This accidental growth of lactic bacteria probably did not lead to the production of homogeneous fermented milks with stable and defined flavors but had the undeniable advantage of preventing the development of pathogen microorganisms.

Later, Romans used vinegar regularly to preserve their foods. Romans were very interested in agriculture and livestock farming, but also on food processing and preservation. To supply a large city such as Rome with foods was not easy, and vinegar, together with other preserving agents such as brines or ice, was used regularly to preserve foods because of its acidifying effect.

Heat treatments were applied extensively to preserve foods much later in human history. Nicolas Appert showed at the beginning of nineteenth century that the application of a thermal treatment to foods contained in tightly closed cans was able to preserve them from spoilage, but about 50 years later, it was Pasteur who gave sense to this procedure by discovering that microbes were responsible for the deterioration of foods and heat acted killing them, and hence preserving foods. Pasteur used heat for the first time with the purpose of inactivating the spoilage microorganisms of wine and beer. The technique was rapidly extended to other liquid foods such as milk and developed in the actual pasteurization treatments. Bigelow and Esty [1] studied in 1920 the *heat inactivation kinetics* of microorganisms and stated the basis for the modern food sterilization processes. Soon after this, Esty and Meyer [2] investigated the effect of acid pH on heat resistance showing that acid pHs lowered the *D* values of microorganisms. Some years later, Sognefest et al. [3] corroborated the important influence of pH on bacterial heat resistance. All these findings led to the introduction in the food canning industries of an acidification step, prior to the thermal treatment.

Customers nowadays are demanding high-quality products combined with food safety. To meet this demand, the industry has developed minimally processed food products using a combination of different technologies in order to reduce the impact of severe heat treatments, which guarantee food safety but produce a negative effect on the sensory and nutritional properties of the product. *Acidification* of the food allows reducing the intensity of thermal treatments and is one of the most frequent procedures used. On the other hand, acidification does not seem to reduce consumer acceptability of canned foods [4].

Present objectives of acidification of canned foods are to inhibit the growth of *Clostridium botulinum* in order to assure food safety and to sensitize spoilage microorganisms to heat especially *bacterial spores*.

This chapter will start with a short discussion on the heat treatments that are applied to canned foods, attending to their pH. Afterward, the chapter will focus on the data related to the heat resistance of bacteria especially bacterial spores, which are the target of heat sterilization treatments, and paying attention to recent relevant data. In Section 23.3, the inhibitory effect of acid pH on bacterial growth will be exposed, as an introduction to its destructive effect at high (heating) temperatures, which will be dealt with in next Section (23.4). The effect of the type of acid on bacterial heat resistance will be described in Section 23.5 and the effect of pH on the recovery of the survivors to the heat treatment in Section 23.6. In Section 23.7, the theories that try to explain the increased sensitization of bacteria to heat when it is applied at acid pHs will be discussed. Finally, Section 23.8 will show the existing models, databases, and software developed to predict bacterial behavior at acid heat treatments.

23.2 CANNED FOODS AND pH

In their natural form, many foods, such as meat, fish, and vegetables, are slightly acid. Most of the fruits are rather acid and only a few foods like egg white are alkaline.

For canned foods, the distinctive pH value is 4.6, which is the minimum pH of growth for *C. botulinum*, the most heat resistant of the food pathogen microorganisms [5].

Low-acid canned foods (with pH values > 4.6) have to be exposed at least to a heat treatment able to reduce *C. botulinum* spore counts in 12 log cycles (2.52 min at 121°C or an equivalent combination of time/temperature), which is known as the botulinum cook [6]. Acid or acidified canned foods (with pH values < 4.6) can be heat treated at temperatures around 100°C for a few minutes at atmospheric pressure without the need of usage of retorts to apply the treatments. Such a simplification in the way to apply these treatments has minimized the efforts invested in the technology, which is essentially the same that was used initially, that is, open water baths heated at 100°C or lower temperatures.

Acid or acidified foods with pH values between 4.0 and 4.5 are usually heat treated to reach acceptable levels of risk of flat sour spoilage caused by *Bacillus coagulans*. Below pH 4, the risk of bacterial spore spoilage disappears and sterilization treatments are not necessary [7].

However, in the last few years, several spoilage incidents related to acid foods such as heat-treated fruit juice products have been reported [8–13]. In most of the cases, the microorganism responsible for the spoilage has been an acidophilic spore-forming bacterium, belonging to a species of the genus *Alicyclobacillus* such as *A. acidoterrestris*. These microorganisms are able to grow at pHs as low as 2.0 in spite of being spore-forming bacteria [10].

Moreover, it has been postulated that the growth of some mesophilic bacilli that are able to grow at pH values close to 4.5 could lead to local increases of the pH of the food that would allow for the development of *C. botulinum* [14,15]. Furthermore, it has been observed that the minimum pH at which *C. botulinum* develops and synthesizes toxin depends on the anaerobic conditions [16] and on the acidulant used to lower the pH [17].

Odlaug and Pflug [6] consider that for a canned food with pH lower than 4.6 to be risky of having the botulinic toxin, the following four situations should occur simultaneously: (1) food with high numbers of *C. botulinum* spores; (2) concurrence of other microorganisms because of pour processing or posttreatment contamination; (3) food composition and storage conditions that enable growth of *C. botulinum* and toxin formation; and (4) growth of other microorganism(s) that create(s) favorable conditions for growth and toxin formation. Fortunately, in the opinion of the same authors [6], the coincidence of the four situations happens very rarely.

The appropriate pH in the cans is obtained by the use of brines with known acid concentrations or tablets of known acid composition that are added to cans of specified volumes. The contents of the cans must be then conveniently stirred to ensure that the pH is below 4.6 in the center of all the food particles [5].

Acids used in canning to lower the pH below 4.6 are usually citric, lactic and malic, but also glucono-δ-lactone [18]. Dall'Aglio et al. [19] stated that while citric and other acids gave an acid taste (or other characteristic tastes related to the acid) to the food product, glucono-δ-lactone could be added to the brine to decrease the pH at the desired level without giving such an acid taste to the preserve. However, in most cases, the choice of *acidulant* used to decrease the pH is based on economic and convenience (ease to apply) criteria, more than on solid scientific data.

23.3 EFFECT OF pH ON BACTERIAL GROWTH

When the pH is shifted from the optimum for growth for a specific microorganism, the growth rate decreases, reaching a value where growth stops completely. Moderately, *acid pHs* cause an increase in the duration of the lag phase, a lower growth rate during the exponential growth phase and a lower maximum population density.

Bacteria deal with wide-ranging changes in the pH of their growth medium by maintaining their internal pH near a fixed optimal value. The proton, more than any other ion, is involved in almost every physicochemical as well as biochemical reaction. The difference in proton concentration across the cytoplasmic membrane (proton electrochemical gradient) together with the difference in electrical charge at both sides (membrane potential) creates a form of potential energy called protonmotive force, which cells can use for a variety of purposes including active transport, maintenance of cell's turgor, maintenance of cell's internal pH, turning flagella, driving a reverse flow of electrons through the respiratory chain to produce reducing power such as reduced nicotinamide adenine dinucleotide (NADH) or generation of adenosine triphosphate (ATP). However, extreme environmental pH values cause dysfunction of this system, leading to an acidification or alkalinization of the bacterial cytoplasm, blockage of the metabolic pathways, and finally growth prevention and even cell death [20].

Typically, bacteria are more sensitive to an environmental stress when they are under the influence of other stresses, that is, the minimum pH for growth will be higher if cells are cultured on a medium containing high concentrations of sodium chloride.

While much of the effect of acid pH can be accounted for by pH, it is well-known that different *organic acids* vary considerably in their inhibitory effects. Apart from the free proton concentration itself (i.e., the pH value), microorganisms are affected by the concentration of nondissociated weak acid, which is also dependent on the pH value. It has been shown that both the ionized and the non-ionized form of an organic acid can contribute to its inhibitory effect, although the undissociated form is generally more inhibitory [21].

Some weak acids (e.g., acetic and lactic acids) are lipophilic in their undissociated form and pass readily through the plasma membrane, entering the bacterial cell. In the cytoplasm at a pH of ~7, the acid molecules dissociate, releasing protons that cannot pass across the lipid bilayer (only by active transport) and accumulate in the cytoplasm, thus lowering the internal pH [5]. Acidified cytoplasm in turn inhibits metabolism and consequently growth. Moreover, the dissociated form of these acids may have inhibitory effects for the cell at high concentrations. Other weak acids do not pass through the bacterial membrane and hence do not acidify the cytoplasm or as proposed for sorbic acid [22] could act as membrane-active compounds. The mechanism for strong acids, such as hydrochloric or phosphoric, is to decrease considerably the external pH, providing a high environmental concentration of protons that determines the acidification of the cytoplasm of the cell.

Some acids are very effective inhibitors of microbial growth and are intentionally added to many foods as preservatives. This is the case of benzoic and sorbic acids [18]. Others such as acetic, fumaric, propionic, or lactic simply prevent or delay the growth of pathogenic and spoilage bacteria [18].

Hsiao and Siebert [23], after studying 11 physical and chemical properties of 17 organic acids commonly found in food systems, attributed the differences in their inhibitory effect to fundamental properties of them including polar groups, number of double bonds, molecular size, and solubility in nonpolar solvents. However, these same authors found different patterns of acid resistance when studying the effects of the same acids on six test bacteria. This could mean that although the effect of different acids can be predictable for some known bacterial species (in function of their molecular properties), the susceptibility of a new bacterium should have to be investigated before any conjecture can be established.

Still, acetic and lactic are commonly found as the most inhibitory acids for many bacteria [17,24–26] and this could be related to their low molecular weight and high *dissociation constant* (pK_a) values. Probably, these are the most important properties of weak acids [27], because they determine the capability of acids to enter bacterial cells in an undissociated form. Acids are only able to enter the cell when they are in the undissociated form. Once inside the cell, they can dissociate and play their role [28]. The amount of undissociated acid present in a system depends on its pH, the pK_a of the acid and its concentration. At a given pH value, acids with the higher pK_a would show the higher proportion of undissociated acid [27]. When an acid has several acid groups, the values of $pK_{a1} \ldots pK_{an}$ are the dissociation constants of the protons and the value pK_{a1} (the lowest pK_a value) corresponds to the first proton, that is, to the balance of undissociated (the form in which the acid may enter the cell)/dissociated acid. Acids with a higher $pK_{a(1)}$ value, such as acetic or lactic, would have higher proportions of the undissociated molecule at higher pH values, hence being more effective.

Prior exposure of bacterial cells to an acidic environment renders them more resistant to acid. More injury has been observed in unadapted cells of *Sighella flexneri*, *Escherichia coli* O157:H7, *Salmonella* Typhimurium, *Salmonella* Senftenberg, and *Staphylococcus aureus* than in acid-adapted or acid-shocked cells [29–32]. The tolerance to an acid environment was shown to be dependent on the *type of acid* used, and *E. coli* O157:H7 acid-adapted and acid-shocked cells were more tolerant than unadapted cells to culture medium acidified with lactic acid, but all three types of cells behaved similarly in medium acidified with acetic acid [30]. Indeed, cellular changes have been described in yeast and in enteric bacteria as a function of *stress adaptation* to weak organic acids [33,34]. The involvement of a multidrug resistance pump has been established in yeast for the resistance development against weak organic acids [33]. Holyoak et al. [35] showed that this pump transports preservatives, sorbic acid, benzoic acid, and acetic acid from the cytosol to the extracellular environment. Enteric bacteria show an "orchestrated response" to proton stress, overlapping arrays of acid stress response proteins [34]. Various systems designed to alleviate acid stress through consumption of protons by amino acid decarboxylases or its elimination by specific membrane transport systems have been described in Gram-negative bacteria [36]. Protein synthesis has also been found to be necessary for the development of acid stress response in Gram-positive bacteria [37].

Bacterial cells surviving exposure to acid stress may exhibit increased resistance to an unrelated stress that is subsequently applied [29,31,32,38]. Cross protection against heat as a result of acid stress has been documented for *E. coli* O157:H7 [30], *Salmonella* Typhimurium [31,39], *Salmonella* Senftenberg [31], *Listeria monocytogenes* [40,41], *S. flexneri* [29], *S. aureus* [32], and *Enterococcus faecium* [37]. The extent of increased heat tolerance was also shown to be dependent on the type of acid [37] and on the quantity added [42]. Conversely, bacterial cells exposed to sublethal heat treatments have been shown to enhance their tolerance to organic acids [43].

Cells of the different bacterial species show different sensitivities to internal acidification and different permeability to acids, which result in different sensitivity to the accumulation of dissociated weak acids. Molds, yeasts, and lactobacilli are frequently selected by their ability to grow at low pH values.

Probably, *C. botulinum* has been the microorganism which more attention has attracted, because of safety implications. Previdi and Gola [44] showed that *C. botulinum* grew slower when the growth medium was acidified with citric acid than with hydrochloric or glucono-δ-lactone, although the minimum pH for growth was the same for all three acidulants. Smelt et al. [17] found that lactic and acetic acids were more inhibitory for *C. botulinum* than citric or hydrochloric acids. Yamamoto et al. [45] found that adipic acid inhibited the growth of several species of *Clostridium* more than citric or hydrochloric.

Lactobacillus brevis was able to grow in buffered peptone water up to pH 3 when it was acidified with citric, hydrochloric, phosphoric, or tartaric acids, but only to pH 3.7 with lactic acid and to pH 4 when acetic acid was used [25]. Other studies have focused their attention on the effect of different acids on *L. monocytogenes* [46–48], *Listeria innocua* [49,50], *S. flexneri* [51], *E. coli* [26,52,53], and *Salmonella* [24]. On the other hand, the inactivation of *S. flexneri* at low pH depended on temperature, and was enhanced with an increase in incubation temperature [29].

In the case of *bacterial spores*, the picture is more complicated, as they have first to germinate and then to outgrow. *Acid pHs* are known to activate spores [54], stimulating them to germinate. However, then the germinating spore will need an adequate pH to proceed with outgrowth.

After vegetative growth, when spore-forming bacteria find nutrient depletion in the medium, they undergo the process of sporulation [55]. However, for this process to occur successfully, sporulating bacteria need adequate conditions in the medium. Usually, these conditions are similar to those during growth, but with narrower margins for the different parameters of growth, such as temperature and pH [55]. Moreover, these conditions during sporulation influence strongly the subsequent heat resistance of the developing spores [56,57]. For example, the D values of several *B. cereus* strains dropped clearly with sporulation pH, decreasing by about 65% per pH unit [58]. z Values were not significantly modified.

As an example of the effect of pH on bacterial growth, Figure 23.1 shows the behavior of a population of *B. coagulans* STCC 4522 spores plated on agar acidified at different levels with different acids The same behavior can be seen with all acids: all the spores germinated and grew at neutral pH and almost all of them at pH 6, but at pH 5.5 and lower the level of inhibition increased

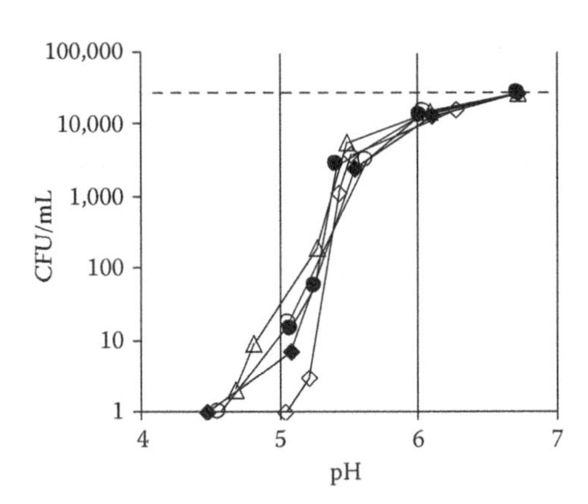

FIGURE 23.1 Growth of *B. coagulans* STCC 4522 spores plated on nutrient Agar acidified to various pHs with different acids: acetic (◇); citric (●); lactic (◆); malic (△); hydrochloric (○). The dashed line represents the concentration of the inoculum.

dramatically. At pH 6 and higher, no differences could be detected between the different acids, but at lower pH values, some differences started to show, although still not significant. Acetic acid appeared to have the stronger inhibitory effect as it inhibited growth at a pH as high as 5. The other acids allowed growth of this microorganism at pH value around 4.5. Once more, the stronger inhibitory effect of acetic acid on bacterial growth is shown.

23.4 EFFECT OF pH ON BACTERIAL HEAT RESISTANCE

Heating medium pH has a great influence on the *heat resistance* of bacteria, and at present it is considered the most important factor determining *heat resistance*, not only because its effects but also because it is easily applicable at an industrial level.

The influence of the pH of the heating medium has been widely investigated. *Acid pH* lowers in general the heat resistance of bacteria. Soon after discovering the kinetics of inactivation of bacteria exposed to heat [1], Esty and Meyer [2] observed in 1922 that *C. botulinum* showed its maximum heat resistance at pHs between 6.3 and 6.9, and Williams [59] gave in 1929 the highest heat resistance values for *B. subtilis* between pH 6.8 and 7.67. Sognefest et al. [3] found that between pH 4.5 and 9 the sterilizing treatment required was less the lower the pH. Other authors, however, have found no effect of pH on the heat resistance of *C. sporogenes* [60], although the pH range was narrow, and more recently on different species of genus *Alicyclobacillus* [61–64]. And still others have found that *C. perfringens* [55] and some strains of *Salmonella* [65] were most heat resistant at pH 5, and Morton et al. [66] have published more recently that toxigenic strains of *C. butyricum* were most heat resistant at pH 4.4, although nontoxigenic strains of this same species were less heat resistant at progressively more acidic pH values.

For the majority of the researchers, maximum heat resistance is shown at neutral pHs [67], and an *acidification* of the heating medium leads to a decrease in the heat resistance of the *spores* of genera *Bacillus* [28,67–82] and *Clostridium* [27,69–72,83–88] as well as ascospores of fungi [89] and vegetative cells of bacteria [90–94].

Different authors have found different relationships between the log *D* value of microorganisms and the pH of the heating medium. Several researchers have found linear relationships: Jordan and Jacobs [95] with *E. coli*; Cerny [67] with *B. subtilis* and *B. stearothermophilus*; Pirone et al. [96] with *C. butyricum*; Mazas et al. [97] with *B. cereus*; Fernández et al. [98] with *B. stearothermophilus* and *C. sporogenes*; Palop et al. [81] with *B. licheniformis*. Other authors found a pH interval that was more effective in decreasing heat resistance. For example, Montville and Sapers [73] found that the greater effect of pH was from 4 to 5.5, while Xezones and Hutchings [83] and Cook and Gilbert

[68] placed this range between 5 and 6. The range of minimum influence is usually located at pH values close to neutrality [72,73,83,86].

Similar findings led to other researchers to draw quadratic polynomial equations to adjust their data: Davey et al. [99] with *C. botulinum*; Mafart and Leguérinel [100] with *C. botulinum, C. sporogenes*, and *B. stearothermophilus*; Couvert et al. [101] with *B. cereus*; and Mañas et al. [94] with *S. typhimurium*. Mafart et al. [102] compared all these models used to describe the effect of pH on heat resistance (*D values*) with data obtained by other authors for *B. stearothermophilus, C. botulinum*, and *C. sporogenes* spores, concluding that linear models fitted better at low heating temperatures and low pHs, while quadratic models would fit better at high temperatures and moderately acidic media. Actually, changes of the pH range of maximum influence on heat resistance, related to the temperature of treatment, were already described for *B. stearothermophilus* spores [71].

Probably, differences in species behavior are also important in determining the overall magnitude of the effect of pH on heat resistance. These differences include the absence of effect in *A. acidocaldarius* in a pH range from 7 to 4 [63], small differences such as the one found in *C. sporogenes* [86] which reduced its D_{110} value only in one-third (from 15.9 to 10.6 min) in the pH interval from 7 to 5 and, probably the greatest effect described in literature for *B. stearothermophilus* ATCC 15952 [103] of more than 20 times in D_{115}. As a summary of all these results, Figure 23.2 shows D_{115} values of these and other spore-forming bacterial species in function of the pH of the phosphate buffer used as heating medium.

Perhaps the smaller effect of acid pH on the heat resistance of *B. coagulans* spores, when compared to other *Bacillus* and *Clostridium* spores, could explain why this is one of the species more frequently isolated from acid or low-acid canned foods [105].

One remarkable exception to the generalized decrease in heat resistance caused by acid pHs is the genus *Alicyclobacillus*, which shows the same heat resistance in media of different pH [63,64]. Although there are data in the literature reporting an influence of the heating medium pH for *A. acidoterrestris* [106], temperature and specific properties of different heating media seem to play a bigger role than pH in contributing to the heat sensitivity of this microorganism [12,107]. As stated before, this microorganism has a quite unusual range of pH for growth (2–6) for a spore-forming bacterium [10]. Again, this lack of effect could account for the frequent isolation of *Alicyclobacillus* species in heat-treated fruit juices [11,12].

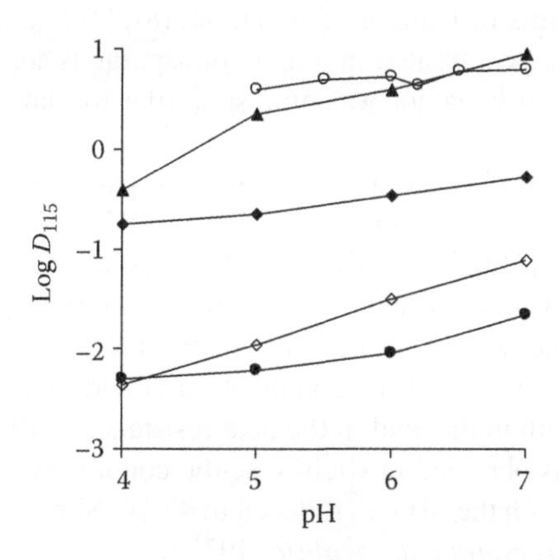

FIGURE 23.2 D_{115} values of different species of spore-forming bacteria in function of the pH of the heating phosphate buffer. (Data taken from Palop, A. et al., *Int. J. Food Microbiol.*, 29, 1, 1996; Palop, A. et al., *Int. J. Food Microbiol.*, 46, 243, 1999; Cameron, M.S. et al., *Appl. Environ. Microbiol.*, 39, 943, 1980; López, M. et al., *Int. J. Food Microbiol.*, 28, 405, 1996; Palop, A., Estudio de la influencia de diversos factores ambientales sobre la resistencia de *Bacillus subtilis, B. licheniformis* y *B. coagulans,* PhD dissertation, Universidad de Zaragoza, Zaragoza, Spain, 1995.) *B. licheniformis* (●); *B. coagulans* (◆); *C. sporogenes* (○); *B. stearothermophilus* (▲); *B. subtilis* (◇).

Finally, to complicate even more the picture, the magnitude of the effect of the pH depends not only on the bacterial species [69,76,81,82] but also on the crop [71].

Other factors apart from the bacterial species and crop such as the sporulation temperature, usually not taken into account by researchers, could be of importance in this regard. Sala et al. [80] found that the range of pH that influenced most the heat resistance of *B. subtilis* STCC 4524 spores was different when the sporulation temperature was 32°C or 52°C.

The effect of pH can be also influenced by the overacting effects of other factors such as sodium chloride or water activity of the heating medium. Hutton et al. [108] with *C. sporogenes* and *C. botulinum* and Periago et al. [109] with *B. stearothermophilus* spores found interacting effects of pH and NaCl concentration in the sense of decreasing heat resistance. Even the food composition itself may have its effect, although in Brown and Thorpes' opinion [72], this last effect could always be attributable to modifications in the pH or the water activity. There are, however, many data in literature [60,82,86,88,110] that seem to show that heat resistance of microorganisms in foods is different from the one they show in a buffer, even of its same pH. Hence, the most generalized opinion [65,77,111] is that heat resistance determinations developed in order to fix thermal treatments for canned foods should be performed in the same foods that are going to be sterilized.

As a consequence, in spite of the high number of investigations carried out to show the effect of the *pH* on the heat resistance, at present only a few general conclusions can be drawn and many aspects still remain unexplained [27]. The variety of strains studied, the diversity of methods used to estimate heat resistance, and the possible influence of uncontrolled factors make it difficult to discern whether the differences found in heat resistance are genetically determined or can be attributed to interferences caused by the methodology. To this respect, it is important to point out that pH varies with temperature, moreover in a different way for each food produce [4,71,77,112]. Hence, the pH determined at room temperature may differ largely from the actual pH at the temperature of treatment. This variation may be in some cases as big as 0.6 pH units (at 130°C) [112]. However, most of the work on the influence of the pH on heat resistance has been done measuring the pH at room temperature, without any knowledge of the pH at the actual temperature of treatment, assuming a constant pH value at all temperatures. This pH variation during heating could be responsible for some of the contradictions found in literature.

Condón et al. [112] developed a thermoresistometer able to measure *pH* at the heating temperature, up to 130°C. With this instrument, Palop et al. [81,82] found that when *B. licheniformis* and *B. coagulans* spores were heat treated in tomato or asparagus adjusted to different pHs (at the temperature of treatment), their behavior was almost similar to that they showed in buffer of the same pHs.

The inactivation of nonspore-forming pathogen microorganisms by a heat treatment is essential for the safe preparation of many foods. Since *E. coli* O157:H7 was isolated from apple cider responsible from an outbreak [113], the study of the heat resistance of this microorganism at acid pH has become of interest for many researchers. Splittstoesser et al. [114] found that an acidification of apple cider reduced the heat resistance of this strain. Dock et al. [115] and Steenstrup and Floros [116] found synergistic effects of the combination of addition of sorbate and benzoate with acidification of apple cider with malic acid in the heat resistance of this strain. The maximum heat resistance of salmonellae was obtained at slightly acidic conditions, around pH 6 [94,117], with a decrease of more than 50% when the pH was reduced to 4 [94]. Similar decreases were observed for *L. monocytogenes* [118] and *Yersinia enterocolitica* [93].

Growth temperature of vegetative bacteria seems to influence the effect of pH of the heating medium on heat resistance. Hence, the magnitude of the influence of pH on heat resistance increases as the growth temperature increases [93,94]. Moreover, the pH of maximum heat resistance for *Y. enterocolitica* changed from 7 for cells grown at 37°C to 5 for cells grown at 4°C [93]. Pagán et al. [93] found that the outer membrane of the cells grown at higher temperatures was more easily disrupted than for cells grown at lower temperatures.

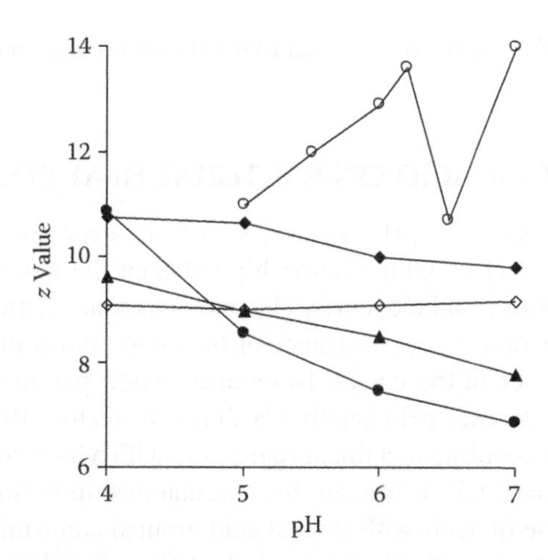

FIGURE 23.3 *z* Values of different species of spore-forming bacteria in function of the pH of the heating phosphate buffer. (Data taken from Palop, A. et al., *Int. J. Food Microbiol.*, 29, 1, 1996; Palop, A. et al., *Int. J. Food Microbiol.*, 46, 243, 1999; Cameron, M.S. et al., *Appl. Environ. Microbiol.*, 39, 943, 1980; López, M. et al., *Int. J. Food Microbiol.*, 28, 405, 1996; Palop, A., Estudio de la influencia de diversos factores ambientales sobre la resistencia de *Bacillus subtilis, B. licheniformis* y *B. coagulans,* PhD dissertation, Universidad de Zaragoza, Zaragoza, Spain, 1995.) *B. licheniformis* (●); *B. coagulans* (♦); *C. sporogenes* (○); *B. stearothermophilus* (▲); *B. subtilis* (◇).

pH of the heating medium also influences the *z values* of bacteria, and it is precisely this effect the one that most controversy has arisen. Some authors have found an increase in *z* value at acid pHs [67,72,81,82,119], which means that at high temperatures, pH may even have no effect on heat resistance [71,88]. However, other authors found a decrease [75,86,87,120], and still others were unable to find any effect [74,76,83–85,103,121].

Figure 23.3 shows *z* values of different spore-forming bacterial species in function of the pH of the buffer used as heating medium. As it can be seen that *z* value remained approximately constant for *B. subtilis* spores (z=9°C) and for *A. acidocaldarius* (z=7°C). It increased from 7.8°C to 9.6°C for *B. stearothermophilus* spores, from 8.9°C to 10.5°C for *B. coagulans* spores, and from 7°C to 11°C for *B. licheniformis* spores when acidifying from pH 7 to 4. A decrease from 14°C to 11°C was observed for *C. sporogenes* PA3679, with a rather odd peak at pH 6.5.

Although the differences in the effect of pH on *z* values could be related to the genetics of the microorganisms, that is, different species would show different behaviors, it has also been proposed that the temperature at which spores were formed could play an important role in their behavior [80]. In their experiments, Sala et al. [80] found that *z* values of a spore suspension of *B. subtilis* sporulated at 32°C did not change in the pH range 4–7, while those of other spore suspension of the same strain, but sporulated at 52°C, increased significantly when the heating medium was acidified in the same pH range.

Sporulation temperature is known to play a key role on the *heat resistance* of *bacterial spores* [122]. Spores sporulated at higher temperatures are more heat resistant [123–126], and Sala et al. [80] showed that sporulation temperature was also related to the effect of pH on *z* values.

Also the medium in which the spores are heat treated can be important for *z* values. In this regard, Condón and Sala [77] found that the *z* value of *B. subtilis* increased as the pH of the buffer was heated but was approximately constant in different foods.

Regarding vegetative cells, Steenstrup and Floros [116] also found an important increase in *z* values of *E. coli* O157:H7 when apple cider was acidified with malic acid. However, Condón et al. [91] did not find any change in *z* value of *Aeromonas hydrophila* related to the pH of the menstruum in which the microorganism was heated. Neither Pagán et al. [93] with *Y. enterocolitica* nor Mañas et al. [94] with *S. typhimurium* found that pH caused any change on the heat resistance with treatment temperature.

23.5 EFFECT OF TYPE OF ACID ON BACTERIAL HEAT RESISTANCE

The *type of acid* used to decrease the pH also affects bacterial *heat resistance*.

Leguérinel and Mafart [127] found a relationship between the lowest pK_a value of nine weak *organic acids* (acetic, L-glutamic, adipic, citric, glucono-δ-lactone, lactic, malic, malonic, and succinic acids) and their effects on the heat resistance of *B. cereus spores* in tryptone salt broth. These authors measured the influence of these acids by creating a new parameter, the z_{pH} value, which is the distance of pH from a reference pH (usually pH 7) that leads to a 10-fold reduction of D value. The relationship was fitted according to a linear regression with a high coefficient of correlation (r_0) ($z_{pH}=-3.37\ pK_a + 21.23$; $r_0=0.965$). It was an inverse relationship between the z_{pH} and the lowest pK_a value of the acid (in case of acids with several acid groups) confirming that higher pK_a led to a higher effect on decreasing heat resistance. The sorted relation of acids (in terms of higher z_{pH} value) was as follows: L-glutamic > malonic > malic > citric > glucono-δ-lactone > lactic > adipic > succinic > acetic.

These results are in agreement with those of other authors [4,28,128,129], which showed that lactic and acetic acids are slightly more effective in reducing heat resistance than other acids. The magnitude of this effect was the same regardless of the temperature of treatment [4].

However, other authors did not find differences in the heat resistance of their bacterial strains when they were heated in the presence of different acidulants. This is the case of Brown and Martínez [130] who found that *C. botulinum* showed the same heat resistance in mushrooms acidified with citric acid or glucono-δ-lactone. Neither Ocio et al. [88] found any effect on *C. sporogenes* spores heated in the same medium with the same acidulants, although the same microorganism was less heat resistant in asparagus acidified with citric acid, but at low heating temperatures [119]. Fernández et al. [111] found again no effects with *B. stearothermophilus* spores.

Palop et al. [129] related the influence of the type of acid used to lower the pH of the heating medium with the sporulation temperature, showing that spores of *B. subtilis* and *B. coagulans* sporulated at high temperatures (52°C) were less heat resistant in asparagus and tomato acidified with lactic or acetic acid than with citric or hydrochloric acids, while the same spores sporulated at 35°C showed similar heat resistance with all acids. The effect was constant along all heating temperatures tested. Hence, a similar interesting effect of that of the sporulation temperature on the effect of the pH on z values [80] mentioned earlier was also shown for the type of acid [129]. Both effects together show the relevance of sporulation temperature in determining the heat resistance of bacterial spores.

Most of these data have been obtained in foods requiring low concentration of acidulants to adjust the pH at the required value because of their natural low pHs. Hence, when determining heat resistance, small differences among acidulants could be undetected because of the small amount of acidulant used. For example, Lynch and Potter [28] were unable to find any influence of the acidulant on the heat resistance of *B. coagulans* and *B. stearothermophilus* when a frankfurter emulsion slurry was adjusted to pH 4.5, but there was such an influence when the pH was adjusted to pH 4.2.

Supporting these findings, Palop [104] showed different degrees of sensitization to heat when at a constant pH value (pH 4) the concentration of different acids was increased. Figure 23.4 shows the D_{99} *values* of *B. coagulans* STCC 4522 heat treated in 1% peptone water acidified to pH 4 with increasing concentrations of different acids. The change in heat resistance caused by an increase in acid concentration depended on the type of acid used. D_{99} decreased significantly when lactic and acetic acid concentrations were increased (from 0.77 to 0.18, more than four times, when the concentration of acetic acid increased from 0.085 to 1 M), and also increased significantly when the concentration of hydrochloric acid was increased. The concentration of all other acids tested did not seem to have an important effect on heat resistance, although there was a trend to increase heat

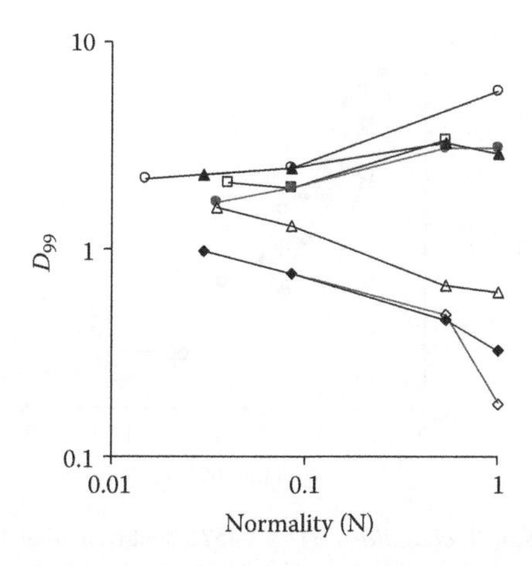

FIGURE 23.4 Heat resistance (D_{99} values) of *B. coagulans* STCC 4522 spores heat treated in 1% peptone water acidified to pH 4 with increasing concentrations of different acids: acetic (◇), ascorbic (□), citric (●), lactic (◆), malic (△), hydrochloric (○), and glucono-δ-lactone (▲).

resistance at increasing concentrations of glucono-δ-lactone and to decrease heat resistance with malic acid. The effects were also shown for *B. subtilis* and *B. licheniformis* spores [104].

Again these findings would point out that high pK_a values lead to lower heat resistances, and that pK_a, together with acid concentration, are important parameters, not only to determine bacterial growth (see Section 23.3) but also for heat resistance. The surprisingly great increase in heat resistance caused by the increase in concentration of hydrochloric acid could be related to the protective effect that would cause the addition of salt (sodium chloride) to the heating medium [65], because NaOH was used in all cases to adjust pH to 4.

However, Leguérinel and Mafart [127] showed that at neutral pH the heat resistance of *B. cereus* spores was the lowest at the lowest acetic acid concentration (0.01 M), while there were no differences in heat resistance at lower pH values (up to pH 4.35) caused by the acid concentration. The authors attributed this protective effect of the acid concentration at neutral pH to the dissociated form of the acid, which accounts for 99.4% of the total concentration of the acid at pH 7, but only to 29.1% at pH 4.35. Perhaps at pH 4, the lower proportion of dissociated acid present (that would leave spores unprotected) could account for the opposite effect found by Palop [104] using acetic acid at this pH.

23.6 EFFECT OF pH ON THE RECOVERY OF SURVIVORS TO HEAT TREATMENT

After a heat treatment, microorganisms are more sensitive to the environmental stresses. Heat causes injuries of different severity, before the microbial cell is irreversibly inactivated. Some of the damages can be repaired by the cells, depending mainly on the severity of heat injury and on *recovery* conditions [131]. These conditions include the temperature of incubation after heat treatment [132,133] and the physicochemical characteristics of the recovery medium [108,134,135]. In this regard, the *pH* is a very important factor.

Microbial *heat resistance* is estimated from the number of survivors of the heat treatment. Hence, any factor influencing cell damage and repair mechanisms might also influence the estimated heat resistance values. Figure 23.5 shows survival curves of *B. coagulans* STCC 4522 spores heat treated at 108°C in pH 7 McIlvaine buffer, plated on Nutrient Agar acidified to different pH values, and incubated for 24 h at 37°C. The heat treatment was exactly the same in all cases, but spores cultured on acid media appeared less heat resistant. $D_{108°C}$ decreased from 3.2 min when spores were incubated at pH 6.22 to 1.3 min when incubated at pH 5.52, and at this pH, there was growth only in the plates corresponding to the first times of treatment, that is, when thermal injury was not important.

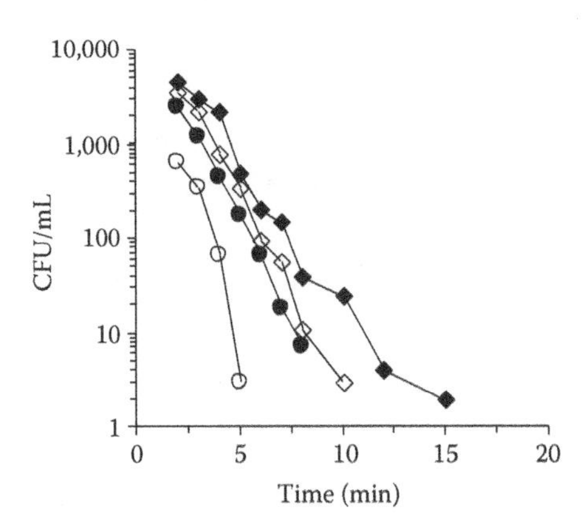

FIGURE 23.5 Survival curves of *B. coagulans* STCC\ 4522 heat treated at 108°C in pH 7 McIlvaine buffer and incubated for 24 h at 37°C in agar acidified to pH 6.22 (♦), 5.98 (◊), 5.85 (●), and 5.52 (○).

At pH 5.05, no growth was observed at any of the heating times. However, as shown in Figure 23.1, the same microorganism was able to grow up to pH 4.5 before any heat treatment was applied, depending also on the type of acid used. It was also observed that the time needed for the colonies to develop was longer at acid pHs and at longer heating times.

Similar results have been shown by Leguérinel et al. [136] with *B. cereus* spores, and by Cook and Brown [137], Cook and Gilbert [138], Mallidis and Scholefield [139], López et al. [140], and Fernández et al. [141] with *B. stearothermophilus* spores. The last authors [141] did not find differences when the recovery medium was acidified with citric acid or with glucono-δ-lactone.

Other parameters of the recovery medium that have influence on the apparent heat resistance of microorganisms are the sodium chloride concentration and the anaerobic conditions [134]. An increase in the salt concentration has been shown to decrease the apparent heat resistance of bacterial spores [134,136,142]. Leguérinel et al. [136] found that there were no interactions between the sodium chloride concentration and the pH of the recovery medium in the apparent heat resistance of *B. cereus* spores, so that their effects could be modeled separately. However, Periago et al. [143] found that the effects of both factors were interacting in the heat resistance of *B. stearothermophilus*. A possible explanation for this controversy, apart from the different bacterial species used in both investigations, could be that the later authors [143] were combining the effects of the heating and of the recovery medium in their study.

The level of inhibition of bacterial growth in an acid medium depends also on the thermal history of the culture. Oscroft et al. [144] found that the efficacy of different acids for preventing growth of several species of *Bacillus* depended on the intensity of the previous thermal treatment, but always citric acid was less effective. Palop et al. [145] reported that a heat treatment as mild as 10 s at 100°C lead to an increase of almost 1 unit in the minimum pH required for *B. coagulans* spores to grow: peptone water acidified to pH 5.2 prevented growth of heat-treated *B. coagulans* spores, but it was necessary to decrease pH to 4.4 to prevent growth of unheated spores. The minimum pH required for heated spores to be able to grow increased with the intensity of heat treatment: even pH 5.6 was able to prevent growth of spores after a heat treatment of 2 min at 100°C.

It has been shown through this chapter that *acid pH* hampers microbial growth and decreases microbial heat resistance; however, the most interesting and somehow unexplored advantage of pH probably comes from the combination of both inhibitory effects. Figure 23.6 shows the D_{95} values of *B. licheniformis* STCC 4523 spores heat treated in 1% peptone water acidified to pH values from 7 to 5.5 with citric acid and then incubated in nutrient agar acidified to different pHs also with citric acid. *D valu*es reduced from 9 to 4 min when the heating medium pH was reduced from 7 to 5.5, and the recovery medium pH was kept constant at pH 7. A decrease of D_{95} from 9 to 3 min was observed when the recovery medium pH was decreased from 7 to 5.5, and the heating medium pH was kept

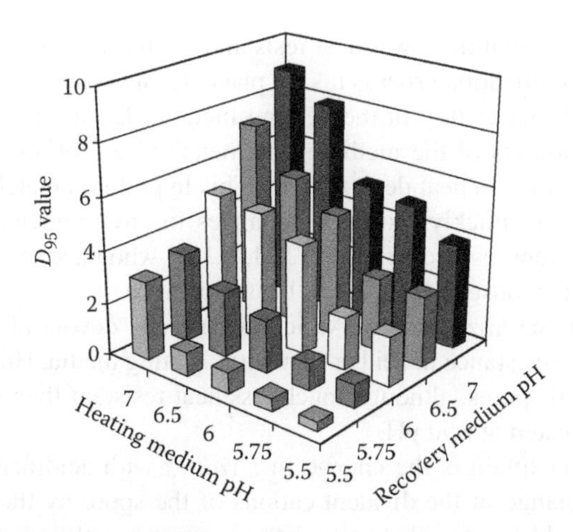

FIGURE 23.6 Effect of heating medium pH and recovery medium pH on the heat resistance (D_{95} value) of *B. licheniformis* STCC 4523 spores.

constant at pH 7. At lower pHs of the recovery medium, no survivors were found. However, when both heating and recovery medium pHs were reduced simultaneously from 7 to 5.5, an important synergistic effect was observed: the D_{95} value was decreased almost 30 times from 9 to 0.34 min. The effects of the acid pH on the heating medium and those of the acid pH in the recovery medium act together in reducing the heat resistance of bacteria so that microorganisms heat treated at acid pH and recovered in an acid medium will show even lower D values.

The combination of *heat treatments* with *acid pHs* has been long used by the food canning industry and offers important advantages to the food processor. In the food industry, the heating medium and the recovery medium are obviously the same. Thus, the intensity of heat treatments can be reduced as heat resistance of bacteria is lower at acid pHs. Moreover, the pH minimum for growth after heat treatment is higher [145].

23.7 MECHANISM OF ACTION OF ACID pH AT HIGH TEMPERATURES

The mechanisms that make bacteria and especially *bacterial spores* resistant to heat are not fully known, and so mechanisms of sensitization of bacterial spores to *acid pH* at high temperatures still remain unrevealed. It has been proposed that there are three main factors that determine the *heat resistance* of bacterial spores: mineralization, protoplast dehydration, and sporulation temperature [122]. Minerals play a key role in the stabilization of proteins and other heat sensitive structures of the spore. It is suggested that minerals, especially calcium, would replace water in the dehydrated protoplast of the spore, contributing to reaching a high degree of immobilization of the molecules and structures contained in it [146]. In this way, they would be less sensitive to heat. Sporulation temperature would be implicated by determining changes in the degree of mineralization of the spores: it has been found that spores obtained at higher temperatures were more mineralized and also more dehydrated [122].

Gould and Dring [147] tried to explain the decrease in heat resistance of bacterial spores heated in acid media by means of a protonization of carboxyl groups of the cortex that would lead to its collapse and to the subsequent protoplast rehydration, and thus to heat sensitization. Supporting this hypothesis, Okereke et al. [148] observed through electron microscopy that heating led to a decrease in the thickness of the cortex and to an increase in the protoplast volume. However, in the opinion of Warth [149], should the cortex collapse by protonization, bacterial spores would be extremely unstable. Warth [149] proposed that the osmotic pressure created by the cortex is caused not only by carboxyl groups but also by the whole structure that includes divalent cations [150,151].

It has also been proposed that the lower heat resistance of bacterial spores at acid pHs could be a consequence of a demineralization process taking place during heat treatment [152,153]. Marquis and Bender [154] showed that acidity in the heating medium led to an exchange of the divalent cations of the spore by protons of the medium, and hence to a protonization of spore proteins, which would be more sensitive to heat denaturation. This hypothesis matches the observation that calcium bonds are easy and quickly broken by changes in environmental conditions [155] and other observations of different researchers [105,122,156,157] who have seen that spores exposed to demineralization treatments loose partially their heat resistance.

If this is the case, demineralized spores, which are already devoid of most of their minerals, would show the same heat resistance in acid or in neutral heating media. However, Palop et al. [105] showed that demineralized spores, although much less heat resistant than native spores, were still less heat resistant when heated at acid pH.

Even more difficult to explain is the change in *z values* with acidification. If heating at acid pH values causes an exchange of the divalent cations of the spore by the protons of the heating medium [152–154], it could happen that at higher temperatures the inactivation process would speed up relatively more than the exchange process. This would explain the decreasing effect of acidification at high temperatures, that is, the increase in *z* values with acidification. Indeed, it has been shown that demineralization speed is different for the different bacterial species [154], that this process is more efficient at temperatures around 60°C than at room temperature [152], and that it needs time to be completed but it is not always successful [158]. However, other situations such as the decrease in *z* values remain still unexplained, and it should not be disregarded that changes in *z* value could simply reflect changes in the activation energy of the denaturation reaction of the molecule(s) responsible for the death of the microorganism as a consequence of the acidification of the heating medium. In this regard, it is assumed that DNA denaturation leads to cell death [159], but denaturation of DNA is not easy, and before it occurs, other cellular structures are damaged impairing microbial growth. At present, mechanisms responsible for inactivation are not fully understood, but several target structures such as some enzymes [160] or the germination system [161] could be implicated.

23.8 MODELING THE EFFECT OF pH ON HEAT RESISTANCE OF BACTERIA

Predictive microbiology is gaining interest as a tool used to guarantee the production of safe foodstuffs. One of the main objectives in this area is to build, estimate, assess, and validate mathematical or probabilistic models with which it is possible to describe the behavior (growth or inactivation) of food-borne microbes in specific environmental conditions (including factors as NaCl, pH, a_w, or temperature). The successful application of predictive microbiology depends on the development and validation of appropriate *models* [162]. Predictive models quantify the effects of interactions between two or more factors and allow interpolation of combinations that have not been tested explicitly [163]. Mathematical and probabilistic models are very useful, not only for hazard analysis and critical control points but also for making decisions in scenarios when there is uncertainty.

As indicated in previous sections, *pH* reduces the thermal resistance of microorganisms and the prediction of the effect of changes of pH during the production process or in formulation on the thermal-resistance parameters (*D* value) or on the number of surviving microorganisms is important. Some models have been developed using thermal resistance data as affected by variations in pH of the food. According to Reichart and Mohacsi-Farkas [164], Fernández et al. [98] and Periago et al. [109] second-order polynomial models give good description of the thermal destruction parameter in vegetative and sporulated bacteria. Davey et al. [99] developed a model based on a quadratic polynomial equation to describe the joint effect of pH and temperature on *D values*. Mafart and Leguérinel [100] proposed an extension of the Bigelow's equation and developed a linear regression model with temperature and the square of the pH. Various other models have been developed relating to pH and temperature on the *heat resistance* of microorganisms [92,94,165,166].

Even Velliou et al. [42] have made an attempt to quantify the effect of an acid treatment into the induced heat tolerance of *E. coli* cells.

In many cases, the models developed only account for the combined effect of two factors: temperature and pH. However, few studies in which various environmental factors are involved have been carried out. Gaillard et al. [167] developed a model that considered the combined effect of temperature, pH, and water activity on the *D* values of *B. cereus* spores. This model is an extension of the first version of the Mafart's model [100].

Periago et al. [109] proposed a quadratic response surface model represented by a polynomial to describe the combined effect of pH, NaCl, and temperature on the *D* value of *B. stearothermophilus* spores. This model had 10 coefficients and predictions derived from the model were close compared with *D* values published in the scientific literature. Nevertheless, in some cases, significant differences were found, indicating the need of obtaining models for specific foods. This model was improved by Tejadillos et al. [168] by using statistics regressions techniques. The new model used only three predictor factors and the variables could easily be computed from the basic variables of pH, temperature, and NaCl concentration. Esnoz et al. [169] developed another model to describe this same combined effect using low-complexity, black-box models based on artificial neural networks. The neural networks models gave better predictions in all experiments tried. When the neural networks were evaluated using new experimental data on the combined effect of pH, temperature, and NaCl, good predictions were also obtained, providing fail-safe predictions of *D* values of *B. stearothermophilus* spores in all cases. The authors concluded that the use of this nonlinear modeling technique made it possible to describe the interacting effects of environmental factors more accurately than the classical predictive models.

All these models have the common fact that they use *D* or *k* parameters as kinetic rates, assuming a logarithmic linear relationship between the number of microorganism survivors and time after heat treatment. Nevertheless, in most cases, this assumption is not valid because of the presence of shoulders and tails in the survivor curve. To overcome this inconvenience, Fernández et al. [170] developed a model based on the Weibull distribution function. This function has been used before by diverse authors to describe the death of a microorganism that undergoes an inactivation treatment when the survivor curve does not follow a first order inactivation kinetic [171]. The authors [170] concluded that Weibull parameters can be used to describe the effect of external factors on heat resistance. The overall model developed can describe the combined effect of pH and temperature with a low number of parameters.

All the information generated through recent years to predict the behavior of microorganisms under different environmental conditions has been used to build huge *databases* and computer programs that facilitate its usage. Some of these programs are free, while other are commercial and need to be registered before use. As an example of free available program, the Pathogen Modeling Program (ars.usda.gov/services/docs.htm?docid=6786) was created by the United States Department of Agriculture. This predictive microbiology application was designed as a research and instructional tool for estimating the effects of multiple variables on the growth, inactivation, or survival of food-borne pathogens. As an example of commercial program, the Food Micromodel (www.arrowscientific.com.au/predictive_micro_sw.html) offers the advantage that has been validated in foods. This is a software package that will predict the growth, survival, and thermal death of the major food pathogens and food-spoilage organisms in a wide range of foods. It is the result of a research project, initiated and funded by the Ministry of Agriculture, Fisheries and Food of the United Kingdom.

These two independent data sets constitute thousands of microbial growth and survival curves that are the basis for numerous microbial models used by the industry, academia, and government regulatory agencies.

In 2003, these two data sets were unified in a common database and are now publicly and freely available. The database, known as ComBase (www.combase.cc), follows a structure developed by the Institute of Food Research of the United Kingdom. Many European research institutions have also added their data to ComBase, and data have been also compiled from scientific literature at the Institute of Food Research. In 2006, the Australian Food Safety Center for Excellence also joined the ComBase Consortium; at present, ComBase contains up to 40,000 records containing full bacterial growth and survival curves.

In all these data sets, factors such as pH, NaCl, or a_w can be entered as independent variables to obtain the variation in the dependant variable, usually the inactivation or growth kinetic parameter.

23.9 CONCLUSIONS

Heat treatments have been applied successfully in food canning factories for several decades. Although the mechanisms that make bacteria less heat resistant remain somehow unknown, the knowledge acquired in the last few years regarding the behavior of bacteria in acid media has allowed for the combination of moderately *acid pHs* with low-intensity heat treatments, which results in foods of higher quality with the same level of safety.

Technical innovations regarding pH-assisted thermal processes are rather few, and acid foods (with pH values below 4.6) are still heat treated at temperatures close to 100°C in open water baths.

At present, it is well-known that the majority of the microorganisms show their maximum *heat resistance* at neutral pHs, decreasing with increase in acidification. The magnitude of the effect of pH will depend on the microbial species and other factors such as food composition, *type*, and concentration *of acid*, or even the sporulation temperature in the case of *bacterial spores*.

pH influences not only heat resistance but also bacterial growth and most importantly influences very much the growth of the survivors to a heat treatment. Hence, the combined effect of the pH of the heating medium and of the *recovery* medium may lead to synergistic reductions of the viability of microbes.

The huge amount of data on the effect of pH on microbial growth and heat resistance has allowed for the development of large *databases* and mathematical *models* that enable to predict the behavior of microorganisms under certain conditions. These data sets are available for food microbiologists, manufacturers, risk assessors, and legislative officers.

REFERENCES

1. WD Bigelow, JR Esty. The thermal death point in relation to time of typical thermophilic organisms. *Journal of Infectious Diseases* 27: 602–617, 1920.
2. JR Esty, KF Meyer. The heat resistance of spores of *B. botulinus* and allied anaerobes. *Journal of Infectious Diseases* 31: 650–653, 1922.
3. P Sognefest, GL Hays, E Wheaton, HA Benjamin. Effect of pH on thermal process requirements of canned foods. *Food Research* 13: 400–416, 1948.
4. JJ Powers. Effect of acidification of canned tomatoes on quality and shelf life. *Critical Reviews in Food Science and Nutrition* 7: 371–396, 1976.
5. DA Corlett Jr., MH Brown. pH and acidity. In: JH Silliker, RP Elliott, AC Baird-Parker, FL Bryan, JHB Christian, DS Clark, JC Olson, TA Roberts, eds. *ICMSF, Microbial Ecology of Foods, Volume I: Factors Affecting Life and Death of Microorganisms*. New York: Academic Press, 1980, pp. 97–117.
6. TC Odlaug, IJ Pflug. *Clostridium botulinum* in acid foods. *Journal of Food Protection* 41: 566–573, 1978.
7. AC Hersom, ED Hutland. *Canned Foods*. Edinburgh, U.K.: Longman, 1980, pp. 80–115.
8. G Cerny, W Hennlich, K Poralla. Spoilage of fruit juice by bacilli: Isolation and characterization of the spoilage microorganisms *Zeitschrift fuer Lebensmittel Untersuchung und Forschung* 179: 224–227, 1984.
9. DF Splittstoesser, JJ Churey, CY Lee. Growth characteristics of aciduric sporeforming bacilli isolated from fruit juices. *Journal of Food Protection* 57: 1080–1083, 1994.
10. CA Wisse, ME Parish. Isolation and enumeration of sporeforming, thermoacidophilic, rod-shaped bacteria from citrus processing environments. *Dairy, Food and Environmental Sanitation* 18: 504–509, 1998.
11. MNU Eiroa, VCA Junqueira, FL Schimdt. *Alicyclobacillus* in orange juice: Occurrence and heat resistance of spores. *Journal of Food Protection* 62: 883–886, 1999.

12. Y Smit, M Cameron, P Venter, RC Witthuhn. *Alicyclobacillus* spoilage and isolation—A review. *Food Microbiology* 28: 331–349, 2011.
13. CE Steyn, M Cameron, RC Witthuhn. Occurrence of *Alicyclobacillus* in the fruit processing environment—A review. *International Journal of Food Microbiology* 147: 1–11, 2011.
14. ML Fields, AF Zamora, M Bradsher. Microbiological analysis of home-canned tomatoes and green beans. *Journal of Food Science* 42: 931–934, 1977.
15. TJ Montville. Metabiotic effect of *Bacillus licheniformis* on *Clostridium botulinum:* Implications for home-canned tomatoes. *Applied and Environmental Microbiology* 44: 334–338, 1982.
16. LA Smoot, MO Pierson. Effect of oxidation-reduction potential and chemical inhibition of *Clostridium botulinum* 10755A spores. *Journal of Food Science* 44: 700–704, 1979.
17. JPPM Smelt, GJM Raatjes, JS Crowther, CT Verrips. Growth and toxin formation by *Clostridium botulinum* at low pH values. *Journal of Applied Bacteriology* 52: 75–82, 1982.
18. JD Dziezak. Preservatives: Antimicrobial agents. A means towards product stability. *Food Technology* 40: 104–111, 1986.
19. GF Dall'Aglio, S Gola, A Versitano, S Quintavalla. L'impiego del glucono-δ-lattone nella preparazione di conserve acide. *Industria Conserve* 63: 123–126, 1988.
20. IR Booth, RG Kroll. The preservation of foods by low pH. In: GW Gould, ed. *Mechanisms of Action of Food Preservation Procedures*. London, U.K.: Elsevier, 1989, pp. 119–160.
21. T Eklund. The antimicrobial effect of dissociated and undissociated sorbic acid at different pH levels. *Journal of Applied Bacteriology* 54: 383–389, 1983.
22. M Stratford, PA Anslow. Evidence that sorbic acid does not inhibit yeast as a classic 'weak acid preservative'. *Letters in Applied Microbiology* 27: 203–206, 1998.
23. CP Hsiao, KJ Siebert. Modeling the inhibitory effect of organic acids on bacteria. *International Journal of Food Microbiology* 47: 189–201, 1999.
24. KC Chung, JM Goepfert. Growth of *Salmonella* at low pH. *Journal of Food Science* 35: 326–328, 1970.
25. BJ Juven. Bacterial spoilage of citrus products at pH lower than 3.5. *Journal of Milk and Food Technology* 39: 819–822, 1976.
26. JH Ryu, Y Deng, LR Beuchat. Behavior of acid-adapted and unadapted *Escherichia coli* O157:H7 when exposed to reduced pH achieved with various organic acids. *Journal of Food Protection* 62: 451–455, 1999.
27. JC Blocher, FF Busta. Bacterial spore resistance to acid. *Food Technology* 37: 87–99, 1983.
28. DJ Lynch, NN Potter. Effects of organic acids on thermal inactivation of *Bacillus stearothermophilus* and *Bacillus coagulans* spores in frankfurter emulsion slurry. *Journal of Food Protection* 51: 475–480, 1988.
29. GL Tetteh, LR Beuchat. Survival, growth, and inactivation of acid-stressed *Shigella flexneri* as affected by pH and temperature. *International Journal of Food Microbiology* 87: 131–138, 2003.
30. JH Ryu, LR Beuchat. Influence of acid tolerance responses on survival, growth, and thermal cross-protection of *Escherichia coli* O157:H7 in acidified media and fruit juices. *International Journal of Food Microbiology* 45: 183–193, 1998.
31. A Álvarez-Ordóñez, A Fernández, A Bernardo, M López. A comparative study of thermal and acid inactivation kinetics in fruit juices of *Salmonella* enterica serovar Typhimurium and *Salmonella* enterica serovar Senftenberg grown at acidic conditions. *Foodborne Pathogens and Disease* 6: 1147–1155, 2009.
32. G Cebrián, N Sagarzazu, R Pagán, S Condón, P Mañas. Development of stress resistance in *Staphylococcus aureus* after exposure to sublethal environmental conditions. *International Journal of Food Microbiology* 140: 26–33, 2010.
33. P Piper, Y Mahe, S Thompson, R Pandjaitan, C Holyoak, R Egner, M Muhlbauer, P Coote, K Kuchler. The pdr 12 ABC transporter is required for the development of weak organic acid resistance in yeast. *EMBO Journal* 15: 4257–4265, 1998.
34. JP Audia, CC Webb, JW Foster. Breaking through the acid barrier: An orchestrated response to proton stress by enteric bacteria. *International Journal of Medical Microbiology* 291: 97–106, 2001.
35. CD Holyoak, D Bracey, PW Piper, K Kuchler, PJ Coote. The *Saccharomyces cerevisiae* weak-acid-inducible ABC transporter Pdr12 transports fluorescein and preservative anions from the cytosol. *Journal of Bacteriology* 181: 4644–4652, 1999.
36. J Kieboom, T Abee. Arginine-dependent acid resistance in *Salmonella enterica* serovar Typhimurium. *Journal of Bacteriology* 188: 5650–5653, 2006.
37. A Fernández, A Álvarez-Ordóñez, M López, A Bernardo. Effects of organic acids on thermal inactivation of acid and cold adapted *Enterococcus faecium. Food Microbiology* 26: 497–503, 2009.
38. A Álvarez-Ordóñez, A Fernández, A Bernardo, M López. Efficacy of trisodium phosphate in killing acid-adapted *Salmonella* Typhimurium. *Journal of Food Safety* 31: 250–256, 2011.

39. GH Leyer, EA Johnson. Acid adaptation induces cross-protection against environmental stresses in *Salmonella typhimurium*. *Applied and Environmental Microbiology* 59: 1842–1847, 1993.
40. JM Farber, F Pagotto. The effect of acid shock on the heat resistance of *Listeria monocytogenes*. *Letters in Applied Microbiology* 15: 197–201, 1992.
41. PJ Taormina, LR Beuchat. Survival and heat resistance of *Listeria monocytogenes* after exposure to alkali and chlorine. *Applied and Environmental Microbiology* 67: 2555–2563, 2001.
42. EG Velliou, E Van Derlinden, AM Cappuyns, E Nikolaidou, AH Geeraerd, F Devlieghere, JF Van Impe. Towards the quantification of the effect of acid treatment on the heat tolerance of *Escherichia coli* K12 at lethal temperatures. *Food Microbiology* 28: 702–711, 2011.
43. WL Hsiao, WL Ho, CC Chou. Sub-lethal heat treatment affects the tolerance of *Cronobacter sakazakii* BCRC 19388 to various organic acids, simulated gastric juice and bile solution. *International Journal of Food Microbiology* 144, 280–284, 2010.
44. MP Previdi, S Gola. Growth of *Clostridium botulinum* in two different media acidified with glucono-δ-lactone, citric acid or hydrochloric acid. *Industria Conserve* 65: 39–41, 1990.
45. Y Yamamoto, N Ono, K Higashi, H Yoshii. Effects of adipic acid on growth and thermal resistance of spores of anaerobic sporeforming bacteria. *Nippon Shokuhin Kogyo Gakkaishi* 36: 551–556, 1989.
46. N Chen, LA Shelef. Relationship between water activity, salts of lactic acid, and growth of *Listeria monocytogenes* in a meat model system. *Journal of Food Protection* 55: 574–578, 1992.
47. SM George, LCC Richardson, MW Peck. Predictive models of the effect of temperature, pH and acetic and lactic acids on the growth of *Listeria monocytogenes*. *International Journal of Food Microbiology* 32: 73–90, 1996.
48. PS Fernández, SM George, CC Sills, MW Peck. Predictive model of the effect of CO_2, pH, temperature and NaCl on the growth of *Listeria monocytogenes*. *International Journal of Food Microbiology* 37: 37–45, 1997.
49. PC Houtsma, BJ Kusters, JC de Wit, FM Rombouts, MH Zwietering. Modelling growth rates of *Listeria innocua* as a function of lactate concentration. *International Journal of Food Microbiology* 24: 113–123, 1994.
50. PC Houtsma, ML Kant Muermans, FM Rombouts, MH Zwietering. Model for the combined effects of temperature, pH, and sodium lactate on growth rates of *Listeria innocua* in broth and Bologna-type sausages. *Applied and Environmental Microbiology* 62: 1616–1622, 1996.
51. LL Zaika, E Moulden, L Weimer, JG Philips, RL Buchanan. Model for the combined effects of temperature, initial pH, sodium chloride and sodium nitrite concentrations on anaerobic growth of *Shigella flexneri*. *International Journal of Food Microbiology* 23: 345–358, 1994.
52. MRS Clavero, LR Beuchat. Survival of *Escherichia coli* O157:H7 in broth and processed salami as influenced by pH, water activity, and temperature and suitability of media for its recovery. *Applied and Environmental Microbiology* 62: 2735–2740, 1996.
53. KA Presser, DA Ratkowsky, T Ross. Modelling the growth rate of *Escherichia coli* as a function of pH and lactic acid concentration. *Applied and Environmental Microbiology* 63: 2355–2360, 1997.
54. RW Berg, WE Sandine. Activation of bacterial spores. A review. *Journal of Milk and Food Technology* 33: 435–441, 1970.
55. AD Russell. *The Destruction of Bacterial Spores*. New York, NY: Academic Press, 1982, pp. 5–142.
56. SE Craven. The effect of the pH of the sporulation environment on the heat resistance of *Clostridium perfringens* spores. *Current Microbiology* 22: 233–237, 1990.
57. A Palop, P Mañas, S Condón. Sporulation temperature and heat resistance of *Bacillus* spores: A review. *Journal of Food Safety* 19: 57–72, 1999.
58. M Mazas, M López, I González, A Bernardo, R Martin. Effect of sporulation pH on the heat resistance and the sporulation of *Bacillus cereus*. *Letters in Applied Microbiology* 5: 331–334, 1997.
59. OB Williams. The heat resistance of bacterial spores. *Journal of Infectious Diseases* 44: 421–465, 1929.
60. M Rodrigo, A Martínez, T Sánchez, MJ Peris, J Safón. Kinetics of *Clostridium sporogenes* PA 3679 spore destruction using computer-controlled thermoresistometer. *Journal of Food Science* 58: 649–652, 1993.
61. K Yamakazi, C Isoda, H Tedzuka, Y Kawai, H Shimano. Thermal resistance and prevention of spoilage bacterium, *Alicyclobacillus acidoterrestris*, in acidic beverages. *Journal of the Japanese Society for Food Science and Technology* 44: 905–911, 1997.
62. M Murakami, H Tedzuka, K Yamakazi. Thermal resistance of *Alicyclobacillus acidoterrestris* spores in different buffers and pH. *Food Microbiology* 15: 577–582, 1998.
63. A Palop, I Alvarez, J Raso, S Condón. Heat resistance of *Alicyclobacillus acidocaldarius* in water, various buffers, and orange juice. *Journal of Food Protection* 63: 1377–1380, 2000.

64. G Ceviz, Y Tulek, AH Con. Thermal resistance of *Alicyclobacillus acidoterrestris* spores in different heating media. *International Journal of Food Science and Technology* 44: 1770–1777, 2009.
65. CR Stumbo. *Thermobacteriology in Food Processing*. New York: Academic Press, 1965, pp. 15–127.
66. RD Morton, VN Scott, DT Bernard, RC Wiley. Effect of heat and pH on toxigenic *Clostridium butyricum. Journal of Food Science* 55: 1725–1727, 1990.
67. G Cerny. Dependence of thermal inactivation of microorganisms on the pH-value of media. II. Bacteria and bacterial spores. *Zeitschrift fuer Lebensmittel Untersuchung und Forschung* 170: 180–186, 1980.
68. AM Cook, RJ Gilbert. Factors affecting the heat resistance of *Bacillus stearothermophilus* spores. II. The effect of sporulating conditions and nature of the heating medium. *Journal of Food Technology* 3: 295–302, 1968.
69. NN Mazokhina, LP Naidenova, LI Rozanova, TV Dashevskaya. Heat resistance and pH effect on microorganisms, causing spoilage of canned foods. *Acta Alimentaria* 2: 385–391, 1973.
70. MFF Leitao, CA Ordorico, C Ciampi, DG Quat. The thermal resistance of *Bacillus stearothermophilus* and *Clostridium* PA 3679 spores in banana puree. *Col Institute of Technology Aliment* 8: 313–327, 1977.
71. KL Brown, RH Thorpe. The effect of pH on the heat resistance and recovery of bacterial spores. *CFPRA Technical Memorandum 185*. Chipping Campdem, 1978.
72. KL Brown, RH Thorpe. The relationship between pH and the heat resistance and recovery of bacterial spores. *CFPRA Technical Memorandum 204*. Chipping Campdem, 1978.
73. TJ Montville, GM Sapers. Thermal resistance of spores from pH elevating strains of *Bacillus licheniformis. Journal of Food Science* 46: 1710–1712, 1981.
74. M Rodrigo, A Martínez, J Sanchís, J Trama, V Giner. Determination of hot-fill-hold-cool process specifications for crushed tomatoes. *Journal of Food Science* 55: 1029–1032, 1990.
75. CG Mallidis, P Frantzeskakis, G Balatsouras, C Katsabotxakis. Thermal treatment of aseptically processed tomato paste. *International Journal of Food Science and Technology* 25: 442–448, 1990.
76. R Behringer, HG Kessler. Influence of pH value and concentration of skimmilk on heat resistance of *Bacillus licheniformis* and *B. stearothermophilus* spores. *Milchwissenschaft* 47: 207–211, 1992.
77. S Condón, FJ Sala. Heat resistance of *Bacillus subtilis* in buffer and foods of different pH. *Journal of Food Protection* 55: 605–608, 1992.
78. A Azizi, S Ranganna. Spoilage microorganisms of canned acidified mango pulp and their relevance to thermal processing of acid foods. *Journal of Food Science and Technology* 20: 241–245, 1993.
79. M Nakajo, Y Moriyama. Effect of pH and water activity on heat resistance of *Bacillus coagulans* spores. *Journal of the Japanese Society for Food Science and Technology* 40: 268–271, 1993.
80. FJ Sala, P Ibarz, A Palop, J Raso, S Condón. Sporulation temperature and heat resistance of *Bacillus subtilis* at different pH's. *Journal of Food Protection* 58: 239–243, 1995.
81. A Palop, J Raso, R Pagán, S Condón, FJ Sala. Influence of pH on heat resistance of *Bacillus licheniformis* in buffer and homogenised foods. *International Journal of Food Microbiology* 29: 1–10, 1996.
82. A Palop, J Raso, R Pagán, S Condón, FJ Sala. Influence of pH on heat resistance of *Bacillus coagulans* in buffer and homogenized foods. *International Journal of Food Microbiology* 46: 243–249, 1999.
83. H Xezones, IJ Hutchings. Thermal resistance of *Clostridium botulinum* (62A) spores as affected by fundamental food constituents. *Food Technology* 19: 113–115, 1965.
84. JAM Lowick, PJ Anema. Effect of pH on the heat resistance of *Cl. sporogenes* spores in minced meat. *Journal of Applied Bacteriology* 35: 119–121, 1972.
85. TC Odlaug, IJ Pflug. Thermal destruction of *Clostridium botulinum* spores suspended in tomato juice in aluminium thermal death time tubes. *Applied and Environmental Microbiology* 34: 23–29, 1977.
86. MS Cameron, SJ Leonard, EL Barret. Effects of moderately acidic pH on heat resistance of *Clostridium sporogenes* spores in phosphate buffer and in buffered pea puree. *Applied and Environmental Microbiology* 39: 943–949, 1980.
87. MH Silla, J Torres, A Arranz, MJ Peris. Citric acid lowers heat resistance of *Clostridium sporogenes* PA 3679 in HTST white asparagus purée. *International Journal of Food Science* 28: 603–610, 1993.
88. MJ Ocio, T Sánchez, PS Fernández, M Rodrigo, A Martínez. Thermal resistance characteristics of PA3679 in the temperature range of 110–121°C as affected by pH, type of acidulant and substrate. *International Journal of Food Microbiology* 22: 239–247, 1994.
89. LR Beuchat. Influence of organic acids on heat resistance characteristics of *Talaromyces flavus* ascospores. *International Journal of Food Microbiology* 6: 97–105, 1988.
90. B Sanz, P López, ML García, PE Hernández, JA Ordóñez. Heat resistance of enterococci. *Milchwissenschaft* 37: 724–726, 1982.
91. S Condón, ML García, A Otero, FJ Sala. Effect of culture age, pre-incubation at low temperature and pH on the thermal resistance of *Aeromonas hydrophila. Journal of Applied Bacteriology* 72: 322–326, 1992.

92. MB Cole, KW Davies, G Munro, CD Holyoak, DC Kilsby. A vitalistic model to describe the thermal inactivation of *Listeria monocytogenes*. *Journal of Industrial Microbiology* 12: 232–239, 1993.
93. R Pagán, P Mañas, J Raso, FJ Sala. Heat resistance of *Yersinia enterocolitica* grown at different temperatures and heated in different media. *International Journal of Food Microbiology* 47: 59–66, 1999.
94. P Mañas, R Pagán, J Raso, S Condón. Predicting thermal inactivation in media of different pH of *Salmonella* grown at different temperatures. *International Journal of Food Microbiology* 87: 45–53, 2003.
95. R Jordan, S Jacobs. Studies on the dynamic of disinfection. XIV. The variation of concentration exponent for hydrogen and hydroxyl ions with the mortality level using standard cultures of *Bact. coli* at 51°C. *Journal of Hygiene Cambridge* 46: 289–295, 1948.
96. G Pirone, S Mannino, M Campanini. Termoresistenza di clostridi butirrici in funzione del pH. *Industria Conserve* 62: 135–137, 1987.
97. M Mazas, M López, I González, A González, A Bernardo, R Martin. Effects of the heating medium pH on the heat resistance of *Bacillus cereus* spores. *Journal of Food Safety* 18: 25–36, 1998.
98. PS Fernández, MJ Ocio, F Rodrigo, M Rodrigo, A Martínez. Mathermatical model for the combined effect of temperature and pH on the thermal resistance of *Bacillus stearothermophilus* and *Clostridium sporogenes* spores. *International Journal of Food Microbiology* 32: 225–233, 1996.
99. KR Davey, SH Lin, DG Wood. The effect of pH on continuous high temperature/short time sterilization of liquids. *American Institute of Chemical Engineering Journal* 24: 537–540, 1978.
100. P Mafart, I Leguérinel. Modelling combined effects of temperature and pH on heat resistance of spores by a linear-Bigelow equation. *Journal of Food Science* 63: 6–8, 1998.
101. O Couvert, I Leguérinel, P Mafart. Modelling the overall effect of pH on the apparent heat resistance of *Bacillus cereus* spores. *International Journal of Food Microbiology* 49: 57–62, 1999.
102. P Mafart, O Couvert, I Leguérinel. Effect of pH on the heat resistance of spores. Comparison of two models. *International Journal of Food Microbiology* 63: 51–56, 2001.
103. M López, I González, S Condón, A Bernardo. Effect of pH of heating medium on the thermal resistance of *Bacillus stearothermophilus* spores. *International Journal of Food Microbiology* 28: 405–410, 1996.
104. A Palop. Estudio de la influencia de diversos factores ambientales sobre la resistencia de *Bacillus subtilis, B. licheniformis* y *B. coagulans*. PhD dissertation, Universidad de Zaragoza, Zaragoza, Spain, 1995.
105. A Palop, FJ Sala, S Condón. Heat resistance of native and demineralized spores of *Bacillus subtilis* sporulated at different temperatures. *Applied and Environmental Microbiology* 65: 1316–1319, 1999.
106. FMS Silva, P Gibbs, MC Vieira, CLM Silva. Thermal inactivation of *Alicyclobacillus acidoterrestris* spores under different temperature, soluble solids and pH conditions for the design of fruit processes. *International Journal of Food Microbiology* 51: 95–103, 1999.
107. MD López, P García, M Muñoz-Cuevas, PS Fernández, A Palop. Thermal inactivation of *Alicyclobacillus acidoterrestris* spores under conditions simulating industrial heating processes of tangerine vesicles and its use in time temperature integrators. *European Food Research and Technology* 232: 821–827, 2011.
108. MT Hutton, MA Koskinen, JH Janlin. Interacting effects of pH and NaCl on heat resistance of bacterial spores. *Journal of Food Science* 56: 821–822, 1991.
109. PM Periago, PS Fernández, MC Salmerón, A Martínez. Predictive model to describe the combined effect of pH and NaCl on apparent heat resistance of *Bacillus stearothermophilus. International Journal of Food Microbiology* 44: 21–30, 1998.
110. M Rodrigo, A Martínez. Determination of a process time for a new product: Canned low acid artichoke hearts. *International Journal of Food Science and Technology* 23: 31–41, 1988.
111. PS Fernández, MJ Ocio, T Sánchez, A Martínez. Thermal resistance of *Bacillus stearothermophilus* spores heated in acidified mushroom extract. *Journal of Food Protection* 57: 37–41, 1994.
112. S Condón, P López, R Oria, FJ Sala. Thermal death determination: Design and evaluation of a thermoresistometer. *Journal of Food Science* 54: 451–457, 1989.
113. PH Sparling. *Escherichia coli* O157:H7 outbreaks in the United States, 1982–1996. *Journal of the American Veterinary and Medical Association* 213: 1733, 1998.
114. DF Splittstoesser, MR McLellan, JJ Churey. Heat resistance of *Escherichia coli* O157:H7 in apple juice. *Journal of Food Protection* 59: 226–229, 1996.
115. LL Dock, JD Floros, RH Linton. Heat resistance of *Escherichia coli* O157:H7 in apple cider containing malic acid, sodium benzoate and potassium sorbate. *Journal of Food Protection* 63: 1026–1031, 2000.
116. LL Steenstrup, JD Floros. Statistical modeling of D- and *z*-value of *E. coli* O157:H7 and pH in apple cider containing preservatives. *Journal of Food Science* 67: 793–796, 2002.

117. CW Blackburn, LM Curtis, L Humpheson, C Billon, PJ McClure. Development of thermal inactivation models for *Salmonella enteritidis* and *Escherichia coli* O157:H7 with temperature, pH and NaCl as controlling factors. *International Journal of Food Microbiology* 38: 31–44, 1997.
118. R Pagán, I Alvarez, P Mañas, FJ Sala. Heat resistance in different heating media of *Listeria monocytogenes* ATCC 15313 grown at different temperatures. *Journal of Food Safety* 18: 205–219, 1998.
119. MH Silla, H Núñez, A Casado, M Rodrigo. The effect of pH on the heat resistance of *Clostridium sporogenes* (PA 3679) in asparagus puree acidified with citric acid and glucono-δ-lactone. *International Journal of Food Microbiology* 16: 275–281, 1992.
120. SA Tait, JE Gaze, A Figueiredo. Heat resistance of *Enterococcus faecalis, Lactobacillus plantarum* and *Bacillus licheniformis* in pH 4 and pH 7 buffers. *CFPRA Technical Memorandum 622*. Chipping Campdem, 1991.
121. F Rodrigo, PS Fernández, M Rodrigo, MJ Ocio, A Martínez. Thermal resistance of *Bacillus stearothermophilus* heated at high temperatures in different substrates. *Journal of Food Protection* 60: 144–147, 1997.
122. TC Beaman, P Gerhardt. Heat resistance of bacterial spores correlated with protoplast dehydration, mineralisation and thermal adaptation. *Applied and Environmental Microbiology* 52: 1242–1246, 1986.
123. PH Khoury, SJ Lombardi, RA Slepecky. Perturbation of the heat resistance of bacterial spores by sporulation temperature and ethanol. *Current Microbiology* 15: 15–19, 1987.
124. JA Lindsay, LE Barton, AS Leinart, HS Pankratz. The effect of sporulation temperature on the sporal characteristics of *Bacillus subtilis. Current Microbiology* 21: 75–79, 1990.
125. S Condón, M Bayarte, FJ Sala. Influence of the sporulation temperature upon the heat resistance of *Bacillus subtilis. Journal of Applied Bacteriology* 73: 251–256, 1992.
126. E Baril, L Coroller, F Postollec, I Leguerinel, C Boulais, F Carlin, P Mafart. Heat resistance of *Bacillus weihenstephanensis* KBAB4 spores produced in a two-step sporulation process depends on sporulation temperature but not on previous cell history. *International Journal of Food Microbiology* 146: 57–62, 2011.
127. I Leguérinel, P Mafart. Modelling the influence of pH and organic acid types on thermal inactivation of *Bacillus cereus* spores. *International Journal of Food Microbiology* 63: 29–34, 2001.
128. JPPM Smelt. Heat resistance of *Clostridium botulinum* in acid ingredients and its significance for the safety of chilled food. *The Netherlands Milk and Dairy Journal* 34: 139–142, 1980.
129. A Palop, J Raso, S Condón, FJ Sala. Heat resistance of *Bacillus subtilis* and *Bacillus coagulans:* Effect of sporulation temperature in foods with various acidulants. *Journal of Food Protection* 59: 487–492, 1996.
130. KL Brown, A Martínez. The heat resistance of spores of *Clostridium botulinum* 213B heated at 121–130°C in acidified mushroom extract. *Journal of Food Protection* 55: 913–915, 1992.
131. DM Adams. Heat injury of bacterial spores. *Advances in Applied Microbiology* 23: 245–261, 1978.
132. GA Prentice, LFL Clegg. The effect of incubation temperature on the recovery of spores of *Bacillus subtilis* 8057. *Journal of Applied Bacteriology* 37: 501–513, 1974.
133. S Condón, A Palop, J Raso, FJ Sala. Influence of the incubation temperature after heat treatment upon the estimated heat resistance values of spores of *Bacillus subtilis. Letters in Applied Microbiology* 22: 149–152, 1996.
134. FE Feeherry, DT Munsey, DB Rowley. Thermal inactivation and injury of *Bacillus stearothermophilus* spores. *Applied and Environmental Microbiology* 53: 365–370, 1987.
135. A Sikes, S Whitfield, J Rosano. Recovery of heat-stressed spores of *Bacillus stearothermophilus* on solid media containing calcium- and magnesium-deficient agar. *Journal of Food Protection* 56: 706–709, 1993.
136. I Leguérinel, O Couvert, P Mafart. Relationship between the apparent heat resistance of *Bacillus cereus* spores and the pH and NaCl concentration of the recovery medium. *International Journal of Food Microbiology* 55: 223–227, 2000.
137. AM Cook, MRW Brown. Relationship between heat activation and percentage colony formation for *Bacillus stearothermophilus* spores: Effect of storage and pH on the recovery medium. *Journal of Applied Bacteriology* 28: 361–364, 1965.
138. AM Cook, RJ Gilbert. Factors affecting the heat resistance of *Bacillus stearothermophilus* spores. I. The effect of recovery condition on colony count of unheated and heated spores. *Journal of Food Technology* 3: 285–293, 1968.
139. CG Mallidis, J Scholefield. Evaluation of recovery media for heated spores of *Bacillus stearothermophilus. Journal of Applied Bacteriology* 61: 517–523, 1986.

140. M López, I González, M Mazas, J González, R Martín, A Bernardo. Influence of recovery condition on apparent heat resistance of *Bacillus stearothermophilus* spores. *International Journal of Food Science and Technology* 32: 305–311, 1997.
141. PS Fernández, FJ Gómez, MJ Ocio, M Rodrigo, T Sánchez, A Martínez. D values of *Bacillus stearothermophilus* spores as a function of pH and recovery medium acidulant. *Journal of Food Protection* 58: 628–632, 1995.
142. I González, M López, M Mazas, J González, A Bernardo. Thermal resistance of *Bacillus cereus* spores as affected by additives in recovery medium. *Journal of Food Safety* 17: 1–12, 1997.
143. PM Periago, PS Fernández, MJ Ocio, A Martínez. A predictive model to describe sensitization of heat treated *Bacillus stearothermophilus* spores to NaCl. *Zeitschrift fuer Lebensmittel Untersuchung und Forschung* 206: 58–62, 1998.
144. CA Oscroft, JG Banks, S McPhee. Inhibition of thermally-stressed *Bacillus* spores by combinations of nisin, pH and organic acids. *Lebensmittel-Winssenschaft und Tehnologie* 23: 538–544, 1990.
145. A Palop, A Marco, J Raso, FJ Sala, S Condón. Survival of heated *Bacillus coagulans* spores in a medium acidified with lactic of citric acid. *International Journal of Food Microbiology* 38: 25–30, 1997.
146. P Gerhardt, RE Marquis. Spore thermoresistance mechanisms. In: I Smith, RA Slepecky, P Setlow, eds. *Regulation of Procaryotic Development*. Washington, DC: American Society for Microbiology, 1989, pp. 43–63.
147. GW Gould, GJ Dring. Heat resistance of bacterial endospores and concept of an expanded osmoregulatory cortex. *Nature* 258: 402–405, 1975.
148. A Okereke, RB Beelman, S Doores, R Walsh. Elucidation of the mechanism of the acid-blanch and EDTA process inhibition of *Clostridium sporogenes* PA 3679 spores. *Journal of Food Science* 55: 1137–1142, 1990.
149. AD Warth. Mechanisms of heat resistance. In: GJ Dring, GW Gould, DE Ellar, eds. *Fundamental and Applied Aspects of Bacterial Spores*. London, U.K.: Academic Press, 1985, pp. 209–226.
150. H Stewart, AP Somlyo, AV Somlyo, H Shuman, JA Lindsay, WG Murrell. Distribution of calcium and other elements in cryosectioned *B. cereus* T spores, determined by high-resolution scanning electron probe X-ray microanalysis. *Journal of Bacteriology* 143: 481–491, 1980.
151. H Stewart, AP Somlyo, AV Somlyo, H Shuman, JA Lindsay, WG Murrell. Scanning electron probe X-ray microanalysis of elemental distribution in freeze-dried cryosections of *B. coagulans* spores. *Journal of Bacteriology* 147: 670–674, 1981.
152. G Alderton, PA Thompson, N Snell. Heat adaptation and ion exchange in *Bacillus megaterium* spores. *Science* 143: 141–143, 1964.
153. G Alderton, JK Chen, KA Ito. Heat resistance of the chemical resistance forms of *Clostridium botulinum* 62A spores over the water activity range 0 to 0.9. *Applied and Environmental Microbiology* 40: 511–515, 1976.
154. RE Marquis, GR Bender. Mineralization and heat resistance of bacterial spores. *Journal of Bacteriology* 161: 789–791, 1985.
155. AB Levine. Biological roles of calcium. In: GJ Dring, GW Gould, DE Ellar, eds. *Fundamental and Applied Aspects of Bacterial Spores*. London, U.K.: Academic Press, 1985, pp. 47–58.
156. G Alderton, N Snell. Base exchange and heat resistance in bacterial spores. *Biochemical and Biophysical Research Communications* 10: 139–143, 1963.
157. RE Marquis, EL Carstersen, SZ Child, GR Bender. Preparation and characterization of various salt forms of *Bacillus megaterium* spores. In: HS Levinson, AL Sorenshein, DJ Tipper, eds. *Sporulation and Germination*. Washington, DC: American Society for Microbiology, 1981, pp. 266–268.
158. WG Murrell, AD Warth. Composition and heat resistance of bacterial spores. In: LL Campbell, HO Halvorson, eds. *Spores III*. Ann Arbor, MI: American Society for Microbiology, 1965, pp. 1–24.
159. GW Gould. Heat-induced injury and inactivation. In: GW Gould, ed. *Mechanisms of Action of Food Preservation Procedures*. London, U.K.: Elsevier, 1989, pp. 11–42.
160. A Palop, GC Rutherford, RE Marquis. Inactivation of enzymes within spores of *Bacillus megaterium* ATCC 19213 by hydroperoxides. *Canadian Journal of Microbiology* 44: 465–470, 1998.
161. SJ Foster, K Johnstone. The trigger mechanism of bacterial spore germination. In: I Smith, RA Slepecky, P Setlow, eds. *Regulation of Procaryotic Development*. Washington, DC: American Society for Microbiology, 1989, pp. 99–108.
162. TA McMeeking, T Ross. Modeling applications. *Journal of Food Protection Supplement* 37–42, 1996.
163. RC Whiting. Microbial modeling in foods. *Critical Reviews in Food Science and Nutrition* 35: 467–494, 1995.

164. O Reichart, C Mohacsi-Farkas. Mathematical modeling of the combined effect of water activity, pH and redox potential on the heat destruction. *International Journal of Food Microbiology* 24: 103–112, 1994.
165. AJ Pontius, JE Rushing, PM Foegeding. Heat resistance of *Alicyclobacillus acidoterrestris* spores as affected by various pH values and organic acids. *Journal of Food Protection* 61: 41–46, 1998.
166. VK Juneja, BS Eblen Predictive thermal inactivation model for *Listeria monocytogenes* with temperature, pH, NaCl, and sodium pyrophosphate as controlling factors. *Journal of Food Protection* 62, 986–993, 1999.
167. S Gaillard, I Leguerinel, P Mafart. Model for combined effects of temperature, pH and water activity of thermal inactivation of *Bacillus cereus* spores. *Journal of Food Science* 63, 887–889, 1998.
168. S Tejadillos, C Armero, PM Periago, A Martínez. A regression model describing the effect of pH, NaCl, and temperature on D values of *Bacillus stearothermophilus* spores. *European Food Research and Technology* 216: 535–538, 2003.
169. A Esnoz, PM Periago, R Conesa, A Palop. Application of artificial neural networks to describe the combined effect of pH and NaCl on the heat resistance of *Bacillus stearothermophilus. International Journal of Food Microbiology* 106, 153–158, 2006.
170. A Fernández, J Collado, LM Cunha, MJ Ocio, A Martínez. Empirical model building based on Weibull distribution to describe the joint effect of pH and temperature on the thermal resistance of *Bacillus cereus* in vegetable substrate. *International Journal of Food Microbiology* 77: 147–153, 2002.
171. MAJS van Boekel. On the use of the Weibull model to describe thermal inactivation of microbial vegetative cells. *International Journal of Food Microbiology* 74: 139–159, 2002.

24 Time–Temperature Integrators for Thermal Process Evaluation

Antonio Martínez Lopez, Dolores Rodrigo, Pablo S. Fernández, M. Consuelo Pina-Pérez, and Fernando Sampedro

CONTENTS

24.1 INTRODUCTION

Consumer demand for new and higher-quality foods has forced the food industry to develop new sterilization and pasteurization systems and to optimize current practices so that such demands can be satisfied without any negative impact on food safety. One of the main areas of current investigation is related to the use of heat in the food preservation, the main objective is the application of the minimum heat levels to destroy or to inhibit the development of the pathogen and spoiling microorganisms and to provide food with a longer shelf life. Technological developments such as treatment with high temperatures for short time periods (HTST or UHT) and the aseptic processing of food containing particles are very interesting because of the potential advantages that they offer from the point of view of nutritive and organoleptic food quality. With the objective of guaranteeing the microbiological safety of food preserved by heat, a strict evaluation of the thermal process is necessary. However, the conventional validation methods of these preservation processes are not always appropriate due to the way in which the food is produced and thermally processed. The time temperature integrators (TTIs) offer an alternative to the thermocouples and the conventional microbiological methods used to quantify the impact of thermal treatment on the microorganisms and others of food components such as vitamins or enzymes. In this chapter, a historical perspective is given on the systems used in the evaluation of the thermal processes and the development and the use of the TTI as a tool for ensuring the safety of thermally processed foods is described.

24.2 CALCULATION AND EVALUATION OF STERILIZATION PROCESSES

The thermal treatment of foods is one of the oldest and most used industrial methods to prolong the shelf life of foodstuffs. Today, its use as preservation technology in its different application forms represents a very significant percentage of all processed or manufactured foods. Apart from the heating system used and the way in which the heat is generated, an evaluation of the impact of the heat on the safety (inactivation of microorganisms, and analysis of risks) and on the quality (nutritive value, and acceptability for the consumer) is essential. Although there is no universal model system to predict the exact number of surviving spores after thermal treatment, the impact of the thermal processes has traditionally been evaluated in four different ways: (1) the method *in situ*, (2) the physicomathematical approach, (3) the inoculated experimental packs, and (4) on the use of TTIs. Traditionally, the first three methods have been used to evaluate the thermal processes, while the TTI is a more recent development.

24.2.1 *In Situ* Method

The *in situ* method is based on the evaluation of certain food quality indexes before and after the thermal treatment. Some of the analyzed parameters are the loss of vitamins [1], sensory appreciation of taste and texture, and the physical evaluation of the color [2]. Using a first-order model for thermal inactivation of these quality indexes, the effect of thermal treatment process can be quantitatively expressed in terms of cooking values "C_T" or lethality values "F_T." These thermal inactivation values are established from the initial and final concentration of the target property in the food and the D_T value (time of decimal reduction), which is defined as the time taken to reduce the level of a thermolabile element of the food by 90% at a constant temperature.

The main advantage of this method is that in some cases, the impact of the thermal process on the chosen parameter can be directly measured and with acceptable accuracy. The disadvantages associated with this methodology are (1) the concentration of the target element that exists naturally in the food may be outside the detection limits of the commonly utilized analytic methods after the thermal process has been applied and (2) the analyses of the studied parameter cannot be applied with routine analysis due to its complexity and high cost.

24.2.2 Physicomathematical Methods

An alternative to the *in situ* method for the evaluation of the thermal processes is the physicomathematical methods developed in recent years. With these methods, the time temperature history of the food obtained by means of mathematical models, under well-known sterilization conditions, is combined with kinetic models allowing us to determine the thermal impact of the treatment on the target element. Stoforos and Merson [3] proposed mathematical solutions for the heating systems for canned solid/liquid products. Models have also been developed for liquid products that contain solid particles in suspension [4,5]. The difficulty of measuring under certain conditions, the required physical parameters such as viscosity values and conductivity at high temperatures, coefficient of heat transfer, and distribution of the residence times of particles in a continuous process, means that the mathematical models do not always accurately predict the time–temperature evolution inside the food, which is necessary to establish the required thermal treatment. In addition, these models must be microbiologically validated.

24.2.3 Inoculated Experimental Packs

Another way that has been used to validate thermal processes is to perform laborious inoculation tests, that is, "inoculated experimental packs" [6]. Rodrigo and Martínez [7] used this methodology to evaluate mathematically obtained sterilization conditions for low-acid artichoke hearts (Table 24.1). The methodology consists of inoculating groups of 50 containers with the reference microorganism (i.e., PA 3679). Each group of 50 containers undergoes different thermal treatment conditions based on the mathematical model. When the sterilization process is finished, the containers are stored at an optimum incubation temperature in order to allow the growth of the inoculated microorganism. The cans are regularly inspected and those that show spoilage symptoms are enumerated. These data allow us to determine whether the mathematically derived sterilization conditions were enough to guarantee the microbiological safety of the food.

Yawger [8] developed a method to evaluate the sterilizing effect of a process determining the reduction in viable spores in a group of inoculated cans (*Bacillus stearothermophilus* or *Bacillus coagulans*). This system is a specific application of the technique of inoculated experimental packs to evaluate the lethality of a thermal process. It differs from the inoculated experimental packs in that it measures the lethality of the process in each inoculated container and not an alteration percentage.

The lethality determined by this system is interpreted as an integrated sterilization (IS) value, which is the equivalent of the lethal value of a process in terms of minutes to the reference temperature in the whole content of the container and not at a single point.

24.2.4 Time Temperature Integrators

The development of the TTIs relates to both consumer demand for high-quality convenience products and the food processing companies that need to save energy and achieve better control of

TABLE 24.1
Inoculated Experimental Packs Data for Artichoke Hearts

	Spoiled Cans/Sterilized		
Total Heating Time (min)	**Inoculated**	**Confidence Limits (P = 0.95)**	**Nonheated (Controls)**
5.0	18/50	11.4–25.8	6/6
10.0	17/50	10.7–25.7	6/6
18.0	0/50	0.0–7.0	6/6
21.5	0/50	0.0–7.0	6/6
35.0	0/50	0.0–7.0	6/6

Source: Extracted from Antonio Martínez, PhD thesis.

the thermal process. Consequently, efforts have been made to optimize the existing systems and develop new heating technologies such as continuous processing retorts, heat exchangers, aseptic processing, microwave or ohmic heating, and combined processes. Because these evaluation methods have serious limitations when applied with these new technologies, a considerable effort has been made to develop TTIs, capable of accurately measuring the impact of thermal treatment on food in terms of quality and safety. In general, these devices are artificial particles made of alginate, plastic, or other carriers, containing a sensor element (microorganism or enzyme) homogeneously distributed inside them. The response of the sensor element is proportional to the intensity of the given thermal process. These devices can be introduced in HTST or UHT systems as well as in cans or retorts.

24.2.5 Biological Lethality (F.BIO) and Physical Lethality (F.PHY) Relationship

The lethality of the preservation process by heat can be determined by physical measures, using thermocouples that record the temperature history of the food during the heat treatment and by biological measures using bacterial spores, or other sensor, immobilized in a TTI. The sterilization values physically and biologically calculated are named physical lethality (F.PHY) and biological lethality (F.BIO) values, respectively. In general, the value of F.BIO is based on the reduction in the number of viable spores after thermal treatment, and it is determined by using a previously established relationship (curve of calibration or equation) among the number of surviving spores (N_F), and the equivalent in minutes of heating (F) at a given temperature [8–12]. Since the spores are sensitive to environmental changes, and there are differences between spores of the same strain, the reduction of the number of viable spores is not constant for the same treatment. However, the value of F.BIO should be constant for a given thermal treatment, since the differences between spore cultures, species of spores, and environmental conditions (different from the heating environment) do not affect the value of an F.BIO that is correctly determined.

The value of F.PHY is calculated by using an appropriate mathematical model and the acquired temperature measurements by using a thermocouple inserted in a cold point of the product (point of the slowest heating) that usually coincides with its geometric center.

Both the F.BIO and F.PHY values should be identical for the same reference temperature. However, several studies mentioned in the literature show differences between both values [13–17]. There are several reasons that could explain these differences. The F.BIO values may be erroneous due to the incorrect use of the TTI, which depends on the biological system and the associated analytic procedures; and the F.PHY values may be erroneous due to the incorrect determination of the required parameters time–temperature history and spore thermal resistance data. Some explanations that may account for the discrepancies among the F.BIO using a TTI system and the associated analytic procedures and the F.PHY value determined by the General Method [18] are mentioned in the following sections.

24.2.5.1 Possible Errors in the Determination of the F.BIO Value

1. Use of a badly calibrated curve
2. Differences between the conditions used to obtain the calibration curve and those used to evaluate the thermal treatment
3. The use of an incorrect lag factor when obtaining the calibration data where the temperature is reached instantly
4. Alteration of the thermal properties of the spores when introducing them in a carrier system
5. Differences between the number of survivors recovered after the calibration experiments and after the nonisothermal treatments
6. Incorrect conversion of the F.BIO values to Fo.BIO values

24.2.5.2 Possible Errors in the Determination of the F.PHY Values by Means of the General Method

1. Errors in the temperature measurements
2. Alteration of the heat transfer characteristics of the product as a consequence of the use of thermocouples as measuring devices
3. Errors in the calculation of the z values
4. Supposition of constant z values in the whole interval of temperature
5. Inexact calculations by means of the Ball formula method

24.2.5.3 Explanations for the Discrepancies between the F.BIO and F.PHY Values

1. The experimental variations can cause differences between the F.BIO and F.PHY. Quite frequently, the thermocouple and the TTI are not located in the same container. Consequently, these values are subject to variations in the heating and cooling rates. Also each thermocouple and each indicator exercises its influence on the results; therefore, it is necessary to get sufficient repetitions for the same experiment.
2. The discrepancy can be caused by the differences in the measurement point inside the same container in which the bioindicator (TTI) is not exposed to a uniform temperature, while the temperature around the thermocouple is uniform.
3. The use of inactivation kinetics obtained by isothermal processes may make inexact predictions in complex systems such as the biological ones in which the temperature changes [19]. Johnson [20] observed an unexpected second inactivation phase in the spores of the *Bacillus cereus* during cooling from 50°C to 40°C after being heated to 90°C, and he concluded that the inactivation data generated at constant temperatures cannot always be used to predict the response of the bacterial spores to changes of temperature. On the contrary, the F.PHY values do not consider the changes in the relationship between the temperature and the rate of inactivation of the spores. Nevertheless, experiments have been done, which show that a good correlation can be obtained between the F.BIO and the F.PHY values. Secrist and Stumbo [21] used containers inoculated with spores of PA 3679 and found a good correlation between the predicted and the experimental values. Rodríguez et al. [22] obtained a good correlation in the predicted and experimental number of survivors after nonisothermal treatment, using spores of *Bacillus subtilis.* Sastry et al. [23] inoculated spores of the *B. stearothermophilus* in mushrooms using alginate gels. Good correlations were obtained in the preliminary studies sterilizing mushrooms in metallic containers.

With the introduction of the aseptic processing, a significant change in the sterilization in comparison with the traditional preservation technologies took place. In this process, the food is pumped through a plate, tubular, or scrapped surface heat exchangers according to the characteristics of the food; it is then cooled and put into aseptic sterile containers. This process is utilized for liquid foods, such as juices, milk, etc., but because of the difficulty in ensuring the lethality reached by the food particle in the sterilization process, it is not often used for foods containing particles larger than 10 mm. On the contrary, as happens in the sterilization process of a container (can or retort pouch), there are no physical methods to determine the temperature reached in the center of the food particle (meat or vegetable pieces) aseptically processed in a heat exchanger. Before authorizing a treatment, the Food and Drug Administration (FDA) requires a demonstration of the sterility reached in the center of the food particle, that is, the cold point of the process. Therefore, mathematical simulation models and validation by means of biological methods (TTI) should be used. At present, some mathematical models have been developed [24,25] that simulate the penetration of heat in the particle. However, for HTST treatments, the discrepancies between the F.PHY and the F.BIO can be greater because of the difficulty in establishing times of residence, temperature transference, etc., which are necessary in order to develop an effective mathematical model.

24.3 GENERAL ASPECTS OF THE TTIs

A TTI can be defined as a small device that shows some irreversible changes with the time–temperature, which is measurable in an easy, exact, and precise way that mimics the changes produced in a temperature sensible factor (microorganism, enzyme, etc.) contained in a foodstuff that suffers the same treatment at variable temperature.

The main advantage of a TTI is its ability to quantify the integrated impact of the time and the temperature in the target attribute without the need for any information on the history of the product real temperature [26].

In accordance with the given definition, a TTI should fulfill some requirements for its structure, behavior, and use: (1) the integrator should contain a calibrated and resistant sensor element for the thermal treatment, and it should experience the same evolution of temperature as the real food; (2) the size of the integrator and its geometry should be similar to the real food, with the sensor element homogeneously distributed in its interior; (3) the carrier should efficiently retain the sensor element so that losses do not take place during the sterilization process; (4) the integrator should be incorporated in the product so that it does not produce distortions during the heat transfer, neither should it modify the profile time–temperature of the food; (5) the integrator should be cheap, easy and quick in its preparation, and easy to analyze and to recover; (6) the integrator should be stable and capable of long-term storage without any loss of functionality; and (7) the integrator should be physically resistant enough to support the heating process without disintegrating.

Besides these requirements, there are some kinetic aspects that an integrator should satisfy. When the sterilizing value of the process is chosen as a concept to express the integrated impact of the time and of the temperature, the TTI should fulfill the following equation:

$$\left(F^{z}_{T_{ref}}\right)_{indicator} = \left(F^{z}_{T_{ref}}\right)_{TTI} \tag{24.1}$$

That is to say, the lethality reached in the integrator should be similar to the response of lethality of the factor used as indicator (microorganism, enzyme, etc.). It is also easy to understand that a system will work as a TTI if its activation energy (E_a) or the z value is the same or similar to that of the factor considered as indicator.

24.4 CLASSIFICATION OF THE TTI BY CONSIDERING THE SENSOR ELEMENT

The TTIs can be divided into the following [26]: (1) chemical systems, (2) physical systems, and (3) biological systems. The biological systems are further divided into microbiological, enzymatic, and nonenzymatic systems. In general, the sensor element of the TTI can be introduced in a carrier system and locate it in a defined position inside the food, or it can be dispersed in the food, as in an inoculated experimental pack validation process. When a physical contact does not exist between the sensor element and the food, the inactivation kinetic of this sensor can be determined independently of the food type and the local conditions. When the sensor element is immobilized in a device, thermophysical properties of the carrier system should be similar to the thermophysical properties of the food to ensure that the transfer of heat of the carrier is similar to that in the food, and in this case, the inactivation kinetic of the sensor should be determined in its own device. If the sensor element is dispersed in the food, it is necessary to keep in mind that the new chemical atmosphere in which the sensor is immersed requires that kinetic calibration may have to be carried out.

24.4.1 CHEMICAL TTI

Chemical TTIs are based on the detection of chemical reactions to quantify the impact of a thermal process. Hendrickx et al. [27] described a detailed revision on the chemical reactions that can be used as TTIs. Different chemical systems such as thiamine heat inactivation [1,28,29] and color changes produced by sugar and amino groups reduction [30] have been used. Methylmethionine sulfonate (MMS) thermal degradation between 121.1°C and 132°C in citrate tampon (pH 4–6) has been correlated to the reduction of microorganisms in thermally treated liquid foods [31]. Under certain process conditions, it is possible to compare the decrease in *B. stearothermophilus* spores with the reduction of the MMS concentration. Sugar hydrolysis has also been used as an indicator of thermal treatments by Kim and Taub [32]. Measurements of this chemical marker have been correlated to the inactivation of *Clostridium botulinum* spores. Betanin, a natural color tracer and major pigment from red beets, was also tested as a chemical TTI for thermal treatment in scraped surface heat exchanger by Mabit et al. [33]. The results showed that although it was sensitive to the level of thermal treatment received, it was also sensitive to the level of mechanical treatment; therefore, it made it inappropriate for the purpose.

TTIs have also been applied to predict lactobacilli growth and consequently the shelf life of modified atmosphere-packed gilthead seabream (*Sparus aurata*) as a function of packaging and refrigerated storage conditions [34]. Photosensitive compounds such as benzylpyridines are excited and colored by exposure to low wavelengths light. The response of UV activatable TTI was kinetically modeled as a function of activation level and temperature to predict the quality of gilthead seabream fillets during distribution and storage.

In spite of the important role that chemical TTIs seem to be able to carry out in the valuation of thermal processes, there is an important disadvantage associated with them. The high value of z (20°C–50°C) or the low value of its activation energy (E_a: 160–60 kJ/mol) makes them unable to guarantee the microbiological safety in the interval of sterilization temperatures. However, the inclusion of sensors in carrier systems that diminish the z value and make it closer to that of the microorganism has been the object of investigation.

24.4.2 Physical TTI

Physical TTIs are based on diffusion phenomena. The system described by Witousky [35] consists of a colored chemical substance that can melt and be absorbed in wick paper under the effect of humid heat (steam). The TTI response is calculated by measuring the distance reached by the melted compound. Its use as an indicator of thermal processes was established by Bunn and Sykes [36]. However, this TTI presents the disadvantage that it cannot be used in thermal processes where dry heat is applied; nor can it be included inside the product because the TTI is activated by steam.

Another system proposed by Swartzel et al. [37] is based on the ionic diffusion and capacity of a semiconductor metal. The diffusion distance can be exactly calculated by measuring the cellular capacitance change before and after a thermal treatment. These systems use several ionic carriers with at least two different activation energies, from which sterilizing values can be calculated by means of the equivalent point method [38]. However, Maesmans et al. [39] have demonstrated that this method can lead to serious errors.

24.4.3 Biological TTI

24.4.3.1 Protein-Based TTI: Enzymatic and Nonenzymatic (Immunochemical)

Thermal treatment can produce irreversible changes in the tertiary structure of proteins. In enzymes, this can affect their activity. Enzymatic systems use this activity loss to measure the impact of thermal processes. Proteins with thermostability at different temperature intervals and with different activation energies have been used. This can cover a wide range of food safety and food quality attributes. Two types of TTI can be defined based on the detection methods to measure protein activity.

1. Enzymatic TTI, where the enzyme residual activity is a direct response of the protein to the thermal impact. A TTI has been calibrated based on the use of the peroxidase enzyme in an organic solvent [27,40]. De Cordt et al. [41,42] proposed the use of the *Bacillus licheniformis* enzyme amylase as a TTI. TTI systems have been developed with the β-galactosidase, lipase, and nitrate reductase enzymes encapsulated in alginate.
2. Immunochemical methods based on the specific interaction of antigen–antibody. The thermal treatment of proteins can affect binding sites so that the antibody does not join the protein. Kinetics of antigenic capacity loss differs between different proteins and in some cases between different binding sites within the same protein. These immunologic techniques (competitive and noncompetitive enzyme-linked immunosorbent assay [ELISA]) have been used by Brown [43] to study the antigenic inactivation after the thermal treatment.

Although these protein systems can provide quick and accurate enough detection measures, they present inconveniences because enzymes become inactive well before reaching the treatment temperature in HTST systems. It is therefore necessary to increase their thermostability so that they can work properly at high temperatures. The necessary manipulations to increase thermostability, adjusting the values of the activation energy (E_a) or the rate constant (k), imply changes in the environmental conditions, stability, and immobilization of enzymes that cannot always be carried out.

24.4.3.2 Microbiological TTI

The use of microbiological systems to control the efficiency of the sterilization processes constitutes one of the main objectives of the food and pharmaceutical industry. A microbiological TTI consists of a carrier system inoculated with a microorganism (essentially bacterial spores) of well-known thermal resistance and concentration under the specific sterilization conditions to assess. The inactivation level of the microbiological integrator in the sterilization treatment gives an idea of the magnitude of the process and therefore of the level of sterility achieved in it [44]. A microbiological integrator should complete the following general requirements:

1. The microorganism should be stable in number and thermal resistance. Results should be reproducible and of low variability [45].
2. The microbiological system (microorganisms, carrier system, and procedures used) should be calibrated under the specific conditions of sterilization to be assessed [46].
3. Relations between the microbiological integrator and the natural microbiological load of the product should be known so that the validity of the sterilization can be ensured [12].

The selection of microorganisms that can be used as biological indicators for sterilization treatments depends on the specific applications for which they have been designed [46]. The strains of the most-resistant species are generally used for the valuation of the sterilization processes, although under certain conditions microorganisms less resistant but quite similar to the natural microflora of the product or easier to detect are used instead. The *B. stearothermophilus* spores are commonly used as biological indicators in processes of sterilization by wet heat [47]. Spores of the *B. subtilis* have been used for treatments with dry heat and in processes with ethylene oxide, while spores of the *Bacillus pumilus* are used for the sterilization by means of ionic radiations [44].

The z values of spores used to control food microbiological security are similar to the z value of the *C. botulinum* (z = 10°C), which is used as a reference in the sterilization of canned products with low acidity. For these foods, spores of *B. stearothermophilus*, *B. subtilis* 5230, *B. coagulans*, and *Clostridium sporogenes* have been used as microbiological indicators in the sterilization by wet heat [45,46]. To determine the impact of the thermal treatment, it is necessary to use spores previously calibrated and validated against well-known physical parameters.

Selected strains of lactic acid bacteria have been tested as a commercial TTI tag for meat products under modified atmosphere [47]. The growth of the TTI microorganisms causes a pH drop in the tag leading to an irreversible color change of the medium chromatic indicator that has been related with quality degradation.

Microbiological control systems are divided into (1) qualitative systems that integrate time–temperature evolution and indicate whether or not the process has reached the established value (negative or positive growth), for which no conclusions can be derived about the extension of the thermal treatment, and (2) quantitative systems that estimate the recount reduction and the intensity of the process that can be determined by plate count of survivors.

24.5 SUPPORTS USED IN THE DEVELOPMENT OF TTIs

To immobilize the sensing element (enzymes, microorganisms, etc.), different carriers or supports have been used.

Pflug [48] developed a hollow plastic bar into which he introduced *B. stearothermophilus* endospores to validate sterilization processes of a great variety of foods like green beans [49], corn, and peas [50]. Rodríguez and Texeira [51] proposed using an aluminum tube with better heat transfer and mechanical resistance than plastic, as a support to evaluate sterilization processes of low-viscosity and quickly heated foods. Another method used is the inoculation of a small glass bulb with an endospore suspension, which is placed in the center of a food piece subject to the conditions of the process [52]. All these methods have one particular disadvantage: the chemical environment of endospores is not that of food. Strips of filter paper impregnated with an established quantity of bacterial endospores in foods heated in conventional systems have also been used [53,54]. The evaluation of the continuous sterilization processes has required the development of another type of integrator, since integrators with a plastic, metallic, or glass support did not suffer the same temperature dragging or evolution as real foods. In this sense, Segner et al. [55] inoculated the food piece with bacterial endospores. However, the endospores were not homogeneously distributed within the whole piece of food, so it did not fulfill one of the necessary requirements for a TTI. Another alternative that has turned out to be the most appropriate is the elaboration of artificial devices that simulate real foods and that contain homogeneously distributed endospores inside [16,23,56–58].

For the simulation of real foods, the devices used are prepared in a gelificable support; therefore, this technique is also known as immobilization in gel. It involves adding some type of pureed gel to the food in the mixture, of which the sensor (microorganism, enzyme, etc.) can be immobilized, and the so-called artificial particles similar in size and shape to those of real foods can be made. This makes it possible to ensure that conditions in the sensing element are as close as possible to the pieces or particles of the real food to be treated, with regard to its chemical environment, particle dynamics during the process, heat penetration, density, etc. [59]. Another advantage of this method is that as microorganisms are homogeneously distributed in the artificial particle, the result of the thermal treatment will be an IS value.

Different types of gelificable supports have been used, such as albumin gels [60], carrageenan gels [61,62], polyacrylamide gels (PGA) [57,63], and recently polysaccharide gels [64]. However, the most commonly used gels are alginate gels [58].

Dallyn et al. [65] described the use of an integrator consisting of the immobilization of bacterial endospores in alginate pellets. This TTI had enough mechanical strength to be used in scratch surface heat exchangers. Later on, Bean et al. [56] inoculated endospores of *B. stearothermophilus* in 1.6–3.2 mm diameter alginate pearls, and they observed that the size of the beads was too small to evaluate sterilization processes of foods that contained large pieces. Other larger (8–24 mm) and cube-shaped alginate artificial particles were developed by Brown et al. [16] who observed discrepancies between the experimental sterilizing values and those calculated by means of a mathematical model of temperature transfer. Another approach was carried out by Sastry et al. [23] who developed an indicator inoculating bacterial endospores suspended in an alginate gel and introducing it inside a mushroom. The lack of homogeneity in the distribution of endospores observed in this integrator hindered the interpretation of results, and in some cases, the mushroom stalk came off the cap during the heating process. Recently, Wang et al. [64] compared hardness, springiness, and water retention of konjac glucomannan gel (g-KGM) as a novel carrier material for TTIs in aseptic processing with those of sodium alginate gel (s-SA). Even though significant differences appear in springiness at a temperature range of 100°C–140°C and in water retention, heat transfer studies performed on g-KGM alone as well as on g-KGM embedded with α-amylase as an integrator, indicated that g-KGM was suitable for industrial ultrahigh sterilization tests.

Other problems present in TTIs developed in an alginate matrix are that some devices cannot be stored for long [16], and the endospores can get lost by lixiviation during the heating. Ocio [58] developed a TTI with alginate and mushroom puree to which she added 4% starch. She observed that this allowed for the freezing of the particle for a long period of time and its later unfreezing, and maintaining mechanical characteristics similar to those of the particle that did not contain starch and that had been stored for 7 days in refrigeration. Later on, Rodrigo [66] used an integrator similar to the one developed by Ocio [58] to evaluate a thermal process under pilot plant conditions. This author [66] did not find significant differences at a 5% level of significance between the experimental lethality values and those calculated by a mathematical model.

24.6 ALGINATE TTI ELABORATION

Alginate and food puree TTIs containing immobilized microorganisms are probably the most versatile and the ones that approach best the physicochemical characteristics of a real food particle (Figure 24.1). Therefore, the production of this type of artificial particles is described later with further details. A TTI based on alginate with immobilized spores inside requires a series of previous studies related to the calibration of the sensor (microorganism, enzyme, etc.) and the stability of the support or carrier (mechanical resistance, capacity of retention of the sensing element inside, etc.)

24.6.1 Thermoresistance of the Sensor Element

Based on the example of the *B. stearothermophilus* spores or microbiological α-amylase, which are much used in this type of TTI, it is first necessary to consider the thermoresistance characteristics of the microorganisms in the medium that will act as carrier.

An important aspect in heat inactivation kinetic studies is the form of heating and the method used for analyzing experimental data. Thermobacteriology studies have conventionally been carried

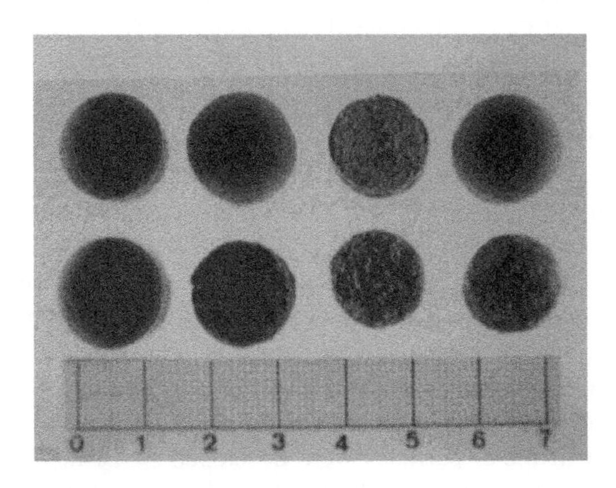

FIGURE 24.1 Alginate–mushroom–starch particles.

out in isothermal conditions. Determination of kinetic parameters in these conditions is relatively simple, and it produces conservative sterilization or pasteurization processes. However, during heat treatment, microorganisms are subjected to conditions that often differ substantially from isothermal experimental conditions. An alternative is to apply nonisothermal heating methods. These methods offer the advantage of subjecting the spores to dynamic temperature conditions, such as those that occur in real thermal sterilization processes. Tucker [67] reviewed the various nonisothermal methodology used in accelerated methods for the stability study of pharmaceutical products. These methods can also be used for microbial inactivation studies. Tucker [67] discussed the following temperature increase programs: (1) uncontrolled temperature increase, (2) linear program, (3) linear program followed by an isothermal period, (4) polynomial program, and (5) hyperbolic program.

With nonisothermal heating processes, it is possible to obtain much information (experimental data) with a single experiment, with consequent savings in material, time, and labor costs. Moreover, nonisothermal studies provide a wealth of data for subsequent development of predictive models.

Experimental data in microorganism inactivation studies are generally analyzed by means of two successive linear regressions. In the first, the logarithm of the concentration of the thermolabile factor remaining after heat treatment is plotted against time, giving the D_T value. In the second regression, the logarithm of D_T is plotted against treatment temperature, giving the z value. This methodology generally provides high confidence ranges, due to the small number of degrees of freedom [68]. Arabshahi and Lund [69] presented a method for analyzing kinetic data for inactivation of thiamine that provided smaller confidence ranges than those obtained by using two linear regressions. To do this, they calculated the activation energy, Ea, from the original experimental data by applying nonlinear regressions in a single step. A possible drawback of this methodology is the high correlation that may exist between the parameters estimated, D_T and z or k_T and Ea, causing convergence and accuracy problems. Nevertheless, the convergence can be improved if the equations are transformed by taking natural logarithms or reparameterizing.

Inactivation predictive models can be an excellent tool for ensuring the microbiologic safety of the thermal process. There is growing interest in predictive microbiology because of its many potential applications, such as estimating the effect of changes or of errors in estimating specific microorganism parameters [70], for example, D_T or z. The successful application of these models depends on developing and validating them in real conditions [71].

The parameters that define the inactivation of microorganism by heat in nonisothermal conditions by analyzing experimental data obtained in real substrates by means of single-step nonlinear regressions are of great value for the development of inactivation models that relate the D_T value to environmental factors of importance in the canning industry, such as pH, sodium chloride concentration, anaerobiosis, or water activity. These models can be incorporated into a hazard analysis and critical control point plan, to safeguard the sterilization process against any eventuality during the manufacturing process that might affect the D_T value of the microorganisms, potentially causing spoilage of the food product.

24.6.2 Preparation of the Carrier

A desirable characteristic that an integrator should satisfy is its mechanical and storage stability. Alginate has been frequently used as carrier to immobilize bacterial spores. Alginate is obtained from a marine alga and their molecules are composed by two monomers mannuronic and guluronic acids (Figure 24.2). Rodrigo [66] prepared an alginate–mushroom–starch mixture to produce cylindrical devices that support the lyophilization process without losses of mechanical properties or microorganisms spores. The mixture must be gelified (Figures 24.3 and 24.4). For this, two procedures can be used.

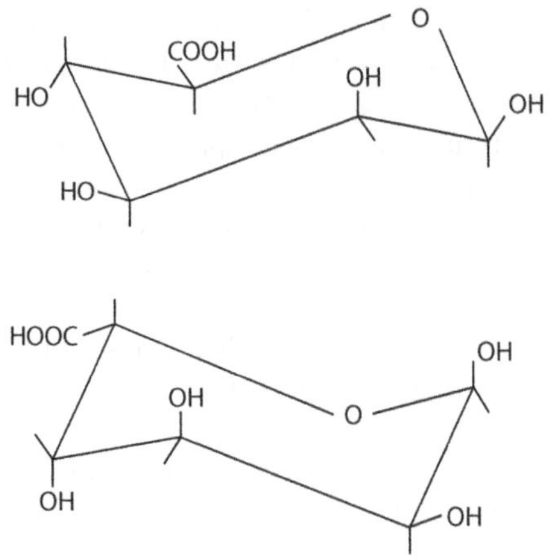

FIGURE 24.2 Structure of mannuronic (above) and guluronic (bottom) acids.

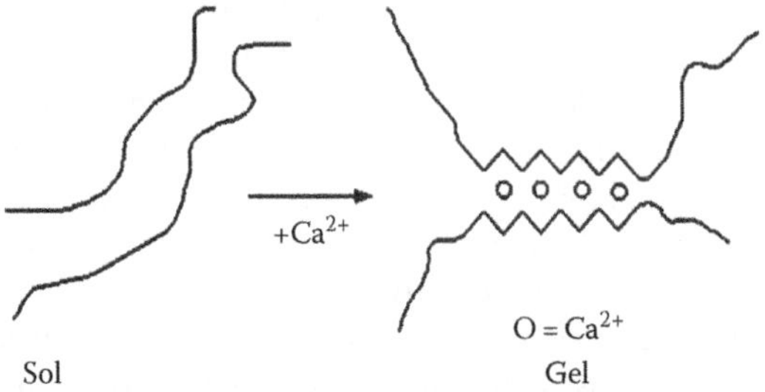

FIGURE 24.3 Gelification process.

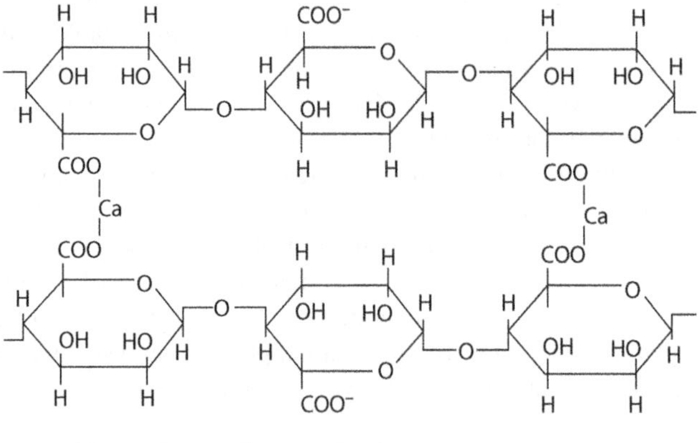

FIGURE 24.4 Structure of the calcium alginate molecule.

1. Internal gelification: In this method, a salt is added to the alginate solution. The salt liberates the calcium in controlled quantities, since a quick contact would cause a precipitation of the gel and the cationic diffusion in it would not be good. Sulfate, carbonate, and calcium orthophosphate acid are the salts that are most commonly used for this purpose.
2. Gelification by diffusion: In this technique, the alginate solution contacts a calcium ion solution. The contact of both solutions can take place in two ways: directly between them or through dialysis membranes (Figure 24.5). Particles formed by alginate gelification are the most interesting particles for the evaluation of thermal processes of real foods containing particles.

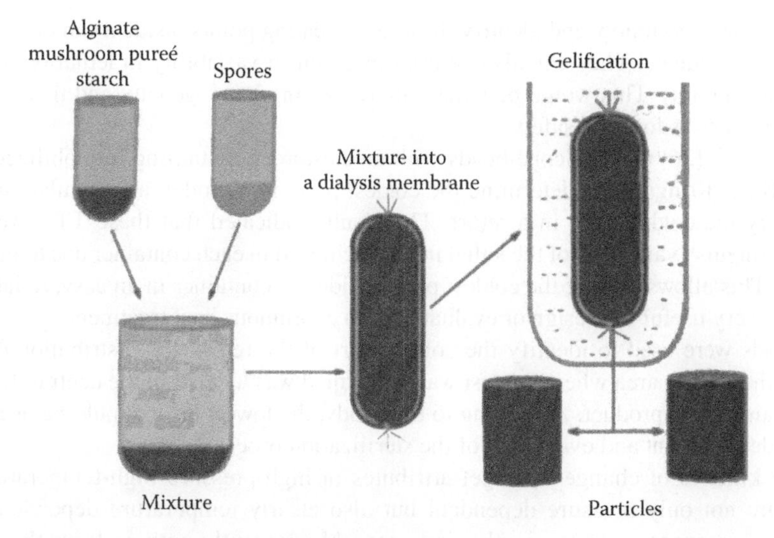

FIGURE 24.5 Gelification process using a dialysis membrane.

24.6.3 Recovering of Sensor Element and Calculating Impact of the Process

After its use, spores immobilized inside the gelified alginate particle must be recovered in order to quantify the thermal treatment impact. For that, devices are introduced in sterile masticator bags containing sodium citrate and α-amylase. After 7 min of treatment serial dilutions are made and spores are plated by duplicate in a suitable culture medium. Counts are made after several hours of incubation at an optimum temperature, depending on the microorganisms. When the sensor is other than a microorganism, the same procedure could be followed, but the analytical methodology can vary as a function of the sensor element.

Table 24.2 shows that there is a good distribution of spores inside the gelified particle and that the extraction procedure is reproducible.

24.7 APPLICATION EXAMPLES OF TTI

From a food safety evaluation point of view, thermal process design should be based on the heat impact achieved in the coldest spot (or slowest heating point) of the food. TTIs can be used as

TABLE 24.2
***B. stearothermophilus* Spore Counts (Forming Colony Units [FCU]) in Three Different Particles Coming from Five Gelified Pieces**

Gelified Piece	Counts (FCU/TTI)		
	TTI 1	TTI 2	TTI 3
1	5.55×10^6	6.40×10^6	5.85×10^6
2	5.75×10^6	5.25×10^6	5.90×10^6
3	4.35×10^6	5.35×10^6	6.65×10^6
4	4.90×10^6	4.30×10^6	4.35×10^6
5	4.35×10^6	6.85×10^6	6.30×10^6

wireless instruments to follow and identify those slow heating points inside a can or jar when other systems are not adequate. They can also be used to establish variability in lethality among different parts inside a retort. This would be a measure of heating heterogeneity and therefore a way to identify potential risks for food safety.

Van Loey et al. [26] used silicone beads with an elastomer containing immobilized *B. subtilis* α-amylase (BSA) 10 mg/mL to determine the coldest point in a model (a particulate food system) and to identify the coldest spot in a retort. The results indicated that these TTIs were sensitive enough to distinguish variations of the lethal impact achieved in each container due to their situation in the retort. This allows finding the coldest point inside the container in an easy, reliable, and fast way, which is very useful for design or evaluation of a continuous heat treatment.

When beads were used to identify the coldest part of the retort, the distribution of lethalities showed that the coldest area where the test was performed was located in the central, lowest part of the box containing the product. According to this study, the lowest layer should be used as a reference for the development and evaluation of the sterilization process.

Since the kinetics of change of target attributes in high-pressure high-temperature (HPHT) processing are not only pressure dependent but also clearly temperature dependent, a control of temperature in space and time is also indispensable. Recently authors from the same group [72] have studied the potential of the readout of a protein system, ovomucoid (1 g/L–0.1 M in MES-NaOH buffer pH 6.2), as an extrinsic, isolated indicator for temperature heterogeneities in a HPHT pilot scale, vertically oriented vessel.

TTIs containing *B. stearothermophilus* spores as sensor element have also been used. Tejedor [73] developed a TTI with alginate and food puree as supporting elements to establish sterilization conditions for a food containing vegetables and tuna fish. The integrator was appropriate to validate the level of safety achieved with the thermal processes in this substrate. The integrated lethalities calculated with a mathematical model of temperature transference and those obtained empirically using the TTI were very similar (Table 24.3).

Tucker [74] used a *Bacillus amyloliquefaciens* α-amylase TTI injected in the center of silicone devices to validate pasteurization processes of products containing large particles. A feasibility study was conducted on an industrial ohmic plant using 10–12 mm whole strawberries as the yogfruit product. The technique developed and demonstrated on continuous pasteurization processes can be applied to almost any process for foods that contain solid particles. Furthermore, Lambourne and Tucker [75,76] developed *B. amyloliquefaciens* and *B. licheniformis* α-amylase TTIs encapsulated in silicone tubes and in silicone cube particles. Another form of encapsulation

TABLE 24.3
Comparison between Experimental and Calculated *B. stearothermophilus* Counts in TTI

			TTI Counts	
Retort Temperature	F_0 (min)	Time (min)	Experimental	Calculated
112	0	0	3.95×10^7	3.95×10^7
	1.10	33.6	9.76×10^6	8.09×10^6
	2.58	44.4	1.96×10^5	1.52×10^5
	3.89	53.2	3.81×10^3	5.74×10^3
115	0	0	4.57×10^7	4.57×10^7
	1.17	25.8	1.33×10^7	2.15×10^6
	2.58	35.4	3.18×10^4	6.11×10^4
	4.44	40.6	2.40×10^3	5.86×10^3

was also employed using polytetrafluoroethylene (PTFE) tubes. The tubes produced some problems with leakage. Devices were employed in the validation of different pasteurization processes. Those devices were found to be a reliable and alternative method for measuring the thermal process delivered to products where conventional probe-based validation techniques were not suitable. The industrial application trials using *B. amyloliquefaciens* and *B. licheniformis* α-amylase were successful in all cases and showed that the pasteurization treatments being applied were in general substantial, and overprocessing was taking place in most cases.

Same authors [77] studied the application of the highly thermostable amylase produced from the extracellular medium of a *Pyrococcus furiosus* fermentation as a sterilization TTI. Inactivation kinetics under isothermal and nonisothermal processing conditions were obtained and industrial measurements validated the ability of the device as a TTI for sterilization processes.

Fernández et al. [78] performed a study to evaluate the lethalities achieved in glass jars containing baby food in a static industrial retort as a function of their position in the crates, distribution within the retort, and variability within the jar. As sensor element, highly heat-resistant spores of *Bacillus sporothermodurans* were used, which were suspended in spheric alginate beads of 4 mm in diameter. The following experimental design was used: 10 inoculated alginate beads were mixed within the content of each glass jar and over 20 jars containing beads were distributed in the four crates of the retort, at different heights and positions (central or corners) within each crate. After the sterilization process, the number of survivors in each individual bead was obtained, and a statistical study was performed.

The results indicated that one of the crates presented a significantly lower level of inactivation than the other ones and that the bottom part of this crate was the coldest spot of the retort. No significant differences were found between the central part and the corners of each crate or within jars. These data allowed an efficient establishment of the variability of inactivation within the product, the coldest spot of each crate, and the coldest spot of the whole retort.

Alicyclobacillus acidoterrestris spores inoculated in alginate beads were characterized under isothermal and nonisothermal conditions simulating industrial heating processes applied to tangerine vesicles. The results showed that it was a suitable microorganism to be used for validation of thermal treatments applied to acidic foods, such as citric fruit juices [79].

24.8 CONCLUSIONS

Thermal processing of food remains as one of the most widely used methods of food preservation. Consumer pressure for more convenient and nutritional foods and the need of saving energy in the industry has been the engine driving the development of new heating and filling systems (heat exchangers heaters and aseptic processing). The main inconvenience that arises with the development of these new heating technologies for low-acid particulated foods is the difficulty to establish a sterilization process that ensures the food safety against pathogenic organisms, such as *C. botulinum*. TTIs can help in developing safe sterilization or pasteurization processes and are essential for their validation. Those systems allow a fast, easy, and correct quantification of the thermal processes impacts in terms of food safety without knowing the actual temperature history of the product. TTIs (chemical, enzymatic, and microbiological) have also demonstrated their ability to validate processes that are carried out in conventional still retorts. In those cases, their usages overcome the need of performing heavy and expensive inoculated experimental packs. Nevertheless, a good knowledge of the heat resistance of the sensor element under isothermal and nonisothermal conditions is required and a study to assess about the mechanical properties of the carrier is advisable in order to ensure that the TTI behaves as the real food under the same treatment conditions.

NOMENCLATURE

A_W	Water activity
D_T	Decimal reduction coefficient at temperature T
E_a	Activation energy
F_o	Treatment time at 121.1°C needed to reach a preestablished number of decimal reductions in a microbial population with z = 10°C
F_T, C_T	Time at a given reference temperature that produces the same reduction in the thermolabile element (microorganism or chemical) during the temperature evolution that undergoes the food
$F^z_{T_{ref}}$	Treatment time at reference temperature needed to reach a preestablished number of decimal reductions in a microbial population with a particular z value
k_T	First-order reaction constant at temperature T
TTI	Time temperature integrator
z	Inverse negative of the slope of the thermal destruction curve

REFERENCES

1. A Mulley, C Stumbo, W Hunting. Kinetics of thiamine degradation by heat. A new method for studying reaction rates in model systems and food products at high temperatures. *Journal of Food Science* 40(5): 985–988, 1975.
2. KI Hayakawa, GE Timbers. Influence of heat treatment on the quality of vegetables: Changes in visual green colour. *Journal of Food Science* 42(2): 778–781, 1977.
3. NG Stoforos, RL Merson. Estimating heat transfer coefficients in liquid/particulate canned foods using only liquid temperature data. *Journal of Food Science* 55(2): 478–483, 1990.
4. JW Larkin. Use of a modified Ball's formula method to evaluate aseptic processing of foods containing particulates. *Food Technology* 43: 124–131, 1989.
5. C Skjöldebrand, TA Ohlsson. A computer simulation program evaluation of the continuous heat treatment of particulate food products Part 1: Design. *Journal Food Engineering* 20(2): 149–165, 1993.
6. National Canners Association. *Laboratory Manual for Food Canners and Processors*. Vol. 1. Westport, CT: Avi Publication, Co., 1968.
7. M Rodrigo, A Martínez. Determination of a process time for a new product canned: Low acid artichoke hearts. *International Journal of Food Science & Technology* 23: 31–41, 1988.
8. ES Yawger. The count reduction system of process lethality evaluation. In: IJ Pflug, ed. *Microbiology and Engineering of Sterilization Processes*, 3rd edn. St. Paul, MN: Environmental Sterilization Services, 1979, pp. 221–232.
9. ES Yawger. Bacteriological evaluation for thermal process design. *Food Technology* 32(6): 59–63, 1978.
10. TE Odlaug, RA Caputo, CC Mascoli. Determination of sterilization (F)-values by microbiological methods. *Developments in Industrial Microbiology* 22: 349–356, 1981.
11. IJ Pflug. Measuring the integrated time-temperature effect of a heat sterilization process using bacterial spores. *Food Process Engineering*, AICHE Symposium Series 78(218): 68–79, 1982.
12. IJ Pflug. *Microbiology and Engineering of Sterilization Processes*, 5th edn. St. Paul, MN: Environmental Sterilization Services, 1982, pp. 1–26.
13. L Michiels. Methode biologique de determination de la valeur steri-lisatrice des conserves apertisees. *Industries Alimentaires et Agricoles* 96: 1349–1353, 1972.
14. DR Thompson, RR Willardsen, FF Busta, CE Allen. *Clostridium perfringens*, population dynamics during constant and rising temperatures in beef. *Journal of Food Science* 44: 646–650, 1979.
15. RA Caputo, CC Mascoli. The design and use of biological indicators for sterilization-cycle validation. *Medicine Development Diagnostic Industrial* 2(8): 23–27, 1980.
16. KL Brown, CA Ayres, JE Gaze, ME Newman. Thermal destruction of bacterial spores immobilized in food/alginate particles. *Food Microbiology* 1: 187–198, 1984.
17. MR Berry, JG Bradshaw. Comparison of sterilization values from heat penetration and spore count reduction in agitating retorts. *Journal of Food Science* 51: 477–482, 1986.
18. WD Bigelow, GS Bohart, AC Richardson, CO Ball. Heat penetration in processing canned food. *Bulletin No 16-L*, Washington, DC: Research Laboratory, National Canners Association, 1920.

19. DR Thompson. The challenge in predicting nutrient changes during food processing. *Food Technology* 36(20): 97–102, 1982.
20. KM Johnson. Effect of initial heat treatment on the expression of temperature sensitivity by outgrowing spores of *Bacillus cereus* at elevated germination and growth temperatures. PhD dissertation, Food Science Department, University of Minnesota, Morris, MN, 1983.
21. JL Secrist, CR Stumbo. Application of spores resistance in the newer methods of process evaluation. *Food Technology* 10: 543–545, 1956.
22. AC Rodríguez, AA Texeira, GH Smerage. Kinetic modeling of heat-induced reactions affecting bacterial spores. *Abstracts of the 47th Annual IFT Meeting*, June 16–19, Las Vegas, NV, 1987, pp. 182–182.
23. SK Sastry, SF Li, P Patel, M Konanayakam, P Bafna, S Doores, B Beelman. A bioindicator for verification of thermal processes for particulate foods. *Journal of Food Science* 53(5):1528–1531, 1536, 1988.
24. SK Sastry, CA Zuritz. A review of particle behaviour in tube flow: Applications to aseptic processing. *Journal of Food Process Engineering* 10: 27–52, 1987.
25. DM Dignan, MR Berry, IJ Pflug, TD Gardine. Safety considerations in establishing aseptic processes for low acid foods containing particulates. *Food Technology* 43(3): 118–121, 1989.
26. A Van Loey, T Haentjens, Ch Scout, E Hendrickx. Enzymic time-temperature integrators for the quantification of thermal processes in terms of food safety. In: FAR Oliveira, JC Oliveira, eds. *Processing Foods*. London, U.K.: CRC Press, 1999, pp. 13–36.
27. M Hendrickx, C Silva, F Oliveira, P Tobback. Optimization of heat transfer in thermal processing of conduction heated foods. In: RP Singh, MA Wirakartakusumah, eds. *Advances in Food Engineering*. London, U.K.: CRC Press, 1992, pp. 221–235.
28. A Mulley, C Stumbo, W Hunting. Kinetics of thiamine degradation by heat: Effect of pH and form of the vitamin on its rate of destruction. *Journal of Food Science* 40(5): 989–992, 1975.
29. A Mulley, C Stumbo, W Hunting. Thiamine: A chemical index of the sterilization efficacy of thermal processing. *Journal of Food Science* 40(5): 993–996, 1975.
30. GC Favetto, J Chiriffe, OC Scorza, CA Hermida. Time-temperature integrating indicator for monitoring the cooking process of packaged meats in the temperature range of 85–100 degrees Celsius. U.S. Patent, 4,834,017, 1989.
31. MF Berry, RK Singh, PE Nelson. Kinetics of Methylme-thionine sulfonium in butter solutions for estimating thermal treatment of liquid foods. *Journal of Food Process Preservation* 13: 475–488, 1989.
32. HJ Kim, IA Taub. Intrinsic chemical markers for aseptic processing of particulate foods. *Food Technology* 47: 91–99, 1993.
33. J Mabit, R Belhamri, F Fayolle, J Legrand. Development of a time temperature integrator for quantification of thermal treatment in scraped surface heat exchangers. *Innovative Food Science & Emerging Technologies* 9: 516–526, 2008.
34. T Tsironi, A Stamatiou, M Giannoglou, E Velliou, PS Taoukis. Predictive modelling and selection of time temperature integrators for monitoring the shelf life of modified atmosphere packed gilthead seabream fillets. *LWT-Food Science and Technology* 44: 1156–1163, 2011.
35. RJ Witousky. A new tool for the validation of the sterilization of parenterals. *Bulletin of Parenteral Drug Association* 11(6): 274–281, 1997.
36. JL Bunn, IK Sykes. A chemical indicator for the rapid measurement of F values. *Journal of Applied Bacteriology* 51: 143–147, 1981.
37. KR Swartzel, SG Ganesan, RT Kuehn, RW Hamaker, F Sadeghi. Thermal memory cell and thermal system evaluation. U.S. Patent 5921981, 1991.
38. KR Swartzel. Arrhenius kinetics as applied to product constituent losses in ultra high temperature processing. *Journal of Food Science* 47(6): 1886–1891, 1982.
39. G Maesmans, M Hendrickx, S De Cordt, P Tobback. Multicomponent time temperature integrators in the evaluations of thermal processes. *Journal Food Process Preservation* 17: 369–389, 1993.
40. Z Weng, M Hendrickx, G Maesmans, P Tobback. Thermostability of soluble and immobilized horseradish peroxidase. *Journal of Food Science* 56(2): 567–570, 1991.
41. S De Cordt, K Vanhoof, J Hu, G Maesmans, M Hendrickx, P Tob-back. Thermostability of immobilized alfa-amylase from *Bacillus licheniformis*. *Biotechnology and Bioengineering* 40: 396–402, 1992.
42. S De Cordt, M Hendrickx, G Maesmans, P Tobback. Immobilized alfa-amylase for *Bacillus licheniforiís*: A potential enzymic time-temperature integrator for thermal processing. *International Journal of Food Science and Technology* 27: 661–673, 1992.
43. HM Brown. The use of chemical and biochemical markers in the retrospective examination of thermally processed formulated meats. *Technical Memorandum No 625, Campden & Chorleywood Food Research Association*, Chipping Campden, U.K., 1991.

44. LG Zechman. Heat inactivation kinetics of *Bacillus stearothermophilus* spores. PhD dissertation, University of Minnesota, Morris, MN, 1988.
45. A Jones, IJ Pflug. *Bacillus coagulans*, FRR B666, as a potential biological indicator organism. *Journal of Parenteral Science Technology* 35: 82–87, 1981.
46. IJ Pflug, TE Odlaug. Biological indicators in the pharmaceutical and medical device industry. *Journal of Parenteral Science Technology* 40: 242–248, 1986.
47. M Ellouze, JC Augustin. Applicability of biological time temperature integrators as quality and safety indicators for meat products. *International Journal of Food Microbiology* 138: 119–129, 2010.
48. IJ Pflug. Method and apparatus for sterility monitoring. U.S. Patent, 3960670, 1976.
49. A Jones, IJ Pflug, R Blanchett. Effect of fill weight on the F-value delivered to two styles of green beans processed in a Sterlimatic retort. *Journal of Food Science* 45: 217–220, 1980.
50. IJ Pflug, A Jones, R Blanchett. Performance of bacterial spores in a carrier system in measuring the F_0-value delivered to cans of f food heated in a steritort. *Journal of Food Science* 45: 940–945, 1980.
51. AC Rodriguez, AA Texeira. Heat transfer in hollow cylindrical rods used as bioindicator units for thermal process validation. *Transactions of the ASAE* 31: 1233–1236, 1988.
52. AC Hersom, DT Shore. Aseptic processing of foods comprising sauce and solids. *Food Technology* 35: 53–62, 1981.
53. G Smith, IJ Pflug, P Chapman. Effect of storage time and temperature and the variation among replicate tests (on different days) on the performance of spore disks and strips. *Applied and Environmental Microbiology* 32: 257–263, 1976.
54. AD Russell. *The Destruction of Bacterial Spores*. London, U.K.: Academic Press, 1982, pp. 1–333.
55. WP Segner, TJ Ragusa, CL Marcus, EA Soutter. Biological evaluation of a heat transfer simulation for sterilizing low-acid large particulate foods for aseptic packaging. *Journal of Food Processing and Preservation* 13: 257–274, 1989.
56. PG Bean, H Dallyn, HMP Ranjith. The use of alginate spore beads in the investigation of ultra-high temperature processing. *Proceedings of the International Meeting on Food Microbiology and Technology*, April 20–23, Parma, Italy. Medicina viva Servizio Congress S.r.l., 1979, pp. 281–294.
57. UA Rönner. A new biological indicator for aseptic sterilisation. *SIK Publication* 516: 43–47, 1990.
58. MJ Ocio. Desarrollo de un integrador tiempo temperatura microbiológico, utilizando esporas de *Bacillus stearothermophillus*, para validar procesos térmicos. PhD dissertation, Facultad de Veterinaria, Universidad de Murcia, Spain, 1995.
59. JE Gaze, LE Spence, GD Brown, SD Holdsworth. Application of an alginate particle technique to the study of particle sterilisation under dynamic flow. Bulletin No 1341, Campden & Chorleywood Food Research Association, University of Gloucestershire, U.K., 1990, pp. 30–35.
60. A Fink, G Cerny. Microbiological principles of short-time sterilization of particulate food. *Proceedings of International Symposium on Progress in Food Preservation Processes*, April 12–14, Ceria, Brussels, Belgium, Vol. 2, 1988, pp. 185–190.
61. F Godia, C Casas, B Castellano, C Solá. Immobilized cells: Behaviour of carrageenan entrapped yeast during continuous ethanol fermentation. *Applied Microbiology and Biotechnology* 26: 342–346, 1987.
62. JC Ogbonna, CHB Pham, M Matsumura, H Kataoka. Evaluation of some gelling agents for immobilization of aerobic microbial-cells in alginate and carrageenan beads. *Microbiology Technology* 3: 421–424, 1989.
63. U Rönner. A new biological indicator for aseptic sterilization. *Food Technology International Europe* 90: 43–46, 1990.
64. JP Wang, L Deng, Y Li, XM Xu, Y Gao, N Hilaire, HQ Chen, ZY Jin, JM Kim, LF He. Konjac glucomannan as a carrier material for time temperature integrator. *Food Science and Technology International* 16: 127–134, 2010.
65. H Dallyn, WC Fallon, PG Bean. Method for the immobilization of bacterial spores in alginate gel. *Laboratory Practice* 26: 773–775, 1997.
66. F Rodrigo. Establecimiento de las condiciones de utilización de un ITT microbiológico, que se utiliza como elemento sensor de esporas de *Bacillus stearothermophillus*. PhD dissertation, Facultad de Ciencias Biológicas, Universidad de Valencia, Valencia, Spain, 1977.
67. I Tucker. Nonisothermal stability testing. *Pharmaceutical Technology* 9(5): 68–78, 1985.
68. DB Lund. Considerations in modelling processes. *Food Technology* 37: 92–94, 1983.
69. A Arabshahi, DB Lund. Considerations in calculating kinetic parameters from experimental data. *Journal of Food Process Engineering* 7: 239–251, 1985.
70. BW McNab. A literature review linking microbial risk assessment predictive microbiology and dose response modelling. *Dairy Food and Environmental Sanitation* 17: 405–416, 1997.

71. TA McMeekin, T Ross. Modelling applications. *Journal of Food Protection* 59(Supplement): 37–42, 1996.
72. T Grauwet, I Van der Plancken, L Vervoort, A Matser, M Hendrickx, A Van Loey. Temperature uniformity mapping in a high pressure high temperature reactor using a temperature sensitive indicator. *Journal of Food Engineering* 105: 36–47, 2011.
73. W Tejedor. Bases para la optimización del proceso de esterilización de una conserva de vegetales y pescado. PhD dissertation, Facultad de Farmacia, Universidad de Valencia, Valencia, Spain, 2001.
74. G Tucker. Application of time temperature integrators for validation of pasteurization processes. Bulletin No 77, Campden & Chorleywood Food Research Association, University of Gloucestershire, U.K., 1999.
75. T Lambourne, G Tucker. Time temperature integrators for validation of thermal processes. Bulletin No 132, Campden & Chorleywood Food Research Association, University of Gloucestershire, U.K., 2001.
76. G Tucker, E Hanby, H Brown. Development and application of a new time-temperature integrator for the measurement of P-values in mild pasteurization processes. *Food and Bioproducts Processing* 87:23–33, 2009.
77. GS Tucker, HM Brown, PJ Fryer, PW Cox, FL Poole, HS Lee, MWW Adams. A sterilization time-temperature integrator based on amylase from the hyperthermophilic organism *Pyrococcus furiosus*. *Innovative Food Science & Emerging Technologies* 8: 63–72, 2007.
78. PS Fernández, A Martínez, F Romero. Estudios de vida útil y validación aplicados al control de calidad. In: *Nuevas Perspectivas del Control de Calidad en la Industria Alimentaria*. Universidad de Murcia, Spain, 2003.
79. MD López, P García, M Munoz-Cuevas, PS Fernández, A Palop. Thermal inactivation of *Alicyclobacillus acidoterrestris* spores under conditions simulating industrial heating processes of tangerine vesicles and its use in time temperature integrators. *European Food Research and Technology* 232: 821–827, 2011.

Index

C

G

H

N

R